Occupational Ergonomics

Theory and Applications

OCCUPATIONAL SAFETY AND HEALTH

A Series of Reference Books and Textbooks

*Occupational Hazards · Safety · Health
Fire Protection · Security · Industrial Hygiene*

1. Occupational Safety, Health, and Fire Index, *David E. Miller*
2. Crime Prevention Through Physical Security, *Walter M. Strobl*
3. Fire Loss Control, *Robert G. Planer*
4. MORT Safety Assurance Systems, *William G. Johnson*
5. Management of Hotel and Motel Security, *Harvey Burstein*
6. The Loss Rate Concept in Safety Engineering, *R. L. Browning*
7. Clinical Medicine for the Occupational Physician, *edited by Michael H. Alderman and Marshall J. Hanley*
8. Development and Control of Dust Explosions, *John Nagy and Harry C. Verakis*
9. Reducing the Carcinogenic Risks in Industry, *edited by Paul F. Deisler, Jr.*
10. Computer Systems for Occupational Safety and Health Management, *Charles W. Ross*
11. Practical Laser Safety, *D. C. Winburn*
12. Inhalation Toxicology: Research Methods, Applications, and Evaluation, *edited by Harry Salem*
13. Investigating Accidents with STEP, *Kingsley Hendrick and Ludwig Benner, Jr.*
14. Occupational Hearing Loss, *Robert Thayer Sataloff and Joseph Sataloff*
15. Practical Electrical Safety, *D. C. Winburn*
16. Fire: Fundamentals and Control, *Walter M. Haessler*
17. Biohazards Management Handbook, *Daniel F. Liberman and Judith Gordon*
18. Practical Laser Safety: Second Edition, Revised and Expanded, *D. C. Winburn*
19. Systematic Safety Training, *Kingsley Hendrick*
20. Cancer Risk Assessment: A Quantitative Approach, *Samuel C. Morris*
21. Man and Risks: Technological and Human Risk Prevention, *Annick Carnino, Jean-Louis Nicolet, and Jean-Claude Wanner*
22. Fire Loss Control: A Management Guide. Second Edition, Revised and Expanded, *Peter M. Bochnak*
23. Computer Systems for Occupational Safety and Health Management: Second Edition, Revised and Expanded, *Charles W. Ross*
24. Occupational Hearing Loss: Second Edition, Revised and Expanded, *Robert Thayer Sataloff and Joseph Sataloff*
25. Protective Clothing Systems and Materials, *edited by Mastura Raheel*
26. Biohazards Management Handbook: Second Edition, Revised and Expanded, *edited by Daniel F. Liberman*
27. Occupational Ergonomics: Theory and Applications, *edited by Amit Bhattacharya and James McGlothlin*

ADDITIONAL VOLUMES IN PREPARATION

Occupational Ergonomics

Theory and Applications

edited by
Amit Bhattacharya
University of Cincinnati Medical School
Cincinnati, Ohio

James D. McGlothlin
National Institute for Occupational Safety and Health
Cincinnati, Ohio

Marcel Dekker, Inc.   New York • Basel • Hong Kong

Library of Congress Cataloging-in-Publication Data

Occupational ergonomics : theory and applications / edited by Amit
  Bhattacharya, James McGlothlin.
      p. cm. — (Occupational safety and health ; 27)
   Includes index.
   ISBN 0-8247-9419-2 (alk. paper)
   1. Human engineering. 2. Industrial hygiene. 3. Industrial safety.
  I. Bhattacharya, Amit. II. McGlothlin, James D. III. Series: Occupational
  safety and health (Marcel Dekker, Inc.) ; 27.
  TA166.O26 1996
  620.8'2—dc20   96-5423
                                                                         CIP

The publisher offers discounts on this book when ordered in bulk quantities. For more information, write to Special Sales/Professional Marketing at the address below.

This book is printed on acid-free paper.

Copyright © 1996 by MARCEL DEKKER, INC. All Rights Reserved.

Neither this book nor any part may be reproduced or transmitted in any form or by any means, electronic or mechanical, including photocopying, microfilming, and recording, or by any information storage and retrieval system, without permission in writing from the publisher.

MARCEL DEKKER, INC.
270 Madison Avenue, New York, New York 10016

Current printing (last digit):
10 9 8 7 6 5 4 3 2 1

PRINTED IN THE UNITED STATES OF AMERICA

Foreword

There are many contemporary books on the subject of ergonomics. Many of these treat the workplace as an application area for ergonomics. In this book, the occupational issues are central to ergonomics, both in the methods used to study how a person is affected by work, and in the many examples of how ergonomic principles can be used to reduce adverse human outcomes in the workplace.

The book is a testimony to the scope of occupational ergonomics today. Not only are a large variety of contemporary methods, applications, and problems presented, but these are addressed by authors who have quite different technical expertise, training, and experience. I view this as a major strength in that such a varied approach avoids the gaps in coverage and sometimes biased views that are created when only a few authors with similar experience attempt to write about such an ecletic field as occupational ergonomics.

The editors have done an excellent job in assuring that each chapter can be understood by readers with disparate backgrounds. People who have medical or life science backgrounds should be able to learn a little about biomechanics. People with an engineering background will be able to learn some human physiology and anatomy. Those with production and operating systems backgrounds will learn how ergonomics can improve organizational effectiveness.

If only a few of the principles presented in this text are implemented in future workplaces, the resulting benefits will, I believe, be great and will serve well as the basis for succeeding editions.

Don B. Chaffin
Director, Center for Ergonomics
Department of Industrial Operations Engineering
University of Michigan
Ann Arbor, Michigan

Preface

Over the last five to seven years the demand of ergonomic needs in the workplace has increased exponentially. OSHA's new ergonomic guidebook for the meat packing industry and the Americans with Disabilities Act were published in 1990. Because of these new developments, as well as the newly developed interest among all relevant professionals to become "ergonomically literate," it is an opportune time to put all the relevant ergonomic information into book form to help meet this important need.

This book fills a critical gap between existing ergonomic books, which are either very narrow in their treatment of the subject matter or very applied. This book is designed to address the broader field of ergonomic practice from its fundamentals to applications to current and emerging issues in the workplace. The chapters on fundamentals are thorough, yet they focus on only those aspects that have direct bearing on the workplace. Professionals with expertise in their respective fields who read this text will be better prepared to understand the ergonomic problems at their workplaces and will learn how to apply the fundamentals of ergonomics to actual workplace ergonomic problem solving. At the same time, new graduate students entering this field will be provided with a good understanding of the fundamentals and shown how these principles translate into problem-solving vehicles in the workplace.

After an introduction to the principles and fundamentals of ergonomics (Part I), there follows the practical applications of those fundamentals in solving ergonomic problems and insight into methods used to assess ergonomic risks and their control (Part II). Part III addresses the issue of connecting the physical ergonomic risk factors to the development of medical incidence and therefore the topic of how a medical surveillance program should be designed to monitor and control ergonomic problems at a work site. In Part IV, the reader is given the opportunity to review ergonomic case studies from various industries. As the field of ergonomics is growing rapidly, we have included in Part V a series of currently important topics such as cumulative trauma disorder, vibration white finger disorder, NIOSH's new lifting formula, OSHA's ergonomic management program, and the impact of the Americans with Disabilities Act (ADA) on ergonomics. Finally, as the field of ergonomics has significant impact at both the national and international levels, Part VI covers the current trends in this field from around the world. The book concludes with appendixes that include software sources and ergonomic checklists, and information on video training materials and other background materials.

Thanks are due Mrs. Mary Mersmann for word processing and indexing assistance. Above all, we are especially grateful for the patience and understanding of our wives, Prakriti Bhattacharya and Nancy McGlothlin, and children, Aleena and Abhirup Bhattacharya and Malia and Jaime McGlothlin.

Amit Bhattacharya
James D. McGlothlin

Contents

Foreword   Don B. Chaffin   *iii*
Preface   *v*
Contributors   *xi*

PART I: Principles of Ergonomics

1. Anthropometry   1
   James F. Annis and John T. McConville

2. Physiology of Body Movement   47
   Stephan Konz

3. Physiological Aspects of Neuromuscular Function   63
   Thomas R. Waters and Amit Bhattacharya

4. Biomechanical Aspects of Body Movement   77
   Angshuman Bagchee and Amit Bhattacharya

5. Biomechanical Models in Ergonomics   115
   Kevin P. Granata and William S. Marras

6. Psychophysical Methodology and the Evaluation of Manual Materials Handling and Upper Extremity Intensive Work   137
   Sheila Krawczyk

7. Instrumentation for Occupational Ergonomics   165
   Robert G. Radwin, David J. Beebe, John G. Webster, and Thomas Y. Yen

8. Occupational Heat Stress   195
   Thomas E. Bernard

9. Physical Work Capacity: Principles and Applications   219
   Ashraf M. Genaidy

10. Worker Participation: Approaches and Issues   235
    Alexander L. Cohen

PART II: Application of Ergonomic Principles

11. Job Analysis   259
    Katharyn A. Grant

12. Workstation Evaluation and Design 279
    David R. Clark

13. Tool Evaluation and Design 303
    Andris Freivalds

14. Manual Materials Handling 329
    Thomas R. Waters and Vern Putz-Anderson

15. Manual Materials Assist Devices 351
    Jeffrey C. Woldstad and Roderick J. Reasor

16. Evaluating Physical Qualifications of Workers and Jobs 367
    Richard J. Wickstrom

17. Office Ergonomics 387
    Mary Brophy and Christin Grant

18. The Human Factors Aspects of Shiftwork 403
    Debra K. Dekker, Donald I. Tepas, and Michael J. Colligan

19. ErgoMOST: An Engineer's Tool for Measuring Ergonomic Stress 417
    Kjell B. Zandin, David L. Gardner, Edward J. Gill, and James R. Wilk

20. The Design and Evaluation of a Musculoskeletal and Work History Questionnaire 431
    Grace Kawas Lemasters and Margaret R. Atterbury

21. Fall Prevention in Industry Using Slip Resistance Testing 463
    Mark S. Redfern and Timothy P. Rhoades

PART III: Medical Surveillance for Ergonomics Programs

22. Record-Based ("Passive") Surveillance for Cumulative Trauma Disorders 477
    Shiro Tanaka

23. Active Surveillance of Work-Related Musculoskeletal Disorders: An Essential Component in Ergonomic Programs 489
    Norka Saldaña

PART IV: Ergonomic Case Studies

24. Development and Implementation of an Ergonomics Process in the Automotive Industry: Reactive and Proactive Processes 501
    Bradley S. Joseph and Glenn Jimmerson

25. Ergonomic Control Measures in the Health Care Industry 519
    Arthur R. Longmate

26. Ergonomic Case Studies in Industry: Health Care 537
    Roger C. Jensen

27. Injuries and Ergonomic Applications in Construction 545
    Hongwei Hsiao and Ronald L. Stanevich

Contents

28. An Ergonomic Analysis and Abatement Recommendations to Reduce Musculoskeletal Stress in Warehousing Operations: A Case Study 569
    Donald S. Bloswick and Emil Golias

PART V: Current Topics

29. Upper Extremity Cumulative Trauma Disorders: Current Trends 581
    Daniel J. Habes
30. Occupational Human Vibration 605
    Michael J. Griffin
31. Revised NIOSH Lifting Equation 627
    Thomas R. Waters and Vern Putz-Anderson
32. OSHA's Ergonomic Program 655
    Gregg LaBar
33. The Americans with Disabilities Act: Implications for the Use of Ergonomics in Rehabilitation 669
    Jerry A. Olsheski and Robert E. Breslin
34. Legal Aspects of Ergonomics 685
    James J. Montgomery
35. Real-Time Exposure Assessment and Job Analysis Techniques to Solve Hazardous Workplace Exposures 699
    James D. McGlothlin, Michael G. Gressel, William A. Heitbrink, and Paul A. Jensen
36. Ergonomics and Concurrent Design 719
    Peter M. Budnick, Donald S. Bloswick, and Don R. Brown

PART VI: International Perspective on Ergonomics

37. Overview of Ergonomic Research and Some Practical Applications in Sweden 733
    Lennart Dimberg
38. Overview of Ergonomic Needs and Research in India 749
    Rabindra Nath Sen
39. Overview of Ergonomic Needs and Research in China 761
    Tian Lin Li, Zun Yong Liu, and Xiou Fen Zhang
40. Overview of Ergonomic Needs and Research in Taiwan 765
    Chi-Yuang Yu and Eric Min-yang Wang

Appendixes

Appendix A. Biomechanical Modelling of Carpet Installation Task 773

Appendix B. Ergonomics Checklists 783

Appendix C. Electronic Sources of Information 807

Appendix D. Ergonomics Software Sources 809

Appendix E. Information Sources from the National Institute for Occupational Safety and Health 815

Appendix F. Ergonomics Journals 817

Glossary 819

Index *827*

Contributors

James F. Annis Consultant, Anthropology Research Project, Inc., Yellow Springs, Ohio

Margaret R. Atterbury Greater Cincinnati Occupational Health Center, Cincinnati, Ohio

Angshuman Bagchee Research Associate, Biomechanics-Ergonomics Research Laboratory, Department of Environmental Health, University of Cincinnati Medical School, Cincinnati, Ohio

David J. Beebe Assistant Professor, Department of Biomedical Engineering, Louisiana Tech University, Ruston, Louisiana

Thomas E. Bernard College of Public Health, University of South Florida, Tampa, Florida

Amit Bhattacharya Professor, Environmental Health and Industrial Engineering, Biomechanics-Ergonomics Research Laboratory, Department of Environmental Health, University of Cincinnati Medical School, Cincinnati, Ohio

Donald S. Bloswick Associate Professor, Department of Mechanical Engineering, University of Utah, Salt Lake City, Utah

Robert E. Breslin Director, Industrial Rehabilitation Programs, Center for Occupational Health, University of Cincinnati Medical Center, Cincinnati, Ohio

Mary Brophy School of Public Health, State University of New York, Albany, New York

Don R. Brown Assistant Professor, Department of Mechanical Engineering, University of Utah, Salt Lake City, Utah

Peter M. Budnick Research Assistant Professor, Department of Mechanical Engineering, University of Utah, Salt Lake City, Utah

David R. Clark Associate Professor, Industrial and Manufacturing Systems Engineering Department, GMI Engineering & Management Institute, Flint, Michigan

Alexander L. Cohen Consultant, Occupational Human Factors, Cincinnati, Ohio

Michael J. Colligan National Institute for Occupational Safety and Health, Cincinnati, Ohio

Debra K. Dekker Senior Associate, COMSIS Corporation, Silver Spring, Maryland

Lennart Dimberg Manager, Health Promotion and Health Care, Volvo Aero Corp., Trollhättan, Sweden

Andris Freivalds Professor, Department of Industrial and Manufacturing Systems Engineering, The Pennsylvania State University, University Park, Pennsylvania

David L. Gardner Vice President—Consulting, H. B. Maynard and Company, Inc., Pittsburgh, Pennsylvania

Ashraf M. Genaidy Associate Professor, Department of Industrial Engineering, University of Cincinnati, Cincinnati, Ohio

Edward J. Gill Vice President—Marketing, H. B. Maynard and Company, Inc., Pittsburgh, Pennsylvania

Emil Golias Senior Industrial Hygienist, Health Response Team, Occupational Safety and Health Administration, Salt Lake City, Utah

Kevin P. Granata Biodynamics Laboratory, The Ohio State University, Columbus, Ohio

Christin Grant Center for Ergonomics, The University of Michigan, Ann Arbor, Michigan

Katharyn A. Grant Industrial Engineer, Division of Biomedical and Behavioral Science, National Institute for Occupational Safety and Health, Cincinnati, Ohio

Michael G. Gressel Engineering Control Technology Branch, National Institute for Occupational Safety and Health, Cincinnati, Ohio

Michael J. Griffin Professor, Human Factors Research Unit, Institute of Sound and Vibration Research, University of Southampton, Southampton, England

Daniel J. Habes Industrial Engineer, Applied Psychology and Ergonomics Branch, National Institute for Occupational Safety and Health, Cincinnati, Ohio

William A. Heitbrink Engineering Control Technology Branch, National Institute for Occupational Safety and Health, Cincinnati, Ohio

Hongwei Hsiao Acting Section Chief, Division of Safety Research, National Institute for Occupational Safety and Health, Morgantown, West Virginia

Paul A. Jensen Engineering Control Technology Branch, National Institute for Occupational Safety and Health, Cincinnati, Ohio

Roger C. Jensen Senior Ergonomist, UES, Inc., Dayton, Ohio

Glenn Jimmerson Automation Systems Specialist, Advanced Manufacturing Technology Department, Ford Motor Company, Dearborn, Michigan

Bradley S. Joseph Corporate Ergonomist, Occupational Health and Safety, Industrial Hygiene, Ford Motor Company, Dearborn, Michigan

Stephan Konz Department of Industrial and Manufacturing Systems Engineering, Kansas State University, Manhattan, Kansas

Sheila Krawczyk Senior Research Associate, Liberty Mutual Research Center for Safety and Health, Hopkinton, Massachusetts

Gregg LaBar Managing Editor, *Occupational Hazards*, Penton Publishing, Cleveland, Ohio

Grace Kawas Lemasters Professor, Department of Environmental Health, University of Cincinnati Medical School, Cincinnati, Ohio

Tian Lin Li Professor, Department of Work Physiology, Institute of Occupational Medicine, Chinese Academy of Preventive Medicine, Beijing, People's Republic of China

Contributors

Zun Yong Liu Associate Professor, Department of Work Physiology, Institute of Occupational Medicine, Chinese Academy of Preventive Medicine, Beijing, People's Republic of China

Arthur R. Longmate Staff Ergonomics Engineer, Safety & Industrial Hygiene, Johnson & Johnson, New Brunswick, New Jersey

William S. Marras Professor and Director, Biodynamics Laboratory, The Ohio State University, Columbus, Ohio

John T. McConville Consultant, Anthropology Research Project, Inc., Yellow Springs, Ohio

James D. McGlothlin Senior Occupational Safety and Health Specialist, Engineering Control Technology Branch, National Institute for Occupational Safety and Health, Cincinnati, Ohio

James J. Montgomery Partner, Montgomery, Rennie & Jonson, Cincinnati, Ohio

Jerry A. Olsheski Assistant Professor, School of Applied Behavioral Sciences and Educational Leadership, Ohio University, Athens, Ohio

Vern Putz-Anderson Chief, Psychophysiology and Biomechanics Section, National Institute for Occupational Safety and Health, Cincinnati, Ohio

Robert G. Radwin Professor, Department of Industrial Engineering, University of Wisconsin—Madison, Madison, Wisconsin

Roderick J. Reasor* Assistant Professor, Department of Industrial and Systems Engineering, Virginia Polytechnic Institute and State University, Blacksburg, Virginia

Mark S. Redfern Associate Professor, Departments of Otolaryngology and Industrial Engineering, University of Pittsburgh, Pittsburgh, Pennsylvania

Timothy P. Rhoades Principal Research Engineer, Applied Safety and Ergonomics, Inc., Ann Arbor, Michigan

Norka Saldaña Director of Safety and Industrial Hygiene, Johnson & Johnson Shared Services, Caguas, Puerto Rico

Rabindra Nath Sen University Professor, Ergonomics Laboratory, Department of Physiology, University of Calcutta, Calcutta, India

Ronald L. Stanevich Branch Chief, Division of Safety Research, National Institute for Occupational Safety and Health, Morgantown, West Virginia

Shiro Tanaka Medical Officer, Division of Surveillance, Hazard Evaluations, and Field Studies, National Institute for Occupational Safety and Health, Cincinnati, Ohio

Donald I. Tepas Professor of Industrial Psychology and Director, Ergonomics Laboratory, Department of Psychology, University of Connecticut, Storrs, Connecticut

Eric Min-yang Wang Associate Professor, Department of Industrial Engineering, National Tsing Hua University, Hsinchu, Taiwan, Republic of China

Thomas R. Waters Research Physiologist, Applied Psychology and Ergonomics Branch, National Institute for Occupational Safety and Health, Cincinnati, Ohio

---

Present affiliation: Senior Industrial Engineer, Management Engineering Services, Eastman Chemical Company, Kingsport, Tennessee

John G. Webster Professor, Department of Electrical and Computer Engineering, University of Wisconsin—Madison, Madison, Wisconsin

Richard J. Wickstrom Physical Therapist and Ergonomist, Disability Control, Inc., Cincinnati, Ohio

James R. Wilk Consultant, H. B. Maynard and Company, Inc., Pittsburgh, Pennsylvania

Jefffrey C. Woldstad* Department of Industrial and Systems Engineering, Virginia Polytechnic Institute and State University, Blacksburg, Virginia

Thomas Y. Yen Research Assistant, Department of Industrial Engineering, University of Wisconsin—Madison, Madison, Wisconsin

Chi-Yuang Yu Associate Professor, Department of Industrial Engineering, National Tsing Hua University, Hsinchu, Taiwan, Republic of China

Kjell B. Zandin Senior Vice President, H. B. Maynard and Company, Inc., Pittsburgh, Pennsylvania

Xiou Fen Zhang Harbin Institute of Industrial Hygiene and Occupational Disease, Harbin, People's Republic of China

---

**Present affiliation*: Department of Industrial Engineering, Texas Tech University, Lubbock, Texas

1
Anthropometry

James F. Annis and John T. McConville
Anthropology Research Project, Inc., Yellow Springs, Ohio

I. INTRODUCTION

Anthropometry involves the systematic measurement of the physical properties of the human body, primarily dimensional descriptors of body size and shape. Anthropologists have been measuring humans for hundreds of years, but for only the last 50 years or so have the dimensions been used in an organized fashion to improve the design and sizing of the things we use in everyday life. Often the problem with the application of anthropometry to a design problem will be the lack of certain necessary measurements or the need to accommodate a wide range in size and shape variability into a single, often inflexible design. Applied anthropometry—that is, the use of anthropometric data in the design and construction of a wide variety of items from clothing to spacecraft—is a relatively new discipline whose practitioners are still learning to cope with the exponential character of technology and its impact on the kinds of information needed to describe the physical and biological characteristics of our species. It grew out of physical anthropology, which traditionally studied body size and function with the goal of resolving our ancestry and identifying the existing varieties of *Homo sapiens*. Today, we are still trying to learn what to measure to satisfy an even younger discipline called ergonomics.

The original impetus for the development of applied anthropometrics, and perhaps ultimately ergonomics, was the need to improve the effectiveness and efficiency of equipment used in combat during World War II. The military tie was so strong in the United States that it has been recognized only relatively recently that civilian industry could also benefit from the proper use of anthropometry in the design of products and workstations. Today industry has embraced the concept so eagerly that the word *ergonomics*, which derives from the Greek *ergon* (work) and *nomos* (natural laws of), has become popularly used in advertisements. A concern for ergonomics currently spawns industrial action committees, comprising members of both management and labor, whose purpose is to improve the human–machine interface to achieve a healthier, safer, and more efficient workforce. Anthropometric data are a necessary and basic tool for attaining this end as well as improving the design of a wide variety of products.

II. BACKGROUND AND SIGNIFICANCE TO OCCUPATIONAL ERGONOMICS

There appears to be two major divisions of ergonomics. The first deals with the worker, the machine the worker uses, and the environment in which the worker operates. The objective of this branch of ergonomics is to create the best possible situation on the job relative to the welfare of the worker's physical and mental health, the efficiency of production, and the quality of the product produced. Second, there are the characteristics of the manufactured product(s) that interact with the human user. In reviewing the literature we have found no less than 20 different definitions of ergonomics. One of the most inclusive examples is as follows:

> The ability to apply information regarding human character, capacities, and limitations to the design of human tasks, machines, machine systems, living spaces, and environment so that people can live, work, and play safely, comfortably, and efficiently.

All of the definitions of ergonomics mention work or the workplace, but, more important from our perspective, they also mention the human operator in the equation. The ways in which work is related to the individual vary greatly throughout industry, yet it is very difficult to think of any work situation in which the application of anthropometry or anthropometric principles could not make the work environment healthier, safer, and more efficient. At the same time, ergonomists and designers are not trying to make their products either cheaper or more expensive but rather better suited to the limitations of the human user. Anthropometry cannot be separated from these ergonomic processes, because they cannot be carried out without a knowledge of human dimensionalities.

The relationship of anthropometry to occupational ergonomics is both straightforward and complex. All the tools used in manufacturing, all the workspaces in which the manufacturing is done, and virtually all of the items produced by the manufacturing process interact with the human body or human bodyspace. In the most automated of manufacturing environments, humans must still make and repair the machines and robots, and the products that come off the assembly line must be designed for human users. Software is still written by humans and entered onto the computer disk or tape through a keyboard operated by the human hand. It is hard to think of an exception. Ultimately there's a human in the loop; hence human dimensions are likely to be needed for some time to come.

Obviously, physical anthropologists are prepared to include almost any measurement that describes the shape, size, or function of the body or its parts as components that make up the area of anthropometry. The data discussed most frequently in this chapter are those used as dimensional descriptors of general body size, those used for specific types of working postures, and those describing reach capabilities. Data for related areas of investigation, such as mass distribution and segmental moments of inertia, range of joint motion (ROJM), strength, and biomechanical aspects of the human body, are discussed elsewhere in this book. In one way or another all of these data are at some point in the design process useful to the occupational ergonomist.

III. RELEVANT CONCEPTS AND TERMINOLOGY

A. Anatomic Concepts

In performing anthropometric measurements, some knowledge of human anatomy is essential, because almost all measurements are defined in terms of some body part or some specific location on a specified part. Subjects being measured are directed to assume specific predefined positions. The standard reference point is the *anatomical position*, in which the person stands erect with arms at the sides and the palms of the hands facing forward. From this posture the descriptive terms that define the body's principal axes and the resultant planes are derived. From these planes, too, the basic terms used to describe the relative position or location of relevant points on the body structure are developed. The most commonly used terms are shown diagrammatically in Figure 1. Typically, the principal axes, X, (front-to-

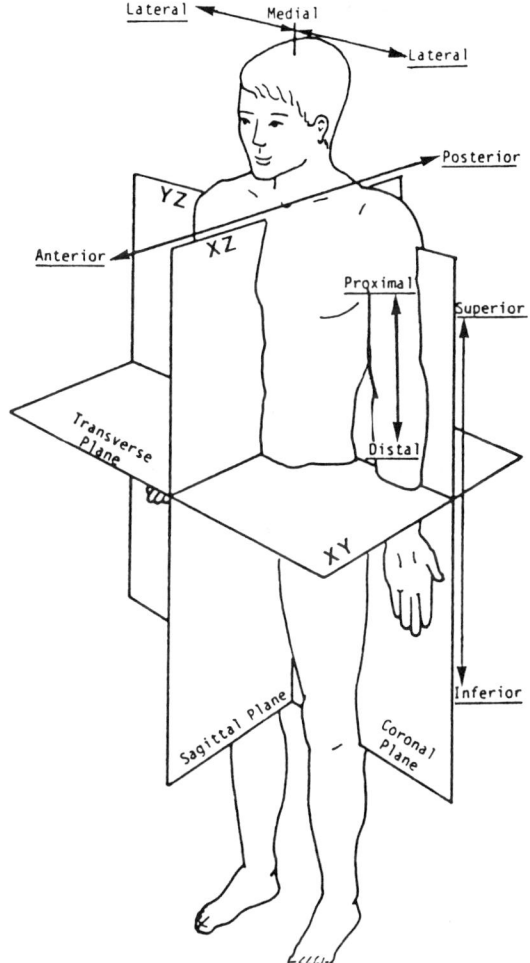

Figure 1 Terminology used to define position and location on the body.

back), Y (side-to-side), and Z (head-to-foot), divide the body into three planes: sagittal, which divides the body into right and left parts (XZ); coronal, which divides the body front to back (YZ); and transverse, which divides the body cross-sectionally (XY).

The relative positions of particular structures or features in or on the body are defined as follows:

Anterior/posterior Structures nearer the front or ventral side are *Anterior* ($+X$) to those located nearer to the back or dorsal surface, which are *posterior* ($-X$).

Medial/lateral Structures located nearer to the center of the body or to the midsagittal plane relative to others are *medial* to those located away from the central body on the left ($+Y$) or right ($-Y$) side, which are *lateral*.

Superior/inferior Structures located nearer the head are located *superior* ($+Z$) to those below, which are termed *inferior* ($-Z$). For example, the heart is located superior to the kidneys.

Proximal/distal On the limbs, parts that are near the trunk are *proximal* whereas those farther from the body central are *distal*. A finger is distal to the elbow.

Frequently the points between which a given dimension is measured are actually drawn on the subject's skin by the anthropometrist. Such points are called *landmarks*. Some landmarks are simply determined by a certain feature found in the topography of the body surface, e.g., the tip of a finger; others must be palpated and marked in relation to skeletal architecture. Twenty-four of the 25 landmarks shown in Figure 2 are of the latter variety.

In a recent survey of U.S. Army personnel (ANSUR) [1], nearly 100 landmarks were used to define the basic group of 132 dimensions measured. Approximately 70 of the landmarks were actually drawn on the subjects' skin by landmarking specialists (e.g., landmarks shown on Figure 2), and another 30 landmarks were located by observation by the measurer, e.g., distal tip of thumb, inferior tip of earlobe.

B. Types of Measurements

Most commonly, anthropometry refers to traditional dimensional descriptors of body size. Except for weight, these measurements basically provide the straight-line or curvilinear distance between two points obtained under static prescribed conditions. Basic categories of static anthropometric dimensions include lengths, depths, breadths, and distances that are basic descriptors of body size. Surface contour measurements, such as arcs and circumferences, are more complex because they contain elements of three-dimensional shape in one plane. In most major surveys, the use of other types of anthropometric measurements is limited. Simple forms of reach, e.g., functional leg length, overhead reach, grip reach, and thumb tip reach, may be measured.

1. Static Measurements

Traditional static dimensions may be briefly described as follows:

Height   Typically the distance along the Z axis from the floor or seated surface to a specific point on the body.

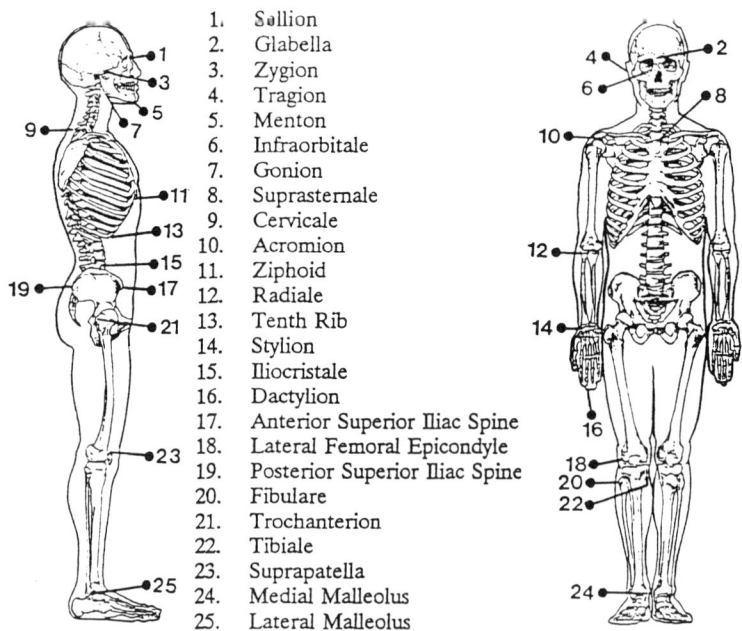

1. Sellion
2. Glabella
3. Zygion
4. Tragion
5. Menton
6. Infraorbitale
7. Gonion
8. Suprasternale
9. Cervicale
10. Acromion
11. Ziphoid
12. Radiale
13. Tenth Rib
14. Stylion
15. Iliocristale
16. Dactylion
17. Anterior Superior Iliac Spine
18. Lateral Femoral Epicondyle
19. Posterior Superior Iliac Spine
20. Fibulare
21. Trochanterion
22. Tibiale
23. Suprapatella
24. Medial Malleolus
25. Lateral Malleolus

Figure 2   Selected skeletal landmarks used to define traditional anthropometric measurements.

Length Usually used to name the distance between two landmarks that are found on a single segmental part of the body. In some cases the term *distance* is used. Some lengths or distances describe the entire segment, whereas others describe a portion of a segment along the longitudinal axis of the part. Some lengths are contours or complex distances measured on the body surface, e.g., Sleeve Length. Distances between linkage centers used in computer models and drafting board manikins are called link lengths.

Depth The distance between two landmarks found on the anterior and posterior surfaces of the body along the X axis.

Breadth The distance between two points found on the right and left (lateral) sides of the body.

Arc Curvilinear or surface contour distance between two points on the body, often on the head or face.

Circumference Closed curvilinear contour that provides the distance completely around the body part. In most cases the circumference is located perpendicular to the longitudinal axis of the body part.

Reach Specialized arm-hand distance in a particular posture or condition.

In most cases, dimension names include the above designations, but in some cases no such descriptive label is used, e.g., Span, Stature, or Scye. The idea of the defining labels used to describe a particular type or class of measurements begs the question of standardization of measurement definitions, a matter over which anthropologists have often been unable to agree. One text that deals with standards was published in 1988 [2]. This book concentrates on nutritional/health assessment measurements, but a number of applied dimensions are discussed. At a conference of practicing anthropologists held at Wright-Patterson Air Force Base in the late 1960s, agreement on a list of 29 dimensions fundamental to applied users was barely achieved [3]. It would be helpful to the user if dimensions with the same name were always measured in the same way. This is very often not the case. Dimensions with the same name from any two surveys may have well been measured differently; hence the difference observed between the two values may in part be due to procedures used and not to population or sampling differences.

2. Dynamic Measurements

Dynamic measurements such as isometric strength and range of joint motion (ROJM) have traditionally been measured in separate surveys. The measurement of ROJM is probably the simplest of such measurements in anthropometrics. Traditionally, planar ROJM measurements are made using a goniometer, which in its simplest form is nothing more than a 180° or 360° protractor with extended arms, one of which is movable to track the segment. A related measurement of this type is the reach envelope, which at some point in the measuring process always involves movement of the arm-shoulder complex and sometimes movement of the trunk or torso. In most cases, however, these measurements are obtained statically, that is, they are expressed as the change or difference in location across the movement, i.e., begin-end delta.

Too few truly dynamic anthropometric measurements are made on humans, and most of those that may be thought to qualify as dynamic contain static elements. For example, isotonic strength testing in which muscle lengths change may be said to be dynamic. Recently, isokinetic devices, in which the speed of the movement produced is controlled, have been developed for this type of testing. Strength and biomechanics are discussed elsewhere in this book.

C. Equipment

1. Traditional Static Measurements

The basic tools of the anthropometrist are the anthropometer, a variety of calipers, and a tape measure. The most commonly used instruments are shown in Figure 3.

Typically, anthropometers are precision instruments made up of four interconnecting sections of tubular metal that are engraved in millimeter (mm) intervals. Current models are square in cross section and are capable of measuring Stature or other heights from the floor and seated surfaces as well as straight lengths and distances up to 210 centimeters (cm) when completely assembled. The heights (starting with 0 mm from the floor or seated surface) are read using a movable slide housing that contains an adjustable perpendicular blade, which is placed in alignment with, or lightly on, the desired measuring point. The slide housing contains a window with centerline that enables the user to read the distance to the nearest 0.5 mm on the engraved scale. Typically, only the nearest whole millimeter is recorded.

The upper two sections of the anthropometer may be used as a beam caliper, as they are equipped with a millimeter scale on the side opposite the main scale that starts with 0 mm at the top fixture (see Figure 3). The beam caliper, which is capable of measuring distances of up to 95.0 cm, is used for measuring whole body depths and breadths as well as many straight linear distances between landmarked points. Anthropometers may be purchased individually or as part of a set that includes a sliding and a spreading caliper and a tape measure.[1] These tools have made up the professional anthropometrist's basic measuring kit over a long time period; however, they are slowly being displaced by more modern automated systems. A wide variety of special application calipers and other instruments are also available for anthropometric measuring, although most of these will not be needed by the industrial ergonomist. Other, less complex items requiring a minimum of equipment or shop skills may be homemade. Of the latter, the most frequently used are special tables for seated measurements with adjustable buttock plates and foot rests. Many of the seated workspace dimensions are obtained using such tables. Static measuring of various regions of the body such as the head, hands, or feet frequently requires the use of stabilizing or referencing surfaces to help control the repeatability of the measurement. A foot box is often used, for example, in

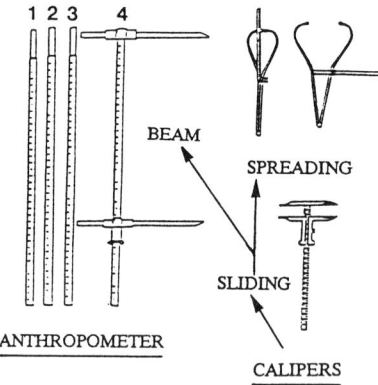

Figure 3 Basic equipment used to perform traditional anthropometric measurements.

---

[1]The best anthropometer currently on the market is the GPM, manufactured in Switzerland and imported into this country by Seritex, Inc., 450 Barell Avenue, Carlstadt, NJ 07072. A catalog is available.

measuring various foot lengths and breadths. To obtain head and face dimensions referenced to fixed surfaces, a device called a headboard has been used in a number of military surveys [4,5]. A subject places the top (vertex) and back (occiput) of the head firmly against the two perpendicular surfaces, and the anthropometrist measures the distance from the two reference surfaces to defined landmark locations using a depth/height gauge. An automated form of this device was developed for use in the 1988 survey of U.S. Army personnel [6,7]. The automated headboard provided three-dimensional (3-D) coordinates for 26 selected landmark locations on over 8000 soldiers.

An industrial ergonomist may find it necessary to perform measurements of employees on the job. In this case an effort should be made to incorporate defined and reproducible controls into the measuring methods. A number of simple devices can be designed and built by users to improve reliability and accuracy or to provide special dimensional values to suit a specific need. If the anthropometrist wishes to compare any data collected to those existing in some database, however, care should be exercised to duplicate the original procedure. In some cases this caveat should extend to the equipment used. The final report that describes the methods and summary statistics for the ANSUR survey [1] contains a description of a number of devices used for special application measurements as well as the procedures used to measure over 132 body dimensions. The ANSUR database represents the largest, most recent, and most comprehensive anthropometric survey of Americans in existence. Thus, we recommend that the reader follow the procedures developed for this survey until comparable data are available on civilians (see Sec. III.E).

2. Looking Ahead

To date, most anthropometric data have been obtained using manually operated instruments or devices such as those discussed above. Usually these measurements are performed one at a time and recorded by hand. These data provide a two-dimensional description of the body worthy of an earlier time. But biologists have for centuries wanted to be able to quantitatively describe the three-dimensional form of the human body. Early in this century, researchers began exploring techniques to obtain large quantities of digital information about the body surface size and structure through the use of stereophotogrammetric cameras [8]. A slow and tedious digitization process was required to obtain the 3-D data from the resulting photographs. Today, the use of lasers, video, cameras, and other devices in combination with graphics software makes it possible to rapidly collect large quantities of high-density digital information on limited areas of the body. In the near future it will likely be possible to scan the entire body and generate 3-D coordinates on up to 500,000 points on the body surface in just a few seconds. Before the use of true 3-D shape becomes routine, work remains to be done on methods used to summarize the massive quantity of 3-D coordinates and improve the ability of optoelectronic devices to resolve the many cracks and crevasses on the surface of the body. This is one of the reasons that traditional anthropometry will continue to be needed for some time to come.

With the onset of the ability to collect high-density digital information in three dimensions, we now can talk about body shape in quantitative ways, for both surface contours and internal organs. Scientists and physicians can now examine 3-D full-color images of a person's internal structures using a variety of computerized scanning techniques. It will be some time before people outside the medical field will know exactly how to use it on the job, but currently such information is invaluable in the early detection or confirmation of cumulative trauma disorders and other afflictions that result from poor ergonomics in the workplace. When high-density 3-D coordinate data for the whole body surface become state-of-the-art, simple point-to-point measurement may be viewed as rudimentary. More important, the engineer, the ergonomist, and the designer will have at their fingertips any human dimension they may need to resolve a given design problem.

D. Statistical Considerations

Fortunately, values for most traditional static anthropometric dimensions are normally distributed. This characteristic permits the use of a number of simplifications that the reader may find helpful. The classic shape of the normal distribution curve is shown in Figure 4 in conjunction with increments of standard deviation (SD) relative to the mean ($\bar{x}$). The mean ± 1 SD can quickly be seen to include 68% of the population sampled, while the mean ± 2 SD and the mean ± 3 SD include the variability exhibited by approximately 95% and 99.8%, respectively, of the population. The relationships between the mean ± increments of the SD and selected percentile levels are given in Table 1. As can be seen, the ubiquitous 5th to 95th percentile range is approximately equivalent to the mean ± 1.65 SD.

Two fundamental questions to be resolved in the application of anthropometric data to design are (1) What database should be selected? and (2) What statistical value(s) should be used? The theoretical answer to the first question is simple: The database of choice is the most recent survey containing the desired dimensions obtained on a large sample of the target user population. In practical fact, this desirable database usually does not exist. Hence, the database will probably have to be selected from the most appropriate of those available.

The choice of a statistical approach depends on the problem to be solved. If the design requires only a single dimension, the designer could attempt to accommodate the entire range of variability in the user population for that variable. Such simple design problems are in reality rare. Usually, a number of body dimensions are required, and it is here that the difficulty begins.

Probably the most common error in thinking about the use of anthropometry is sizing and design is that if an individual is small (or large) for a given dimension, then that person is small (or large) for all other dimensions. This is seldom, if ever, true—a fact that can be demonstrated in at least two ways.

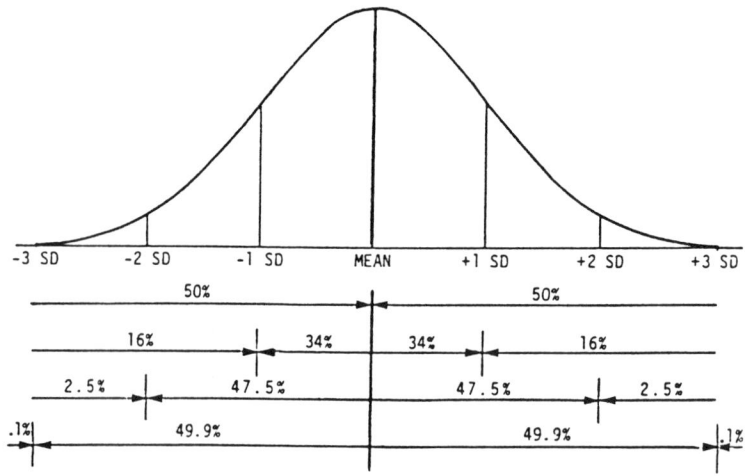

Figure 4 The normal distribution curve and relative proportions of the population represented by multiples of the standard deviation.

Table 1 Using the Mean and Standard Deviation to Estimate Percentile Values for Normally Distributed Data

Percentile value	Formula
99.5	Mean + (2.58 × SD)
99	Mean + (2.32 × SD)
97.5	Mean + (1.95 × SD)
97	Mean + (1.88 × SD)
95	Mean + (1.65 × SD)
90	Mean + (1.28 × SD)
80	Mean + (0.84 × SD)
75	Mean + (0.67 × SD)
70	Mean + (0.52 × SD)
50	Mean
30	Mean − (0.52 × SD)
25	Mean − (0.67 × SD)
20	Mean − (0.84 × SD)
10	Mean − (1.28 × SD)
5	Mean − (1.65 × SD)
3	Mean − (1.88 × SD)
2.5	Mean − (1.95 × SD)
1	Mean − (2.32 × SD)
0.5	Mean − (2.58 × SD)

One common misconception is that the mean, which lies near the center of the most densely populated portion of the distribution, is the best value to use to satisfy size requirements of the largest number of people. This is true only if the problem is univariate and if the design will include no adjustability. The "average person" has long been dear to designers' hearts, but, as has been demonstrated by Daniels [9], this individual probably does not exist. In 1952, Daniels analyzed a sample of 4063 U.S. Air Force flyers to determine how many men with average Stature (mean ± 0.3 SD) would also be of average size for a successively inputted series of clothing design dimensions for which the same criterion for average was used. A total of 1055 men met the average criterion for Stature, but not a single individual was found who also had average values for all 10 clothing dimensions. We recently conducted a similar analysis on male and female subsamples taken from the ANSUR survey [10]. To reduce overall variability, samples of 2074 white males and 1438 white females were used in the analysis. The results of the analysis are shown in Table 2.

As can be seen in Table 2, no men remained in the sample after screening for eight dimensions, and only one woman of the 309 with average Stature was average for all 10 dimensions measured. The rate of dropout would vary with a different list of dimensions, of course, but in all cases a very small group of individuals would remain after a few rounds of screening. As can be seen from Table 2, for example, only 5–6% of the original subjects remained after only the second round of selection. The "average person" is a statistical concept; such a person does not in fact exist.

A second misconception involves use of percentile values in designs. Probably the most frequently specified design limits are the 5th and 95th percentiles. These percentiles are often used to designate the smallest and largest individuals for whom an item or workspace will

Table 2 The "Average"[a] Man and Woman (values in mm)

Dimension	Men			Women		
	Mean ± 0.3 SD	n[b]	Percent[c]	Mean ± 0.3 SD	n[b]	Percent[c]
Stature	1745–1784	486	23.4	1614–1652	309	21.5
Chest Circumference	978–1018	115	5.5	891–930	86	6.0
Sleeve Length	876–898	60	2.9	789–809	35	2.4
Crotch Height	823–849	28	1.4	749–773	21	1.5
Vert. Trunk Circ.	1624–1668	18	0.9	1527–1566	14	1.0
Hip Circumference	968–1004	10	0.5	951–987	6	0.4
Neck Circumference	374–385	3	<0.1	310–319	3	0.2
Waist Circumference	849–899	1	0.1	771–821	2	0.1
Thigh Circumference	580–608	0	0.0	563–590	2	0.1
Crotch Length	632–656	0	0.0	603–636	1	<0.1

[a]Average defined as the mean ± 0.3 SD.
[b]Number of original sample remaining.
[c]Percentage of original sample remaining.
Source: Ref. 10.

be designed, but just as with "average," a person with a 5th percentile value for one dimension will be 5th percentile for very few others. Thus, if a group of 5th percentile (or 95th percentile) dimensions are used in a design, one can be quite sure that only a very few, if any, individuals will be totally accommodated.

Another pitfall encountered in the use of percentiles for design purposes is that, except for the 50th percentile, they are not additive. This was demonstrated by McConville and E. Churchill [11] and by Robinette and T. Churchill [12], who examined the effect of combining percentile body segments. When seven 5th and 95th percentile height segments were each added together, the total was significantly below the 5th percentile Stature and well above the 95th percentile Stature, respectively.

A similar analysis of the more recent ANSUR data reveals similar results [10]. As can be seen in Table 3, the segmental totals are over 9 cm (approximately 3.5 in.) different from the direct percentile value. The segments are depicted by number in Figure 5. Certainly the cumulative error is significant, even for the most lenient design requirements. As with the earlier example, the error is negative (sum of segments is less than directly computed percentiles) at the 5th percentile level and equivalently positive at the 95th percentile level. Although the cumulative errors are large, the error associated with each segment may result in a poor design when minimum or maximum values used in conjunction with preferred clearances are mixed.

E. Sources of Data

Most engineers, ergonomists, and designers have a favorite source for the anthropometric data they use in solving design problems. The large number of textbooks, handbooks, and data collections that include some anthropometric data on American adults would lead one to assume that there are ample data to solve almost any design or sizing problem. Unfortunately, this is far from the truth. For the most part, available data are simple univariate summary statistics that suffer from the drawbacks discussed above. Often, the authors say little about

Anthropometry

Table 3 5th and 95th Percentile Models: Males and Females[a] (values in cm)

Segment	5th percentile		95th percentile	
	Male	Female	Male	Female
Floor to Lateral Malleolus	5.84	5.23	7.64	6.97
Lateral Malleolus to Lateral Femoral Epicondyle	39.63	36.28	47.62	44.05
Lateral Femoral Epicondyle to Greater Trochanter	38.67	36.27	46.90	43.97
Greater Trochanter to Iliac Crest	11.97	10.22	17.08	15.24
Iliac Crest to Tenth Rib	2.39	3.22	7.52	8.10
Tenth Rib to Seventh Cervical	36.21	33.02	43.43	39.79
Seventh Cervical to Tragus of Ear	8.67	8.04	12.35	11.56
Tragus of Ear to Top of Head	12.18	11.42	14.04	13.25
Sum for Segments	155.56	143.70	196.58	182.93
ANSUR Stature	164.69	152.78	186.65	173.73
Difference	−9.13	−9.08	+9.93	+9.20

[a]The data presented here were derived from ANSUR [1] and originally presented in Ref. 10. N for the females = 2208, N for the males = 1774.

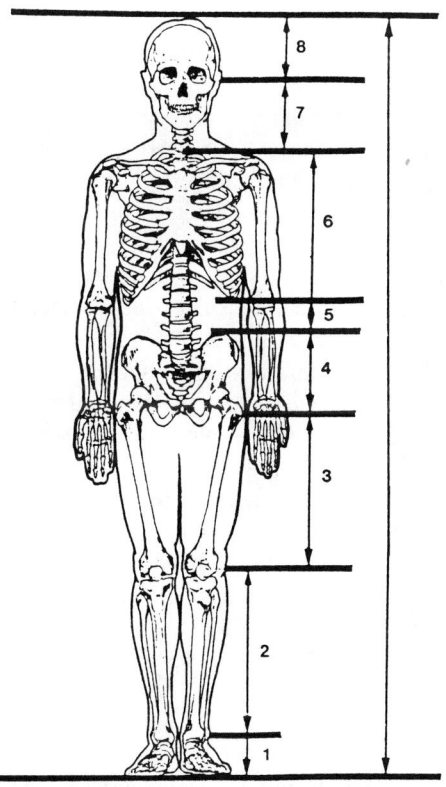

Figure 5 Segmental heights used in Table 3.

the source of the data, offering little description of the makeup and size of the sample represented.

The lack of a current and comprehensive database representing the U.S. civilian population is a serious shortcoming. So far as we know, only one nationwide survey of U.S. civilian men and women is available that includes applied workspace dimensions, and its data are now over 30 years old.

Fortunately, the status of anthropometry on U.S. military personnel is much better. In fact, data tables presented in many textbooks and handbooks in this country were compiled from U.S. military samples. Data that appear in popular texts authored by Europeans are also occasionally based on U.S. military data [13,14]. For the immediate future at least, users may be forced to employ military data for civilian problem solving.

1. Civilian Data

In 1939–1940 more than 10,000 women and more than 100,000 children were measured in major surveys conducted by the Department of Agriculture to obtain dimensional data relevant to clothing sizing and pattern design [15, 16]. These surveys were well planned and executed, with excellent data analysis and reporting, but they are clearly dated for purposes of describing the current civilian population. In addition, few of the clothing dimensions selected for measurement in the surveys are of relevance to ergonomists. The data from these surveys formed the basis of a series of recommended sizing standards for women's and children's clothing. These standards are now considered defunct. No such clothing survey has ever been conducted for adult civilian males in the United States.

The U.S. Department of Health, Education and Welfare initiated a cycle of national health surveys (HES) in 1960 [17]. The first of the series included 18 anthropometric measurements on a nationwide sample of 3091 men and 3581 women between the ages of 18 and 79. The measurements included 12 ergonomically useful dimensions, such as Knee Height, Sitting Height, and Buttock-Knee Length that are relevant in seated workspace layouts. Although these surveys continue to be carried out at 10-year intervals, the most recent anthropometric data collected include only Stature and Weight plus a series of skinfolds, girths, and bony breadths to be used for nutritional assessment. The currently ongoing health and nutrition (examination) survey (HANES) is the third in the series (HANES III).

There have been quite a number of small specialized anthropometric surveys on U.S. civilians. Some results are proprietary and thus not widely available. Nonproprietary data may be found in journals and occasionally in books, but because the samples are small they are typically composed of some selected subset of the population. One example can be found in the excellent two-volume set on work ergonomics produced by the Human Factors group at Kodak [18]. Some of the anthropometry presented was collected on Kodak employees. Care should be exercised in using such data, however, because local population demographics or employee selection processes may be reflected in them. For example, selection criteria can have an effect on the correlation coefficient (r) of Stature and Weight. The r value for a sample of airline stewardesses measured in 1971 [19] is 0.729, whereas for U.S. Army women measured in 1988 [1], $r = 0.529$, and for U.S. civilian women as measured in the HES [17], $r = 0.205$. The data show that although both stewardesses and Army women are subsets of the general female civilian population, they are clearly selectively sized subsets of that population.

2. Military Data

The anthropometric database for U.S. military populations is far more current and comprehensive than that for the U.S. civilian population. Since the end of World War II, each of the military services has conducted one or more major body size surveys of their populations. Such surveys are designed to obtain body size information for a wide variety of needs—sizing and design of clothing and personal protective equipment, workspace layout, and modeling, among others. They usually include some 70–150 body measurements made on relatively large samples of 500 or more. Thus, they are an excellent source of anthropometric design data for the ergonomist if they are used with knowledge of their limitations.

The chief problem associated with the use of military data in civilian applications is fairly obvious. Military samples are truncated at the ends of the distribution for most variables, and very few military personnel are over 50 years of age. Military selection criteria limit Stature to approximately 155–203 cm for men and 145–203 cm for women. And, because of existing requirements for fitness levels in the military as well as height and weight limits, the physique of military personnel reflects less body fat and a greater lean body mass on the average than will be found in the civilian population. The pros and cons of applying military anthropometry to the design problems of a civilian workforce have been discussed by a number of authors [20,21]. Despite the potential problems, military data are often the only source for body measurements of interest.

The majority of the military data have been published as technical reports prepared by the various sponsoring agencies and are available through the National Technical Information Service (NTIS). These reports vary in completeness from a presentation of simple univariate statistical summaries accompanied by a description of the survey methodology and measurement techniques used [22] to detailed data analyses that supplement summary statistics with correlation matrices, simple and multiple regressions, selected bivariate frequency tables, and subset data analysis [5]. ANSUR, the most recent military survey [1], is modeled on the Clauser et al. 1972 report [5] and is reasonably comprehensive. It is a particularly useful survey in that both Army men and women were measured in the same survey, which was conducted by a measurement team trained to perform the measurements in as nearly as possible exactly the same way on both sexes. In addition, considerable effort was put into the sampling strategy so that special databases could be constructed to match the demographics (age and racial/ethnic proportions) of the U.S. Army for some time to come [23].

In the 1970s, raw data tapes of several military surveys and a few civilian surveys that were suitable for mainframe computers [24] were made available through NTIS to interested parties so that researchers could obtain the complete data sets to conduct their own analysis. Under contract to the U.S. Air Force [25,26] and through arrangement with other agencies, Anthropology Research Project, Inc. (ARP) has collected raw data files in its data bank for some 50 surveys—military and civilian, American and foreign, male and female. A summary listing of these surveys by name, number of subjects, and number of variables is provided in Table A.1 of the Appendix: Ergonomic Tables and Figures. This database and associated analysis software are available to support a variety of ergonomic design functions including sizing of personal items and workstation design.

Hundreds of different anthropometric dimensions have been measured over the years. Of these, nearly 300 are listed in the *NASA Anthropometric Source Book,* Volume II [27]. This data book provides the mean, standard deviation, coefficient of variation, and percentile breakouts for 295 traditional static dimensions. The data derive from 61 foreign and domestic military and civilian surveys of males and females. Although somewhat outdated, this work is still one of the most comprehensive presentations of anthropometric data available.

In sum, there is a fairly impressive array of anthropometric data available on the U.S. military and a rather more limited amount on civilians. Because data will not be available for every sample of interest or for every dimension required, the user must be knowledgeable about the strengths and limitations inherent in the available data and use them appropriately.

3. Future Needs

At no time in history has the workforce been so diverse, and going on into the 21st century we expect that trend to continue. As diversity due to age, sex, race, and ethnicity increases, so too does the complexity of industrial and military design problems. A given job may be performed in the morning by an athletic 20-year-old man, 6 ft. 8 in. tall, weighing 220 lb and in the afternoon by a 50-year-old woman, 5 ft tall, weighing 100 lb. Such problems already confront today's ergonomists; in the near future they will become routine.

Fortunately, the advent of scanning techniques combined with computer graphics technology has expanded our ability to rapidly describe the three-dimensional shape of the body in digital form. So far as is known, a full body surface scanner is not yet available. However, a system that uses a low-power He-Ne laser combined with a color video camera to digitize over 150,000 coordinates in 3-D space in about 17 sec is in use in our laboratory at the present time. This scanner is capable of digitizing a volume as large as 14 in. on a side. Currently, scan data on the heads of over 1000 children and youths are being analyzed. Ultimately such instrumentation will enable the derivation of custom dimensions upon demand, both simple point-to-point descriptors and sufficient 3-D coordinates to define the human shape.

F. Applications for Anthropometry

To ensure that the 1988 ANSUR survey would include dimensions that could be used for a variety of purposes for some time to come, over 360 dimensions that had been previously measured in 34 domestic and foreign surveys were examined for their potential utility [28]. Thirteen categories of measurements based upon area of application were identified:

Basic body size descriptors
Key dimensions for specifying and directing the sizing and design of personal items
Clothing and personal equipment pattern and/or construction dimensions
Dimensions used to develop manikins or test dummies
Dimensions useful for the design and sizing of load-carrying equipment
Head and face dimensions used primarily in the design of headgear, optical, and auditory devices
Hand, finger, wrist, and forearm measurements used in the design and sizing of gloves and in the construction of hand forms
Foot and ankle dimensions needed for the design and sizing of shoes and boots
Dimensions central to the design and layout of single- and multiperson workstations
Dimensions used to design and specify anthropometrically compatible aircraft and other vehicles
Dimensions used for developing the link or skeletal system for most three-dimensional kinematic anthropomorphic analogs
Dimensions useful for the development of 3-D anthropomorphic dummies, 3-D computer models of humans, and evaluation of workstations
Dimensions recommended for use in standards

These categories are not mutually exclusive, and in many instances dimensions appear in two or more categories.

Only 30–40 dimensions common to workspace or workstation design are likely to be of interest to the practicing ergonomist. Approximately this number of traditional dimensions have been selected, principally from the HES [17] and the 1988 ANSUR [1] surveys, for discussion in this chapter. These dimensions are illustrated in the Appendix in Figures A.1–

A.4. For detailed descriptions of the measurement procedures used in the two surveys, the reader is referred to the referenced publications.

IV. ANTHROPOMETRIC CRITERIA FOR ERGONOMIC APPLICATION

Three major factors related to body size variability that the ergonomist must consider in specifying the dimensions of a workstation or design are sex, age, and race or ethnicity. It is likely that in many cases all three areas of concern will apply to the problem at hand. To examine the range of anthropometric variability within these subgroups as briefly and efficiently as possible, the presentation here is restricted principally to those dimensions for adults most relevant to workspace sizing and design.

A. General Body Size Variability

Probably the most frequently used body size descriptors are Stature and Weight. Not only are they more readily available than most other body size data, but, taken together, these two variables have a higher correlation with a larger number of dimensions than any other pair of measurements [15]. The range of variability in Stature and Weight for both sexes is shown in the Appendix in Figures A.5 [1,17,22,29-31] and A.6 [1,5,17,19,29,32] for a number of military and civilian anthropometric surveys going back to the 1960s.

When military and civilian samples such as those presented in Figures A.5 and A.6 are examined for the range of the 1st to 99th percentile values in Stature and Weight, the differences are surprisingly small. A notable exception is the much wider distribution of civilian women's Weight values, which reflect the many heavier women found in the civilian population. The data also reveal the selective hiring requirements for such special populations as female flight attendants [19] and law enforcement officers [30]. As expected, the distributions in the nationwide civilian surveys, i.e., HANES [29] and HES [17], are wider than in the smaller surveys. There is no clear evidence for secular change (people getting larger over time) in either variable when data from the older surveys are compared to those in the more recent ANSUR survey [1]. Evidence of body size difference for other dimensions will be found in association with much of the discussion to follow.

B. Sexual Variation

Women today occupy many of the workplaces that had been until recently the sole province of men. This phenomenon presents a wide range of new challenges to designers and ergonomists, who must now be concerned about a wider range of users. Size and shape differences between the sexes have been, until recently, poorly understood or ignored in design. Historically, the assumption has been that the 50th percentile female is essentially equivalent to the 5th percentile male. Although this is a reasonable assumption for height and weight, it does not necessarily hold true for many other variables. A study conducted by Robinette et al. [33] demonstrated that women are not simply scaled-down versions of men. A more recent analysis of ANSUR [1] data, performed at ARP, compared differences in the mean values for 130 dimensions for samples that were closely matched in stature and weight (and age). The data for 30 dimensions selected to cover the range of differences observed in the analysis are presented in Table A.2 in the Appendix. The female mean as a percentage of the male mean was found to range from a low of 86.8% (Neck Circumference) to a high of 110.2% (Hip

Breadth). Basically the data reflect differences in proportion caused by male muscularity and female pelvic structure as well as variations in the amount and location of body fat. ANSUR data are particularly useful for this type of analysis, because both sexes were measured within the same time period by the same measuring team and using identical techniques. Table A.3 in the Appendix shows a comparison of female and male proportions for civilians taken from the HES survey [17]. Like the military samples of Table A.2, the two samples were matched on the basis of ±2.5 cm in Stature and ±5 kg in Weight. When men and women are nearly the same in height and weight, women are slightly larger on the average for a number of important workspace dimensions, particularly hip-related dimensions that are important in seated workspace/workstation design.

C. Age Variation

It is common knowledge that adult body size and shape alters with aging, but there are few anthropometric data to help the ergonomist quantify these changes [34]. No large sample of civilians has been specifically measured to track this phenomenon. Ideally, such a study should be longitudinal—that is, measurements should be repeated periodically on the same individuals over their adult lifetime. Some short-term longitudinal data that include dimensions that would be useful to the ergonomist have been collected [35–39], usually over 5–10-year periods. The largest, longest ongoing longitudinal study, covering over 60 years, is the Fels Institute study of human growth [40], and unfortunately it does not include applied dimensions.

By far the most anthropometric data available in the United States are cross-sectional, that is, measured on a sample of subjects at one point in time. Although not an ideal source, cross-sectional data from a large random sample containing subjects of a wide range of ages can nevertheless reflect differences in body size and shape between persons of different ages in the population at one point in time. These are the people the ergonomist/designer must fit into workspaces.

Such data may be examined by breaking out groups of subjects from contiguous age brackets. This approach is illustrated in the Appendix in Tables A.4 and A.5, which give the data for men and women from the HES survey [17] ranging from less than 24 to more than 75 years of age. The data were analyzed in terms of changes in the mean values observed for six age brackets referenced to the youngest group, those of ≤ 24 years of age. The greatest change observed for either sex occurs in the apparent loss in Stature and associated components of standing body height. Also of some significance is the increase in Seat Breadth and Elbow-Elbow Breadth, particularly among women. The rather large increase in the distance between women's elbows with age is believed to reflect the change occurring in the architecture of the shoulder girdle, e.g., a forward rotation of the glenoid cavity of the scapula.

Military data are basically limited in demonstrating age-related changes in body size, because the oldest individuals on active duty are typically in their early fifties. ANSUR [1] data were examined for the same basic dimensions listed in Tables A.4 and A.5 for the HES samples. Briefly, soldiers show the same increases in mass-related dimensions through their mid-to-late forties. Changes in other applied dimensions prior to the beginning of the sixth decade appeared not to be significant for design purposes.

D. Racial/Ethnic Variation

A major component in the diversity to be found in the workplace of the 21st century will be due to racial/ethnic variation. Anthropologists have never fully agreed on a universal definition of race, but it is herein reserved to identify the major classic subdivisions: white, black, and Asian. Ethnic difference or ethnicity refers to those populations that can be identified in association with a geographic or national boundary or a social or religious population comprising individuals belonging to one of the races or to an admixture of races. In the United States, for example, Hispanics constitute a large and growing population that, strictly speaking, is an ethnic group. Depending on the demographic distribution of racial/ethnic groups sampled in a given survey, the label "Other" has traditionally been used to designate individuals in a total sample who have not been sampled in sufficient numbers to make up a statistically useful subsample. Considering the major surveys used in the chapter, the "Other" group for the HES database [17] is believed to be principally composed of Hispanics, Asian Americans, and Native Americans, whereas the "Other" subsample from ANSUR [1] is known to include mainly the latter two groups.

The means and standard deviations for the dimensions measured in the HES survey were broken out by the available racial/ethnic groups and are presented in the Appendix in Tables A.6 and A.7 for men and women, respectively. The same values for similar dimensions measured on the men and women of the ANSUR survey are given in Tables A.8 and A.9. Although comparisons between the two databases, even for dimensions with the same name, should be viewed with caution because differences in measuring techniques may confound the results, the data reveal some instructive points. Black/white mean values for Buttock-Knee Length, for example, differ by 1–1.5 cm for men and 2 cm for women for both civilian and military personnel. For both of these groups, the difference in Sitting Height as a proportion of Stature can be explained by the length of the legs. In short, African Americans of a given stature have longer legs than white Americans. Hispanic values given with the ANSUR data in Tables A.8 and A.9 reflect overall smaller body sizes than either the black or white sample. These and other findings of racial/ethnic body size variation in the U.S. population should be taken into consideration in plotting effective designs for both civilian and military workspaces.

E. Specialized Data

Although traditional anthropometry has many uses, its applications often baffle engineers and ergonomists faced with real-life design problems. Traditional anthropometry had its origins in the laboratory, where lightly clothed subjects stand or sit immobile in the strictly controlled postures required to obtain reliable measurement results. In the workplace, people rarely stand with their heads in the Frankfort plane[2] or sit at display consoles with hips and knees at perfect 90° angles. They slump, stoop, or stretch to reach controls. They may be required to wear bulky clothes or protective headgear on the job, which further complicates the problems. For these and other reasons the engineer/designer may not find the anthropometry needed to solve a particular problem.

Some data for a number of specialized measurements of use to the ergonomist or engineer are available, particularly for military personnel. The following sections provide some data and suggested sources of information that can supplement the more traditional dimensional variables.

---

[2]The Frankfort plane is an arbitrary reference position of the head used for anthropometry in which a landmark just above the tragus of the right ear (tragion) is horizontally aligned with a landmark on the lowest border of both right and left orbital ridges (infraorbitale).

1. Clothed Anthropometry

Commonly, anthropometric dimensions are measured on the bare skin or over minimal clothing. Most ergonomists and engineers are concerned with dimensions on individuals arrayed in possible combinations of clothing. Many jobs require, at least occasionally, that the worker wear restrictive multilayered clothing. The effect of such heavy clothing on the ability to perform required tasks was probably first recognized by the military. As a result, almost all available clothed anthropometric data were obtained from military ensembles. The first data of this type were published in 1946 [41]. Summaries of subsequent studies may be found in a variety of handbooks and government technical reports, but most of the studies were conducted on fairly small samples of subjects.

Results of a typical clothed anthropometry study are summarized in the Appendix in Table A.10 [42]. The four ensembles tested in this study are listed at the bottom of the table. Ensemble 1 (ground soldier—hot weather) may be similar to normal clothing worn in the work environment, whereas Ensemble 4 (extremely cold weather, chemical protection added) is something like what the handler of hazardous material might wear. In general, body size can be seen to increase while overhead reach and mobility decrease from Ensemble 1 to Ensemble 4, as chemical protective clothing and/or cold weather layers were added. Forearm-Forearm Breadth and Hip Breadth, for example, increase by two-thirds, while overhead reach capability is reduced by about 20% in Ensemble 4. As can be seen, increases in body depths and breadths can be quite large, and mobility is substantially reduced in overhead reach. For example, Forearm-Forearm Breadth and Hip Breadth, Sitting were increased by approximately 66.7%, and the overhead reaching capability was reduced by about 20% in Ensemble 4.

Anthropology Research Project, Inc. recently completed a pilot study of five different U.S. Army ensembles on a sample of five civilian male and five civilian female subjects [43]. The data are currently being analyzed; however, indications are that the range of variability is not substantially different from that presented in Table A.10, except that the replicability of the measurements was found to be very labile even under strict controls for many measurements. A companion study [44] on the effects of the same ensembles on more than 36 planar joint motions was undertaken at the same time. Although different subjects were used in the studies, the data provide evidence of the combined effects of complex clothing ensembles on body size and mobility. Although the number of subjects was small in both studies, the data were collected under what were perhaps the best documented and best controlled studies of this type in recent times. The full technical reports of these studies should be available through the National Technical Information Service (NTIS) in 1995.

2. Working Postures

Recognizing the limitations of measurement data collected from subjects sitting and standing in stylized postures, anthropologists at the Aeromedical Laboratory at Wright-Patterson Air Force Base in the 1950s collected data on men positioned in a few common working postures [45]. Investigators from the same laboratory later repeated and expanded upon the earlier study [46]. Thirteen of 26 dimensions were measured, and the remainder were derived photogrammetrically. All were related to common working postures. Data for the 13 directly measured dimensions and the age, Stature, and Weight of the sample are summarized in Table A.11 in the Appendix , and the associated measurements are illustrated in Figures A.7a and A.7b.

Using the mean Stature as a reference, interesting relationships can be computed. For example, a man on his knees with his torso erect may require overhead clearance averaging more than 74% of his Stature. A standing man should be able to grasp an object that is at a height equal to approximately 122% of his Stature. An average-sized man lying on his back

Anthropometry

may be able to grip an object about 750 mm (29.5 in.) above the floor or at a distance of about 42% of his Stature. The same man lying on the floor will have a Horizontal Length about 25 mm (1 in.) more than his Stature.

These data are presented principally because of their unusual nature and because no comparable data have been collected in the intervening years. The measurements should be repeated and expanded upon using a current workforce population.

3. Reach

Traditional anthropometry provides little help for the ergonomist who must position equipment or controls within the reach of a worker. Like many of the special applied anthropometric measurements, interest in reach and reach envelope measurements was spurred by problems encountered in designing complex aircraft cockpits. One of the first to investigate reach envelopes was Dempster [47]. Some 20 years later, Kennedy [48] conducted a three-dimensional reach study on a relatively small sample of civilian men ($n = 20$) and women ($n = 30$) who were believed to be representative of the USAF population. Examples of these data are shown in Figures 6 and 7.

The measurements were made on seated subjects in seats designed to simulate those found in an aircraft cockpit. The seat was unpadded, with a pan angle of 6° and a back angle of 103°. Investigators used a special reach apparatus that permitted reaches to be controlled in defined positions over most of the sphere that encompasses the seated worker. The reach task was to grasp a small knob between the thumb and index finger and push it away along the test azimuth until the arm was extended to the fullest extent possible with the shoulders still in contact with the seat back. The push rod could be positioned at 15° intervals relative to the seat reference point (SRP). The illustrations at top left and right in Figures 6 and 7 give the 5th, 50th, and 95th percentile envelopes in the midcoronal (0 cm, *YZ*) and midsaggital (0 cm, *XZ*) planes. The illustrations provide the envelopes at 46 cm in front and behind the SRP (*YZ*) and 46 cm to the right and left of the SRP (*XZ*). The effect of arm position on reach is clearly demonstrated.

Many additional reach envelopes, including horizontal views, are provided in this and other publications [49–53]. In many cases the data are from the same source, and they are almost entirely military.

More traditional two-dimensional reach data obtained from ANSUR men and women [1] are provided in Table A.12 in the Appendix. The values for some of the reaches were derived mathematically from other measurements completed in the survey. The reaches are illustrated in Figures A.3 and A.4. In general, the female reaches average more than 90% of the male values, with Functional Leg Length, Wrist-Thumbtip Length, and Wrist-Center of Grip having the nearest equivalence to male values. Men retain about 68% of their overhead reach when seated, compared to 64% in women. These and other evidence of sexually dimorphic traits are reflected in the reach data.

4. Linkage and Computer Models

Drafting board manikins and the majority of current computer man-models are based on the concept of the anthropometric links outlined by Braune and Fischer [54] in their classic biomechanical study. This concept was later refined and expanded by Dempster [47], who defined the link as a straight-line distance between adjacent centers of joint rotation on a segment of the body. Although it has not been possible to measure the internal links as defined by Dempster, it is possible to measure a series of surface dimensions that approximate, to varying degrees, the internal link lengths. These anthropometric link lengths are summarized for the ANSUR [1] males and females in the Appendix in Table A.13 and illustrated in

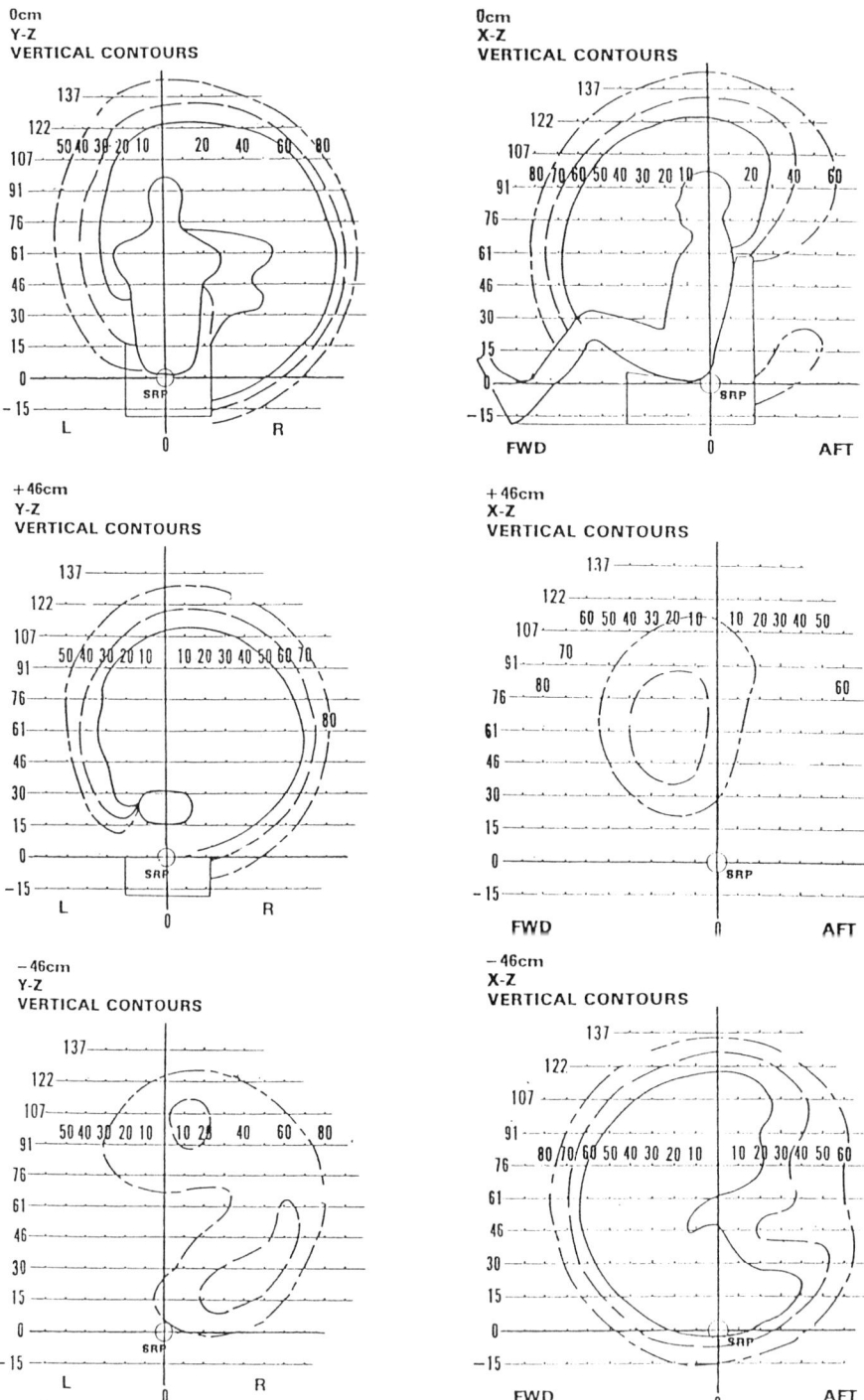

Figure 6 Selected vertical reach envelopes for seated men in the coronal plane (*YZ*) and sagittal plane (*XZ*) relative to the seat reference point (SRP).

Anthropometry

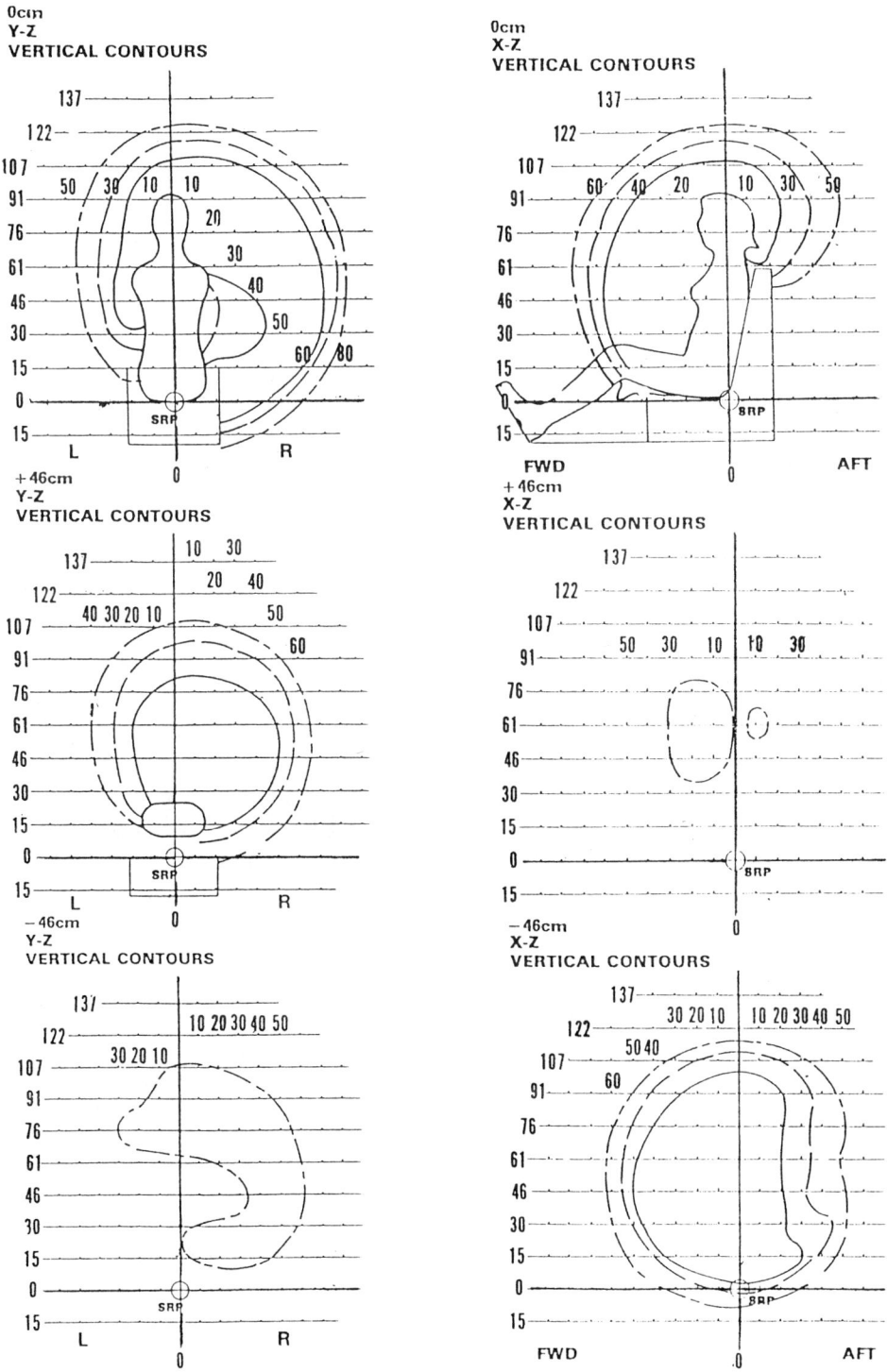

Figure 7 Selected vertical reach envelopes for seated women in the coronal plane (*YZ*) and sagittal plane (*XZ*) relative to the seat reference point (SRP).

Figure 8. The links are either directly measured, as in Biacromial Breadth and Acromiale-Radiale Length, or derived by subtraction, where Thigh Link Length, for example, is Trochanteric Height minus Lateral Femoral Epicondyle Height.

Regression equations for predicting link lengths from Stature and Weight are given in Table A.14, again based on the ANSUR [1] survey data. It is reiterated that these anthropometric link lengths are approximations from surface landmarks and not the internal center-to-center link lengths.

F. Anthropometry and Seated Work

As automation encroaches, the number of jobs requiring standing work postures will decline and the attention of occupational ergonomists will be increasingly focused on the seated workstation. With each passing year, more workers will be seated at some form of video display terminal (VDT) or at least in front of a control panel, which may include video displays. All too frequently, seated postures assumed over extended periods lead to severe discomfort and a variety of job-related disorders. The hand, wrist, arm, and back disorders that sometimes result from either stasis or repetitive movements associated with postures used in badly fitted workstations are discussed elsewhere in this text.

A critique of the current recommended standards for VDT design [55] is beyond the scope of this chapter. Suffice it to say that the univariate statistics used as a basis for most such standards are not adequate to the task of designing workstations for an increasingly diverse population of users. Among available workspace dimensions, the importance of Popliteal Height and Buttock-Popliteal Length to the chair are clear, but the relationship of Thigh

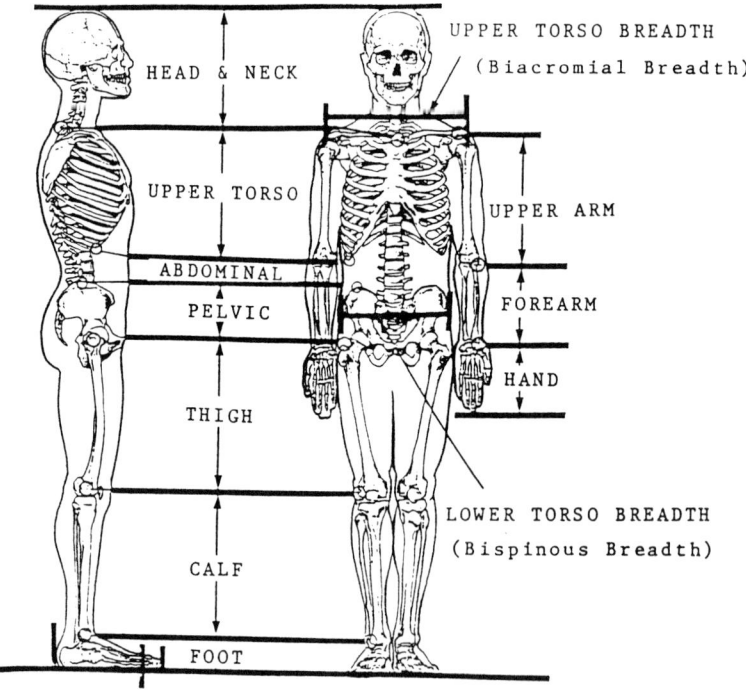

Figure 8 Links and related landmark centers on the skeleton.

Clearance Height and Elbow Rest Height to the preferred work surface height is considerably more difficult to deal with. Statistically speaking, this is because the correlation of these two dimensions with other workspace dimensions is quite poor. It is very easy to forget that the work surface has some thickness, that handbook values for Popliteal Height and Sitting Knee Height frequently do not include shoes, and that the Buttock-Knee Length does not take much clothing into account.

With the above caveats in mind, let us reexamine some concepts introduced earlier in this chapter and then look briefly at a more promising way of selecting dimensional values for seated work designs. Faulty thinking related to the "average man" concept and the cumulative error introduced by summing percentile values for body segments were discussed in Sec. III.D. Perhaps the best way to avoid these pitfalls is through the use of regression equations, which usually produce less cumulative error. Frequently one may know or can easily find statistics for Stature (or Weight) of a target population, but no data exist for the seated work dimensions needed for a given user population. As noted, the best overall pair of variables in a multiple regression equation with two independent variables are Stature and Weight. This is so because Stature relates well to many segmental heights and lengths, whereas Weight is volumetric and correlates well with complex measurements such as arcs, contours, and circumferences. Tables A.15 and A.16 in the Appendix give a series of Stature and Weight regression equations for a group of workspace dimensions selected from the ANSUR [1] and HES surveys [17], respectively. The equations are included here because they are not readily available but are invaluable in estimating reliable values for dimensions when the Stature and Weight ranges desired in a design are known. The principal errors that may be incurred through their use arise from the fact that regression equations are population-specific. That is, in this case, the value computed for a given dimension will be accurately predicted only to the extent that target populations resemble the ANSUR or HES databases. Because these two surveys represent the best available databases covering both civilian and military populations, the regressed values may still be superior to inputting a series of percentile values or taking a value for a given dimension from a table of univariate data from an unknown source. Any value within the range for Stature and Weight of either ANSUR [1] or HES [17] data can be inputted in the appropriate equation, and the standard error of the estimate (SE_{est}) can be used in the calculations much like the standard deviation is used with the mean. That is, by adding and subtracting increments of the SE_{est} to/from the predicted value, various levels of confidence limits can be established for the variable. If one wanted to establish the range of adjustability for the seat pan of a chair, the height and weight for a very small person and that for a very large person could be inputted into a regression equation for predicting Popliteal Height. For example, the lower limit could be established by a person who is 1524 mm (60 in.) tall and weights 45.4 kg (100 lb), and the upper limit might be a person who is 1980 mm (78 in.) tall and weighs 95 kg (210 lb). By inputting these values into the Popliteal Height (PH) equations (Table A.16) for HES women and men, the following results are obtained:

PH(women) = (0.301 × 1524) − (0.347 × 45.4) − 62.7 = 380.3 mm

PH(men) = (0.321 × 1980) − (0.372 × 95.0) − 88.6 = 511.6 mm

As noted above, the predicted range of 380.3–511.6 mm (15–20 in.) can now be expanded by using SE_{est} to ensure inclusion of a high percentage of the population. The value can also be adjusted to account for various heel heights. If designing for a population with characteristics of age and weight per height that are similar to those of the military, the ANSUR regressions can be used.

G. General Guidelines

The compilation of general guidelines to use when applying anthropometric data to ergonomic design problems can be dangerous. Nevertheless, the statements below offer an approach that should be of relatively universal usefulness.

1. Identify the specific dimension(s) you need for the problem at hand. If more than one major design dimension is involved, you have a multivariate problem to solve.
2. Determine the maximum error the design can accept for the dimension(s) identified. If the allowable error is ±10% or greater, or if multiple dimensions are not involved, the database selection is less critical.
3. Identify the demographics of your user population with regard to such variables as age range, sex, and race/ethnicity. Determine whether there are any special anthropometric characteristics of the target population that could affect your sizing or design options.
4. Seek out the availability of relatively recent anthropometry on the user population (the database should not be older than one generation). If satisfactory data cannot be found, and if it is not feasible to create the needed database, as is likely the case, use the best available database that includes the dimension(s) you require. For example, as mentioned earlier, the most comprehensive listing of foreign and domestic databases currently available in Volume II of the *NASA Anthropometric Source Book* [27]. Use the HES data [17] if the design cannot tolerate the shortcomings of a military database. The best current source of military data is ANSUR [1].
5. If the problem is multivariate (most are) and the allowable error is small, try to locate multiple regression equations for the selected database that permit the use of independent input variables (a) for which you know the target population range and (b) that have the best correlation with your list of design variables. Remember that regression equations are population-specific, so some error will be incurred if the user population is very different.
6. In the final analysis, if data on the required dimensions do not exist, collect data on at least 30 individuals representative of the target population. Be consistent with your procedures, record the method used, and add to the database should the design effort prove to be successful.

APPENDIX: ERGONOMIC TABLES AND FIGURES

Table A.1 Databank Holdings

Survey date	Surveyed population	Sample size	Variables
	U.S. Military Populations		
	Males		
1950	U.S. Air Force pilots	4000	146
1959	U.S. Army aviators	500	46
1964	U.S. Navy aviators	1529	98
1965	U.S. Air Force ground personnel	3869	161
1966	U.S. Army ground personnel	6682	73
1966	U.S. Navy enlisted	4095	73
1966	U.S. Marines enlisted	2008	73
1967	U.S. Air Force flyers	2420	189
1970	U.S. Army flyers	1482	88

Table A.1 Continued

Survey date	Surveyed population	Sample size	Variables
1981	U.S. Navy flyers	1087	108
1988	U.S. Army personnel	5500	225
	Females		
1946	U.S. Women's Army Corps	7563	65
1968	U.S. Air Force women	1905	139
1977	U.S. Army women	1331	151
1981	U.S. Navy flyers	351	112
1988	U.S. Army personnel	3500	225
	Foreign Military Populations		
1960	Turkish armed forces	912	151
1961	Greek armed forces	1071	151
1961	Italian armed forces	1342	151
1961	Korean military flyers	264	132
1964	Vietnamese military forces	2129	51
1967	German Air Force	1466	152
1969	Iranian military	9414	74
1970	Latin-American armed forces	1985	76
1970	Royal Air Force aircrew	2000	64
1972	Royal Air Force Head Study	500	46
1972	Royal Australian Air Force	482	18
1973	French military flyers	174	118
1974	Royal New Zealand Air Force aircrew	238	63
1974	Canadian military forces	565	33
1975	British Army Survey	1537	61
1975	English Guardsmen	100	61
1976	English Transport Corpsmen	161	61
1976	United Kingdom Gurkhas	36	61
1976	Hong Kong Chinese military	73	47
1977	Australian personnel	2945	32
1981	Israeli aircrewmen	360	63
1985	Dutch military	1010	40
1985	Canadian aircrew	519	32
	U.S. Civilian Populations		
	Adult males		
1961	Air traffic controllers	678	65
1962	Health Examination Survey (HES)	3091	18
1962/81	Matched Health Examination Survey (HES) (ages 18–65)	2761	70
1974	Law enforcement officers	2989	23
1975	Health & Nutrition Examination Survey (HANES) (ages 18–74)	6563	11
1980	Health & Nutrition Examination Survey (HANES II) (ages 18–75)	5921	13
1981	U.S. miners	270	44
1984	Hispanic Health & Nutrition Examination Survey (HHANES) (ages 18–65)	1619	21

(continued)

Table A.1 Continued

Survey date	Surveyed population	Sample size	Variables
	Adult females		
1962	Health Examination Survey (HES)	3581	18
1971	Airline stewardesses	423	73
1975	Health & Nutrition Examination Survey (HANES) (ages 18–74)	10123	11
1980	Health & Nutrition Examination Survey (HANES II) (ages 18–75)	6598	13
1981	U.S. miners	86	44
1984	Hispanic Health & Nutrition Examination Survey (HHANES) (ages 18–65)	2037	21
	Children and Youths		
	Males		
1975	Health & Nutrition Examination Survey (HANES) (ages 1–18)	3571	13
	Females		
1975	Health & Nutrition Examination Survey (HANES) (ages 1–18)	3533	13

Table A.2 Comparison of Dimensional Values of Military Men and Women Matched for Stature and Weight[a] (weight in kg, age in yrs, all others in cm)

Dimension	Men ($n = 91$) Mean	SD	Women ($n = 153$) Mean	SD	Percent (F/M × 100)
Age	25.6	0.67	25.7	0.6	—
Stature	168	1.5	167.2	1.3	—
Weight	65.3	2.67	64.7	2.8	—
Neck Circumference	36.5	1.28	31.7	1.1	86.8
Biceps Circumference	31.6	1.36	28.2	1.5	89.2
Chest Below Bust Circum.	87.2	3.15	78.4	3.1	89.9
Scye Circumference	41.9	1.59	37.9	1.4	90.5
Shoulder Circumference	111.6	3.17	103.8	3	93
Bideltoid Breadth	46.5	1.48	43.7	1.5	93.8
Axillary Arm Circumference	31.3	1.24	29.4	1.4	93.9
Bimalleolar Breadth	6.9	0.29	6.5	0.3	94.2
Chest Breadth	30	1.4	28.5	1.4	95
Waist-to-Hip Length	17.1	1.61	16.4	2	95.9
Biacromial Breadth	38.4	1.37	36.9	1.3	96.1
Radiale-Stylion Length	25.2	0.91	24.4	0.8	96.8
Sleeve Outseam	57	1.78	55.5	1.6	97.4
Shoulder-Elbow Length	35.1	1.06	34.3	1	97.7
Head Circumference	55.9	1.22	54.6	1.2	97.7
Hand Length	18.4	0.64	18.1	0.6	98.4

Table A.2 Continued

Dimension	Men (n = 91) Mean	SD	Women (n = 153) Mean	SD	Percent (F/M × 100)
Acromion-Radiale Length	32.3	0.91	31.8	1	98.5
Knee Height	47.2	1.26	46.7	1.3	98.9
Chest Height	121.3	1.89	120.3	2.3	99.2
Crotch Height	79	2.22	78.4	1.8	99.2
Cervicale Height	194.5	1.41	194.4	1.5	99.9
Acromial Height	136.8	1.8	136.8	1.8	100
Shoulder Length	14.6	1.06	14.6	1	100
Waist Height, Navel	100.3	2.27	100.7	2	100.4
Calf Circumference	35.6	1.47	35.9	1.6	100.8
Buttock Height	83.6	2.1	84.8	2	101.4
Waist Circumference	79	4.01	80.4	5.5	101.8
Waist Breadth	28.7	1.29	29.8	1.9	103.8
Buttock Circumference	91.8	2.37	99.1	3.6	108
Hip Breadth	32.2	0.93	35.5	1.5	110.2

[a]Individuals matched within ±5 kg in weight and ±2.5 cm in stature.
Source: Ref. 1.

Table A.3 Comparison of Dimensional Values of Civilian Men and Women Matched for Stature and Weight (weight in kg, age in yrs, all others in cm)

Dimension	Men (n = 198) Mean	SD	Women (n = 154) Mean	SD	Percent (F/M × 100)
Age	46.1	15.5	41.1	14.3	—
Stature	167.1	1.45	166.4	1.46	—
Weight	69.4	2.88	68.6	2.66	—
Biacromial Breadth	38.7	1.81	36.1	1.63	93.3
Buttock-Knee Length	56.9	1.90	59.1	2.04	103.9
Buttock-Popliteal Length	47.4	2.47	50.2	2.48	105.9
Chest Girth[a]	96.9	4.05	89.6	3.96	92.5
Elbow-Elbow Breadth	40.9	2.56	39.5	2.81	96.6
Elbow Rest Height	23.7	2.93	23.5	2.83	99.2
Knee Height	51.9	1.88	51.8	1.80	99.8
Popliteal Height	42.0	1.78	41.4	1.94	98.6
Right Arm Girth[a]	30.4	1.86	29.5	1.79	97.0
Seat Breadth	34.3	1.73	39.5	2.81	115.2
Sitting Height	84.2	2.50	84.5	2.56	100.4
Sitting Height Erect	88.2	2.50	87.4	2.41	99.1
Thigh Clearance Height	14.0	1.11	14.2	1.30	101.4
Waist Girth[a]	87.0	7.04	78.2	5.35	89.9

[a]Original name.
Source: Ref. 17.

Table A.4 Mean Changes in Workspace Dimensions with Aging Civilian Men (values in cm)

Dimension	Age group						
	≤24	25–34	35–44	45–54	55–64	65–74	≥75
Biacromial Breadth	40.0	+0.1	−0.1	−0.4	−0.8	−1.3	−2.6
Buttock-Knee Length	58.9	+1.0	+0.2	+0.7	−0.3	−0.6	−1.1
Buttock-Popliteal Length	49.2	+0.6	+0.0	−0.1	−0.2	−0.4	−0.8
Elbow-Elbow Breadth	40.3	+1.4	+2.4	+2.9	+2.8	+2.7	+1.7
Elbow-Rest Height	24.8	+0.1	+0.1	−0.3	−0.8	−1.7	−2.9
Knee Height	54.2	+0.6	0.0	−0.2	−0.6	−1.0	−1.8
Popliteal Height	44.3	+0.4	−0.5	−0.8	−1.0	−1.3	−2.1
Seat Breadth	34.8	+0.8	+1.0	+1.2	+0.9	+0.5	+0.1
Sitting Height	91.4	+0.6	+0.1	−0.4	−1.7	−3.0	−4.6
Stature	174.3	+1.6	−0.4	−1.1	−2.8	−4.3	−7.1
Thigh Clearance Height	14.4	+0.3	+0.1	−0.1	−0.5	−0.8	−1.1

Source: Ref. 17.

Table A.5 Mean Change of Workspace Dimensions with Aging in Civilian Women (values in cm)

Dimension	Age group						
	≤24	25–34	35–44	45–54	55–64	65–74	≥75
Biacromial Breadth	35.3	+0.3	+0.5	+0.3	0.0	−0.5	−0.8
Butt-Knee Lgth	56.6	+0.3	+0.6	+0.0	0.0	−0.2	−0.7
Buttock-Popliteal Length	47.8	+0.2	+0.2	+0.2	+0.2	0.0	−0.6
Elbow-Elbow Brdth	35.6	+1.2	+3.3	+4.5	+6.1	+6.1	4.5
Elbow-Rest Ht	22.9	+0.7	+1.0	+0.5	−0.3	−1.6	−2.1
Knee Height	50.0	0.0	0.0	−0.5	−0.7	−1.0	−0.7
Popliteal Ht	40.6	−0.7	−1.0	−1.2	−1.5	−1.7	−1.2
Seat Breadth	35.1	+1.0	+2.0	+2.2	+2.5	+2.2	+1.0
Sitting Height	85.3	+0.3	+0.3	−0.5	−1.5	−3.8	−4.8
Stature	162.1	−0.3	−0.8	−2.3	−3.6	−5.9	−6.9
Thigh Clrnc Ht	13.5	+0.2	+0.5	+0.5	+0.2	0.0	−0.3

Source: Ref. 17.

Table A.6 Selected Workspace Dimensions for U.S. Civilian Men by Race/Ethnicity (weight in kg, all other in cm)

Dimension	White (n = 2669)		Black (n = 358)		Other (n = 64)	
	Mean	SD	Mean	SD	Mean	SD
Biacromial Breadth	39.6	2.10	39.9	2.17	39.5	2.45
Buttock-Knee Length	59.1	2.89	60.1	2.87	56.5	2.44
Buttock-Popliteal Length	49.2	3.02	50.8	2.93	47.7	3.44
Chest Girth[a]	99.7	8.19	96.3	8.76	96.7	7.75
Elbow-Elbow Breadth	42.2	4.62	40.5	4.54	40.5	4.64
Elbow-Rest Height	24.4	2.83	21.7	3.05	25.2	3.12
Knee Height	54.1	2.86	55.1	2.99	52.1	2.27
Popliteal Height	43.9	2.64	45.0	2.62	42.7	24.4
Right Arm Girth[a]	30.7	3.18	31.1	3.56	28.9	3.03
Seat Breadth	35.6	2.74	34.0	2.79	34.3	2.40

Table A.6 Continued

Dimension	White (n = 2669)		Black (n = 358)		Other (n = 64)	
	Mean	SD	Mean	SD	Mean	SD
Sitting Height	86.8	3.60	84.0	3.30	84.5	3.39
Sitting Height, Erect	90.9	3.54	87.7	3.44	88.7	3.23
Stature	173.5	6.93	172.2	6.46	168.7	5.30
Thigh Clearance Ht	14.3	1.68	14.3	1.77	13.5	1.50
Waist Girth[a]	89.2	11.39	85.4	11.69	86.5	10.62
Weight	75.3	12.54	73.1	13.20	69.0	10.35

[a]Original name.
Source: Ref. 17.

Table A.7 Selected Workspace Dimensions of U.S. Civilian Women by Race/Ethnicity (weight in kg, all others in cm)

Dimension	White (n = 3051)		Black (n = 469)		Other (n = 62)	
	Mean	SD	Mean	SD	Mean	SD
Biacromial Breadth	35.3	1.88	36.3	1.94	34.8	2.47
Buttock-Knee Length	56.6	2.95	58.6	3.43	54.9	2.99
Buttock-Popliteal Length	47.7	2.91	49.7	3.40	46.6	3.37
Chest Girth[a]	87.8	7.84	89.5	9.47	90.9	9.00
Elbow-Elbow Breadth	38.8	5.23	39.7	5.99	39.8	5.61
Elbow-Rest Height	23.4	2.82	21.2	2.79	22.4	3.60
Knee Height	49.5	2.63	50.8	2.95	48.8	2.50
Popliteal Height	39.6	2.52	40.8	2.89	39.0	2.00
Right Arm Girth[a]	28.5	4.06	29.7	5.08	28.4	4.24
Seat Breadth	36.7	3.63	36.5	4.30	36.0	3.85
Sitting Height	82.2	3.72	79.4	3.66	79.7	3.42
Sitting Ht, Erect	85.1	3.51	82.6	3.62	82.5	3.21
Stature	160.4	6.51	160.3	6.92	156.5	5.67
Thigh Clearance Ht	13.7	1.77	14.1	2.18	12.8	1.76
Waist Girth[a]	76.0	11.70	80.4	13.48	81.5	13.05
Weight	63.1	13.15	67.9	16.95	62.0	13.91

[a]Original name.
Source: Ref. 17.

Table A.8 Selected Workspace Dimensions for Military Men by Race/Ethnicity (values in cm)

Dimension	Black (n = 1465)		White (n = 1979)		Hispanic (n = 1075)		Other (n = 985)	
	Mean	SD	Mean	SD	Mean	SD	Mean	SD
Acromial Height, Sitting	57.8	2.82	60.7	2.73	58.8	2.58	59.2	2.80
Bideltoid Breadth	49.2	2.62	49.2	2.52	48.8	2.46	48.9	2.76
Buttock-Knee Length	63.0	2.97	61.5	2.76	59.7	2.72	59.7	3.21
Buttock-Popliteal Length	51.3	2.63	49.9	2.48	48.3	2.42	48.3	2.82

(*continued*)

Table A.8 Continued

Dimension	Black (n = 1465)		White (n = 1979)		Hispanic (n = 1075)		Other (n = 985)	
	Mean	SD	Mean	SD	Mean	SD	Mean	SD
Elbow Rest Height	20.7	2.48	24.0	2.34	23.2	2.31	23.6	2.49
Eye Height, Sitting	77.0	3.23	80.4	3.15	77.4	3.06	78.3	3.21
Forearm-Forearm Breadth	54.1	4.33	54.9	4.20	54.5	4.26	54.0	4.48
Hand Circumference	21.6	1.00	21.4	0.91	20.9	0.91	21.0	1.04
Hip Breadth, Sitting	36.0	2.64	37.0	2.41	36.3	2.40	36.1	2.53
Knee Height, Sitting	57.0	2.76	55.8	2.65	54.3	2.56	54.1	2.98
Popliteal Height	44.4	2.54	43.3	2.36	42.0	2.33	41.8	2.67
Sitting Height	89.0	3.34	92.6	3.22	89.6	3.15	90.6	3.35
Span	186.4	8.31	181.7	7.56	177.3	7.65	177.8	8.60
Stature	175.6	6.79	176.5	6.50	170.9	6.40	171.7	7.07
Thigh Clearance Height	17.1	1.34	16.7	1.17	16.6	1.20	16.6	1.31
Waist Depth	22.1	2.45	22.7	2.55	22.5	2.56	22.1	2.55
Waist Height, Sitting	22.8	1.51	23.9	1.47	22.9	1.47	23.3	1.46

Source: Ref. 1.

Table A.9 Selected Workspace Dimensions for Military Women by Race/Ethnicity (values in cm)

Dimension	Black (n = 1360)		White (n = 1387)		Hispanic (n = 337)		Others (n =403)	
	Mean	SD	Mean	SD	Mean	SD	Mean	SD
Acromial Height, Sitting	54.3	2.64	56.7	2.61	54.7	2.42	55.1	2.85
Bideltoid Breadth	43.4	2.27	43.1	2.24	43.1	2.07	43.4	2.42
Buttock-Knee Length	60.1	2.79	58.1	2.70	56.9	2.66	57.1	3.36
Buttock-Popliteal Length	49.3	2.53	47.4	2.41	46.4	2.39	46.6	3.00
Elbow Rest Height	20.4	2.34	23.4	2.19	22.3	2.11	22.5	2.48
Eye Height, Sitting	72.4	2.97	75.1	3.10	72.6	2.93	73.1	3.14
Forearm-Forearm Breadth	46.5	3.38	47.0	3.48	46.6	3.28	46.7	3.66
Hand Circumference	18.8	0.84	18.5	8.11	18.2	0.75	18.4	0.92
Hip Breadth, Sitting	38.1	2.70	38.8	2.80	38.1	2.48	37.8	2.94
Knee Height, Sitting	52.5	2.53	51.0	2.45	49.8	2.34	50.1	3.01
Popliteal Height	39.7	2.37	38.6	2.23	37.5	2.06	37.8	2.53
Sitting Height	83.6	3.08	86.6	3.21	84.0	3.01	84.7	3.29
Span	170.9	7.60	164.6	7.28	162.5	7.22	163.9	8.77
Stature	163.0	6.22	163.3	6.34	158.7	5.92	160.0	6.93
Thigh Clearance Height	16.2	1.21	15.7	11.60	15.5	1.08	15.5	1.26
Waist Depth	20.5	2.44	20.1	2.47	20.6	2.30	20.2	2.52
Waist Height, Sitting	22.1	1.42	23.3	1.37	22.4	1.40	22.4	1.44

Source: Ref. 1.

Table A.10 The Effect of Various Military Ensembles[a] on Selected Workspace Dimensions—Mean Change from Nude Value (values in cm)

Dimension	Ensemble 1	Ensemble 2	Ensemble 3	Ensemble 4
Stature	4.7	6.5	7.5	8.9
Chest Depth	6.5	9.1	12.6	15.6
Thumbtip Reach	1.7	0.8	6.9	4.3
Overhead Reach, Sitting	−6.6	−9.1	−21.1	−24.1

Table A.10 Continued

Dimension	Ensemble 1	Ensemble 2	Ensemble 3	Ensemble 4
Sitting Height	1.9	2.7	4.6	6.0
Eye Height, Sitting	−1.2	−1.6	−0.5	−1.9
Knee Height, Sitting	3.4	4.3	4.1	5.8
Popliteal Height, Sitting	−2.5	−1.7	−0.6	−1.6
Buttock-Knee Length	4.7	3.9	6.7	8.5
Buttock-Popliteal Length	3.1	1.0	4.7	2.2
Shoulder Breadth, Sitting	1.5	6.6	12.5	15.0
Forearm-Forearm Breadth, Sitting	13.0	20.2	27.0	31.7
Hip Breadth, Sitting	14.3	16.7	18.2	22.6
Head Length	9.0	9.0	12.3	15.3
Head Breadth	8.8	8.8	13.7	14.8
Foot Length	3.4	8.4	6.4	11.7
Foot Breadth	1.1	5.8	3.0	6.3
Heel Breadth	1.7	4.3	3.6	3.7
Hand Length	–	0.0	2.4	1.4
Hand Breadth	–	0.8	–	–

[a]Ensemble 1: Ground soldier—hot weather.
Ensemble 2: Ground soldier—hot weather, plus chemical protection.
Ensemble 3: Ground soldier—extremely cold weather.
Ensemble 4: Ground soldier—extremely cold weather, plus chemical protection.
Source: Ref. 42.

Table A.11 Selected Body Dimensions of Men in Common Working Positions (values in mm, all others indicated)

Dimension	Mean	SD	Range	Percentile 5th	Percentile 50th	Percentile 95th
Overhead Reach Breadth	377	17.3	340–422	345	378	404
Max. Overhd Reach Ht.	2240	85.9	1913–2410	2116	2235	2377
Overhead Reach Ht, Fist	2141	82.9	1836–2314	1996	2146	2225
Horizontal Length	1778	59.7	1582–1895	1676	1773	1877
Horizontal Length, Knee Bent	1484	53.6	1349–1621	1400	1466	1575
Arm Reach, Supine	747	32.6	630–836	671	742	818
Bent Knee Ht, Supine	502	22.9	432–559	462	500	536
Squatting Height	1117	49.3	1034–1224	1036	1107	1194
Kneeling Height	1304	45.0	1199–1392	1224	1295	1382
Bent Torso Height	1309	70.1	1120–1443	1176	1321	1420
Max. Squatting Breadth	562	54.1	447–716	478	559	653
Kneeling Leg Length	672	33.5	569–744	617	673	729
Bent Torso Breadth	448	22.4	401–518	414	445	485
Age (yr)	2.9	9.0	18–61	–	–	–
Stature	1752	57.9	1575–1859	–	–	–
Weight (kg)	73	9.6	154.9–104.8	–	–	–

Source: Ref. 46.

Table A.12 Selected Reach and Related Length Data from Military Subjects (values in cm)

Dimension	Sex	Mean	SD	Percentile				
				1st	5th	50th	95th	99th
Finger tip Reach	M	87.5	4.21	78.7	80.9	87.3	94.6	98.0
	F	80.0	3.94	71.8	73.8	79.9	86.8	89.6
Functional Grip	M	75.1	3.68	67.3	69.3	75.0	81.3	84.0
Reach	F	68.6	3.39	61.5	63.2	68.5	74.4	77.0
Functional Leg	M	108.2	5.10	96.9	100.2	108.0	116.9	120.3
Length	F	101.2	4.91	89.8	93.3	101.1	109.4	112.8
Index Finger Reach	M	86.2	4.14	77.4	79.6	86.1	93.2	96.5
	F	78.9	3.87	70.8	72.8	78.8	85.5	88.3
Overhead Fingertip	M	223.1	9.75	200.6	207.3	223.0	239.3	245.0
Reach	F	206.2	9.24	185.0	191.5	205.9	221.7	227.8
Overhead Fingertip	M	143.4	5.90	129.3	133.8	143.3	153.2	156.7
Reach, Sitting	F	132.7	5.59	119.7	123.3	132.7	141.8	145.4
Vertical Grip Reach	M	210.7	9.24	189.3	195.8	210.6	226.0	231.3
	F	134.7	8.71	174.6	180.8	194.5	209.4	215.1
Vertical Grip Reach,	M	131.0	5.45	117.8	122.1	130.9	14.01	143.2
Sitting	F	121.2	5.13	109.2	112.7	121.3	129.6	132.9
Wrist-Center of Grip	M	7.0	.49	6.0	6.2	6.9	7.8	8.2
	F	6.6	.49	5.6	5.9	6.6	7.5	7.9
Wrist-Wall Length	M	74.8	3.73	66.4	68.8	74.8	81.1	84.1
	F	67.9	3.43	60.6	62.5	67.9	73.7	76.2

Source: Ref. 1.

Table A.13 Link Lengths for Military Men and Women (values in cm)

Link	Males		Females		Derivation
	Mean	SD	Mean	SD	
Head	13.1	—	11.4	—	Vertex–Tragion
Neck	10.6	1.12	9.8	1.07	Tragion–Cervicale
Upper Torso (vertical)[a]	39.8	2.18	36.4	2.06	Cervicale–10th Rib
Upper Torso (horizontal)[b]	39.7	1.80	36.3	1.74	Biacromial Breadth
Abdominal	4.8	1.57	5.6	1.47	10th Rib–Iliocristale
Lower Torso (horizontal)	23.1	1.98	22.0	2.05	Bispinous Breadth
Pelvic	14.5	1.56	12.7	1.52	Iliocristale–Trochanterion
Thigh	42.7	2.51	4.01	2.35	Trochanterion–Lateral Femoral Epicondyle
Calf	43.5	2.45	40.0	2.36	Lateral Femoral Epicondyle–Lateral Malleolus
Foot (vertical)	6.7	0.55	6.1	0.53	Lateral Malleolus Height
Foot (horizontal)	27.0	1.31	24.4	1.22	Foot Length
Upper Arm	34.1	1.72	31.2	1.67	Acromion–Radiale
Forearm	27.0	1.57	24.3	1.55	Radiale–Stylion
Hand	19.4	0.98	18.1	0.97	Hand Length

[a]Also called Thorax Link.
[b]For Clavicle Link, divide Biacromial Breadth by 2.
Source: Ref. 1.

Table A.14 Stature (S) and Weight (W) Regression Equations for Selected Link Lengths on ANSUR Males (M) and Females (F) (weight in kg, stature in cm)

Dependent variable			Slope stature		Slope weight		Intercept	SE_{est}	R^2
Lateral Malleolus Height	(M)	=	$0.032 \times S$	+	$0.006 \times W$	+	6.2	4.8	0.218
	(F)	=	$0.033 \times S$	+	—[a]	+	6.4	4.9	0.159
Foot Length	(M)	=	$0.117 \times S$	+	$0.022 \times W$	+	46.3	91.3	0.514
	(F)	=	$0.110 \times S$	+	$0.028 \times W$	+	47.9	8.8	0.478
Ball of Foot Length	(M)	=	$0.084 \times S$	+	$0.017 \times W$	+	35.3	7.9	0.426
	(F)	=	$0.084 \times S$	+	$0.028 \times W$	+	29.0	70.2	0.463
Lateral Femoral	(M)	=	$0.342 \times S$	+	—[a]	−	99.3	13.0	0.757
Epicondyle Height	(F)	=	$0.332 \times S$	+	—[a]	−	76.8	13.1	0.720
Trochanterion Height	(M)	=	$0.648 \times S$	−	$0.034 \times W$	−	182.8	23.8	0.751
	(F)	=	$0.601 \times S$	+	—[a]	−	117.5	24.2	0.714
Illiocristale Height	(M)	=	$0.722 \times S$	−	$0.016 \times W$	−	181.8	20.1	0.847
	(F)	=	$0.694 \times S$	+	—[a]	−	141.6	19.5	0.837
Waist Height (Omph)	(M)	=	$0.736 \times S$	−	$0.040 \times W$	−	200.5	20.1	0.844
	(F)	=	$0.736 \times S$	−	$0.045 \times W$	−	188.7	19.1	0.847

(continued)

Table A.14 Continued

Dependent variable		Slope stature		Slope weight		Intercept	SE_{est}	R^2
Tenth Rib Height	(M)	$0.698 \times S$	+	$0.018 \times W$	−	118.5	18.8	0.865
	(F)	$0.714 \times S$	+	—[a]		118.7	17.6	0.869
Cervicale Height	(M)	$0.900 \times S$	+	$0.028 \times W$	−	82.8	10.6	0.972
	(F)	$0.805 \times S$	+	$0.017 \times W$	−	77.7	10.2	0.970
Suprasternale Height	(M)	$0.843 \times S$	+	$0.030 \times W$	−	64.8	11.4	0.963
	(F)	$0.849 \times S$	+	$0.023 \times W$	−	68.0	10.1	0.967
Biacromial Breadth	(M)	$0.083 \times S$	+	$0.053 \times W$	+	210.0	14.9	0.312
	(F)	$0.104 \times S$	+	$0.050 \times W$	+	162.8	14.6	0.296
Acromion-Radiale Length	(M)	$0.192 \times S$	+	$0.015 \times W$	−	8.5	10.2	0.647
	(F)	$0.195 \times S$	+	$0.012 \times W$	−	13.3	10.5	0.604
Radiale-Stylion Length	(M)	$0.156 \times S$	+	$0.008 \times W$	−	10.3	11.2	0.485
	(F)	$0.352 \times S$	+	—[a]		20.3	11.5	0.444
Hand Length	(M)	$0.084 \times S$	+	$0.012 \times W$	+	36.6	7.3	0.435
	(F)	$0.086 \times S$	+	$0.015 \times W$	+	30.6	7.4	0.416
Wrist-Index Finger Length	(M)	$0.081 \times S$	+	$0.010 \times W$	+	29.9	6.7	0.457
	(F)	$0.085 \times S$	+	$0.012 \times W$	+	23.6	6.5	0.453
Wrist-Thumbtip Length	(M)	$0.055 \times S$	+	$0.007 \times W$	+	22.7	5.3	0.373
	(F)	$0.056 \times S$	+	$0.009 \times W$	+	21.4	5.3	0.357

[a]This independent variable does not add significantly.
Source: Ref. 56.

Anthropometry

Table A.15 Stature (S) and Weight (W) Regression Equations for Selected Workspace Dimensions on ANSUR Males (M) and Females (F) Soldiers (weight in kg, stature in cm)

Dependent variable			Slope stature		Slope weight		Intercept	SE_{est}	R^2
Acromial Height, Sitting	(M)	=	$0.240 \times S$	+	$0.057 \times W$	+	132.5	21.7	0.463
	(F)	=	$0.269 \times S$	+	$0.044 \times W$	+	89.6	21.2	0.454
Bideltoid Breadth	(M)	=	$-0.039 \times S$	+	$2.009 \times W$	+	395.8	13.8	0.716
	(F)	=	$-0.026 \times S$	+	$2.030 \times W$	+	331.9	13.1	0.662
Buttock-Knee Length	(M)	=	$0.287 \times S$	+	$0.079 \times W$	+	50.6	16.2	0.705
	(F)	=	$0.254 \times S$	+	$1.033 \times W$	+	92.2	17.4	0.654
Buttock-Popliteal Length	(M)	=	$0.274 \times S$	+	$0.038 \times W$	−	9.7	16.5	0.616
	(F)	=	$0.263 \times S$	+	$0.060 \times W$	+	15.6	17.6	0.555
Elbow Rest Height	(M)	=	—[a]	+	$0.060 \times W$	+	183.6	26.4	0.060
	(F)	=	$0.047 \times S$	+	$0.042 \times W$	+	117.7	26.2	0.044
Eye Height, Sitting	(M)	=	$0.359 \times S$	+	$0.014 \times W$	+	149.7	23.5	0.529
	(F)	=	$0.391 \times S$	+	—[a]	+	101.6	22.1	0.560
Forearm-Forearm Breadth	(M)	=	$-0.172 \times S$	+	$3.051 \times W$	+	572.6	27.2	0.611
	(F)	=	$-0.118 \times S$	+	$3.069 \times W$	+	432.5	21.0	0.633

(*continued*)

Table A.15 Continued

Dependent variable		Slope stature			Slope weight		Intercept	SE_{est}	R^2
Hand Circumference	(M)	$0.029 \times S$	+		$0.043 \times W$	+	128.8	7.5	0.393
	(F)	$0.003 \times S$	+		$0.041 \times W$	+	106.7	6.9	0.333
Hip Breadth, Sitting	(M)	$-0.026 \times S$	+		$2.006 \times W$	+	250.8	12.2	0.764
	(F)	$-0.052 \times S$	+		$2.085 \times W$	+	292.6	15.8	0.664
Knee Height, Sitting	(M)	$0.343 \times S$	+		$0.030 \times W$	−	66.7	12.7	0.793
	(F)	$0.332 \times S$	+		$0.033 \times W$	−	45.8	13.4	0.742
Popliteal Height	(M)	$0.354 \times S$	−		$0.040 \times W$	−	155.7	12.5	0.748
	(F)	$0.352 \times S$	−		$0.073 \times W$	−	138.9	13.0	0.699
Sitting Height	(M)	$0.377 \times S$	+		$0.019 \times W$	+	236.7	23.8	0.551
	(F)	$0.414 \times S$	+		—[a]	+	176.7	22.9	0.570
Span	(M)	$0.951 \times S$	+		$0.055 \times W$	+	110.9	47.2	0.688
	(F)	$0.986 \times S$	+		$0.030 \times W$	+	47.4	50.1	0.620
Thigh Clearance	(M)	$-0.046 \times S$	+		$1.007 \times W$	+	164.0	6.8	0.707
	(F)	$-0.041 \times S$	+		$1.027 \times W$	+	147.5	7.5	0.618
Waist Depth	(M)	$-0.121 \times S$	+		$2.022 \times W$	+	263.5	14.0	0.699
	(F)	$-0.148 \times S$	+		$2.078 \times W$	+	273.5	15.0	0.638
Waist Circumference	(M)	$-0.347 \times S$	+		$7.075 \times W$	+	863.1	41.3	0.771
	(F)	$-0.424 \times S$	+		$9.028 \times W$	+	907.1	48.2	0.661

[a]This independent variable does not add significantly.
Source: Ref. 56.

Table A.16 Stature (S) and Weight (W) Regression Equations for the HES Males (weight in kg, stature in mm)

Dependent variable			Slope stature		Slope weight		Intercept	SE_{est}	R^2
Biacromial Breadth	(M)	=	0.069 × S	+	0.648 × W	+	228.2	18.1	0.268
	(F)	=	0.097 × S	+	0.538 × W	+	164.8	16.1	0.309
Right Arm Girth[a]	(M)	=	-0.109 × S	+	2.412 × W	+	314.3	15.6	0.768
	(F)	=	-0.134 × S	+	2.853 × W	+	319.5	17.3	0.832
Chest Girth[a]	(M)	=	-0.217 × S	+	6.300 × W	+	896.1	36.3	0.810
	(F)	=	-0.184 × S	+	5.384 × W	+	832.0	35.9	0.804
Waist Girth[a]	(M)	=	-0.511 × S	+	8.622 × W	+	1125.6	55.6	0.765
	(F)	=	-0.450 × S	+	7.901 × W	+	985.4	55.8	0.786
Sitting Height	(M)	=	0.335 × S	+	0.607 × W	+	237.4	24.8	0.540
	(F)	=	0.416 × S	+	0.096 × W	+	144.5	26.7	0.519
Sitting Height, Erect	(M)	=	0.379 × S	+	0.385 × W	+	219.5	23.3	0.597
	(F)	=	0.416 × S	+	0.149 × W	+	171.1	23.3	0.586
Knee Height, Sitting	(M)	=	0.318 × S	+	0.243 × W	−	28.2	17.1	0.649
	(F)	=	0.309 × S	+	0.347 × W	−	19.8	16.1	0.646
Popliteal Height	(M)	=	0.321 × S	−	0.372 × W	−	88.6	16.8	0.604
	(F)	=	0.301 × S	−	0.347 × W	−	62.7	17.4	0.554
Thigh Clearance	(M)	=	-0.020 × S	+	0.973 × W	+	105.6	12.1	0.489
	(F)	=	-0.002 × S	+	0.938 × W	+	80.9	13.1	0.494
Buttock-Knee Length	(M)	=	0.260 × S	+	0.734 × W	+	85.7	17.8	0.628
	(F)	=	0.240 × S	+	1.189 × W	+	108.5	18.5	0.646
Buttock-Popliteal Length	(M)	=	0.232 × S	+	0.438 × W	+	58.8	24.2	0.378
	(F)	=	0.203 × S	+	0.930 × W	+	94.7	23.0	0.439
Seat Breadth	(M)	=	-0.005 × S	+	1.816 × W	+	226.7	16.2	0.666
	(F)	=	-0.009 × S	+	2.195 × W	+	241.9	21.8	0.657
Elbow-Elbow Breadth	(M)	=	-0.189 × S	+	3.383 × W	+	494.6	24.5	0.721
	(F)	=	-0.185 × S	+	3.483 × W	+	464.7	25.3	0.776
Elbow-Rest Height	(M)	=	0.042 × S	+	0.646 × W	+	119.5	28.4	0.104
	(F)	=	0.084 × S	+	0.463 × W	+	66.3	27.6	0.101

[a]Original name.
Source: Ref. 18.

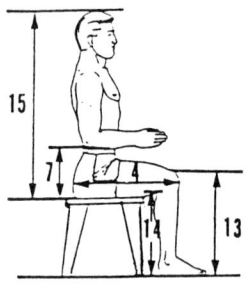
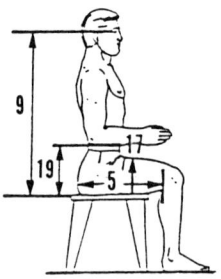
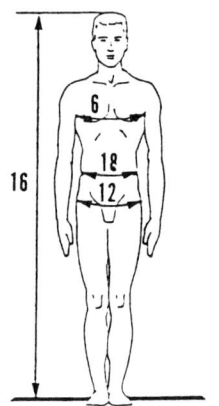
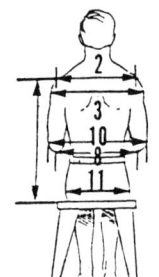

1. Acromial Height, Sitting
2. Biacromial Breadth
3. Bideltoid Breadth
4. Buttock-Knee Length
5. Buttock-Popliteal Length
6. Chest Circumference
7. Elbow-Rest Height
8. Elbow-Elbow Height
9. Eye Height, Sitting
10. Forearm-Forearm Breadth
11. Hip Breadth, Sitting
12. Hip Circumference
13. Knee Height, Sitting
14. Popliteal Height
15. Sitting Height
16. Stature
17. Thigh Clearance Height
18. Waist Circumference
19. Waist Height, Sitting

Figure A.1 Selected traditional dimensions useful in workspace design and evaluation.

Anthropometry

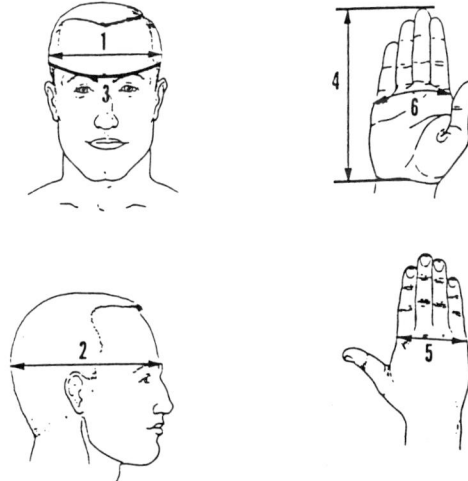

1. Head Breadth
2. Head Length
3. Head Circumference
4. Hand Length
5. Hand Breadth
6. Hand Circumference

Figure A.2 Basic head and hand dimensions.

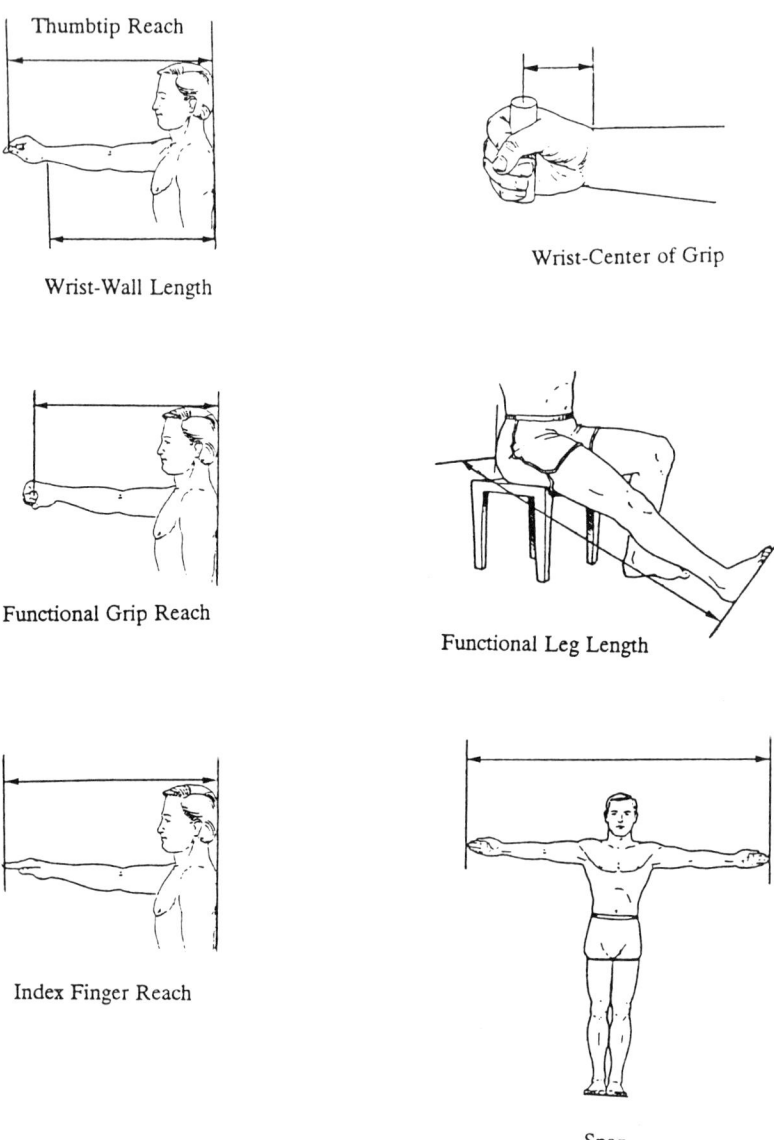

Figure A.3 Selected reaches from the wall and other functional dimensions.

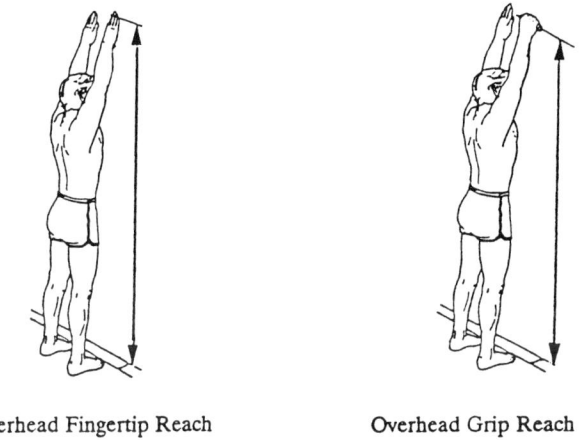

Overhead Fingertip Reach

Overhead Grip Reach

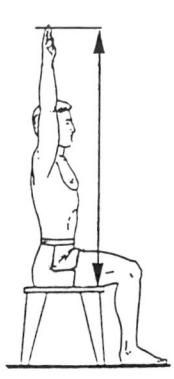

Overhead Reach, Sitting

Vertical Grip Reach, Sitting

Figure A.4 Selected overhead reach dimensions.

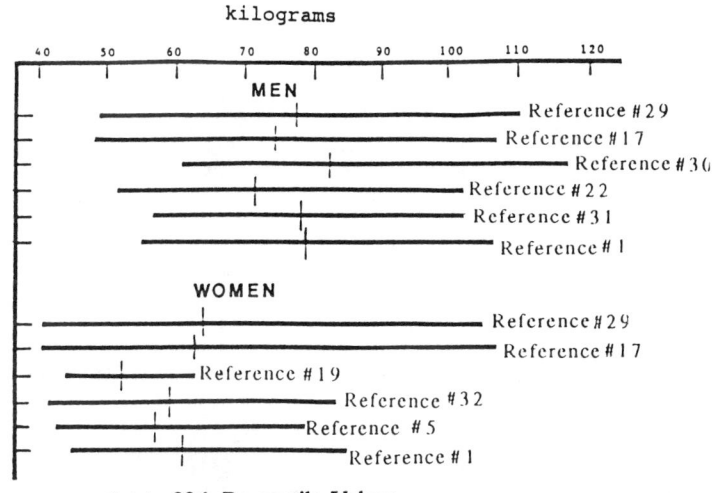

Figure A.5 Range of weight for men and women in selected U.S. anthropometric surveys.

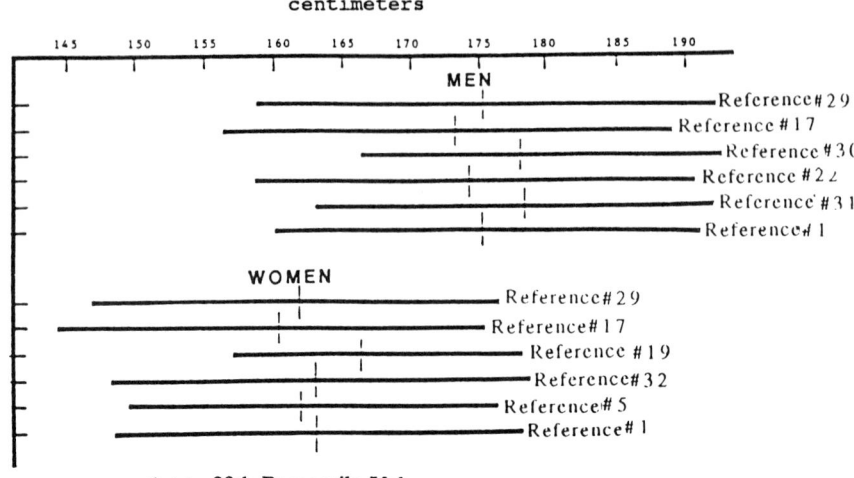

Figure A.6 Ranges of stature for men and women in selected U.S. anthropometric surveys.

(a)

Figure A.7 Visual description of workspace measurements documented in Table A.14.

Anthropometry

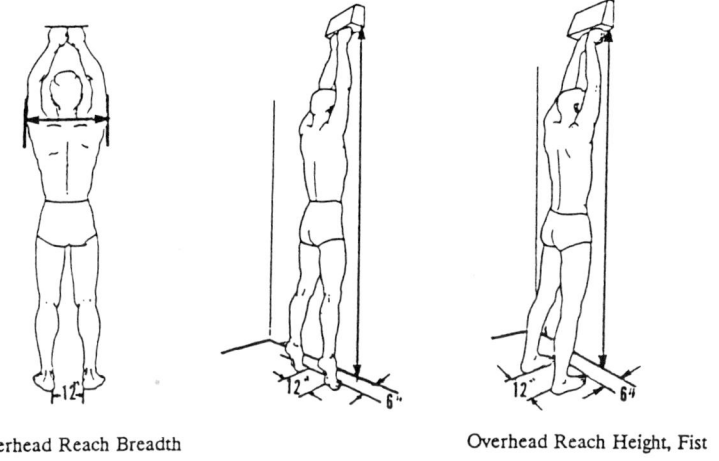

Overhead Reach Breadth

Overhead Reach Height, Fist

Maximum Overhead Reach Height

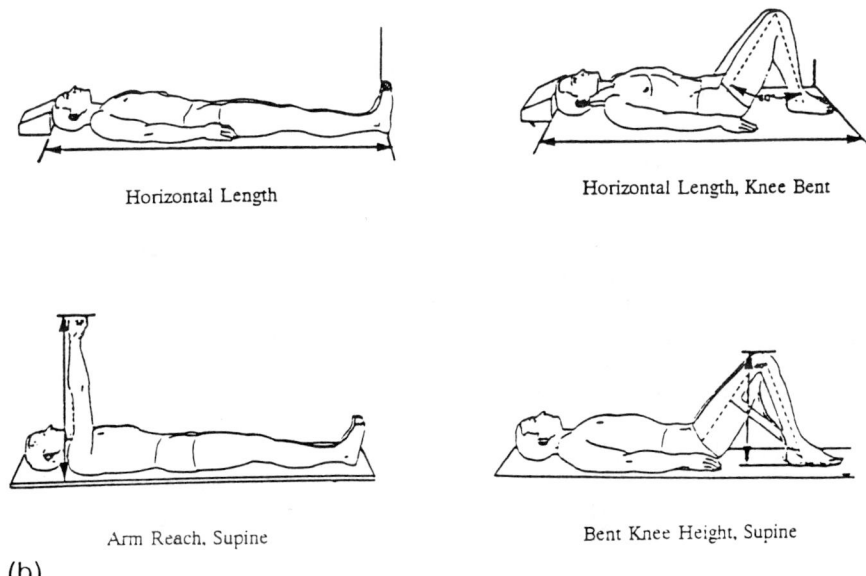

Horizontal Length

Horizontal Length, Knee Bent

Arm Reach, Supine

Bent Knee Height, Supine

(b)

Figure A.7 Continued

REFERENCES

1. C. C. Gordon, B. Bradtmiller, C. E. Clauser, T. Churchill, J. T. McConville, I. Tebbetts, and R. A. Walker, *1987-1988 Anthropometric Survey of U.S. Army Personnel: Methods and Summary Statistics*, Techn. Rep. NATICK/TR-89-044, U.S. Army Natick Research, Development and Engineering Center, Natick, MA, 1989.
2. T. G. Lohman, A. F. Roche, and R. Martorell, *Anthropometric Standardization Reference Manual*, Human Kinetics Books, Champaign, IL, 1988.
3. H. T. E. Hertzberg, The Conference on Standardization of Anthropometric Techniques and Terminology, *Am. J. Phys. Anthropol.* 28(1):1-16 (1968).
4. H. T. E. Hertzberg, E. Churchill, C. W. Dupertuis, R. M. White, and A. Damon, *Anthropometric Survey of Turkey, Greece and Italy*, Macmillan, New York, 1963.
5. C. E. Clauser, P. E. Tucker, J. T. McConville, E. Churchill, L. L. Laubach, and J. A. Reardon, *Anthropometry of Air Force Women*, AMRL-TR-70-5 (AD 743 113), Aerospace Med. Res. Lab., Wright-Patterson Air Force Base, OH, 1972.
6. J. F. Annis and C. C. Gordon, *The Development and Validation of an Automated Headboard Device for Measurement of Three-Dimensional Coordinates for the Head and Face*, Techn. Rep. NATICK/TR-88/048 (AD A201-186), U. S. Army Natick Research, Dev. Eng. Center, Natick, MA, 1988.
7. J. F. Annis, An automated device used to develop a new 3-D data base for head and face anthropometry, in *Advances in Industrial Ergonomics and Safety,* Vol. 1, (A. Mital, Ed., Taylor & Francis, Philadelphia, 1989.
8. R. E. Herron, Biostereometric measurement of body form, *Yearb. Phys. Anthropol.* 16:80-121 (1972).
9. G. S. Daniels, *The "Average Man"?*, Techn. Note WCRD 53-7, Wright Air Dev. Center, Wright-Patterson Air Force Base, OH, 1952.
10. J. F. Annis and J. T. McConville, Application of anthropometric data in sizing and design, in *Advances in Industrial Ergonomics and Safety,* Vol. II, B. Das, Ed., Taylor & Francis, Philadelphia, 1990.
11. J. T. McConville and E. Churchill, *Statistical Concepts in Design*, AMRL-TR-76-29, Aerospace-Medical Res. Lab., Wright-Patterson Air Force Base, OH, 1976.
12. K. Robinette and T. Churchill, *Design Criteria for Characterizing Individuals in the Extreme Upper and Lower Body Size Ranges*, AMRL-TR-79-33 (AD A072 353), Aerospace Medical Res. Lab., Wright-Patterson Air Force Base, OH, 1979.
13. S. Pheasant, *Bodyspace Anthropometry, Ergonomics and Design*, Taylor & Francis, London, 1986.
14. E. Grandjean, *Fitting the Task to the Man*, 4th ed., Taylor & Francis, London, 1988.
15. R. O'Brien and W. C. Shelton, *Women's Measurements for Garment and Pattern Construction*, Misc. Publ. No. 454, U.S. Gov. Printing Office, Washington, DC, 1941.
16. R. O'Brien, M. A. Girshick, and E. P. Hunt, *Body Measurements of American Boys and Girls for Garment Pattern Construction*, Misc. Publ. 366, U.S. Department of Agriculture, U.S. Gov. Printing Office, Washington, DC, 1941.
17. H. Stoudt, A. Damon, R. MacFarland, and J. Roberts, *Weight, Height and Selected Body Dimensions of Adults, United States 1960-62*, Public Health Service Publ. No. 1000, Ser. 11, No. 8, U.S. Gov. Printing Office, Washington, DC, 1965.
18. S. H. Rodgers (Ed.) *Ergonomic Design for People at Work,* Vol. I and II, Van Nostrand Reinhold, New York, 1986.
19. C. C. Snow, H. M. Reynolds, and M. A. Allgood, *Anthropometry of Airline Stewardesses*, Dept. Transportation Rep. No. FAA-AM-75-2, FAA Office of Aviation Medicine, Civil Aeromed. Inst., Oklahoma City, OK, 1975.
20. J. T. McConville and C. E. Clauser, *Anthropometric Resources vs. Civilian Needs*, AMRL-TR-78-111 (AD A061-390), Aerospace Med. Res. Lab., Wright-Patterson Air Force Base, OH, 1978.
21. W. S. Marras and J. Y. Kim, Anthropometry of industrial populations, *Ergonomics* 36(4): 371-378, (1993).

22. R. M. White and E. Churchill, *The Body Size of Soldiers: U.S. Army Anthropometry—1966*, Techn. Rep. 72-51-CE (AD 743 465), U.S. Army Natick Laboratories, Natick, MA, 1971.
23. B. Bradtmiller, J. Ratnaparkhi, and I. Tebbetts, *Demographic and Anthropometric Assessment of U.S. Army Anthropometric Data Base*, Techn. Rep. NATICK/TR-86/004 (AD A164-637), U.S. Army Natick Res., Dev. Eng. Center, Natick, MA, 1985.
24. E. Churchill, P. Kikta, and T. Churchill, *The AMRL Anthropometric Data Bank Library*, Vols. I–V, AMRL-TR-77-1 (AD A047 314), Aerospace Med. Res. Lab., Wright-Patterson Air Force Base, OH, 1977.
25. J. T. McConville, E. Churchill, C. E. Clauser, and M. Alexander, The Aerospace Medical Research Laboratory's Anthropometric Data Bank: a resource for designers, *Preprints 1977 Sci. Program*, Aerospace Med. Assoc., 1977 Annu. Sci. Meeting, Las Vegas, NV, 1977.
26. J. C. Robinson, K. M. Robinette, and G. F. Zehner, *User's Guide to Accessing the Anthropometric Data Base at the Center for Anthropometric Research Data*, 2nd ed., AL-TR-1992-0036, Armstrong Lab., U.S. Air Force Systems Command, Wright-Patterson Air Force Base, OH, 1992.
27. E. Churchill, J. T. McConville, L. L. Laubach, T. Churchill, P. Erskine, and K. Downing, *Anthropometric Source Book*, Vol. II, *A Handbook of Anthropometric Data*, NASA Ref. Publ. 1024 (NTIS No. N79-13711/3/XPS), National Aeronautics and Space Administration, Washington, DC, 1978.
28. C. E. Clauser, J. T. McConville, C. C. Gordon, and I. O. Tebbetts, *Selection of Dimensions for an Anthropometric Data Base*, Vol. I, *Rationale, Summary and Conclusions*, Techn. Rep. NATICK/TR-86/053 (AD A179 566), U.S. Army Natick Res., Dev. Eng. Center, Natick, MA, 1986.
29. A. Engle, R. S. Murphy, K. Maurer, and E. Collins, *Plan and Operation of the HANES I Augmentation Survey of Adults 25 to 74, United States 1974–75*, DHEW Publ. PHS 78-1314, Ser. 1, No. 14, U.S. Public Health Service, NCHS, Hyattsville, MD, 1978.
30. J. I. Martin, R. Sabeh, L. L. Driver, T. D. Lowe, R. W. Hintz, and P. A. C. Peters, *Anthropometry of Law Enforcement Officers*, Rep. No. NELC, Techn. Docu. 442, TD 442, Law Enforcement Standards Laboratory, Nat. Bureau of Standards, Naval Electronics Laboratory Center, San Diego, CA, 1975.
31. H. G. Grunhofer and G. Kroh (Eds.), *A Review of Anthropometric Data of German Air Force and United States Air Force Personnel*, AGARD-AG-205, Advisory Group for Aerospace Res. Dev., Neuilly sur Seine, France, 1975.
32. L. L. Laubach, J. T. McConville, E. Churchill, and R M. White, *Anthropometry of Women of the U.S. Army—1977, Report No. 1—Method and Survey Plan*, Techn. Rep. NATICK/TR-77/021 (AD A043 715), U.S. Army Natick Res. Develop. Command, Natick, MA, 1977.
33. K. Robinette, T. Churchill, and J. T. McConville, *A Comparison of Male and Female Body Sizes and Proportions*, AMRL-TR-79-69 (AD A074 807), Aerospace Med. Res. Labor., Wright-Patterson Air Force Base, OH, 1979.
34. J. F. Annis, H. W. Case, C. E. Clauser, and B. Bradtmiller, Anthropometry of an aging work force, *Exp. Aging Res.* *17*(3): 157–176 (1991).
35. J. S. Friedlaender, P. T. Costa, Jr., R. Bosse, E. Ellis, J. G. Rhoads, and H. W. Stoudt, Longitudinal physique changes among healthy white veterans at Boston, *Hum. Biol.* 49(4):541–558 (1977).
36. A. Damon and H. W. Stoudt, The functional anthropometry of old men, *Hum. Factors* (vol?):485–491 (1963).
37. B. Bell, C. L. Rose, and A. Damon, The Normative Aging Study (NAS), the Veterans Administration longitudinal study of healthy aging, *Gerontologist* 6: 179–184 (1966).
38. W. C. Chumlea, P. J. Garry, W. C. Hunt, and R. L. Rhyne, Distributions of serial changes in stature and weight in a healthy elderly population, *Hum. Biol.* 60(6): 917–925 (1988).
39. N. W. Shock, R. C. Greulich, R. Andres, D. Arenberg, P. T. Costa, Jr., E. G. Lakatta, and J. D. Tobin, *Normal Human Aging: The Baltimore Longitudinal Study of Aging*, Vols. I and II, (NIH Publ. AM-84-2450). U.S. Govt., Printing Office, Washington, DC, 1984.

40. Anonymous, Fels study reaches 60, *Vitalsigns* 15(1):(1989) Wright State Univ. School of Medicine, Dayton, OH.
41. F. E. Randall, A. Damon, R. S. Benton, and D. I. Patt, *Human Body Size in Military Aircraft and Personal Equipment*, Techn. Rep. AAF-TR-5501 (ATI 25 419), Air Materiel Command, Wright Field, Dayton, OH, 1946.
42. R. F. Johnson, *Anthropometry of the Clothed U.S. Army Ground Troop and Combat Vehicle Crewman*, Techn. Rep. NATICK/TR-84/034, U.S. Army Natick Res., Dev. Electronics Center, Natick, MA, 1984.
43. S. P. Paquette, H. W. Case, J. F. Annis, T. L. Mayfield, S. E. Kristensen, and D. N. Mountjoy, *The Effect of Multilayered Military Ensembles on Body Size*, NATICK TR/-, U.S. Army Natick Res. Dev. Engin. Center, Natick, MA, 1993.
44. S. P. Paquette and R. A. Maulucci, *The Effects of Clothing on Body Size and Range of Joint Motion*, Techn. Rep. NATICK/TR- Natick Res. Dev. Eng. Center, Natick, MA, 1993.
45. H. T. E. Hertzberg, I. Emanuel, and M. Alexander, *The Anthropometry of Working Positions*, WADC Techn. Rep. 54-520, Wright Air Development Center, Wright-Patterson Air Force Base, OH, 1956.
46. M. Alexander and C. E. Clauser, *Anthropometry of Common Working Positions*, AMRL-TR-65-73, Aerospace Med. Res. Lab., Wright-Patterson Air Force Base, OH, 1965.
47. W. T. Dempster, The anthropometry of body action, *Ann. N. Y. Acad. Sci.* 63(4):559–585 (1955).
48. K. W. Kennedy, *Reach Capability of Men and Women: A Three-Dimensional Analysis*, AMRL-TR-77-50, Aerospace Med. Res. Lab., Wright-Patterson Air Force Base, OH, 1978.
49. H. W. Stoudt, *Arm-Leg Reach and Workspace Layout*, in *Anthropometric Source Book*, Vol. I, *A Handbook of Anthropometric Data*, NASA Ref. Publ. 1024 (NTIS No. N79-13711/3/XPS), Natl. Aeronautics and Space Administration, Washington, DC, 1978, Chap. 5.
50. W. E. Woodson, B. Tillman, and P. Tillman, *Human Factors Design Handbook*, 2nd ed., McGraw-Hill, New York, 1992.
51. R. D. Huchingson, *New Horizons for Human Factors Design*, McGraw-Hill, New York, 1981.
52. A. Damon, H. W. Stoudt, and R. A. McFarland, *The Human Body in Equipment Design*, Harvard Univ. Press, Cambridge, MA, 1966.
53. J. S. Roebuck, Jr., K. H. E. Kroemer, and W. G. Thomson, *Engineering Anthropometry Methods*, Wiley, New York, 1975.
54. W. Braune and O. Fischer, *The Center of Gravity of the Human Body as Related to the German Infantryman*, Leipzig, 1889 (ATI 138 452; available from Defense Documentation Center).
55. The Human Factors Society, *American National Standard for Human Factors Engineering of Visual Display Terminal Workstations*, ANSI/HFS Standard 100-1988, Human Factors Society, Santa Monica, CA, 1988.
56. J. Cheverud, C. C. Gordon, R. A. Walker, C. Jacquish, L. Kohn, A. Moore, and N. Yamashita, *1988 Anthropometric Survey of U.S. Army Personnel: Correlation Coefficients and Regression Equations,* Techn. Rep. NATICK/TR-90/036, U.S. Army Natick Res. Dev. Eng. Center, Natick, MA, 1990.

2
Physiology of Body Movement

Stephan Konz
Kansas State University, Manhattan, Kansas

I. INTRODUCTION

This chapter gives some scientific background for analysis of physical work. The potential problem is excessive stress on the cardiovascular system. Excessive stress on the muscular system (usually during manual material handling) is covered elsewhere in this volume.

Excessive stress can result from a "high" metabolic requirement on a "normal" person. It also can be from a "medium" or even "low" metabolic requirement on people with lower levels of cardiovascular fitness (e.g., some female workers, older workers, or handicapped workers). The goal is to design jobs so that "everyone" can do them. This requires knowledge concerning the energy requirements of various tasks and the ability of people to do physical work.

II. BACKGROUND AND SIGNIFICANCE TO OCCUPATIONAL ERGONOMICS

Hard physical work still occurs in some industries. This chapter describes how the energy required can be estimated and the response of the cardiovascular system to the energy requirements can be anticipated. Recommendations are given for work limits.

III. RELEVANT CONCEPTS AND TERMINOLOGY

A. Metabolism

Metabolic rate can be estimated from (1) tables and formulas and (2) measurements.

1. Tables and Formulas

Metabolic rate is divided into three components: basal, activity, and digestion.

$$TOTMET = BSLMET + ACTMET + DIGMET$$

where TOTMET is the total metabolic rate, BSLMET is the basal metabolic rate, ACTMET is the activity metabolic rate, and DIGMET is the digestion metabolic rate (also called specific dynamic action).

This chapter is a modification of part of Chapter 11, Work physiology and biomechanics, in Stephan Konz, *Work Design: Industrial Ergonomics*, 3rd ed., Publishing Horizons, Worthington, OH, 1990.

Metabolic rate can be expressed in many different units (see Table 1). The standard power unit is the watt (W). The energy content of food is given in kilocalories (kcal, commonly spoken of as calories).

Basal Metabolic Rate. Basal metabolic rate maintains body temperature, functions and blood circulation.

$$\text{BSLMET} = \text{BSMET} \times \text{WT}$$

where
BSLMET = basal metabolic rate, W
BSMET = 1.28 W/kg for males
      = 1.16 W/kg for females
WT = body weight, kg

The reason BSMET is lower for females is that females tend to have a higher percentage of body fat than males, and fat has a limited metabolism.

Konz [1] gives a more complex formula that also considers the effect of age from 5 to 70. Children have a higher basal metabolic rate per kilogram body weight as they have a higher surface area-to-volume ratio (and thus require more heat to maintain body temperature), and growth takes energy.

Activity Metabolic Rate. Activity metabolic rate provides the energy for activities. See Table 2 for some examples. Working in a contorted posture (e.g., kneeling, bent over) increases metabolic rate [3].

Pandolf et al. [4] give walking metabolism (without a load) as a function of terrain and velocity.

$$\text{WLKMET} = C[2.7 + 3.2(v - 0.7)^{1.65}]$$

where
WLKMET = walking metabolism (total), W/kg of body weight
C = terrain coefficient
  − 1.0 for treadmill, blacktop road
  = 1.1 for dirt road
  = 1.2 for light brush
  = 1.3 for hardpacked snow
  = for softer snow = 1.3 + 0.082 × (foot depression, cm)
  = 1.5 for heavy brush

Table 1 Energy Conversions

1 J	=	1. N-m			
1 J	=	238.9 kcal			
1 W	=	1. J/s			
1 W	=	0.85885 kcal/h	1 kcal/h	=	1.163 W
1 W	=	3.413 Btu/h	1 Btu/h	=	0.293 W
1 W	=	0.00134 hp	1 hp	=	746 W
1 W	=	6.116 kg-m/min	1 kg-m/min	=	0.1635 W
1 W/m²	=	0.01718 met[a]	1 met[a]	=	58.2 W/m²
1 W/m²	=	10.764 W/ft²	1 W/ft²	=	0.0929 W/m²
1 kcal	=	3.968 Btu	1 Btu	=	0.252 kcal
			1 Btu	=	776.65 ft-lb

[a]A met is a measure of metabolic rate

Physiology of Body Movement

Table 2 Activity Cost for Various Activities[a]

Cost (W/kg)	Activity
0.4	Crocheting, eating, reading aloud, sewing by machine, sitting quietly, writing
0.6	Playing cards, standing relaxed, typing with electric typewriter
0.7	Paring potatoes, doing office work (standing), sewing with foot-driven machine, standing at attention, playing the violin
0.8	Dressing and undressing, knitting a sweater
0.9	Singing in a loud voice
1.0	Driving a car, tailoring
1.2	Dishwashing, typing rapidly
1.4	Washing floors
1.5	Cello playing, light laundry
1.6	Riding a walking horse, sweeping bare floor with broom
1.7	Golf, organ playing, painting furniture with brush
1.9	Sweeping with hand (push) carpet sweeper
2.7	Doing heavy carpentry
3.0	Cleaning windows
3.1	Cleaning with upright vacuum cleaner
3.5	Dancing the waltz
4.1	Ice skating
4.5	Weeding a garden
5.0	Horseback riding (trot)
5.1	Playing Ping-Pong
5.8	Dancing rhumba, playing tennis
6.6	Sawing wood (hand saw)
7.9	Playing football
8.9	Fencing

[a]For total energy cost, add the basal metabolism and, if appropriate, the cost of digestion. See the text for equations for walking, carrying, and running.
Source: Ref. 2.

$\quad\quad\quad$ = 1.8 for swamp
$\quad\quad\quad$ = 2.1 for sand
$\quad\quad v$ = velocity, m/s (for $v > 0.7$ m/s (2.5 km/h, 1.56 mi/h)

Pandolf et al. [5] give the following equation for standing or walking very slowly:

$$\text{WLKMET} = 1.5\,\text{WT} + 2(\text{WT} + \text{WTL})(\text{WTL}/\text{WT})^2 + C(\text{WT} + \text{WTL})(1.5v^2 + 0.35vG)$$

where

WLKMET = metabolic rate (total) for walking slowly, W
$\quad$ WT = body weight, kg
$\quad$ WTL = weight of a load on the shoulders, kg
$\quad\quad v$ = velocity (speed) of walking, m/s ($v < 1$ m/s)
$\quad\quad C$ = coefficient for terrain (see above)
$\quad\quad G$ = grade, %

The first term (1.5 WT) is the metabolic cost of standing without a load and is equal to 1.5 W/kg. The second term is the cost of load bearing while standing. Walking on the level is $C(\text{WT} + \text{WTL})(1.5\,v^2)$. The cost of climbing a grade is $C(\text{WT} + \text{WTL})(0.35\,vG)$.

The metabolic cost of walking a specific distance is minimized when $v = 1.13$ m/s (4.05 km/h). Stride length (L) divided by stature height (h) varies linearly, with L/h ranging from 0.67 at $v = 0.8$ m/s to 0.9 at $v = 1.7$ m/s [6].

The metabolic cost of carrying (walking with a load) depends upon the load location. Soule and Goldman [7] reported that loads on the head use 1.2 times the energy of carrying a kilogram of your own body; in the hands, loads require 1.4–1.9 times as much; on the feet, loads required 4.2–6.3 times as much. Legg and Mahanty [8] found that it takes 6.4 times as much energy to carry 1 kg on the feet as on the back.

As a general guideline for carrying, minimize the load's moment arm, along both the frontal and transverse axes.

The cost of running [9] is

$$\text{RUNMET} = -142/\text{WT} + 11 + 0.04V^2$$

where
RUNMET = running metabolism (total), W/kg
V = velocity, km/h
WT = weight, kg

When estimating metabolic rate from tables or formulas, remember that most people do not work continuously; they take microbreaks. The duration of each task element can be estimated from videotape or occurrence sampling.

Digestion Metabolic Rate. Digestion and the transformation of food within the body are called specific dynamic action. For the typical carbohydrate–fat–protein mixture of the U.S. diet, and not worrying about time since the meal,

$$\text{DIGMET} = 0.1(\text{BSLMET} + \text{ACTMET})$$

If you wish to make DIGMET constant in a laboratory experiment, have the subject standardize what is eaten at meals and control the time between the meal and the experiment.

In 1932, Klieber [10] determined the following formula for a variety of animals (dove, rat, pigeon, hen, dog, sheep, human, cow, and steer):

$$\text{DAYMET} = 70 \text{ WT}^{3/4}$$

where
DAYMET = metabolism during a day, kcal
WT = body weight, kg

The exponent could have been predicted as 2/3 because surface area increases by 2 and volume increases by 3. Peters [11] says that the difference is due to the differing shape of the animals as the size increases.

2. Measured Metabolism

In a laboratory situation where its accurate determination is important, metabolic rate can be measured indirectly by measuring oxygen consumption (indirect calorimetry). This is akin to measuring the fuel consumption of your car by measuring its oxygen intake through the carburetor. A clip is placed on the subject's nose to prevent nose breathing, and a tube connected to an oxygen supply is inserted in the mouth. The body's conversion efficiency of fuel to energy is estimated. A recent development is the use of instantaneously reacting sensors placed in the air flow of the exhaust air, eliminating the need for the noseclip.

$$\text{TOTMET} = 60 \text{ ENERGY} \times \text{OXUPTK}$$

where
TOTMET = total metabolism, W
ENERGY = energy equivalent of 1 liter of oxygen, W-h/liter
[This depends upon the respiratory quotient (RQ), which in turn depends upon the proportion of fat vs carbohydrate metabolized during the exercise.]
= 5.36 for RQ = 0.83 (rest)
= 5.66 for RQ = 0.86 (exercise up to 60% of maximum rate)
= 6.40 for RQ = 1.0 (100% of maximum oxygen uptake)
OXUPTK = oxygen uptake (VO_2), liters of oxygen/min

The metabolic rate of a laboratory task can be controlled to be constant by having the subject walk on a treadmill, pedal a bicycle ergometer, or walk up and down steps to a metronome beat.

B. Cardiovascular System

This section describes the anatomy of the system, the response to exercise, three major factors in cardiovascular performance, and cardiovascular limits.

1. Anatomy

Figure 1 shows an "engineer's view" of the cardiovascular system. The heart has two pumps; the right side pumps to the pulmonary circulation, and the left side pumps to the systemic circulation. The blood has transport and storage functions.

The pulmonary circulation starts at the right atrium (Latin: room) and goes through a valve to the right ventricle (the pump itself). Systole is contraction of the heart muscle (the aortic valve is open); diastole is relaxation of the heart muscle (the aortic valve is closed). When heart rate increases, it does so by reducing the diastolic time. When the blood returns from the lungs (with more oxygen and less carbon dioxide), it enters the systemic circulation.

After passing through the left atrium and ventricle, the blood enters the artery. Oxygen is removed and carbon dioxide added in the muscles. In addition, the blood picks up new "fuel" at the intestines, and this fuel may be removed at the muscles (if they are working). When muscles work, they add waste products to the blood, which transports them to the liver (for possible biotransformation), kidney (elimination of water-soluble compounds in the urine), and intestine (feces).

2. Response to Exercise

The cardiovascular system responds to exercise in five ways, by altering (1) heart rate, (2) stroke volume, (3) artery–vein oxygen concentration differential, (4) blood distribution and (5) by going into oxygen debt.

Heart Rate. Heart rate (with some exceptions) is a good estimator of metabolic rate. The exceptions are the following:

1. Heart rate is increased by emotions (the effect is relatively greater when the person is sedentary, i.e., heart rate due to exercise is low).
2. Heart rate is increased by vasodilation (heat stress for unacclimatized people; acclimatized people sweat instead of vasodilating).
3. Heart rate may not increase as much as predicted at very heavy metabolic rates, as other cardiovascular system responses may be increased also.

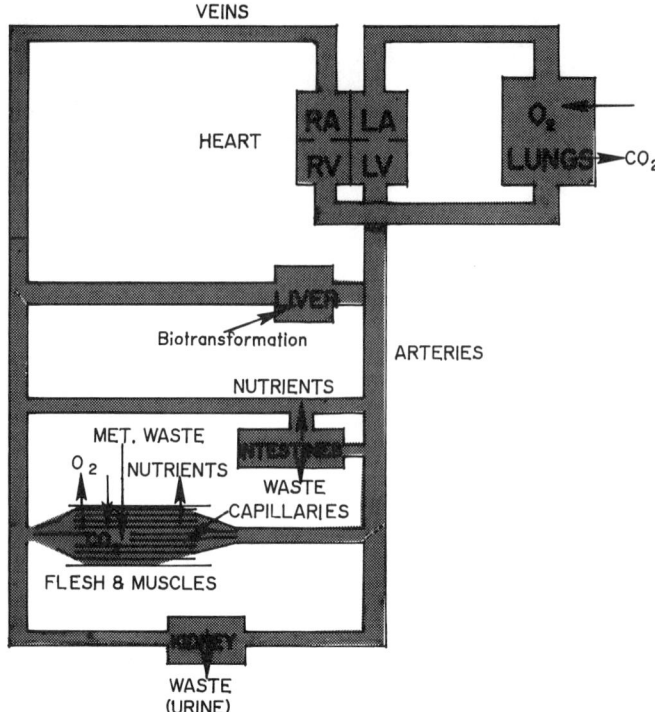

Figure 1 Engineer's view of the cardiovascular system shows a pulmonary circuit and a systemic circuit. The heart has two pumps. The pulmonary pump (right ventricle) sends blood to the lung, where O_2 is added and CO_2 is removed. The blood returns to the left atrium. The systemic pump (left ventricle) sends blood out into the arteries. After passing through various organs and the capillaries, the blood returns in the veins. In the "marsh" of the capillaries, the blood gives up oxygen and nutrients to the flesh and obtains CO_2 and metabolic wastes. Nutrients are added to the blood from the intestines. Wastes are biotransformed in the liver and reinserted into the blood; the modified wastes are removed by the kidney (urine) and intestines (feces). (From Ref. 2.)

Nonetheless, for light and medium work loads, heart rate is a good predictor of metabolic rate. A predictor equation [12] is

INCHR = K + 0.12 INCMET

where
  INCHR = increase in heart rate, beats/min
  K = constant
    = 2.3 for arm work (cranking)
    = −11.5 for leg work (walking) or arm + leg work
    [The 13.8 difference in the coefficient for the legs is due to venous pooling in the legs.]
  INCMET = increase in metabolism, W

There are four ways to measure heart rate.

1. *Light* is used in two ways. With one technique, a light is shined on an artery in the earlobe. A photocell on the other side of the earlobe sees the light during "ebb tide"

between beats. With another, a device placed on the finger detects reflected light. Both the earlobe and finger models are sensitive to body movement and so work best with a stationary person.

2. *Sound* is used by physicians with a stethoscope. (Women instinctively comfort their babies by carrying them on their left arm so the baby's head lies against the left chest and the baby can easily hear the mother's heart beat.)
3. *Palpation* is the detection, with the fingers, of the surge of blood that follows each beat. Common locations are the arteries in the wrist and neck. Count the number of beats in 10 s and multiply by 6, or count for 15 s and multiply by 4. In working situations, this is done after the work stops—that is, the "recovery pulse" is measured.
4. *Electronics* is the most common and the only practical technique for a nonstationary person. Usually, electrodes are attached to the chest and the EKG waveform is recorded. To minimize muscle noise, place the electrodes above and below the heart in a triangle (sternum and about 5 cm outside each nipple); if the signal will be sent by radio, the third (ground) electrode is not needed. To improve the signal, wipe the skin with a solvent, add electrode paste, and avoid hairs. Typically the unit is about the size of a pack of cigarettes and is belt-mounted. Depending on the unit, the output can be displayed on the unit, stored on a small tape recorder, or sent by radio or by "hard wire."

Heart rate can also be estimated subjectively, rather than being measured, by asking the individual to rate "perceived exertion." See Table 3. This concept has been validated by many experimenters for many tasks. Ljunggren [14] reported that it is not even necessary to have the individual's own perceived exertion; this exertion can be estimated by an observer!

The standard prediction of maximum heart rate [15] is

HRMAX = 220 − AGE

where HRMAX is maximum heart rate in beats per min and AGE is age in years.

Table 3 Borg's New Rating of Perceived Exertion (RPE) Scale[a]

Vote	Subjective description
6	No exertion at all
7	Extremely light
8	
9	Very light
10	
11	Light
12	
13	Somewhat hard
14	
15	Hard
16	
17	Very hard
18	
19	Extremely hard
20	Maximal exertion

[a]This scale can be used to predict heart rate. Use the words for guidance, but then vote with a number. Heart rate = 10 × Vote.
Source: Ref. 13.

The standard deviation of the prediction is 10 beats/min. Thus a 50th percentile individual at age 30 would have a maximum heart rate of 190. Maximum heart rate is relatively unaffected by physical fitness. When exercising for cardiovascular fitness, initially exercise at 60–70% of HRMAX; an experienced exerciser can aim for 75–85% of HRMAX.

Figure 2 shows how the heart rate (primarily determined by aerobic oxygen supply) responds to constant-intensity exercise. The heart rate cost of a task can be determined in three ways. The simplest is to subtract the individual's basal heart rate from the peak. For example, Joe's peak of 110 and basal of 70 give a task cost for him of 40 beats/min. Determining the basal is more complicated than it might seem. People subconsciously increase their heart rates just before exercising. To get a good estimate of the basal heart rate, measure it 5 or 10 min before exercise or after the work when heart rate has returned to basal—say 5 min after light work and 10–15 min after moderate or heavy work. A problem with the first method is that it assumes that the peak represents the heart rate during the work, i.e., that the top of the curve is flat.

A second way is to consider the area under the curve (area B in Figure 2). However, this assumes that the heart rate instantaneously returns to basal. The best way is to consider work cost as area B + C. For accurate work, the area under the curve should be calculated with a planimeter, but for most purposes it can be estimated from a series of rectangles.

Stroke Volume. The second method of adjusting oxygen supply to the body is stroke volume. The amount of blood pumped by the left ventricle can be predicted as

$$SV = STROVB + 0.000050(TOTMET - 200), \quad TOTMET < 500$$

where

SV = stroke volume, liters/beat
$STROVB$ = basal male stroke volume, liters/beat
(female = 0.9 × male)
= (SI × DBSA) / 1000

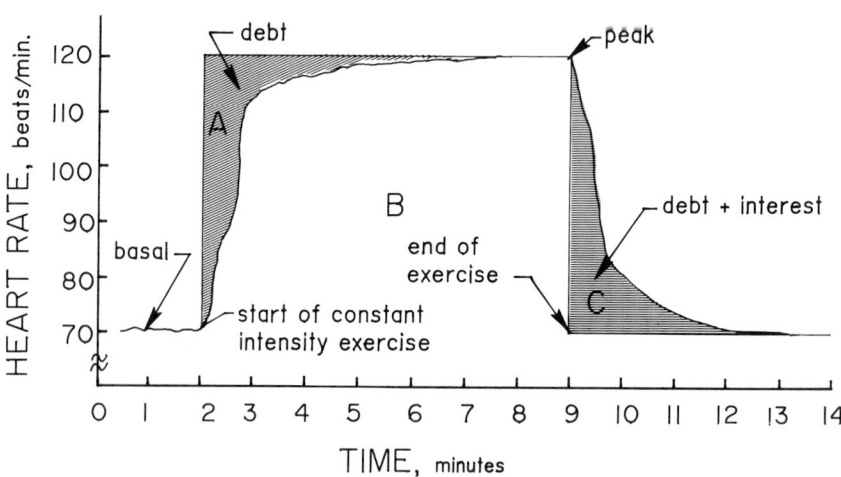

Figure 2 Aerobic response (and thus heart rate) lags the exercise onset. The deficit (area A) is replaced by anaerobic oxygen. The anaerobic supply in the blood is composed of alactate (energy equivalent of about 1.9 liters of oxygen) and lactate (about 3.1 liters) [2, p. 398]. During recovery (area C), the anaerobic oxygen in the blood is replaced, but the replacement process itself uses oxygen ("interest"), so area C is larger than area B. For task cost, use area B + C. (From Ref. 2.)

where

SI = stroke index, mL/ (beat–m²)
  = $53.45 - 0.194 \text{ AGE}$
AGE = age, yr
$DBSA$ = DuBois surface area, m²
  = $0.007184 \text{ HT}^{0.725} \times \text{WT}^{0.425}$
HT = height, cm
WT = weight, kg
$TOTMET$ = total metabolism, W

For TOTMET over 500 W, add (0.000025)(TOTMET − 500) to the SV equations.

Stroke volume depends upon body posture, exercise, and physical fitness. When the person is lying down, stroke volume may be 0.12 liter/beat, but for sitting and standing a value of 0.08 is more typical. Exercise with the legs improves venous return; stroke volume may increase to about 0.40 liter. Exercise with only the arms tends to permit venous pooling in the legs, and stroke volume changes little.

Maximum stroke volume is primarily a function of physical fitness. Maximum SV = 0.135 liter for excellent cardiovascular fitness, 0.120 for good, 0.100 for fair, 0.090 for poor, and 0.085 for very poor. See Table 3 for maximum oxygen uptake levels for these fitness levels. World class athletes may have a maximum SV as high as 0.200 liter/beat, giving them a resting heart rate in the 50s. Stroke volume peaks at about 40% of maximum oxygen consumption.

For blood pressure, see Box 1. For cardiac output, see Box 2.

A-V Differential. The arterial-venous oxygen differential is the third method of adjusting oxygen supply to the body.

$$OXUPTK = CO \times AVDIF$$

Box 1 Blood Pressure

Blood pressure commonly is measured with a sphygmomanometer. The cuff is applied to the upper arm while listening to the sound of the blood flowing in the artery below the cuff. After the cuff is tightened, there is no sound because there is no blood flow in the artery. As the cuff is loosened, blood begins to flow turbulently and a noise is heard. This first sound is called *systolic* blood pressure. As the cuff is loosened further, the turbulence decreases until the flow becomes laminar and the sound disappears. The disappearance of the sound indicates *diastolic* blood pressure, the minimum or base pressure exerted by the blood on the artery walls.

Pulse pressure = systolic pressure − diastolic pressure

Blood pressure can be estimated [16] as

$SBP = 101.3 + 0.68 \text{ AGE}$

$DBP = 63.7 + 0.36 \text{ AGE}$

where
SBP = systolic blood pressure, mm Hg
DBP = diastolic blood pressure, mm Hg
AGE = age, yr

Life insurance studies indicate that blood pressures below 110/70 are optimal for a long lifespan. Resting diastolic below 90 is satisfactory, between 90 and 100 is suspicious, and above 100 is poor.

Box 2 Cardiac Output

Cardiac output is the volume of blood/min from the left ventricle:

$$CO = HR \times SV$$

where
CO = cardiac output, liters/min
HR = heart rate, beats/min
SV = stroke volume, liters/beat

Cardiac output, for a resting young man, is about 5 liters/min; maximum for a normally sedentary young man is about 25; and maximum for a world class athlete is about 35.

Basal, activity, and skin cardiac output can be predicted from formulas.

Basal cardiac output is [17]

$$COBASL = CI \times DBSA$$

where
$COBASL$ = basal cardiac output, liters/min
CI = cardiac index, liters/(min-m^2)
    = $4.29 - 0.029\ AGE + 0.003\ AGE^2$ ($5 < AGE < 70$)
AGE = age, yr
$DBSA$ = DuBois surface area, m^2

Activity cardiac output is [15]

$$COACT = CLMW \times TOTMET$$

where
$COACT$ = activity cardiac output, liters/min
$CLMW$ = conversion from liters of blood to watts
    = 0.166 for TOTMET < 700 W
    = 0.114 for TOTMET > 700 W
$TOTMET$ = total metabolism, W

Activity cardiac output increases primarily in the muscles—up to 18 times basal flow.

Skin cardiac output is [18]

$$COSKIN = (CSKIN \times ISKINT) + (CCORE \times ICORET)$$

where
$COSKIN$ = skin cardiac output, liters/min
    = 0.4 for basal conditions
$CSKIN$ = skin coefficient, liters/(°C-min)
$ISKINT$ = increase in skin temperature, °C
$CCORE$ = core (hypothalamus) coefficient, liters/(°C-min)
$ICORET$ = increase in core temperature, °C

Rowell [4] says that for tasks with TOTMET < 0.2 VO$_2$MAX, the increase in COSKIN comes from a decrease in COBASL + COACT rather than an increase in cardiac output. That is, for lighter work, the skin blood flow is just a redistribution of blood rather than an increase.

In a person who needs to lose heat, vasodilation brings hot central blood to the surface, where heat is transferred to the environment by radiation and convection; COSKIN can increase to up to 4 times COBASL. This vasodilation circulation bypasses the muscles as the blood flows from the small arteries (arterioles) to the small veins (venules) through bypasses (arteriovenous anastomoses). This flow through a "high resistance" naturally increases the heart rate. Thus the common observation is that heart rate increases in heat.

However, for persons who are heat-acclimatized, the sweating ability is enhanced. Because they can sweat more and thus lose heat by evaporation, they do not need to lose as much heat by vasodilation. Because their blood does not use the high-resistance paths, heart rate is lower for acclimatized people than for nonacclimatized people. Thus, although heat stress can affect heart rate, it does not affect the oxygen consumption.

where
OXUPTK = oxygen uptake, VO_2 at standard temperature (0°C) and pressure (760 torr), dry (STPD), liters O_2/min
CO = cardiac output, liters blood/min
AVDIF = arterial-venous oxygen differential, liters O_2/liter blood

While resting, the arterial oxygen content is 19 mL per 100 mL of blood and the venous oxygen content is 15 mL/100 mL. That is, for every 100 mL of blood passing the muscles, the muscles get 4 mL of oxygen. However, in an emergency (fleeing from a tiger), the muscles can get 13 mL O_2/100 mL of blood; that is, the venous oxygen content drops to 6 mL. In highly trained athletes, the AVDIF can be 16. The coronary blood supply, even under normal circumstances, has an AVDIF of 17; thus more oxygen for the heart must come from more blood, not from an increase in the AVDIF.

Blood Distribution. The fourth way of getting more oxygen to a muscle is through redistribution. During exercise, the capillary density increases from 200 per square millimeter at rest to 600 mm^{-2}. Muscle blood flow can change from 2 mL per 100 mL of tissue to 14. Blood stored in the lungs can increase from 500 mL to 1500 mL. As exercise increases, the kidneys and intestines gradually use less blood and send their blood to the skin and muscles. If food is present in the stomach, cramps may result.

Oxygen Debt. If you are "underdeposited," you go into debt. If the aerobic supply of oxygen (i.e., from the lungs) is not sufficient, the muscles draw upon the anaerobic oxygen stored in the blood. However the anaerobic supply is limited, and any that is borrowed must be repaid—with interest. See Figure 2.

3. Cardiovascular System: Gender, Age, and Training

Gender. For physical work, the average female has a maximum oxygen uptake (VO_2MAX) 15-30% below that of males [19]; this is due to a higher percentage of body fat and a lower hemoglobin concentration in the blood.

As fat tissue contains little blood, blood volume for an adult male averages 75 mL/kg (it is 65 for a female and 60 for a child) [15]. For the same age and body weight, a female has a lung volume about 10% lower than that of a male. Females also have lower hemoglobin content than males (13.9 vs 15.3), lower hematocrit (the relative amount of plasma and corpuscles) (42 vs 47), and lower arterial oxygen content (16.7 mL/100 mL vs 19.2/100). Thus for submaximal work (oxygen uptake of 1.5 liters/min), females need 9 liters of cardiac output to transport 1 liter of oxygen whereas men require only 8 liters.

Although these figures help explain the difference in athletic performance, cardiovascular differences between men and women should have little importance for most industrial work because most industrial tasks should not be designed to require maximum cardiovascular output.

Age. Our body's physical performance peaks at around age 25. After that it is all downhill.

For the specific index of VO_2MAX, Dehn and Bruce [20] reported, from three studies, that it declines by 1.04, 0.94, and 0.93 mL/(kg-min) per year for males; in their own study, the decline was 1.32 for habitually inactive males and 0.65 for active males. Astrand et al. [21] reported declines, over 21-year period, for Swedish physical education instructors, of 0.64 for males and 0.44 for females. Ilmarinen [22] says that the decline in VO_2MAX after age 20-25 is 1-2% per year, but there are large individual variations. Schacher et al. [23] combined gender, age, and body fat to predict treadmill VO_2MAX:

MALE VO_2MAX = 66.734 - 0.315 AGE - 0.678 PFAT

FEMALE VO_2MAX = 58.094 - 0.356 AGE - 0.494 PFAT

where AGE is age in years and PFAT is percent body fat.

Jackson et al. [24] emphasize that most of the decline is due to physical activity level and percent body fat, not aging. The aging effect was just 0.27 mL/(kg-min) per year.

$$\text{MALE VO}_2\text{MAX} = 47.9 - 0.27 \text{ AGE} + 3.41 \text{ SRPA} - 0.20 \text{ PFAT}$$
$$- 0.09 \text{ SRPA} \times \text{PFAT}$$

where SRPA is a measure of self-reported physical activity.

What does this decline in capability mean for jobs? Henschel [25] gives a good summary: "Cardiovascular capacity to perform light to moderate physically exhausting work is not grossly age-dependent up to age 65, although capacity for hard, exhausting work is strongly age-dependent, with maximum capacity between 20 and 25."

Training. Fitness has "dimensions" of cardiovascular fitness (endurance), muscle strength, and flexibility. Specific training techniques are more appropriately found in books on kinesiology or athletics. However, two general statements are

1. To train the cardiovascular system to increase cardiac output, train with large muscle groups.
2. To strength train specific muscles, train with the specific muscles.

Some exercises strengthen weak muscles (work hardening), and other exercises stretch tight muscles and ligaments. If the job loads the muscles dynamically, relax and stretch them. If the job loads the muscles statically, the exercise should move them.

4. Cardiovascular Limits

In seeking to determine cardiovascular limits, the first question is, How to you determine an individual's work capability? The second is, What proportion of the capability should be used?

Capability. The capability of the cardiovascular system is determined through a test to determine the maximum oxygen uptake, VO_2MAX, which is measured in milliliters per kilogram per minute. VO_2MAX is a product of cardiac output and A-V differential. Table 4 shows categories of fitness for U.S. males. (Females typically have VO_2MAX 15-30% below that of males [19]) so their critical values would be about 75% of male values. A male under 30 having a value of 33 mL/(kg-min) would be considered to be in fair shape. A female under 30 having a value of 33 would compare with 33/0.75 = 44 and be in good shape.

Wisner [26] gives many values of VO_2MAX for various countries. For developing countries, he emphasizes that the low values of aerobic power of many people are due to poor nutrition and parasitoses.

Table 4 Maximum Oxygen Consumption VO_2MAX, mL/(kg-min), for U.S. Males of Various Ages and Degrees of Cardiovascular Fitness

Cardiovascular fitness	Age, yr			
	Under 30	30–39	40–49	50+
Very poor	<25.0	<25.0	<25.0	—
Poor	25.0–33.7	25.0–30.1	25.0–26.4	25.0
Fair	33.8–42.5	30.2–39.1	26.5–35.4	25.0–33.7
Good	42.6–51.5	39.2–48.0	35.5–45.5	33.8–43.0
Excellent	51.6+	48.1+	45.1+	43.1+

Source: Ref. 2.

How is VO$_2$MAX determined? One possibility (a maximal test) is to put a person on a treadmill or bicycle ergometer and exercise the person to exhaustion. (The VO$_2$MAX determined on a treadmill will be higher than on a bicycle, which in turn will be higher than for lifting [7].) However, this requires trained staff and equipment. Another test is to record the distance a person can run/walk in 720 s (12 min):

$$VO_2MAX = -10.3 + 35.3 \text{ DIST}$$

where
VO$_2$MAX = maximum oxygen uptake, mL/(kg-min)
DIST = distance run in 720 s, miles

A submaximal test is the step test. There are two 5-min sessions of stepping up and down on a 40-cm step with 15 cycles/min for the first test and 25 cycles/min for the second. The heart rate is measured from the pulse from 30 to 60 s after work stops, from 90 to 120 s, and from 150 to 180 s [27].

Testing the capabilities of individuals for screening purposes has become controversial in the United States because of antidiscrimination laws.

Proportion of Capability. Endurance work (lasting 4–8 h) can be carried out at a rate of 33–50% of a person's VO$_2$MAX. The general concept is to avoid the use of anaerobic metabolism. Rohmert [28] says that very fit individuals can work at 50% of their VO$_2$MAX for 8 h a day for 6 days a week. Jorgensen [29] recommends 50% for trained workers and 33% for untrained. If the task is primarily upper body work, the maximum watts should be about 30% less; see Eastman Kodak [30] for a more extensive discussion. Mital et al. [31] base their lifting guidelines on 21–23% of uphill treadmill aerobic capacity or 28–29% of bicycle aerobic capacity; note that lifting has static components as well as dynamic.

Eastman Kodak [30] recommends (their Figures 11.1 and 15.1) a sustained maximum of 33% VO$_2$MAX for an 8-h shift, 30.5% for a 10-h shift, and 28% for a 12-h shift. Mital et al. [31] reduce their 28–29% of bicycle aerobic capacity for 8 h to 23–24% for 12 h.

For periods of less than 8 h of work performance, see Figure 3.

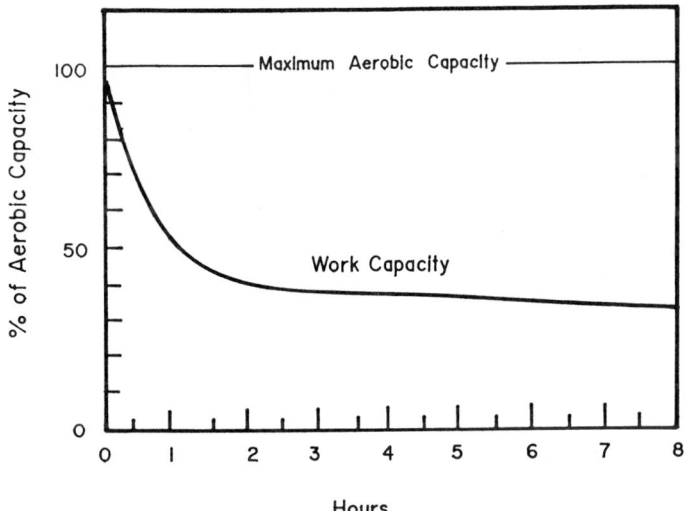

Figure 3 Work capacity declines with working time. (From Ref. 2.)

Assuming you wish to exclude only a small percentage of the population, the above percentages translate to about 350 W, 5 kcal/min, and 100–120 heart beats/min. Wisner [26] states that "it is more or less universally admitted that heartbeat rates should not exceed 110 beats/min during the working day. During more intensive work periods, 130 beats/min should not be exceeded."

The fact that people can work hard does not mean that they should. Jobs that require high metabolic rates should be considered prime candidates for mechanization, as motors are a far more efficient power source than muscles. Consider fork lifts, balancers, manipulators, and horizontal transfer rather than vertical transfer. After all, although scientists may be interested in physiology, engineers should be interested in productivity.

Two administrative solutions are job rotation and part-time work. In job rotation, people shift jobs periodically within the day. In part-time work, the job is given to several people who work part of the day; for example, employing Joe for 4 h in the morning and Pete for 4 h in the afternoon or scheduling enough people that the entire job is done in part of a shift (e.g., package handling for delivery services is often done within 2–3 h/day).

When setting work standards, industrial engineers usually use fatigue allowances. See Chapter 28 in Konz [2] for some tables of allowances.

IV. PHYSIOLOGICAL CRITERIA FOR ERGONOMIC APPLICATIONS

The cardiovascular limits for heavy work are given in terms of aerobic capacity, which is usually expressed in terms of oxygen capacity (VO_2MAX) or heart rate. In summary,

For an 8-h shift, heart rate should not average over 110 beats/min.
For parts of a shift, heart rate should not average over 130 beats/min.

REFERENCES

1. S. Konz, *Work Design: Industrial Ergonomics*, 2nd ed., Publishing Horizons, Worthington, OH, 1984.
2. S. Konz, *Work Design: Industrial Ergonomics*, 3rd ed., Publishing Horizons, Worthington, OH, 1990.
3. A. Freivalds and C. Bise, Metabolic analysis of support personnel in low-seam coal-mines, *Int. J. Ind. Ergon. 8*: 147–155 (1991).
4. K. Pandolf, M. Haisman, and R. Goldman, Metabolic energy expenditure and terrain coefficients for walking on snow, *Ergonomics 19*: 683–690 (1976).
5. K. Pandolf, B. Givoni, and R. Goldman, Predicting energy expenditure with loads while standing or walking very slowly, *J. Appl. Physiol. Respir. Environ. 43*:(4): 577–581 (1977).
6. R. Alexander, Stride length and speed for adults, children and fossil hominids, *Am. J. Phys. Anthropol. 63*: 23–27 (1984).
7. R. Soule and R. Goldman, Energy cost of loads carried on the head, hands or feet, *J. Appl. Physiol. 27*: 687–690 (Nov. 1980).
8. S. Legg and A. Mahanty, Energy cost of backpacking in heavy boots, *Ergonomics 29*(3): 433–438 (1986).
9. W. van der Walt and C. Wyndham, An equation for prediction of energy expenditure of walking and running, *J. Appl. Physiol. 34*(5): 559–563 (1973).
10. M. Klieber, *The Fire of Life*, Wiley, New York, 1961.
11. R. Peters, *The Ecological Implications of Body Size*, Cambridge Univ. Press, Cambridge, U.K., 1989.

12. R. Andrews, The relationship between measures of heart rate and measures of energy expenditure, *Am. Inst. Ind. Eng. Trans. 1*(1): 2–10 (1969).
13. G. Borg, Psychophysical scaling with applications in physical work and the perception of exertion, *Scand. J. Work Environ. Health 16*(Suppl. 1): 55–58 (1990).
14. G. Ljunggren, Observer ratings of perceived exertion in relation to self ratings and heart rate, *Appl. Ergon. 17*(2): 117–125 (1986).
15. P. Astrand and K. Rodahl, *Textbook of Work Physiology*, 3rd ed., McGraw-Hill, New York, 1986.
16. A. Roozbahar, G. Bosker, and M. Richardson, A theoretical model to estimate some ergonomic parameters from age, height and weight, *Ergonomics 22*(1): 43–58 (1979).
17. A. Guyton, *Textbook of Medical Physiology*, Saunders, Philadelphia, 1961.
18. J. Stolwijk, Mathematical model of thermoregulation, in *Physiological and Behavioral Temperature Regulation*, J. Hardy, A. Gagge, and J. Stolwijk, Eds., Thomas, Springfield, IL, 1970, Chap. 48.
19. J. Vogel, J. Patton, R. Mello, and W. Daniels, An analysis of aerobic capacity in a large United States population, *J. Appl. Physiol. 60*: 549 (1986).
20. M. Dehn and R. Bruce, Longitudinal variations in maximum oxygen intake with age and activity, *J. Appl. Physiol. 33*: 805–807 (1972).
21. I. Astrand, P. Astrand, I. Hallbeck, and A. Kilbom, Reduction in maximal oxygen uptake with age, *J. Appl. Physiol. 35*(5): 649–654 (1973).
22. J. Ilmarinen, Job design for the aged with regard to decline in their maximal aerobic capacity. Part II. The scientific basis for the guide, *Int. J. Ind. Ergon. 10*: 65–77 (1992).
23. C. Schacherer, A. Rowe, and A. Jackson, Development of prediction models for physical work capacity: practical and theoretical implications, *Proc. Hum. Factors Soc.* 1992, pp. 674–678.
24. A. Jackson, E. Beard, L. Wier, and J. Stuteville, Multivariate model for defining changes in maximal physical working capacity of men, ages 25 to 70 years, *Proc. Hum. Factors Soc.* 1992, pp. 171–174.
25. A. Henschel, Effect of age on work capacity, *Am. Ind. Hyg. Assoc. J. 31*(4): 430–436 (1970).
26. M. Sharp, E. Harman, J. Vogel, J. Knaik, and S. Legg, Maximum aerobic capacity for repetitive lifting: comparison of three standard exercise modes, *Eur. J. Appl. Physiol. 57*: 753–760 (1988).
27. W. Tuxworth and H. Shahnavaz, The design and evaluation of a step test for the rapid prediction of physical work capacity in an unsophisticated industrial work force, *Ergonomics 20*(2): 181–191 (1977).
28. W. Rohmert, Problems in determining rest allowances. Part 2. Determining rest allowance in different human tasks, *Appl. Ergon. 4*(2): 43–58 (1979).
29. K. Jorgensen, Permissible loads based on energy expenditure measurements, *Ergonomics 28*(1): 365–369 (1985).
30. Eastman Kodak, *Ergonomic Design for People at Work*, Vol. 2, Van Nostrand Reinhold, New York, 1986.
31. A. Mital, A. Nicholson, and M. M. Ayoub, *A Guide to Manual Material Handling*, Taylor & Francis, London, 1993.

3
Physiological Aspects of Neuromuscular Function

Thomas R. Waters
National Institute for Occupational Safety and Health, Cincinnati, Ohio

Amit Bhattacharya
University of Cincinnati Medical School, Cincinnati, Ohio

I. INTRODUCTION

The mechanical motions of the body segments necessary to accomplish a task are skillfully controlled by a myriad of neuromuscular components and a series of well-orchestrated neural events encompassing both the central and peripheral nervous systems. Some of these issues are discussed in the following sections of this chapter.

The maintenance of ergonomically efficient posture requires the optimal orientation of various interconnected human body segments to produce minimal biomechanical loadings of the joints. The neural components of movements are the planning and programming units and the performance units. Planning and programming are the functions of the precortical centers (cerebral cortex, bilateral movement), the basal ganglia, the cerebellum, and the thalamus. The premotor and sensory regions provide the input for planning to the basal ganglia and the cerebellum. The performance units include the motor cortex and spinal cord, with the smoothing function performed by the cerebellum. In order for a person to perform a motor act smoothly, all the somatosensory systems have to work in harmony to help provide "accurate" information regarding the position of body segments, muscle tensions, and joint motions to the higher centers. For a detailed discussion of this topic, readers should refer to comprehensive texts of physiology and neuroanatomy [1-5].

II. BACKGROUND AND SIGNIFICANCE TO OCCUPATIONAL ERGONOMICS

A worker who performs a simple manual task, such as picking up a box from a table, does so without much difficulty. The underlying physiological chain of events required to perform this seemingly simple task, however, is very complex and involves a number of physiological systems including the muscular, sensory, and central and peripheral nervous systems. Prior to the actual event of reaching for the box, the worker uses his or her visual system and the higher brain centers to make a judgment regarding its weight, size, and location with respect to the orientation and location of his or her body. This information is synthesized by the brain, and a movement plan is developed that will generate a series of complex control commands. These control commands are relayed to the appropriate muscle groups (via motor neurons) to allow for smooth movement of the body segment and load unit.

At the muscle–bone level, when the appropriate commands are received by the muscle, the muscle contracts and applies a pulling force on the tendon attached to the bone segment. If the tensile force generated by the muscle is strong enough to create a moment that will overcome the moment created by the weight of the segment and any external load applied to the segment, then the segment and load will be moved.

To ensure smooth movement, a complex system of sensory elements monitors the condition of the musculoskeletal system. These sensors measure the joint forces, muscle forces, and muscle length and provide continuous feedback to the central nervous system, which then modifies the commands sent to the muscles. The sensors consist of various muscle stretch (change-of-length detector) and tension-monitoring receptors, which balance reciprocally acting pairs of muscles to control the position of a segment at any point in time (e.g., extensor versus flexor muscles around a joint). This positional control system is crucial to coordinated muscle movement. The sensory units are integrated into a highly evolved system of reflexes that provide a short-latency connection between the sensory and motor components of the system that results in stereotypical motor responses to sensory stimulus. This arrangement allows for a rapid response to perturbations in the status of the system (e.g., the knee jerk response to stretching the patella ligament of the knee by striking it with a hammer).

III. RELEVANT CONCEPTS AND TERMINOLOGY

A. Major Components of the Neuromuscular System

Table 1 lists the five main elements of the neuromuscular system and their purpose. As shown in Figure 1, the five component groups of the neuromuscular system are organized into a highly complex command and control system that integrates physiological and mechanical functions into a single system capable of performing work.

1. Cerebral Cortex, Cerebellum, and Subcortical Centers

In humans, the programming center for a movement is made up of the cerebral cortex of the brain and various components of the subcortical centers (basal ganglia, brainstem nuclei, and brainstem reticular formation). These systems send descending commands (efferent) to the motor neurons and finally to the muscles (Figure 2). During contraction of the muscles, various receptors in the muscles (muscle spindles), tendons (Golgi tendon organ), and joints send real-time information (afferent) about the status of the body segment movement to the higher centers for processing. The role of the cerebellum is to smooth the movement of the body segment. It receives information from afferent systems (vestibular, proprioception, and visual) as well as commands from the higher centers. Input signals from the cerebral cortex are

Table 1 Major Components of the Neuromuscular System

Component	Purpose
Cerebral cortex, cerebellum, and subcortical centers	Central processing unit for coordination of muscular activity
Motor neurons	Neural circuitry
Muscle fibers	Force-generating neuromechanical actuator
Visual, vestibular, and somatic sensory receptors	Feedback sensors for detection of position, stretch, pressure, tension, etc.
Bones, joints, ligaments, and tendons	Mechanical support and linkage connectors

Neuromuscular Function

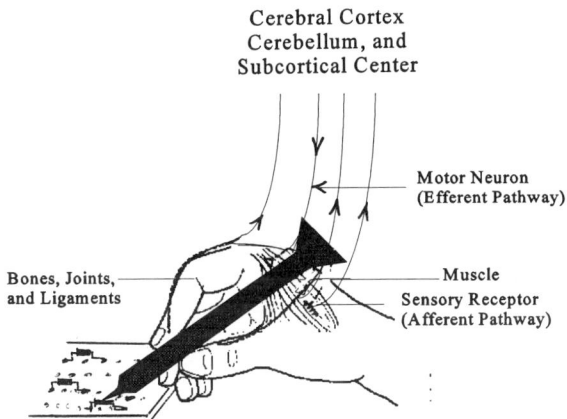

Figure 1 Components of the neuromuscular system showing the direction of flow of neural signals. A complex system of interneurons (not shown) also provides reflex pathways in which afferent sensory signals can directly affect motor signals without traveling to the higher centers of the brain. The final common pathway for motor function is the motor neuron, which connects directly to the muscle fibers. (Adapted from Ref. 3.)

conveyed via brainstem nuclei to the cerebellum and tell the brain what the muscles should be doing. Based on its knowledge of the status of various afferent systems (as it relates to motor coordination) and the nature of "expected" controlling commands from the higher centers, the cerebellum is capable of producing smooth movement. It is not clear how the cerebellum processes this information. However, although its exact mechanism of action is not clear, it certainly plays an important role in minimizing the error signal between what the muscles are doing and what they "should" be doing. The cerebellum does not initiate muscle contractions directly but controls the activity in the descending motor commands. People with cerebellum damage have jerky uncoordinated movements, poor balance, and unsteady gait [1].

2. Motor Neurons

The motor neuron, illustrated in Figure 2, is a nerve cell that links the central nervous system and the appropriate muscle fiber or fibers. The motor neuron consists of a cell body containing the cell nucleus, an elongated segment or axon, and the endplate or neuromuscular junction. When a command signal of sufficient strength is received at the body of the motor neuron, the cell membrane is depolarized (there is a drop in voltage due to a change in cell

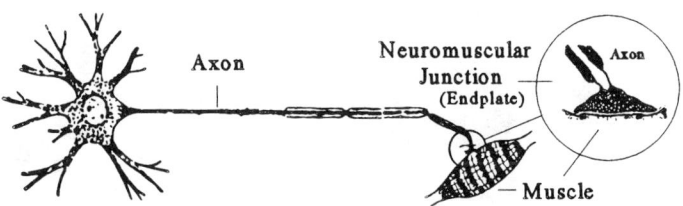

Figure 2 Illustration of a motor neuron showing the axon and the neuromuscular junction (endplate). When the cell body is stimulated, an action potential is transmitted to the endplate, where a neurotransmitter (acetylcholine) is released into the neuromuscular junction. The neurotransmitter causes the muscle fibers to contract.

membrane permeability. This depolarization initiates an action potential along the nerve axon. The action potential is a propagating electric signal that travels toward the endplate. When the action potential reaches the endplate, acetylcholine, a chemical transmitter, is released into the neuromuscular junction. This chemical transmitter then depolarizes the endplate membrane of the muscle fiber to a threshold value (to +30 mV from a resting potential of −70 mV). This action electrically triggers the muscle fiber contraction.

Motor neurons and muscles are organized into motor units that consist of a single motor neuron, its axon, and all the muscle fibers innervated by it. The number of muscle fibers innervated by a single motor unit varies from a few (e.g., muscles that move fingers) to several hundred (e.g., back muscles). The number of muscle fibers innervated by one motor unit is dependent on the function of the muscle rather than its size. Muscles that cause large and strong body motions usually have more muscle fibers under the control of a motor unit than those required to perform fine precision movements.

3. Muscles

There are over 600 muscles in the human body accounting for about 45% of the total body weight. Muscles are composed of one of three kinds of fibers, depending upon the function of the muscle. The three types are skeletal, smooth, and cardiac. Skeletal muscle is connected to the bones of the body, and when contracted it causes the body segments to move. Smooth muscle is found in the stomach, intestinal tracts, and walls of blood vessels. Cardiac muscle is the contractile tissue found in the heart that pumps the blood for circulation. In this chapter, we focus our attention on the skeletal muscles.

A single skeletal muscle, as shown in Figure 3, consists of hundreds to tens of thousands of muscle fibers, totaling about a quarter of a billion in an average person. Each muscle fiber consists of a single cylindrical muscle cell with a diameter of 10–90 μm and a length

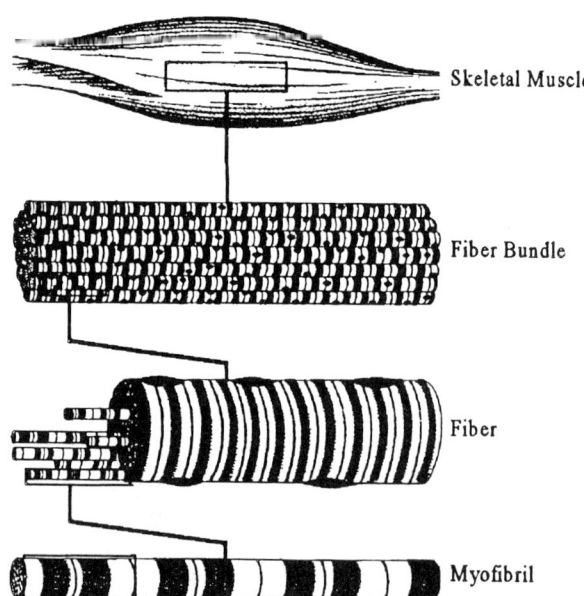

Figure 3 Illustration of a skeletal muscle showing the fiber composition. Each fiber is innervated by one motor neuron that controls the contraction of the muscle fiber. (Adapted from Ref. 6.)

of up to 30 cm. Muscle fibers are further divided into individual contractile elements called myofibrils. Myofibrils are the basic contractile element of the muscle, and the forces and movement in a muscle cell are generated by special protein molecules and contractile proteins in the myofibrils. Each muscle fiber contains hundreds of myofibrils that actively contract in the presence of calcium, which is released when a contraction is initiated. Contraction is accomplished by a complex chemical process whereby adjacent filaments of molecular proteins, actin and myosin, are pulled toward each other in a sliding motion that results in shortening of the muscle cell and the development of tensile force. Generally, muscle fibers are shorter than the muscle they make up, but some fibers run the entire length of the muscle.

The amount of tension or force generated by a contracting muscle is dependent on its precontraction length, the velocity of the contraction, and the direction of muscle movement during the contraction (i.e., whether the muscle is lengthening or shortening). When a muscle is shortening during a contraction, the activity is defined as *concentric*. Conversely, when a muscle is lengthening during a contraction, the activity is defined as *eccentric*. Figure 4 graphically illustrates the relationship between muscle force, length, velocity of contraction, and direction of movement.

For a concentric contraction, there is an optimal precontraction muscle length that will produce a maximum tension force when the muscle is stimulated. This length is called the *resting length* of the muscle. If the precontraction muscle length is at or below 60% of the resting length, the muscle will not produce any tension when stimulated. Therefore, the ability of a muscle to produce an optimal force is strongly dependent upon the position of the body segment to which the muscle is attached. For example, optimal biceps muscle tension is generated when the elbow joint angle (subtended between the forearm and the upper arm) is in the region of 90–100°. Moreover, as the velocity of a concentric contraction increases, the force decreases.

Muscles are usually attached to bones in a paired arrangement, with the *agonist* muscles performing the main movement of the body segment and the *antagonist* muscles acting as the

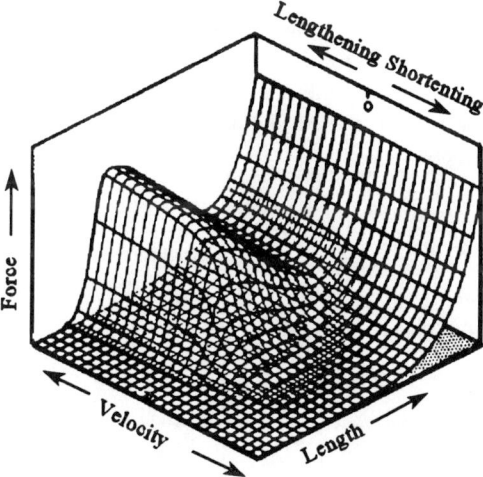

Figure 4 Graphical illustration of the effects of muscle length, contraction velocity, and direction of muscle movement on muscle force (assuming a fixed excitation level). The surface plot shows the magnitude of force developed as a function of muscle length, velocity, and movement direction. The ridge of the peak corresponds to the resting length of the muscle. (Adapted from Ref. 7.)

controller. *Fixator* muscles provide support to the proximal joints, and *synergist muscles* help prevent undesirable movements of other joints when the agonist muscle passes over more than one joint.

4. Visual, Vestibular, and Sensory Receptors

In order to move a body segment(s) and/or maintain whole-body upright balance during static and dynamic conditions, the brain must receive feedback information regarding the position and movement of the body. This type of information is provided by a series of biological sensors located in the joints, muscles, and tendons and under the skin, as well as the visual and vestibular systems. The output from these sensory components is transmitted by afferent neurons to the spinal cord and brain, where they are processed and used to alter the motor signals. Feedback from these sensory receptors is essential for smooth, coordinated movement because the sensory input provides the cues necessary to alter the timing of the motor program that controls the motor function. For new, unlearned tasks, the visual and vestibular systems are especially important in generating the coordinated motor patterns for smooth motion. The flowchart shown in Figure 5 illustrates how the sensory system provides feedback information to alter the motor signals sent to the muscles.

The visual and vestibular systems provide information about the spatial orientation of the body and the movement of the head. The *visual system* provides information regarding the orientations (horizontal and vertical) of objects in three-dimensional space. The *vestibular system*, illustrated in Figure 6, provides information regarding the position and movements of the head and their relationship to gravitational forces. The three orthogonally placed semicircular canals provide information regarding acceleration of the head in three-dimensional space. The utricle and the saccule are position sensors that provide information regarding the position of the head in space.

The actions of the vestibular system are not consciously felt unless one is required to perform a motor task under dim light and/or walk or stand on an uneven surface. The literature indicates that the role of the vestibular system for motor task performance is not as critical when the other afferents from the visual and proprioception systems are intact.

Proprioceptive sensors provide information about the status of the muscles themselves, such as how hard the muscles are pulling and how fast they are being stretched. Muscle stretch receptors (muscle spindle) and the Golgi tendon organs (tension-monitoring receptors) transmit information regarding muscle length and tension to the controlling centers in the cortex and subcortical units as well as to the cerebellum.

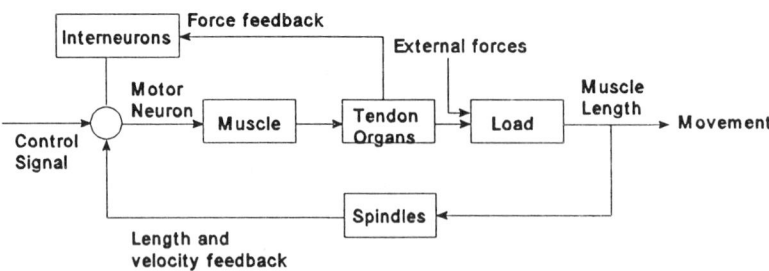

Figure 5 Flowchart of muscle control system. Feedback loops from tendon organs and muscle spindles provide sensory feedback for control of muscular function. Descending pathways from the brain provide control signals, where they are integrated with the sensory input and sent to the muscles via the motor neurons. (Adapted from Ref. 7.)

Neuromuscular Function

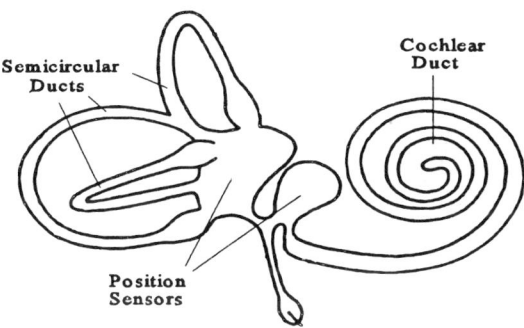

Figure 6 Vestibular sensory receptors. Semicircular ducts provide sensory input regarding angular acceleration. The position receptors provide information about the orientation of the head. (Adapted from Ref. 7.)

Kinesthetic sensors measure joint movements, and somatic sensors in the skin provide information about temperature, pressure, and pain.

5. Bones, Joints, Tendons, and Ligaments

The bones, joints, tendons, and ligaments serve as an integrated support frame for the body and provide an attachment point for the muscles. The bones are connected by a complex arrangement of connective tissue to form a custom-fitted joint that provides a flexible yet stable connection with a wide range of motion. As the muscles contract, they transmit their forces to the bones through the tendons, thereby causing the bones to move around the joints.

There are three classes of joints: fibrous, cartilaginous, and synovial [4]. Most body movements take place around the synovial joints. There are six types of synovial joints: hinge, pivot, ellipsoid, ball-and-socket, plane, and saddle joints. Examples of these joints are found in the elbow, neck, wrist, shoulder, between the bony arches of the vertebrae, and at the base of the thumb. The type of motion and the maximum range of motion of the body segment possible at the synovial joint is dependent upon the shape of the articulating bones, the strength and the orientation of the surrounding ligaments, and the size and strength of the muscles.

The muscles are attached to bones by collagen fibers called tendons. Tendons have strong tensile characteristics but do not possess contracting properties. Tendons are instrumental in transmitting forces generated by the contracting muscles. For example, finger motions are caused by the force transmitted by the long tendons attached to the contracting muscles in the lower arm (between the elbow and the wrist).

Movements of the body segments around the articulating joint surfaces have been given specific names. In general, when the angle between two attached bone segments decreases during a movement, the movement is called *flexion*, and when the angle between the two attached bone segments increases, the movement is called *extension*. When the movement of a limb is away from the midline of the body, the movement is called *abduction*, and when the movement is toward the midline of the body, the movement is called *adduction*. *Pronation* and *supination* movement occur when a body segment undergoes a rotation about its long axis. Pronation or medial rotation is a movement toward the center of the body, whereas supination or lateral rotation defines a movement away from the center of the body. Table 2 provides some examples of body movements and the corresponding muscles producing those movements.

Table 2 Table of Segment Movements and Muscles Used in Making Those Movements

Body movement	Agonist	Antagonist
Elbow flexion	Biceps brachii	Triceps brachii
Shoulder abbduction	Deltoid	Pectoralis
Spinal extension	Erector spinae	Rectus abdominus
Hip flexion	Psoas, iliacus, rectus femoris, pectineus, tensor fascia	Gluteus maximus and hamstring

B. Motor Control

1. Motor Programs

Learned tasks, such as touching your nose, standing on one foot, or hitting a baseball, are controlled by overall motor plans that are stored in the central nervous system. These programs provide the proper sequence of motor activity needed to coordinate a specific planned motor function. It is important to note that the motor program must not only provide motor signals to initiate a specific movement but must also provide the motor signals needed to maintain posture. Complex, multijoint movement patterns are difficult to learn, because they are composed of numerous nested subprograms of varying complexity. On the other hand, some neural patterns are repeated so frequently that they may function without external timing cues or sensory input. These programs are sometimes referred to as central pattern generators. Motor activity is monitored by sensory receptors, and the status of the system is relayed to the central nervous system, where the motor program may be altered. Additional neural networks within the spinal cord may modify the motor signals through a feedback system of reflexes.

2. Reflexes

The neuromuscular system has the capability of rapidly adjusting motor function as a result of sensory input through highly developed neural circuits or reflex pathways. These reflex pathways consist of efferent motor neurons, afferent sensory neurons, and interneurons that link and integrate the signals between the sensory and motor neurons. A reflex can be thought of as a negative feedback control system that results in a stereotyped motor response to a sensory stimulus. A number of reflexes have been identified, such as the antigravity reflex for assistance in standing, the stretch reflex for maintaining constant muscle length (demonstrated by the knee tendon jerk test), the flexor and crossed-extensor reflexes that provide a matched response of leg withdrawal to painful stimuli and a resultant extension of the opposite leg, and the reciprocal inhibition reflex that turns the antagonist muscle off during a motor function. It is the network of interneurons in the spinal cord that integrates the control signals from the brain and the afferent signals from the sensory receptors to produce a flow of triggering signals to the muscles. The motor neurons, however, are the final common pathway for muscle activation.

3. Static and Dynamic Contractions

Muscles perform two basic types of tasks: dynamic body segment movement and the holding of body segments in static postures. Concentric contractions involve shortening of the muscles, causing a movement such as lifting a box. Eccentric contractions involve lengthening of an actively contracting muscle, such as the controlled lowering of a weight on a glass surface. In this case, the active muscle is controlling the rate and the motion under the action of the external force of gravity. A static contraction is performed when a posture is maintained without any movement. This type of contraction can cause rapid muscle fatigue due to poor blood circulation and the buildup of metabolites and waste products in the muscle. Most body movements or sustained postures require the use of more than one muscle.

4. Recruitment and Rate Coding

To allow for a wide range of muscle force levels, two physiological mechanisms are used to control the amount of force generated by a muscle: recruitment and rate coding.

Recruitment is defined as the number of motor units involved in the muscle contraction, and *rate coding* is defined as the frequency of discharge of the motor neurons. For slow contractions, such as postural control, the small, slow, fatigue-resistant motor units are recruited first, and the larger, fast, fatigueable motor units are recruited as the desired movement becomes faster or more powerful. Rate coding is used to modulate the force level by increasing the discharge rate as more force is needed. In general, for powerful contractions, all of the motor units are recruited before maximum discharge rates are reached.

C. Types of Movements

When the decision has been made to move one or more body segments, the neuromuscular system must choose which muscles to turn on, when they should be turned on, and how much force each muscle should develop. Since a wide range of activation sequences, muscle combinations, movement speeds, and motion trajectories could be used to achieve a desired movement, a predictable set of movement strategies must exist in order to optimize the efficiency of the movement. In general, complex movements can be resolved into simpler individual movements that are controlled by the agonist and antagonist muscles, which work together to achieve a specified movement. In a typical movement, for example, the agonist muscle contracts first to initiate the rotation of the segment about the joint at the proper velocity, and a short time later the antagonist muscle contracts to slow or stop the motion of the segment. In broader terms, the accuracy and speed of a movement dictate the selection and sequencing of the muscles, but postural constraints and system equilibrium must also be considered.

A number of goal-directed movement strategies have been proposed. Some of these strategies are based on the method of excitation control. For example, speed-dependent movements are modulated by the excitation amplitude, whereas speed-independent strategies are modulated by the excitation duration [8]. In some movements, coactivation of the agonist and antagonist muscles is needed to increase joint stiffness to stabilize the joint when it is exposed to high inertial loading. Significant coactivation has been shown in the spine for dynamic movements [9]. The stretch–shorten cycle has been proposed as another strategy for controlling movement. The stretch–shorten cycle is a linked eccentric–concentric contraction designed to use the additional force generated by a muscle when it is contracting eccentrically (lengthening), which is helpful in high-performance movements. Also, synergistic muscle action (cooperating agonists) may be used to assist in the development of the proper segment velocities.

D. Assessing Neuromuscular Function

1. Electromyography

When a muscle is activated, it creates an electric discharge (myoelectric signal) that can be measured directly from the muscle or through an electrode attached to the surface of the skin. Measurement and recording of these signals is called electromyography (EMG). These electric signals provide information about the intensity and duration of the contraction. Although the activity of individual motor units can be measured with needle and fine wire electrodes, surface electrodes are typically used to measure the activity of whole muscle groups. In this case, the measured signal is the summation of all the active motor units within the recording area of the electrodes. Under certain conditions, EMG activity can be used to measure the magnitude of the force being developed in a selected muscle group. Care must be taken, however, in determining muscle force from EMG activity, due to the length–tension and velocity–tension characteristics mentioned previously. For more information about making EMG measurements, the National Institute for Occupational Safety and Health has published a users' manual [10].

2. Evoked Potentials

Evoked potentials are electric signals initiated in a neural pathway by the application of an external stimulus (electric pulse) to a motor or sensory component to determine the functional status of the pathway of interest. Measurement of the amplitude or conduction velocity of an evoked potential along a neural pathway is useful in evaluating the function of the sensory or motor system. For example, visual or auditory evoked potentials can be used to assess the function of the optic or auditory nerve, and somatosensory evoked potentials can be used to assess the function of the various peripheral nerves. Multiple stimulations are required to measure evoked potentials from the skull because the signals must be averaged to remove the random electroencephalographic (EEG) activity of the brain.

3. Postural Stability

When a person is standing upright, the neuromuscular system provides motor control to the musculature of the supporting limbs to maintain a stable posture. These fine motor contractions result in a small natural oscillation or swaying of the body. Techniques for measuring the extent of this natural swaying are discussed in Chapter 4. It has been shown that the natural swaying of the body is modified in clinical and neurological disorders, such as in humans exposed to neurotoxic industrial chemicals and by man made drugs and physically fatiguing tasks [11–14]. Excessive body sway could very well interfere with the safe performance of tasks and may jeopardize safety in the workplace. Successful application of this method has been illustrated for workers exposed to chemicals that affect the neurological system and may impair their balance, making them sway more than members of an unexposed population [14].

4. Muscular Strength Measurement

Strength is a measure of the maximum force that can be produced by a single muscle or by a series of muscles under prescribed conditions. Owing to the voluntary nature of the test, it is usually thought of as the maximum voluntary contraction level. Standardized tests have been developed for measuring static (isometric) strength as well as dynamic (isokinetic and isoinertial) strength. Recall that muscle force is a function of the resting length of the muscle, the velocity of contraction, and the direction of motion. Thus, muscle strength measures depend upon the position of the body and the type of motion occurring at the time of measurement. Maximum strength values have been published for a wide range of test conditions. For more information on assessment of strength, see the chapter on physical work capacity in this book or refer to studies reported in the literature [15,16].

5. Kinesiology

In its broadest sense, the term *kinesiology* refers to the study of movement. In the context of this chapter, however, kinesiology is the measurement and analysis of human motion and the way in which it relates to the musculoskeletal system that generates those movements. Kinesiological measurements are important to understand musculoskeletal functions such as gait, posture, and static and dynamic muscle action. Kinesiology, which is closely related to biomechanics, is useful for identifying abnormalities in musculoskeletal function (see Chapter 4). For example, movement requirements can be analyzed and used to assess work demands in ergonomics, motion patterns can be compared to normative patterns to assist the clinician in diagnosing neuromuscular dysfunction, and kinematic measurements may be used to improve athletic mechanics.

Measurement techniques are available for either two- or three-dimensional analysis. Two-dimensional kinesiological measurements may be made with a simple goniometer (a device designed to measure the relative rotation of a given joint) or with a video camera and markers placed at the joints of interest. The linear translation, velocity, and acceleration of the given joint as well as the relative angle between the segments and the rotation velocity and acceleration of the segments can then be obtained from a frame-by-frame analysis of the videotape. Three-dimensional measurements are usually complex and require sophisticated methods for acquiring and analyzing the kinesiological data. A full description of kinesiology is beyond the scope of this chapter. For more information, refer to studies reported in the literature [15,17,18].

6. Tremor

The muscular contraction needed to move a body part or maintain it in a fixed position is accompanied by small, arrhythmic, involuntary oscillations in the muscle forces that are not visible to the untrained eye. These muscle oscillations are referred to as *physiological tremor*. When muscles become fatigued or damaged, the physiological tremor increases, and large-amplitude monorhythmic oscillations are observed. Musculoskeletal function can thus be assessed by measuring the amplitude and frequency of these oscillations. Findley and Capildeo [19] have edited a book that describes methods of assessing movement disorders from tremor measurements. In addition, Galinsky et al. [20] describe a portable device that can be used to measure tremor in a field environment.

E. Metabolic Considerations

As mentioned previously, the basis of muscular contraction is the transformation of chemical energy derived from food taken into the body into useful mechanical energy in the form of muscular contractions. To achieve this transformation, high-energy phosphate compounds, such as adenosine triphosphate (ATP) and phosphocreatinine (PCr), provide the chemical energy needed for muscular contractions. These high-energy-yielding compounds are crucial in both aerobic metabolism (metabolic processes that use oxygen) and anaerobic metabolism (metabolic processes that occur without oxygen).

1. Functional Fiber Types

Because a whole muscle is required to operate across a wide range of exertion conditions, it is composed of a mixture of different functional types of muscle fibers. Based on the amount of time it takes a muscle fiber to reach its peak tension, two types of muscle fibers have been identified: slow (Type I, or red) and fast (Type II, or white). The slow fibers, which take 80–100 ms to reach peak tension, have more myoglobin and rely on aerobic metabolism. They do not fatigue easily, and therefore a task can be maintained over long periods (e.g., endurance running activities). The fast fibers, which usually take about 40 ms to reach peak tension, have higher concentrations of glycolytic enzymes and glycogen and rely on anaerobic metabolism. The fast fibers fatigue easily, but they are best suited for strong and quick body movements (e.g., weight-lifting activities). The exact reasons behind the nature of the functional behavior of these fibers are still not clear. The literature suggests that the type of nerve fibers that innervate these muscle fibers dictate their functional behavior [2]. The slow muscle fibers are innervated by small-diameter, low-conduction velocity nerve fibers, whereas the fast muscle fibers are innervated by large-diameter, high-conduction velocity nerves. The slow fibers are always active at low levels of contraction, providing a sustained tonic muscle activity such as that required for maintaining posture of the body. The fast fibers are only active during strong movements of the body segments. Most body motions require a combination of slow and fast motor activities. For example, a hand–wrist manipulative task of using a screwdriver above head level may require slow unit activity of the shoulder muscles to stabilize the posture of the arm, but fast units of the hand wrist muscle are needed to turn the screwdriver. A muscle may consist of anywhere between 10 and 90% of any one type, but most muscles have a relatively even distribution (soleus muscle may range as high as 70% Type I).

2. Aerobic Metabolism and Bloodflow

Muscular exertions requiring aerobic metabolism need ample blood flow to carry oxygen to the tissues and to carry away metabolic by-products. Therefore, it is essential that blood flow be maintained to muscles with high workloads. When a muscle is contracted at high tonic levels, such as when a sustained static posture is required, the muscle contractions may inhibit adequate blood flow, thereby reducing the capability for aerobic metabolism. From an ergonomic perspective, it is important to limit static postures so that physiologic function is not compromised.

Maximum aerobic power, which is defined as the highest oxygen uptake an individual can attain during exercise, is a measure of the capability of the cardiovascular system to provide oxygen to the muscles for aerobic metabolism [3]. Maximum aerobic power has been measured as high as 7.4 liters/min for a male and 4.5 liters/min for a female cross-country skier [3]. The mean for an industrial population would be about 3.0 liters/min for men and about 2.0 liters/min for women [3,16]. Refer to Chapters 2 and 9 in this book for more information on maximum aerobic power, physical work capacity, and cardiovascular capacity.

3. Anaerobic Metabolism

During activities requiring strong muscular contractions, anaerobic metabolism plays a crucial role in providing energy to the muscles. Energy is provided to the muscles by the anaerobic breakdown of ATP, PCr, and glycogen in a lactic phase. It is difficult to measure an individual's capacity for anaerobic metabolism, but it is known that there is limited capability to sustain high workloads due to the limited supply of energy-yielding substrates. High sustained workloads will result in the buildup of high concentrations of lactate in the muscle, which is removed by the circulatory system. When a muscle has depleted its stores of anaerobic substrate and the workload is higher than about 50% of the maximum voluntary contraction, the muscle may begin to lose strength and become fatigued.

4. Oxygen Debt

During the recovery period following exercise, the amount of oxygen consumed in excess of the resting value is called *oxygen debt*. The higher the exercise level or workload, the higher is the level of oxygen debt incurred. In an exhaustive workload, the energy demand is not met adequately by aerobic metabolism; therefore, anaerobic energy production provides the necessary energy, causing lactic acid to build up. In other words, the more strenuous the workload, the longer it takes to achieve preexercise-level metabolism. Therefore, the concept of oxygen debt is critical in the design of a work (i.e., exercise) and rest regimen so that a task can be performed without experiencing fatigue. The literature provides recommendations of work–rest regimens using the principles of oxygen consumption and oxygen debt [3,16].

5. Local Muscle Fatigue

Static as well as dynamic muscular contractions can result in local muscle fatigue. Local muscle fatigue occurs when the endurance time for the muscle is exceeded. The endurance time for a muscle is dependent on the amount of force developed by the muscle as a percentage of the maximum force attainable by the muscle. For example, a muscle can sustain a force of about 15% of its maximum indefinitely without becoming fatigued, but it can sustain 50% of its maximum force for only about 1 min [3]. Similarly, a muscle can sustain a repetitive contraction rate of about 30 contractions per minute if the force is about 60% of maximum but can sustain a rate of only about 10 contractions per minute if the force is about 80% of maximum [3].

6. Whole Body Fatigue

When the metabolic demands of dynamic and sustained activity exceed the energy-producing capacity of a worker, muscle contraction is affected and whole body fatigue is usually experienced. Physiologists generally recommend that energy expenditure not exceed about 50% of maximum aerobic power for 1 hr of work, about 40% for 2 hr of work, and about 33% for 8 hr of continuous work [21]. These values are designed to prevent fatigue, which is believed to increase a worker's risk of musculoskeletal injury. Intervals of heavy, continuous work should be separated by light duty jobs, so that recovery can occur. For more information on whole body fatigue, refer to Astrand and Rodahl [3] and McArdle et al. [22].

REFERENCES

1. A. J. Vander, J. H. Sherman, and D. S. Luciano, *Human Physiology: The Mechanism of Body Function*, McGraw-Hill, New York, 1970.
2. A. C. Guyton, *Function of the Human Body*, Saunders, Philadelphia, 1969.
3. P. Astrand and K. Rodahl, *Textbook of Work Physiology*, McGraw-Hill, New York, 1986, pp. 19,115,334.
4. B. Tyldesley and J. I. Grieve, *Muscles, Nerves and Movement: Kinesiology in Daily Living*, Blackwell Scientific, London, 1989.
5. A. M. Burt, *Textbook of Neuroanatomy*, Saunders, Philadelphia, 1993.
6. J. Dudel, in *Fundamentals of Neurophysiology*, R. F. Schmidt, Ed., Springer-Verlag, New York, 1978, Ch. 5, pp. 129–137.
7. V. B. Brooks, *The Neural Basis of Motor Control*, Oxford Univ. Press, New York, 1986.
8. R. M. Enoka, *Neuromechanical Basis of Kinesiology*, Human Kinetics, Champaign, IL 1988.
9. W. S. Marras, Toward an understanding of dynamic variables in ergonomics, *Occup. Med.: State Rev.* 7(4):655–677 (1992).
10. DHHS(NIOSH), *Selected Topics in Surface Electromyography for Use in the Occupational Setting: Expert Perspectives*, Dept. of Health and Human Services, Natl. Institute for Occupational Safety and Health, Cincinnati, OH, 1992.
11. A. Bhattacharya, R. Shulka, K. N. Dietrich, J. Miller, A. Bagchee, R. Bornschein, C. Cox, and T. Mitchell, Functional implications of postural disequilibrium due to lead exposure, *Neurotoxicology* 14(2-3):179–190 (1993).
12. R. Seliga, A. Bhattacharya, P. Succop, R. Wickstrom, D. Smith, and K. Willeke, Effect of work-

load and respirator wear on postural stability, heart rate and perceived exertion, *Am. Ind. Hyg. Assoc. J.* 52(10):417–422 (1991).
13. A. Bhattacharya, R. Shukla, R. Bornschein, K. Dietrich, and J. Kopke, Postural disequilibrium quantification in children with chronic lead exposure, *Neurotoxicology* 9(3):327–340 (1988).
14. D. Sack, D. Linz, R. Shukla, C. Rice, A. Bhattacharya, and R. Suskind, Health status of pesticide applicators: postural stability assessments, *J. Occup. Med.* 35(12):1196–1202 (1993).
15. D. B. Chaffin and G. B. J. Andersson, *Occupational Biomechanics*, Wiley, New York, 1984.
16. Eastman Kodak Co., *Ergonomic Design for People at Work, Vol. 2*, Van Nostrand Reinhold, New York 1986, p. 472.
17. P. Allard, I. A. F. Stokes, and J. P. Blanchi (Eds.), Three-dimensional analysis of movement, *Human Kinetics*, Champaign, IL 1995.
18. W. J. Vincent, *Statistics in Kinesiology*, Human Kinetics, Champaign, IL 1995.
19. L. J. Findley and R. Capildeo (Eds.), *Movement Disorders: Tremor*, Oxford Press, New York, 1984.
20. T. L. Galinsky, R. R. Rosa, and D. D. Wheeler, Assessing muscular fatigue with a portable tremor measurement system suitable for field use, *Behav. Res. Methods, Instrum. Comput.* 22(6):507–516 (1990).
21. T. Waters, V. Putz-Anderson, A. Garg, and L. J. Fine, Revised NIOSH equation for the design and evaluation of manual lifting tasks, *Ergonomics* 36(7):749–776 (1993).
22. W. D. McArdle, F. I. Katch, and V. L. Katch, *Exercise Physiology: Energy, Nutrition, and Human Performance*, Lea & Febiger, Philadelphia, 1981.

4
Biomechanical Aspects of Body Movement

Angshuman Bagchee and Amit Bhattacharya
University of Cincinnati Medical School, Cincinnati, Ohio

INTRODUCTION

A significant number of the principles of ergonomics are derived from biomechanics, which can be defined as the systematic study of the human body as governed by the laws of physics. Laws of classical mechanics, in the form of fluid mechanics, statics and dynamics, and solid-state mechanics are applied to the scientific study of human posture and movement. Several definitions of the term *biomechanics* have been proposed by various authors. Fung [1] describes it as the mechanism of the living system. It is simply the application of the laws of mechanics to the biological system. The study of biomechanics dates back to the medieval ages, where the references to physical laws that govern human body movements can be found in the works of Leonardo da Vinci (1452–1519). Advances in the last two decades in the field of biomechanics have helped us understand several important physiological functions, like the role of the neuromuscular system in human movement, fluid mechanical principles governing the blood flow and respiratory systems, tissue biomechanics, and the finite-element modeling of bone structures.

The application of biomechanics to the occupational field, however, did not gain momentum until after World War II. Occupational biomechanics, a subfield of biomechanics, can be defined as the application of biomechanical principles to characterize and evaluate the effect of work task demand on the kinematic and kinetic responses of the workers [2,3].

II. BACKGROUND AND SIGNIFICANCE TO OCCUPATIONAL ERGONOMICS

With the advent of the industrial revolution, the emphasis shifted toward mass production. Occupational injuries and diseases were either neglected as an inevitable side-effect of a mass production system or were attributed to several other factors not related to the workplace. In the 1700s, Ramazzini suggested that the cause of such occupational diseases was the use of excessive force and unnatural postures [4]. It was not until recently that the principles of biomechanics were applied to a systematic study of human performance in the work environment.

Applied biomechanics forms the primary structure of occupational ergonomics, dealing with characterization of the loading of the musculoskeletal system. From the simple lever system that helps in quantitating the loading of the musculoskeletal system due to manual lifting of a weight, to complex measurements of interdiscal pressure to estimate the loading of the spinal unit, biomechanics helps us explain and improve a number of ergonomic problems and is a rapidly expanding field of study. Occupational biomechanics is applied in the determination of the bone–muscle–joint loading of the worker due to his or her interaction with tools, equipment, and the workplace. It also provides scientific guidelines for developing new tools that will reduce musculoskeletal disorders and for developing and modifying workstations to reduce worker discomfort. Progress in computer technology has allowed the development of computer models for predicting musculoskeletal loading associated with the performance of certain tasks. It allows, through the use of these models, the development of safe weight-lifting limits, chair design, and workplace layouts to conform to the specific working population. Since occupational biomechanics combines engineering concepts and the laws of physics with medicine, a multidisciplinary approach is required with expertise from a number of other fields, including bioinstrumentation, kinesiology, physiology, engineering, occupational therapy, rehabilitation engineering, and several other allied fields. This chapter outlines the major concepts of biomechanics that are relevant to occupational ergonomics. Appropriate references are cited throughout the text for in-depth information on specific topics.

III. RELEVANT CONCEPTS AND TERMINOLOGY

A. Principles of Applied Mechanics

Mechanics is the branch of physics that deals with the motion and deformations resulting from forces and moments acting upon a body. Applied mechanics is limited to the application of these principles to rigid bodies, deformable bodies, or fluids. Though the human body segments are somewhat deformable, for simplification they are assumed to be rigid bodies acted upon by point forces and moments. Depending on the state of the object under the influence of disturbing forces, the branch of mechanics is further subdivided into *statics*, which deals with bodies that are in a state of static equilibrium ("at rest"), and *dynamics*, which deals with bodies that are in a state of motion due to the action of the acting forces. The principles of statics are useful in determining the loading on the musculoskeletal system and the forces acting at the different body segment joints and muscles, while the principles of dynamics are useful in determining the forces generated when the body is in motion, as in running or diving.

Newtonian mechanics is based on the three dimensions of length, width, and height, along with the measurement of time. Quantities that require only magnitude for their description are known as scalars. Some examples of scalar quantities are mass, time, temperature, and speed. Vector quantities require the following for their specification: (1) magnitude, (2) direction, and (3) point of application. Examples of vector quantities include velocity, force, and momentum. In a three-dimensional case, a velocity vector **v** may be expressed as **v** = a**i** + b**j** + c**k**, where **i**, **j**, and **k** are three orthogonal unit vectors representing the three-dimensional axis system (Figure 1). The quantities a, b, and c are magnitudes (real numbers). Although scalar quantities can be added and multiplied algebraically, vector quantities require special mathematical methods for performing these operations. Some of the mathematical operations with vector quantities are described below. Refer to Özkaya and Nordin [5] for a detailed treatment of the subject as applied to biomechanics. It is customary to use boldface letters or letters with a bar or arrowhead over them to signify vectors. Boldface letters—**u**, **v**, **w**, etc., are used here to denote vector quantities.

1. Vector Addition

Two vectors **u** = a_1**i** + b_1**j** + c_1**k** and **v** = a_2**i** + b_2**j** + c_2**k** can be added by adding the individual components along the **i**, **j**, and **k** directions as follows:

$$\mathbf{w} = \mathbf{u} + \mathbf{v} = (a_1 + a_2)\mathbf{i} + (b_1 + b_2)\mathbf{j} + (c_1 + c_2)\mathbf{k}$$

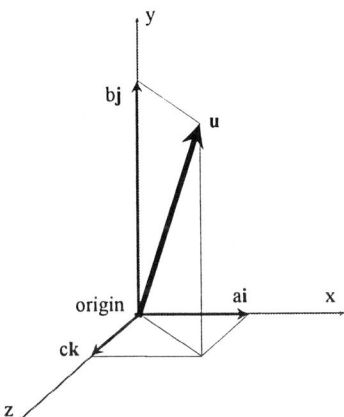

Figure 1 Representation of a vector quantity.

2. Vector Product

Two different kinds of products can be formed with vectors. The *dot* or *scalar product*, which has a magnitude but no direction, is formed as follows:

$$w = \mathbf{u} \cdot \mathbf{v} = |\mathbf{u}| \, |\mathbf{v}| \cos \theta = (a_1 \times a_2 + b_1 \times b_2 + c_1 \times c_2)$$

where θ is the angle between the two vectors $\mathbf{u}$ and $\mathbf{v}$.

The *cross* or *vector product* between the two vectors, resulting in another vector with both magnitude and direction, is given by

$$\begin{aligned} w = \mathbf{u} \times \mathbf{v} &= |\mathbf{u}| \, |\mathbf{v}| \sin \theta \\ &= (b_1 c_2 - b_2 c_1) \, \mathbf{i} + (a_1 c_2 - a_2 c_1) \, \mathbf{j} + (a_1 b_2 - a_2 b_1) \, \mathbf{k} \end{aligned}$$

Figure 2 shows the vectors $\mathbf{u}$ and $\mathbf{v}$ and their cross product $\mathbf{w}$. The direction of the resultant vector $\mathbf{w}$ is perpendicular to the plane formed by the vectors $\mathbf{u}$ and $\mathbf{v}$, shown as plane P in Figure 2. If $\mathbf{u}$ and $\mathbf{v}$ lie in the plane of this page, $\mathbf{w}$ would then be sticking into or out of the page. The exact direction of the resultant vector $\mathbf{w}$ can also be given by the *right-hand*

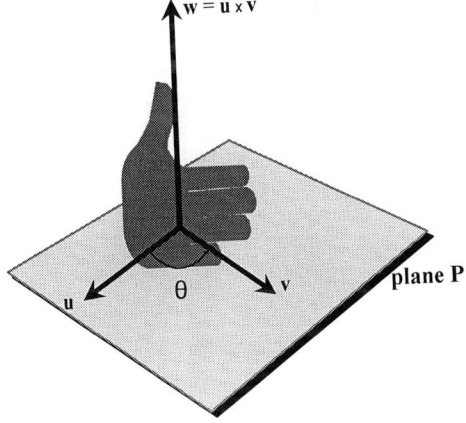

Figure 2 Right-hand rule used in calculating cross product.

rule, whereby the direction of the vector **w** is along the thumb if the rest of the fingers curl up from **u** toward **v** (see Figure 2). Note that the vector product is not commutative, i.e., **u** × **v** ≠ **v** × **u**.

3. Resolving Vectors into Components

Vector quantities can be resolved into their component vectors along any set of directions. Vectors are usually resolved into components along mutually perpendicular directions that form the *axis system*. An axis system is important to specify the location of any vector in space. In the above example, **u** = a**i** + b**j** + c**k** may be resolved into components along the orthogonal axis system formed by x, y, and z (**i**, **j**, and **k** are unit vectors along x, y, and z), as shown in Figure 1. This kind of coordinate system is the most widely used and is known as a *rectangular* or *Cartesian coordinate system*. In a two-dimensional system, the z axis is absent. For example, **q** = x**i** + y**j** represents a two-dimensional vector.

4. Forces

To understand the concepts of mechanics as applied to the human body, it is important to be familiar with several terms associated with forces and moments. *External* forces act on the outside of the object; examples of such forces are gravity, inertia, and ground reaction forces. *Internal* forces act within the structure of the body in reaction to the external forces and are called *stresses*. If we consider the forces in the body segments due to external forces like the gravitational pull of a load being carried in the hand, then the muscle forces generated can be assumed to be internal forces opposing the load. A force is a vector quantity that requires direction, magnitude, and point of application for it to be specified.

The application of forces on a rigid body may or may not cause the body to move or change direction. Figure 3 shows a force vector **F** acting at the forearm. The rotation of the arm occurs about the instantaneous center of rotation located at the elbow joint. This point can be considered as the origin, with the x axis along the forearm and the y axis orthogonal to it. The direction of the force vector can be specified with respect to this coordinate system. Such a coordinate system is termed the local or joint-based coordinate system. In a metric system of measurement, forces are expressed in newtons (N), where 1 N is defined as the force required to maintain a body of mass 1 kg at a constant acceleration of 1 m/sec^2. Force in the U.S. customary unit system is expressed in pounds (lb).

5. Newton's Laws of Motion

The basis of mechanics is formed primarily by Newton's laws of motion (Newtonian mechanics). Briefly these laws can be described as follows:

First Law. A body tends to remain in its inertial state of rest or motion unless and until acted upon by an external disturbing force.

Second Law. The net rate of change of momentum of a body is equal to the external forces acting on the body. Mathematically, this can be stated as

$$\mathbf{F} = \frac{\Delta(m\mathbf{v})}{t}$$

where $\Delta(m\mathbf{v})$ denotes the change in the momentum of the body. For a constant mass, the equation becomes

$$\mathbf{F} = \frac{\Delta(m\mathbf{v})}{t} = m\mathbf{a}$$

Biomechanical Aspects of Body Movement

F denotes the net force (algebraic sum of all forces) acting on the rigid body, **a** is the acceleration achieved by the body as a result of the force, and *m* is the mass of the body. This forms the basis for the governing equations discussed later.

    Third Law. For every action of force, there is always an equal and opposite reaction force produced. Thus, forces work in pairs. As an example, ground reaction forces are produced due to the action of body force on the ground.

6. Moments

Moments are formed by forces acting in a certain manner on a rigid body that causes the body to rotate about an axis. For example, a push applied to the edge of the door causes the door to rotate about its hinges. Similarly, torque is a moment that causes the body to twist as a result of a force. Moment is a vector quantity and is obtained by forming a cross product of the distance vector with the force vector, as follows:

$$\mathbf{M} = \mathbf{r} \times \mathbf{F} = |\mathbf{r}|\,|\mathbf{F}|\sin\theta$$

where **r** is the distance vector from the axis of rotation to the force vector **F**. If d is the perpendicular distance of the force vector **F** from the axis, the above expression simplifies to a moment vector with magnitude $M = dF$ and the direction given by the right-hand rule. In the metric system, moments are expressed in newton-meters (N-m); in the U.S. customary system it is expressed in pound-feet (lb-ft) or pound-inches (lb-in.).

7. Governing Equations

Newton's second law of motion states that the net resultant force on the body is equal to the rate of change of momentum for the body. Mathematically,

$$\Sigma\mathbf{F} = m\mathbf{a}$$

where $\Sigma\mathbf{F}$ denotes the vector sum of all the forces acting on the rigid body and **a** denotes the resulting acceleration of the body due to the action of the forces. A similar expression can be stated for the moments acting on the body,

$$\Sigma\mathbf{M} = I\alpha$$

where $\Sigma\mathbf{M}$ denotes the vector sum of all external moments acting on the rigid body, α is the resulting angular acceleration, and I is the moment of inertia of the rigid body about the axis of rotation. Moment of inertia is the resistance to change in angular velocity, $I = \Sigma mr^2$.

8. Equilibrium

Equilibrium is a special state of the body in which the body either is at rest (static equilibrium) or is moving with a constant velocity (dynamic equilibrium). In the latter case, a constant velocity signifies zero acceleration (**a** = 0; α = 0). In cases of either static or dynamic equilibrium, the net acceleration of the body is zero, and the above governing equations can be written as

$$\Sigma\mathbf{F} = 0 \quad \text{and} \quad \Sigma\mathbf{M} = 0$$

where the linear and angular acceleration of the body are both zero. Though a true condition of equilibrium is not always achieved, many biomechanical systems can be approximated to be in a state of equilibrium to simplify the evaluation of forces and moments. Care should be taken, however, to understand the factors and events that may contribute to violate the assumption of equilibrium and place the calculated forces and moments in error. For example, a running activity is highly dynamic, and thus principles of static equilibrium may not be applicable in calculating the joint forces and moments for such an activity.

B. Statics

Statics is concerned with bodies that are in a state of rest or under a static equilibrium. The forces acting upon the body in a static state are balanced; in other words, the net resultant of the forces acting on the body is zero. The principles of statics can be applied to stationary or near-stationary body segments.

As discussed earlier, forces acting on a body can be expressed as vectors. These forces acting at various segments of the human body are both two- and three-dimensional in nature. In most cases, the degrees of freedom for these segments are limited to a single plane. For example, the elbow joint is formed by the upper and lower arm, which move relative to each other in a single plane when performing simple activities such as lifting. Thus, in most cases, the forces can be approximated to be two-dimensional. Proper placement of the local coordinate system can yield values of forces and moments that are biomechanically useful and easier to interpret. Forces acting on the body segment, being vector quantities, can be resolved into components along the axes, as illustrated in the following example. The reader is encouraged to consult further references for a detailed treatment of vectors [6].

Example Problem 1

Figure 3 shows the force **F** generated by the biceps brachii of the forearm. The magnitude of the biceps force vector **F** is 40 N, and it is making an angle of 60° with the forearm. We wish to calculate the force component along the forearm (stabilizing force), F_S, and the force component perpendicular to the forearm, F_T (tangential, so called because it is acting at a tangent to the circular path in which the forearm will rotate due to the force).

Solution

The forces F_T and F_S can be calculated by resolving the force **F** into its components (see Figure 3) along the direction of the forearm (F_S) and along a direction perpendicular to the forearm (F_T), using the following set of equations [6]:

$$F_T = |F| \sin \theta = 40.0 \times \sin 60° = 34.64 \text{ N}$$

$$F_S = |F| \cos \theta = 40.0 \times \cos 60° = 20.0 \text{ N}$$

For F_T component of the force **F** in the above example is called the "rotational" force, for it has the effect of rotating the forearm about the elbow joint. The F_S component is the

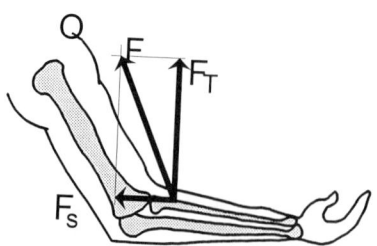

Figure 3 Resolution of muscle force into orthogonal components.

"stabilizing" component; it compresses or pulls apart the joint, depending on the direction of the force.

Application of force give rise to moment or torque as described earlier and may cause the body segment to rotate. Thus, moment can be calculated by multiplying the force by the perpendicular distance of the force vector from the pivotal axis. In the above example, if the distance of the point of application of the tangential force F_T from the elbow (pivotal point) is $d = 5$ cm, then the moment would be calculated as $M = d\,F_T = 34.64$ N $\times$ 5 cm $=$ 173.2 N-cm $= 1.732$ N-m. Note that the direction of the moment vector M is given by the right-hand rule discussed earlier and would be along an axis perpendicular to the plane of the forces, coming out of the plane of the paper in the above example. Also note that the component F_S will not produce any moment as it passes through the pivotal axis, and the corresponding d is zero.

In a condition of static equilibrium, the body remains stationary under the action of external forces (F_1, F_2, F_3, F_4, F_5, . . .) and moments (M_1, M_2, M_3, M_4, M_5, . . .). Using the equation stated earlier for equilibrium, the net algebraic sum of the forces and moments acting on the body at any time is zero and can be expressed by the following equations:

$$\Sigma F = F_1 + F_2 + F_3 + F_4 + \cdots = 0$$

$$\Sigma M = M_1 + M_2 + M_3 + M_4 + \cdots = 0$$

The above sum of vectors can be simplified for a two-dimensional case by equating all the components of forces and moments to zero:

$$\sum_{i=1}^{i=n} (F_x)_i = 0, \quad \sum_{i=1}^{i=n} (F_y)_i = 0, \quad \sum_{i=1}^{i=n} (M_z)_i = 0$$

Here, the forces $(F_x)_i$ and $(F_y)_i$ lie in a plane while $(M_z)_i$ is along an axis perpendicular to the plane of the forces. Using the above equations, any set of forces and moments can be added algebraically to obtain the unknown forces and moments in a system at equilibrium. We will be applying the concept of equilibrium when solving for forces in a lever system.

C. Lever System

A lever system operates on the application of the governing laws of mechanics discussed above. It is the simplest form of machine system existing in nature. One example of this system is the seesaw system shown in Figure 4. The shortest distance of the load and effort forces from the instantaneous center of rotation (or fulcrum) are called the *load arm* and *effort arm*, respectively. Taking the moments about the fulcrum F, and applying the principles of static equilibrium, we have

$$\Sigma M_z = 0$$

Load $\times$ Load arm $-$ Effort $\times$ Effort arm $= 0$

or

Load $\times$ Load arm $=$ Effort $\times$ Effort arm

In terms of the quantities shown in Figure 4,

$$L\,d_L = E\,d_E$$

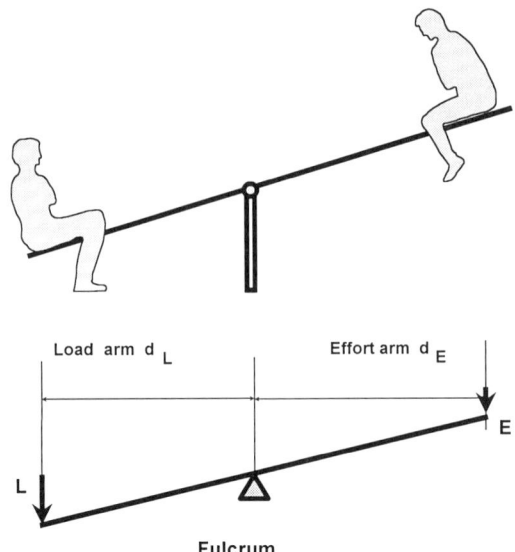

Figure 4 Lever system.

$$E = \frac{d_L}{d_E} L$$

Thus,

$$\text{Effort} = \frac{\text{load arm}}{\text{effort arm}} \times \text{Load}$$

The amount of force, or effort, needed to balance the load can be varied by changing the ratio of the load arm to the effort arm. Hence, the placement of the fulcrum with respect to the load and the effort is critical in determining the value of the force needed to balance the load. *Mechanical advantage* is defined as the ratio of the load to the effort required to balance it. From the above equation, this is also equal to (effort arm) / (load arm). A mechanical advantage value greater than unity means that the effort required to balance the load is *smaller* than the load itself. Most of the lever systems found in the human body actually work at a mechanical disadvantage (mechanical advantage < 1.0), requiring greater effort than the balanced load. On the other hand, having a smaller effort arm is anatomically advantageous, because the muscle has to move very little to achieve a large movement of the load.

Several examples of lever systems can be found in the human body. There exist three classes of lever systems, based on the positioning of the load and effort with respect to the fulcrum. These are termed (1) first class; (2) second class; and (3) third class lever systems. The system in which the fulcrum lies between the load and the effort, as shown in Figures 4 and 5a, is known as a first class lever system. An example of this kind of lever system in the human body exists at the neck joint as shown in Figure 5a. A second class lever system has the load situated between the effort and the fulcrum. An example of the second class of lever system can be found in the human body at the ankle joint, where standing on the toes would require the forces through the Achilles tendons to balance the body weight with the

Biomechanical Aspects of Body Movement

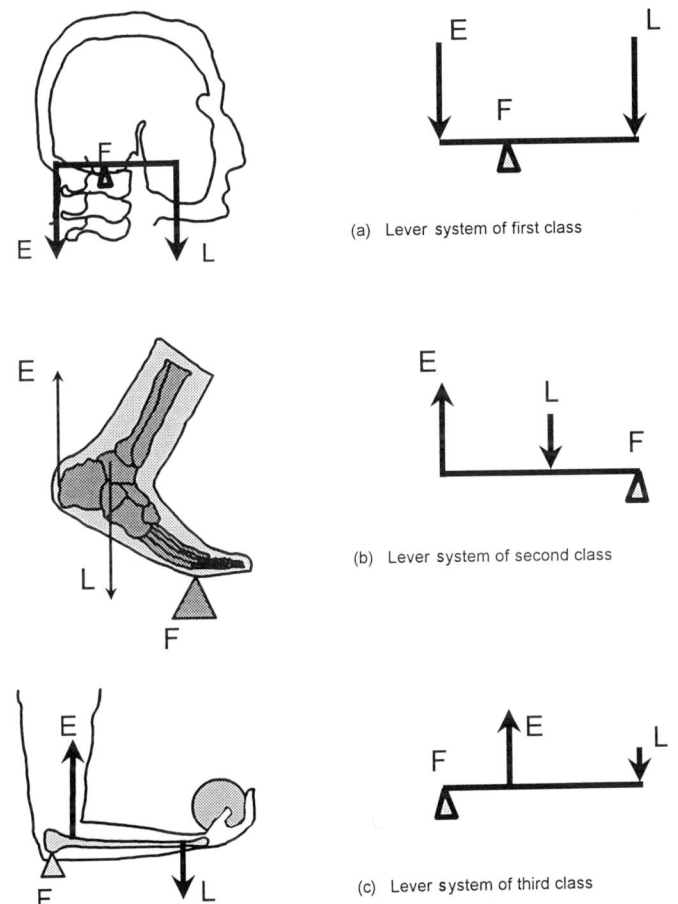

Figure 5 Three classes of lever systems. Examples and their schematic representation.

fulcrum at the big toe, so as to produce a second class lever system as shown in Figure 5b. A third class lever system is the type most commonly found in the human body. Here, the effort lies between the load and the fulcrum. The effort required to counterbalance the load is always greater than the load. Figure 5c shows a third class lever system. An example of the third class of lever system is found in the forearm, where the effort is exerted by the biceps brachii muscles as shown in Figure 5c.

The principles of a lever system are useful in determining the muscle forces required to perform several activities of daily life. They also provide better understanding of the importance of a reduced load arm in order to reduce the effort (muscle force) required for performing the task. Several examples are presented below to illustrate the principles of levers as they apply to the various body segments.

Example Problem 2

Calculate the lifting force required by the operator to lift the wheelbarrow shown in Figure 6. The load **L** of the wheelbarrow is 70 kg. The perpendicular distances of the load vector

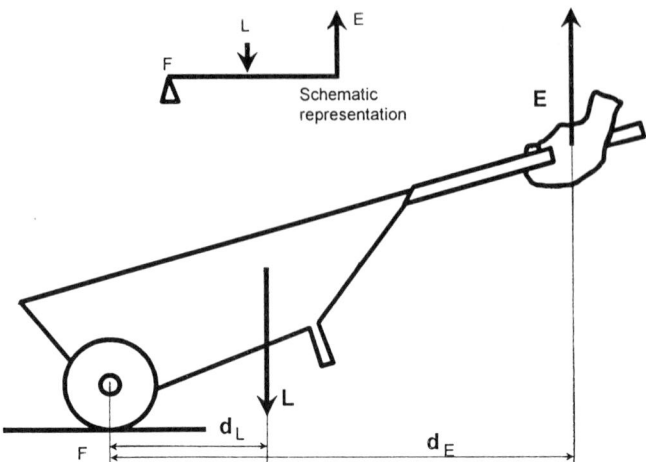

Figure 6 Lever system applied to a wheelbarrow.

L and the effort vector **E** from the instantaneous center of rotation (fulcrum) are 30 cm and 120 cm, respectively. What class of lever system is illustrated in this example? Also, calculate the mechanical advantage of the system.

Solution

Using the equation for a lever system described earlier:

Effort = load arm

$$\text{Effort} = \frac{\text{load arm}}{\text{effort arm}} \times \text{load} = \frac{30}{120} \times 70 = 17.5 \text{ kg}(= 1/1.7 \text{ N})$$

The mechanical advantage is defined as follows:

$$\text{Mechanical advantage (MA)} = \frac{\text{load}}{\text{effort}}$$

Since load × load arm = effort × effort arm, we have

$$\text{MA} = \frac{\text{effort arm}}{\text{load arm}} = \frac{120}{30} = 4.0$$

Example Problem 3

Figure 7 shows a worker performing a task on a table that requires bending of the torso, head, and arms. The balancing force can be assumed to be provided by the extensor muscles of the back, shown by the vector **E** acting at a distance of d_e (= 5 cm) from the instantaneous center of rotation (fulcrum) situated between the fifth lumbar and first sacral (L_5S_1) vertebrae. The approximate location of the gravitational forces (loads) due to the head, trunk, and arms from the fulcrum are also shown. The values of loads and their load arm distances for a 61-kg (600-N) man are given below:

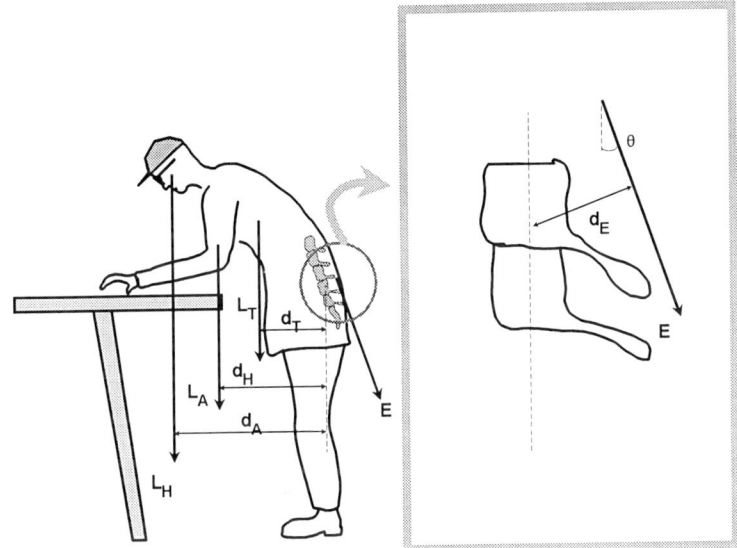

Figure 7 Calculation of back muscle forces.

Load	Load arm
L_H = 8.4% of bw = 50.4 N	55.0 cm
L_A = 5.1% of bw = 30.6 N	30.0 cm
L_T = 50% of bw = 300.0 N	13.0 cm

where bw = body weight (600N).

Calculate the muscle force **E** required for this lever system. What is the total weight (or force) experienced by the L_5S_1 vertebrae? Assume that the component of **E** acting through L_5S_1 is given by $E \cos \theta$, where θ is the angle of inclination of the vector **E** to the vertical direction (assume $\theta = 30°$).

Solution

Summing up the load moment about the fulcrum:

$$L_H d_H + L_A d_A + L_T d_T = 50.4 \times 55 + 30.6 \times 30 + 300 \times 13$$
$$= 7590 \text{ N-cm}$$

Equating the net moment about the fulcrum to zero, we have

$$E \times d_E = 7590$$

Therefore,

$$E = 7590 / d_E = 7590 / 5 = 1518 \text{ N}$$

In terms of body weight, this effort is 1518 / 600.0 = 2.53 bw.

The total loading on the spinal column is the net sum of forces acting on L_5S_1:

$$L_H + L_A + L_T + E \cos \theta = 50.4 + 30.6 + 300 + 1518 \times \cos 30°$$

$$= 1695.6 \text{ N } (2.83 \text{ times bw})$$

Thus, even though only 63% of the total body weight is physically present above L_5S_1, the actual loading is about three times the body weight.

Example Problem 4

Assume that, in the above case, the person is holding a load weighing 1 kg (10 N) in his hands. Assume that the distance of this load from his L_5S_1 is about 60 cm. What are the total muscle force and the net loading on L_5S_1?

Hint. Add the moment due to the load weighing 10 N to the equation above and re-evaluate the muscle force. The weight of the load should be included when calculating the net loading on L_5S_1.

Answer. 1638 N (2.73 bw); 1809.5 (3.0 bw).

Example Problem 5

Figure 8 shows three postures, commonly found at the workplace: (a) standing upright, (b) sitting upright, (c) sitting and performing a task at arms' reach. The gravitational force of the upper body is assumed to be acting along the line G. The distance of the line G from the fifth lumbar and first sacral vertebrae (L_5S_1) is shown to be (a) 8 cm in a standing posture, (b) 8.5 cm in a seated posture, and (c) 10 cm in a seated posture with arms extended for stenographic work. The weight of the head, trunk, and arms constitute 62% of the body weight. The balancing moment is provided by the back extensor muscles. The distance of the line of back extensor muscle force **F** acting on the vertebrae (d_F) is assumed to be 4 cm. Calculate the resultant loading **R** on the L_5S_1 vertebrae for a woman weighing 600 N (61 kg). What are the biomechanical implications of these postures based on your calculations?

Hint. Use the following equation for calculating the net loading **R** on L_5S_1.

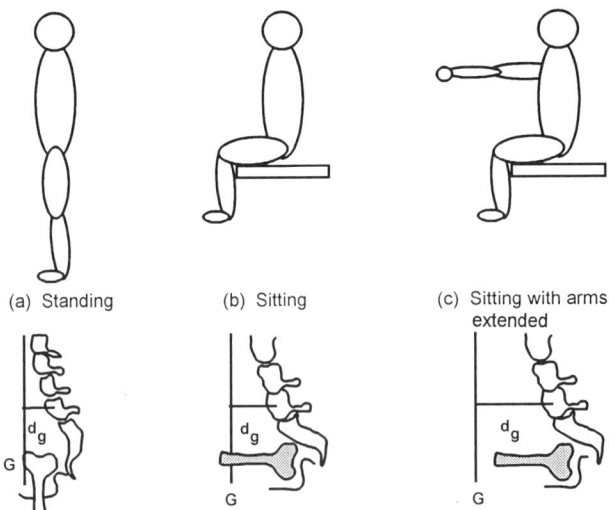

Figure 8 Influence of posture in the loading of the spinal column. (Modified from Ref. 8.)

$$R = G + F_m$$

where F_m is the back muscle force, R is the net loading on L_5S_1, and G is the gravitational force.

$$R = G + \frac{d_g}{d_F} G = \frac{d_g + d_F}{d_F} G$$

here, d_g is the distance of the force G and d_F is the distance of the back muscles from the instantaneous center of rotation at L_5S_1.

Answer. (a) 1116 N; (b) 1164 N; (c) 1302 N.

E. Principles of Statics Applied to Specific Human Body Joints

In this section, several numerical examples of the application of principles of statics are presented for various body segments. The aim is to illustrate the application of simple principles of statics in evaluating loading of the musculoskeletal system. A comprehensive discussion of the biomechanics of all body segment movements can be found elsewhere [3,6–8] and is beyond the scope of this chapter.

1. Wrist and Hand

The wrist and hand are primary body segments for performing the activities of daily life. The movement is achieved through the articulation of several small bone structures in a highly complex interaction, generating forces and movements in three-dimensional planes. A detailed discussion of the bone articulation can be found in Ref. 3.

When gripping or supporting an object, forces are applied at the distal joints of the fingers. Forces are transmitted through a series of bone joint articulations of the hand. These joints are the distal interphalangeal joint (DIP) between the distal and proximal phalanges, the proximal interphalangeal joint (PIP) between the middle and proximal phalanges, the metacarpophalangeal joint (MCP) between proximal carpal and the metacarpal bones, and the carpometacarpal joint (CMC) between metacarpal and carpal bone, as illustrated in Figure 9. Since each of these joints has ligamental tissues attached to the bones for supporting the structure, the external forces are balanced in part by each of the four major joints mentioned above and as illustrated in the following example.

Example Problem 6

Figure 9 shows the hand pressing on a surface with 1 kg force ($\approx$ 10 N). Calculate the moment at each of the joints, using the distance of the joints and the effort arm tabulated below (distance of the muscle force at each joint from its center of rotation, or fulcrum, shown in Figure 9).

Joint	Distance from load	Effort arm
DIP	2.0 cm	0.50 cm
PIP	5.5 cm	0.75 cm
MCP	10.5 cm	1.00 cm
CMC	20.0 cm	1.25 cm

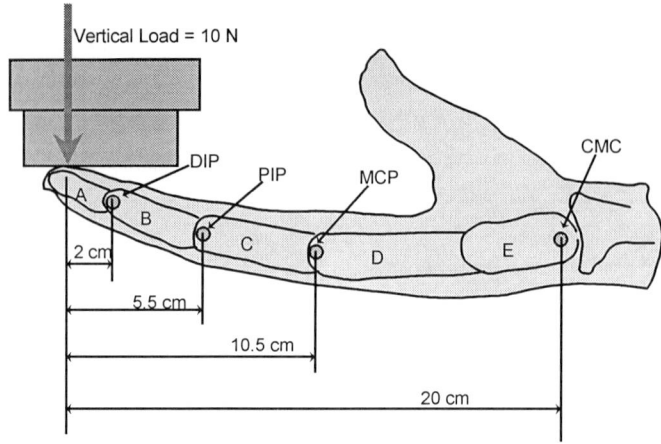

Figure 9 Forces acting on the hand. (Modified from Ref. 3.)

Solution

The moment about each of the joints can be calculated by multiplying the distance of the load from the fulcrum and the load. The values can be tabulated as follows.

Joint	Moment (Load × Load arm)
DIP	20 Ncm (= 0.2 Nm)
PIP	55 Ncm (= 0.55 Nm)
MCP	105 Ncm (= 1.05 Nm)
CMC	200 Ncm (= 2.0 Nm)

Example Problem 7

Using the effort arm distances listed in Example Problem 6, calculate the muscle force that will be required, if it is assumed that each of the joints is acting alone to balance the moment due to the load.

Solution

The muscle force (effort) required at each joint can be calculated by using the principles of statics, assuming that the net moment about the particular joint is zero. Thus, if $\mathbf{E}$ is the muscle effort acting at the effort arm d_F, then $\mathbf{E} = \mathbf{M}/d_F$, where $\mathbf{M}$ is the moment calculated in the previous example. The following table lists the muscle forces.

Joint	Muscle effort (moment / effort arm)
DIP	40 N
PIP	73.3 N
MCP	105 N
CMC	160 N

As we can see in the above examples, the moment due to the external load increases as we go from fingertip to the wrist, due to an increase in the moment arm. The muscle force needed to balance the moment would thus be excessively high if only one of the joints were to act alone. In reality, the moment at each of the joints is counterbalanced by smaller moments produced by the flexor profundus tendons in each of the individual joints [3].

The grip force required to hold an object depends on the size and weight of the object and the type of grip. Figure 10 shows two different kinds of grips and the tangential forces produced in the synovium tendons of the hand. The tendon, in turn, wraps around the bones of the wrist to create a pulley system as depicted in Figure 11, where the person is gripping a tool while flexing his hand. This flexion produces a normal force on the tendon synovium itself that is given by [2]

$$\mathbf{F}_N = \frac{\mathbf{F}_T e^{\mu\phi}}{R}$$

where $\mathbf{F}_N$ = normal force on the tendons
$\mathbf{F}_T$ = tangential force produce along the tendon
R = radius of the pulley system
ϕ = angle subtended by the tendon on the pulley
μ = coefficient of friction between the tendon and the pulley

For the purpose of simplification, the value of μ can be assumed to be zero [6]. The above equation is further simplified to

$$\mathbf{F}_N = \mathbf{F}_T / R$$

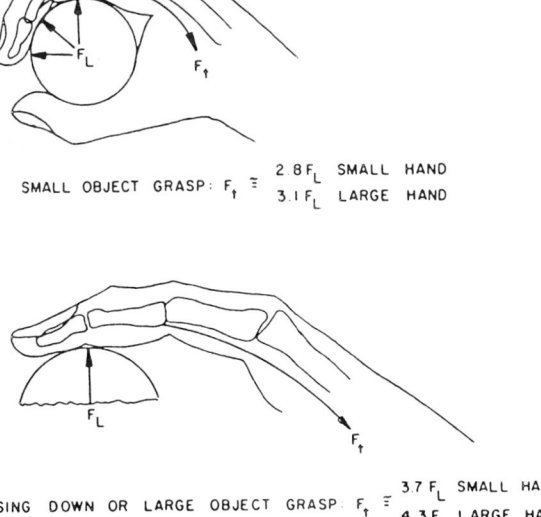

Figure 10 Forces experienced by wrist during gripping. (Reproduced from Ref. 2 by permission of the author.)

The reaction force on the bones due to the tendons, $\mathbf{F}_R$, can be calculated by resolving the forces, yielding

$$\mathbf{F}_R = 2\mathbf{F}_T \sin(\phi/2)$$

Example Problem 8

Figure 11 shows the forces acting on the hand of a person operating a screwdriver. Calculate the tendon force $\mathbf{F}_T$ and the reaction force $\mathbf{F}_R$. For the purpose of this example, the grip force $\mathbf{P}$ acting at the finger may be assumed to be 1 N. Assume a small hand in operation on a small grip size. Also assume that the subtended angle (wrist angle) is 30°.

Solution

Using the equation shown in Figure 10,

$$\mathbf{F}_T = 2.8\ \mathbf{P} = 2.8 \times 1 = 2.8\ \text{N}$$

Using the equation for $\mathbf{F}_R$, we have

$$\mathbf{F}_R = 2\ \mathbf{F}_T \sin \phi/2 = 2 \times 2.8 \times \sin 30°/2 = 1.45\ \text{N}$$

Example Problem 9

Repeat the above analysis for an extension of the wrist as shown in Figure 12, where the person is using a paint roller. Assume that the grip pressure force $\mathbf{P}$ required for holding is 5 N. Assume a wrist angle of 25° and a small grip size. Assume a factory worker with large hand size.

    Hint.   Use the above methodology.
    Answers. $\mathbf{F}_T = 15.5\ \text{N}$; $\mathbf{F}_R = 6.7\ \text{N}$.

Several studies have been conducted to investigate the range of motion of the hand and wrist and the analysis of the forces acting on the adjacent structures [2,3]. These values act as guidelines for determining untoward postures of the hand and the extent of forces that may be detrimental when applied at extreme angles of flexion and extension of the wrist. One of

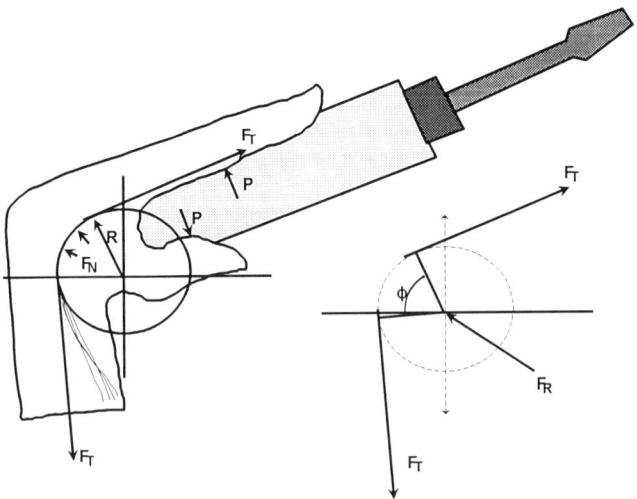

Figure 11 Forces acting on the wrist during flexion.

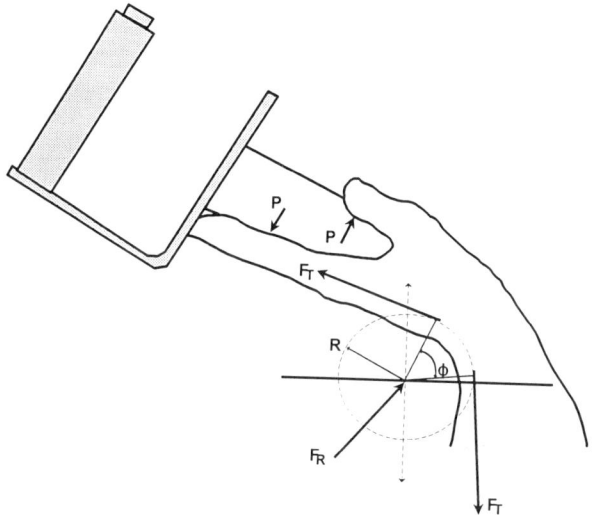

Figure 12 Forces acting on the wrist during extension.

the most common musculoskeletal cumulative trauma disorders is the carpal tunnel syndrome, which causes inflammation of the nerves at the wrist joints due to excessive and repeated loading of the median nerve at the wrist joint structure. Additional issues related to carpal tunnel syndrome are given elsewhere in this book.

2. Elbow

Weight lifting is a common activity we perform every day. Most of the time we simply grab the weight in one hand and flex our arms to raise it up. Lifting can be performed using one or both hands. It can be pointed out here that one-arm lifting would load only one side of the body framework. This will produce an asymmetric loading of the spinal column and skeletal system as a whole. In arm lifting, the elbow joint acts as the instantaneous center of rotation (fulcrum).

As was described earlier, depending on the muscles involved in supporting the weight, the arm uses either a second or third class lever system. This is illustrated further in Figure 13. In Figure 13a, the principal muscle applying the force is the brachoradialis, attached to the radial bone near the wrist, as shown. The combined load of the arm and the weight is shown as **L**. Since the load held by the hand (point of application of the load) lies between the effort and fulcrum, it can be termed a second class lever system. In Figure 13b, the muscle applying the force is the biceps brachii. This is a third class lever system [7,9].

Example Problem 10

Calculate the muscle force (effort) needed in Figures 13a and 13b if the value of the load **L** is 90 N and the load arm (perpendicular distance of the application of **L** from the fulcrum) is 18 cm. The effort arm (perpendicular distance of the force vector **E** from the fulcrum) can be assumed to be 4 cm for (a) and 2 cm for (b). Calculate the resultant force **R** on the fulcrum F, if it is given that **E** acts at an angle of 15° in Figure 13a and 60° in Figure 13b, measured from the left horizontal.

Hint. Calculate the value of **E** using the equation $\mathbf{E} = [(\text{load arm})/(\text{effort arm})] \times \text{load}$. Resultant force **R** can be calculated by resolving the forces **L** and **E** along the two axes.

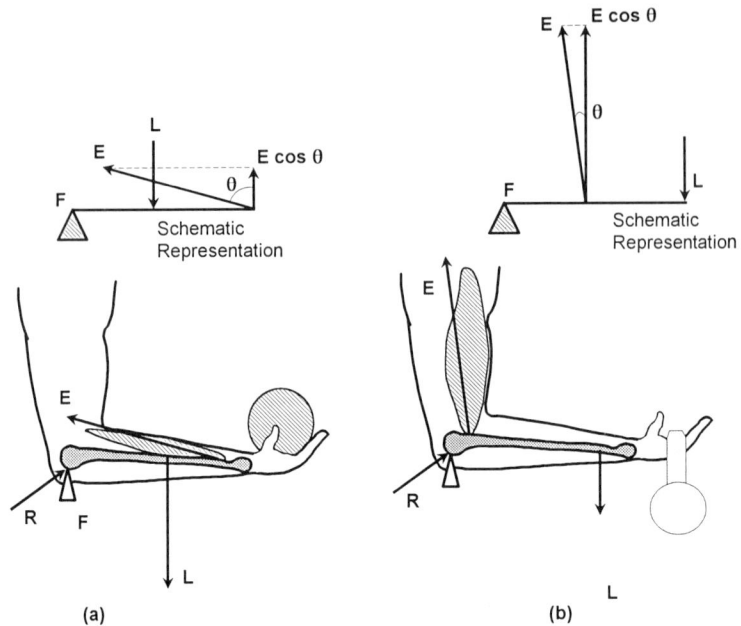

Figure 13 Application of different muscle forces may determine what lever system should be applied for calculating the resultant forces.

Answer. (a) 405 N; (b) 810 N. Resultants: (a) 391.48 N at −2.2° to right horizontal; (b) 733.4 N at −56.5° to right horizontal.

3. Shoulder

Movement at the shoulder level is one of the most intricate to quantify. The motion around the shoulder joint involves a number of muscles acting together at several possible joints to achieve the wide range of movement of the arm with respect to the torso. The combined range of motion is achieved by four articulations: the glenohumeral, sternoclavicular, acromioclavicular, and scapulothoracic [3]. The near-global range of motion exceeds the individual range of motion at any of these articulations. The main range of motion occurs at the glenohumeral joint formed by the humeral head and the glenoid fossa of the scapula. It forms an open-ended ball-and-socket joint, formed by a pear-shaped depression at the glenoid fossa and the humeral head capable of performing rotational, translational, and rolling motion with respect to the fossa. Several muscles act at the four articulations, making it difficult to calculate the precise forces experienced by each muscle group. Simple assumptions can be made to calculate the major forces and reactions at these joints, as shown in the following example problem.

Example Problem 11

The main joint forces at the shoulder act at the glenohumeral joint. The major muscle group involved in the articulation is the deltoid muscle. Figure 14 shows a construction worker performing a ceiling operation, which requires him to operate a tool with stretched arms as shown. Assume the weight of the tool to be 0.5 kg (5 N), acting at a distance of 80 cm from the instantaneous center of rotation O at the shoulder. Assume that the weight of the arm is 0.05 times the body weight and acts at a distance of 35 cm from O. For a man weighing 100

Biomechanical Aspects of Body Movement

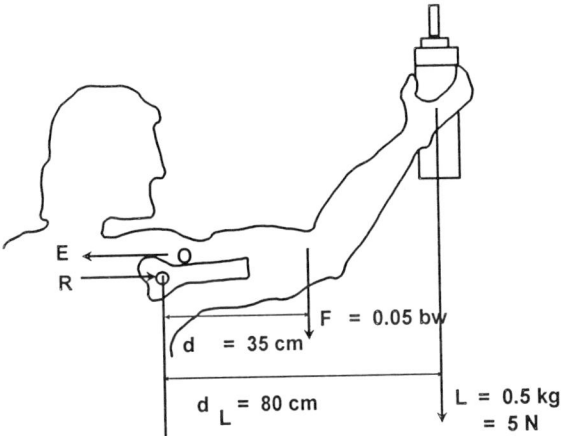

Figure 14 Forces at shoulder during an overhead task performance.

kg (1000 N), calculate the effort **E** applied by the deltoid muscles with an effort arm of 3.5 cm.

Solution

Equating the moments about the point O to zero, we have

$$-(E \times 3.5) + (0.05 \times 1000 \times 35) + (5 \times 80) = 0$$

$$E = 2150/3.5 = 614.3 \text{ N } (0.6 \text{ bw})$$

Since the only horizontal force acting at the joint is the muscle effort **E**, the horizontal reaction force **R** on the joint is equal to **E**, or 614.3 N.

Example Problem 12

Suppose the worker in the above example has to apply an upward thrust to the tool for driving the nails into the roof, increasing the effective downward force to 50 N (instead of 5 N). Recalculate the deltoid force **E**.
    Hint. Use the above equation for calculating **E** (=**R**).
    Answer. E = 1642.9 N (1.64 times bw)].

Example Problem 13

At a Stockholm construction site, several complaints were received concerning strains and aches in the upper extremities due to the excessive use of bolt guns (Hilti Model DX450). A bolt gun support was developed to eliminate the loading on the upper extremities by transferring the loading to the waist belt. Figure 15 shows the biomechanical configuration of the loading (left) without bolt-gun support, and (right) with bolt-gun support. Calculate the shoulder abduction force being provided by the deltoid muscles, which are assumed to be acting at a distance of 3 cm from the instantaneous center of rotation (fulcrum), to support the gun in Figure 15a. The weight of the gun (W_g) and the weight of the arm (W_a) and their distances from the fulcrum O are shown in the figure for an 800-N worker.
    Hint. Calculate the moments about O due to the arm and gun, and equate them to the moment created by the shoulder deltoid muscles.
    Answers. 373.3 N.

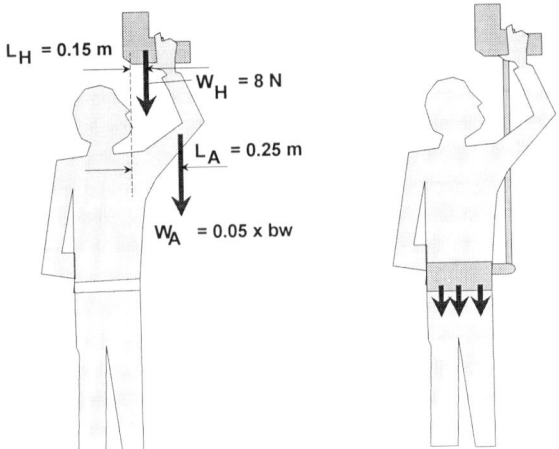

Figure 15 Modification of a heavy tool to reduce the loading of the shoulder.

4. Knee

The human knee acts as a pulley system as shown in Figure 16. The patellar tendon and the quadriceps each apply balancing forces to maintain the flexion of the knee. In a static case of a lifting task, forces generated in the knee can be calculated, as illustrated in the following example problem.

Example Problem 14

Figure 17 shows a person lifting a weight. The superincumbent weight (above the knee joint) of the person with the package can be assumed to be 900 N. Also, assume that d_w, the distance of the line of the center of gravity from the instantaneous center of rotation (fulcrum), is 5 cm. The corresponding distance, d_p, of the patellar force vector **P** is 4 cm. Assuming a static posture as shown, calculate the patellar tendon forces.

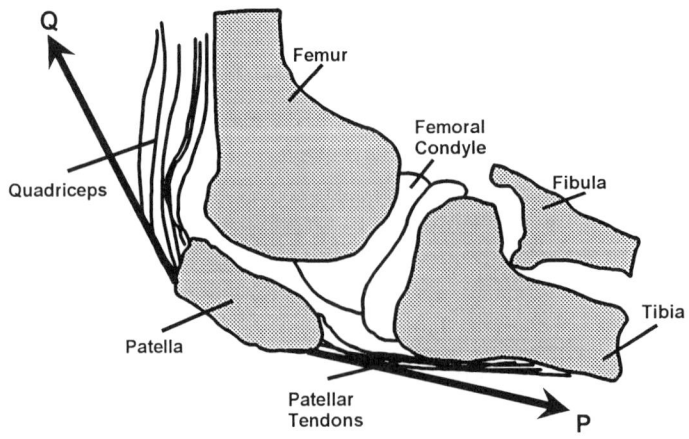

Figure 16 Anatomy of a knee joint.

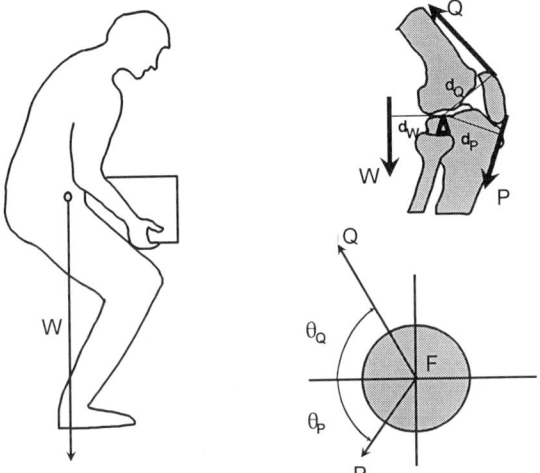

Figure 17 Forces at knee during manual materials handling.

Solution

Using the equilibrium of moments about the fulcrum,

$$W \times d_W = P \times d_P$$

Hence,

$$P = \frac{d_W}{d_P} W = \frac{5}{4} \, 900 = 1125 \text{ N}$$

Assuming the patella to be a perfect pulley, the forces **P** and **Q** are equal in magnitude.

Example Problem 15

In the above example, the anatomical measurements reveal the approximate angles of the force vectors **P** and **Q** with the left horizontal to be θ_P and θ_Q, with the values of 60° and 50°, respectively (see Figure 17). Calculate the resultant force on the patella.

    Hint. The resultant force on the patella can be calculated by resolving the vectors **P**, **Q**, and resultant **R** into x and y components (assume that the x axis is along the right horizontal).

    Answers. 1290.5 N acting at an angle of 5° to the right horizontal.

The examples above assume that the knee is acting as a perfect pulley, with equal tension on the ligaments carrying the forces. However, this approximation is not valid for extreme angles of knee flexion [10].

F. Dynamics

Dynamics is concerned with bodies that are in motion. Apart from the system of forces acting on the body, there exist the inertial forces due to the motion of the individual segments of the body. The forces that cause the motion may vary with time, as do the position and orientation of the segments. Such time-dependent measurements are termed temporal quantities and play a vital role in dynamics. Dynamics can be further subdivided into kinetics, which is concerned with the study of forces and moments that cause the motion, and kinematics, which deals with the measurement of motion itself. Newton's second law of motion can then be applied to the body segments for evaluating the forces and moments. As mentioned in the laws governing the principles of mechanics, for a system of forces **F** and moments **M** acting on a body, the law can be mathematically stated as

$$\Sigma \mathbf{F} = m\mathbf{a}, \quad \Sigma \mathbf{M} = I_o \alpha$$

Here, **a** (in m/s^2) and α (rad/s^2) denote the resulting linear and angular acceleration of the body, m (kg) is the mass of the body, and I_o (kg-m^2) is the moment of inertia of the body about the axis of rotation. Thus, both kinematic and kinetic measurements are required to fully describe the motion of the body and the forces that cause the motion.

In the case of a static loading, the body is assumed to be at rest. Hence, the inertial forces are absent in a static loading, and the right-hand sides of the above equations are equal to zero (no linear or angular accelerations). Thus, static loading constitutes a special case of dynamic loading. Since, in a real-life scenario, the worker is involved in various activities that may include dynamic motion of the body segments, inertial forces play a very important role in determining the total musculoskeletal loading on the human body. The importance of dynamic motion of the segments on the loading of the body segment joints is further explained in another chapter of this book.

1. Kinematics

Kinematics describes the geometry of motion. It is concerned with the temporal change in the position of the body segments. Both linear and angular displacement of the body segments are important in characterizing motion. For simplification, we assume that the body segments behave as rigid links, and the displacement, velocity, and acceleration of a fixed point on this rigid link are evaluated.

If linear displacement is denoted as x, velocity is defined as the rate of change of x in a particular direction. Acceleration is defined as the rate of change of velocity, or the rate of change of displacement. Mathematically,

$$\text{Velocity} = \mathbf{v} = \frac{d}{dt}(x)$$

$$\text{Acceleration} = \mathbf{a} = \frac{d}{dt}(\mathbf{v}) = \frac{d}{dt}\left(\frac{d}{dt}x\right) = \frac{d^2}{dt^2}(x)$$

Similarly, if angular displacement is defined as θ, the angular velocity and angular acceleration are defined as follows:

$$\text{Angular velocity} = \omega = \frac{d}{dt}\theta$$

$$\text{Angular acceleration} = \alpha = \frac{d}{dt}\left(\frac{d}{dt}\theta\right) = \frac{d^2}{dt^2}\theta$$

Since linear and angular displacement, velocity, and acceleration are vector quantities, they require magnitude, direction, and point of application for their specification. These quantities must be specified with respect to a global reference axis system or a local joint-based axis system. Special passively illuminated markers can be placed at body segments and captured in a video image. This video can be digitized using specialized instrumentation to ob-

Biomechanical Aspects of Body Movement

tain the movement of each body segment in space. Figure 18a shows the digitized image of the lower limb of a child walking over an obstacle. Figure 18b shows how the angle between the ankle and lower leg (shin) changes during this walk [11]. Additional information needed to measure kinematic information is included in another chapter of this book.

2. Kinetics

Kinetics is concerned with the measurement of forces and moments that cause motion. For simplification, the human body is assumed to be made up of individual link segments, each of which behaves as a rigid body, that interact with each other to transmit the forces, moments, and the motion caused by the action of these forces. The rigid link segments behave in accordance with Newton's laws of motion.

Application of Newton's second law of motion yields the set of equations for dynamic equilibrium stated at the beginning of this section. The calculation of ground reaction forces plays an important role in understanding the result of body movement during the performance of tasks. As we will see later in this chapter, the principles of dynamics can be applied to complex link-segment models of the human body in order to evaluate the joint forces and moments, which play an important part in understanding the functions and limitations that govern human movement.

G. Spinal Unit

1. Structure

The spinal unit forms the integral part of the framework of bone that provides structural support to the upper part of the body. The complex movement pattern allowed by the spinal column is key to the maintenance of this structural support, while allowing the necessary range

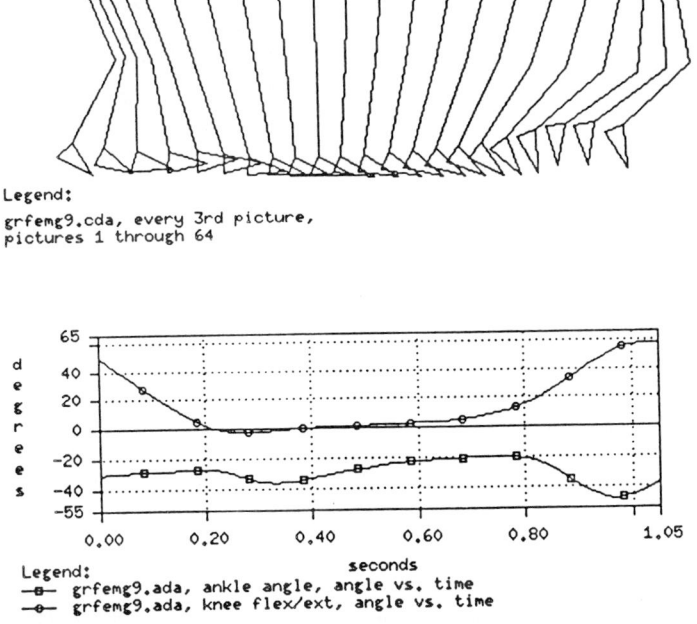

Figure 18 Use of videography for digitization of human movement and measurement of segmental displacement and angles.

of motion of the upper body. The spinal unit bears a major portion of the weight of the head, arm, and trunk (HAT). During simple activities like walking or standing erect to perform a task, the spinal column has to not only support the gravitational forces of the HAT but also support the forces and moments created due to motion of the upper body parts. Such forces may often exceed the limitations of biomechanical loading that can be endured by the spinal unit and can result in long-term or acute injury. Understanding the loading on the spinal column and its behavior under the action of these forces is important in deriving the criteria of safe manual materials handling practices.

The spinal column, also known as the vertebral column, is a complicated structure. The structure can be simplified by assuming it to consist of fairly rigid bone segments (vertebrae) connected by means of ligaments and intervertebral disks. The column consists of 33 vertebrae linked to one another through intricate connections of tissues, ligaments, and intervertebral disks to form a strong but flexible supporting column for the head and trunk. The vertebral column can be subdivided into several regions. The cervical region consists of seven vertebrae and is followed by the thoracic region, which contains 12 vertebrae. The lumbar region follows, with five vertebrae. This is followed by the five fused vertebrae of the sacral region and three to four fused coccygeal segments. Normally the column appears symmetrical in the frontal plane and has characteristic curvatures in the sagittal plane. These curvatures provide additional natural shock absorbance and flexibility at the intervertebral joints. There exist anterior convexity in the cervical and lumbar regions and posterior convexity in the thoracic and sacral regions. The curvatures in the thoracic and sacral regions are primarily due to the wedge-shaped structure of the vertebrate unit, while the curves in the cervical and lumbar regions are due to wedge-shaped intervertebral disks. In the event of the loading of the complete spinal unit, the lumbar and cervical regions are flattened to a greater extent [12]. Deviations from these natural curvatures of the spine may generate excessive stresses in the intervertebral units and can cause permanent deformations, depending on the extent of the loading.

Integrated vertebral movement of the spine is the result of coordinated action of the soft tissues interlacing the vertebrae. In order to describe the biomechanics of spinal movement, it is important to understand the structure of the spinal units or vertebrae. Figure 19 shows side and top views of the vertebral units including the ligaments found in the region. Figure 19a shows vertebrae with the intervertebral disks between them. The structure and function of the intervertebral disk will be discussed later. Figure 19b shows the major ligaments associated with the vertebral units. These ligaments allow smooth movement of the spinal column within the feasible ranges of motion with minimum resistance and expenditure of energy and provide vital support in the event of a traumatic shock to the spinal column. Without the fine coordination of these soft tissues, smooth spinal motion would not be possible. Each of the components of the spinal column has its own physical properties that combine to provide a quantitative picture of the spinal column's functional capabilities.

2. Intervertebral Disk

The intervertebral disk is present between two vertebrae as discussed earlier and is shown in Figure 19a. It is subjected to a variety of loadings during the movement of the column. In fact, when a person is standing upright, the forces to which a disk is subjected are much greater than the weight of the portion of the body above it. This is due to the fact that the line through the center of gravity of the HAT lies a little in front of the spinal column, which creates moments about the instantaneous center of rotation situated somewhere in the middle of the vertebral unit. This is further illustrated in the following example problem.

Biomechanical Aspects of Body Movement

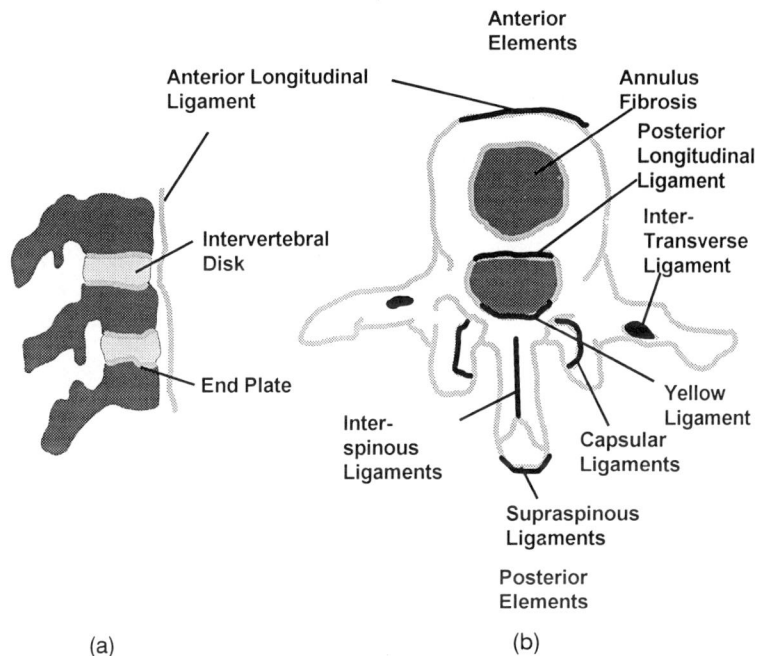

Figure 19 Elements of a spinal unit.

Example Problem 16

Figure 20 shows a single vertebral unit under the action of the gravitational load **W** of the head and trunk acting at a distance d_w (=6 cm) in front of the instantaneous axis of rotation (fulcrum) on the vertebrae. The muscle force exerted is shown as **F** and acts at a distance d_F. Calculate the ratio of the forces **F** and **W**. What is the combined compression force on the vertebral unit?

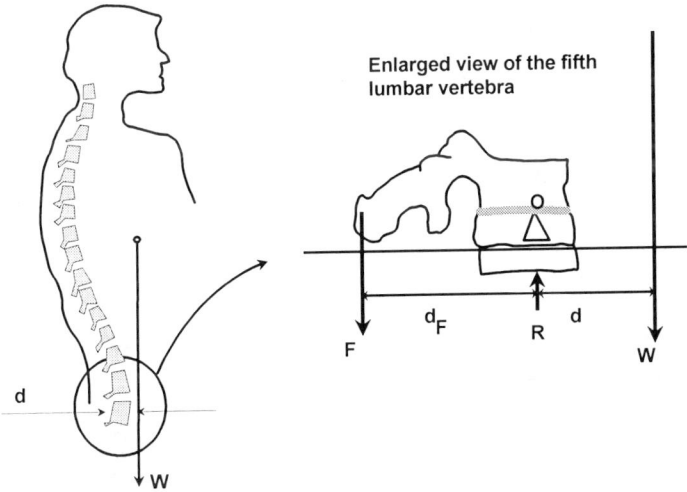

Figure 20 Forces experienced by the fifth lumbar vertebra due to the body weight. (Adapted from Ref. 12.)

Solution

Using the laws of static equilibrium and applying the equation for a lever system, we have that the net moment about the instantaneous center of rotation (fulcrum) is zero:

$$\mathbf{W} \times d_w + (-\mathbf{F} \times d_F) = 0$$

$$\mathbf{W} \times d_W = \mathbf{F} \times d_F$$

Therefore

$$\frac{\mathbf{F}}{\mathbf{W}} = \frac{d_W}{d_F}$$

For $d_W = 6$ cm and $d_F = 3$ cm, we have

$$\frac{\mathbf{F}}{\mathbf{W}} = \frac{6}{3} = 2.0$$

To calculate the compressive load on the vertebral unit, we add the forces experienced by the unit:

$$\mathbf{R} = \mathbf{F} + \mathbf{W} = 2.0\mathbf{W} + \mathbf{W} = 3.0\mathbf{W}$$

Thus, the vertebral unit experiences three times the weight of the body mass above it.

If the weight of the head and trunk in the above example equals 0.6 times the weight of the person, then the compressive force on the vertebral column would be 3.0 × 0.6 or 1.8 times the weight of the body. Studies have estimated the loading of the spine to be as much as three times the body weight when in a seated posture.

The intervertebral disk consists of three main parts. These are the nucleus pulposus, the annulus fibrosus, and the cartilaginous endplates. The nucleus pulposus is a gelatinous center with embedded fibrous content, with a high water content (70–90%). The water content decreases with age, which in effect changes the viscoelastic properties of the disk. The nucleus pulposus is more dorsally placed in the lower back and has a larger swelling capacity in the cervical and lumbar regions. The annulus fibrosis forms the outer peripheral ring of the disk and consists of fiber running in a helical pattern with alternating layers of fibers criss-crossing each other. The fibers run at about 30° to the disk plane and cross each other in adjacent layers at an angle of about 120°. The fibers are attached directly to the vertebral units on the periphery and to the endplates in the interior of the structure. This allows relative movement of the different spinal substructures in a smoother manner. The cartilaginous endplates form the upper and lower surfaces of the intervertebral disks and are formed out of hyaline cartilage tissues.

Since the disk is the major load-bearing component of the spine, several studies have been performed to establish the behavior of the disk under various loading conditions. In general, the disk is subjected to loads (both compressive and tensile) and shear stresses or a combination of loading. The forces may be exerted suddenly (jerk loading) or over a longer period of time. The loading can be sustained, causing a creep failure, or cyclic, causing a fatigue failure. In cases of creep or fatigue failure, the level of force required to cause the failure is often much lower than that required for failure under normal conditions. The deformation curve for the disk under normal loading condition is sigmoidal, implying that the disk offers little resistance at low loads, but its stiffness increases dramatically with increase in the applied loading. This provides flexibility of movement at low-loading conditions and provides supportive action at higher loading conditions [12]. It has been demonstrated that the disk is very stable at high levels of compressive loading. This type of loading causes the disk material to bulge out, which in turn introduces compressive and shear stresses in the disk material. The degenerated disk behaves in a slightly different fashion. The load is transmitted from one endplate to the other primarily via the annulus. This causes the endplates to sustain the major portion of the loading at the disk's outer edges. At this outer layer of the disk, the tangential forces are small but the annulus fibers are exposed to twice as much stress. Unlike the nondegenerated disk, the axial stress becomes compressive.

Torsional forces introduce similar internal stresses in the disk structure. These generated stresses are not evenly distributed throughout the disk. The magnitude of these shear forces is usually higher at the periphery than in the central parts of the disk. Experiments indicate that a relatively high level of shear force is required to cause any failure in the disk structure in the horizontal plane [12].

Due to the fact that the instantaneous center of rotation (fulcrum) for a vertebral unit lies somewhere in the middle of the unit, both compressive and tensile loads are generated in different parts of the disk. Measurement of the tensile strength of the disk have been made along various directions to the plane of the disk [12]. The tensile strength is the highest along a 15° direction to the disk plane. The anterior regions are stronger than the posterior regions. The strength distribution provides most protection at areas prone to herniation under severe loading. Unequal distribution of the tensile strength along the cross section of the disk shows that the disk is highly anisotropic and is more adaptive in resisting loads in certain directions than in others. The disk has been found to be stiffer in compression than in tension. This has been attributed to the buildup of fluid pressure within the nucleus pulposis under compressive loading. The tensile stress can be divided into its normal and shear components along and perpendicular to the plane of the disk. Although the normal component can be resisted by the alternating fibers, there exists no provision for resisting shear stresses. Another factor contributing to the vulnerability of the disk to tensile loading is Poisson's effect, whereby the cross-sectional area is reduced in tensile loading whereas it is increased due to the bulging in compressive loading.

Experimental findings suggest that bending and torsional loads, and not compressive loads, are the most potentially damaging to the disk. Bending of the spinal column produces both compressive and shear forces on the material of the disk. The disk material is pushed out in the direction of the curve as shown in Figure 21. Excessive bending may cause this

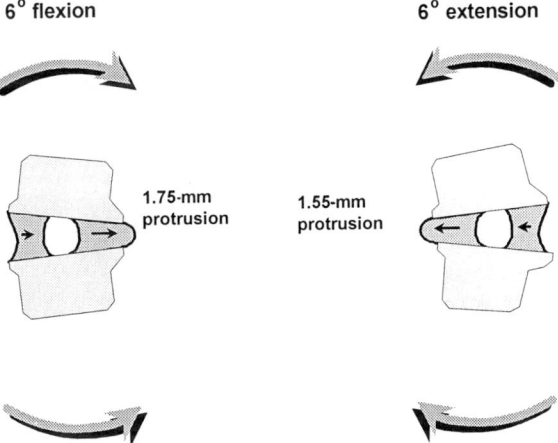

Figure 21 Protrusion of the discal material during flexion or extension of the spine. (Adapted from Ref. 12.)

bulge to impinge against nerve roots, causing painful physiological conditions. The loading in occupational tasks is often a complex combination of the different loading conditions defined above. Experimental values of the stiffness coefficients of the intervertebral disk may be found in several references [12]. The stress values are important in determining safe limits of musculoskeletal loading and are used as guidelines for loading limits in manual materials handling.

3. Intradiscal Pressure

Measurement of the fluid pressure in the intradiscal cavity is probably the most effective way of studying the loading characteristics in vivo. In this method the intradiscal pressure of the nucleus pulposus acts as a load transducer and indicates the magnitude of axial loading on the spinal column. Figure 22 shows the level of intradiscal pressure for various lumbar support during seated postures. Figure 23 indicates the measured intradiscal pressure for various daily life postures. Increased pressure indicates a greater muscular effort in maintaining the posture and hence a larger stress on the spinal column. As is evident from Figure 23, a seated posture may actually produce higher disk pressures than a standing posture. This is attributed to the fact that the line of gravity for the upper body acts farther away from the instantaneous center of rotation.

4. Biomechanical Aspect of Back Pain

Back injury and pain can have a variety of causes. The phenomenon of back pain is most prevalent in the occupational setting. It is the most frequently reported form of musculoskeletal disorder. Several physical, phychosomatic, biomechanical, and anatomical factors have been shown to cause back pain, alone or in combination with one another.

As the disk material does not have any nerve endings, simple compressive forces are not responsible for pain. During the application of compressive loads on the spinal column,

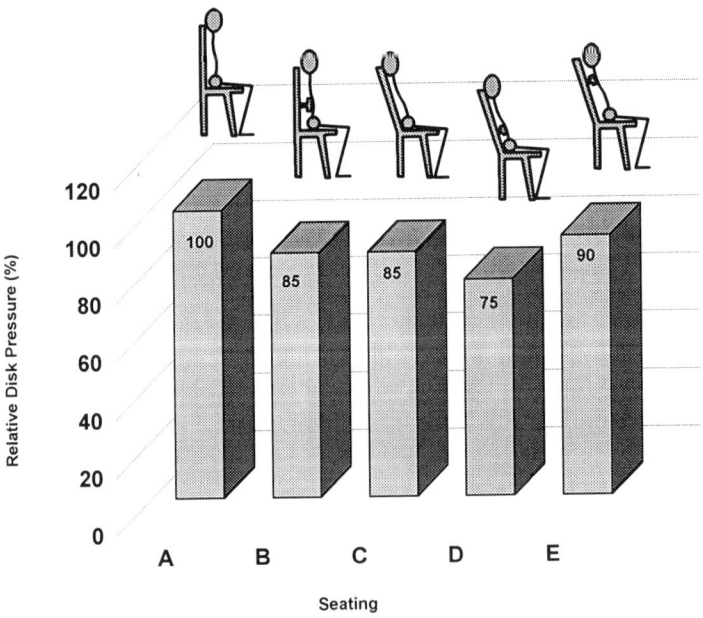

Figure 22 Relative disk pressure experienced during sitting with various inclinations of the back support and presence of a lumbar support. (Modified from Ref. 8.)

Biomechanical Aspects of Body Movement

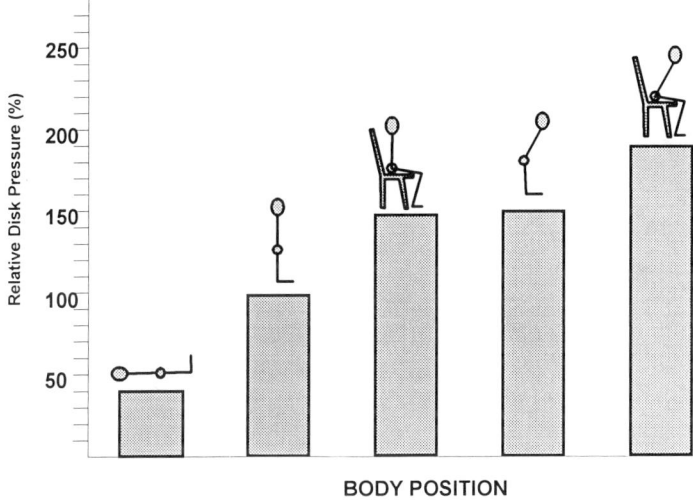

Figure 23 Relative disk pressure during various postures.

the disk material can protrude outside and impinge upon the spinal nerves that surround the disk and the vertebral units. This phenomenon is depicted in Figure 24. In the static condition of the loading of the spine, an increase in the lumbosacral angle can cause an increase in the natural lumbar curvature (lordosis). This is termed *swayback*. About three-fourths of all back pain cases are due to excessive lordosis [13]. Pain is generated due to facet impingement and irritation, as depicted diagrammatically in Figure 25.

H. Center of Gravity

Upright stability in human beings is possible due to a complex mechanism of the neuromusculoskeletal system that acts in a systematic manner to control the orientation of various

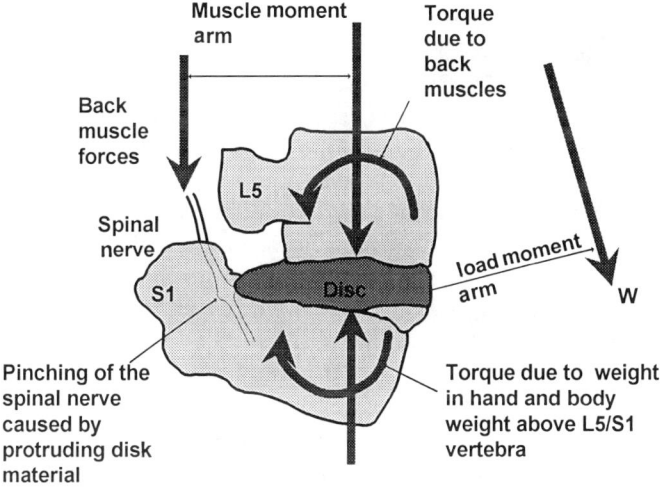

Figure 24 Pinching of the spinal nerve. (Adapted from Ref. 13.)

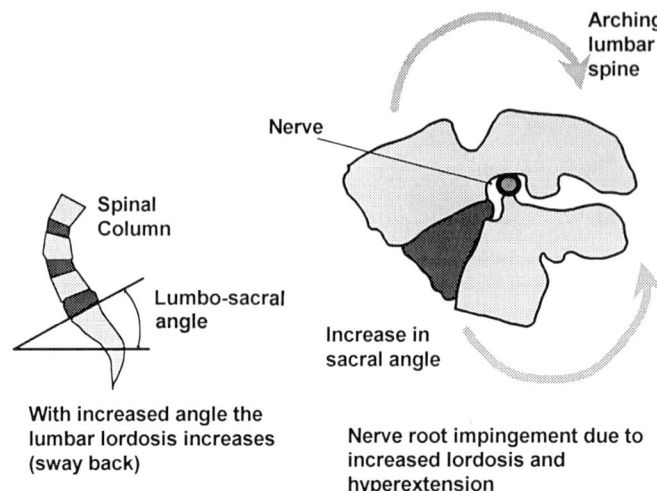

Figure 25 Back ache caused by "swayback." (Adapted from Ref. 13.)

body segments above its base of support to achieve the upright stance. To attain upright stability the body constantly tries to maintain its center of gravity (COG) at a comfortable location between the feet. Three major joints of the body—ankle, knee, and hip—help move the COG of the body appreciably. Like many other body postures, the same location of the COG is achieved by several possible movement strategies of the three joints. The muscle groups moving these joints are in turn activated by motor neurons. Commands to the motor neurons come from both the spinal cord and subcortical levels, depending on the kind and duration of excitation. A combination of reflex action and voluntary movements helps activate these muscles. At least one pair of muscles are acting at each of these three joints to achieve fine movement. A result of that is a continuous effort to correct the positioning of the COG, creating a slight swaying of the body in both the anterior–posterior and medial–lateral directions.

The physics of upright stability dictates that the line of projection through the COG must lie within the base of support for the body to maintain its position in a static stability. Several factors, including muscle strength, age, reaction time, and environmental conditions, affect the actual limit within which the COG must be maintained at all times for stable upright support [14]. In reality, the limits within which the person's COG must be maintained for him or her to hold the upright posture without falling may actually be smaller than the outer periphery of the feet but depends heavily on the stance or support. This is illustrated in Figure 26, where the base of support is estimated for three different stances. Studies have indicated the presence of a closed-loop feedback system for maintaining the COG within the limits of basal support [15] or performing a movement strategy to modify or increase the base of support, either by taking a step (staggered feet stance) or by grabbing an external support.

In static postures, the projection of the COG can be estimated by the center of pressure (COP) measurement under the feet [16]. Under normal circumstances, the center of pressure is continuously shifting as the body is trying to maintain balance, creating a bodily sway as discussed above. The locus of the COP can be experimentally determined by using force platforms [16,17]. As the subject stands on a force platform, the forces and moments generated by the plantar surface of the subject's feet are collected via an amplifier and analog-to-digital board interfaced to a personal computer. These forces and moments are further ana-

Biomechanical Aspects of Body Movement

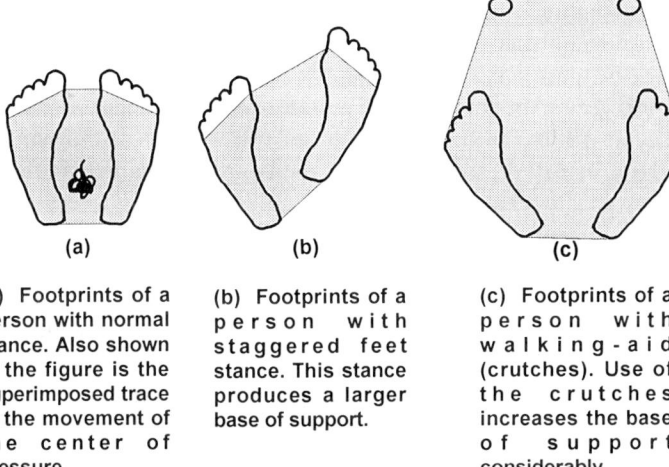

(a) Footprints of a person with normal stance. Also shown in the figure is the superimposed trace of the movement of the center of pressure.

(b) Footprints of a person with staggered feet stance. This stance produces a larger base of support.

(c) Footprints of a person with walking-aid (crutches). Use of the crutches increases the base of support considerably.

Figure 26 Base of support during various upright stances. (a) Footprints of a person with normal stance. Also shown in the figure is the superimposed trace of the movement of the center of pressure. (b) Footprints of a person with staggered feet stance. This stance produces a larger base of support. (c) Footprints of a person with walking aid (crutches). Use of the crutches increases the base of support considerably.

lyzed to determine the time-variant location of the COP. The trace of the locus of the COP is plotted for the entire test period. The length and spread of COP sway, with respect to the stability boundary described earlier, are important indirect measures of lack of stability. It forms a tool for evaluating workplaces and tasks that present a very high potential for loss of balance for the worker and possible fall or injury due to a momentary loss of balance [18].

I. Link-Segment Model

The skeletal system along with the muscles, jointly called the musculoskeletal system, is the primary source of human body movement. The skeletal system consists of the bones that form the hard interior structure of the body and provides support to many of the body organs. The bones are also responsible for articulation, and a combination of these articulations give rise to human movement. Movement of each individual bone is accomplished by the musculature attached to it. The musculature is in turn activated by the nervous system. As we saw in Chapter 3, the nervous system along with the musculoskeletal system forms a closed-loop feedback system to achieve complex movement patterns such as walking or playing a piano.

To apply the laws of physics and engineering in a simplified manner to describe human movement, the body is divided into segments [19]. This simplified approach assumes that the body segments are formed of rigid links joined to each other by pin joints that allow relative motion similar to the body joint that is being studied. So far we have discussed the individual body segments (elbow, shoulder, wrist, lower extremities, etc.) and how each of these segments behaves when acted upon by external loading of the joints. In the workplace, a worker might be involved in a task that creates a simultaneous complex loading on several body segments. In order to evaluate the musculoskeletal loading characteristics of the entire body it is essential to combine the loadings of the various body segments in a composite link-segment model.

In a link-segment modeling approach, the human body is modeled as a kinematic chain of linkages. Very few of these linkages are closed-loop, where the movement of any one link in the system can be used in calculating the movement of other links in the system. Most of the kinematic linkages in the human body are an open-loop system, where the same movement of an individual link can be obtained by several alternative movement patterns of the other links in the system. Forces in the human body are transmitted via these open-loop link systems by the nervous system, which controls the link segments. Because of the open-loop nature of the link segments, a given body movement can be achieved by the nervous system in several alternative movement patterns of the kinematic chain, though the end result is the same. It is important to understand this limitation of the mathematical link-segment model in reproducing human movement. Most of the time, a careful examination of the physically possible range of motion of these link segments can narrow down the choice of alternative movement patterns. Figure 27a shows the link segments that model the human leg. The upper leg (thigh, lower leg, and foot can be assumed to be three separate rigid body segments, attached to each other through pin joints. Each of these segments can be separately consid-

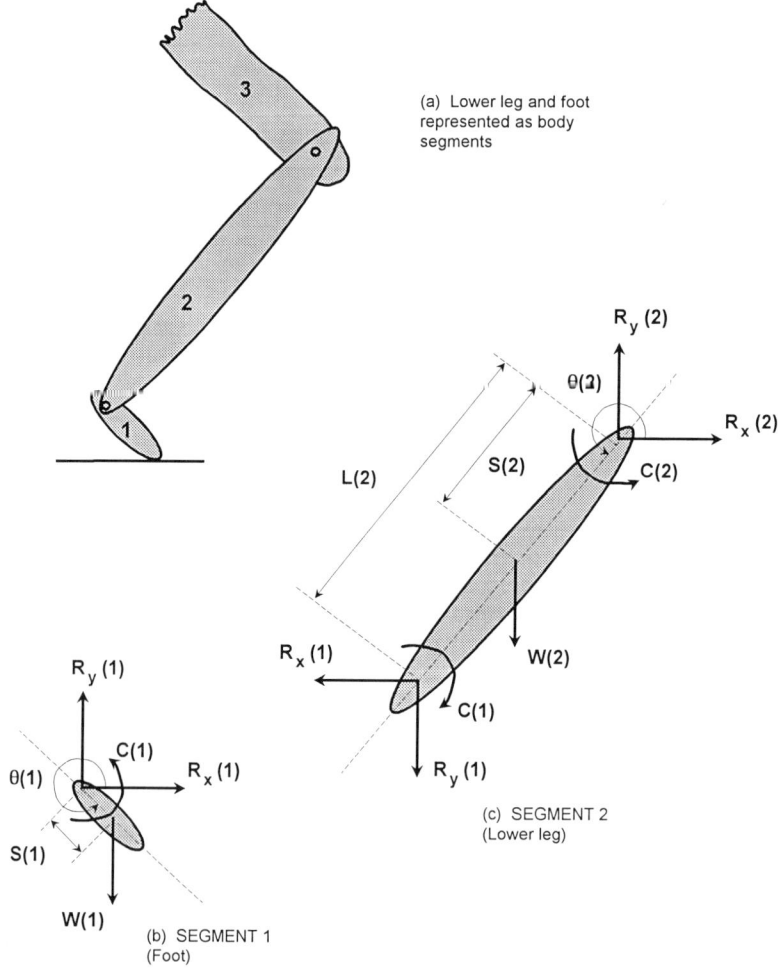

Figure 27 Line-segment modeling of the lower leg. (Modified from Ref. 3.)

ered for analysis of the forces and moments exerted on it. Figures 27b and 27c show the separated segments for ease of analysis. These are called *free-body diagrams*. In a state of either static or dynamic equilibrium, the net algebraic sum of the forces and moments acting on the segment must be equal to zero (in dynamic equilibrium, the body has a constant linear and/or angular velocity). Within each of these segments, there may be internal muscle forces acting to keep the segment in a state of equilibrium.

Several assumptions are made to simplify the analysis and apply the existing equation of motion to the body segments.

1. Each body segment has a fixed mass located at its COG.
2. Bone joints are approximated as pure hinge joints (with no loss of motion).
3. The location of the COG does not shift with motion.
4. Mass moment of inertia (denoted by I) of each body segment about its COG is constant during the movement.
5. At each joint, a resultant force acts that is equal to the joint reaction force plus the result of all muscle forces (agonists and antagonists) acting across the joint.
6. Only the moment or torque generated by the resultant muscle force (not individual muscles) around the joint is used for analysis.
7. Each muscle crosses only one joint.

An axis system can be placed on the segmental joint (where it connects and interacts with other segments). Forces acting at the joint can be resolved into components acting along the x and y axes. The equation of motion can be formulated for the three body segments as shown below.

Foot Segment. From the free-body diagram shown in Figure 27b, summation of the horizontal and vertical forces yields

$$\Sigma \mathbf{F}_x = 0 \quad \text{and} \quad \Sigma \mathbf{F}_y = 0$$

or

$$\mathbf{R}_x(1) = \mathbf{M}(1) \times \mathbf{A}_x(1)$$

and

$$\mathbf{R}_y(1) - \mathbf{W}(1) = \mathbf{M}(1) \times \mathbf{A}_y(1)$$

where
$\mathbf{R}_x(1), \mathbf{R}_y(1)$ = horizontal and vertical components of the resultant muscle and reaction forces at the ankle joint
$\mathbf{A}_x(1), \mathbf{A}_y(1)$ = horizontal and vertical components of the linear acceleration of the center of mass (CM)

Summation of the moments about the CM yields

$$\Sigma \mathbf{M}_{CM} = I(1)\,\alpha(1)$$

or

$$\mathbf{C}(1) + \mathbf{R}_x(1)[S(1)\sin\theta_1] - \mathbf{R}_y(1)[S(1)\cos\theta_1] = I(1)\,\alpha(1)$$

where
$\mathbf{C}(1)$ = resultant muscular torque at the ankle joint
$I(1)$ = moment of inertia of the foot about the COG of the foot (perpendicular to the sagittal plane, assuming that the motion is limited to the sagittal plane)
$\alpha(1)$ = angular acceleration of the foot

θ_1 = angle of the foot segment with respect to the right horizontal in the counterclockwise direction
$S(1)$ = distance between the proximal joint and the CM or COG of the foot segment

Similar to segment 1 (foot segment), we can formulate the three equations for segment 2 (lower leg) shown in Figure 27c.

$$\mathbf{R}_x(2) - \mathbf{R}_x(1) = M(2) \times \mathbf{A}_x(2)$$

$$-\mathbf{R}_y(1) + \mathbf{R}_y(2) - W(2) = M(2) \times \mathbf{A}_y(2)$$

$$-\mathbf{C}(1) + \mathbf{C}(2) + \mathbf{R}_x(1)\,[L(2) - S(2)]\sin\theta_2 - \mathbf{R}_y(1)\,[L(2) - S(2)]\cos\theta_2$$
$$-\mathbf{R}_y(2)\,S(2)\cos\theta_2 + \mathbf{R}_x(2)\,S(2)\sin\theta_2 = I(2)\,\alpha(2)$$

Similarly, the equation of motion can be formed for each of the connecting segments. Because of the number of variables involved, the equations become highly complex for a three-dimensional analysis. A generalized set of two-dimensional equations for static and dynamic cases is given below to illustrate the process.

Generalized Equations for a Two-Dimensional Static Model. Note: Here, $\mathbf{A}_x$, $\mathbf{A}_y$, and the horizontal force may be assumed to be zero.

$$\Sigma \mathbf{R}_j(y) = \mathbf{R}_{j-1}(y) + \mathbf{W}_j$$

where
$\mathbf{R}_j(y)$ = resultant muscular and reaction force at jth joint (vertical)
$\mathbf{R}_{j-1}(y)$ = resultant muscular and reaction force at the previous $(j-1)$th joint.
$\mathbf{W}_j$ = weight of the jth body segment

For the torque at the joint,

$$\mathbf{C}_j = \mathbf{C}_{j-1} + L_j \cos\theta_j\, \mathbf{W}_j + L_{j-1}\cos\theta_j\, \mathbf{R}_{j-1}(y)$$

where
$\mathbf{C}_{j-1}$, $\mathbf{C}_j$ = torques at the $(j-1)$th and jth joints
L_j = distance from joint j to the CM or COG of the link
L_{j-1} = link length from joint j to the adjacent joining joint $(j-1)$
θ_j = angles of the links at each joint j with respect to the right horizontal

Generalized Equations for a Two-Dimensional Dynamic Model. For the dynamic model the orthogonal components of the linear acceleration, $\mathbf{A}_x$ and $\mathbf{A}_y$, need to be calculated for each joint. The equations are as follows:

$$\mathbf{A}_x(j) = 0$$

$$\mathbf{A}_x(j+1) = -L_i[\dot{\theta}_j^2 \cos\theta_j + \ddot{\theta}_j \sin\theta_j] + \mathbf{A}_x(j)$$

$$\mathbf{A}_y(j+1) = -L_i[\dot{\theta}_j^2 \sin\theta_j - \ddot{\theta}_j \cos\theta_j] + \mathbf{A}_y(j)$$

where
L_i = length of segment i
θ_j = angular position of joint j with respect to the right horizontal, measured counterclockwise
$\dot{\theta}_j$ = angular velocity of segment i at joint j
$\ddot{\theta}_j$ = angular acceleration of segment i at joint j

Biomechanical Aspects of Body Movement

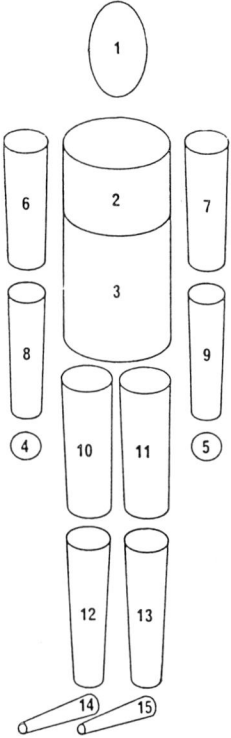

Figure 28 Three-dimensional modeling of a human body. (Adapted from Ref. 23.)

As in the above analysis, equations can be formed for each individual body segment, and the forces and moments at each individual joint can be calculated. The moment of inertia of the segment is calculated based on the fact that each body segment can be assumed to be a simple conical volume [23], as shown in Figure 28.

The above equations can be programmed into a computer for calculation of the joint forces and moments. The anthropometric measurements can be fed into the model that includes the segmental length, weight, diameter/circumference, and segment orientation with respect to each other. The displacement, velocity, and acceleration of the segments can be obtained from kinematic data collected through videographic techniques. Several commercial packages are available for performing such sophisticated analyses. A simple application of the above link-segment modeling can be found in models that calculate the joint forces and moments as well as the loading on the L_5S_1 vertebral unit, for manual materials handling tasks [2, 23].

IV. BIOMECHANICAL CRITERIA FOR ERGONOMIC APPLICATIONS

Why is it essential to study the biomechanical criteria for ergonomic applications? This chapter has presented a number of biomechanical principles that can be applied to human body movement and form the basis for developing criteria to be used in applied ergonomics. For example, the limitation of the forces produced in the fingers at various flexions or extensions of the wrist may help us design hand tools that require optimum forces for a worker to operate them without excessive strain. Such interventions, on the basis of biomechanical criteria, reduce the risk of injury to the body due to impact loads or cumulative trauma by giving due consideration to the physical limitations of the human body. Determination of the criteria for cumulative trauma disorders due to incorrect postures, loading, and frequency of task performance have been based on the biomechanical analysis of the static as well as dynamic loading of the body segments. Such standards are already under preparation [20].

Several other important applications of biomechanical analysis may be cited. The lifting guidelines suggested by the National Institute of Occupational Safety and Health (NIOSH) use the biomechanical criteria of a maximum allowable compressive loading on the L_5S_1 vertebral units (3.4 kN, 770 lb) in setting the maximum load that can be safely lifted by a person. This recommendation includes several other biomechanical, physiological, and psychophysical factors for evaluating this loading of the human body that are based on the physical limitations due to the strength and posture of the body [21]. These include the frequency of lifting (which increases the dynamic loading as well as the metabolic demand), forward and upward reach (static loading), travel of the load, control at the endpoint of the loading, asymmetry of lift, coupling of the load, etc. Another application of biomechanical analysis is found in the design of equipments, tools, and workstations to optimally utilize the midrange of motion of the body segments, since it has been shown that forces and moments can be applied maximally at the midrange of motion of the segments. Existing databases have documented the maximum forces that can be applied at various limb positions and body postures. These databases may be used to control the amount of work output depending on the posture of the worker, without inflicting any form of harmful musculoskeletal loading. For example, the maximal two-handed upward vertical force (including lifting) that can be applied repetitively by a 41–50-year-old male in a seated position (without backrest) is 11 kg when the person has to stretch out his arms [22]. Any job requiring the person to perform such a task should be designed on the basis of this criterion for avoiding worker discomfort and possibility of injury. Such databases are limited in their scope of population size and type and need to be expanded to include the specific working population under consideration.

At present, lack of extensive data for all segments of the human population prevents us from more accurate analysis of safe limits of the loading of the body segments. With the help of further research, further advances in computer capabilities, and sophisticated equipment for measurement, better and more realistic biomechanical models can be generated to enable us to set forth safe and acceptable limits of physical work and human output.

REFERENCES

1. Y. C. Fung, *Biomechanics*, Springer-Verlag, New York, 1981.
2. D. B. Chaffin and G. B. Andersson, *Occupational Biomechanics*, Wiley, New York, 1984.
3. M. Nordin and V. H. Frankel, *Basic Biomechanics of the Musculoskeletal System*, 2nd ed., Lea and Febiger, Philadelphia, 1989.
4. E. R. Tichauer, *The Biomechanical Basis of Ergonomics*, Wiley-Interscience, New York, 1978.
5. N. Özkaya and M. Nordin, in *Fundamentals of Biomechanics: Equilibrium, Motion, and Deformation*, D. Leger, Ed., Van Nostrand Reinhold, New York, 1991.
6. B. LeVeau, *Biomechanics of Human Motion*, Saunders, Philadelphia, 1976.
7. B. A. Gowitzke and M. Milner, *Scientific Basis of Human Movement*, 3rd ed., Williams and Wilkins, Baltimore, 1988.
8. Eastman Kodak Co., Human Factors Section, *Ergonomics for People at Work*, Van Nostrand Reinhold, New York, 1986.

9. B. Tyldesley and J. I. Grieve, *Muscles, Nerves, and Movement: Kinesiology in Daily Living*, Blackwell Scientific, London, 1989.
10. M. I. Ellis, B. B. Seedhom, V. Wright, and D. Dowson, An evaluation of the ratio between the tensions along the quadriceps tendon and the patellar ligament, *Eng. Med.* 9(4): 189–194 (1980).
11. Peak Performance Technologies, *User's Reference Manual*, Englewood, CO, 1994.
12. A. A. White and M. M. Panjabi, *Clinical Biomechanics of the Spine*, Lippincott, Philadelphia, 1978.
13. R. Cailliet, *Low Back Pain Syndrome*, 3rd ed., F. A. Davis Co., Philadelphia, 1981.
14. A. Bagchee and A. Bhattacharya, Method for estimating fall potential in clinical, occupational, and environmental exposure cases (Abstract), *J. Biomech.* 26(3): 305 (1992).
15. M. L. Root, W. P. Orien, and J. H. Weed, Normal and abnormal function of the foot, in *Clinical Biomechanics*, Vol. II, Clinical Biomechanics Corp., Los Angeles, 1977.
16. T. Shimba, An estimation of center of gravity from force platform data, *J. Biomech.* 15:533–550 (1987).
17. A. Bhattacharya, R. Morgan, R. Shukla, et al., Non-invasive estimation of afferent inputs for postural stability under low levels of alcohol, *Ann. Biomed. Eng.* 15: 533–550 (1987).
18. A. Bhattacharya, The role of worker's balance in falls at the workplace, Regional Conf. on Ergonomics, Safety and Health in Construction, July 19–20, 1994.
19. W. T. Dempster, Space Requirements of the Seated Operator: Geometrical, Kinematic, and Mechanical Aspects of the Body with Special Reference to the Limbs, Wright Air Development Center Tech. Rep. No. 55-159, Dayton, OH, Wright Patterson Air Force Base, Nat. Tech. Info. Service No. AD-087892, 1955.
20. ANSI, American National Standard Institute: Control of Cumulative Trauma Disorders, Draft ANSI Z - 365, 1993
21. T. R. Waters, V. Putz-Anderson, A. Garg, et al., Revised NIOSH equation for the design and evaluation of manual lifting tasks, *Ergonomics* 36(7): 749–776 (1993).
22. Materials Handling Research Unit, University of Surrey (England) (Preparers), *Force Limits in Manual Work*, IPC Science and Technology Press, Surrey, England, 1980.
23. E. P. Hanavan, A Mathematical Model of the Human Body, Wright-Patterson Air Force Base, AMRL-TR 64-102, 1964.

5
Biomechanical Models in Ergonomics

Kevin P. Granata and William S. Marras
The Ohio State University, Columbus, Ohio

I. INTRODUCTION

Why should an ergonomist be interested in biomechanical models of the human body? In order to answer this question one must consider what can be gained from biomechanical modeling. The goal of biomechanical modeling in an ergonomics context is to gain insight into the how the joints of the body are strained during work as well as to provide a framework for understanding how the musculoskeletal system behaves during work task performance. Biomechanical models provide qualitative and quantitative answers that explain how the body is loaded during work. Quantification is important because it allows one to address the issue of how much is too much loading on a joint. Through quantification of joint loading one can not only assess the risk associated with existing jobs but also predict whether someone is likely to suffer an injury with a proposed design of a workplace. Qualitative analyses allow comparison of biomedical loads between two or more sets of conditions. Models of musculoskeletal stresses may allow ergonomists to determine which of several workstation designs will result in the least amount of structural loading on a specific joint. Thus, biomechanical models used in ergonomics provide both a means to quantitatively assess work and a means to predict whether a proposed workstation can be expected to lead to injuries.

One must recognize that our current knowledge of joint biomechanics is extremely limited in that we are able to provide only gross approximations and estimates of forces supplied by the muscles or imposed on the body. Although the complexity of joint motion and force development can be appreciated from studies of anatomy, physiology, and kinesiology, quantitative descriptions of their mechanical behavior are still very simplistic in comparison. Through the application of biomechanical modeling, we can test our hypotheses, simulate joint dynamics, and predict the loads developed within a joint. By developing validated biomechanical simulations, complex anatomical structures and motions can be not only described but also explained.

II. BACKGROUND AND SIGNIFICANCE TO OCCUPATIONAL ERGONOMICS

The objective of biomechanical analyses is to explain how the body works from a loading perspective. Application of these analyses also provides insight into how, why, and when a joint may fail. A biomechanical model is simply a mathematical representation of the musculoskeletal system and permits one to quantitatively describe the loading effects of a task. A biomechanical model that is capable of accurately simulating the forces developed within a joint may be employed to investigate the relation between joint load and exertion parameters such as joint angle, torsion, velocity, applied load, and impact. When these joint loads are compared to the structural strengths or tolerance limits of the joint, one may use this information to explain why an injury occurred during a particular exertion. When the probabilistic nature of joint tolerances is considered, these models could help explain why some individuals will suffer an occupationally related injury under certain conditions whereas others may not. In addition, common injuries that occur for unknown reasons may be examined to determine the causative factors, thereby indicating a possible means for injury prevention.

Models are used to help us organize and test our knowledge and understanding of how different workplace or loading factors are related to the probability of failure of a joint or structure. Perhaps one of the greatest values of biomechanical modeling is that it provides a mechanism to test hypotheses generated from epidemiological studies of injuries in the workplace. For example, studies [1] have shown that twisting motions in the workplace are a contributing factor in occupationally related low back disorder risk. Biomechanical models [2] can explain how spine loading patterns differ during twisting and bending motions. Thus, biomechanical models could provide a cross-validation for suspected risk factors. In addition, once we understand the mechanism of occupationally related injury we can use these models to determine how to change the work environment to minimize the risk of injury. In this manner, biomechanical models can help us determine how much exposure to a risk factor is too much, so that low-risk work environments can be created.

Occupationally related low back pain provides a classic example of how biomechanical modeling can be employed to investigate why and how an injury occurred as well as to suggest possible solutions. The epidemiological literature demonstrates that, in general, occupations associated with manual materials handling industries are subject to high incidence rates of low back disorders (LBDs). The studies clearly indicate that lifting and parameters associated with lifting result in increased risk of injury. Biomechanical models of the trunk and low back have been developed to qualitatively and quantitatively describe trunk and spinal loads as a function of the lifted weight, the distance it is held from the body, and associated trunk moment, trunk posture, position, and motion [3–6]. These models have been successful in describing why people who perform certain lifting tasks are more susceptible to injury than those who perform different tasks. Furthermore, the models have provided estimates of the direction and magnitude of loads on the spine. These data can be employed to determine how the spine fails under typical lifting conditions. Although we are far from a complete understanding of the biomechanical cause of occupationally related low back pain, significant progress has been made in the recognition and explanation of many of the causative parameters.

Biomechanical models are developed to understand or interpret the functional behavior and joint loadings that occur during a task or exertion of interest. Therefore, the conditions under which the model is developed and operates must be realistic and accurate for the model to be useful for ergonomic purposes. For example, static models [3,4,7] that assume symmetric lines of action and neglect motion are not intended to represent the biomechanics of common, complex, dynamic motions such as many of those tasks observed in the workplace. Similarly, analyses that neglect significant anatomic structures and physiologic principles can misrepresent the biomechanical loads imparted by and upon the joints. The fact that a computer model "works" does not guarantee that it is correct or useful. Inaccurate or unrealistic model output may lead to misdiagnosis of the joint load, potential, injury, and risk, thereby permitting conclusions that may exacerbate the problem.

Biomechanical Models

Unfortunately, as a biomechanical model becomes more accurate and realistic, it often becomes more complex. Simplistic biomechanical models are often easy to use and can be employed by those who are not necessarily experts in ergonomics. Results from simple models are also easier to interpret than their complicated counterparts. Furthermore, highly complex (and often more accurate) models representing realistic tasks require significantly more input data to achieve those ends. Often the more complex models cannot be used by many whose primary concern is the design of the workplace. On the other hand, when applied and interpreted correctly, simple biomechanical models are an invaluable source of information. These models are particularly attractive to nonsophisticated users. However, the danger with these simple models is that the user may not be aware of the model limitations or of assumptions inherent to the model and thus may misapply the model or use it to lead them to an incorrect solution to a manual materials handling problem. Therefore, there is often a trade-off between criteria of accuracy and realism versus simplicity and ease of use. In general, for ergonomic purposes, a model need only be as complex as is necessary to accurately and reasonably describe the nature of loads occurring at a joint of interest due to a particular work task.

It is important to recognize that the fundamental goal of a biomechanical model is to quantitatively describe the resultant internal loads experienced by a joint as a consequence of an occupationally related task. As mentioned in an earlier chapter, both external and internal loads can be imposed on the body as a result of work. External loads are those associated with events outside the body such as the forces of gravity acting upon a box that one is attempting to lift. Internal loads are associated with forces and moments within the body used to counterbalance the external load. These typically are generated by muscles and ligaments within the body. However, the distance between a joint's center of rotation and the external load is typically much greater than the distance from the joint's center of rotation to the muscle attachment. Therefore, muscles, tendons, and ligaments are at a mechanical disadvantage, and the internal forces are typically much larger, often orders of magnitude larger, than the applied external force. Holding a 10-lb weight, i.e., an external load, in front of one's body at arm's length could require internal forces within the back muscles in excess of 150 lb. Biomechanical models help us to place the contributions of both the internal and external forces in perspective relative to occupationally related risk of injury.

Biomechanical models used for ergonomic purposes can be found for both the back and the wrist. The anatomy of the back and that of the wrist are drastically different; therefore, these models have developed along different paths.

A. Models of the Trunk and Back

Several mathematical models of the lumbar spine have been developed to determine the loads on the spine during lifting exertions [3–10]. These models have been designed to enhance the knowledge of low back mechanics, improve the ability to diagnose the cause and risk of injury during manual materials handling, and appraise recommendations designed to reduce lifting injuries.

One must recognize that loading of the spine can occur in several different ways. Compression, shear, and torsion loading can be imposed on the spine during the performance of a task. Realistically, most lifting tasks result in some combination of these loadings on the spine. In addition, loading can occur at various levels of the spine. The spine is a series of 33 vertebrae, and each vertebra could be loaded differently during an occupational task. Hence, the system has the potential to be represented in a very complex model. Most current models used in ergonomics assume that the spine consists of a single rod or a single or double joint so the models can be solved.

Once estimates of loading are derived for a particular task, these loadings are typically compared to spine tolerances. Tolerances of the vertebral end plates have been determined in vitro by Evans and Lissner [11], Sonada [12], and Jagger et al. [13] for pure compressive loads. Brinckmann et al. [14] determined compressive tolerance of the spine as a function of combinations of loading level and frequency of loading for a population of cadaver specimens. Shirazi-Adl [15] used finite-element modeling to determine tolerances of the disk during complex loadings of the spine.

1. Biomechanical Risk Factors

One of the major challenges for any biomechanical model of the spine is to determine which components of the workplace need to be considered in order to accurately account for the risk associated with the occupational demands of the work. As mentioned earlier, biomechanical models can be extremely complex and difficult to use if they comprehensively include all biomechanical components of work. Similarly, if the models are too simplistic they may not be very useful for the control of occupationally related low back disorders. Thus, it is imperative to identify those components that are necessary to include in an accurate biomechanical model that can be applied to the workplace.

Marras and associates [1,16] performed an industrial in vivo study to assess the contribution of various biomechanical workplace factors to the risk of suffering a work-related low back disorder. Over 400 manual materials handling jobs were observed in 48 different industries. The medical records in these industries were examined so that specific jobs historically categorized as presenting low, medium, or high risk of occupationally related low back disorder could be identified. One hundred fourteen job-dependent characteristics were recorded for each job. Among these characteristics, workplace factors such as the weight of the object lifted, the distance of the object from the worker, work heights, and cycle times were recorded. In addition, workers wore a triaxial goniometer on the back [lumbar motion monitor (LMM)] that documented the three-dimensional angular position, velocity, and acceleration characteristics of the lumbar spine associated with the various jobs. A multiple logistic regression model indicated that a combination of five trunk motion and workplace factors was best associated with the risk of low back disorder. These factors were lifting frequency, load moment, trunk lateral velocity, trunk twisting velocity, and trunk sagittal angle. Increases in the magnitude of these factors significantly increased the risk of low back disorder. This study indicated that by quantitatively considering the combination of these five factors it is possible to predict the probability of membership in the high-risk catagory 10 times more effectively than chance. As this is the only study to comprehensively examine the biomechanical factors associated with occupationally related low back disorders, these results provide valuable information as to which factors need to be included in biomechanical models of the trunk. The study suggests that if all five of the risk factors could be accounted for in a biomechanical model, one would be able to gain an appreciation for how back injuries are associated with spine loading.

2. Static Models

Biomechanical models of lifting have progressed from static two-dimensional analyses to more recent attempts at understanding dynamic three-dimensional stresses on the spine. The newer models endeavor to accurately and realistically represent the mechanical loading and behavior of the lower back while refraining from as much unnecessary complexity as possible.

Biomechanical Models

A simple two-dimensional lifting model can be developed by representing the body as a stick figure as shown in Figure 1. If the spine is assumed to be a rigid rod attached to the pelvis by a single hinge, all motion and trunk mechanics are centered about that point. We can allow the back muscles to be represented by a single tensioned cable strung between the upper spine and the posterior aspect of the pelvis. When weight is placed in the stick man's hands, the only thing preventing the hinged spine from falling forward is the tension in the back muscle equivalent. The external moment tending to cause the stick man to bend forward can be described by the two-dimensional vector product of the lifted external load and the moment arm from the back joint to the weight. Similarly, the internal, restorative moment that supports the trunk in an upright posture against the external load can be described by the vector product of the back muscle equivalent and the moment arm from the back joint to the muscle. If we assume that there is no motion and know the moment arm distances, we can apply the laws of static equilibrium to determine the tensile force in the modeled back muscle required to hold a given external load. Thereby we can solve for the tensile force in the back muscle equivalent.

This simple analysis can be extended to determine the compressive load imposed on the spine by the stick man when he holds the weight. In this model, compression is simply represented as the sum of vertical forces. Clearly, the weight held by the stick man adds compression on the spine. Note also that the back muscle equivalent adds a significant vertical force that must be supported by the spine. Therefore, in this example, the compressive force is the sum of the weight in the hand, body segment weights, and, approximately, the back muscle equivalent tensile force.

The analysis described here is neither accurate nor realistic. It simply provides "ballpark" estimates of the loads that act on the spine. The biomechanical model is two-dimensional, whereas the human body is three-dimensional. Great numbers of muscles in the trunk have

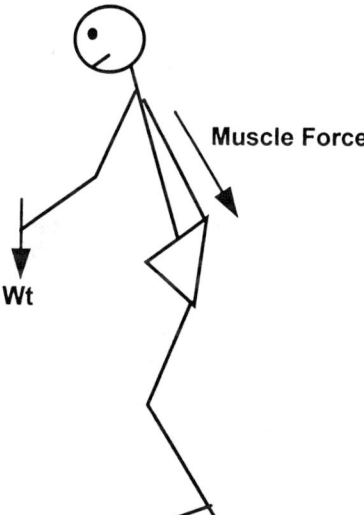

Figure 1 A rudimentary lifting model representing sagittally symmetric, two-dimensional exertions. The spine is modeled as a rigid rod with a single hinge at its base. Back muscles are represented by a single tensioned cable. Although the model is neither accurate nor realistic, it may provide a "ballpark" estimate of spinal loading.

been neglected, each with unique endpoints, i.e., insertions and origins. Those muscles exist in three-dimensional space and often wrap around the skeletal structures of the trunk. Hence, such a model is unable to evaluate shear or torsional loads applied to the spine. The spine is not a rigid rod with a frictionless hinge at its base. Instead, it is a multielement structure with complex geometry and passive nonlinear resistance to bending and twisting motions. No passive forces were included in the model. The modeled exertion was represented as a static, sagittally symmetric lift. Lifting velocity and acceleration influence trunk mechanics and spinal load. Lifts are seldom performed directly in front of the body. Instead, most occur in combination with twisting and lateral bending and with the weight held in an asymmetric posture, i.e., off to one side. Thus, when using such a model one must consider how model predictions may be limited by inappropriate assumptions about the workplace conditions.

Clearly, the simple stick man model described here is useful for its designed purpose, i.e., to demonstrate the fundamental concepts of lifting models, but it is greatly limited in its usefulness for accurate biomechanical analyses of lifting exertion and concomitant spinal loads. One of the original, groundbreaking lifting models [3] is very similar to the stick man model described here. It has served as a simple analytical tool for job and workplace evaluation. To overcome its limitations, more advanced models have been developed. The single-extensor stick man model has served as the ancestral foundation for the more advanced models.

Models have also been developed that attempt to account for the three-dimensional nature of trunk loading. Chaffin and Erig [17] extended the single-equivalent stick man representation to three-dimensional space. Their model was able to accommodate asymmetric lifts but was still limited to static evaluations and was not sensitive to complex loadings of the spine due to the activities of the multiple internal muscle forces involved in a lift. Schultz and Andersson [7] developed a model that accounts for the three-dimensional architecture of the trunk. As shown in Figure 2, this model assumes that the trunk can be viewed as a cross section through a level of the lumbar spine. Schultz and Andersson described the relationship

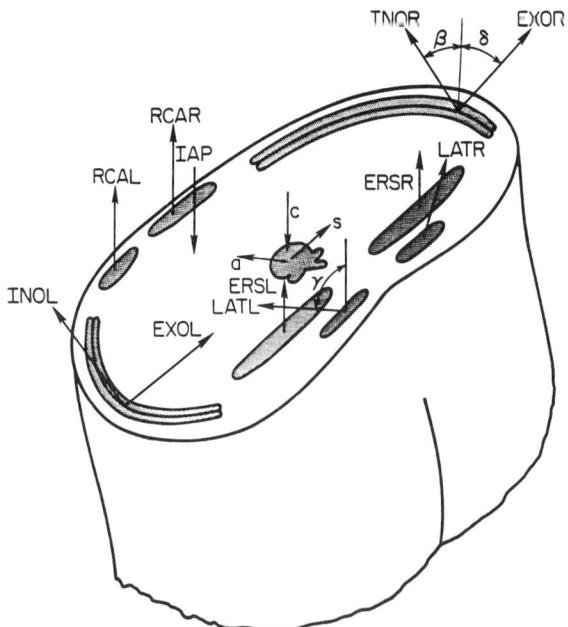

Figure 2 Cross-sectional view of the human trunk at the lumbro-sacral junction. (Adapted from Ref. 7.)

between the 10 trunk muscles, intra-abdominal pressure (IAP), and three-dimensional spine loading in six force and moment equations. Since there were 11 unknown muscle and IAP values and only six equations available to derive a solution, the model resulted in a statically indeterminate problem. This static model was solved either by accepting simplifying assumptions about the muscle activities (i.e., no coactivation) or through the application of linear programming techniques. However, because of the limited number of functional equations in the model, the linear programming approach was not able to predict coactivation of the trunk muscles that had been observed even under static conditions [18]. More recent models have attempted to include coactivation in the solution by using adjustments to the linear programming model [19] and double optimization functions [20].

Biomechanical models of spinal loading require accurate kinetic assessment of muscle forces and realistic representation of lifting kinematics. Static modeling may represent static exertions, but most lifting exertions include a dynamic component.

3. Lifting Dynamics

The lifting models of Chaffin [3], Shultz and Andersson [7], and Gracovetsky and Farfan [21] attempted to estimate spinal load from equations of static equilibrium. Consequently, the models were representations of static posture lifting. Schultz and Andersson [7] indicated that their model of static lifting could be employed to model quasi-dynamic movement. They implied that motion can be represented as a series of static postures linked together in time to give the illusion of motion. This concept is similar to connecting a series of still photographs to produce a motion picture. However, the laws of physics demonstrate that forces associated with static exertions are not the same as those associated with dynamic exertions. Furthermore, physiology demonstrates that the contractile velocity of a muscle influences its ability to generate force. Static modeling cannot satisfy these conditions. The significance of considering these dynamic factors was confirmed via industrial biomechanical assessments [1,16].

Evidence has shown that static lifting models underpredict spinal loading during dynamic lifting exertions. Marras and Sommerich [6] concluded that "spine compression increases directly with [isokinetic] trunk velocity" during lifting exertions. Similar conclusions were reached by Granata and Marras [22], Frievalds et al. [23], McGill and Norman [8], and Goel et al. [24], who predicted a 30–40% increase in dynamic lifting moment and spinal compression compared to static exertions.

In addition to the mechanical factors, lifting dynamics also influence muscle coactivity. Muscle coactivity is important because it is responsible for the complex loading of the spine. Because many of the trunk muscles are not oriented in a vertical direction, they can apply compression, shear, and torsional forces on the spine. The number of active trunk muscles and the magnitude of the force in each muscle are also significantly affected by lifting velocity. Measures of muscle coactivity [25,26] have demonstrated that the activity in muscles other than prime movers, i.e., synergistic and antagonistic muscles, increases with lifting velocity. The relation between dynamic motion and muscle coactivity directly affects the loads on the lumbar spine [27]. Therefore, dynamic modeling must be employed to accurately represent spinal loads during lifting exertions.

Quasi-dynamic models of isokinetic lifting have been developed [5,27,28] that incorporate the behavior of muscle coactivity patterns. The models are labeled quasi-dynamic because lifting exertions are represented as smooth, continuous trunk motions, typically isokinetic, but the influences of mass dynamics are not considered. McGill and Norman [8] concluded that isokinetic models overpredict peak dynamic lifting moments by an average of 25%. To accurately represent the biomechanics of lifting, fully dynamic modeling must be employed [29].

4. Trunk Mechanics

Both external and internal forces contribute to spinal loading. In fact, internal (i.e., muscle) forces typically contribute far more to spinal stress than external forces. Moment arm distances of the muscles are small compared to almost any external moment arm distance. Because muscles are at a severe mechanical disadvantage when attempting to offset external moments, extremely large forces are required to overcome this handicap. Estimates of spinal loading must therefore include both the external and the much larger internal (muscle) forces.

Chaffin [3] and Chaffin and Baker [30] developed an early low back lifting model to predict spinal loads under static two-dimensional sagittally symmetric conditions. The model computed the force in a single extensor muscle required to offset reactive trunk moments generated by external forces. It did not account for the muscle synergy or flexor antagonism evident in multiple-muscle systems [18,31–33]. Hof and Van Den Berg [34] demonstrated that the forces exerted by synergistic and antagonistic muscles significantly affect the total joint torque. Hence, multiple-muscle models must be employed to accurately represent lifting biomechanics.

One of the first multiple-muscle models intended to overcome these limitations was the model of Schultz and Andersson [7]. This model used 10 muscle equivalents to predict static sagittally symmetric and asymmetric [19] spinal loading on the lumbar spine. Optimization techniques were employed to estimate muscle forces from an indeterminate system of equations. Gracovetsky and Farfan [21] implemented a similar analysis with a much more comprehensive anatomic model of the low back. In addition, Jager and Luttmann [35] and Goel et al. [24] used optimization procedures to predict loading during dynamic and asymmetric lifting exertions. However, optimization-based predictions of muscle coactivation do not agree well with observed muscle coactivation during dynamic lifting activities [26,36].

Deterministic models, including the single-equivalent-muscle models, by their nature cannot account for muscle force variability within and between subjects performing a lifting exertion. The models assume that internal forces can be completely described from the applied external forces and trunk kinematics. Muscle tensile forces in these models are determined from a system of equations representing lifting mechanics and possible constraints introduced by the optimization objective functions [37]. Variability within the system is not possible in deterministic modeling, because muscle kinetics are reduced to a single, "unique," optimum solution [38,39]. Therefore, deterministic modeling represents the optimal solution to the objective function of the model, but not necessarily the realistic muscle activity and spinal load. It is important to be able to account for this variability in muscle recruitment during an occupational activity. Variation in muscle force can significantly influence spinal loading. Therefore, to accurately model lifting mechanics, the variation in muscle coactivity must be considered.

There is little question that a significant amount of antagonistic coactivation exists and is the rule rather than the exception under typical manual materials handling conditions. This coactivity has been noted in response to several workplace factors in several studies. Marras et al. [18], Gracovetsky et al. [31], Zetterburg et al. [32], and McGill and Norman [33] measured antagonistic muscle activity and concluded that coactivity contributes to trunk stability. Rajulu [40] and Marras and Mirka [26,41] also measured significant antagonistic coactivity as a function of external load, trunk position, velocity, and acceleration.

Optimization methods have often been suggested as a means to overcome the statically indeterminate nature of the trunk's multiple-muscle system and permit analytic solutions to more complex biomechanical models. However, these optimization schemes typically predict unrealistic muscle coactivity levels, including an absence of antagonistic activity [27,37]. In an attempt to improve upon this situation, muscle coactivity has been predicted using nonlinear optimization techniques [24,42], but with limited success. The predicted muscle activity amplitudes do not agree with measured values.

It is extremely important to account for muscle coactivity because this coactivity significantly influences spinal load [27,34]. Trunk flexor muscles of the abdomen have a greater mechanical advantage than the extensor muscles of the back. Minimal abdominal activity requires substantial back muscle tension to offset the antagonistic moments. Therefore, muscle cocontraction influences the activity in the prime movers necessary to achieve static or dynamic equilibrium. If optimization techniques inaccurately predict coactivity, then muscle activity in the prime movers will also be incorrectly estimated, and subsequent predictions of spinal loads will be inaccurate. Furthermore, if the coactivity of all trunk muscle equivalents is not monitored, then damaging components of torsion and shear loads remain unpredicted [15,43]. The assumption of insignificant coactivity and antagonism is valid in a very limited and artificial set of circumstances and ensures underestimation of spinal loading. Therefore, ignoring muscle coactivity through deterministic optimization risks inaccurate estimation of spinal loading.

5. Electromyography-Assisted Models

To avoid estimation errors associated with deterministic prediction of muscle force, biomechanical models of lifting exertions have been developed that are assisted or guided by biological input. The most common biological input used in biomechanical modeling involves electromyographic (EMG) measures of muscle activity. Electromyography-assisted models employ measured myoelectric activity to determine individual muscle forces. Thus, they avoid the problem of neglected muscle coactivity associated with statistically indeterminate models. Relative weighted EMG values recorded during lifting exertions are used to predict the tension developed in muscles and their resultant trunk moments and spinal loads. By their nature, EMG-assisted models accurately represent the neuromuscular control system of the trunk through direct measurement of muscle activity [42,44]. The advantages of EMG-assisted models are that (1) they are not limited by the constraints of an optimization objective function; (2) muscle coactivity and antagonistic forces are accurately represented via measurement; (3) coactive variability within and between subjects is directly measured; (4) physiologic coefficients predicted by the model can be used for instantaneous validity checking; and (5) accuracy can be documented via direct comparison of measured and predicted lifting moments.

McGill and Norman [9] developed a dynamic model using EMG measures to estimate force in 12 trunk muscles. Myoelectric activity was measured on one half of the body, and contralateral muscles were presumed to behave similarly to their measured counterparts. Therefore, the model represents sagittally symmetric lifting only and disregards muscle-induced lateral shear and torsion at the base of the spine. A more recent publication by McGill [45] employs bilateral placement of electrodes in an attempt to predict lateral bending moments. Muscle force F is represented by

$$F = \text{gain} \times \frac{\text{EMG}(t)}{\text{EMG}_{\text{Max}}} \times \text{Area} \times F(\text{Vel}) \times F(L) \qquad (1)$$

where Area is the cross-sectional area of the muscle, Gain represents the physiologic muscle force per unit area assigned to 35 or 50 N/cm^2, and $F(\text{Vel})$ and $F(L)$ are modulation factors representing theoretical muscle force–velocity [46] and length–strength [47] relations, respectively. Muscle activity is represented by normalized EMG levels, i.e., EMG/EMG$_{\text{Max}}$, where

EMG$_{Max}$ is a constant. Unfortunately, normalizing by a constant value of EMG$_{Max}$ may have limited applicability. Marras et al. [18,25] demonstrate that maximum EMG levels change as a function of trunk angle and subsequent muscle length. Thus, the normalization constant must be a function of trunk position. Muscle length directly influences the relation between measured EMG and joint torque [48], due in part to the physiologic length–strength relation of muscle [47]. Isokinetic velocity and acceleration also significantly affect measured EMG levels [26,36,41,49]. Therefore, dynamic effects on myoelectric activity and muscle force must be included in any EMG-assisted low back model.

An EMG-assisted model developed by Reilly and Marras [28] incorporated intramuscular EMG data from five left–right pairs of trunk muscles. Bilateral muscle measures allowed prediction of relative asymmetric moments, although experimental data were collected only in the sagittal plane. A similar model by Marras and Sommerich [5] was validated under asymmetric moments, although experimental data were collected only in the sagittal plane. A similar model by Marras and Sommerich [5] was validated under asymmetric exertions, i.e., in a vertical bending plane rotated from the sagittal. Myoelectric inputs were normalized as a function of both trunk bending angle and asymmetry and were modified for length and velocity artifact via empirically derived regression equations [5,26]. Time-dependent kinetic, kinematic, and EMG data were approximated by a three-segment, straight-line representation of the data profiles. Good correlations were achieved between predicted and measured trunk moments (Figure 3). However, approximating the dynamic data profiles by straight-line segments reduced the true power of the EMG-assisted model by artificially representing the measured muscle activity.

The Marras–Sommerich [5] model was the first EMG-assisted lifting model to allow validation of predicted results. Straight-line segment representations of the measured trunk extension moments were compared with predicted moments to test the model accuracy. A validity check was performed by requiring a physiologically reasonable gain, i.e., muscle force per unit area, predicted from each trial. Although the output gain factor was checked for validity, it was allowed to vary with each lifting task, implying that the subject's strength capacity changes with each exertion. Clearly, a subject's muscle strength per unit area does not change from trial to trial. A dynamic EMG-assisted lifting model proposed by Granata and Marras [10] treated gain as a subject-dependent input constant. This introduced possible

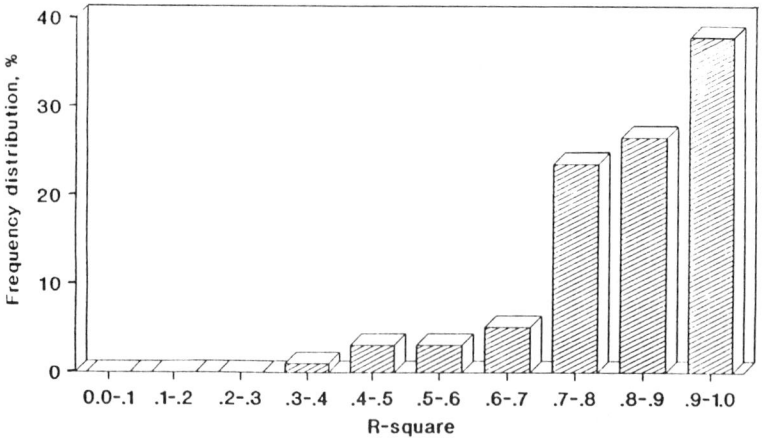

Figure 3 Association (R^2) between predicted and measured trunk torque (total of 98 trials). (From Ref. 5.)

intertask variability into the magnitude of predicted extension moment with each lifting task. The model sampled and processed dynamic data at each point in time (100 Hz) throughout small time windows, requiring no interpolation or linearity assumptions, although the entire lifting motion was not represented. As shown in Figure 4, the processed EMG represented dynamic changes in muscle forces and measured the coactivation in the multimuscle system. The model was tested under isometric, isokinetic, and isoinertial conditions at three separate angles of asymmetry. Coactive bilateral EMG data were normalized as a function of trunk angle and asymmetry.

Models extending the concept of EMG-assisted modeling to trunk loading during axial twisting exertions have also been published. Pope et al. [50] endeavored to model static twisting exertions, and McGill [51] attempted to model dynamic twisting exertions. Both models suffered from physiologically unrealistic gain values. They required muscle force capacity far in excess of accepted values [52–54]. Marras and Granata [2] demonstrated that the muscle area employed in EMG-assisted modeling must represent the maximum area of the muscle as opposed to the area of the muscles that is found at the transverse plane defined by the lumbosacral junction. This modification permitted successful simulation of static and isokinetic torsional exertions.

In order to allow the biomechanical modeling of typical daily lifting activities, a model must be capable of simulating spinal loading during unconstrained, free-dynamic coupled motion. It is necessary that muscles represented by the model be permitted to change length, direction, and velocity so that appropriate modulation of the muscle myoelectric activity can occur. Recently, EMG-assisted models have been developed that successfully treat these three measures as dependent variables [2,29]. These models represent the trunk musculature as a

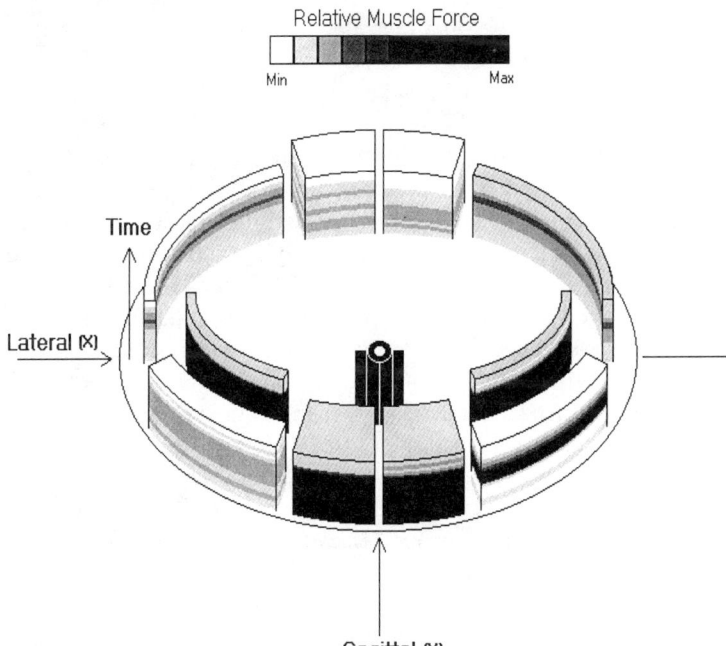

Figure 4 Relative muscle force as a function of time (vertical axis) in each of the 10 modeled trunk muscles viewed from the posterior aspect. Note the measured coactivity, especially in the antagonistic musculature. (Adapted from Ref. 2)

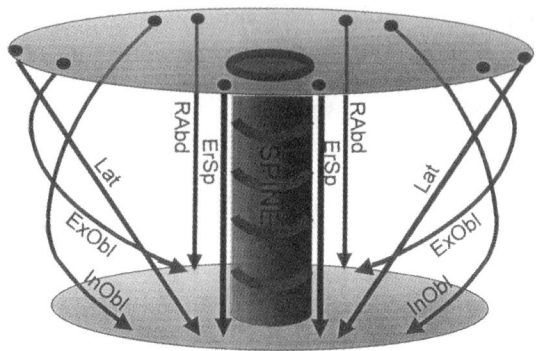

Figure 5 Vectors representing internal muscle forces from the dynamic model of Granata and Marras [22] and Marras and Granata [2].

series of force vectors as shown in Figure 5. These vectors can change in orientation and magnitude as the trunk moves through space and can thus more accurately account for the loading contributions of the trunk musculature.

Interpreting electromyographic data collected during dynamic motions can be difficult because of the motion artifact related to physiologic and analytic phenomena [26,36,49,55–57]. Electromyography-assisted models must appropriately handle EMG dynamics. This has been accomplished by modulating the EMG on either a purely theoretical [9,45] or purely empirical [5,10] basis without examining the relation between the two. Analysis of muscle length and velocity modulation factors appropriate for free-dynamic lifting exertions [58] demonstrated that empirically developed relations agree with the theoretical relations reported in the physiological literature.

As can be seen from this discussion, EMG-assisted models are the only models available at this time that have the potential to accurately assess the loads on the lumbar spine associated with the biomechanical risk factors identified via industrial biomechanical studies [1,16]. This is true because these models are the only ones that are capable of accurately assessing muscle coactivity. Granata and Marras [22] recently assessed the influence of coactivation upon spinal compression estimates. Their study concluded that during free-dynamic lifting tasks a single-equivalent-muscle model would underestimate the compression on the spine by 40% compared to a full 10-muscle, EMG-driven coactivation model. Hence it is extremely important to consider the activities of the complete set of trunk muscles when attempting to assess loads on the lumbar spine.

Data necessary as input to an EMG-assisted model are dynamic myoelectrical data and trunk kinematics. Dynamic kinetic data are also necessary if model performance validation and quality assurance measures are desired. Because electromyographic activity must be measured along with kinematic information in order to run an EMG-driven model, most occupational tasks of interest are simulated in the laboratory.

The major disadvantage of EMG-assisted models is that they require the measurement of myoelectric activity. This measurement can be tedious and laborious, making it difficult to assess tasks on a routine basis. Mirka and Marras [59] have developed a means to estimate expected EMG activity given the parameters of a lifting exertion. They developed a

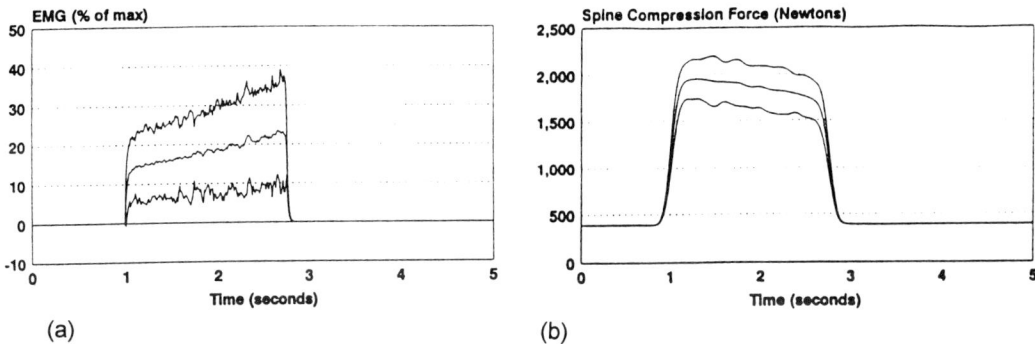

Figure 6 Mean and standard deviation of expected EMG activity in erector spinae (a) and spinal compression (b) muscle during an extension motion of the trunk.

stochastic EMG model that would simulate the range of expected EMG activity from each of 10 trunk muscles given the load lifted, dynamic characteristics of the trunk, and position profile of the trunk. Figure 6a shows the mean and standard deviation of expected EMG activity in the erector spinae muscle for a given trunk lift. Mirka and Marras used these stochastically generated EMG activities as input to an EMG-assisted model. These estimated muscle activities agreed well with measured muscle activities during a validation study. By running the model over many trials the authors were able to predict the expected loading on the spine over repeated lifting without measuring muscle activity. Figure 6b shows the expected range of spinal loads that would be expected via 50 simulations of a lifting activity using the stochastically generated EMG as input to the EMG-assisted model described above. Hence, future multiple-muscle models may be developed that can enjoy the advantages of EMG-assisted models without the need to actually measure muscle activity for a particular task.

B. Models of the Hand and Wrist

As with the back, it is important that we understand how the structures of the hand and wrist are loaded during work so that we can develop a better understanding of how the risk of injury changes as working conditions change. Over the past decade we have seen a significant increase in repetitive trauma illnesses associated with work. Illnesses such as carpal tunnel syndrome, tendinitis, and tenosynovitis have become common in industry, whereas they were unheard of in most industries just a few years ago. Thus, it is imperative that we begin to understand those occupationally related factors that are associated with an increased risk of hand and wrist disorders so that we can control exposure to these risk factors on the job.

Compared to the back, biomechanical modeling efforts have been sparse for the hand and wrist. In addition, few biomechanical models have been used for ergonomic purposes. This dearth of models of the hand and wrist may be due, in part, to the high degree of complexity in the hand and wrist structure compared to the back. This complexity is a function of the elaborate anatomic construction of the wrist. The hand and wrist complex is unique in that most of the muscles that supply the internal force during gripping and finger action are located in the forearm rather than in the hand or wrist itself. Thus, internal forces must be transmitted from the forearm muscles through tendons to the fingers and hand joints. For that reason, biomechanical models have traditionally represented the wrist as a pulley system as shown in Figure 7. The modeling issues, therefore, have centered on including the appropriate

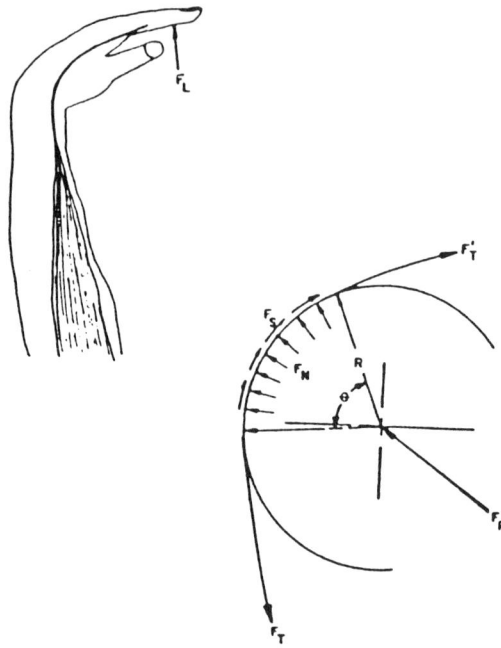

Figure 7 Armstrong and Chaffin's [64] biomechanical model of a flexor tendon wrapping around the flexor retinaculum, $F_R = 2F_T \sin(\theta/2)$, where F_T is the tendon force and F_R is the resultant reaction force exerted against the tendon. (Adapted from D. Chaffin and G. Andersson, *Occupational Biomechanics*, John Wiley and Sons, New York, 1984, p. 219. Reprinted by permission of John Wiley and Sons.)

anatomical representation of the hand and wrist and including enough information that solutions to the models could be found.

In an attempt to reduce the degree of complexity associated with biomechanical modeling of the hand and wrist, most of the existing models focus on specific segments of the hand (i.e., finger or wrist joint) and the forces exerted during specific postural conditions. For the most part, the vast majority of these models were developed for the purpose of understanding the implications of hand injury or hand surgery; very few were developed specifically for ergonomic purposes. Few models have been concerned with the assessment of cumulative trauma disorders (CTDs), which have been identified as a major risk factor by ergonomists for many occupationally related wrist injuries. However, we can still gain an appreciation for the state of ergonomic biomechanical modeling through these injury-based models.

As was the case for the back, the early models of the hand and wrist attempted to evaluate the structure and loading in the hand and wrist while the wrist assumed static postures. Landsmeer [60] was one of the first to formulate a model of the structures in the fingers. Through anatomical investigations he was able to describe, with a series of three two-dimensional models, how the tendons bridge a finger joint. His models consisted of two interphalangeal joints that formed a biarticulating system, which was traversed by both a flexor tendon and an extensor tendon. Thus, the model could permit both agonist and antagonist actions of the musculoskeletal system. This model was able to explain how changes in the state of equilibrium between the two tendons were able to generate variations in finger positions. Chao et al. [61] later developed a three-dimensional model of a finger joint in selected isometric hand positions. The joint and tendon orientations were derived from biplanar X-ray analyses.

Coordinate systems defined at each joint helped identify the constraint forces and moments. These authors formulated a pylon concept that explained the relationship between applied force, passive tendon force, active tendon force, and the finger joint. This formulation resulted in a statically indeterminate problem. Electromyographic and physiological assessments were used to solve the problems. This study did provide an understanding of the functional anatomy of the hand. The model predicted the average forces in the finger tendons in selected positions. Both the Landsmeer and Chao et al. models relied on simple static equilibrium conditions for their solutions. These models, although simple in their conception, required detailed anatomical input for their success. The complexity also limited their use to static hand postures. Another difficulty of these hand and wrist models was that there was no means to validate them. Furthermore, simplifying assumptions were used to solve the indeterminacy of the system, generating results that neglected coactivity. Analyses by Ferguson et al. [62] demonstrated that not only muscle coactivity in the hand and wrist are significant, but their neglect may cause joint and tendon loads to be underpredicted by as much as 30%.

The model of Chao et al. [61] was further developed by An et al. [63] to generate a normative model of the human hand that computed the tendon locations and excursions under various functional configurations of the hand. In this model the tendon orientations and their force and moment parameter contributions to the analytical model were added to the governing equilibrium equations.

Armstrong and Chaffin [64] developed a two-dimensional static model of the wrist that calculated the resultant force exerted by a tendon on adjacent wrist structures. This model represented the extrinsic finger flexor tendons as frictionless pulley–belt mechanisms passing through the carpal tunnel. The model permitted one to quantitatively determine how wrist size and wrist position affect forces on the tendons and their adjacent structures. It assumed that a single equivalent tendon force produced a reaction force in the carpal tunnel during flexion or extension. It was intended to determine when loading of the carpal canal reached the point where the tendon sheaths would become inflamed and/or when significant pressure would develop around the median nerve. Goldstein et al. [65] showed that a sufficient magnitude of tensile force applied to the tendons can cause residual strain in these tendons. This was the first model intended for ergonomic purposes. As with previous finger models, their wrist model did not attempt to account for the dynamic nature of many occupational tasks. Marras and Schoenmarklin [66] and Schoenmarklin et al. [67] have identified rapid motion as a risk factor for CTDs under some work situations. Schoenmarklin and Marras [68] considered the effects of incorporating acceleration into the Armstrong–Chaffin [64] model. As shown in Figure 8, acceleration significantly increases the friction in the wrist. Thus, the model emphasizes the importance of including wrist motion in ergonomic models.

In order to deal with the problem of indeterminacy, researchers have attempted to find solutions to the models through optimization techniques. Penrod et al. [69] used the principle of minimal muscle effort to create a linear optimization program that calculated the contributions of tendons passing through the wrist to externally applied static moments. Chao and An [70] also used linear programming methods to solve the redundancy problem associated with hand modeling. Again, the appropriateness of these methods depends upon whether coactivation muscle is present in the task of interest.

Many of the more recent models have been concerned with more than simply predicting the forces occurring in the muscles and tendons of interest in the hand and wrist. Recently, ergonomic biomechanical models have attempted to predict hand postures when grasping handles in a power grip orientation. Buchholz and Armstrong [71] developed a three-dimensional kinematic model of the hand that predicts joint angles of the fingers and the hand. They assumed that the hand and fingers are a series of ellipses that wrap around a handle. How-

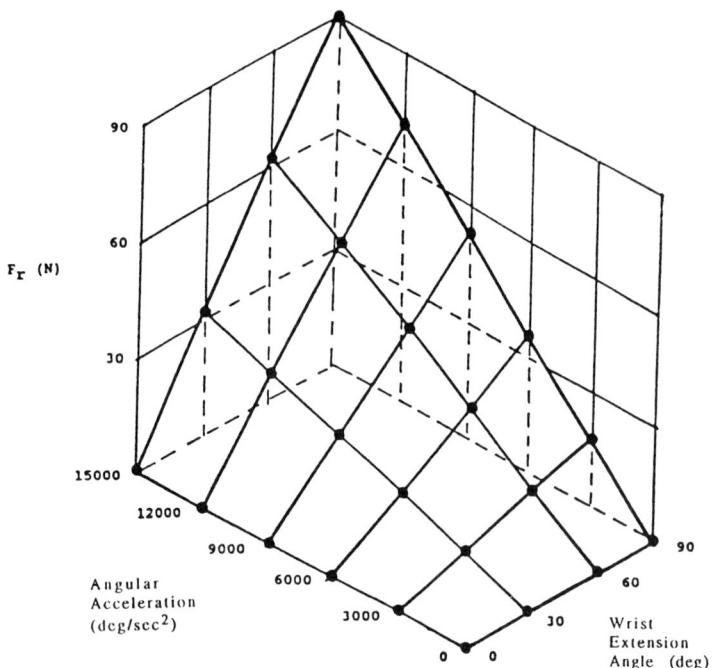

Figure 8 The resultant reaction force (F_R) exerted by the carpal bones or flexor retinaculum against a flexor tendon as a function of wrist angle and acceleration. (From Ref. 68.)

ever, no estimates of internal loading are provided in this model. Moore et al. [72] developed a model of the wrist based upon the relationship between hand forces and tendon contributions described by An et al. [73]. This model computes the frictional work between tendons and their adjacent structures during power grasps.

In an attempt to consider the coactive nature of loading on the wrist, Schoenmarklin [74] developed an EMG-assisted model of the wrist. Figure 9 shows the anatomical relationship among the tendons passing through the wrist. This model monitors the wrist flexor and extensor muscles and conditions the EMG signal for velocity artifact and muscle length changes. By considering both flexor and extensor muscle forces, the model is able to estimate the amount of friction associated with dynamic wrist motion activities.

In conclusion, biomechanical modeling of the hand and wrist can be considered still in its infancy compared to that of other parts of the body. Progress is occurring slowly but steadily. Upper extremity models have the additional burden of attempting to describe phenomena that are not well understood such as tendon sheath inflammation. However, these models are crucial in the understanding of the CTD process and the proper design of the workplace.

IV. CRITICAL REVIEW OF CURRENT STATUS

As is evident from this review, our knowledge of the human body and how its loading is associated with the risk of injury is still in a very rudimentary stage. It is much better developed with respect to the back than with respect to the hand and wrist. We can use the development of the models in these fields as a lesson on how to approach industrial biomechani-

Biomechanical Models

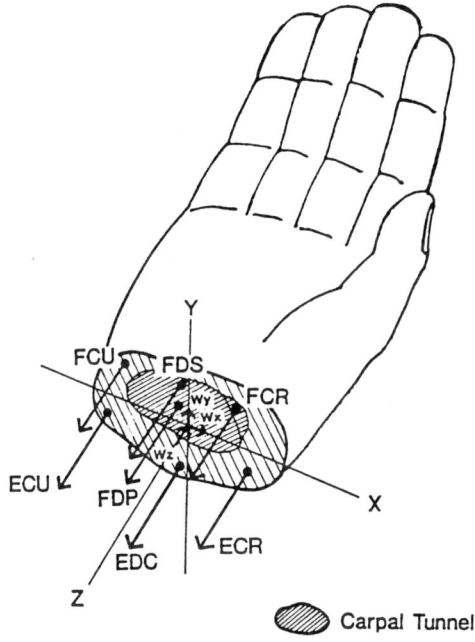

Figure 9 Cross-sectional view of the wrist from the EMG-driven model developed by Schoenmarklin [74].

cal problems in the workplace. For both the back and the hand, our representation of available knowledge began with models of the structures under static (isometric) conditions. These models were relatively straightforward in their development, and some could be validated under idealized, isometric conditions. However, there is only limited evidence that the application of such models has made a difference in industrial injury rates. Thus, although these models have been a useful starting point for the consolidation of knowledge of the biomechanics of a particular part of the body, their usefulness has been limited because many of them do not have the realism necessary to describe and understand the risk factors inherent in many industrial jobs.

In the case of both the back and the wrist, we have seen that once we begin to relax the assumption that the body is subject to isometric forces in the workplace, our ability to account for risk via modeling improves significantly [1,16,67]. This situation can be appreciated via a review of possible biomechanical variables that could be included in our models. Table 1 shows the trade-off between the degree of realism associated with various test conditions (under which biomechanical models have been developed) and the components of the physical sciences that are accounted for in the models [75]. This table shows that the physical conditions associated with isometric models (shown in the left-hand column) are the most limited in their ability to relate to realistic physical conditions. The only factors associated with force on a structure are the mass handled or force applied. The models do not have the ability to consider the effects of acceleration or jerk forces that could surely become significant when considering internal loadings of the body.

As we consider more of the physical factors associated with the variables in each column, more realistic conditions can be accounted for. For example, in Table 1 as we move through the columns from left to right, we find more factors such as acceleration and jerk

Table 1 Independent and Dependent Variables in Several Techniques Used to Measure Motor Performance

Variables	Isometric (static) Indep.	Isometric (static) Dep.	Isokinetic Indep.	Isokinetic Dep.	Isoacceleration Indep.	Isoacceleration Dep.	Isojerk Indep.	Isojerk Dep.	Isoforce Indep.	Isoforce Dep.	Isoinertial Indep.	Isoinertial Dep.	Free dynamic Indep.	Free dynamic Dep.
Displacement, linear/angular	const.*		C	X	C	X	C	X	C	X	C	X		X
Velocity, linear/angular	0	const.*	0	C	X	C	X	C	X	C	X	X		
Acceleration, linear/angular	0		0		const.*		C	X	C	X	C	X	X	
Jerk, linear/angular	0		0		0		const.*		C	X	C	X	X	
Force, torque	C	X	C	X	C	X	C	X	const.*	C	X		X	
Mass, moment of inertia	C		C		C		C		C	const.*		C	X	
Repetition	C	X	C	X	C	X	C	X	C	X	C	X	C	X

C = variable can be controlled; 0 = variable is not present (zero); X = can be dependent variable. * = set to zero.
The boxed constant variable provides the descriptive name.
Source: Ref. 75.

forces included in a model based on a given technique of measurement, and a greater appreciation for the collection of variables influencing the loading on a particular joint can be gained. Once valid models have been developed that can account for and predict the variables that are listed as dependent variables in the free-dynamic (far right) column in this table, our knowledge of biomechanical loading in the body will be complete. However, it is expected that these models will be extremely complicated and difficult to apply. Therefore, the goal of current biomechanical modeling efforts should be to develop models that are not unnecessarily complex but include enough of the dependent variables that reasonable control of workplace risk can be attained.

V. FUTURE CONCERNS

Current biomechanical models consist of isolated models of a particular part of the body. As shown in this review, models exist for only a few portions of the body. It is expected that our knowledge of joint loading for each particular part of the body will improve to the point where we can accurately assess the loads on a joint in isolation. However, there are two further challenges that must be addressed.

First, we must work toward developing a whole body model that can accurately reflect imposed loads on several joints of the body simultaneously. We know that the body acts in a coordinated fashion to accomplish many routine tasks. Thus, the load imposed upon one joint of the body can affect joints in other parts of the body. It is also evident that if one joint becomes compromised the human can compensate by changing the loading sequence experienced by the other joints of the body. A real challenge for the future will be to develop isolated joint models so that they can accurately account for the chain of events that occurs throughout the body, thereby treating the body as a true system. Only then can we truly begin to understand the limitations that might be imposed upon the body through injury and appreciate the implications of legislative acts such as the Americans with Disabilities Act.

Finally, human loadings are known to be subject to temporal demands. Greater effort must be placed upon understanding how the imposed loads of a joint are altered by age. We must also increase our understanding of how loading of a joint changes throughout the workshift. In addition, we must begin to develop models that can describe how tissues repair themselves and how this process changes as a function of age. Only when these elements are in place can we begin to truly understand and quantify the risk associated with occupational demands.

REFERENCES

1. W. S. Marras, S. A. Lavender, S. E. Leurgans, S. L. Rajulu, W. G. Allread, F. A. Fathallah, and S. A. Fersuson, The role of dynamic three-dimensional trunk motion in occupationally-related low back disorders: the effects of workplace factors, trunk position and trunk motion characteristics on risk of injury, *Spine 18*: 617–628 (1993).
2. W. S. Marras and K. P. Granata, A biomechanical assessment and model of axial twisting in the thoraco-lumbar spine, *Spine 20*: 1440–1451 (1995).
3. D. Chaffin, A computerized biomechanical model. Development of and use in studying gross body actions, *J. Biomech. 2*: 429–441 (1969).
4. DHHS, *Work Practices Guide for Manual Lifting,* Dept. Health and Human Services, Nat. Institute for Occupational Safety and Health (NIOSH), Publ. No. 81-122, 1981.

5. W. S. Marras and C. M. Sommerich, A three-dimensional motion model of loads on the lumbar spine. I. Model structure, *Hum. Factors 33*: 123-137 (1991).
6. W. S. Marras and C. M. Sommerich, A three-dimensional motion model of loads on the lumbar spine. II. Model validation, *Hum. Factors 33*: 139-149 (1991).
7. A. Schultz and G. Andersson, Analysis of loads on the lumbar spine, *Spine 6*: 76-82 (1981).
8. S. M. McGill and R. W. Norman, Dynamically and statically determined low back moments during lifting, *J. Biomech. 8*: 877-885 (1985).
9. S. M. McGill and R. Norman, Partitioning the L4-L5 dynamic moment into disc, ligamentous, and muscular components during lifting, *Spine 11*: 666-678 (1986).
10. K. P. Granata and W. S. Marras, An EMG-assisted model of loads on the lumbar spine during asymmetric trunk extensions, *J. Biomech. 26*: 1429-1438 (1993).
11. F. G. Evans and H. R. Lissner, Biomechanical studies on the lumbar spine and pelvis, *J. Bone Joint Surg. 41A*: 218-290 (1959).
12. T. Sonada, Studies on the compression, tension, and torsion strength of the human vertebral column, *J. Kyoto Prefect Med. Univ. 71*: 659-702 (1962).
13. M. Jager, A. Luttmann, and W. Laurig, Lumbar load during one-handed bricklaying, *Int. J. Ind. Ergon. 8*: 261-277 (1991).
14. P. Brinckmann, M. Biggemann, and D. Hilwed, Fatigue fracture of human lumbar vertebrae, *Clin. Biomech. 3*: S1-S23 (1988).
15. A. Shirazi-Adl, Stress in fibers of a lumbar disc, analysis of the role of lifting in producing disc prolapse, *Spine 14*: 96-103 (1989).
16. W. S. Marras, S. A. Lavender, S. E. Leurgans, F. A. Fathallah, S. A. Fersuson, W. G. Allread, and S. L. Rajulu, Biomechanical risk factors for occupationally-related low back disorders, *Ergonomics 28*: 377-410 (1995).
17. D. B. Chaffin and M. Erig, Three-dimensional biomechanic static strength prediction model sensitivity to postural and anthropometric inaccuracies, *IIE Trans. 23*: 215-227 (1991).
18. W. S. Marras, A. I. King, and R. L. Joynt, Measurements of loads on the lumbar spine under isometric and isokinetic conditions, *Spine 9*: 176-188 (1984).
19. A. Schultz, G. Anderson, R. Ortengren, K. Haderspeck, A. Nachemson, and S. Gotegorg, Loads on the lumbar spine, *J. Bone Joint Surg. 64*: 713-720 (1982).
20. J. C. Bean, D. B. Chaffin, and A. B. Schultz, Biomechanical model calculation of muscle forces: a double linear programming method, *J. Biomech. 21*: 59-66 (1988).
21. S. Gracovetsky and H. Farfan, The optimum spine, *Spine 11*: 543-573 (1986).
22. K. P. Granata and W. S. Marras, The influence of trunk muscle coactivity upon dynamic spinal loads, *Spine 20*: 913-919 (1995).
23. A. Freivalds, D. B. Chaffin, A. Garg, and K. S. Lee, A dynamic biomechanical evaluation of lifting maximum acceptable loads, *J. Biomech. 17*: 251-262 (1984).
24. V. K. Goel, J. S. Han, J. Y. Ahn, T. Cook, J. N. Weinstein, J. Winterbottom, D. McGowan, and D. Dawson, Loads on the human spine during dynamic lifting with knees straight, *Adv. Bioeng. 20*: 33-36 (1991).
25. W. S. Marras, P. E. Wongsam, and S. L. Rangarajulu, Trunk motion during lifting: the relative cost, *Int. J. Ind. Ergon. 1*: 103-113 (1986).
26. W. S. Marras and G. A. Mirka, A comprehensive evaluation of trunk response to asymmetric trunk motion, *Spine 17*: 318-326 (1992).
27. W. S. Marras, Predictions of forces acting upon the lumbar spine under isometric and isokinetic conditions: a model—experimental comparison, *Int. J. Ind. Ergon. 3*: 19-27 (1988).
28. C. Reilly and W. S. Marras, Simulift: a simulation model of the human trunk motion, *Spine 14*: 5-11 (1989).
29. K. P. Granata and W. S. Marras, An EMG-assisted model of trunk loading during free-dynamic lifting exertions, *J. Biomech.* (1994).
30. D. Chaffin and W. Baker, A biomechanical model for analysis of symmetric sagittal plane lifting, *AIIE Trans. II*: 16-27 (1970).
31. S. Gracovetsky, H. F. Farfan, and C. Helluer, The abdominal mechanism, *Spine 10*: 317-324 (1985).

32. C. Zetterberg, G. B. Andersson, and A. B. Schultz, The activity of individual trunk muscles during heavy physical loading, *Spine 12*: 1035–1040 (1987).
33. S. M. McGill and R. W. Norman, Potential of the lumbodorsal fascia forces to generate back extension moments during squat lifts, *J. Biomed. Eng. 10*: 312–318 (1988).
34. A. L. Hof and J. W. Van Den Berg, Linearity between the weighted sum of the EMG's of the human triceps surae and the total torque, *J. Biomech. 10*: 529–539 (1977).
35. M. Jager and A. Luttmann, Biomechanical analysis and assessment of lumbar stress during lifting using a dynamic 19-segment human model, *Ergonomics 32*: 93–112 (1989).
36. W. S. Marras and G. A. Mirka, Muscle activities during asymmetric trunk angular accelerations, *J. Orthop. Res. 8*: 824–832 (1990).
37. D. E. Hardt, Determining muscle forces in the leg during normal human walking. An application and evaluation of optimization methods, *J. Biomech. Eng. 100*: 72–78 (1978).
38. K. N. An, B. M. Kwak, E. Y. Chao, and B. F. Morrey, Determination of muscle and joint forces: a new technique to solve the indeterminate problem, *J. Biomech. Eng. 106*: 364–367 (1984).
39. D. R. Pederson, R. A. Brand, C. Chang, and J. S. Arora, Direct comparison of muscle force predictions using linear and nonlinear programming, *J. Biomech. Eng. 109*: 192–199 (1987).
40. S. Rajulu, Decomposition of electromyographic signals for biomechanical interpretation, Ph.D. Dissertation, Ohio State Univ., 1990.
41. W. S. Marras and G. A. Mirka, Electromyographic studies of the lumbar trunk musculature during the generation of low-level trunk acceleration, *J. Orthop. Res. 11*: 811–817 (1993).
42. D. G. Thelen, A. B. Schultz, and J. A. Ashton-Miller, Prediction of dynamic isometric muscle forces from myoelectric signals during complex loadings of the trunk, *Adv. Bioeng. 20*: 505–508 (1991).
43. A. Shirazi-Adl, A. M. Ahmed, and S. C. Shrivastava, Mechanical response of the lumbar motion segment in axial torque alone and in combination with compression, *Spine 11*: 914–927 (1986).
44. W. S. Marras and C. Reilly, Networks: internal trunk loading activities under controlled trunk motion conditions, *Spine 13*: 661–667 (1988).
45. S. M. McGill, A myoelectrically based dynamic three dimensional model to predict loads on lumbar spine tissues during lateral bending, *J. Biomech. 25*: 395–414 (1992).
46. A. V. Hill, The heat of shortening and the dynamic constants of muscle, *Proc. Roy. Soc. Biol. 126*: 136–195 (1938).
47. A. M. Gordon, A. F. Huxley, and F. J. Julian, The variation in isometric tension with sacromere length in vertebrate muscle fibers, *J. Physiol. (Lond.) 184*: 170–192 (1966).
48. J. Vredenbregt and G. Rau, Surface electromyography in relation to force, muscle length and endurance, in *New Developments in Electromyography and Clinical Neurophysiology,* Vol. 1, J. E. Desmedt, Ed., Karger, Basel, Switzerland, 1973, pp. 607–622.
49. B. Bigland and O. C. J. Lippold, The relation between force velocity and integrated electrical activity in human muscles, *J. Physiol. 123*: 214–224 (1954).
50. M. H. Pope, G. B. J. Andersson, H. Broman, M. Svensson, and C. Zetterburg, Electromyographic studies of the lumbar trunk musculature during the development of axial torques, *J. Orthop. Res. 4*: 288–297 (1986).
51. S. M. McGill, Electromyographic activity of the abdominal and low back musculature during the generation of isometric and dynamic axial trunk torque: implications for lumbar mechanics, *J. Orthop. Res. 9*: 91–103 (1991).
52. T. Weis-Fogh and R. M. Alexander, *Scale Effects in Animal Locomotion*, Academic, London, 1977, pp. 511–525.
53. S. M. McGill and R. W. Norman, Effects of an anatomically detailed errector spinae model on L4-S1 disc compression and shear, *J. Biomech. 20*: 591–600 (1987).
54. J. G. Reid and P. A. Costigan, Trunk muscle balance and muscular force, *Spine 12*: 783–786 (1987).
55. P. V. Komi, Measurement of force–velocity relationship in human muscle under concentric and eccentric contractions, *Med. Sport, Biomech. III 8*: 224–229 (1973).
56. P. V. Komi, Relationship beween muscle tension, EMG and velocity of contraction under concentric and eccentric work, *New Dev. Electromyogr. Clin. Neurophysiol. 1*: 596–606 (1973).
57. W. S. Marras, Industrial electromyography, *Int. J. Ind. Ergon. 6*: 89–93 (1990).

58. K. P. Granata, An EMG-assisted model of trunk loading during free-dynamic lifting exertions, Ph.D. Dissertation, Ohio State Univ., 1993.
59. G. A. Mirka and W. S. Marras, A stochastic model of trunk muscle coactivation during trunk bending, *Spine 18*: 1396-1409 (1993).
60. J. M. F. Landsmeer, The coordination of finger joint motions, *J. Bone Joint Surg. 45*: 1654-1662 (1963).
61. E. Y. Chao, J. D. Opgrande, and F. E. Axmear, Three-dimensional force analysis of finger joints in selected isometric hand functions, *J. Biomech. 9*: 387-396 (1976).
62. S. A. Ferguson, F. A. Fathallah, K. P. Granata, J. Y. Kim, and W. S. Marras, Coactivity effects upon carpal tunnel contact forces, *Proc. Hum. Factors Soc.,* Seattle, WA, 1993, pp. 705-709.
63. K. N. An, E. Y. Chao, W. P. Cooney, and R. L. Linscheid, Normative model of the human hand for biomechanical analysis, *J. Biomed. Eng. 9*: 313-320 (1979).
64. T. J. Armstrong and D. B. Chaffin, Some biomechanical aspects of the carpal tunnel, *J. Biomech. 12*: 567-570 (1978).
65. S. A. Goldstein, T. J. Armstrong, D. B. Chaffin, and L. S. Mathews, Analysis of cumulative strain in tendons and tendon sheaths, *J. Biomech. 20*: 1-6 (1987).
66. W. S. Marras and R. W. Schoenmarklin, Wrist motions in industry, *Ergonomics 36*: 341-351 (1993).
67. R. W. Schoenmarklin, W. S. Marras, and S. E. Leurgans, Industrial wrist motions and incidence of hand/wrist cumulative trauma disorders *Ergonomics 37*: 1449-1459 (1994).
68. R. W. Schoenmarklin and W. S. Marras, A dynamic biomechanical model of the wrist joint, *Proc. Hum. Factors Soc.,* 34th Annual Meeting, Orlando, FL, 1990.
69. D. D. Penrod, D. T. Davey, and D. P. Singh, An optimization approach to tendon force analysis, *J. Biomech. 7*: 123-129 (1974).
70. E. Y. Chao and K. N. An, Graphical interpretation of the solution to the redundant problem in biomechanics, *J. Biomech. Eng. 100*: 159-167 (1978).
71. B. Buschholz and T. J. Armstrong, A kinematic model of the human hand to evaluate its prehensile capabilities, *J. Biomech. 25*: 149-162 (1992).
72. A. Moore, R. Wells, and D. Ranney, Quantifying exposure in occupational manual tasks with cumulative trauma disorder potential, *Ergonomics 34*: 1433-1453 (1991).
73. K. N. An, E. Y. Chao, W. P. Cooney, and R. L. Linscheid, Forces in the normal and abnormal hand, *J. Orthop. Res. 3*: 202-211 (1985).
74. R. W. Schoenmarklin, Biomechanical analysis of wrist motion in highly repetitive, hand-intensive industrial jobs, Ph.D. Dissertation, Ohio State Univ., 1991.
75. W. S. Marras, J. D. McGlothlin, D. R. McIntyre, M. Nordin, and K. H. E. Kroemer, Dynamic measures of low back performance, in *American Industrial Hygiene Association Ergonomics Guide*, 174-ER-93, 1993.

6
Psychophysical Methodology and the Evaluation of Manual Materials Handling and Upper Extremity Intensive Work

Sheila Krawczyk
Liberty Mutual Research Center for Safety and Health, Hopkinton, Massachusetts

I. INTRODUCTION: WHY PSYCHOPHYSICS?

Psychophysical methods are a consistent, reproducible, quick, inexpensive, and convenient way to assess the degree of physical strain on the human body. Psychophysical criteria have also been correlated with physiological criteria and some injury indices. Psychophysical methods utilize the results of the central nervous system integration of various information, including the many signals elicited from the peripheral working muscles and joints, and from the central cardiovascular and respiratory functions. All of these signals, perceptions, and experiences are combined and utilized by means of psychophysical methods.

Various physical stressors found in manual work, such as excessive forces, high rates of repetition, and awkward or sustained postures have been associated with musculoskeletal injuries. There is an absence of quantitative dose–response data that examine the relationship between work parameters and morbidity patterns. Epidemiological studies that examine the relationship between work and morbidity patterns require a lot of time and resources for data collection [1]. Biomechanical methodologies cannot address issues of repetitive work and fatigue [2,3]. Separate physiological criteria are needed for whole body or lower body tasks and upper body work [4–12]. More information is needed about the relationship between risk factors, work parameters, and the development of musculoskeletal injuries so that work can be designed to reduce the risk of the development of musculoskeletal injuries.

Another approach is a psychophysical methodology, in which subjects or workers determine the relationship between work factors and the perception of physical stress, exertion, fatigue, and discomfort on the body. Since localized muscle fatigue may be an early symptom of some use-related musculoskeletal injuries [13–15], psychophysical methodologies may serve as a more sensitive indicator for the risk of the development of musculoskeletal injuries [16].

The psychophysical methodology is an approach that allows for the simultaneous evaluation of the combined effects of different physical stressors. Specifically, the effects of different work task parameters, as well as the combined effects, can be evaluated using psychophysical methods. Psychophysical methods have been used extensively to evaluate manual materials-handling tasks, upper extremity–intensive tasks, and other manual work. A major advantage of psychophysical studies is that the results can be readily applied as guidelines in the workplace. Psychophysical data can provide guidance in the analysis and design of repetitive manual work that is commonly found in manufacturing and production assembly facilities, warehouses, retail trades such as grocery and discount stores, and other workplaces.

II. BACKGROUND

Psychophysics is an old branch of psychology that studies the relationship between sensation and physical stimulus intensity. Weber's law, from the early 1800's, states that

$$\frac{\Delta I}{I} = k$$

where I = intensity of a physical stimulus, ΔI = increment of I producing a just noticeable difference (jnd), k = constant. In essence, this means that the percentage error stays constant. The constant k is a function of the particular parameter being measured. For instance, the constant k is 3% for brightness, 7% for length, and about 2.5% (or 1/40) for weight. In practice, one cannot tell the difference between a 40-g weight and a 40.5-g weight. For one to be able to just notice the difference the second weight would have to weigh 41 g.

Fechner's law, which is based on Weber's law, states that sensation increases as the logarithm of physical stimulus intensity:

$$S = k \log I$$

where S = strength of sensation, k = constant, and I = intensity of a physical stimulus. For instance, in the judgment of loudness of sound, the sound energy increases logarithmically with respect to linear judgments, and that is why the decibel (dB) scale is used.

Over the years, it became clear that the logarithmic-linear relationship between physical stimulus intensity and sensation was an accurate description over only a limited range of physical stimulus intensity. In 1960, Stevens showed that a wider range of physical stimulus intensity could be accurately described if sensation were expressed as a power function of physical stimulus intensity [17].

III. RELEVANT CONCEPTS AND TERMINOLOGY

A. Psychophysical Methodology and Physiological Criteria

1. Psychophysical Power Law

The general form of Stevens' power law is:

$$\psi = k(\phi - \phi_0)^n$$

where ψ = psychological magnitude, k = constant, ϕ = physical magnitude, and ϕ_0 is a constant value corresponding to the threshold of detection [17]. The value of ϕ_0 is usually negligible, but its importance assumes larger proportions when subjective scales are extended downward to very low values [17]. For ranges of stimuli well above the minimum detectable level, the value of ϕ_0 is usually negligible [17].

Thus, neglecting ϕ_0, the psychophysical power law proposed by Stevens states that the psychological magnitude ψ is related to the physical magnitude ϕ by:

$$\psi = k\phi^n$$

Or, as stated in the terms used previously:

$$S = kI^n$$

where S = strength of sensation, k = constant, and I = intensity of a physical stimulus. The constant k is a function of the particular unit of measurement. The exponent n depends mainly on the modality tested. For perception of muscular force n is about 1.6 [18], and for force of handgrip it is 1.7 [17]. The value of n ranges from 0.33 for brightness to 3.5 for electric shock and is determined by magnitude estimation [17]. Using this method, the observer estimates the apparent strength or intensity of his or her subjective impressions relative to a standard. Magnitude estimation is the most useful of four principal methods to construct these ratio scales of apparent magnitude.

The power function is plotted as a straight line in log–log coordinates, with the slope of the line equal to the value of the exponent n. Accordingly, modern psychophysical theory [17] provides that the strength of a sensation is related by a power function to the intensity of its physical stimulus [16–26]. Most psychophysical relations may be described by this power function [17].

2. Borg Scales

The above ratio scaling methods are very good methods to describe how the subjective intensity varies with physical stimulus intensity. One major drawback of the above ratio scaling methods is that they do not provide absolute levels for interindividual comparisons. Although general functions for a group of subjects can be determined. It is difficult to compare subjects with each other because subjects are asked only to make relative comparisons. To overcome the difficulties associated with the ratio scaling methods, a category scale was developed by Borg [18]. Borg's [18] first subjective rating scale had 21 grades. All of the odd scale values from 3 to 19 were anchored with verbal expressions, chosen so that the scale should receive a good interindividual reliability. The scale was presented to the subjects with equal distances between the figures and in the following terms: 3, Extremely light; 5, Very light; 7, Light; 9, Rather light; 11, Neither light nor laborious; 13, Rather laborious; 15, Laborious; 17, Very laborious; and 19, Extremely laborious. Very high correlations have been obtained between heart rate and these ratings during work tests [18]. This indicates the differential value of the scale but not the general validity of the growth function [27]. Since perceived exertion determined by ratio scaling methods increased with an exponent of about 1.6, Borg [18] concluded that an integration of central factors (such as heart rates) and peripheral factors (such as blood lactates, with an exponent of about 2) would explain the psychophysical variation better than any single physiological variable [27].

Another category scale for ratings of perceived exertion (RPE), shown in Table 1, was constructed by Borg [28] to increase linearly with the exercise intensity for work on a cycle ergometer. Since oxygen consumption and heart rate increase linearly with work load, this would be a convenient means of constructing a scale, even if it did violate the true growth of the perceived intensities [27]. The values of the RPE scale, shown in Table 1, grow fairly linearly with work load. The correlation between the ratings and heart rate is also very high [28]. For middle-aged persons at work loads of medium intensity, heart rate should be fairly close to 10 times the RPE value.

The Borg RPE scale has been widely used to study the perception of exertion in laboratory, clinical, and occupational settings. One of the most common uses of the RPE scale is in the clinical diagnosis of patients with coronary and respiratory disturbances. The normal growth pattern and the level of exertion change dramatically in different clinical populations. The RPE scale is also used in rehabilitation, and for the prescription and regulation of exercise intensities, or as a means to evaluate certain training situations [16]. A similar use, of interest to ergonomists, is in the evaluation of different work tasks. Perceived exertion has been used in ergonomic evaluations of heavy aerobic work tasks. However, the value of subjective estimations may be especially evident in job situations where the work tasks con-

Table 1 The Borg Scale for Ratings of Perceived Exertion (RPE)

6	
7	Very, very light
8	
9	Very light
10	
11	Fairly light
12	
13	Somewhat hard
14	
15	Hard
16	
17	Very hard
18	
19	Very, very hard
20	

sist of short-term static work, intermittent or varied work, or upper extremity intensive work for which valid physiological measurements are difficult to obtain. Borg [27] suggests that the RPE scale is the best one for most applied studies of perceived exertion. The RPE scale is one of the most frequently used indices of physical stress [16–20, 22–25, 27–30].

To meet the twofold demands of ratio scaling and level estimations, Borg [27] developed the category ratio (CR) scale, shown in Table 2, so that perceptual ratings would increase as a positively accelerating function. This scale contains some of the category properties of the RPE scale, and also contains ratio properties. The verbal expressions are set so that perceptual intensity increases according to a power function. The number "10" is defined as the strongest effort and exertion a person has ever experienced [16]. Since a person may imagine an intensity that is even stronger, the "absolute" maximum is somewhat higher. By

Table 2 The Borg Category Ratio (CR) Scale of Perceived Exertion, Which Was Constructed as a Category Scale with Ratio Properties

0	Nothing at all	
0.5	Very, very weak	(just noticeable)
1	Very weak	
2	Weak	(light)
3	Moderate	
4	Somewhat strong	
5	Strong	(heavy)
6		
7	Very strong	
8		
9		
10	Very, very strong	(almost max)
•	Maximal	

anchoring the highest number at a well-defined perception, with some degree of "sameness" for different individuals, a good point of reference is obtained [16]. Thus, two individuals working at their respective maximal working capacities will be experiencing the same degree of perceived exertion, even though their physical outputs may be different. Similarly, two individuals working at 50% of their respective working capacities will experience the same amount of perceived exertion, even though their physical outputs may be different [31]. This scale gives psychophysical functions comparable to those obtained with magnitude estimation. Exponents of about 1.6 have been obtained for perceived exertion in cycle ergometer exercise. In addition, close correlation between scale ratings and both blood lactate and muscle lactate levels have been obtained [32]. Borg [27] suggests that this scale may be best suited for subjective symptoms, such as aches and pain.

3. Visual Analog Scales

The visual analog scales (VAS) [19,26,30–31,33–39] is another frequently used measurement of physical stress. One of the most common visual analog scales consists of a horizontal 10-cm line with verbal anchors or descriptions at the endpoints. The subject is instructed to indicate the degree of perceived sensory intensity by placing a vertical line at the appropriate position along the horizontal continuum. The distance from the left end of the VAS to the subject's vertical line is measured in centimeters to obtain a value between 0 and 10. A visual analog scale that has been used for the evaluation of upper extremity intensive work [30,35–38] is shown in Figure 1.

Ulin et al. [30] used this scale and another VAS with the verbal anchors "very uncomfortable work" and "very comfortable work" on the left and right ends, respectively, and Borg's CR scale for university students in the assessment of screw-driving tasks at different heights. She found that the scales compared in sensitivity and that any one could be used with reliable results. She suggests that the VAS may be preferred by subject populations that are not as verbally oriented as university students, or in production situations where workers do not have the time to read and consider all of the verbal anchor points of the Borg scale.

In an automobile assembly plant, Armstrong et al. [39] utilized 10-cm visual analog scales for worker assessments of hand tool mass, tool handle circumference, horizontal work location, vertical work location, and overall ratings. For tool mass, the verbal anchors used were "too heavy," "just right," and "too light" at 0, 5, and 10 cm, respectively, and "very uncomfortable," "somewhat uncomfortable," and "very comfortable" at 0, 5, and 10 cm, respectively. For tool handle circumference, the verbal anchors were "too large," "just right," and "too small" at 0, 5, and 10 cm, respectively. For the horizontal and vertical work locations, the verbal anchors were "very uncomfortable," "somewhat uncomfortable," and "very comfortable" at 0, 5, and 10 cm, respectively. The overall ratings had the verbal anchors

For a normal 8-hour workday:

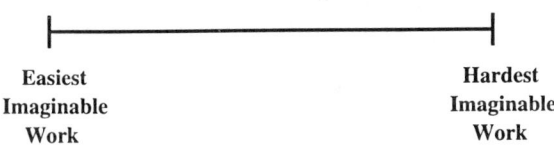

Easiest Imaginable Work   Hardest Imaginable Work

Figure 1 A 10-cm visual analog scale (VAS) with the verbal anchors "easiest imaginable work" and "hardest imaginable work" at the left and right endpoints, respectively, that has been used for the evaluation of manual work.

"very good," "fair," and "very poor" at 0, 5, and 10 cm, respectively. Males and females did not differ significantly in their assessments. The workers' subjective assessments using the 10-cm visual analog scales showed strong correlations with the tool characteristics and the work tasks.

Borg [19] compared different rating methods, among them a 11-cm VAS with the verbal anchors "no exertion at all" and "maximal exertion" on the left and right endpoints of the scale, respectively. For short-time work on a bicycle ergometer, good correlations between heart rate and perceived exertion ratings were obtained independently of which scale was used. In addition, the correlation coefficients between heart rate and perceived exertion ratings were very similar for the different methods of the physical work tests. Similarly, Harms-Ringdahl et al. [33] found that there was no significant difference between the ratings made on a 10-cm VAS and Borg's category scale for ratings of perceived pain (BRPP), which is similar to the CR scale, for the assessment of pain in response to loading of soft tissue structures at the radial side of the elbow joint.

Neely et al. [31] compared a 10-cm VAS with the verbal anchors "MIN" and "MAX" on the left and right endpoints, respectively, with Borg's CR scale, heart rate, and blood lactate levels for leg exertion during an exercise test using a bicycle ergometer. As is usual in these kinds of tests, heart rate correlated well with power. The physiological measures of heart rate and blood lactate levels (an indicator of peripheral strain) were correlated with both the CR and VAS ratings of leg exertion.

Seymour et al. [34] evaluated VASs of different lengths (5, 10, 15, and 20 cm) with the verbal anchors "no pain" and "worst pain imaginable" on the left and right endpoints, respectively. Also evaluated were 10-cm VASs with different verbal anchors on the right endpoints ("troublesome," "miserable," "intense," "unbearable," and "worst pain imaginable"). For dental pain among males and females, high correlations were found between the scores on all of the scales. Scales of length 10 or 15 cm had the smallest measurement error. Considering the different verbal anchors, the scale with the verbal anchor "worst pain imaginable" was found to be the best choice for comparing present pain or worst pain between different groups. Using this scale, no significant difference was found between males and females, or between patients with different dental conditions. Seymour et al. [34] suggest that the use of a 10-cm VAS with the verbal anchor "worst pain imaginable" was the most suitable in the measurement of dental pain. Similarly, Price et al. [26] found that a 15-cm VAS was a valid and reliable measure for both the intensity and the unpleasantness of experimentally induced pain from noxious heat stimuli delivered to the forearm, or for chronic back and/or shoulder pain.

In conclusion, it appears that visual analog scales with a length of approximately 10 cm and verbal anchors suited to the extremes of the intensity of the physical stress of concern may provide results that are well correlated with both physical measures (e.g., tool characteristics, work tasks) and physiological measures.

4. Preferred Maximums

In ergonomic assessments of physical work tasks, the various Borg scales and visual analog scales have been used to rate the physical intensity of the work tasks. The subjects are given a task to perform and then are instructed to rate the work, utilizing the scales as if they were performing the task for a normal 8-hr workday. Thus, a determination of the intensity of the work can be made to compare different work tasks and their parameters, such as weight, frequency, and so forth.

Another way that psychophysics has been applied in the study of work is in the determination of preferred maximums. The subject is given control of one of the task parameters or variables, usually the weight of the object being handled. All of the other task parameters or variables, such as frequency, size, distance, and so on are kept constant. Subjects are instructed to work as hard as they can without straining themselves or without becoming unusually tired, weakened, overheated, or out of breath [40,41]. Individual subjects integrate all of their sensory inputs, monitor their feelings of exertion and fatigue, and adjust the weight accordingly. In this way, a preferred maximum for the particular task is obtained.

This preferred maximum methodology has been utilized extensively in studies of manual materials handling [24,40–58], to determine maximum weights or forces, frequencies, and so on for various lifting, lowering, pushing, pulling, and carrying tasks. When weight is the variable that is adjusted, the resulting preferred maximum may be referred to as the maximum acceptable weight of lift (MAL) [42]. When weight and/or frequency are the variables adjusted, the resulting preferred maximum workload may be calculated for comparison purposes as the maximum acceptable workload (MAWL), which is equal to the load in kg × frequency in actions/minute × distance lifted in meters [43].

This preferred maximum methodology has also been utilized in upper extremity intensive work. Krawczyk et al. [35,37] had subjects adjust the weight for a repetitive upper extremity transfer task. The results are shown in Figure 2. As frequency and distance increased, the preferred weight decreased. Similarly, Kim and Fernandez [59] and Marley and Fernandez [60,61] had subjects determine the maximum acceptable frequency for hand-held pneumatic drilling tasks. Likewise, Putz-Anderson and Galinsky [62] had subjects psychophysically determine maximum work durations for repetitive elevated arm movement tasks. Snook et al. [63] had subjects psychophysically determine maximum forces for wrist flexion and extension.

Researchers [42,44] have found significant differences in psychophysical studies between male industrial workers and students, and between female industrial workers and housewives, respectively. Thus, in psychophysical studies, it is especially important that the subject population be the same type as the target population in which the psychophysical results are to be

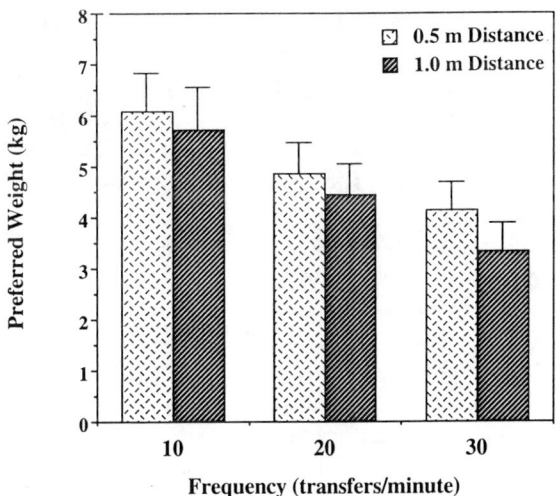

Figure 2 Mean preferred weights for each of the frequency and distance combinations for an upper extremity transfer task, over an 8-hr workday, for 20 industrial subjects. 95% confidence intervals are shown by error bars.

applied. For instance, if the target population is industrial workers, then the experimental subject population should also be industrial workers.

5. Lower Body Versus Upper Body Work

For whole body or lower body tasks, where the large muscle groups in the legs are used, perceived exertion ratings and physiological criteria, such as heart rate and oxygen consumption, are highly correlated [5,19,24,28,31,64–66]. Some physiological measures, such as heart rate, do not respond equivalently for whole body or lower body tasks and upper body work. Physiological methodologies where heart rate, measures of oxygen consumption, and energy expenditure criteria are monitored are not as sensitive to upper extremity work [4–12]. For instance, heart rate tends to increase more for arm work than for whole body or lower body tasks. Thus, physiological fatigue criteria that have been developed from leg or whole body exertions cannot validly be applied to tasks performed by the arms. Separate physiological fatigue criteria are needed for tasks that involve mainly arm work.

Snook and Irvine [4] found that there were no significant differences in heart rate for arm work in a laboratory setting and at a shoe factory. However, there was a significant heart rate difference between arm lifts and leg lifts performed in the laboratory. They concluded that fatigue criteria that have been developed from leg tasks cannot validly be applied to tasks performed mainly by the arms; energy expenditure rate guidelines for arm tasks should be lower than those for leg tasks. Based on their results, the mean heart rate should not exceed 99 beats/minute for arm tasks and 112 beats/minute for leg tasks.

Gamberale [5] examined the relationships between perceived exertion and physiological indicators of exertion during different exercises. Borg's RPE scale was used to determine perceived exertion at different work loads for lifting weights with the arms, working with a wheelbarrow, and also exercising on a bicycle ergometer. In addition, heart rate, oxygen uptake, and blood lactate concentration were measured. There was a linear relationship between RPE and heart rate independent of the kind of work producing the physical exertion. For all three exercises, a closer relationship to heart rate was obtained when the subjects rated their overall feeling of exertion, rather than the exertion on the arms or on the legs. However, the highest level of perceived exertion in relation to heart rate was found in the exercise of lifting weights with the arms. This exercise also yielded the highest level of blood lactate concentration in relation to oxygen uptake. The blood lactate concentrations in the exercise of lifting weights were higher than in the exercise of pushing the wheelbarrow, even though the oxygen uptake was lower. At a given level of oxygen uptake, the exercise of lifting weights was more anaerobic in character than the exercise of pushing the wheelbarrow and of working on the bicycle ergometer. These results suggest that the higher the blood lactate concentration an exercise produces as compared with oxygen uptake, the higher will be the level of the overall perception of exertion as compared with heart rate. In addition, the higher the blood lactate concentration an exercise produces as compared to oxygen uptake, the higher will be the perceived exertion on the most involved muscle groups as compared with the overall perceived exertion.

Similarly, Borg et al. [6] examined perceived exertion using the RPE and the CR scales related to heart rate and blood lactate concentration during arm and leg exercise. The arm and leg exercise was performed on bicycle ergometers, one of which was specially adapted for arm exercise. The responses obtained were at least twice as high for arm cranking as for cycling. The largest difference was found for blood lactate concentration, and the smallest for RPE and heart rate. The incremental functions were similar in both exercises, with approximately linear increases in RPE and heart rate, and positively accelerating functions for CR and blood lactate concentration.

When perceived exertion on the CR scale was set as the dependent variable and a simple combination of heart rate and blood lactate was used as the independent variable, a linear relationship was obtained for both kinds of exercise, as had been previously found for cycling, running, and walking. Thus, for exercise of a steady-state type with increasing loads, the incremental curve for perceived exertion can be predicted from a simple combination of heart rate and blood lactate [6].

Mermier et al. [7] also compared upper and lower body activities and found significantly different results for upper body versus lower body tasks. Three tasks were used: exercising on a cycle ergometer, lifting (waist to shoulder height), and vacuuming. Their goal was to estimate ventilation by using ventilation-on-heart-rate regressions established during exercise testing to estimate ventilation in the field. For men and women, ventilation increased more steeply relative to heart rate for the exercises involving the upper body (lifting and vacuuming) compared with the lower body exercise (cycling). The regression coefficient describing the increase of ventilation with heart rate was approximately 30% greater with upper body exercise. The differences in the mean regressions for upper and lower body exercise tended to be greater in women than in men. However, these physiological criteria were consistent in that ventilation-on-heart-rate regression slopes derived from tests in which progressively increasing workloads were used were comparable to those obtained during variable and nonprogressive protocols.

From these studies, it can be concluded that the psychophysical indices are correlated with the physiological indices; however, the relationships are different for arm work than for whole body or lower body work. The physiological indices are clearly more sensitive for upper body or arm work than for whole body or lower body tasks. Thus, different physiological guidelines should be used for arm work compared with whole body or lower body work.

B. Psychophysical Consistency and Reproducibility

1. Borg Scales and Visual Analog Scales

The Borg scales, VASs and variations of these scales have been shown to be consistent and reproducible for measurement and comparison of physical stimuli [4,34–38,40,45–53, 56,59,60–61,67–68] regardless of which scale was used [19,25–26,30–33]. Borg [19] compared four different rating methods: the RPE scale, graded from 6 to 20; the original Borg scale; graded from 1 to 21; a 9-point graded scale; and an 11-cm visual analog scale. Good correlations between heart rates and perceived exertion ratings were obtained independently of which scale was used. The correlations between the different ratings were also satisfactory. Similarly, Aristila et al. [25] compared three variations of the Borg scale and found good correlation of the perceived exertion ratings with heart rate and excellent reproducibility. Krawczyk et al. [35,37] found that the VAS perceived exertion ratings for an upper extremity transfer task were consistent throughout an 8-hr workday, as shown in Figure 3. Harms-Ringdahl et al. [33] compared Borg's category scale for ratings of perceived pain (BRPP) and a 10-cm VAS and found no significant difference between the two scales. There was also no significant difference between the first and second time a scale was used by the same subject. Ulin et al. [30] found equivalent results for Borg's CR scale and two different 10-cm visual analog scales with different sets of verbal anchors in the evaluation of pneumatic screw-driving tasks. Seymour et al. [34] compared visual analog scales of different lengths (5, 10, 15, and 20 cm) and 10-cm visual analog scales with different verbal anchors on the right endpoints. High correlation was found between the scores on all of the scales. No significant difference was found between scores recorded by males and females or between those given by patients with two different dental conditions. Likewise, Price et al. [26] found that 15-cm

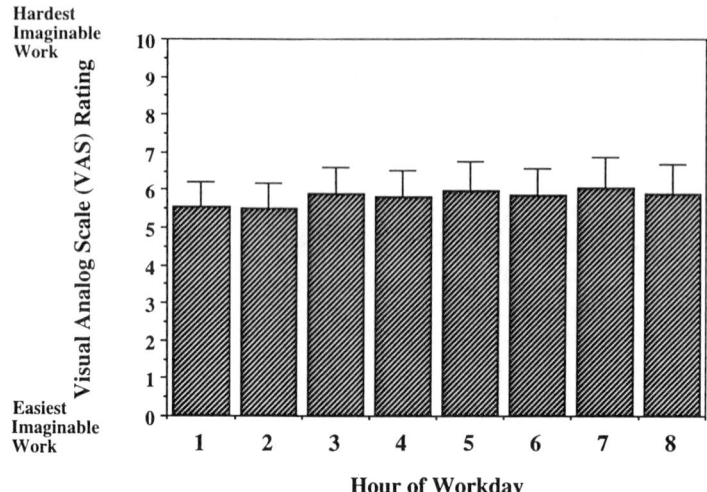

Figure 3 Overall mean VAS-perceived exertion ratings combined for all frequencies and distances at the end of each hour throughout an 8-hr workday, for an upper extremity transfer task, for 20 industrial subjects. 95% confidence intervals are shown by error bars.

VASs can be used as a valid and reliable measure for both the intensity and the unpleasantness of experimentally induced pain or chronic (back and/or shoulder) pain.

2. Manual Materials Handling

Psychophysical methodologies have been shown to be consistent for manual materials-handling tasks. Ciriello et al. [45] found that subjects were able to determine their maximum acceptable weights and forces for various manual materials-handling tasks during the first 40 min of testing, and that they remained consistent throughout the 4-hr duration of the test session. Ljungberg et al. [24] found that it took subjects only 5-10 min to select a weight for lifting tasks that remained consistent over one hour. Snook and Irvine [40] found that subjects were able to determine their maximum acceptable frequency of lifting in 40 min.

Psychophysical methodologies have been shown to be reproducible for manual materials handling tasks. Snook and Irvine [40] replicated an experiment three times and showed only insignificant differences among the three replications. Griffin et al. [46] found only a 7% decrease, which was not significant, between test and retest one week later for acceptable weights of lifting. Similarly, Foreman et al. [47] found a 7% decrease in the rating of acceptable dynamic lifting strength between two days, and the difference was not statistically significant. Fernandez et al. [48,49] concluded that the psychophysical approach was reproducible as subjects arrived at the same estimates of lifting capacity in repeated trials. Legg and Myles [52,53] found that maximum acceptable load for lifting determined two times a day for five days did not change significantly. When this load was used over an 8-hr workday, soldiers performed the lifting and lowering tasks without metabolic, cardiovascular, or subjective evidence of fatigue.

Conversely, Mital [54] found a larger decrease in the maximum acceptable weight of lift throughout an 8-hr workday when compared to estimates made at 25 min. Much of this decrease in the maximum acceptable weight took place by the completion of the second hour. After two hours, the slope of the decrease in acceptable weight had leveled off.

The differences in results from these manual materials-handling studies may be due to a frequency effect. Ciriello and Snook [55] reported the tendency of the psychophysical method to produce overestimates of maximum acceptable weights and forces for tasks with very high frequencies. This was verified with experiments of 4-hr duration by Karwowski and Yates [50–51] and 8-hr duration by Fernandez et al. [48,49]. These studies [48–51] found that selected lifting weights did not differ significantly with time at low frequencies, but that at high frequencies there were larger, sometimes significant decreases. Since Mital [54] combined all frequencies in his results, the decrease with time may have been due to this frequency effect.

3. Upper Extremity Intensive Work

Psychophysical methods have been shown to be consistent for upper extremity intensive work. Krawczyk et al. [35,37] found that subjects could accurately predict their perceived exertion for a normal 8-hr workday for a repetitive upper extremity transfer task after one hour. After subjects performed the same task throughout a full 8-hr workday, the difference in the perceived exertion between the first two hours and the last two hours of performing the repetitive upper extremity transfer tasks was only 8%, as shown in Figure 3. Thus, the perceived exertion rating made in the first two hours, where subjects were supposed to imagine what it would be like to perform the task "for a normal 8-hr workday," closely agreed with what they thought eight hours later, after they had actually performed the transfer task for eight hours. Likewise, the difference in the maximum preferred weights between the first two hours and the last two hours of performing the repetitive upper extremity transfer tasks was only 5%, as shown in Figure 4.

Thus, perceived exertion and preferred weight were both consistent when determined at hourly intervals throughout an 8-hr workday. In addition, these results verified that a psychophysical determination for a "normal 8-hr workday" does not require a *full* 8-hr workday.

Other studies have also shown that perceived exertion is consistent over time. Krawczyk and Armstrong [35,36] found no significant differences in perceived exertion at 30, 60, 90, and 120 min for a similar repetitive upper extremity transfer task performed over a 2-hr time period. Likewise, in another study, Krawczyk et al. [35,38] found no significant differences

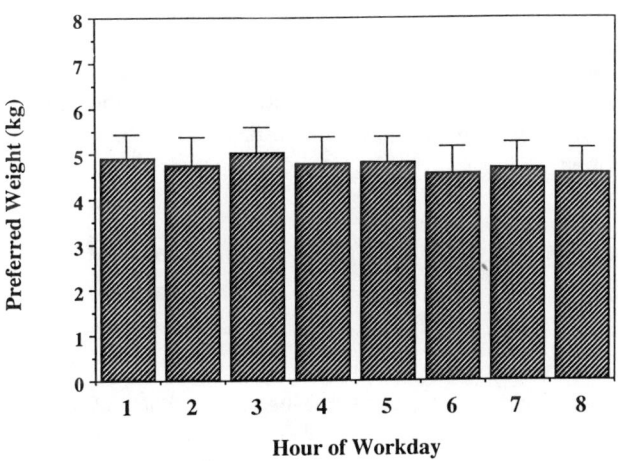

Figure 4 Overall mean preferred weights combined for all frequencies and distances at the end of each hour throughout an 8-hr workday, for an upper extremity transfer task, for 20 industrial subjects. 95% confidence intervals are shown by error bars.

in perceived exertion determined after 30 and 60 min for combinations of transferring and screw-driving tasks.

Psychophysical methods have been shown to be reproducible for upper extremity intensive work too. Ulin et al. [30] used Borg and two different VASs to rate perceived exertion for pneumatic screw-driving tasks and found equivalent results regardless of which scale was used. Kim and Fernandez [59] and Marley and Fernandez [60,61] found no significant differences in the maximum acceptable frequency for three experimental replications of handheld pneumatic drilling tasks.

4. Preferred Maximums: Initially Heavy Versus Initially Light

In the determination of maximum acceptable preferred weights, when subjects started with a heavy container weight, Krawczyk et al. [35,37] found that the resultant preferred weight for an upper extremity transfer task was greater than that determined when starting with a light weight. This is shown in Figure 4, where hours 1, 3, 5, and 7 were started with a heavy container weight and hours 2, 4, 6, and 8 were started with a light weight. This difference was statistically significant. However, it is not of practical significance since this difference was only about 3%. The perceived exertions shown in Figure 3 also reflect this: hours 1, 3, 5, and 7 were slightly greater than hours 2, 4, 6, and 8. Legg and Myles [52,53] found this same trend for manual materials-handling tasks. The difference was 15%; however, it was not statistically significant. To eliminate the effect of the initial weight, the resultant preferred weights determined from an initially heavy container weight and an initially light container weight should be averaged together.

IV. PSYCHOPHYSICAL CRITERIA FOR ERGONOMIC APPLICATION

A. Psychophysical Criteria and Injury Indices

Psychophysical measures of physical stress have been correlated with health outcomes, disability, and compensation [69,70]. Psychophysical criteria have been used in developing recommendations for permissible workloads. After numerous studies investigating different manual materials-handling tasks, Snook [56] concluded that designing the job to fit the worker, using psychophysically determined guidelines, can reduce up to one-third of industrial back injuries. Other studies [69,71–72] have shown that overexertion injuries would be reduced if manual materials-handling tasks were designed to match acceptable levels of perceived exertion. Snook [70] reviewed a number of studies, suggesting that the setting of maximum permissible workloads in industry would have a significant effect upon low-back disability and low-back compensation.

B. Psychophysical Studies in Manual Materials Handling and Upper Extremity Intensive Work

1. Overview

The psychophysical methodology is an approach that allows the evaluation of the combined effects of different physical stressors to be evaluated simultaneously. Psychophysical methodologies have been used extensively in the evaluation and design of manual materials handling tasks [5,22,24,40–58,64–68,73–78] such as lifting, lowering, pushing, pulling, carrying, and also walking tasks. The National Institute for Occupational Safety and Health (NIOSH) [79–81] equation for the design and evaluation of manual lifting tasks utilizes biomechanical, physiological, and psychophysical criteria. Researchers at the Liberty Mutual

Insurance Company [56–57] have composed extensive tables of psychophysically determined maximum acceptable weights and forces for manual materials handling tasks to serve as guidelines that are consistent with worker capabilities and limitations. The variables in these studies included task frequency, distance, height and duration, object size and handles, extended horizontal reach, and combination tasks. The guidelines are intended to assist industry in the control of low back pain through reductions in initial episodes, length of disability, and recurrences [82].

Likewise, utilizing psychophysically determined guidelines for upper extremity intensive work may reduce the risk of the development of upper extremity cumulative trauma disorders (CTDs) or repetitive stress injuries (RSIs), such as carpal tunnel syndrome (CTS). Psychophysical methodologies have been used to assess upper extremity intensive work and different hand tools, to derive guidelines for the design of upper extremity work [4,10–11,30,35–39,59–62,83–96].

At present, the psychophysical methodology may be the most appropriate way to evaluate upper extremity intensive work. Unfortunately, there are no definitive data concerning what constitutes excessive quantities of the work-related risk factors: force, repetition, awkward postures, mechanical stresses, vibration, and cold temperature [97,98]. More information is needed about the relationship between risk factors, work parameters, and the development of CTDs so that work can be designed to reduce the risk of the development of CTDs. Epidemiological studies that examine the relationship between work and morbidity patterns require a lot of time and resources to collect these types of data [1]. Biomechanical methodologies cannot address repetitive work and fatigue issues [2,3]. As discussed earlier (Section III.A.5.), physiological methodologies, where heart rate, measures of oxygen consumption, and energy expenditure criteria are monitored, are not as sensitive for upper extremity work [4–12].

In the absence of validated epidemiological, biomechanical, or physiological methods of assessing upper extremity intensive work, an alternative approach is a psychophysical methodology, in which subjects or workers determine the relationship between work factors and the perception of physical stress, exertion, fatigue, and discomfort on the body. Since localized muscle fatigue may be an early symptom of use-related CTDs [13–15], psychophysical methodologies may serve as a more sensitive indicator for the risk of the development of CTDs [16].

2. Task Variables

Weight. An increase in perceived exertion with increased weight has been found in manual materials-handling tasks (whole-body exertions) [5,22,24,52,64,74,99], as well as a decrease in working endurance time [77]. Similarly, an increase in perceived exertion with increased weight or force on the upper extremity has been found by researchers examining one-handed lifts [10,11,35–37], repetitive arm elevations [12], screw driving [93,95], drilling [59], and gripping tasks [91]. Decreased psychophysically determined work durations [62] and increased electromyographic activities [100] have also been found in the upper extremities with increased weight or force.

When the variable adjusted by the subject is weight, how does the preferred maximum weight for an 8-hr workday compare to maximum strength? Pytel and Kamon [58] found that for males and females, the maximum acceptable load selected for repetitive lifting (manual materials handling, whole-body exertions) was 22% of the experimental maximum load that an individual was able to lift once without risk of injury. In comparison, Krawczyk et a. [35,37] found that for males and females, the maximum preferred weights for a repetitive upper extremity transfer task ranged from 3.3 to 6.1 kg, as shown in Figure 2. The smallest preferred weight of 3.3 kg occurred at a repetition rate of 30 transfers/minute and a transfer distance of 1.0 m, and the largest maximum preferred weight of 6.1 kg occurred at 10 transfers/minute and 0.5 m distance. These maximum preferred weights were 16-29% of the measured "upper extremity strength."

As frequency increases, preferred weight decreases in studies of maximum preferred weights or forces, when weight or force is the variable adjusted by the subject. This is shown in Figure 2. Numerous studies have shown a frequency and weight tradeoff: as frequency increased, psychophysically determined acceptable weight or force decreased [35,37,41,45, 48-51,55,56,66,67,74]; or, as weight or force increased, psychophysically determined acceptable frequency decreased [11,40,56,59,74,91,92].

Some upper extremity studies have shown that the effect of weight was greater than the effect of frequency [10,35,36,100]. For manual materials-handling tasks, the total psychophysically determined maximum acceptable workload (load in kg × frequency in actions/minute × distance lifted in meters) was affected more by weight than by frequency, which results in a higher workload performed at higher frequencies [24,43,56,75]. This result is in accordance with the frequency effect discussed earlier (Section III.B.2.) for manual materials-handling tasks: the tendency of the psychophysical method to produce overestimates of maximum acceptable weights and forces for tasks with high frequencies. Similarly, the results of an 8-hr study [35,37] of preferred maximum weights for the upper extremity, showed that subjects tolerated a higher perceived exertion rating with higher task frequency, as shown in Figure 5, even though they were always supposed to be working at their maximum level.

Frequency. An increase in perceived exertion with increased task frequency has been found in manual materials handling tasks [64,66,74,76], as well as a decrease in working endurance time [76,77]. Similarly, an increase in perceived exertion with increased task frequency for upper extremity tasks has been found by researchers for one-handed lifts [10,35-37], as shown in Figure 5, and for screw-driving tasks [96]. Decreased psychophysically determined work durations [62] and increased electromyographic activities [100] have also been found in the upper extremities with increased task frequency.

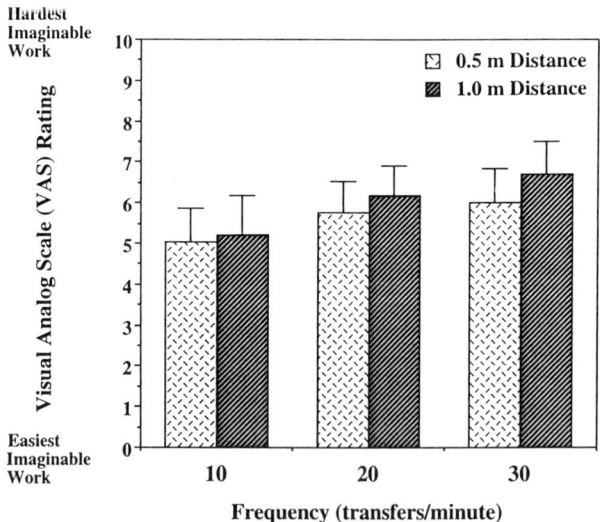

Figure 5 Mean VAS perceived exertion ratings for each of the frequency and distance combinations for an upper extremity transfer task, over an 8-hr workday, for 20 industrial subjects. 95% confidence intervals are shown by error bars.

Increasing frequency obviously does not increase the static strength required for the task. However, the rate of energy expenditure will increase as transfer frequency increases, and to a lesser degree as weight increases. Increases in the rate of energy expenditure have been shown to increase perceived exertion for whole-body exertions [5,19,24,64–66,73]. However, remember (Section III.A.5.) that physiological methodologies are not as sensitive for upper extremity work [4–12].

Energy expenditure rates do not appear to be the limiting factor in upper extremity intensive work. By process of elimination, this provides more credence that inadequate recovery time for the upper extremity, and the resulting localized discomfort and fatigue, may be a limiting factor. For the upper extremity, a frequency threshold may exist [35,36]. Below this threshold significant increases or decreases in perceived exertion do not occur, and above this threshold there is inadequate recovery time for the upper extremity, resulting in greater perceived exertion, localized discomfort and fatigue, and presumably greater risk of the development of CTDs. Indeed, morphological tissue changes (resembling peritendinitis crepitans) may occur in over exercised limbs and have been experimentally induced by over-exercising the hind limb in rabbits [101]. Thus, time and load characteristics have been correlated with a subsequent cumulative trauma injury. Likewise, an accumulation of strain was found to occur in tendinous tissues of the upper extremity during physiological loading in human cadaver hands [102].

Distance. Distance is a task variable that consists of yet another whole series of variables. The NIOSH [81] equation for the design and evaluation of manual lifting tasks has three distances: the vertical travel distance between the origin and the destination of the lift, the vertical distance of the hands from the floor, and the horizontal distance of the hands from midpoint between the ankles. The Liberty Mutual Insurance Company [56,57] tables of psychophysically determined maximum acceptable weights and forces for manual materials-handling tasks also have many aspects of distance: the vertical distance of lifting or lowering; the distance away from the body (box width); the distances of floor level to knuckle height, knuckle height to shoulder height, and shoulder height to arm reach for lifting and lowering; the distances of close to the body and extended horizontal reach for lift; the distance of pushing, pulling, or carrying; and the vertical distance from the floor to the hands for the pushing, pulling, and carrying tasks. Usually these distances are significant and should be considered, but for some tasks they may not be significant. This will depend on the distance considered and the task involved. For instance, no significant differences in maximum acceptable weight were observed among six different box sizes in the carrying tasks [57]. Thus, one should refer to the psychophysical tables [56,57] to determine what is appropriate for the application involved.

Similarly, there are many aspects of distance for upper extremity intensive work. For upper extremity work, the distance component may be significant in that it acutely affects posture. In a repetitive upper extremity transfer task [35,37], as distance increased, perceived exertion increased and preferred weight decreased for distances of 50 and 100 cm, as shown in Figures 5 and 2, respectively. The distance was lateral, producing a side-to-side motion in front of the subjects. The 50-cm distance neither challenged nor exceeded the reach envelope (where one can reach) for all subjects, while the 100-cm distance challenged the reach envelope. Thus, the different distances had some effect on the trunk posture and the resultant moments produced about the shoulder, elbow, and wrist. However, in a similar repetitive upper extremity transfer task, Krawczyk and Armstrong [35,36] found no significant distance effect on perceived exertion for distances of 25, 51, and 76 cm. Again, these transfer distances required a lateral displacement, producing a side-to-side transfer motion right in front of the subject. However, none of these distances challenged or exceeded the reach envelope for all subjects. Thus, postural changes were not required at the different distances.

In a similar study [11] of upper extremity one-handed lifts in the horizontal plane that considered distances measured from in front of the body, subjects stood in front of a 91-cm-high work table and were required to move dumbbells distances of 38 and 63 cm towards themselves. This task required some trunk flexion, and the distances were found to be significantly different for a 2.3-kg load, while no significant difference was found between the two distances for a 4.5-kg load. In another study where the distance parameter required postural changes, Ulin et al. [94] found that ratings of perceived exertion increased with increasing horizontal distance away from the body for screw-driving tasks. In a checkstand configuration study [89], design variations (89- and 76-cm heights, and 8- and 23-cm distances), which ultimately affected the transfer distance and body posture, had significant effects on some comfort ratings. Likewise, postural discomfort has been shown to increase as a function of horizontal and vertical distance from the body for automobile assembly tasks [39]. Other upper extremity studies have found significant distance effects when the distances studied were vertical work heights [30,62].

Another factor which was shown importance is the direction of the transfer distance. Increased electromyographic activities [100] have been found in the upper extremity with different directions of movements from outward points within the reach (20–230° measured from the frontal plane of the subjects) to a fixed point near the body for a constant 38-cm transfer distance.

3. Design Applicability

Psychophysical methodologies are quick, relatively inexpensive, and convenient. Psychophysical criteria have also been shown to be consistent and reproducible, and well correlated with physiological criteria and some injury indices. In addition, a major advantage of psychophysical studies is that the results can be readily applied as guidelines in the workplace, such as in manufacturing and production assembly facilities, warehouses, and retail trades, for example, grocery and discount stores.

The results of preferred maximum studies can be directly applied, such as the Liberty Mutual Insurance Company [56,57] tables of psychophysically determined maximum acceptable weights and forces for manual materials-handling tasks, and the preferred weight results for the upper extremity transfer tasks shown in Figure 2. In this case, work tasks can be designed within these work parameters. The objective would be to modify the weight, frequency, distance, and other task parameters singularly or in combination to fit both the psychophysically determined acceptable task parameters (Liberty Mutual Insurance Company tables or Figure 2) and then given requirements for the particular work task. For instance, if for a given transfer task the frequency were fixed, say, at a particular assembly line speed, weight and/or distance could be modified appropriately. Conversely, if weight were held constant, frequency and/or distance could be modified. For example, Figure 2 shows that the average maximum preferred weight should be 6.1 kg for a one-handed upper extremity transfer task at a work pace of 10 transfers/minute and a distance of 0.5 m. Conversely, for a one-handed upper extremity transfer task of 6.1 kg at a distance of 0.5 m, the frequency should be ≤10 transfers/minute.

The NIOSH [81] equation for the design and evaluation of manual lifting tasks utilizes biomechanical, physiological, and psychophysical criteria, and thus provides another method to directly apply psychophysical criteria as workplace guidelines. For studies of perceived exertion, at least a relative rating with respect to different work tasks and sometimes an absolute measure with respect to heart rate (Borg RPE scale) can be obtained. In addition, Section

IV.C. shows how discomfort analyses may be used to determine the effects of specific task and workstation attributes.

For both manual materials handling tasks and upper extremity intensive work, weight or force has been shown to have the most significant effect [10,35,36,100]. Thus, weight and force may be the most important work parameter to consider in the analysis and design of work. For example, to decrease the weight for the worker, the following questions should be considered: Does the object to be lifted need to be that heavy? Could the object be made out of a lighter material? Could the object be made less bulky, so that it could be lifted closer to the body? Could it be lifted in parts? Does the object need to be manually lifted and/or transferred? Could the workstation be modified to partially or fully support the weight of the object? Could a mechanical assist (e.g., hoist, articulating arm) be used to transfer or support the object? Weight or force guidelines are shown in the Liberty Mutual Insurance Company tables and in Figure 2.

The second most important parameter to consider may be frequency or work pace [35,36]. As frequency increases, perceived exertion increases and preferred weight decreases, as shown in Figures 5 and 2, respectively. The Liberty Mutual Insurance Company tables and Figure 2 show some frequency guidelines. Recalling that there may be a frequency threshold for upper extremity transfer tasks [35,36], if the task frequency were 30 or more transfers/minute, a decrease in perceived exertion would occur if the frequency were decreased to 20 transfers/minute. However, further decreases in frequency below 20 transfers/minute may not have a significant effect on perceived exertion.

Distance may also be an appropriate parameter to consider in the analysis and design of work. As distance increases, perceived exertion increases and preferred weight decreases, as shown in Figures 5 and 2, respectively. The Liberty Mutual Insurance Company tables and Figure 2 give some distance guidelines. For the upper extremity, some studies have shown the importance of different distances, especially if the different distances produce postural changes, whereas lateral, side-to-side distances within the reach envelope may have only a minimal effect on perceived exertion [35,36]. Thus, decreases in distance within the reach envelope may not significantly decrease perceived exertion.

In conclusion, psychophysical data can provide guidance in the analysis and design of manual materials-handling tasks, upper extremity intensive tasks, and other manual work. Further studies of work-related musculoskeletal injuries and health surveillance will be required to determine the effect of decreased perceived exertion, discomfort, and fatigue, and to verify the effectiveness of psychophysically determined guidelines in reducing the risk of work-related musculoskeletal injuries. Until psychophysically determined guidelines are validated as effective in helping to reduce the risk of the development of work-related musculoskeletal injuries, they should be used in conjunction with both an active and passive surveillance program. If it is found that workers are adversely affected while using psychophysically determined guidelines, appropriate workplace and medical interventions should be implemented.

4. Work Enlargement

Psychophysical studies can be used to verify the positive effects of work enlargement. Krawczyk et al. [35,38] used a psychophysical methodology to examine combination tasks consisting of upper extremity transferring and screw-driving components. Five combination tasks were performed with different proportions of transferring and screw driving: 100% transfer, 75% transfer and 25% screw drive, 50% transfer and 50% screw drive, 25% transfer and 75% screw drive, and 100% screw drive. The left hand always performed the transferring component and the right hand always performed the screw-driving component of the combination tasks. The positive effect of using both extremities to perform the different components of the combination tasks was verified using VAS perceived exertion ratings and body part discomfort analyses.

Overall VAS perceived exertion decreased as the work was enlarged to utilize both upper extremities. Figure 6 shows that the lowest mean overall VAS perceived exertion rating was for the 50% transferring and 50% screw-driving task. The workload of this task was evenly distributed between the left and right upper extremities, since the transferring was done with the left hand and the screw driving was done with the right hand. By allowing different extremities to perform the task components, the frequency for each extremity was effectively decreased. Consequently, this task allowed the maximum amount of physiological recovery time that could be provided simultaneously for both the left and right upper extremities. Most subjects reported a preference for the balanced nature of this combination task.

The overall VAS perceived exertion increased as the combination task utilized more of one upper extremity than the other and involved more of either the transferring or screw driving task components, as shown in Figure 6. Thus, the left upper extremity (transferring component) or the right upper extremity (screw-driving component) would become the limiting factor of the work. Even though the less varied tasks affected only one extremity, this limiting factor affected *overall* VAS ratings.

The highest overall perceived exertions were for the tasks that required only transferring or only screw driving, as shown in Figure 6. These tasks allowed the least amount of physiological recovery time for the upper extremity (left or right) that was responsible for performing the task. When there is not enough recovery time, the body part experiences fatigue and discomfort. Body part discomfort surveys showed a proportional increase in frequency and severity of discomfort (using a discomfort VAS, shown in Section IV.C., Figure 8) in the respective upper extremity that was required to perform a greater proportion of the combination tasks.

In addition to the overall VAS perceived exertion, the subjects used separate visual analog scales to rate the perceived exertion of the transferring and screw-driving components. As shown in Figure 7, as the transferring proportion of the task increased, the transferring perceived exertion increased. Likewise, as the screw-driving proportion of the task increased, the screw-driving perceived exertion increased. Psychophysical methods are usually used to

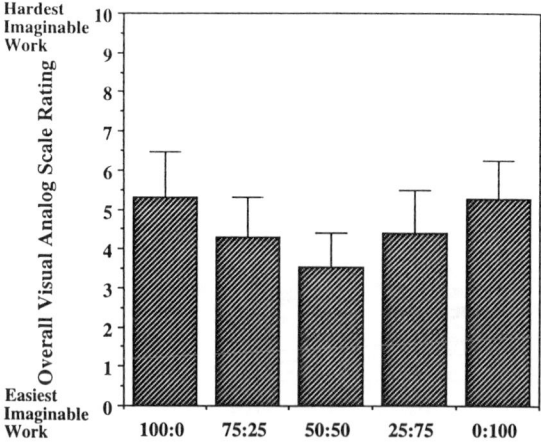

Figure 6 Mean overall VAS perceived exertion ratings for an upper extremity combination task involving transferring (performed with the left hand) and screw-driving (performed with the right hand) components, for 24 industrial subjects. 95% confidence intervals shown by error bars.

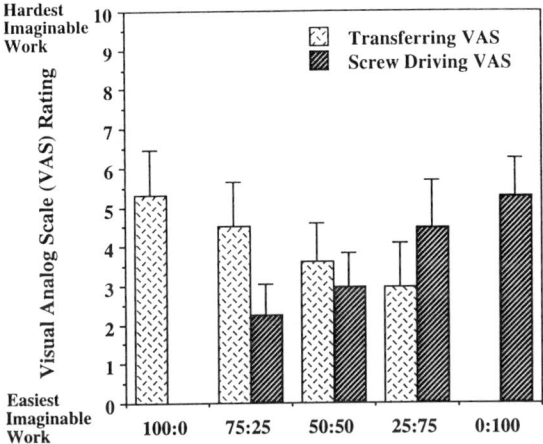

Figure 7 Mean transferring and screw driving visual analog scale (VAS) perceived exertion ratings for an upper extremity combination task involving transferring (performed with the left hand) and screw driving (performed with the right hand) components, for 24 industrial subjects. 95% confidence intervals shown by error bars.

combine the effects of physical stressors such as weight and frequency [5,22,24,35-37, 40-58,64-68,73-78]. However, in these upper extremity combination tasks that required two components (transferring and screw driving), subjects were able to do the opposite: to separate the effects of physical stressors. Subjects were able to discriminate between the transferring and screw-driving parts of the tasks, and to give a psychophysical rating accordingly.

The body part discomfort surveys revealed that the less varied work, which utilized more of one upper extremity than the other and involved more of one particular task component (transferring or screw driving), produced greater body part discomfort severity concentrated in fewer body parts. Conversely, the more varied work produced decreased body part discomfort severity, and the discomfort was more evenly distributed throughout the different body parts. Similarly, Westgaard and Jansen [103] found that production workers had significantly higher symptom scores of self-reported musculoskeletal complaints for some body parts than did the group with more varied work tasks.

C. Discomfort Analyses

Psychophysical methodologies utilize an individual's perception of physical strain on the human body. From this perception, individuals may rate how hard they feel that they are working using Borg or VASs, or they may adjust a work variable to achieve a maximum level of normal workday exertion. When carried out over a period of time, this perception will incorporate feelings of exertion and also feelings of fatigue and discomfort.

Another way to assess individual perception of exertion over time is by tracking discomfort directly. Both overall and body part discomfort can be monitored for both frequency (how often) and severity (how much) [30,35-39,65,83-90,95,96,99,104-107,109-112]. The frequency of overall and body part discomfort can be determined simply by asking individuals if they have any discomfort, either overall or in particular body parts or regions of the body. Likewise, the severity of discomfort can be quantified using various rating scales for overall discomfort or discomfort in particular body parts or regions of the body. A VAS such as the one shown in Figure 1, with the verbal anchors "easiest imaginable work" and "hardest imaginable work" on the left and right endpoints, respectively, can be used to determine overall discomfort. As explained earlier, the results from the VAS shown in Figure 1 were comparable to the results from the VAS with "very uncomfortable work" and "very comfortable work" anchors [30]. Thus, it appears that as long as the two verbal anchors represent extremes of some work-related physical intensity, the results will be comparable. For body part discomfort, separate VASs can be used for each body part or region, with the anchors "no discomfort at all" and "worst imaginable discomfort" [35,38], as shown in Figure 8. A simple numerical scale may also be used to quantify discomfort severity. In this case, the individual picks a number to represent his or her discomfort, say, from 0 to 7 or from 0 to 9 [104], where the numbers are verbally anchored by comfort and discomfort descriptions. This can be done for overall discomfort and also separately for each body part or region. Specific types of discomfort, such as pain, soreness, stiffness, numbness, and tingling, may also be quantified [63]. The relative severity of body part discomfort may be determined by using a body part discomfort survey that shows a body part diagram and asks the individual to rate the different body parts from most discomfort to least discomfort [105,106].

Discomfort has been associated with poor work performance [84,85,105–107]. Corlett and Bishop [105,106] found that relieving the discomfort of workers increased work output and reduced production costs. Many upper extremity intensive tasks cause discomfort, fatigue, and pain [30,35–38,88–90,95,96,108]. Discomfort analyses have been shown to be beneficial in the evaluation of various work apparatus [39,65,95,109–111], tasks [35–38,83–87,96,99,104,112], and workstations [30,39,88–90,95,96,105–107]. For instance, for automobile assembly tasks, postural discomfort has been shown to increase as a function of horizontal and vertical distance from the body [39]. Since localized muscle fatigue may be an early symptom of use-related upper extremity cumulative trauma disorders [13–15], discomfort analyses may serve as a more sensitive indicator for the risk of the development of some work-related musculoskeletal disorders.

Krawczyk et al. [35–38] used both VAS perceived exertion ratings and body part discomfort data to verify that the overall VAS perceived exertion ratings were a reflection of

Figure 8 A body part discomfort survey with 10-cm VASs for each body part, with the verbal anchors "no discomfort at all" and "worst imaginable discomfort" on the left and right endpoints, respectively, that has been used for the evaluation of manual work. Using this body part discomfort survey, discomfort frequency and severity can be assessed for particular body parts or regions of the body.

the discomfort in the upper extremities due to the performance of upper extremity intensive work, and that perceived exertion ratings may be an appropriate measure to use for overall assessment of the tasks. The tasks were repetitive upper extremity transfer tasks and pneumatic screw-driving tasks. Hagberg [12] found that increases in ratings or perceived exertion with work load indicated an importance of local factors, that is strain on muscles and tendons. Likewise, Wiker et al. [86,87] found that global reports of discomfort or fatigue were strongly related to the severity of symptoms experienced in specific muscle groups. Similarly, Yoshitake [104] found that general fatigue correlated well with symptoms of fatigue in some part of the body. Nevertheless, Kuorinka [83] concluded that local discomfort ratings were more sensitive to differences in work methods, and thus, more reliable than general discomfort ratings for ergonomic purposes.

Body part discomfort analyses may provide specific information about the relationship between task and workstation attributes (e.g., upper extremity intensive, height adjusted, standing) and their effects on the body. In a checkstand configuration study [89], subjects were found capable of distinguishing specific symptoms related to different checkstand designs.

Body part discomfort analyses may reveal additional aspects of a job that may be important for task analysis and redesign. For example, Krawczyk et al. [35-37] used body part discomfort analyses in two different studies of upper extremity transfer tasks and verified a positive effect of table height adjustment with respect to lower back, torso, and buttocks discomfort. In one study [37], where the height of the horizontal conveyor was adjusted so that the vertically oriented container handle was at elbow height for each subject, the reported lower back, torso, and buttocks discomfort was minimal, with a cumulative response rate of less than 14%. Whereas, in the other study [36], when the table height was fixed at 91 cm for each subject, the lower back, torso, and buttocks region was reported to be among most uncomfortable body regions. A full 75% rated this as one of the five more uncomfortable body regions (of a total of eleven). On average, the lower back, torso, and buttocks region was the fourth most uncomfortable body part—ranked just after the dominant upper extremity body parts and just ahead of the lower legs and feet. This was despite the fact that in the former study with minimal lower back, torso, and buttocks discomfort, the experimental workday was a full 8-hr duration. And, in the latter study with much more lower back, torso, and buttocks discomfort, the experimental task was carried out for only two hours each day, with lower weights and shorter distances too. These studies certainly suggest that the discomfort of the lower back can be decreased or minimized by adjusting the workstation height appropriately for each worker. These discomfort analyses show the relative importance of the effects of different workplace design attributes that may be addressed to minimize body part discomfort. Similar results have been shown in other upper extremity intensive work. Ayoub [88] found that all cashiers experienced some type of body pain. The areas that caused discomfort were the back, feet, legs, shoulders, neck, knees, hands, elbows, and wrists. Back pains occurred among 90% of the cashiers.

In Krawczyk et al.'s [35-38] studies of upper extremity intensive work, the subjects were standing while performing the work tasks. Even though the experimental tasks were mainly upper extremity intensive, the standing component taxed the lower extremities. This was evident since in addition to upper extremity discomfort, there was also a substantial number of discomfort responses in the lower legs and feet, and the number of discomfort responses for the lower legs and feet increased throughout the workday. Similarly, in a study of seven supermarkets, Ryan [90] found that the checkout department had the highest rates of discomfort or pain symptoms for almost all body areas. The lower back, lower limbs, and feet were the body areas with the highest rates of discomfort or pain symptoms. Ryan [90] found a positive and significant correlation between the proportion of time spent standing and symptoms in the lower limb and foot, especially in the checkout department, where 90% of the time was spent standing in one place. Rys and Konz [110] also found that body comfort decreases as the duration of the standing work increases, regardless of the type of floor. They found the greatest increases in discomfort in the lower body parts. None of the upper parts of the body had comfort significantly affected by floor surface.

In conclusion, discomfort analyses may serve as a sensitive indicator for the risk of the development of work-related musculoskeletal disorders. Overall and body part discomfort analyses may provide specific information about aspects of the work apparatus, tasks, and workstations that need to be addressed. Minimizing work-related discomfort may improve worker performance, increase output, decrease production costs [84,85,105–107], and increase worker satisfaction.

V. CONCLUSION

Psychophysical methods are a consistent, reproducible, quick, inexpensive, and convenient way to assess the degree of physical strain on the human body. Psychophysical criteria have been correlated with physiological criteria and some injury indices. Psychophysical methods utilize the results of the central nervous system integration of various information, including the many signals elicited from the peripheral working muscles and joints, and from the central cardiovascular and respiratory functions. All of these signals, perceptions, and experiences are combined and utilized with psychophysical methods.

The effects of different work task parameters, as well as the combined effects, can be evaluated using psychophysical methods. A major advantage of psychophysical studies is that the results can be readily applied as guidelines in the workplace. Psychophysical data can provide guidance in the analysis and design of repetitive manual work tasks, such as those found in manufacturing and production assembly facilities, warehouses, retail trades (e.g., grocery and discount stores), and other workplaces. The objective should be to modify the weight, frequency, distance, and other task parameters singularly or in combination to fit both the psychophysically determined acceptable task parameters and the given requirements for a particular work task. In addition, body part discomfort analyses may contribute specific information about the relationship between task and workstation attributes (e.g., upper extremity intensive, height adjusted, standing) and their effects on the body.

Health surveillance is required to verify the effectiveness of psychophysically determined guidelines in reducing the risk of work-related musculoskeletal injuries. Further studies of work-related musculoskeletal injuries and health surveillance is required to determine the effect of decreased perceived exertion, discomfort, and fatigue, and to verify the effectiveness of psychophysically determined guidelines in reducing the risk of work-related musculoskeletal injuries. Until psychophysically determined guidelines are validated as effective in helping to reduce the risk of the development of work-related musculoskeletal injuries, they should be used in conjunction with both an active and passive surveillance program. If it is found that workers are adversely affected while using psychophysically determined guidelines, appropriate workplace and medical interventions should be implemented.

REFERENCES

1. C. H. Hennekens and J. E. Buring. *Epidemiology in Medicine*. Little, Brown and Company, Boston, Massachusetts/Toronto, Ontario, Canada, 1987, pp. 258–271.

2. M. F. Tracy, Biomechanical methods in posture analysis, in *Evaluation of Human Work*, J. R. Wilson and E. Nigel Corlett, Eds., Taylor & Francis, New York, 1990, pp. 571-604.
3. T. J. Armstrong and D. B. Chaffin, Some biomechanical aspects of the carpal tunnel, *J. of Biomechanics* 12:567-570 (1979).
4. S. H. Snook and C. H. Irvine, Psychophysical studies of physiological fatigue criteria, *Human Factors* 11(3):291-300 (1969).
5. F. Gamberale, Perceived exertion, heart rate, oxygen uptake and blood lactate in different work operations, *Ergonomics* 15(5):545-554 (1972).
6. G. Borg, P. Hassmén, and M. Lagerström, Perceived exertion related to heart rate and blood lactate during arm and leg exercise. *Eur. J. Appl. Physiol.* 65:679-685 (1987).
7. C. M. Mermier, J. M. Samet, W. E. Lambert, and T. W. Chick, Evaluation of the relationship between heart rate and ventilation for epidemiologic studies. *Arch. of Env. Health* 48(4):263-269 (1993).
8. R. B. Andrews, Estimation of values of energy expenditure rate from observed values of heart rate, *Human Factors* 9(6):581-586 (1967).
9. E. Asmussen, and I. Hemmingsen, Determination of maximum working capacity at different ages in work with the legs or with the arms, *Scand. J. of Clin. and Lab. Invest.* 10:67-71 (1958).
10. A. Garg, Physiological responses to one-handed lift in the horizontal plane by female workers, *Am. Ind. Hyg. Assoc. J.* 44(3):190-200 (1983).
11. A. Garg and U. Saxena, Maximum frequency acceptable to female workers for one-handed lifts in the horizontal plane, *Ergonomics* 25(9):839-853 (1982).
12. M. Hagberg, Work load and fatigue in repetitive arm elevations, *Ergonomics* 24(7):543-555 (1981).
13. K. N. Baidya and M. G. Stevenson, Local muscle fatigue in repetitive work, *Ergonomics* 31(2):227-239.
14. R. H. T. Edwards, Muscle fatigue and pain, *Acta Med. Scand. Suppl.* 711:179-188 (1986).
15. F. Valencia, Local muscle fatigue: A precursor to RSI? *Med. J. of Australia* 145:327-330 (1986).
16. G. A. V. Borg, Psychophysical scaling with applications in physical work and the perception of exertion, *Scand. J. of Work, Environ. and Health.* 16(Suppl. 1):55-58 (1990).
17. S. S. Stevens, The psychophysics of sensory function, *Am. Scientist* 48:226-253 (1960).
18. G. A. V. Borg, *Physical Performance and Perceived Exertion, Studia Psychologica et Paedagogica, Series altera, Investigationes XI*, CWK Gleerup, Lunds Sweden, 1962.
19. G. A. V. Borg, Perceived exertion: A note on "history" and methods, *Med. and Sci. in Sports*, 5(2):90-93 (1973).
20. H. Eisler, Subjective scale of force for a large muscle group, *J. of Exp. Psychol.* 64(3):253-257 (1962).
21. J. C. Stevens and W. S. Cain, Effort in isometric muscular contractions related to force level and duration, *Perception & Psychophysics.* 8(4):240-244 (1970).
22. C. E. Baxter, H. Stålhammar, and J. D. G. Troup, A psychophysical study of heaviness for box lifting and lowering, *Ergonomics* 29(9):1055-1062 (1986).
23. F. I. Gamberale, I. Holmér, A. S. Kindblom, and A. Nordström, Magnitude perception of added inspiratory resistance during steady-state Exercise, *Ergonomics* 21(7):531-538, (1978).
24. A.-S. Ljungberg, F. Gamberale, and Å. Kilbom, Horizontal lifting: Physiological and psychological responses, *Ergonomics* 25(8):41-757 (1982).
25. M. Arstila, H. Wendelin, I. Vuori, and I. Välimäki, Comparison of two rating scales in the estimation of perceived exertion in a pulse-conducted exercise test, *Ergonomics* 17(5):577-584 (1974).
26. D. D. Price, P. A. McGrath, A. Rafii, and B. Buckingham, The validation of visual analog scales as ratio scale measures for chronic and experimental pain, *Pain* 17:45-56 (1983).
27. G. A. V. Borg, Psychophysical bases of perceived exertion, *Med. and Sci. in Sports and Exercise* 14(5):377-381 (1982).
28. G. A. V. Borg, Perceived exertion as an indicator of somatic stress, *Scand. J. of Rehabil. Med.* 2(3):92-98 (1970).

29. B. J. Noble, Clinical application of perceived exertion. *Med. and Sci. in Sports and Exercise* *14*(5):406–411 (1982).
30. S. S. Ulin, C. M. Ways, T. J. Armstrong, and S. H. Snook, Perceived exertion and discomfort versus work height with a pistol-shaped screwdriver, *Am. Ind. Hyg. Assoc. J.* *51*(11):588–594 (1990).
31. G. Neely, G. Ljunggren, C. Sylven, and G. Borg, Comparison between the visual analogue scale (VAS) and the category ratio scale (CR-10) for the evaluation of leg exertion. *Int. J. Sports Med.* *13*:133–136 (1992).
32. B. J. Noble, G. A. V. Borg, I. Jacobs, R. Ceci, and P. Kaiser, A category-ratio perceived exertion scale: Relationship to blood and muscle lactates and heart rate. *Med. and Sci. in Sports and Exercise* *15*(6):523–528 (1983).
33. K. Harms-Ringdahl, A. M. Carlsson, J. Ekholm, A. Raustorp, T. Svensson, and H.-G. Toresson, Pain assessment with different intensity scales in response to loading of joint structures, *Pain* *27*:401–411 (1986).
34. R. A. Seymour, J. M. Simpson, J. E. Charlton, and M. E. Phillips, An evaluation of length and end-phrase of visual analogue scales in dental pain, *Pain* *21*:177–185 (1985).
35. S. Krawczyk, Psychophysical determination of work design guidelines for repetitive upper extremity transfer tasks over an eight hour workday, Ph.D. diss., The University of Michigan, Ann Arbor, Michigan, 1993.
36. S. Krawczyk and T. J. Armstrong, Perceived exertion over time of hand transfer tasks: Weight, frequency, and distance, in *Designing for Everyone*, Y. Quéinnec and F. Daniellou, Eds., Taylor & Francis, London, 1991, pp. 167–169.
37. S. Krawczyk, T. J. Armstrong, and S. H. Snook, Preferred weights for hand transfer tasks for an eight hour workday, in *Arbete Och Hälsa Vetenskaplig Skriftserie, Proceedings of National Institute of Occupational Health International Scientific Conference on Prevention of Work-Related Musculoskeletal Disorders PREMUS*, M. Hagberg and Å. Kilbom, Eds., Stockholm, Sweden, 1992, pp. 157–159.
38. S. Krawczyk, T. J. Armstrong, and S. H. Snook, Psychophysical assessment of simulated assembly line work: Combinations of transferring and screw driving tasks, *Proceedings of the Human Factors and Ergonomics Society 37th Annual Meeting*, Seattle, Washington, 1993, pp. 803–807.
39. T. J. Armstrong, L. Punnett, and P. Ketner, Subjective worker assessments of hand tools used in automobile assembly, *Am. Ind. Hyg. Assoc. J.* *50*(12):639–645 (1989).
40. S. H. Snook and C. H. Irvine, Maximum frequency of lift acceptable to male industrial workers, *Am. Ind. Hyg. Assoc. J.* *29*(November-December):531–536 (1968).
41. S. H. Snook, C. H. Irvine, and S. F. Bass, Maximum weights and work loads acceptable to male industrial workers, *Am. Ind. Hyg. Assoc. J.* *31*:579–588 (1970).
42. A. Mital, Patterns of differences between the maximum weights of lift acceptable to experienced and inexperienced materials handlers, *Ergonomics* *30*(8):1137–1147 (1987).
43. L. M. Nicholson and S. J. Legg: A psychophysical study of the effects of load and frequency upon selection of workload in repetitive lifting, *Ergonomics* *29*(7):903–911 (1986).
44. S. H. Snook and V. M. Ciriello, Maximum weights and work loads acceptable to female workers, *J. of Occup. Med.* *16*(8):527–534 (1974).
45. V. M. Ciriello, S. H. Snook, A. C. Blick, and P. L. Wilkinson, The effects of task duration on psychophysicall-determined maximum acceptable weights and forces, *Ergonomics* *33*(2):187–200, (1990).
46. A. B. Griffin, J. D. G. Troup, and D. C. E. F. Lloyd, Tests of lifting and handling capacity: Their repeatability and relationship to back symptoms, *Ergonomics* *27*(3):305–320, (1984).
47. T. K. Forcman, C. E. Baxter, and J. D. G. Troup, Ratings of acceptable load and maximal isometric lifting strengths: The effects of repetition, *Ergonomics* *27*(12):1283–1288, (1984).
48. J. E. Fernandez and M. M. Ayoub, The psychophysical approach: The valid measure of lifting capacity, in *Trends in Ergonomics/Human Factors V*, F. Aghazadeh, Ed., North-Holland, Elsevier, Amsterdam, 1988, pp. 837–845.
49. J. E. Fernandez, M. M. Ayoub, and J. L. Smith, Psychophysical lifting capacity over extended periods, *Ergonomics* *34*(1):23–32 (1991).

50. W. Karwowski and J. W. Yates, The effect of time in the psychophysical study of the maximum acceptable amounts of liquid lifted by females, *Proceedings of the Human Factors Society 28th Annual Meeting*, San Antonio, TX, 1984, pp. 586-590.
51. W. Karwowski and J. W. Yates, Reliability of the psychophysical approach to manual lifting of liquids by females, *Ergonomics* 29(2):237-248 (1986).
52. S. J. Legg and W. S. Myles, Maximum acceptable repetitive lifting workloads for an 8-hour work-day using psychophysical and subjective rating methods, *Ergonomics* 24(12):907-916 (1981).
53. S. J. Legg and W. S. Myles, Metabolic and cardiovascular cost, and perceived effort over an 8 hour day when lifting loads selected by the psychophysical method, *Ergonomics* 28(1):337-343 (1985).
54. A. Mital, The psychophysical approach in manual lifting: A verification study, *Human Factors* 25(5):485-491 (1983).
55. V. M. Ciriello and S. H. Snook, A study of size, distance, height, and frequency effects on manual handling tasks, *Human Factors* 25(5):473-483 (1983).
56. S. H. Snook, The design of manual handling tasks, *Ergonomics* 21(12):963-985 (1978).
57. S. H. Snook and V. M. Ciriello, The design of manual handling tasks: Revised tables of maximum acceptable weights and forces, *Ergonomics* 34(9):1197-1213 (1991).
58. J. L. Pytel and E. Kamon, Dynamic strength test as a predictor for maximal and acceptable lifting, *Ergonomics* 24(9):663-672 (1981).
59. C. H. Kim and J. E. Fernandez, Psychophysical frequency for a drilling task, *Int. J. of Ind. Ergonomics* 12:209-218 (1993).
60. R. J. Marley and J. E. Fernandez, A psychophysical approach to establish maximum acceptable frequency for hand/wrist work, in *Advances in Industrial Ergonomics and Safety III*, W. Karwowski and J. W. Yates, Eds., Taylor & Francis, New York, 1991, pp. 75-82.
61. R. J. Marley and J. E. Fernandez, Psychophysical frequency and sustained exertion at varying wrist postures for a drilling task, *Ergonomics* 38(2):303-325 (1995).
62. V. Putz-Anderson and T. L. Galinsky, Psychophysically determined work durations for limiting shoulder girdle fatigue from elevated manual work, *Int. J. of Ind. Ergonomics* 11:19-28 (1993).
63. S. H. Snook, D. R. Vaillancourt, V. M. Ciriello, and B. S. Webster, Pyschophysical studies of repetitive wrist flexion and extension, *Ergonomics* 38(7):1488-1507 (1995).
64. S. S. Asfour, M. M. Ayoub, A. Mital, and N. J. Bethea, Perceived exertion of physical effort for various manual handling tasks, *Am. Ind. Hyg. Assoc. J.* 44(3):223-228 (1983).
65. C. G. Drury and J. M. Deeb, Handle positions and angles in a dynamic lifting task, part 2. Psychophysical measures and heart rate, *Ergonomics* 29(6):769-777 (1986).
66. A. Garg and J. Banaag, Maximum acceptable weights, heart rates and RPEs for one hour's repetitive asymmetric lifting, *Ergonomics* 31(1):77-96 (1988).
67. S. H. Snook, Psychophysical acceptability as a constraint in manual working capacity, *Ergonomics* 28(1):331-335 (1985).
68. S. H. Snook and C. H. Irvine, Psychophysical studies of physiological fatigue criteria, *Human Factors* 11(3):291-300 (1969).
69. G. D. Herrin, M. Jaraiedi, and C. K. Anderson, Prediction of overexertion injuries using biomechanical and psychophysical models, *Am. Ind. Hyg. Assoc. J.* 47(6):322-330 (1986).
70. S. H. Snook, Psychophysical considerations in permissible Loads, *Ergonomics* 28(1):327-330 (1985).
71. S. H. Snook, R. A. Campanelli, and J. W. Hart, A study of three preventive approaches to low back injury, *J. of Occup. Med.* 20(7):478-481 (1978).
72. D. H. Liles, S. Deivanayagam, M. M. Ayoub, and P. Mahajan, A job severity index for the evaluation and control of lifting injury, *Human Factors* 26(6):683-693 (1984).
73. M. G. Wardle and D. S. Gloss, A psychophysical approach to estimating endurance in performing physically demanding work, *Human Factors* 29(6):745-747 (1978).
74. F. Gamberale, A.-S. Ljungberg, G. Annwall, and Å. Kilbom, An expermental evaluation of psychophysical criteria for repetitive lifting work, *Applied Ergonomics* 18(4):311-321 (1987).
75. A. Garg and U. Saxena, Effects of lifting frequency and technique on physical fatigue with special

reference to psychophysical methodology and metabolic rate, *Am. Ind. Hyg. Assoc. J.* *40*(10):894–903 (1979).

76. A. M. Genaidy and S. Al-Rayes, A psychophysical approach to determine the frequency and duration of work-rest schedules for manual handling operations, *Ergonomics 36*(5):509–518 (1993).

77. A. M. Genaidy and S. S. Asfour, Effects of frequency and load of lift on endurance time, *Ergonomics 32*(1):51–57 (1989).

78. S. H. Snook and V. M. Ciriello, Maximum weights and work loads acceptable to female workers, *Occup. Health Nursing* (August):11–20 (1974).

79. V. Putz-Anderson and T. Waters, Revisions in NIOSH guide to manual lifting: Proceedings of a national strategy for occupational musculoskeletal injury prevention—Implementation issues and research needs, The University of Michigan, Ann Arbor, Michigan, 1991.

80. National Institute for Occupational Safety and Health (NIOSH), scientific support documentation for the revised 1991 NIOSH lifting equation: Technical contract reports, *Psychophysical Basis for Manual Lifting Guidelines* by M. M. Ayoub (purchase order number 88-79313, 1988), U.S. Department of Health and Human Services, Public Health Service, Centers for Disease Control, NIOSH, Cincinnati, OH, 1989.

81. T. R. Waters, V. Putz-Anderson, A. Garg, and L. J. Fine, Revised NIOSH equation for the design and evaluation of manual lifting tasks, *Ergonomics 36*(7):749–776 (1993).

82. S. H. Snook, Approaches to the control of back pain in industry: Job design, job placement and education/training. *SPINE: State of the Art Reviews 2*(1):45–59 (1987).

83. I. Kuorinka, Subjective discomfort in a simulated repetitive task, *Ergonomics 26*(11):1089–1101 (1983).

84. R. W. Schoenmarklin and W. S. Marras, Effects of handle and work orientation on hammering: I. Wrist motion and hammering performance, *Human Factors 31*(4):397–411 (1989).

85. R. W. Schoenmarklin and W. S. Marras, Effects of handle and work orientation on hammering: II. Muscle fatigue and subjective ratings of body discomfort, *Human Factors 31*(4):413–420 (1989).

86. S. F. Wiker, D. B. Chaffin, and G. D. Langolf, Shoulder postural fatigue and discomfort: A preliminary finding of no relationship with isometric strength capability in a light-weight manual assembly task, *Int. J. of Ind. Ergonomics 5*:133–146 (1990).

87. S. F. Wiker, D. B. Chaffin, and G. D. Langolf, Shoulder posture and localized muscle fatigue and discomfort, *Ergonomics 32*(2):211–237 (1989).

88. M. A. Ayoub, Ergonomic deficiencies: I. Pain at work, *J. of Occup. Med. 32*(1):52–57 (1990).

89. P. Harber, D. Bloswick, J. Luo, J. Beck, D. Greer, L. F. Peña, Work-related symptoms and checkstand configuration: An experimental study, *Am. Ind. Hyg. Assoc. J. 54*(7):371–375 (1993).

90. G. A. Ryan, The prevalence of musculo-skeletal symptoms in supermarket workers, *Ergonomics 32*(4):359–371 (1989).

91. J. B. Dahalan and J. E. Fernandez, Psychophysical frequency for a gripping task, *Int. J. of Ind. Ergonomics 12*:214–230 (1993).

92. J. E. Fernandez, J. B. Dahalan, M. G. Klein, and R. J. Marley, Using the psychophysical approach in hand-wrist work, *Proceedings of M. M. Ayoub Institute for Ergonomics Research*, Occupational Ergonomics Symposium, Lubbock, TX, April 1993. pp. 63–70.

93. R. Örtengren, T. Cederqvist, M. Lindberg, and B. Magnusson, Workload in lower arm and shoulder when using manual and powdered screwdrivers at different working heights, *Int. J. of Ind. Ergonomics 8*:225–235 (1991).

94. S. S. Ulin, T. J. Armstrong, S. H. Snook, and A. Franzblau, Effect of tool shape and work location on perceived exertion for work on horizontal surfaces, *Am. Ind. Hyg. Assoc. J. 54*(7):383–391 (1993).

95. S. S. Ulin, T. J. Armstrong, S. H. Snook, and W. M Keyserling, Examination of the effect of tool mass and work posture on ratings of perceived exertion for a screw driving task, *Int. J. of Ind. Ergonomics 12*(1–2):105–115 (1993).

96. S. S. Ulin, T. J. Armstrong, S. H. Snook, and W. M. Keyserling, Perceived exertion and dis-

comfort associated with driving screws at various work locations and at different work Frequencies, *Ergonomics* 36(7):833–846 (1993).
97. K. H. E. Kroemer, Cumulative trauma disorders: Their recognition and ergonomics measures to avoid them, *Appl. Ergonomics* 20(4):274–280 (1989).
98. T. J. Armstrong, R. G. Radwin, D. J. Hansen, and K. W. Kennedy, Repetitive trauma disorders: Job evaluation and design, *Human Factors* 28(3):325–336 (1986).
99. B. G. Coury and C. G. Drury, Optimum handle positions in a box-holding task, *Ergonomics* 25(7):645–662 (1982).
100. H. Strasser and E. Keller, Local muscular strain dependent on the direction of horizontal arm movements, *Ergonomics* 32(7):899–910 (1989).
101. O. Rais, Heparin treatment of peritenomyosis (peritendinitis) crepitans acuta, *Acta Chirurgica Scand. Suppl.* 268:(1961).
102. S. A. Goldstein, T. J. Armstrong, D. B. Chaffin, and L. S. Matthews, Analysis of cumulative strain in tendons and tendon sheaths, *J. of Biomech.* 20(1):1–6 (1987).
103. R. H. Westgaard and T. Jansen, Individual and work related factors associated with symptoms of musculoskeletal complaints. II. Different risk factors among sewing machine operators, *Br. J. of Ind. Med.* 49:154–162 (1992).
104. H. Yoshitake, Relations between the symptoms and feeling of fatigue, *Ergonomics* 14(1):175–185 (1971).
105. E. N. Corlett and R. P. Bishop, A technique for assessing postural Discomfort, *Ergonomics* 19(2):175–182 (1976).
106. E. N. Corlett and R. P. Bishop, The ergonomics of spot welders, *Appl. Ergonomics* (March):23–32 (1978).
107. V. Bhatnager, C. G. Drury, and S. G. Schiro, Posture, postural discomfort, and performance, *Human Factors* 27(2):189–199 (1985).
108. L. Lannersten and K. Harms-Ringdahl, Neck and shoulder muscle activity during work with different cash register systems, *Ergonomics* 33(1):49–65 (1990).
109. C. G. Drury and B. G. Coury, A methodology for chair evaluation, *Appl. Ergonomics* 13(3):195–202 (1982).
110. M. Rys and S. Konz, Floor mats, *Proceedings of the Human Factors Society 34th Annual Meeting*, Orlando, Florida, 1990, pp. 575–579.
111. S. S. Ulin, S. H. Snook, T. J. Armstrong, and G. D. Herrin: Preferred tool shapes for various horizontal and vertical work Locations, *Appl. Occup. and Environ. Hyg.* 7(5):327–337 (1992).
112. J. M. Deeb, C. G. Drury, and K. L. Begbie, Handle positions in a holding task as a function of task height, *Ergonomics* 28(5):747–763 (1985).

7
Instrumentation for Occupational Ergonomics

Robert G. Radwin
University of Wisconsin—Madison, Madison, Wisconsin

David J. Beebe
Louisiana Tech University, Ruston, Louisiana

John G. Webster and Thomas Y. Yen
University of Wisconsin—Madison, Madison, Wisconsin

I. INTRODUCTION

Ergonomists are concerned with measurement of a variety of physical, physiological, biophysical, and environmental factors. These include physical phenomena such as human motion (e.g., joint angles, force, acceleration), physiological processes related to energy expenditure and metabolism (e.g., heart rate, oxygen consumption, body temperature), and biopotentials associated with muscle contraction (e.g., electrocardiograms, electromyograms) and environmental exposures (illumination, acoustic noise, whole body and hand–arm vibration). An understanding of the principles, operation, and use of biomedical instruments is necessary for acquiring accurate measurements of any of these phenomena.

Many measurements in industrial ergonomics involve the use of electronic or mechanical instruments. Too often a lack of the basic fundamentals of measurement, electronic principles, and sensors can lead to unwitting measurement errors. Ergonomics practitioners and researchers should therefore be familiar with how electronic instruments operate and be aware of possible sources of error that can affect measurement systems.

Figure 1 shows a block diagram of a generalized instrumentation system. Most instrument systems include (1) a sensor, (2) an amplifier, (3) a signal conditioning stage, and (4) a display and storage device. Sensors are transducers that measure physical, chemical, or thermal events as electric signals. Because these signals are often very small, they have to be amplified in order to be accurately measured. The signals usually contain noise not related to the activity being measured, and because these unrelated signals are amplified along with the desired signal, signal conditioning is necessary to remove them from the wanted signals. The display presents the measured information so an observer can access the data. The signal may also be stored for further processing or viewed at a later time. Often the storage, processing, and display device is a digital computer.

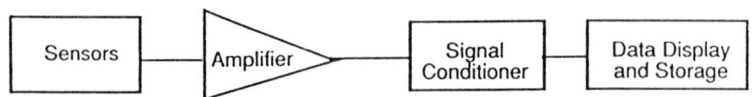

Figure 1 Block diagram of generalized instrumentation system.

II. RELEVANT CONCEPTS AND TERMINOLOGY

This section reviews the fundamentals of physics, electricity, and electronics that govern the operation and use of various instruments for measuring physical and physiological phenomena.

A. Measurement Principles

1. Accuracy

Accuracy is the difference between the quantity as measured and its true value. This difference may arise from a variety of sources. It can be a gross measurement error (attributed to misreading or improper use of the instrument), a systematic error (caused by a particular instrument's shortcomings or environmental factors), or a random error (due to random variations in the measurement itself). Random variations can often be reduced by making multiple measurements and taking an average. When the true value of a measurement is not known, error is sometimes determined by taking the standard deviation of a large number of repeated measurements. Accuracy is sometimes expressed in terms of percent error of full scale and usually varies over the range of the instrument.

2. Precision

Precision is the ability of an instrument to reproduce the same measurement over and over again. The precision of an instrument is often qualified by the number of significant figures the instrument is capable of resolving (see below). It is important to realize that higher precision does not necessarily imply better accuracy.

3. Resolution

Resolution is the smallest change that an instrument is capable of detecting. Resolution is determined by the manufacturer based on the instrument's accuracy and precision. The resolution of an analog meter is limited by the smallest division on the meter scale. A digital instrument's resolution is the quantity that the least significant digit represents.

4. Range

Minimum range is limited by the minimum quantity the instrument can resolve. The maximum range is limited by the greatest input that will not damage the instrument or the greatest quantity that an instrument can measure within a certain accuracy.

5. Calibration

Calibration references a measurement against a more accurate reference standard, usually traceable to the National Bureau of Standards. Typically, the reference standard should be at least an order of magnitude more accurate than the instrument being calibrated. Since accuracy depends on the quantity measured, calibration should be performed over the complete range of the instrument.

Instrumentation

6. Reading an Analog Scale

Errors can arise from improper readings off an analog display. An anthropometeric study found that considerable error was attributable to the precision used for reading the instrument rather than the actual precision limits of the instrument. Untrained subjects tended to measure body dimensions consistently to the nearest 1/4 in. (6 mm) or less, irrespective that the rule they were provided had 1/16-in. (1-mm) precision [1]. Measurements should be recorded from analog scales using all the digits that one is sure of plus one digit estimated nearest to the true value. For instance, if an analog scale is divided into tenths of a unit, the measurement may be recorded using all digits up to a tenth of a unit, plus a last digit that is the best estimation read by the observer.

B. Instrument Dynamics

1. Response Characteristics

Because physiological and physical measurements are seldom static, the dynamic features of instrumentation must be taken into consideration. System response characteristics should be considered for all sensors, amplifiers, signal conditioners, and display devices relative to the signal being measured. The response to a step input is often used for characterizing instruments. The specific response depends on the ordinary differential equations that describe the instrument input–output relationships. Figure 2 shows typical dynamic characteristics. Most instruments can be classified as zeroth-order, first-order, or second-order systems.

A zeroth-order system output is always proportional to its input. Consequently the step response for zeroth-order systems has no time delays, oscillations, or overshoot. Zeroth-order systems can be described by their static sensitivity, which is the ratio of the output signal divided by the input signal. Therefore the output is proportional to the input signal times the static sensitivity. A simple potentiometer is an example of a zeroth-order instrument.

The output of a first-order instrument requires a finite time to reach its final level for a step input, but no overshoot occurs. First-order systems are characterized by their static sensitivity and time constant. The time constant is defined as the time required for the sys-

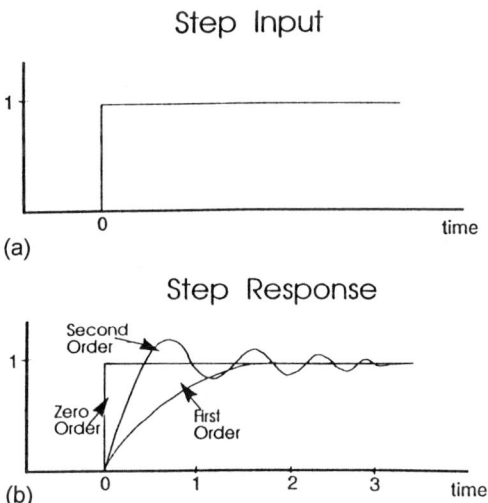

Figure 2 (a) Step input and (b) step response for zeroth-, first-, and second-order systems.

tem to reach 63% of its final value for a step input. The smaller the time constant, the faster the output rises. A mercury thermometer is an example of a first-order instrument.

For a step input, second-order systems take a finite time to reach the final value, may overshoot, and can yield oscillations. Many higher order instruments can be treated as second-order systems by making some simplifying assumptions. A simple spring scale is an example of a second-order instrument. Second-order systems can be modeled by a mass, spring, and viscous damping element, as shown in Figure 3a, and their characteristics may be described by their static sensitivity, natural frequency, and damping ratio. An excited second-order system oscillates or resonates as its natural frequency. Inputs that are near or at the system's natural frequency should be avoided because they cause the instrument to oscillate uncontrollably. The damping ratio is a dimensionless quantity that is greater than 1 when the system is overdampened and less than 1 when the system is underdampened, in which case the output for a step input will overshoot the final value (see Figure 3b).

2. System Step Response

Often the actual mass, spring, and viscous damping elements are not known. Second-order system characteristics are sometimes obtained empirically by their step response using the following parameters (see Figure 3c).

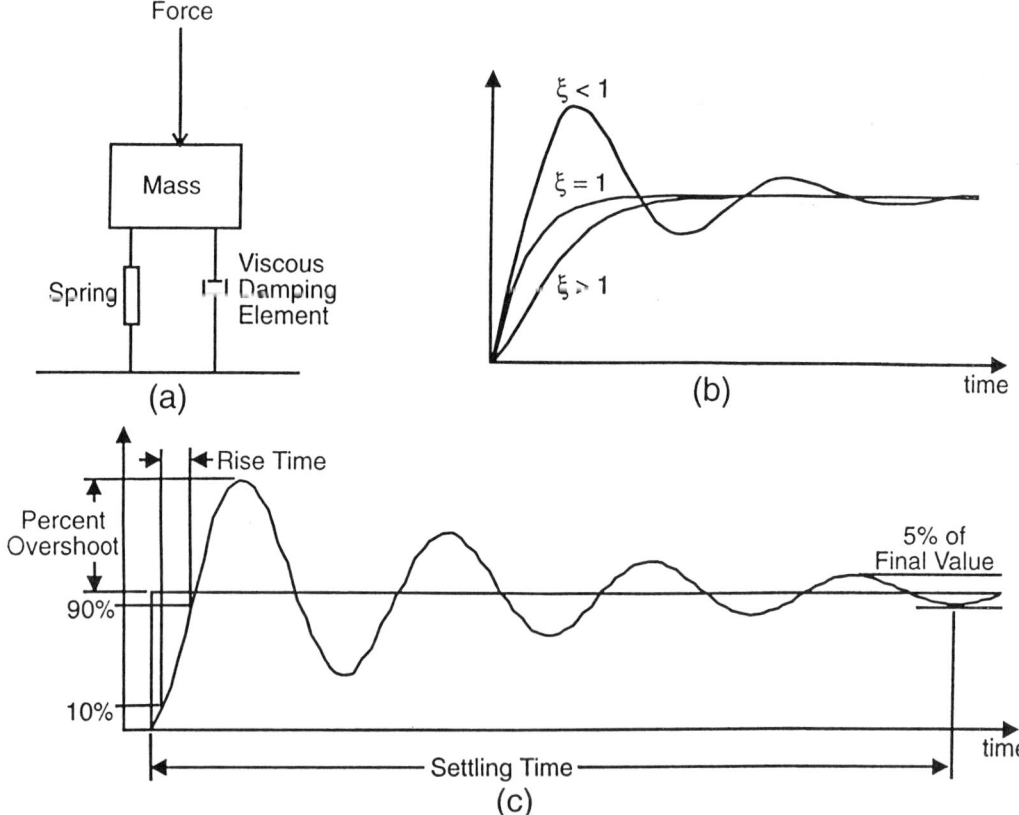

Figure 3 (a) Generalized second-order system model consisting of mass, spring, and viscous damping element. (b) Second-order system response based on damping ratio ζ. (c) Step response parameters describing a second-order signal.

Percent overshoot. The amount by which the peak output exceeds the final value is expressed as a percentage of the final value. Percent overshoot is a convenient measure of relative stability.

Settling time. Settling time is the total time needed for the response to settle within a certain percentage (e.g., 5%) of the final value.

Rise time. Rise time is sometimes defined as the transition time for the response to change from 10% to 90% of its final value.

3. Instrument Specifications

Additional instrument specifications describe operating characteristics that are important to consider when selecting an instrument for a particular application. Among others, the following specifications should be considered.

Input and Output Range. These ranges are usually specified as the minimum and maximum voltages for the input and output. This is particularly important to consider when interconnecting instruments or instrument components such as sensors and amplifiers.

Overload Range. The maximum input that will not damage the instrument is the overload range. The overload range should be sufficiently large that any anticipated input does not exceed that level.

Input and Output Impedance. Impedance is the resistance to the flow of current. The most fundamental equation in electric circuits is Ohm's law, which states that voltage across an electrical conductor is directly proportional to the current passing through it such that $V = IR$, where V is voltage (volts), I is electric current (amperes), and the constant R is electrical resistance in ohms (Ω). In direct current (dc) circuits, impedance corresponds to resistance. The impedance for direct current is not necessarily the same as for alternating current (ac). Impedance is a function of resistance, inductance, and capacitance for ac circuits and is expressed as ohms.

Resistors offer the same opposition to current flow for dc and ac circuits, and therefore have the same impedance. Inductors and capacitors build up voltage that opposes the flow of ac. This opposition to alternating current, called reactance, must be combined with resistance, resulting in a complex quantity that has properties of both magnitude and phase. Consequently, impedance is a function of ac frequency. The reactance produced by inductance is proportional to frequency, whereas the reactance produced by capacitance is inversely proportional to frequency. The dc impedance for a capacitor is infinite, and its ac impedance is $1/2\pi f C$, where f is frequency (Hz). The dc impedance for an inductor is zero, and its ac impedance is $2\pi f L$. A circuit's total impedance is determined by the complex combination of the resistance and reactance for all of the circuit components.

In order to transfer maximum electric power from one device to another, the two impedances must be matched. To prevent an output component from being loaded down by having the input component draw current and changing its output, the impedance of the output component should be much less (usually many orders of magnitude less) than the impedance of the input component.

Frequency Response. The frequency response describes the characteristics of an instrument as a function of the input frequency. Frequency response is often presented as a plot of the frequencies that shows that input signals are attenuated, or amplified, or where resonances occur. Frequency is usually plotted on the x axis using a logarithmic scale, and the response ratio of the output/input is plotted on a decibel scale such that

$$\text{Response (dB)} = 20 \log_{10}\left(\frac{\text{output}}{\text{input}}\right)$$

It is often desirable for the input signal to be within a frequency where the response is flat so that the output signal is unaffected by the input signal frequency (does not change with frequency).

Nonlinearity. Nonlinearity is a measure of how much an instrument output varies from its input. It is sometimes expressed as the maximum deviation from a linear output in terms of a percentage of the reading or a percentage of full scale, whichever is greater.

Repeatability. Repeatability is a measure of the variation of the output for the same input over time.

Signal-to-Noise Ratio. The ratio between the peak or rms amplitude of the signal and the noise is usually expressed as a fraction (e.g., 1000/1) or in decibels as SNR = $20 \log_{10}$ (signal/noise).

Stability. Stability is a measure of the output drift as a function of time, temperature, humidity, shock, or vibration.

II. SENSOR PRINCIPLES

Sensors can be electromechanical, electrochemical, or electrothermal. Electromechanical sensors measure phenomena associated with kinetics and kinematics, such as motion and force. Electrochemical sensors include electrodes for measuring gas concentrations or bioelectrical phenomena. Electrothermal sensors are used for measuring temperature.

A. Electromechanical Sensors

Electromechanical sensors convert mechanical inputs into electrical outputs. Mechanical parameters usually include displacement, velocity, acceleration, and force. Electromechanical sensors are based on resistive, piezoelectric, capacitive, and inductive transduction.

1. Sensors Based on Electrical Resistance

The resistance of a cylindrical electrical conductor is $R = \rho L/A$, where L is its length, A is its cross-sectional area, and ρ is the resistivity of the particular material (Ω-m). The higher the resistivity of a material, the more the material acts as an insulator. Resistors are often made from carbon composites because of their very high resistivity, or from high-resistivity metal wire, which is usually wound in coils to increase the length. The resistivity of metals increases with increasing temperatures because of the increased number of collisions that electrons make, thus increasing their electrical resistance.

Mechanical actions usually affect sensors based on electrical resistance either by changing L or A (for metals) or by changing ρ (for semiconductors), consequently changing the sensor's resistance and the voltage across it. The simplest displacement sensor is a variable resistor, or potentiometer. The electrical resistance is proportional to the position of the slider. Metal strain gauges are made from lengths of very fine wire (<25 μm diameter). When the wire is stretched, its resistance changes, mainly because of changes in its cross-sectional area A and length L. Resistive sensors are sensitive to temperature changes; this must be taken into account when measurements are being made.

Silicon strain gauges are made of diffused resistors integrated into a silicon substrate. Silicon and metal strain gauges are used in a similar manner, and both provide a very linear response within the elastic limits of the material to which they are fastened. Silicon strain gauges exhibit even higher temperature effects than metal because deformation affects ρ. These temperature effects can often be controlled using special compensation circuits.

The gauge factor

$$\frac{\Delta R/R_0}{\Delta L/L_0}$$

specifies a strain gauge's sensitivity to mechanical deformation. In this proportion, $\Delta R/R_0$ is the fractional change in resistance due to strain and $\Delta L/L_0$ is the fractional change in strain gauge length. The gauge factor for metal strain gauges is typically between 2 and 5, whereas the gauge factor for silicon can be as high as 170. This mechanical deflection, usually a fraction of 1%, causes a small but measurable change in resistance.

Strain gauges are usually used in pairs. Two gauges can be placed on the top of a cantilever beam and two on the bottom. The folded cantilever beam permits four gauges to be placed on the top surface, which eases manufacture and increases performance [2]. When strain gauges are arranged in a Wheatstone bridge circuit, their sensitivity can be increased. The circuit in Figure 4 uses two strain gauges in adjacent legs of the bridge. The output voltage of the bridge, V_o, is given by the equation

$$V_o = V_e \left(\frac{R_2}{R_2 + R_4} - \frac{R_1}{R_1 + R_3} \right)$$

When the beam bends, gauge 1 on the top of the beam elongates and increases ΔR, while gauge 2 on the bottom is compressed and decreases ΔR. If two fixed resistors in the opposite two legs of the bridge have resistance identical to that of the undeformed strain gauges, R_0, they do not deform. This arrangement is known as a half-bridge circuit. If ΔR is much less than R_0, then

$$V_o \approx -\frac{V_e}{2R_0} \Delta R$$

and the output is proportional to the change in resistance. A full bridge configuration in which all four legs contain strain gauges (R_1 and R_3 on top of the beam and R_2 and R_4 on the bottom) yields even higher sensitivity with an output given by $V_o \approx \Delta R/R_0$.

The bridge circuit configuration increases sensitivity by $2(1 + v)$ over that of a single strain gauge configuration, where v is Poisson's ratio. Bridge circuits also provide temperature compensation, which prevents output shifts due to temperature effects on the transducers. When maximum output is desired for any strain gauge, a low-modulus material (such as aluminum) is often used to increase strain per unit force. When four strain gauges are used

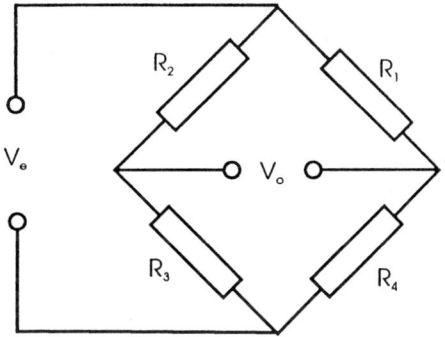

Figure 4 Wheatstone bridge circuit.

in a Wheatstone bridge, they yield an electrical output that is insensitive to bending stresses due to the force being applied off center or at an angle, and they can be temperature-compensated [2]. Strain gauges are often used in load cells and as force sensors in scales and pressure transducers.

2. Piezoelectric and Piezoresistive Sensors

Piezoelectric materials, such as quartz and barium titanate, produce a change in charge distribution when subjected to a mechanical stress. Piezoresistive sensors produce changes in resistance when stressed. Piezoelectric and piezoresistive load cells require minute deformations of their atomic structure within a block of crystalline material. Typical uses for piezoelectric and piezoresistive sensors include accelerometers and load cells.

Because piezoelectric sensors operate on changes in charge distribution, their most notable drawback is their inability to respond to static loads. Quartz is a naturally occurring piezoelectric material, and deformation of its crystalline structure changes its electrical characteristics such that the electric charge across its surface is altered. The charges collect on metal electrodes deposited onto the surface of piezoelectric material. Special amplifiers called charge amplifiers are used that output a voltage proportional to a charge at its input.

Piezoelectric force sensors operate on the principle that the charge across a piezoelectric material is proportional to its deformation, which is proportional to the applied force. Piezoelectric accelerometers operate the same way, except that a small mass M is mounted on top of the piezoelectric material. When the accelerometer moves with an acceleration a, the mass exerts against the piezoelectric material a force $F = Ma$ that is proportional to its acceleration. Piezoelectric device sensitivity is expressed in terms of coulomb charge per acceleration or force unit. Small-mass devices have low charge sensitivity.

Piezoresistive sensors are not limited to measuring changing signals and are suitable for static loads. Piezoresistive sensors are usually configured in a Wheatstone bridge like strain gauges. Consequently they require an external excitation voltage and a sensitive instrumentation amplifier.

3. Electrocapacitive Sensors

Capacitors have the capacity to store an electric charge. The simplest capacitor consists of two parallel metal plates separated by an electrical insulator. When the voltage across a capacitor changes, opposite polarity charges accumulate on the plates, resulting in an electric current flowing through the capacitor that is proportional to the voltage change. The current through a capacitor is $I = C\, dV/dt$, where C is capacitance measured in farads (F). The capacitance of a plate capacitor is described by the equation $C = \varepsilon A/x$, where x is the plate separation distance, ε is the dielectric constant of the insulator, and A is the plate area. These plates are constructed such that when an applied force changes x, the capacitance C changes. Applications for electrocapacitive sensors include their use in transducers for measuring minute displacement changes and for microphones.

4. Inductive Transducers

An inductor consists of multiple turns of insulated wire wound as a coil. When current flows through the coil a magnetic field is produced that is proportional to the current passing through the coil. If the current changes, the magnetic field changes, and a voltage is induced across the ends of the coil. The voltage across an inductor is $V = dI/dt$, where L is inductance measured in henrys (H). The inductance of a coiled wire is proportional to the square of the number of turns of wire, n, such that $L = n^2 G\mu$, where G is a geometric form factor and μ is the effective permeability of the material the wire is looped around.

Inductors are often used as displacement sensors where changes in displacement result in changes in inductance. The most common sensor using this principle is the linear voltage displacement transducer (LVDT). LVDTs have better resolution than potentiometers and provide very high motion sensitivity at the expense of more complex signal-processing circuitry. Usually an ac signal is fed through a primary coil. The coil is wrapped around a movable core along with a secondary coil, so that a voltage is induced across its output. The amplitude of the ac voltage depends on the position of the core relative to the coils. The core can be moved in and out along the axis of the two coils. The ac induced voltage is converted to direct current by using a rectifying circuit. LVDTs make good linear displacement transducers because of their high resolution, almost zero friction, and no wear.

B. Electrochemical Sensors

Electrodes convert ionic bioelectrical potentials into an electric current or voltage that can be amplified or measured. The most common types of electrochemical sensors used in ergonomics include electrodes for measuring oxygen concentrations in expired air and electromyogram and electrocardiogram electrodes.

1. Biopotential Electrodes

Needle or wire electrodes are invasive and are used for measuring ionic activity in a localized region is tissue such as muscle or nerve. Surface electrodes are noninvasive and measure potentials externally at the surface of the skin, such as for EMG, ECG, and EEG measurements. Figures 5 and 6 show examples of both types of electrodes.

Biopotentials are usually measured as the potential difference between two electrodes. All electrodes involve an interface between a metal and an electrolyte. Ions of opposite charge migrate from the metal to the electrolytic solution and from the solution into the electrode,

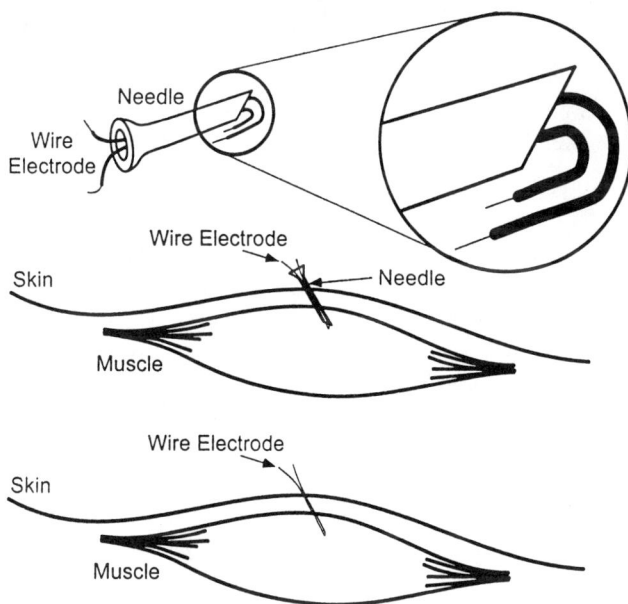

Figure 5 Needle/wire electrode shown placed in a muscle.

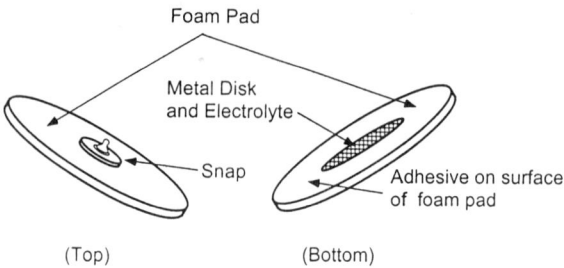

Figure 6 Typical foam pad surface electrodes.

so that an electrical double layer is formed and a potential difference is developed across the interface. This electrode potential occurs independently of the specific bioelectrical activity present. When the potential difference between two electrodes is measured, the electrode potential cancels out. Small, random fluctuations in electrode potential sometimes do not cancel and cause drift in the output voltage.

Surface electrodes usually consist of a metal disk mounted in a plastic receptacle. Silver-silver chloride (Ag-AgCl) electrodes, consisting of a silver chloride–coated silver disk, are common, although other metals are sometimes used. The space between the disk and the skin is filled with an electrolyte gel or paste such as a mixture of sodium chloride and glycerol. Disposable electrodes are available that already contain the gel.

2. Ion-Selective Sensors

The Nernst equation describes the relationship between ion concentrations and the voltage across a membrane E_k such that

$$E_k = \frac{RT}{nF} \ln \left[\frac{\text{ion concentration inside}}{\text{ion concentration outside}} \right]$$

where
R = the universal gas constant
T = the temperature (kelvins)
n = the ion valence
F = Faraday's constant

When an ion-selective membrane separates the solution from another (i.e., the ion concentrations are different), a voltage will appear across the membrane. This principle can be used for measuring the concentrations of different ions, depending on the composition of the membrane. pH electrodes use glass membranes to measure H^+ concentration. The principle is also used for CO_2 and O_2 concentration measurements.

C. Electrothermal Sensors

Although mercury column thermometers are widely used for measuring temperature and are reliable and inexpensive, and they have a high thermal capacity, so they may alter the temperature being measured. They also respond slowly, so rapid changes in temperature cannot be measured. Perhaps the greatest disadvantage is that they do not produce an electric signal that can be used for continuous measurements and recordings. A variety of electronic sensors are available for measuring temperature that overcome these limitations.

Instrumentation

Thermistors are ceramic materials that, unlike metals, decrease in resistance with increasing temperatures. Transistor-based temperature sensors use the temperature dependence of voltage across a diode junction. Themocouples operate on the principle that when two dissimilar metals are brought together, a voltage is observed across the junction. The magnitude of that voltage is dependent on temperature.

D. Photosensors

Photosensors can detect radiant energy or light. Photoconductive cells, or photoresistors, contain elements that change conductivity as a function of incident electromagnetic radiation. Materials such as cadmium sulfide are photoconductive. Photosensors are often used for detecting when an incident beam of light is broken or for quantifying the amount of incident light such as used in a light meter.

III. AMPLIFICATION

Electric signals produced by most sensors are usually too small to measure directly (a body surface electrocardiogram signal measured at the electrode is usually less than 1 mV). In order for signals to be observed, processed, or recorded, they need to be amplified to a magnitude of 1 V or more. The characteristics of an amplifier describe how an input signal appears on its output.

A. Gain

The amplifier output voltage V_o is the product of the input voltage V_{in} multiplied by the amplitude factor (gain) A such that $V_o = AV_{in}$. An amplifier having a gain of 100,000 will amplify a 50-µV EMG signal to 0.5 V.

B. Impedance

Instrumentation amplifiers should have high input impedance (at least 10 MΩ) so they provide minimal loading of the signal being measured.

C. Frequency Response

Frequency response specifies the upper and lower frequency of the range in which a signal can be amplified without a loss in gain.

D. AC and DC Coupling

When a sensor is connected to an amplifier directly, it is said to be dc-coupled. That is, the complete signal is input to the amplifier. If a dc offset voltage is present, that offset will be amplified along with the ac signal. When a sensor is coupled through a series capacitor, it is said to be ac-coupled, and only the alternating signal is input to the amplifier because capacitors have infinite impedance to dc signals. Examples for both ac- and dc-coupled amplifier outputs are shown in Figure 7.

E. Saturation

The maximum output of any instrument is limited to its power supply voltage. When this maximum output is reached, no further increase is possible, regardless of what input is ap-

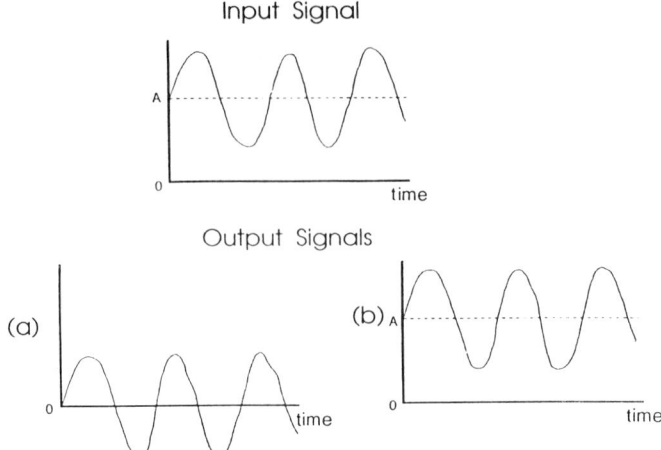

Figure 7 Signal output for (a) ac-coupled amplifier, (b) dc-coupled amplifier.

plied. An instrument in this state is said to be saturated. Just as a saturated sponge cannot hold any more water, a saturated instrument is unaffected by any further increase in the input signal. Figure 6 illustrates a saturated and an unsaturated amplifier.

F. Zero Suppression

Direct current amplifiers may have an offset adjustment to compensate for drift in the output.

G. Differential Amplifiers

Differential amplifiers amplify the difference between the signals at the two inputs. If the same signal is sent to both inputs of a differential amplifier, the output is expected to be zero. This requires identical impedance characteristics at each input. In the real world this is usually not the case. The common mode rejection ratio (CMRR) is a measure of an amplifier's ability to reject common mode signals (i.e., signals that are equally applied to both inputs of a differential amplifier). The CMRR for a differential amplifier is determined by measuring the output voltage for the same voltage at both inputs of a difference amplifier. The common mode gain (CMG) is the ratio between the output voltage V_o and the input voltage V_i such that CMG = V_o/V_i. The CMRR is the ratio between the amplifier gain (differential gain) A and the CMG such that CMRR = A/CMG. Since many unwanted or modifying inputs are common mode (the same signal appears on both amplifier inputs), a large CMRR is desirable. The CMRR should be greater than 10,000 for instrumentation amplifiers. Differential amplifiers are often used for biopotential recordings such as EMGs to eliminate electrical noise common to both inputs, such as line current interference.

The intrinsic mismatch between the inputs gives rise to a voltage between the inputs called the offset voltage. Although offset voltages are typically only a few millivolts, they can give rise to significant errors if the signal one is attempting to measure is also small. Many amplifiers have adjustments for eliminating offset voltages.

H. Temperature

The environment in which instruments are used can affect their performance. Most electric components are affected by temperature. The resistance of a conductor, for instance, increases with temperature. The offset voltage of an amplifier is usually temperature-dependent, giving rise to measurement errors if large temperature changes occur.

I. Noise

Any unwanted signal at an amplifier's input is amplified along with the desired signal. Noise can come from a variety of sources, including capacitive coupling between electrode leads and ac wiring, induced currents from ac magnetic fields, skin–electrode movement artifact, and popcorn noise from the amplifier itself. Fortunately, the effect of noise can often be minimized by taking simple corrective precautions. For example, the use of shielded cabling can eliminate capacitive coupling.

IV. SIGNAL CONDITIONING

A. Signals

Bioelectric signals are sometimes simply recorded as time-varying signals. Often complex signals have to be reduced to a form related to the quantity being measured using circuitry for rectification, integration, differentiation, and root mean square.

1. Time-Variant Signals

Signals measured from electronic sensors are usually represented as a voltage changing with time. These time signals are usually displayed on an oscilloscope or printed on a strip chart recorder, or they may be converted into digital form and displayed on a computer screen. A simple sinusoidal signal, as shown in Figure 8, is simply described by its amplitude and period. When the signal is more complex or random, time plots are inadequate for extracting useful information. Since physiological signals are often complex, signal-processing methods are needed for displaying and interpreting the information the signals contain. Common signal-conditioning functions include rectification, logarithm, integration, differentiation, and root mean square (see Figures 8 and 9).

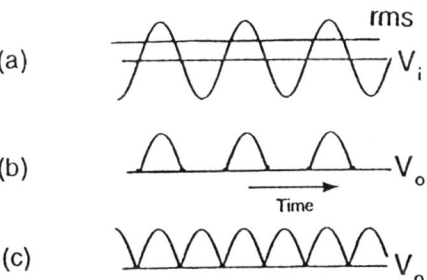

Figure 8 (a) Simple sinusoidal signal; (b) half-wave rectified sine wave; (c) full-wave rectified sine wave.

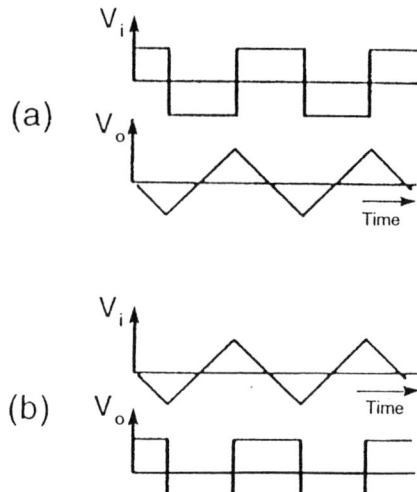

Figure 9 (a) Differentiation; (b) integration.

2. Rectification

A half-wave rectifier has an output signal equivalent to the input signal when the input is a positive value and an output of zero when the input is negative. A full-wave rectifier is a circuit that produces the absolute value $|x|$ of the input signal. Figures 8b and 8c show a half-wave and full-wave rectified output for a sine wave.

3. Logarithmic Functions

Logarithm circuits depend on the logarithmic relationship between a transistor base-emitter voltage and its collector current. Antilog circuits can be obtained by interchanging the circuitry. Log and antilog circuits may be used for multiplication, division, raising a voltage to a power, or taking a root.

4. Differentiation

The output of a differentiator circuit is proportional to the derivative of the input signal (see Figure 9a). Differentiators may be used for transforming displacement data into velocity or acceleration. Differentiators tend to accentuate the high-frequency noise, which can often mask the desired output.

5. Integration

An integrator outputs the integral of the input signal (see Figure 9b). Since differentiation tends to enhance noise in a signal, integration is often a more desirable approach. For example, velocity can be obtained by differentiating a displacement signal or integrating an acceleration signal. A less noisy signal would be obtained by integrating the output of an accelerometer. Often integration occurs over a specified time period. If, on the other hand, the signal being integrated has a low frequency drift, differentiation may be more desirable because over integration time these drifts accumulate into large errors.

6. Root Mean Square

The root mean square (rms) of a signal is the square root of the average of the square of the signal, such that

Instrumentation

$$\text{Root mean square} = \left[\frac{1}{T}\int_0^T f^2(t)\, dt\right]^{1/2}$$

For a simple sine wave,

$$f(x) = A \sin \frac{2\pi}{T} t, \qquad \text{rms} = \left[\frac{1}{T}\int_0^T A^2 \sin^2 \frac{2\pi t}{T}\, dt\right]^{1/2} = \frac{A}{\sqrt{2}} = 0.707 A$$

Some rms detectors simply scale the peak amplitude of a sinusoidal signal by 70.7% for determining rms (see Figure 8a). Such circuitry would give false readings for any signal except a sine wave. A true rms detector contains circuitry for determining the root mean square (rms) of an arbitrary function $f(x)$ using circuitry for instantaneously squaring the signal, integrating over a defined time constant, and taking the square root. Some averaging rms detectors simply rectify and integrate the signal without squaring as an approximation. The accuracy of this approximation depends on the magnitude of the signal.

B. Filters

Electronic filters separate unwanted signals from sensor signals according to frequency. Three general types of filters are (1) low-pass, (2) high-pass, and (3) bandpass filters (see Figure 10). Low-pass filters allow low frequencies to pass unattenuated, while high-frequency signals are attenuated and vice versa for high-pass filters. Bandpass filters pass signals unattenuated within a specific frequency range.

In practice, filters are not ideal. Signals at the unwanted frequencies are not completely eliminated; they are only reduced in amplitude. Often the signal-to-noise ratio is great enough that the noise is practically eliminated. Significant attenuation of unwanted signals is greatest when there is a large frequency difference between the signal and the noise. The corner

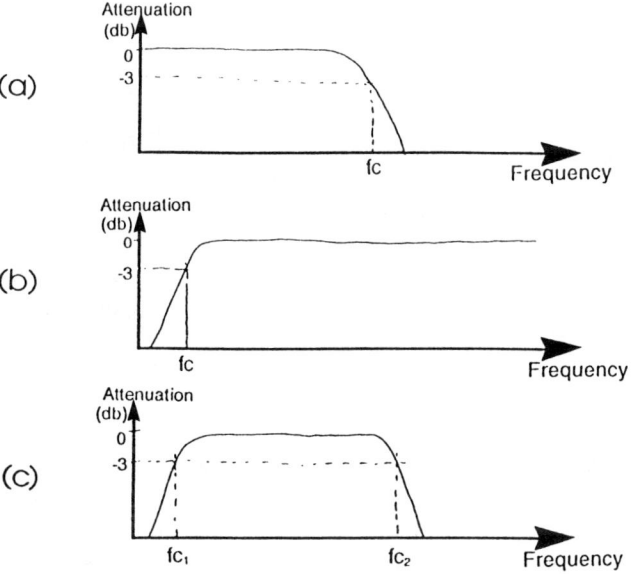

Figure 10 Three general types of filters: (a) Low-pass; (b) high-pass; (c) bandpass.

frequency (f_c) is usually defined as the frequency at which the signal is attenuated by 71% or 3 dB [20 log (0.707)]. Frequencies inside the corner frequency are much less attenuated than 3 dB, where attenuation is insignificant. Attenuation beyond the corner frequency depends on the filter roll-off, which is specified as a slope in terms of either decibels per octave or decibels per decade. An octave is a frequency multiplied by 2, and a decade is a multiple of 10. Therefore, if a filter attenuates –60 dB/octave, then the signal is attenuated –60 dB for each doubling of frequency. The more negative the slope, the sharper the filter and the more noise is reduced beyond the corner frequency.

V. SIGNAL PROCESSING

A. Digital Circuits

Computers represent signals as binary digits containing 0's and 1's. The conversion of analog signals into digital representations accounts for the powerful manipulation and storage capabilities of the computer. Many instruments contain interface circuitry so they can communicate data with a computer and permit the computer to remotely operate the instrument.

1. Digital Devices

Digital electronic devices use two different voltage levels to represent the two logic states of 0 and 1. Computers generally use 0 V and +5 V for representing logic levels of 0 and 1, respectively. This is the convention used for transistor-transistor logic (TTL), named for the type of circuitry. Some digital devices, such as modems and printers, use different voltage level conventions such as –12 V for logical 1 and +12 V for logical 0. Low-power devices now use 0 V and +3 V conventions for 0 and 1. Although less common, some digital devices use current levels instead of voltage levels.

2. Digital-to-Analog Conversion

A digital-to-analog converter (DAC) in its simplest form is a set of resistors each having a different resistance, providing a binary-weighted scaling of input digital voltages. The output voltage is proportional to the current, which is determined by which of the resistors are connected to the reference voltage.

3. Analog-to-Digital Conversion

There are numerous types of analog-to-digital converter (ADC) circuits available, differing in conversion speed and complexity. Voltage-to-frequency converters provide an output train of pulses at a frequency proportional to input current or voltage. A counter accumulates the pulse count for a fixed period of time. High resolution and relatively good accuracy over a wide range of input voltage are possible, but conversions are slow. A simple counter, or servo converter, contains a DAC and a digital counter. When the analog output of the DAC is greater than or equal to the input analog voltage, the counter stops counting and the digitized value is the counter value. Because a simple up-counter converter takes a long time for each conversion, an up-down counter can be used to track the input level. This counting process is faster but still relatively slow, especially for rapidly changing signals. Successive approximation converters are the most popular type of ADC converter because of their conversion speed and simplicity. Successive approximation converters operate at a fixed conversion time independent of the amplitude of the input analog signal. The converter operates by approximating the analog input signal with a binary code and successively revises this code for each bit in the code until the best approximation is achieved.

Analog-to-digital converter range and resolution are limited by the number of bits in the converter. Conversions involving more bits require more time and therefore need faster converters if they are to maintain a nominal conversion rate. Conventional 12-bit ADC converters are capable of operating at conversion speeds of better than 1000 samples per second. An N-bit ADC converter has a range of 2^N levels. Therefore the range for a 12-bit ADC is 2^{12} = 4096 discrete levels. A converter's range is sometimes expressed as its dynamic range computed as $20 \log 2^N$ (dB), or $20 \log 4096 = 72$ dB for a 12-bit converter. An ADC voltage resolution depends on its input voltage limits. For example, a 12-bit converter with a –5 V to +5 V input range provides a resolution of (10 V)/(4096 bits) = 2.44 mV/bit.

B. Sampling Theorem

Analog signals must be sampled at a sufficiently high sampling rate in order to completely represent the data while not accumulating so many data that it is difficult to store and handle them. The Nyquist sampling theorem states that if a continuous band-limited signal contains no frequency component greater than f_c, then the original signal can be completely represented if it is sampled at a rate of at least $2f_c$ samples per second. The frequency $2f_c$ is known as the Nyquist frequency. Although completely represented, signals sampled just as the Nyquist frequency are usually not visually satisfactory. Oversampling, or sampling a signal at rates greater than the Nyquist frequency, is often used for sampling data that are intended for visual display.

If an analog signal is undersampled when digitized by using a sampling frequency less than the Nyquist frequency, aliasing can occur. Aliasing confounds the sampled signal with a signal different from the actual signal and results in unreliable data. This phenomenon is sometimes observed in the cinema when the wheels of a moving vehicle appear to be rotating more slowly than they should, or even rotating backwards, because of the limited number of frames per second at which the event is captured relative to the rotating speed of the wheels. Aliasing is easily prevented by sampling using a sampling rate greater than the Nyquist frequency. Sometimes the signal of interest is much less than the total frequency content of the entire signal. Antialiasing filters are nothing more than low-pass filters that attenuate signals greater than the Nyquist frequency in order to prevent aliasing. These filters also help eliminate interference noise at frequencies above the Nyquist frequency that can corrupt digitized data and are recommended whenever analog data are digitized.

C. Digital Filters

Digital filters can implement all analog filter functions (i.e., low-pass, high-pass, bandpass, integration, differentiation) via software. Two advantages of digital filters are that the filters can be adjusted without the need for special hardware and filtering can be performed after the data have been sampled. This is very useful for extracting noise that wasn't evident while the data were being collected. Digital filters can be subdivided into two categories: (1) finite impulse response (FIR) filters, and (2) infinite impulse response (IIR) filters. Digital filters are implemented using difference equations. The speed of the filter is dependent on the complexity and the number of coefficients in the difference equation. The design of digital filters often involves a trade-off between desired performance and speed. In other words, it is possible to design a digital filter with optimal performance, but often the resulting difference equation is computationally inefficient.

Finite impulse response (FIR) digital filters have $M + 1$ coefficients and have the general equation $y_i = b_0 x_i + b_1 x_{i-1} + b_2 x_{i-2} + \cdots + b_M x_{i-M}$, where x_i is the digitized signal input and y_i is filtered output. For example, if a two-coefficient FIR filter has coefficients $b_0 = 1/2$ and $b_1 = -1/2$, then the filter response is $y_i = 1/2\ x_i + 1/2\ x_{i-1}$, which is simply the average of every input data point with the point preceding it. FIR filters are inherently stable, but sharp roll-offs are difficult to achieve with low-order (containing few coefficients) filters.

An infinite impulse response (IIR) filter uses both input and output values and thus has feedback. The general IIR filter has $N+M+1$ coefficients and has the equation $y_i = a_1 y_{i-1} + a_2 y_{i-2} + \cdots + a_N y_{i-N} + b_0 x_i + b_1 x_{i-1} + b_2 x_{i-2} + \cdots + b_M x_{i-M}$. A simple IIR filter has the equation $y_i = (1-a)y_{i-1} + ax_{i-1}$, where $a = 1 - e^{-T/\tau}$, T = sampling period, $t = 1/2\pi f_c$, and fc is the filter corner frequency. This digital filter behaves similarly to a single-pole Butterworth filter.

Integer filters are a special class of filters designed for maximum computational efficiency at the expense of very specific performance limitations. To achieve computational speed, integer filters use only integer coefficients. This allows the computer to perform multiplication via bit shift operations, which are much faster than standard multiply operations. The use of only integer coefficients results in a limited choice in the placement of poles during the filter design.

Standard design methodologies have been developed for many digital filters; they are discussed and applied by Tompkins [4]. It is relatively easy to implement digital filters without an in-depth knowledge of signal processing theory. However, one must be careful to consider the effects of phase, aliasing, and sampling rate when implementing them.

D. Spectrum Analysis

Spectrum analysis views a time-variant signal as a function of frequency. Real-time spectrum analyzers resolve the frequency content of complex signals using a network of narrowband filters tuned to different frequencies. Simple spectrum analysis may be performed using filter networks containing filters that are successively tuned to multiples of the previous filter. This is the kind of spectrum obtained using octave-band and one-third octave-band analyzers.

Digital computers and analog-to-digital converters make it possible for signals to be sampled, digitized, and mathematically transformed as a function of frequency using the Fourier transform. The fast Fourier transform (FFT) is an efficient algorithm commonly used for calculating Fourier transforms. The major limitation of FFT analyzers is that signals may be sampled for only a limited time window. The FFT algorithm assumes that the signal being transformed is periodic and therefore repeats indefinitely, which may not be appropriate in some cases.

Many FFT analyzers contain a fixed number of points resolution for the spectrum; 400 points is typical. The frequency resolution of an FFT analyzer is inversely proportional to the time window for a fixed number of points spectrum. Sample rates are therefore adjusted in proportion to the time window. Consequently a spectrum for a longer time window will contain a finer frequency resolution than one for a shorter time window, but it will have a smaller bandwidth. Consider a spectrum analyzer that has a 1-kHz sample rate for a 1-s time window and a frequency resolution limited to 1 Hz for a 500-Hz spectrum. If the time window were cut in half to 0.5 s, the sample rate would be doubled to 2 kHz, so the frequency resolution would be limited to 2 Hz for a 1000-Hz bandwidth. An increased spectrum resolution would also be obtainable if the number of points were increased without changing the sample rate.

Digital FFT spectrum analyzers often allow averaging repeated samples of the signal. Although FFT analyzers are extremely useful for ergonomics, extreme care must be taken to ensure that the signal sampled is representative of the actual signal under study and that the signal's spectrum doesn't vary over time. When a signal's statistical properties remain the same over time, the signal is said to be stationary.

E. Interfaces

Interfaces are circuits that convert data from an external instrument or circuit into a form suitable for a computer or convert digital data from the computer to a form accessible to an external device.

1. Serial Interfaces

Serial interfaces transmit and receive digital data one bit at a time. Because data bits are sent one at a time, only two data lines are necessary; one line is used for transmission and the other is used for reception. This makes serial transmission cable small and inexpensive. Since data are sent one bit at a time, however, data transmission speed is slow. The RS-232 serial interface convention is the most widely used.

RS-232 serial interface circuits perform two basic operations. First, they convert an 8-bit parallel data word from the computer into a serial data word to be sent to the serial device. Second, they convert the data from a serial device into an 8-bit parallel data word that is transferred to the computer. Because of the extensive use of asynchronous serial communication, a common device performing the parallel-to-serial and serial-to-parallel conversions was developed that is called a universal asynchronous receiver transmitter (UART). The rate at which serial transmission occurs is called the baud rate. This is the number of bits of information that are transmitted during 1 s. If x bits are transmitted during the time interval T_b, then the baud rate is

Baud rate (bits/s) = x/T_b

For proper interpretation of the received data, the receiver must be operating at the same baud rate as the transmitter.

2. Parallel Interfaces

Parallel communication interfaces send and receive data on multiple signal lines at the same time. If a computer commuincates eight bits of parallel data, eight separate lines are needed. Devices that operate using parallel data interfaces include keyboards, printers, and a variety of test and measurement instruments.

The output bits from a digital communication interface may be used for opening and closing circuits. It is necessary to isolate the interface logic circuit from the load the circuit is controlling. Isolation can be accomplished by using either an electromagnetic or a solid-state relay. Some electromagnetic relays, called reed switches, can be activated with the 5-V signal directly from the digital communication interface and require only a few milliamperes of current. Solid-state relays use optoisolation; they contain a light-emitting diode (LED) and a light-sensitive element such as a photocell, photodiode, phototransistor, or light-activated silicon-controlled rectifier in one encased package.

3. IEEE-488 Interface

Many modern electronic instruments are now being equipped with special circuitry for interfacing them with a computer. Such an interface allows a computer to read data directly from the instrument as well as externally control most of the instrument functions. The IEEE-488 standard makes it possible for instruments from numerous manufacturers to be interfaced with a computer. The standard is based on 8-bit parallel data transmission, and it is usually suitable only over short distances from the computer within the same room. Instruments using the IEEE-488 interface have a standard 24-pin connector, often located on the rear panel of the instrument. Many instruments now have this interface as a standard feature or it is available as an option. Interface boards and software for the IEEE-488 interface are available for most microcomputers.

F. Virtual Instruments

Software like LabVIEW is capable of performing many of the functions of hard-wired instruments, such as generating signals, measuring voltages, processing, and displaying signals using a PC. This software allows the computer to control physical instruments through IEEE-488 and RS-232 interfaces and microcomputer data acquisition boards containing analog-to-digital converters, digital-to-analog converters, timing circuitry, and digital I/O ports. Virtual instruments are created using interactive software that provides a graphical user interface (GUI) for simulating the front panel of an instrument on the computer screen, analogous to the front panel of a physical instrument, and the instrument is operated from the keyboard or mouse. Users write programs using virtual instrument software instead of physically interconnecting electronic components. The software provides icons for analog and digital displays, control switches and knobs, and signal processing operations. Icons can be interconnected to create arbitrary new instruments at the desire of the computer operator. Physical instruments containing serial or parallel interfaces for communicating with the computer are represented by instrument driver icons, are accessible and controllable, and can be programmed remotely from the computer. Such software is a powerful tool for controlling hardware and processing and storing data while taking advantage of the flexibility of a computer. An example of an instrument implemented through software as a virtual instrument is shown in Figure 11a. The icon block diagram of the virtual instrument is shown in Figure 11b.

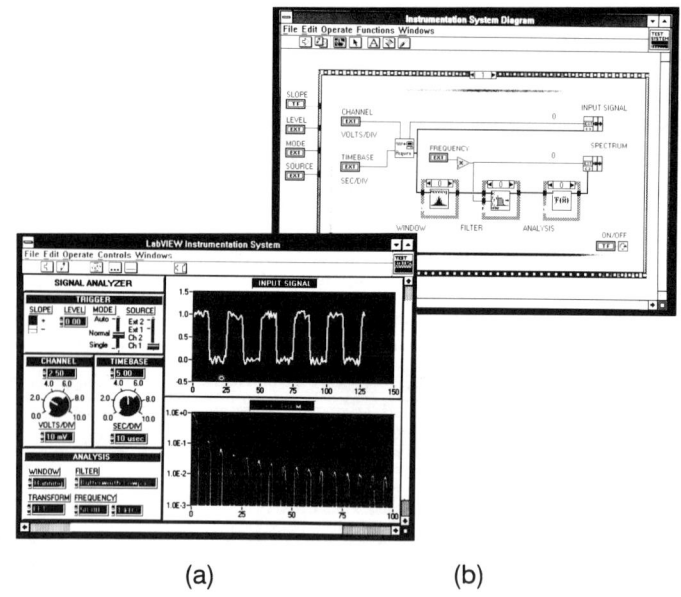

(a)   (b)

Figure 11 (a) A virtual instrument implemented with LabView software. (b) Icon block diagram. (Courtesy of National Instruments.)

VI. INSTRUMENTS IN INDUSTRIAL ERGONOMICS PRACTICE AND RESEARCH

Various instruments are used by ergonomics practitioners and researchers. Often a specific measurement need necessitates selecting an instrument. The following discussion describes some common instruments used in ergonomics and their principles of operation. An understanding of particular instrumentation benefits and limitations will result in more accurate measurements and selection of the instrument most suited for a specific need.

A. Human Kinematics

Kinematic measurements in ergonomics include linear and angular displacements, velocities, and acceleration. Movement is often measured relative to an anatomical landmark such as a body segment, joint, or other anatomical prominence, using a body landmark, or with respect to an external reference.

1. Acceleration

Acceleration is usually measured using devices called accelerometers. An accelerometer consists of a small mass and a sensor that measures the force associated with the acceleration of that mass. Because the mass is constant, the force sensor voltage is proportional to acceleration. Today's accelerometers are small enough and light enough to attach directly to the limbs for measuring body motions. Accelerometers are usually sensitive to movement in only one direction. Triaxial accelerometers can be used to measure movement in any direction by resolving the motion into three orthogonal components. These are simply three accelerometers mounted at right angles to each other.

2. Velocity

Velocity is seldom measured directly but can be calculated through integration of acceleration measurements or differentiation of displacement signals (see below).

3. Displacement

Although accelerometers can be used for measuring displacement by integration, direct-displacement measurements available for displacement are more accurate and include optical methods, electroacoustic instruments, and electrogoniometers.

Accelerometers for Displacement Measurements. Double integration of the acceleration data will measure displacement but constant errors will compound in the integration. Displacement can also be calculated by integrating velocity data once. Piezoelectric accelerometers are usually satisfactory for measuring only changing acceleration signals and are not adequate for constant acceleration. Consequently, constant velocity and displacement cannot be measured using piezoelectric devices.

Optical Motion Analysis. Optical displacement measurement systems capture movements using stroboscopic photographs, cinematography, video, or some other optoelectrical device. In the simplest form, body segments can be illuminated using reflective markers located on the joints or limbs for producing bright points or lines. Movement can be measured simply by using multiple-exposure images from stroboscopic photographs and manually measuring displacement or inputting the data using a computer digitizing tablet.

Computerized motion analysis systems, such as Selspot, contain light-emitting diode (LED) markers affixed to anatomical landmarks that are flashed and picked up by a special camera that locates the x and y coordinates of the specific marker that is flashing and sends the information to a computer. Since the LED markers are illuminated sequentially, the computer is able to track a large amount of markers, typically 30 or more.

Video motion analysis uses a conventional video camera mounted in a fixed location for recording brightly illuminated reflective markers attached to anatomical landmarks. The video signal is recorded on tape for off-line analysis or directly digitized using video capture circuitry into a pixel image for a microcomputer. Either an observer or software is used for identifying the high-contrast illuminated markers, and locations of x and y coordinates of the markers are measured as pixels relative to a calibrated image. Since markers have to be tracked in space, these systems are limited in their number of markers.

Three-dimensional motion is possible by using two cameras located at known angles. Additional cameras are needed for tracking markers when they may become hidden from one camera's view. As many as five or more cameras may be used for motion analysis.

Electrogoniometers. Electrogoniometers are instruments that can directly measure the angular displacement of a joint. They are usually made from potentiometers, angle encoders, or strain gauges attached to body segments. Electrogoniometers range in complexity. The best instruments account for nonlinearities of joint rotation and have little cross-talk between multiarticular joints. Electrogoniometer signals can be directly recorded or sampled by an ADC, and they don't have a problem with visual obstructions, but they do have some limitations. They may interfere with normal movement patterns by limiting the range motion or restricting movement. A commercially available lightweight strain gauge type of electrogoniometer is shown in Figure 12.

B. Force Measurements

1. Mechanical Force Transducers

Simple spring scales are useful for measuring static forces in the field, such as the weight of objects or constant loads. A push-pull spring force transducer with a variety of attachments

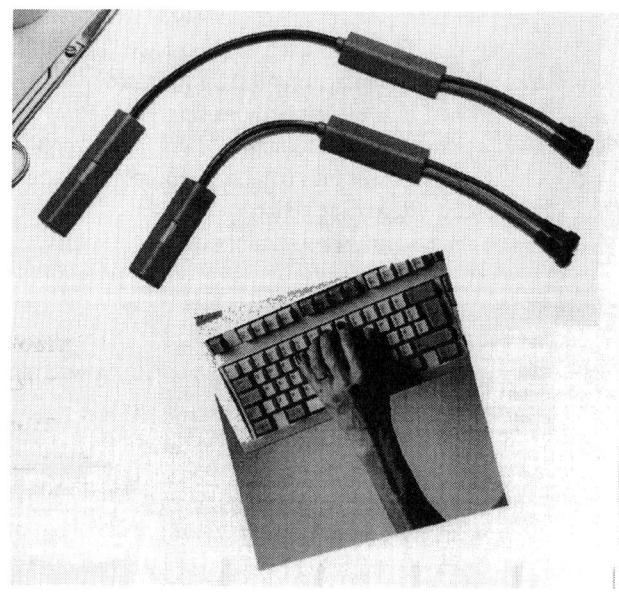

Figure 12 A lightweight strain gauge type of electrogoniometer shown mounted on the wrist. (Courtesy of Penny and Giles).

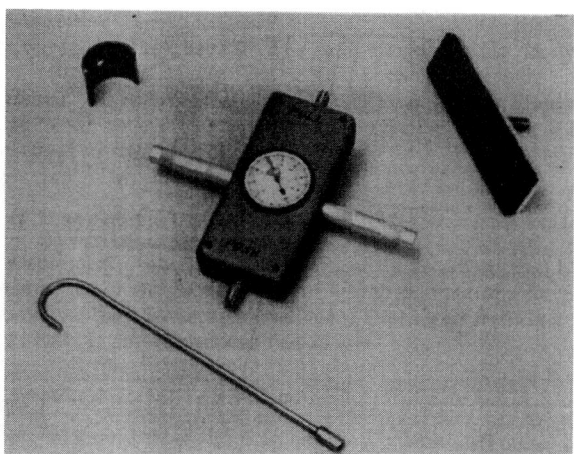

Figure 13 Spring-type force transducer with a variety of attachments. (Courtesy of Eastman Kodak [5].)

is shown in Figure 13. But because spring scales behave as second-order systems with resonant frequencies often less than 1 Hz, they are unsuitable for measuring varying forces. To measure a force that changes with time, electronic load cells are used. Because springs can become permanently deformed when stretched beyond their elastic limits, their calibration should be periodically tested using known loads.

2. Strain Gauge Load Cells

A simple load cell consists of a metal beam with strain gauges bonded on the sides. Commercial load cells of various types are available for full-scale loads of a few grams up to hundreds of thousands of newtons, often in several degrees of accuracy and price ranges.

A difficulty with strain gauge load cells is that the measured force is limited to a single point on the beam or plate. Forces applied at points further away from where the strain gauge is mounted produce greater moments than forces applied at a point closer to the strain gauge. To get consistent and repeatable force measurements, the same point of force application must be maintained. Radwin et al. [6] built a strain gauge instrument that linearly summed forces applied at multiple locations along the length of a beam that was independent of the point of application. Such an instrument is useful for measuring forces that are summed, such as grip force produced by several fingers. This type of strain gauge instrument can be used as a handle for measuring hand force while operating tools as equipment (see Figures 14a and 14b).

3. Piezoelectric Load Cells

Rapidly changing forces can be measured using piezoelectric force sensors. Because of their small size, many commercial piezo load cells have the necessary support electronic circuit built into their already small package. A single piezo load cell is useful over a very wide range of forces because the 1% nonlinearity applies to any calibration range. Piezoelectric load cells can measure forces as great as 16,000 N. Because piezoelectric load cells are very stiff, they are suitable for measuring isometric forces. Because they respond only to changing and impulsive forces, piezoelectric load cells are unsuitable for measuring steady-state forces.

4. Conductive Polymer Force Sensors

A durable and thin conductive polymer sensor was found useful for measuring external forces when conventional force sensors are too large. Because of their small size and durability, these sensors can be attached to the fingers and hands [7]. The conductive polymer sensing elements

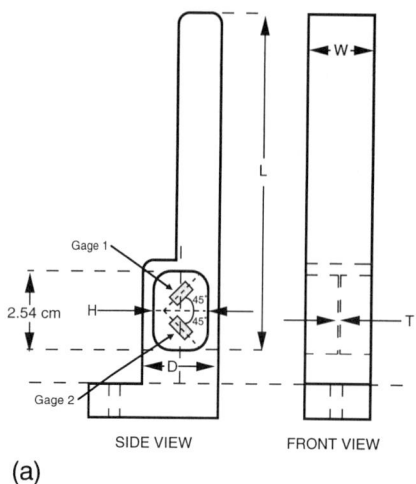 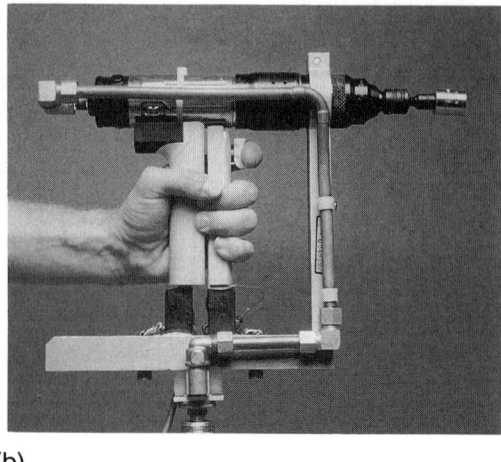

Figure 14 (a) Diagram of a shear stress strain gauge load cell. (b) The shear stress strain gauge load cell mounted on a pistol-grip pneumatic hand tool. (Courtesy of Radwin et al. [6].)

are composed of two conducting interdigitated patterns deposited on a thermoplastic sheet facing another sheet that contains a conductive polyetherimide film. As applied force increases, the two layers compress together, increasing the contact area and decreasing its electrical resistance. A dome for distributing force over the active sensing area is necessary for these elements to operate as force sensors. Without the dome the measurements are erroneous. These sensors are very limited; their useful range is up to 30 N with an accuracy of 1 N; however, there are few alternatives available for directly measuring finger and hand forces. A schematic diagram of this type of force sensor is shown in Figure 15a. A force–time plot for force measurements from four fingers during the lifting of a box is shown in Figure 15b.

5. Force Platforms

Force platforms are used for measuring forces in two or more directions, such as ground reaction force acting on the feet during standing or walking. Ground reaction force contains a vertical component plus two shear components acting along the surface [8]. This measurement is accomplished by using three or more strain gauge piezoelectric, piezoresistive, or capacitative force sensors that are arranged at right angles to each other. A common force plate configuration contains a flat plate supported by four triaxial force transducers (see Figure 16).

C. Electromyogram

Electromyograms (EMGs) are recorded using indwelling electrodes or surface electrodes. Concentric needle electrodes have a central insulated wire surrounded by a 200-µm-diameter cannula. The tip is beveled for insertion. Insulated wire electrodes are inserted through a carrier needle, which is then withdrawn. Both types permit superior localization near the tip, but they are invasive, making them unsuitable for field use. Surface electrodes are more practical for industrial ergonomics applications outside of the laboratory. A popular commercially available surface electrode is the Beckman silver–silver chloride miniature electrode. It is small (about 10 mm in diameter), light (250 mg), and reusable [9].

Being very small in amplitude, ranging from less than 50 µV up to 5 mV, EMG signals have to be amplified. A differential instrumentation amplifier with a gain greater than

Instrumentation

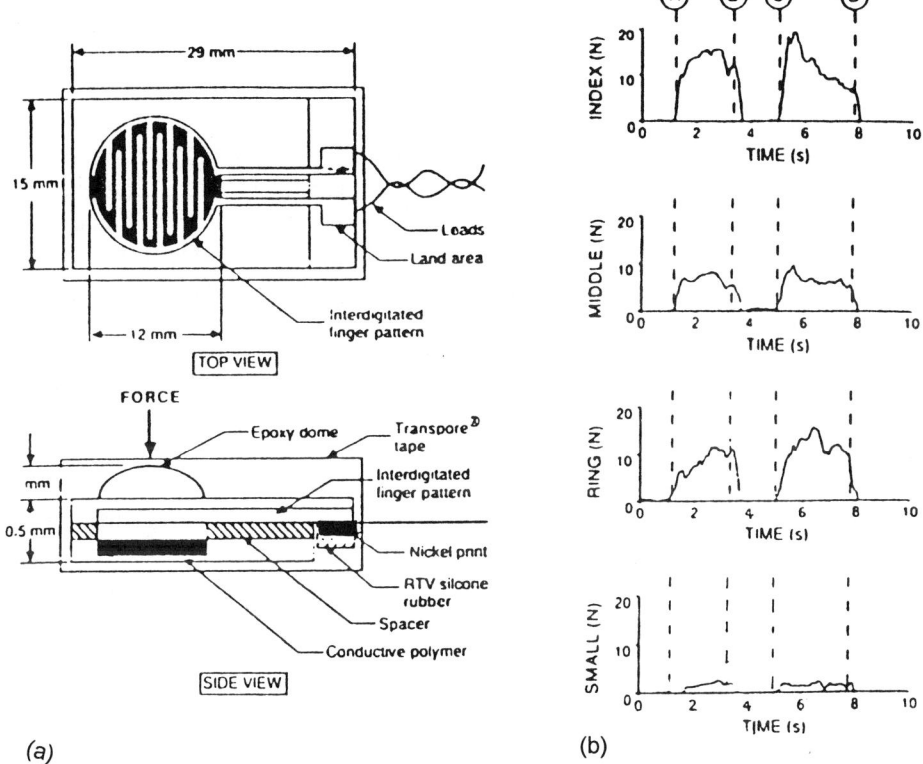

Figure 15 (a) Conductive polymer force sensor; (b) finger forces measured during a box-lifting task using the conductive polymer force sensors. The box was lifted from the floor at A, placed on a table at B, lifted from the table at C, and placed back on the floor at D. (Courtesy of Jensen et al. [7].)

1000 is commonly used. Since the input impedances for differential amplifiers used for measuring electromyograms are never perfectly balanced, objectionable interference from power line sources is inevitable. This can sometimes be minimized by shielding leads and grounding each shield. Lowering skin impedance and using very high input impedance amplifiers is

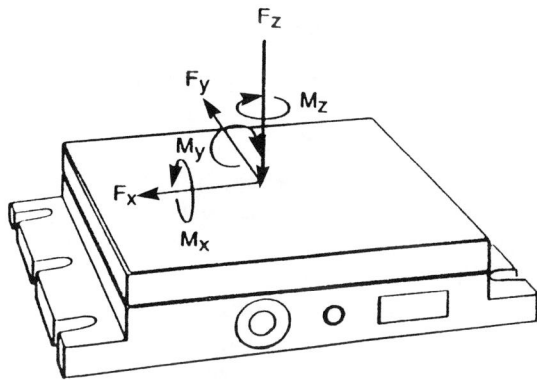

Figure 16 A force platform for measuring the three orthogonal force vectors (F_x, F_y, F_z) and three moments (M_x, M_y, M_z) about a single force vector applied to the surface.

also helpful. Skin impedance is lowered by cleansing the skin with rubbing alcohol and using conductive electrode gel. The EMG amplitude is greatly affected by electrode placement and skin impedance.

The bandwidth for raw EMG signals is less than 500 Hz, with most of the power at frequencies less than 100 Hz. Integrated or rms electromyograms are obtained by passing the EMG signal through a rectifier and an integrator circuit or rms detector. The signal is usually integrated over a specific time period, such as 100 ms, and then integration begins again. The time integral has a linear relationship with muscle tension under isometric conditions. Sometimes the EMG signal is rectified and low-pass filtered so that the output has a deliberate change with frequency. The resulting envelope of the EMG signal is related to muscle activity and has a much lower frequency content than the raw EMG (less than 10 Hz). This enables the signal to be recorded in conjunction with the development of force produced from the muscle activity.

D. Energy Expenditure

Heart rate is sometimes used for approximating energy expenditure. The respiratory demands of physical work, however, are best assessed by measuring the fractions of oxygen and carbon dioxide in expired air. Ventilation rate can be measured by exhaling through a gas meter that mechanically measures the volume of breath that accumulates over a fixed time period, typically 1 min. A face mask is usually worn as the subject performs manual tasks. Oxygen consumption and carbon dioxide production are measured from the collected air.

1. Heart Rate

Heart rate is usually measured using electrocardiograph (ECG) signals and a cardiotachometer. ECG electrodes are similar to EMG electrodes and amplified using similar amplifiers. Interference from contracting muscle EMG signals can sometimes be picked up by ECG electrodes. This problem is sometimes minimized by relocating the electrodes. Pulse rate monitors can continuously monitor heart rate using a photooptical pulse sensor clipped to the earlobe.

2. Ventilation

Rotating vane flowmeters contain a small turbine in the flow path air. The number of revolutions of the rotating turbine is proportional to the volume of air flowing past it. Rotation is measured using electromechanical sensors or the interruption of a light beam using a photosensor.

3. Oxygen and Carbon Dioxide Concentrations

The percentage of expired CO_2 can be measured by using an infrared detector or mass spectrometer. Mass spectrometers are instruments that produce a stream of ions from a substance and separate the ions into a spectrum according to their mass-to-charge ratios. The instrument determines the partial pressure or molar fraction of each type of ion present for O_2, CO_2, N_2, etc. Some systems can measure as many as eight component gases. Mass spectrometers are very versatile but quite expensive. Infrared analyzers are more cost-effective and are usually suitable for ergonomic applications. Most gases absorb varying levels of infrared light (3–30 μm wavelength) at distinct characteristic wavelengths. Hence the concentrations of components of a gas mixture can be measured by the transmission of infrared light. When an instrument measures transmission for a well-defined set of wavelengths selected for particular gases, such as CO_2, it is known as a nondispersive infrared (NDIR) analyzer.

The polarographic P_{O_2} electrode is based on an oxidation reaction that occurs at the Ag-AgCl anode and a reduction reaction that occurs at the glass-coated Pt cathode [3]. A polarizing voltage sufficient to drive the reaction (600–800 mV) produces a current linearly proportional to the number of O_2 molecules in solution. A membrane permeable to O_2 and other gases separates the electrode from the surroundings. The electrode is zeroed using a CO_2/N_2 gas and calibrated using two gases containing known concentrations of O_2. The reaction partially consumes O_2 and is very sensitive to temperature. The amount of oxygen consumed is measured from the difference between the amounts of oxygen inspired and expired. Expired gas collected in a mixing chamber or in a sample bag is analyzed for P_{O_2}. Portable battery-powered measurement systems, such as the Oxylog, are available for measuring oxygen uptake of mobile subjects.

E. Vibration

Since vibration is oscillatory motion, it is most conveniently measured by using an accelerometer. Accelerometers should be selected carefully. It is important that the size and mass of an accelerometer be small enough not to interfere with the measurement by loading. Accelerometers weighing more than 15 g are unsuitable for vibration measurements made by mounting them on a human body. Acceleometer sensitivities are proportional to their mass; the smaller the accerlerometer, the less sensitive it is. It is important that an accelerometer not be excited by frequencies near or at its resonant frequency. This would produce erroneously large measurements. Accelerometer resonant frequencies vary in inverse proportion to their mass. Accelerometers are also influenced by temperature changes, humidity, and other harsh environmental conditions.

1. Whole Body Vibration

The frequencies of interest for whole body vibration are dc to 80 Hz. Piezoresistive accelerometers are best suited for whole body vibration measurements because they operate at dc and low frequencies. Furthermore, piezoresistive accelerometers are self-calibrating because they can be oriented in the direction of gravity for an input equivalent to the acceleration of gravity and then turned 90° to gravity for removing gravity from the accelerometer.

Accelerometers are mounted on the body near bony eminences and surfaces. A triaxial seat disk accelerometer may be inserted between a vehicle seat and a passenger's buttocks. Vibration at the feet of a vehicle passenger can be measured by mounting an accelerometer directly to the floor.

2. Hand-Arm Vibration

Hand-arm vibration usually is transmitted through the handles of manually operated equipment such as powered hand tools. Piezoelectric accelerometers are often used for hand-arm vibration measurements. Triaxial acceleometers can be mounted directly on a tool handle for measuring vibration along orthogonal axes. Accelerometers are usually mounted directly to the handle using a hose clamp or similar strap or by welding a mounting block with a stud for the accelerometer.

3. Vibration Analysis

Human vibration measurements are usually recorded and spectrum analysis is used for identifying the frequency characteristics of vibration exposure. Vibration exposure time is determined by measuring the duration of tool operation during a suitable sample period.

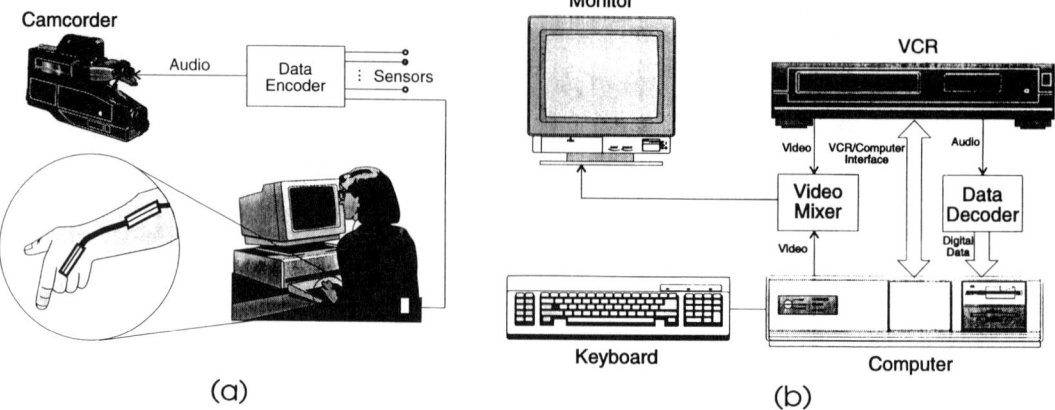

Figure 17 (a) Video-based data collection system. (b) Interactive multimedia data extraction and analysis system.

F. Temperature and Humidity

Electronic thermometers based on thermocouples and thermistors are commercially available and provide a calibrated temperature output. Rise time, size, long-term stability, and sensitivity should all be considered when choosing a temperature sensor. Thermal expansion thermometers, such as the mercury column variety, typically have a rise time of 10 s, whereas a thermistor may have a rise time of 1 s and a thermocouple may have a rise time on the order of 0.05 s.

Electronic sensors are capable of determining relative humidity. Resistive humidity sensors use materials whose resistance decreases when they absorb moisture. Capacitive humidity sensors use materials for which the dielectric constant (and thus capacitance) changes with moisture uptake.

G. Integration of Instrument and Multimedia

The effective integration of measurements using multiple instruments with arbitrary work activities can be overwhelming. A large amount of collected data requires an efficient way to extract meaningful information. Radwin and Yen [10] developed a method for integrating the data collected from transducers with the video image recorded from a camcorder. An example of this system is shown in Figure 17. Analog signals from up to 32 transducers (electrogoniometers, force sensors, EMGs, etc.) are digitized, coded, and recorded directly onto the audio track of a VHS tape in synchronization with video images of the subject performing the operation. The data and video are analyzed using an interactive computer-controlled video-based analysis system. Data segments are extracted from the tape based on the elemental breakpoint for the task as determined off-line by the analyst. The videotape can be viewed in real time, slow, faster, or frame by frame in either the forward or backward direction, allowing flexible breakpoint assignments. Spectral analysis was used for quantifying postural repetitiveness, sustained posture, and postural deviation magnitude [11,12].

REFERENCES

1. J. C. Yoon and R. G. Radwin, The accuracy of consumer-made body measurements for women's mail-order clothing, *Hum. Factors 26*: 557–568 (1994).
2. E. O. Doebelin, *Measurement Systems: Application and Design*, 4th ed., McGraw-Hill, New York, 1990.
3. J. G. Webster, Ed., *Medical Instrumentation: Application and Design,* 2nd ed., Houghton Mifflin, Boston, 1992.
4. W. J. Tompkins, Ed., *Biomedical Digital Signal Processing: C-Language Examples and Laboratory Experiments for the IBM PC*, Prentice-Hall, Englewood Cliffs, NJ, 1993.
5. Eastman Kodak Company, *Ergonomic Design for People at Work,* Vol. 2, Van Nostrand Reinhold, New York, 1986.
6. R. G. Radwin, G. P. Masters, and F. W. Lupton, A linear force-summing hand dynamometer independent of point of application, *Appl. Ergon. 22*: 339–345 (1991).
7. T. R. Jensen, R. G. Radwin, and J. G. Webster, A conductive polymer sensor for measuring external finger forces, *J. Biomech. 24*: 851–858 (1991).
8. D. A. Winter, *Biomechanics and Motor Control of Human Movement,* Wiley, New York, 1990.
9. C. J. De Luca, Electromyography, in *Encyclopedia of Medical Devices and Instrumentation*, J. G. Webster, Ed., Wiley, New York, 1988, pp. 1111–1120.
10. T. Y. Yen and R. G. Radwin, A video-based system for acquiring biomechanical data synchronized with arbitrary events and activities, *IEEE Transactions on Biomedical Engineering 42*(a): 944–948 (1995).
11. R. G. Radwin and M. L. Lin, An analytical method for characterizing repetitive motion and postural stress using spectral analysis, *Ergonomics 36*(4): 379–389 (1993).
12. R. G. Radwin, M. L. Lin, and T. Y. Yen, Exposure assessment of biomechanical stress in repetitive manual work using frequency-weighted filters, *Ergonomics 37* (12):1984–1998 (1993).

8
Occupational Heat Stress

Thomas E. Bernard
College of Public Health, University of South Florida, Tampa, Florida

I. INTRODUCTION

Heat stress is basically the combined effects of climatic conditions (called the environment for simplicity), metabolism, and clothing. The obvious environmental factors are air temperature, humidity, radiant heat, and air movement. Metabolism is associated with the demands to perform work, and the concern in heat stress is the rate of heat generated by metabolic processes. Clothing material, construction, and usage affect the exchange of heat between the body and the environment.

Heat strain includes changes in body temperature, heart rate, and sweating. During exposures to heat stress, the temperature of most body compartments increases, and these changes facilitate the removal of heat from the body. The so-called core temperature is an important measure of heat strain. Blood flow moves heat from the deep body tissues, including working muscles where it is generated, to the skin, where it is dissipated. Heart rate reflects the cardiovascular response to blood flow requirements and is a sensitive indicator of physiological strain due to both work demands and heat stress. Finally, sweating ensues to provide for evaporative cooling of the body.

In summary, as muscles work, they metabolize substrates to supply energy, they generate heat, and they require removal of metabolic by-products. The rate of blood flow through the muscles is in proportion to the metabolic demands. The heat generated in the muscles is distributed by the blood, and the temperature of the body increases. The heat in the core compartment is moved to the skin by additional blood flow that is directly proportional to the metabolic rate and inversely proportional to the difference between the core and skin temperatures. There may be an additional heat load due to hot air and hot surfaces, and this gain is influenced by air temperature, humidity, and motion, by the temperature of the surroundings, and by the clothing worn. Heat is removed from the body primarily by evaporation of sweat, and the rate of evaporation is influenced by the ambient humidity and air motion as well as by the type and configuration of the clothing.

The evaluation of heat stress by the application of heat stress models and heat strain criteria are the primary emphasis of this chapter, which also provides a framework for viewing the principles of recognition and control. The goals of this chapter are to describe a basic model of heat stress, outline heat strain in response to the stress, describe ways that heat stress may be recognized in the workplace, discuss two conventional methods of heat stress assessment in industrial settings, highlight alternative assessment methods based on physiological response, and present an overall framework for the management of heat stress.

II. BACKGROUND AND SIGNIFICANCE TO OCCUPATIONAL ERGONOMICS

Fundamentally, ergonomics is the study of fitting a task to workers to enhance their performance and protect their well-being. To fit a task requires either (1) a model for the stressors that reasonably predicts the resulting strain or risk of disorders or (2) a method of evaluating the strain. By managing the stress or strain, performance is enhanced and the risk of disorders is controlled to within acceptable limits. Heat stress becomes interesting in the application of occupational ergonomics when the factors of environment, work, and clothing come together to influence an individual's capacity for work and risk of suffering heat-related disorders. Occupational heat stress is well enough understood to use either stress models or strain evaluation for design of work.

III. OCCUPATIONAL HEAT STRESS

The management of occupational heat stress requires an understanding of the factors of heat stress and how it may affect a worker. The three factors of environment, metabolic rate, and clothing combine to dictate the rate of heat gain and the amount of heat loss required to protect the worker.

Management of heat stress is directed to two fronts: general controls and job-specific countermeasures. General controls include training, heat stress hygiene practices, and medical surveillance. The job-specific controls are directed toward containing or managing the risk of specific exposures.

A. Model of Heat Stress

A description of heat exchange between the body and the environment is a frequently used means to qualitatively and quantitatively express the factors that affect the level of heat stress. The heat balance equation may be reported in different ways, but they have the same fundamental structure [1,2]:

$$S = (M - W) + C + R + E + K + (C + E)_{resp} \tag{1}$$

Each term in the equation represents a rate of heat flow for a different modality. By convention, positive values mean that the flow is toward the body, and negative, away from the body. In other words, a positive value means that there is a tendency for body temperature to increase, and a negative value means that there is a tendency for it to decrease.

S is the rate of heat storage in the body. As heat is stored, the temperature of the body increases. Over the course of a day, there is no net storage of heat, but over shorter periods of time there will be increases due to heat stress exposures. Under extreme conditions, the rate of storage may reach 500 W.

$M - W$ is the net heat gain due to metabolic rate (M) less the rate of external work accomplished (W). Because external work is only about 10% of the total metabolic rate [3], W is usually ignored. M is the internal source of heat during heat stress and is always positive and proportional to work rate. When no work is being performed, the internal heat generation is about 100 W, but this can easily increase to 500 W during heavy work.

C is the convective heat flow between the body surface and the surrounding air. It is directly proportional to the temperature gradient between the skin and air and to air speed. The proportionality constant is modified by clothing. C is positive when air temperature is greater than mean skin temperature (nominally 36°C in heat stress) and negative with lower air temperature. Whether the convective heat transfer is positive or negative, the value does not often exceed 50 W.

R is the radiant (infrared) heat flow between the body surface and the solid surroundings. It is proportional to the temperature gradient between the skin and the average surface temperature of the surroundings. Like convection, the proportionality constant is modified by clothing. R is positive when the average temperature of the surroundings is greater than skin temperature. The value of radiant heat transfer does not often exceed 50 W.

E is the rate of heat loss from the evaporation of sweat. The skin is wetted by water from the sweat glands, and as the water evaporates the energy required for vaporization is taken from the skin. In this way, the body is cooled. The value of E in Eq. (1) is always negative, but it is often reported without the minus sign. (Sometimes the heat balance equation is reported with a minus sign in front of E to emphasize the fact that evaporation is a loss. This contradicts the sign convention presented above.) Evaporative cooling is generally under physiological control and is adjusted to minimize storage. E can be limited by the environment and clothing as well as by physiological capability. Looking at the environment, the maximum rate of cooling is proportional to air speed (up to 2 m/s) [4] and to the difference in water vapor pressure on the skin and in the air (humidity). The proportionality constant is modified by clothing. When the environmental and clothing conditions are favorable for evaporative cooling, the rate of cooling may be physiological control and can range from 0 to a physiological limit of –600 W.

K represents heat conduction from direct contact between the body and a solid surface (with the possibility of intervening clothing). Large contact surface areas are required before conduction becomes an important avenue of heat gain or loss in industrial evaluations of heat stress. Significant heat flow rates over smaller surface areas are more likely to affect local tissue by inducing extreme discomfort, pain, or a burn [5]. Conduction as a factor in heat stress will not be considered further in this chapter.

$(C + E)_{resp}$ represents the heat exchange in the respiratory track due to convection and evaporation. Although these paths are important under sedentary conditions, they represent a minor pathway of heat exchange for industrial heat stress. For the sake of discussion, this term will be ignored.

Equation (1) is reduced to the form

$$S = M + C + R + E \qquad (2)$$

That is, the rate of heat storage is equal to the internal heat generation plus the gains or losses from convection and radiation plus the loss due to sweat evaporation.

The heat balance expressed by either Eq. (1) or Eq. (2) is a comprehensive means of describing the level of heat stress because the environment, work, and clothing factors can be accounted for, at least in theory. Recall that the major source of heat gain is the metabolic heat generated by performing the work and that the major avenue of heat loss is evaporation. Although radiation and convection are important, they have a much smaller contribution. The International Organization for Standardization (ISO) has stipulated a method to quantify the values in Eq. (2), and this method will be addressed in the evaluation section [2].

Ideally, heat stress is negligible if there is no storage of heat in the body ($S = 0$). In this case, the required level of evaporative cooling (E_{req}), which is under physiological control, is determined from the equation

$$E_{req} = -(M + C + R) \qquad (3)$$

Sometimes the value of E_{req} is greater than the maximum evaporative cooling that can be sustained physiologically or by the combination of clothing and environment. In this case, there is a net gain in heat (positive value for storage rate), and an acceptable increase in core temperature is usually specified ($\Delta T_{re, max}$). The maximum acceptable amount of heat storage (H_{max}) then follows from the equation

$$H_{max} \text{ [W–h]} = 0.75 \times \text{body weight [kg]} \times \Delta T_{re,max} \text{ [°C]} \qquad (4)$$

Then the safe exposure time (t_{max}) can be estimated from the storage rate and the maximum allowable increase in heat storage,

$$t_{max} \text{ [h]} = H_{max}/S \qquad (5)$$

For example, if $S = 200$ W and $H_{max} = 100$ W–h, then t_{max} would be 0.5 h (30 min). This is the evaluation principle behind time-limited heat stress exposures. The greater the rate of storage, the shorter is the safe exposure time for the same storage limit.

B. Description of Heat Strain

Physiological strain comprises the body's responses to heat stress. One response is an increase in body temperature as metabolic rate increases and heat is moved. Another is the cardiovascular response that is necessary to support both the metabolic demands and the demand to move heat from the core to the skin. The sweating response is necessary to support evaporative cooling.

Starting with the generation of internal heat from metabolism, there is an initial tendency to increase body core temperature at the onset of work, even in thermally neutral environments. This is followed by a leveling off once the necessary adjustments are completed. The amount of storage and the related core temperature depend primarily on the individual and the level of metabolism. The controlled level of storage is marked by the work-specific core temperature ($T_{core-work}$). A simple relationship that describes this increase is given by the equation [6].

$$T_{core-work} \text{ [°C]} = 36.5 + 3.0 \, fV_{O_2, max} \qquad (6)$$

where $fV_{O_2, max}$ is the fraction of the individual's maximum aerobic capacity that the work metabolism represents. For a given work demand, $T_{core-work}$ will increase with decreasing fitness levels of the workers, or for a given worker $T_{core-work}$ will increase with increasing demands.

Equation (6) is predictive of $T_{core-work}$ for a wide range of environments. If the environmental conditions begin to limit the ability of the physiological systems to remove the metabolic heat, then T_{core} will increase out of proportion to the work demands [7,8]. Evaluation schemes for long-term exposures to heat stress attempt to limit the increase in core temperature to roughly that dictated by the work, that is, without environmental heat stress [9].

So core temperature can be assessed as a measure of physiological strain in response to heat stress. The World Health Organization (WHO) suggested that long-term exposures to heat stress should be limited in such a way that core temperature does not exceed 38°C [10]. However, WHO recognized that there can be transient responses greater than 38°C that are safe for intermittent periods. For controlled and well-monitored exposures, 39°C is acceptable.

Blood flow from the core tissues to the periphery is responsible for moving the metabolic heat to the skin, where it can be dissipated to the environment. As a normal response to the metabolic demands of work, the cardiovascular system adjusts blood flow to the working muscles in proportion to the work. In addition, cardiac output must increase to meet the needs of heat transport from the deep body tissues to the skin, where the heat can be dissipated from the body. Under some circumstances, the gradient between the core and skin temperatures decreases. When this happens, more blood must be delivered to the skin to maintain the rate of heat transport. As an index of cardiovascular response to work and heat stress, heart rate is a valuable measure [11].

Several guideposts are available to judge whether heart rates are excessive. One is a daily average below 110 beats per minute (bpm) [10]. A second threshold on heart rate is a sustained rate (about 1 min) greater than 90% of the maximum heart rate [12]. (A popular rule of thumb for estimating maximum heart rate is 220 − age.) A third guidepost is recovery heart rate [13,14]. The heart rate at 1 min of recovery ($HR_{rec,\ 1\text{-min}}$) should be less than 110 bpm to avoid excessive cardiovascular strain [15].

Sweating is the remaining component of heat strain to be discussed in this chapter. Sweat rate is adjusted physiologically to a level necessary to support the required amount of evaporative cooling. The rate of sweating is limited either by the environment or by physiological capacity. Acclimatization to heat stress is a physiological adaptation to repeated heat stress exposures that increases the individual's capacity to secrete sweat and thus increases the rate of evaporative cooling [16]. It has the beneficial effects of lowering the levels of core temperature and heart rate for the same level of heat stress.

In terms of physiological strain, sweating is important as a potential cause of dehydration. Sweat losses on the order of 1 liter/h or a total of 5 liters in a day are limits within which most workers can replace the water lost to sweat. If the losses exceed these values or if there is not adequate water replacement, then dehydration can occur. Marked changes in physical work capacity are noted when dehydration exceeds 1.5% of body weight [17].

If the physiological strain, which is a normal response to heat stress, becomes excessive for an individual, then that individual has an increased risk for a heat-related disorder. Major heat-related disorders are described in the following list along with their symptoms and signs. The first three are a progression of events from a simple, subclinical dehydration to a serious medical emergency. The other three are common disorders associated with heat stress.

1. *Dehydration.* Dehydration is an excessive loss of water due to either inadequate replacement or an illness.

Symptoms.   No early symptoms (< 1.5%); fatigue/weakness; dry mouth
Signs.   Loss of work capacity; increased response time

The signs and symptoms start when dehydration exceeds 1.5% of body weight, and they become more severe as the dehydration increases. Dehydration over 5% is incapacitating. The most harmful consequence of dehydration is the increased risk for heat exhaustion and heat stroke.

2. *Heat exhaustion.* Heat exhaustion is marked by lowered work capacity and inability to continue. It is due to a reduced cardiovascular capacity associated with decreased systemic pressure due to vasodilation or advanced dehydration.

Symptoms: Fatigue; weakness; blurred vision; dizziness, headache
Signs: High pulse rate; profuse sweating; low blood pressure; insecure gait; pale face; collapse; normal or slightly elevated body temperature

3. *Heat stroke.* Heat stroke is a medical emergency. It is the result of a fundamental breakdown in the thermoregulatory center that leads to very high core temperatures and subsequent effects on a broad range of organ systems. The usual causes are subnormal tolerance, lack of acclimatization, excessive exposure to heat stress, or drug or alcohol abuse.

Symptoms: Chills; restlessness; irritability

Signs: Euphoria; disorientation; erratic behavior; unconsciousness; convulsions; red face; hot, dry skin (usual); shivering; collapse; body temperature $\geq 40°C$

    4. *Heat syncope*. Heat syncope is a faintness experienced when there is a significant drop in blood pressure, usually occurring when there is a sudden change in posture to an upright position or after prolonged maintenance of an upright posture. Under these circumstances, blood pools in the lower extremities and in the skin, where the blood vessels are dilated due to heat stress. This reduces blood flow to the head.

Symptoms: Blurred vision (grey-out); fainting (brief) (blackout)
Signs: Near-fainting behavior; brief fainting; normal temperature

    5. *Heat cramps*. Heat cramps are muscle cramps associated with work under conditions of severe heat stress. They are sometimes attributed to electrolyte imbalances. The cramps will occur in fatigued muscle or the abdominal muscles, and the cramping may occur during work or after hours.

    6. *Heat rash*. Prolonged, uninterrupted sweating along with inadequate cleaning of the skin may lead to an inflammation of the sweat gland ducts, thus reducing the amount of sweat secretion. The result is a skin rash marked by small red eruptions and itching sensations.

C. Recognition of Heat Stress in the Workplace

Recognition is the first step in the management of industrial hazards, and heat stress is not an exception. There are several perspectives from which someone may recognize that heat stress is an important factor in the performance of work. These perspectives include contributors in the workplace, worker behaviors, medical surveillance, and physiological sampling. Many times recognition is a qualitative judgment, and that is the way it is treated here. If there is reason to believe that sufficient heat stress is present to be a hazard, then evaluation is required.

    The first perspective is recognition of the contributing factors to heat stress hot environments, high work demands, or protective clothing. These are the very factors that are considered in the quantitative evaluation of heat stress. Generally, if the environment is commonly recognized as being hot, heat stress is likely. Sometimes the role of metabolic rate in heat stress is underestimated. If the metabolic rate is moderate or high, heat stress may be present in environments that are judged comfortable by less active people. Clothing is also an important factor. Most people recognize that vapor-barrier clothing may cause significant heat stress in cool environments. To a lesser degree, other kinds of clothing will reduce the rate of evaporative cooling and thus increase the level of heat stress. Any time clothing is worn that is not constructed of lightweight cloth, the potential for heat stress should be considered. A clear sign that heat exposures are significant is when the clothing is soaked with sweat.

    Worker behaviors provide a second perspective on recognizing heat stress. These may be observed during a walk-around or through discussions with employees and supervisors. One behavior is to modify clothing materials or the way clothing is worn to enhance evaporative cooling, and another is to seek shade or other cooler locations. A second group of employee behaviors is directed toward reducing the metabolic rate by increasing the number of breaks, slowing the pace of work, not following safe work practices, and neglecting maintenance. Other features of employee behavior include increased irritability and absenteeism and decreased morale during periods of heat stress exposure. Sluggish decision making, increased number of errors, and decreased quality of work are further indicators of heat-induced behavioral changes.

Medical surveillance is a useful recognition tool. Assuming for the moment that heat-related disorders are not being clinically diagnosed, there are patterns of complaints that may be related to overexposure to heat stress. Looking through first aid or clinic logs or reviewing the types of complaints that employees present, the following may be seen: fatigue, faintness, nausea, headache, skin rashes, and muscle cramps. If temperatures are taken routinely, they may be elevated, and, if asked, employees may report particularly concentrated urine (this is due to dehydration). There may also be a pattern of increased accidents during periods of heat stress exposure [18, 19].

In the face of indicators mentioned in the preceding paragraphs, physiological sampling may provide the final and definitive clues that heat stress is at a level that justifies an evaluation. Remembering that physiological strain associated with heat stress includes increases in sweating, body temperature, and heart rate, these responses can be another recognition tool. Monitoring sweat loss can be troublesome, but an indicator of dehydration is the change in body weight between the beginning and end of a shift. If the weight loss exceeds 1.5% of total body weight measured at the beginning, then significant dehydration from sweating is likely [17]. To assess the potential for excessive body temperature, a reading of oral temperature or other acceptable surrogate for core temperature may be taken at the end of a work cycle. If the temperature corrected for differences from core temperature exceeds 38°C, then the heat stress exposures require evaluation and control [20]. At the same time work is interrupted to take a worker's temperature, recovery heart rate can be noted. This is accomplished by asking the worker to sit down at the end of the work cycle, starting the temperature measurement, and assessing the heart rate after about 1 min [14]. The assessment can be by palpation, where a 10-s count is taken and multiplied by 6. If the recovery heart rate ($HR_{rec,\ 1\text{-min}}$) is less than 110 bpm, then the cardiovascular response is not excessive [15]. If it is greater than 110, and especially if it is greater than 120, then heat stress may affect the well-being of the individual [21].

D. Evaluation of Heat Stress

The evaluation of heat stress requires a characterization of the environment along with an estimate of the metabolic rate and some consideration of clothing. After outlining important environmental measures, two evaluation schemes will be presented. One scheme is empirical, and the other is rational. In both cases, the emphasis is on an assessment of heat stress as a predictor of physiological or heat strain.

1. Environmental Measures

The environmental contributions to heat stress are convection and radiation as well as limits on evaporative cooling. Environmental measures, therefore, need to encompass these contributors in some fashion. Common measures of the environment are described in this section [22].

Dry bulb temperature (T_{db}) is the measure of air temperature using a sensor freely suspended in the air. (The sensor should be shielded from sources of radiant heat so that the temperature is not affected by radiant heat.) The primary value of T_{db} is to help estimate the direction and magnitude of convective heat exchange. Alone, it is a very weak indicator of heat stress.

Psychrometric wet bulb temperature (T_{pwb}) is the value read from a temperature sensor that is covered by a wetted wick over which air is forced at a speed greater than 3 m/s. This temperature is a function of the air temperature decreased in proportion to the amount of evaporative cooling from the wick, which is inversely proportional to the water vapor pressure (absolute humidity) in the ambient air. In other words, as the humidity increases, T_{pwb} approaches T_{db}. The combination of T_{db} and T_{pwb} is used to determine the level of humidity in the air. The value for water vapor pressure is used to determine the maximum amount of evaporative cooling that can be supported by the environment.

Natural wet bulb temperature (T_{nwb}) is similar to T_{pwb} except that the air motion over the wick is simply the ambient air motion; the air is not forced over the wick by a fan or other means. Because the wick is exposed to, and the evaporation rate depends on, the ambient conditions, T_{nwb} is a good index of the ability of the environment to support evaporative cooling.

Globe temperature (T_g) is measured by a temperature sensor in the middle of a black-painted copper sphere. (The sphere was originally 6 in. in diameter, but new instruments use a smaller one.) It is used to estimate the average temperature of the solid surroundings for estimation of the radiant heat exchange. Because the globe is in intimate contact with the surrounding air, T_g also responds to convective heat flow between the globe and the air. Empirically, T_g is a good index of sensible heat exchange (combined effects of convection and radiation).

Air speed (V_{air}) is a measure of the air movement over the body. It affects the rate of convective heat flow as well as the rate of evaporative cooling. It is used to compute these avenues of heat exchange.

The wet bulb globe temperature (WBGT) index [23] is affected by both sensible heat exchange and the ability of the environment to support evaporative cooling. It is the principal index for empirical methods of heat stress evaluation. Under most conditions,

$$\text{WBGT} = 0.7 \, T_{nwb} + 0.3 \, T_g \tag{7}$$

When the workplace is in direct sunlight, then

$$\text{WBGT}_{sunshine} = 0.7 \, T_{nwb} + 0.2 \, T_g + 0.1 \, T_{db} \tag{8}$$

2. Empirical Scheme for Heat Stress Evaluation

A widely accepted method to assess the presence of heat stress in the workplace was first proposed by the National Institute for Occupational Safety and Health (NIOSH) in 1986 [20] and adopted in its current form (1995/6) by the American Conference of Governmental Industrial Hygienists (ACGIH) in 1991 [24]. The goal of this evaluation scheme is to limit the time-weighted average body core temperature over 1-2 hours to 38°C for chronic exposures to heat stress. To accomplish this goal, the authors of the NIOSH criteria document looked for practical methods to assess the environment and to combine the effects of environmental and metabolic rate as well as acclimatization state.

WBGT was selected as the index of the thermal environment because of its simplicity, ruggedness, and predictive value. Investigators then looked for a protective limit on WBGT at various levels of metabolic rate. The protective limit was a value that would limit the increase in core temperature to 38°C or less for about 95% of workers. The threshold then was a map of limiting WBGTs as a function of metabolic rate. Because acclimatization state affects the ability to physiologically control heat stress, separate exposure limits were developed for unacclimatized and acclimatized workers. Both NIOSH and the ACGIH recommend lower thresholds for workers who are not acclimatized.

The exposure limits developed by NIOSH (and adopted by ACGIH) are illustrated in Figure 1. In essence, the lines represent the points at which thermal equilibrium can be maintained at or below 38°C core temperature by most workers. Notice that as the rate of

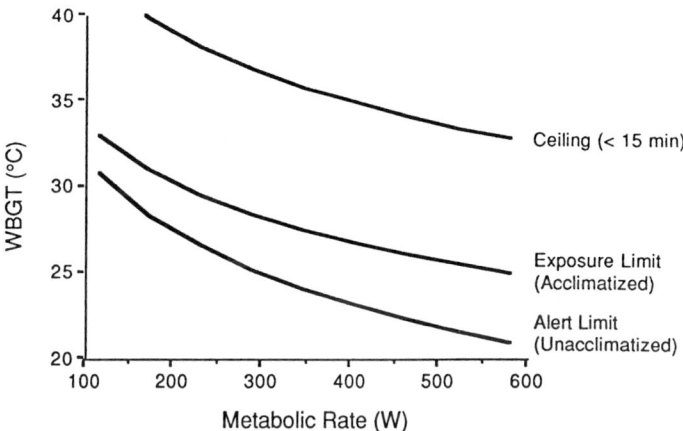

Figure 1 Thresholds for heat stress as a function of metabolic rate and WBGT to limit core temperature to 38°C for acclimatized and unacclimatized workers based on ordinary work clothes as described by NIOSH [20] and the ACGIH [24].

internal heat generation from metabolism increases, the environmental conditions must be cooler to maintain sufficient heat loss.

Clothing effects are not addressed directly by NIOSH other than to say that the thresholds apply to workers wearing customary, single-layer work clothes. A line of thinking initiated by Ramsey [25] and further developed by work sponsored by the Electric Power Research Institute (EPRI) [26, 27] has led to the use of clothing correction factors by the ACGIH. The factors in essence suggest what the effective increase in WBGT is when clothing other than ordinary work clothes is worn. Table 1 is a list of clothing correction factors (CFs) recommended by the ACGIH and values I have estimated for other clothing configurations. To best use these in the evaluation of heat stress, the correction factor is added to the prevailing WBGT in the workplace.

Table 1 WBGT Correction Factors for Clothing Ensembles[a]

	ACGIH	Other sources
Work clothes	0	0
Coveralls	2	
Winter uniform	4	
Particle barrier		4–7
Vapor-transmitting	6	
Vapor barrier		8
Encapsulating suit		11

[a]The correction factors are added to the WBGT measured for the task(s) requiring nonwork clothes, or the correction factor can be subtracted from the limit value if only one clothing ensemble applies for the analysis.

For multiple locations, multiple clothing ensembles, and multiple tasks, the effective WBGT and effective metabolic rate (M) are determined from a time-weighted average (TWA). That is

$$\text{WBGT}_{\text{effective}} = \{\Sigma[(\text{WBGT}_i + \text{CF}_i)t_i]\}/T \tag{9}$$

and

$$M_{\text{effective}} = \{\Sigma M_i t_i\}/T \tag{10}$$

where $T = \Sigma t_i$.

Table 2 illustrates this simple process for determining whether a job is above the threshold. As can be seen, the overall demands are above the threshold. The time in the rest location (Task 3) can be increased to 25 min to bring the work cycle to the threshold.

The time interval encompassed for time weighting is usually 1–2 h. If the work is cyclic in nature and includes the rest place, a TWA of one work cycle up to a 2-h duration is the appropriate time base. If the work is very intermittent or irregular, the selection of the time base requires some discretion, but it should not be more than 2 h. For combinations of $\text{WBGT}_{\text{effective}}$ and $M_{\text{effective}}$ that are above the line in Figure 1 for the appropriate acclimatization state, some workers may have difficulty coping with the heat stress. The further above the line, the greater the problem. Below the line, heat stress should not be a risk factor.

Although the evaluation is for time intervals between 1 and 2 h, the assumption is that the work is performed at these levels over the 8-h day. Shorter times at high levels of heat stress are also safe, but the WBGT-based method described here is not designed to evaluate those exposures. Both the U.S. Navy [28] and EPRI [26] have proposed WBGT-based methods to evaluate short-term exposures. Both methods consider WBGT and metabolic rate (and EPRI considers clothing as well) in recommending a time limit for the exposure. In these cases, heat stress is an important job factor if the work time exceeds the recommended exposure time. Rational methods of heat stress evaluation also provide safe exposure times.

3. Rational Scheme for Heat Stress Evaluation

Rational methods for heat stress evaluation use a different approach to the problem. The starting point is a model of heat exchange between the person and the environment similar to the one presented in Eqs. (1) and (2). For a given scheme, computations or estimations are described to determine values for the different avenues of heat exchange. Generally, if thermal balance can be achieved, then no storage occurs and the exposure is not limited. On the other hand, if $S > 0$, then the exposure is time-limited and the time limit can be determined from

Table 2 Illustration of the Use of the WBGT-Based Threshold and Correction Factors

Task	Clothing	M (W)	Measured WBGT (°C)	Corrected WBGT (°C)	Threshold WBGT (°C)	Time (min)
1	Vapor-transmitting	300	33	39		10
2	Coveralls	400	33	35		20
3	Work clothes	100	20	20		10
Time-weighted average		300	29.8	32.3	28.2	40
Change Task 3 time to 25 min:						
Time-weighted average		245	27.1	28.9	29.2	55

Eq. (5). The time limit can be used to evaluate the level of heat stress with respect to the work time.

A recent rational scheme has been accepted by the ISO [2]. It uses two criteria to prescribe time limits. The first is a maximum amount of heat storage [refer to Eq. (5)], and the second is a limit on the total volume of sweat that is required to maintain thermal balance. The computational details for this method are provided in the Appendix along with an example.

The method requires the collection or estimation of a number of environmental and work parameters including dry bulb, psychrometric wet bulb, and globe temperatures, air speed, metabolic rate, and clothing insulation. If the psychrometric wet bulb is not known, it can be taken as 1°C lower than the natural wet bulb temperature. The ISO method assumes cotton clothing, but blends with polyester can be treated as all cotton. There is also an emerging database on insulation and vapor permeation for noncotton clothing that can be adapted for this method [29].

From these data, heat transfer coefficients can be estimated for convection, radiation, and maximum evaporative cooling. Then the values in Eq. (3) can be determined to estimate the required evaporative cooling. Because the proportion of sweat that is evaporated decreases with increasing rates of sweating and evaporative cooling requirements, a level of sweating efficiency is estimated in order to estimate the total amount of sweating required to support the evaporative cooling.

The next step in the ISO evaluation process is to select the criteria that will be used for the evaluation. The selection is based on the acclimatization state and the acceptable level of heat stress. In the ISO nomenclature, the Warning level represents the highest level at which suitable and healthy workers are not at risk. The Danger level is a limit at which some workers may be at risk. Based on the selection, limits on sweat rate, skin wetness, heat storage, and sweat volume (to avoid dehydration) are set (see Table 3 in the Appendix).

Knowing the requirements and the limits, it is possible to estimate the amount of evaporative cooling that is available and the amount of sweating it will require. Based on these predicted values, time limits for the exposure can be computed to avoid excessive heat storage or sweat loss (dehydration).

If there are a series of locations, then the effect of the total exposure is computed from the time-weighted average of the individual values of required evaporation and maximum evaporation. That is,

$$E_{req} = \Sigma(t_i E_{req,\,i})/T \tag{11}$$

$$E_{max} = \Sigma(t_i E_{max,\,i})/T \tag{12}$$

where $T = \Sigma t_i$. The evaluation is then reentered at the point where skin wetness is computed. To account for recovery periods, this method of multiple locations can be used, where one of the locations is the recovery area.

4. Physiological Evaluation of Heat Stress

An alternative method to evaluate the level of heat stress is to look at the heat strain associated with the work. This concept was introduced earlier as a means of recognition. It is also a valid means of evaluation when the effects are monitored instead of the exposure [20]. Physiological strain for heat stress exposures is well reflected in sweating, core temperature, and heart rate.

Sweating. Sweat losses are interesting in that they provide a way to anticipate dehydration. Over 1–2-h intervals, the average rate of sweating can be assessed as [6]

$$\text{Sweat rate [L / h]} = \frac{BW_{initial} - BW_{final} + \text{ingestion [kg]}}{\text{time [h]}} \tag{13}$$

where BW is body weight in kilograms before and after the measurement interval, and ingestion is the weight of drink and food consumed during the same period. This equation assumes that there is no loss of body weight by excretion. The hourly sweat loss should be less than 1 liter. Water losses greater than 1 liter/h are not easily sustained [30].

There is also a total sweat loss threshold of 5 liters in the workday. Again, it is difficult to replace more than 5 liters of water in 8 h. To monitor total sweat loss means that weight loss by excretion must also be assessed. As a practical measure, total sweat loss is not a routine measure. An estimate of dehydration over the workday, however, is available. This estimate in the percent change in body weight from the beginning to end of the work shift. As mentioned above, a loss of more than 1.5% of body weight (about 1–1.5 kg) is significant [17].

Core Temperature. Core temperature is an abstraction, and there are various means to measure the "core" temperature. In the laboratory, the usual means are rectal, tympanic, and esophageal temperatures. Esophageal is very effective for assessing the physiological responses to heat stress and strain. It is less indicative, however, of the total thermal storage. Tympanic temperature was an early favorite for predicting thermoregulatory response but has yielded to esophageal. A personal preference is to use rectal temperature (T_{re}) as the best indicator of the three for total heat storage for industrial settings.

Unfortunately, the three measures of core temperature are difficult to measure in the field for obvious social and logistical reasons. Therefore, a surrogate measure of core temperature is usually sought. These include oral and ear canal temperatures as well as readings obtained with a surface-mounted device. Each method has its advantages and disadvantages. Oral temperature measurement is the most readily recognized method, with a long history of use for the evaluation of industrial heat strain. Modern devices have electronic means that allow them to respond more rapidly than mercury-in-glass thermometers. Necessary precautions include (1) no eating or drinking for 15 min prior to measurement and (2) mouth closed during the measurement (no talking or mouth breathing). Generally, adding 0.5°C to the value is a good predictor of rectal temperature, although the average steady-state difference may reach 1°C.

An alternative measure is the temperature in the ear canal. The temperature sensor is placed in the ear canal near the tympanic membrane, and care is taken to insulate the ear canal from the outside environment. Insulating techniques have included cotton and hearing protectors. The ear canal temperature values are sensitive to the environmental conditions. For instance, in hot environments or with high radiant heat, the value may be greater than rectal temperature. Temperature in the ear canal is also more reactive to the level of heat stress than rectal temperature, which means that it will rise or fall faster.

A recently proposed noninvasive technique for assessing core temperature is the use of an insulated disk sensor placed on the chest [32]. Its original purpose was to provide an alert once a threshold limit is achieved, and the disk temperature gives an estimate of core temperature based on population data. It has a linear relationship with core temperature that varies with individuals and the clothing worn. Therefore, to enhance the value of individual measures, it is best to reference it to another measure of core temperature.

Once a surrogate method of assessing core temperature is selected, then a sampling strategy can be developed. Basically, the time-weighted average core temperature should be 38°C for the 8-h day, with a ceiling limit of 39°C.

Heart Rate. Heart rate measures are relatively easy to obtain in a reliable and acceptable fashion.

Historically, recovery heart rate was used to assess work demands and heat stress because palpation of the radial artery was a simple field method. Criteria have evolved from the initial work of Brouha in 1960 [13] through efforts by Fuller and Smith [14] to a single sample at 1 min of recovery [15, 21]. A worker is stopped and asked to sit down and the recovery heart rate is noted at the end of 1 min. The original method called for a 30-s count from 30 s to 1 min, but recovery heart rate can be assessed by palpation for 10 s, by electronic meters, or by other means near the end of the minute. If the value is less than 110 bpm, there is no excessive strain.

Data loggers in conjunction with heart rate monitors are now readily available [31]. These enable investigators to record heart rate on a minute-by-minute basis for an entire work shift. Review of the heart rate data for a worker includes an evaluation of the average for the workday as well as an examination of peaks. The average over a workday should be less than about 110 bpm [10]. Peak heart rates (sustained for more than 1 min) should be less than 90% of the individual's maximum heart rate [an estimate for HR_{max} is 195 − 0.67 (age − 25)] [32].

Because the average and peak heart rate evaluations can miss periods of significant cardiovascular strain, ergonomists will visually examine the patterns for heart rate creep or other indicators of failure to achieve complete control [33]. Bernard and Kenney [32] proposed a quantitative method of assessing the heart rate pattern over the course of the day (or lesser periods). The analysis is based on moving-time averages (akin to time-weighted averages on the same time base and continuously updated). This method has been adapted to real-time monitoring.

E. Management of Heat Stress

The management of heat stress is the natural and obligatory next step if the evaluation indicates that heat stress is a significant workplace hazard. Management comes in two stages: general controls and job-specific controls. The following paragraphs outline when it is appropriate to implement general and specific controls and describe what the controls may be.

The controls are developed within the overall health and safety activity at the individual site. There are four important components of the heat stress management program:

Policy Statement The policy statement acknowledges the potential for heat stress exposure, the role of heat stress control in the overall health and safety program, and the availability of health and safety personnel to assist line management.
Delineation of responsibilities The senior line managers have responsibility with some delegation of authority, first line supervision has day-to-day responsibility to ensure compliance and provide a role model, and employees must follow guidelines and practice good heat stress hygiene.
Workplace monitoring Workplace monitoring follows good industrial hygiene practice for monitoring hazards.
Review and evaluation There should be a periodic review of the heat stress program to be assured that it is meeting its goals and to make adjustment as necessary.

The actual embodiment of the controls to manage heat stress are varied and adaptable in infinite combinations with the exercise of common sense and a little imagination. One valuable source is the current practices of workers.

1. General Controls

General controls are those actions required to reduce the risk of heat-related disorders among workers who may be exposed to heat stress in the workplace. General controls are the first stage of countermeasures, and they transcend the actual conditions in which the exposure may occur. Included in general controls are heat stress hygiene practices, training, and medical surveillance.

To make the decision about who might be exposed to heat stress and when, we must refer back to the evaluation step. Anytime workplace conditions combine to place the situation above the threshold for unacclimatized workers in Figure 1, then heat stress is present at a level that should be considered a hazard (albeit a low-risk hazard at the threshold). If the calculation of time limits is used to evaluate the level of stress, then conditions that combine in the evaluation scheme to recommend an exposure time of less than 8 h are the threshold at which to implement general controls. If only occasional samples are taken of the environment, they should be taken at the hottest part of the day during periods of hot weather, in other words, when the highest levels of heat stress are most likely.

Heat Stress Hygiene Practices. Heat stress hygiene practices are those actions that an individual can take to lower the risk of heat-related disorders. They consist of the following:

Self-determination The individual must interrupt a heat stress exposure once extreme discomfort or the initial symptoms of a heat-related disorder are detected.

Fluid replacement Because thermal regulation depends on sweating and the necessary loss of water, the water must be replaced at frequent intervals to maintain acceptable hydration. Sometimes drinking is restricted or impossible (e.g., wearing a respirator), and then allowances for pre- and postexposure hydration must be considered.

Lifestyle and diet Practicing a generally accepted healthy lifestyle (getting adequate sleep, limiting nonwork exposures to heat stress, exercising, not abusing alcohol or drugs, and eating a well-balanced diet) greatly reduces the risk of heat-related disorders.

Health status Those with any chronic disease should inform their physician of occupational exposures to heat stress and follow the recommendations. Those with an acute illness should report the condition to a supervisor, and the heat stress exposures should be restricted or reduced.

Acclimatization Because acclimitization requires at least 3 days, allowances must be made for those workers who are not acclimatized to the heat, and performance expectations should therefore be reduced.

It is clear that heat stress hygiene practices are the responsibility of individual workers but that management must minimize the barriers to the practice of heat stress hygiene.

Training. Training is a fundamental health and safety practice for those who may be exposed to a hazard such as heat stress. Annual training that fits within the training practices of the site should be sufficient. The training should include discussions of heat stress and strain, heat-related disorders, heat stress hygiene practices, and controls used in the facility.

Medical Surveillance. Medical surveillance encompasses physicals and monitoring of sentinel health events. Both preplacement and periodic physicals appropriate to evaluation of an individual's capacity to deal with heat stress are recommended. The physical should include comprehensive medical and work histories, comprehensive physical examination and tests, assessment of drug use, and ability to use personal protection. Good practice dictates that there be a written opinion as to the suitability of exposing the individual to heat stress.

Monitoring of sentinel health events includes monitoring individuals as well as the population. Relevant events are heat-related disorders, patterns of accidents, absenteeism, and chronic fatigue.

2. Specific Controls

Specific controls are those that apply to specific manifestations of heat stress. They follow the traditional hierarchy of engineering controls followed by administrative controls, with the use of personal protection as a last resort. Specific controls are appropriate for workers who are exposed to heat stress above the threshold for acclimatized workers in Figure 1, and only when they are exposed above the threshold.

Engineering Controls. Engineering controls change the conditions so that the level of heat stress is reduced, ideally below the exposure thresholds. Methods of engineering controls include the following.

Reduce the metabolic rate. A very effective means to reduce heat stress is to reduce the amount of internal heat generation. This can be done by spreading the work out, mechanization, and increasing staffing.

Change clothing requirements. A proper balancing of risks from heat stress and from agents requiring protective clothing may lead to the selection of different barrier fabrics with markedly different evaporative cooling capacities. For instance, some vapor-transmitting fabrics cause much less heat stress than vapor-barrier fabrics.

Reduce temperature and humidity. Reductions of air temperature and humidity are frequently achieved through spot or dilution ventilation. This is another method to significantly reduce the level of heat stress in the workplace. The ventilation systems can be temporary or permanent and may include mechanical cooling. Dramatic cost savings have been demonstrated with this method.

Increase air motion. Increasing air speed via fans is a time-honored method to enhance evaporative cooling, but it is of limited value once air speed exceeds 2 m/s. When air temperature is greater than 40°C, increasing air motion may actually increase heat stress.

Control radiant heat. When radiant heat is high, the effects can be reduced through combinations of insulating exterior surfaces and reducing surface emissivity. In addition, shields can be very effective.

Administrative Controls. Administrative controls manage the risk through work practices. They are relatively easy to implement, although they may not be the most cost-effective. Administrative controls include the following.

Planned work time. Limiting the heat stress exposure to a time period that would ensure that most workers are not overexposed is one way to limit the risk. The work time limit can be based on the WBGT or ISO methods of estimating safe exposure times discussed above for evaluation purposes.

Self-determination. Giving employees the opportunity to subjectively control the pace of work and the work time is frequently used as a means of controlling heat stress exposures. Self-pacing is a valuable means of reducing the physiological strain and improving efficiency. Subjective self-limitation, however, may not be reliable. Physiological monitoring to provide objective information on heart rate and body temperature will improve the reliability. Recommendations for personal monitoring include those physiological measures discussed above that can be performed and assessed in real time. The advantage of personal monitoring is that it allows the more heat-tolerant workers more exposure time. In this way, personal monitoring can improve productivity while controlling the risk of heat-related disorders.

Recovery allowances. It is important to provide adequate recovery from heat stress exposures. Including recovery times and locations in the analysis of overall evaluation of heat stress exposure provides insight as to whether or not the recovery allowance is adequate. If the exposure is above the threshold, the recovery is insufficient, and the exposure or recovery conditions or times must be adjusted.

Scheduling work. To the extent possible, scheduling work to times when the heat stress levels may be lower (e.g., night, during outages) is a useful way to control exposures.

Personal Protection. Personal protection for heat stress exposures means providing a microenvironment around the worker that allows a greater loss of heat. There are three types of personal cooling systems that have demonstrated effectiveness:

1. *Circulating air systems*. Venting air from supplied-air hoods or supplying breathing grade air directly under clothing enhances evaporative and convective cooling. Many times, the cooling is sufficient to virtually eliminate heat strain. The major disadvantage is that worker mobility is restricted with the air line.
2. *Liquid cooling systems*. This type of personal cooling is based on circulating cooling liquid (e.g., water) around some portion of the body within enclosed tubes or channels. The rate of cooling depends on the surface area of the body covered. The heat is taken up by a heat sink that is usually composed of ice but could theoretically be another material. The service time depends on the size of the heat sink or the ability to replenish the heat sink. The major disadvantage to these systems is the cost.
3. *Ice cooling garments*. Ice cooling garments cool the body by direct transfer of heat from the body to the heat sink by conduction. These are sometimes referred to as passive systems because there is no mechanical movement of air or liquid.

Under certain conditions of high radiant heat, reflective clothing over parts or all of the body can reduce the level of heat stress. The reflective clothing must be chosen carefully because it will reduce the rate of evaporative cooling.

In choosing personal protection it is crucial that the means be selected to best match the capabilities and requirements of the cooling system to the job mission. The decision should be confirmed by actual trials in the field including physiological measures to demonstrate a reduction in physiological strain.

IV. CRITICAL REVIEW OF CURRENT STATUS

Two approaches to the evaluation of heat stress are described in this chapter. One approach is through the use of heat stress models such as WBGT-based thresholds and required sweat rate methods. The other is monitoring physiological strain, which includes some combination of body temperature, heart rate, and sweat loss.

The use of models of heat stress relies on the predictive ability of the model to indicate when excessive physiological strain is likely or when the risk of a heat-related disorder is significant. The criteria are selected to protect most workers. The principal advantage of current models of heat stress for occupational applications is that they are reasonably and acceptably good at achieving the necessary level of protection.

The disadvantages of heat stress models center on interindividual variability, sensitivity to environmental and work level uncertainty, and evolving means to account for clothing. Because individuals respond differently to heat stress and differ in the attendant risks of heat-related disorders, the criteria are selected to be protective of at least 95% of the healthy working population. Furthermore, because the WBGT is a simple index of environmental conditions, it cannot adequately reflect all possibilities, and this shortcoming is adjusted for in WBGT-based criteria. Many healthy workers are capable of safely working at levels of heat stress greater than the criteria levels, and these workers are more likely to self-select into the more stressful jobs. Experience makes workers and supervisors aware of this fact, which, in turn, makes it difficult for them to accept the criteria.

Finally, the models were developed with the assumption that ordinary cloth work clothes would be worn. Beyond ordinary work clothes, the clothing effects on heat stress are dictated by insulation, permeability, and ventilation [34], which in turn depend on the fabrics, construction, and use. The ACGIH has recommended clothing correction factors for the measured value of WBGT to account for some other clothing ensembles. These values are based on shifts in the upper limits of environmental conditions observed for establishing thermal equilibrium. Some data would suggest that this is a valid approach, but it has not been vigorously evaluated. The heat balance method proposed by the ISO can be modified if the insulation and vapor resistance characteristics for non-cotton clothing are known. Again, some limited data are available to suggest that the method can be extended to non-cotton clothing.

Physiological evaluation examines the individual's response to heat stress. The advantage of monitoring physiological strain is that it removes the need to know well the environmental, work, and clothing factors as they contribute to heat stress. To the extent that a representative sample of the worker population is selected, the exposure to heat stress is sufficiently understood to guide decisions on the need for controls. It was the population response to heat stress in terms of core temperature and recovery heart rate that helped establish the WBGT thresholds, so it makes sense that these can be used to directly evaluate the work environment. By extension, physiological monitoring can be a tool to control exposures for individuals. The disadvantage is that the effectiveness of the controls cannot be estimated before they are first tested in actual or simulated work conditions.

V. FUTURE CONCERNS

Understanding how different protective clothing ensembles affect heat stress is important for evaluating heat stress in many work circumstances. That is, the correction factors for the WBGT-based model of heat stress should be valid for the clothing that is most common as well as the novel fabrics that are emerging for protective clothing. Furthermore, the assumption that adjustments based on changes associated with thermal equilibrium will work for higher levels of heat stress requires confirmation. With respect to the heat balance methods, how non-cotton clothing affects dry and evaporative heat transfer needs to be evaluated and made available to practitioners.

Expanding the use of personal monitoring for evaluation of heat strain is appropriate and practical, and it may be the only way worker protection can be ensured during work with some common forms of protective clothing. Devices for monitoring core temperature and/or heart rate are available, and others can certainly be developed. With supporting data, these can be used to assess population responses to heat stress for the purposes of evaluation and also for improved self-assessment of individual exposures. The successful implementation of cost-effective personal monitoring can reduce the problems of conservative limits based on heat stress models and further reduce the risk of heat-related disorders.

APPENDIX: ISO REQUIRED SWEAT RATE ANALYSIS FOR THE EVALUATION OF HEAT STRESS[1]

The terms and units used in the ISO sweat rate analysis are as follows.

C	Convective heat exchange, W/m²
E_{max}	Maximum rate of evaporative cooling, W/m²
E_p	Predicted rate of evaporative cooling, W/m²
E_{req}	Required evaporative cooling, W/m²

---

[1]Based on Ref. 2.

F_{cl} Clothing reduction factor, dimensionless
F_{pcl} Clothing reduction factor for water vapor permeation, dimensionless
h_c Coefficient of convection, W/(m²-°C)
h_{cf} Coefficient of convection—forced, W/(m²-°C)
h_{cn} Coefficient of convection—natural, W/(m²-°C)
h_r Coefficient of Radiation, W/(m²-°C)
I_{cl} Clothing insulation, m²-°C/W
M Metabolic rate, W
P_a Ambient water vapor pressure, kPa
P_{sk} Water vapor pressure on skin, kPa
Q_{max} Maximum heat storage, W-h/m²
R Radiant heat exchange, W/m²
SA   Body surface area, m²
SR   Sweat rate, liters/h
SV_{max} Maximum sweat loss, W-h/m²
SW_{max} Maximum sweat rate, W/m²
SW_p Predicted sweat rate, W/m²
SW_{req} Required sweat rate, W/m²
T_{db} Dry bulb (air) temperature, °C
T_g Globe temperature, °C
T_{pwb} Psychrometric wet bulb temperature, °C
T_r Mean radiant temperature, °C
T_{sk} Average skin temperature, °C
t Time limit, min
t_d Time limit based on dehydration, min
t_w Time limit based on skin wetness, min
V Adjusted air speed, m/s
V_{air} Air speed, m/s
w_{max} Maximum skin wetness, dimensionless
w_p Predicted skin wetness, dimensionless
w_{req} Required skin wetness, dimensionless
η Sweating efficiency, dimensionless

In the following description of the ISO Required Sweat Rate Analysis, some deviations from the standard are made, and these are noted by double brackets (<< >>).

The first step in the process is to compute the components of heat exchange. For metabolism, the metabolic rate must be divided by the body surface area (SA) (value for standard man is 1.8 m²).

$$M = \frac{\text{metabolic rate}}{\text{SA}}$$

The ambient water vapor pressure (P_a) can be estimated from the following equation or determined from a psychrometric chart. <<This equation is not in the standard but is adopted from the program listing and ASTM Standard E 337.>>

$$P_a = 0.6105 \, \exp\left[\frac{17.27 \, T_{pwb}}{T_{pwb} + 237.3}\right] - 0.0669 \, (T_{db} - T_{pwb})(1 + 0.00115 T_{pwb})$$

Occupational Heat Stress

The mean radiant temperature (T_r) is the average temperature of the solid surroundings. A value for ζ is selected for the effects of natural or forced convection on the globe temperature.

ζ = greater value of $\{0.4(|T_g - T_{db}|)^{0.25}, 2.5\ V_{air}^{0.6}\}$

$T_r = 100\ \{[(T_g + 273)/100]^4 + \zeta(T_g - T_{db})\}^{0.25} - 273$

Compute average skin temperature (T_{sk}):

$T_{sk} = 30.0 + 0.093\ T_{db} + 0.045\ T_r - 0.571\ V_{air} + 0.254\ P_a + 0.00128\ M - 3.57\ I_{cl}$

Because the simple act of working causes some air motion around the body, an adjustment to air velocity (V_{air}) is made.

$V = V_{air} + 0.0052\ (M - 58)$

The increase due to work activity should be limited to 0.7 m/s, which occurs when $M > 193$ W/m². <<Although V is not limited in the standard, there are limits to how much air movement affects convection and evaporative cooling. As a further limit, V should not exceed 3.0 m/s.>>

The coefficient for convection (h_c) can be dominated by natural or forced convection, so each must be estimated and the larger value selected.

$h_{cn} = 2.38\ (|T_{db} - 36|)^{0.25}$

If $V \leq 1$ m/s, then $h_{cf} = 3.5 + 5.2V$. If $V > 1$ m/s, then $h_{cf} = 8.7 V^{0.6}$.

h_c = greater value of $\{h_{cn}, h_{cf}\}$

Now the coefficient for radiation (h_r) for a standing person can be determined.

$h_r = 4.1 \times 10^{-8}\ \dfrac{(T_{sk} + 273)^4 - (T_r + 273)^4}{T_{sk} - T_r}$

The effects of clothing isolation on sensible heat exchange is reflected in the clothing reduction factor (F_{cl}).

$F_{cl} = \dfrac{1}{(h_c + h_r)I_{cl} + 1/(1 + 1.97\ I_{cl})}$

where I_{cl} is the clothing insulation.

Convective heat exchange (C) is then computed as

$C = h_c F_{cl}(T_{db} - T_{sk})$

Radiant heat exchange (R) is then computed as

$R = h_r F_{cl}(T_r - T_{sk})$

The required evaporative cooling (E_{req}) follows from Eq. (3):

$E_{req} = -(M + C + R)$

<<Heat exchange in the respiratory track is given in the ISO standard but is not included here.>>

Clothing also affects the transmission of water vapor, and the clothing reduction factor for water vapor permeation (F_{pcl}) for cotton clothing is

$$F_{pcl} = 1/\{1 + 2.22\ h_c(I_{cl} - [1 - 1/(1 + 1.97\ I_{cl})]/(h_c + h_r)\}$$

The water vapor pressure on the skin (P_{sk}) can be computed from the following equation or determined from a psychrometric chart or tables for saturation pressure at the skin temperature.

$$P_{sk} = 0.6105\ \exp[17.27\ T_{sk}/(T_{sk} + 237.3)]$$

Now the maximum evaporative cooling to the environment (E_{max}) can be estimated.

$$E_{max} = -16.7\ h_c F_{pcl}\ (P_{sk} - P_a)$$

To achieve the necessary rate of evaporation, the skin must reach a level of wetness called the required skin wetness (w_{req}).

$$w = E_{req}/E_{max}$$

$$w_{req} = \text{minimum value } (w, 1.0)$$

<<The standard limits wetness at this stage to 2.0. In practice, w cannot exceed 1.0, and greater values are indirectly eliminated later in the evaluation.>>

At higher levels of wetness, some sweat does not evaporate but rather is lost from the body by dripping. Sweating efficiency (η) is an indicator for the proportion of sweat that is evaporated.

$$\eta = 1.0 - w_{req}^2/2$$

Then the required sweat rate (SW_{req}) can be computed.

$$SW_{req} = -E_{req}/\eta$$

The criteria appropriate for the evaluation are selected from Table 3. Then the evaluation can proceed.

Table 3 ISO Criteria by Acclimatization State and by Alert Level[a]

Criteria	Units	Unacclimatized		Acclimatized	
		Warning	Danger	Warning	Danger
SW_{max}*	W/m²	200	250	300	400
	liters/h	0.52	0.65	0.78	1.04
w_{max}	—	0.85	0.85	1.0	1.0
Q_{max}	W-h/m²	50	60	50	60
SV_{max}	W-h/m²	1000	1250	1500	2000
	liters	2.5	3.25	3.9	5.2
*If $M < 65$, then					
SW_{max}	W/m²	100	150	200	300
	liters/h	0.26	0.39	0.52	0.78

[a]The following descriptions are not given as such in the standard, but they are intended to help in the application of the limits.

Unacclimatized: Workers have not had at least 5 days of 2-h heat stress exposures over the preceding 7 days.
Acclimatized: Workers have had at least 5 days of 2-h heat stress exposures over the past 7 days.
Warning: If work time exceeds recommended time for Warning, controls should be considered. Adequate recovery should be allowed.
Danger: If work time reaches Danger time, controls are necessary.

Occupational Heat Stress

Table 4 List of Work Conditions, Intermediate Values for the ISO Required Sweat Rate Analysis, and the Final Values (Copy of a spreadsheet output designed for the analysis.)

Heat Stress Assessment by Required Sweat Rate Analysis

Work Sheet — Criteria: ISO

Work Situation

Work Situation	Data		Time Limit [min]	
Acclimation State (Code 1 – 3)	1	Acclimated	Warning	194
Clothing (Code or I-Value in m2 °C/W)	1	Work Clothes	Danger	363
Metabolic Rate (Code or M-Value in W)	2	Moderate Work		

			WBGT [°C]	TLV [°C]
Environment (Enter Values)		[°C]	34.7	28.4
	Tdb	45	Not adjusted for clothing	
	Tnwb	26		
	Tg	55	Predicted Values	Warning	Danger
	Vair [m/s]	0.3	Evaporative Limit [min]	194	480
Task Sequence Number		1	Dehydration Limit [min]	300	363
			wp [-]	0.64	0.68
Temperature Units (Enter Code 1 – 2)		1	Ep [W/m2]	239	254
Criteria Basis (Enter Code 1 – 3)		1	SRp [L/h]	0.78	0.86

Intermediate Values (Current Conditions)

			Analysis		
Psychr Wet Bulb [°C]		24.4	Ereq [W/m2]	254.4	254
Pv [kPa]		1.65	Emax [W/m2]	374.4	3.06
M [W]		290			
M [W/m2]		161	wreq [-]	0.68	45
ΔVmet [m/s]		0.54	eff [-]	0.77	26
ΔV [m/s]		0.54	SRreq [L/h]	0.86	55
Vadj [m/s]		0.84			
V' [m/s]		0.84	Limits	Warning	Danger
			SRmax [L/h]	0.78	1.04
Tsk [°C]		37.2	wmax [-]	1.00	1.00
Psk [kPa]		6.33	Qmax [Wh/m2]	50.00	60.00
			SVmax [L]	3.90	5.20
hcn		3.98			
hcf		7.85			
hc		7.85	Condition A (pred=req)	FALSE	FALSE
			wp [-]	0.68	0.68
mrt coef		1.21	Ep [W/m2]	254	254
Tr [°C]		63.3	SRp [L/h]	0.86	0.86
hr		5.68			
			Condition B (w - limit)	FALSE	FALSE
hc + hr		13.53	wp [-]	0.68	0.68
Icl [m2 °C/W]		0.089	eff [-]	0.77	0.77
icl-interm		0.07797	Ep [W/m2]	254	254
Fcl [-]		0.49	SRp [L/h]	0.86	0.86
Fpcl [-]		0.61			
1/RT [W/m2 kPa]		79.95	Condition C (SR - limit)	TRUE	FALSE
			w-interm	1.25	0.94
C [W/m2]		29.9	wp [-]	0.64	0.76
R [W/m2]		72.2	eff [-]	0.80	0.71
Cres [W/m2]		2.3	Ep [W/m2]	239	284
Eres [W/m2]		-11.1	SRp [L/h]	0.78	1.04

If $w_{req} \leq w_{max}$ and $8\,SW_{req} \leq SV_{max}$, then the heat stress is not time-limited.

If $w_{req} \leq w_{max}$, then the following are the predicted values for skin wetness, evaporative cooling, and sweat rate:

$$w_p = w_{req}, \quad E_p = E_{req}, \quad SW_p = SW_{req}$$

If $w_{req} > w_{max}$, then skin wetness is limited, thus limiting evaporative cooling and sweat rate. Under these conditions, the following are the predicted values for wetness, evaporative cooling, and sweat rate:

$$w_p = w_{max}, \quad E_p = w_p E_{max},$$
$$SW_p = -E_p/\eta_p \quad \eta_p = 1.0 - w_p^2/2$$

If $SW_p > SW_{max}$ based on either of the predicted values above, then the predicted sweat rate is greater than can be achieved. Recomputation of the predicted values is as follows.

$$\phi = -E_{max}/SW_{max}$$
$$w_p = (\phi^2 + 2)^{0.5} - \phi \quad \text{<<or } w_{max}, \text{ whichever is less>>}$$
$$\eta_p = 1.0 - w_p^2/2, \quad E_p = w_p E_{max}$$
$$SW_p = -E_p/\eta_p \quad \text{(which is usually } SW_{max})$$

To estimate the sweat volume rate for a standard man (SR),

$$SR\ [L/h] = 0.0026\ SW_p$$

From the predicted values for evaporative cooling rate and sweat rate, time limits may be computed. The time limit based on inadequate evaporative cooling (t_e) is

$$t_e\ [min] = 60 Q_{max}/(E_p - E_{req})$$

The time limit (t_d) based on the possibility of dehydration is

$$t_d\ [min] = 60\ SV_{max}/SW_p$$

The recommended maximum exposure time is the lesser of t_e and t_d.

<<A draft version of the standard provided for an adjustment in sweat rate to account for the transition that occurs. The following is this adjustment.>> If $10 \leq t < 30$ min, then recalculate as follows:

$$E_p^* = E_p(t - 10)/20$$
$$t_e^*\ [min] = 60 Q_{max}/(E_p^* - E_{req})$$
$$t = \text{minimum value } (t_e^*, t_d)$$

If $t < 10$ min, then heat stress is very high. Set $t = 0$.

ACKNOWLEDGMENT

Dr. Francis N. Dukes-Dobos provided a critical review of this manuscript, and I appreciate his valuable comments.

REFERENCES

1. H. S. Belding and T. F. Hatch, Index for evaluating heat stress in terms of resulting physiological strain, *Heating, Piping Air Cond.* 27:129-135 (1955).
2. International Organization for Standardization (ISO), *Hot Environments—Analytical Determination and Interpretation of Thermal Stress Using Calculation of Required Sweat Rate*, ISO 7933, Geneva, 1989.
3. J. Stegemann, *Exercise Physiology*, Year Book Medical Publishers, Chicago, 1981, p. 71.
4. E. Kamon and B. D. Avellini, Wind speed limits to work under hot environments for clothed men, *J. Appl. Physiol.* 46:340-349 (1979).
5. H. Siekmann, Recommended maximum temperatures for touchable surfaces, *Appl. Ergon.* 21: 69-73 (1990).
6. D. McK. Kerslake, *Stress of Hot Environments*, Cambridge Univ. Press, Cambridge, U.K., 1972.
7. M. Nielsen, Die Regulation der Körpertemperatur bei Muskelarbeit, *Skand. Arch. Physiol.* 79:193 (1938).
8. A. R. Lind, A physiological criterion for setting thermal environmental limits for everyday work, *J. Appl. Physiol.* 18: 51-56 (1963).
9. F. N. Dukes-Dobos and A. Henschel, Development of permissible heat exposure limits for occupational work, *Am. Soc. Heating, Refrig. Air Cond. Eng. J.*, 15:57-62 (1973).
10. World Health Organization (WHO), *Health Factors Involved in Working Under Conditions of Heat Stress*, Tech. Rep. Ser. 42, WHO, Geneva, 1969.
11. E. Kamon, Scheduling cycles of work for hot environment conditions, *Ergonomics* 22:427-439 (1979).
12. American College of Sports Medicine, *Guidelines for Exercise Testing and Prescription*, Lea & Febiger, Philadelphia, 1991.
13. L. Brouha, *Physiology in Industry*, Pergamon, London, 1960.
14. F. H. Fuller and P. E. Smith, Evaluation of heat stress in a hot workshop by physiological measurements, *Am. Ind. Hyg. Assoc. J.* 42:32-37 (1981).
15. NIOSH, OSHA, USCG, and EPA, *Occupational Safety and Health Guidance Manual for Hazardous Waste Site Activities*, DHHS(NIOSH), Washington, DC, 1985, pp. 85-115.
16. A. R. Lind and D. E. Bass, Optimal exposure time for development of acclimatization to heat, *Fed. Proc.* 22:704 (1963).
17. J. F. Greenleaf and M. H. Harrison, Water and electrolytes, in *Exercise, Nutrition and Health*, D. K. Layman, Ed., American Chemical Society, New York, 1985.
18. H. S. Belding, T. F. Hatch, B. A. Hertig, and M. L. Riedesel, Recent developments in understanding of effects of exposure to heat, in *Proceedings of 13th International Congress on Occupational Health*, New York, Craftsman, New York, 1961.
19. J. D. Ramsey, C. L. Burford, M. Y. Beshir, and R. C. Jensen, Effects of workplace thermal conditions on safe work behavior, *J. Safety Res.* 14:105-114 (1983).
20. National Institute for Occupational Safety and Health (NIOSH), *Criteria for a Recommended Standard—Occupational Exposure to Hot Environments*, Revised criteria 1986, USDHHS (NIOSH) 86-113, Washington, DC, 1986.
21. T. E. Bernard and W. L. Kenney, Heart rate recovery, Presented at American Industrial Hygiene Conference, 1988.
22. International Organization for Standardization (ISO), *Thermal Environments—Instruments and Methods for Measuring Physical Quantities*, ISO 7726, Geneva, 1985.
23. C. P. Yaglou and D. Minard, Control of heat casualties at military training centers, *AMA Arch. Ind. Health* 16:302-316 (1957).
24. American Conference of Governmental Industrial Hygienists (ACGIH), *Threshold Limit Values and Biological Exposure Indices for 1995-1996*, ACGIH, Cincinnati, 1995.
25. J. D. Ramsey, Abbreviated guidelines for heat stress exposure, *Am. Ind. Hyg. Assoc. J.* 39:491-495 (1978).

26. T. E. Bernard, W. L. Kenney, L. F. Hanes, and J. F. O'Brien, *Heat-Stress Management Program for Power Plants* NP4453L, Electric Power Research Institute, Palo Alto, CA, 1991.
27. W. L. Kenney, WBGT adustments for protective clothing, *Am. Ind. Hyg. Assoc. J. 48*:A576–A577 (1987).
28. Tri-Services Document, *Prevention, Treatment and Control of Heat Injury*, Army TB Med 507, 1980.
29. W. L. Kenney, D. J. Mikita, G. Havenith, S. M. Puhl, and P. Crosby, Simultaneous derivation of clothing-specific heat exchange coefficients, *Med. Sci. Sports Exercise 25*:283–289 (1993).
30. B. McArdle, W. Dunham, H. E. Hollong, W. S. S. Ladell, J. W. Scott, M. L. Thomson, and J. S. Weiner, *The Prediction of the Physiological Effects of Warm and Hot Environments*, Rep. No. RNP 47/391, Medical Research Council, London, 1947.
31. D. P. Humen and D. R. Boughner, Evaluation of commercially available heart rate monitors, *Can. Med. Assoc. J. 131*:585–589 (1984).
32. T. E. Bernard and W. L. Kenney, Rationale for a personal monitor for heat strain, *Am. Ind. Hyg. Assoc. J. 55*:505–514 (1994).
33. Eastman Kodak Company, *Ergonomic Design for People at Work,* Vol. 2, Van Nostrand Reinhold, New York, 1986.
34. U. Reischl, A. W. Spaul, F. N. Dukes-Dobos, and E. G. Hall, Ventilation analysis of industrial protective clothing, in *Trends in Ergonomics/Human Factors*, Proc. Annu. Int. Ind. Ergon. Safety Conference, Miami, FL, June 9–12, 1987, Part A, S. A. Asfour, Ed., North-Holland/Elsevier, Amsterdam, 1987, pp. 421–428.

9
Physical Work Capacity
Principles and Applications

Ashraf M. Genaidy
University of Cincinnati, Cincinnati, Ohio

I. INTRODUCTION

The objective of this chapter is to provide a brief review of physical work capacity. The chapter is organized into two sections: (1) principles and (2) applications of physical work capacity. The principles of physical work capacity (PWC) include definition, components of PWC, factors affecting PWC, assessment procedures of PWC, and normative physical work capacity data. The applications of physical work capacity are discussed with reference to ergonomic job design and task analysis, job-simulated exercise programs, preemployment placement and screening, and essential job functions.

II. BACKGROUND AND SIGNIFICANCE TO OCCUPATIONAL ERGONOMICS

Physical work capacity (PWC) plays a central role in the process of carrying out ergonomic stress analysis in industry. The objective of applying ergonomic principles in the workplace is to maintain a balance between job stress requirements and PWC. If PWC is exceeded, the worker is at risk of overexertion. If the job stress requirements are less than PWC, the worker is underutilized and there is a productivity loss.

Tichauer [1] pointed out that ergonomic stresses can be classified into two distinct categories: (1) traumatogenic (high dose over short duration) and (2) cumulative (low dose over long duration). In most cases, ergonomic stress analysis is usually centered around assessing the effects of traumatogenic stresses on individuals. Once assessed, the stress requirements are compared to PWC. This chapter deals with the application of PWC in this regard. However, the application of PWC to cumulative stresses has not been fully explored by ergonomic researchers. It is believed that future developments will deal with this subject.

III. PRINCIPLES OF PWC

A. Definition

Physical work capacity is the functional capacity of an individual to perform a certain task that requires muscular activity over a period of time. The length of time may vary from few seconds (e.g., strength) to several hours (e.g., endurance).

B. Components of PWC

The components of physical work capacity are muscular strength, muscular endurance, cardiovascular endurance, and joint flexibility (Figure 1). A brief description of each component is given below. Special emphasis is given to the components of PWC that have been used to define human abilities in the workplace.

1. Muscular Strength

Muscular strength is the maximum force exerted by a group of muscles under prescribed conditions. Muscular strength is usually determined over very few work cycles (≤10).

There are two types of muscular strength: static and dynamic. Static strength does not involve any motion during physical exertion. Dynamic strength, however, requires movement during the trajectory of motion.

Static strength, also known as maximum voluntary contraction or isometric strength, is the maximum force exerted by a group of muscles over a single trial. *Dynamic strength* is the maximum load that can be handled by a person one time (one-repetition maximum or 1 RM) or several times without resting between the repetitions (e.g., 10 repetitions or 10 RM) for the task under consideration.

2. Muscular Endurance

Muscular endurance is the ability of a specific group of muscles to continuously perform a task until the individual is no longer able to sustain the task. The task duration may be from a few minutes to several hours depending upon the work intensity. Muscular endurance can be measured in terms of endurance time (the maximum length of time during which an individual is capable of maintaining a workload continuously), oxygen consumption, or heart rate.

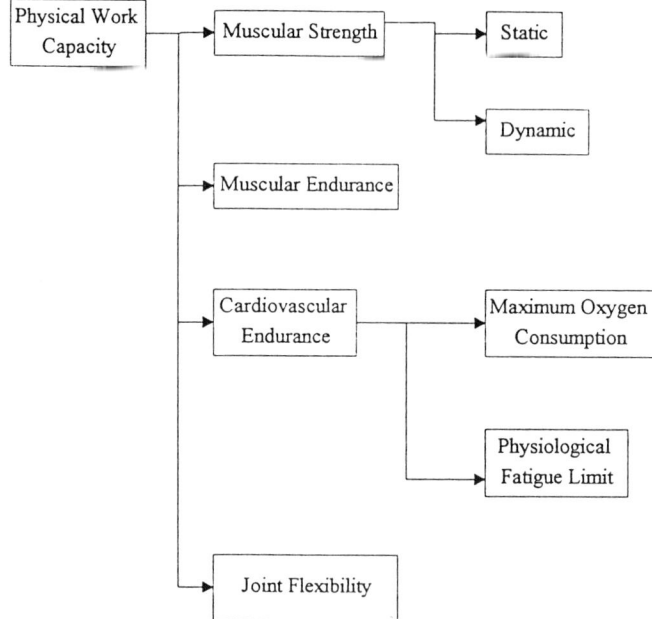

Figure 1 Components of physical work capacity.

3. Cardiovascular Endurance

Cardiovascular endurance is a measure of the ability of the cardiovascular system to perform continuous work until exhaustion. It can be defined for maximal and submaximal loads.

For maximal loads, cardiovascular endurance is known as maximum oxygen consumption or maximum aerobic power [2]. *Maximum oxygen consumption* is the maximal amount of oxygen that a person can consume during physical work while breathing air at sea level [2]. On average, the individual can last 4–5 min at maximal loads.

No standard term has been used for cardiovascular endurance at submaximal loads, although Asfour et al. [3] and Genaidy et al. [4] coined the term *physiological fatigue limit*. This limit defines the ability of the cardiovascular system to sustain submaximal loads for a work duration of a few minutes to several hours without excessive fatigue.

4. Joint Flexibility

Joint flexibility, also known as joint mobility [5], is usually defined relative to a specific posture. Joint flexibility is assessed in terms of the angular deviation of a body segment from the neutral position around a joint. The maximum angular deviation is termed the range of motion.

5. Other Components

From a biomechanics standpoint, industrial operations that require excessive loading of the lumbar spine are evaluated with respect to the *compressive strength* of the lumbar spine [6]. The compressive strength is usually determined around the limits set when testing cadaver specimens around failure points.

Genaidy et al. [7] suggest that from a safety standpoint the evaluation procedure should be made with reference to the *damage load*, which is the limit causing microtrauma to the lumbar spine tissue. The following regression equations can be used to predict the biomechanical tolerance limit of the lower back using the concept of damage load [7]:

$$DL = -805.18 + (0.74554 * CS)$$

$$CS = 7222.41 - (1047.71 * AGE) - (1279.18 * GENDER) + (56.73 * PP)$$

where
- DL = damage load (N)
- CS = compressive strength (N)
- AGE = age group (1 for age group 20–29 years, 2 for 30–39 years, 3 for 40–49 years, and 4 for 50 years and above)
- GENDER = gender (male = 1; female = 2)
- PP = population percentage

Since cumulative trauma disorders are becoming widespread in the workplace, Silverstein et al. [8] determined that any job requiring a hand force of more than 4 kg is classified as a high force job. Moreover, a job is considered highly repetitive if the duration of the work cycle is less than 30 sec or if the basic motions are repeated over 50% of the cycle time.

C. Factors Affecting PWC

Personal, task, and environmental parameters are important factors that affect the physical work capacity of an individual. Some of the most important personal factors are age, gender, body weight, and fitness level.

It is well established that physical work capacity declines as individuals age. The maximum physical work capacity is usually achieved in the age range of 25–35 years. The PWC of an individual who is over 60 years of age is about 50% of the values attained around 25–35 years. On average, the female physical work capacity is about two-thirds of the male capacity. Fitness level can significantly improve the physical work capacity of an individual. The capacity of a very fit person may reach as high as two to three times that of the least fit person.

The characteristics of a work task are an important consideration in deciding on the most appropriate type of physical work capacity for a given job. This can be attributed to a number of factors including (1) the use of different muscle groups, because each muscle group produces a certain output, and (2) a physiological limiting factor for each task (i.e., strength vs. endurance).

It is well established that environmental factors affect a worker's physical work capacity. For example, it was found that lifting capacity decreased by about 12% when subjects handled loads at 32°C wet bulb globe temperature (WBGT) compared to lifting loads at moderate ambient temperatures, defined as ≤27°C WBGT [9]. Thus, the effects of environmental factors should be taken into account when individuals are engaged in outdoor activities in hot or cold environments.

D. Assessment of Physical Work Capacity

The procedures for assessing physical work capacity should be standardized to enable the comparison of the PWCs of different employees performing the same job or to compare their values to normative data in the published literature.

Standard procedures for assessing static strength are reported by Chaffin [10]. Important factors to consider during the assessment procedure are (1) exertion duration, (2) the strength-measuring device, (3) rest periods between physical exertions, (4) body positions, and (5) subject instructions.

Kamon and Ayoub [11] described the methods used to determine maximum oxygen consumption. These methods can be grouped into direct and indirect methods. The direct methods tax the cardiorespiratory system to its limit, whereas indirect methods estimate the maximum oxygen consumption around submaximal loads.

Asfour et al. [12] and Genaidy [13] documented the details of assessing dynamic strength using the concept of one-repetition maximum (1 RM). It should be noted that dynamic tests should be closely monitored because they tax the musculoskeletal system to its limits. Consequently, the individual can be exposed to a higher risk of musculoskeletal injury.

The tests used in assessing the flexibility of various joints are described by Genaidy [13] and Guo et al. [14]. Also, tests for muscular endurance have been documented in studies by Genaidy and coworkers [13–15].

It is recommended that any physical work capacity evaluation test should be preceded by warm-up flexibility exercises for the body parts engaged in the test.

E. Normative Physical Work Capacity Data

1. Static Muscular Strength

Batti'e et al. [16] documented a comprehensive database of isometric strengths for three body parts: arm, back, and leg. The static strengths were collected for men and women of various age groups (20–29, 30–39, 40–49, 50–59, and ≥60 yr).

Mathiowetz et al. [17] reported power and precision grip strength data as a function of gender and age (20–24, 25–29, 30–34, 35–39, 40–44, 45–49, 50–54, 55–59, 60–64, 65–69, 70–74, and ≥75 yr). Three types of precision grip strength were measured: tip pinch (thumb tip to index finger tip), key pinch (thumb pad to lateral aspect of middle phalanx of index finger), and palmar pinch (thumb pad to the pads of index and middle fingers). For each of the hand strengths, the individual is seated with the elbow flexed at 90° and close to the body and the forearm in a neutral position.

The static strength data are summarized in Tables 1–5.

2. Joint Flexibility

Laubach [18] compiled the range-of-motion data published by Barter et al. [19] and Harris and Harris [20]. The results are given in Table 6 for both men and women. It should be pointed out that this database lacks important flexibility measures such as the range of motion of the trunk. Batti'e et al. [21] reported some of these measures, which are presented here in Table 7 for men and women of various age groups (20–29, 30–39, 40–49, 50–59, and ≥60 yr).

3. Muscular and Cardiovascular Endurance

Over the past several decades, the physical work capacity measures for continuous activities have been based on highly dynamic tasks (bicycle ergometer and treadmill), which are inapplicable to industrial tasks. As pointed out earlier, physical work capacity is task-specific, requiring the use of different muscle groups and combined static and dynamic muscular contractions.

Recently, Genaidy et al. [4], Genaidy and Asfour [22], and Asfour et al. [23,24] established endurance time curves for lifting tasks for up to 8 hr. The results are graphically displayed in Figure 2 for various lifting tasks.

Table 1 Isometric Strength (N) Normative Data

Body part	Age (yr)	Male		Female	
		Mean	SD	Mean	SD
Arm	20–29	406	77	222	60
	30–39	387	81	222	60
	40–49	369	89	222	55
	50–59	345	81	206	54
	60+	322	98	229	54
Back	20–29	582	154	368	80
	30–39	558	172	372	142
	40–49	468	126	375	104
	50–59	500	141	287	100
	60+	469	79	237	75
Leg	20–29	1043	262	523	189
	30–39	1003	255	517	178
	40–49	927	249	512	159
	50–59	847	238	460	154
	60+	802	230	475	151

SD = standard deviation.
Source: Ref. 16.

Table 2 Power Grip (N) Normative Data

Age (yr)	Hand[a]	Male Mean	Male SD	Female Mean	Female SD
20–24	R	538	92	313	65
	L	465	97	271	58
25–29	R	537	102	331	62
	L	492	72	282	54
30–34	R	542	100	350	85
	L	491	97	302	79
35–39	R	532	107	330	48
	L	502	97	295	52
40–44	R	520	92	313	60
	L	502	83	277	61
45–49	R	489	102	277	67
	L	449	101	249	57
50–54	R	505	81	293	52
	L	453	76	255	48
55–59	R	450	119	255	56
	L	370	104	210	53
60–64	R	399	91	245	45
	L	342	90	203	45
65–69	R	405	92	221	43
	L	342	88	182	37
70–74	R	335	96	221	52
	L	288	81	185	45
75+	R	292	93	190	49
	L	245	76	167	40

[a]R = right; L = left.
Source: Ref. 17.

4. Manual Materials Handling

As manual handling activities are a major cause of musculoskeletal injuries in the workplace, they have been extensively studied by many researchers. As a result, several comprehensive databases have been reported in the published literature.

Snook and Ciriello [25] published tables on the dynamic strengths of male and female workers performing infrequent (less than once every 5 min) manual materials handling activities (lifting, lowering, carrying, pushing, and pulling). Moreover, they documented endurance limits (more frequent than once every 5 min) for manual handling activities. The endurance limits are recorded in terms of the maximum acceptable weight of load. Other researchers such as Ayoub et al. [26] and Mital [27,28] reported capacity limits for frequent lifting tasks. The limits established by Snook and Ciriello [25], Ayoub et al. [26], and Mital [27,28] are based on the psychophysical approach.

The National Institute for Occupational Safety and Health [29] and Hidalgo [30] developed comprehensive lifting capacity limits for frequent and infrequent activities. These limits are based on biomechanical, physiological, and psychophysical criteria.

5. Upper Extremity Activities

Genaidy et al. [31] published preliminary recommendations for repetition, force, and posture. These recommendations were established on the basis of epidemiological and biomechanical data reported by many investigators. They are presented in Table 8.

Table 3 Tip Pinch (N) Normative Data

		Male		Female	
Age (yr)	Hand	Mean	SD	Mean	SD
20–24	R	80	13	49	9
	L	76	10	47	8
25–29	R	81	20	53	8
	L	78	23	50	8
30–34	R	78	30	56	13
	L	78	21	52	13
35–39	R	80	16	52	11
	L	79	17	53	11
40–44	R	79	18	51	12
	L	79	16	49	13
45–49	R	83	22	59	13
	L	78	18	54	12
50–54	R	81	18	56	10
	L	79	18	51	11
55–59	R	74	15	52	8
	L	67	17	46	6
60–64	R	70	18	45	9
	L	68	17	44	9
65–69	R	76	19	47	9
	L	68	13	47	11
70–74	R	61	12	45	12
	L	59	12	44	10
75+	R	62	15	43	13
	L	62	17	41	11

Source: Ref. 17.

IV. APPLICATIONS OF PWC

A. Role of PWC in Physical Ergonomics Job Design

Genaidy [32] outlined a procedure for the use of physical work capacity within the context of the physical ergonomics job design cycle (Figure 3). This procedure is centered upon the ergonomic index E,

$$EI = \frac{SR}{PWC}$$

where SR signifies stress requirements and PWC is physical work capacity.

The physical ergonomics job design cycle consists of the following steps:

Step 1. Review medical records. This step involves the computation of epidemiological measures on the basis of data collected from medical records. These measures can be prepared for various factors such as the type of operation and personal factors such as age and gender.

Step 2. Identify high-risk operations. In light of epidemiological measures, one can identify operations exposing employees to a high risk of injury and/or illness.

Table 4 Key Pinch (N) Normative Data

Age (yr)	Hand	Male Mean	SD	Female Mean	SD
20–24	R	116	16	78	9
	L	110	16	72	9
25–29	R	119	22	78	9
	L	111	20	74	9
30–34	R	117	21	83	13
	L	117	23	79	16
35–39	R	116	14	74	9
	L	114	17	71	12
40–44	R	114	12	74	13
	L	112	18	70	13
45–49	R	115	18	78	14
	L	110	20	74	13
50–54	R	119	20	74	11
	L	116	19	72	12
55–59	R	108	19	70	11
	L	102	21	65	10
60–64	R	103	24	69	12
	L	99	18	63	11
65–69	R	104	17	67	12
	L	98	16	64	13
70–74	R	86	11	65	13
	L	85	13	61	13
75+	R	91	21	56	10
	L	85	13	51	12

Source: Ref. 17.

Step 3. Perform a task analysis of high-risk operations. Task analysis involves detailed analysis of industrial operations through video recording and observation in order to identify the various elements of each job as well as the risk factors associated with them.

Step 4. Quantify stress requirements in terms of a measurable output. For an existing job, stress requirements are determined on the job. For a new operation that does not yet exist, stress requirements should be designed so that they match the physical work capacity of the potential employees. An example of stress requirements is the maximum load handled on the job for infrequent lifting tasks [33–35].

Step 5. Determine the physical work capacity based on physiological, biomechanical, or psychophysical criteria, as appropriate. The outcome of employee testing should be a measurable output identical to that used in the quantification of stress requirements. An example of physical work capacity is isometric strength measured for a given job position [33–35].

Step 6. Calculate the ergonomic index (EI_{comp}) as the ratio of stress requirements and physical work capacity. The ergonomic index is a unitless measure.

Step 7. Set the desired ergonomic index range in the workplace (EI_{des}). This will be based on how much risk the company is willing to accept. Initially, the occupational health and safety practitioner may be forced to accept the status quo. Later the EI_{des} may be revised as needed (see step 11).

Table 5 Palmer Pinch (N) Normative Data

Age (yr)	Hand	Male Mean	SD	Female Mean	SD
20–24	R	118	25	77	10
	L	114	26	73	13
25–29	R	116	19	79	14
	L	112	19	76	13
30–34	R	110	21	86	22
	L	113	25	81	21
35–39	R	116	18	78	19
	L	115	24	76	15
40–44	R	109	19	76	14
	L	110	22	74	16
45–49	R	107	15	80	13
	L	105	17	78	13
50–54	R	106	24	77	14
	L	107	26	73	13
55–59	R	105	21	71	14
	L	95	20	69	13
60–64	R	97	15	66	14
	L	94	14	64	12
65–69	R	95	13	63	14
	L	94	18	61	15
70–74	R	81	15	64	12
	L	84	15	62	9
75+	R	83	19	53	12
	L	81	17	51	12

Source: Ref. 17.

Step 8. If EI_{comp} exceeds EI_{des}, apply engineering and administrative controls. An example of engineering controls is the reduction of the value of a task variable such as the frequency of handling. An example of administrative controls is the application of a job rotation program.

Step 9. If EI_{comp} is less than EI_{des}, it may be possible to increase the stress requirements in order to improve the work efficiency and match the physical work capacity.

Step 10. If EI_{comp} is within EI_{des}, no correction is needed.

Step 11. Test if EI_{des} results in a minimum risk of musculoskeletal disorder. This can be accomplished by conducting a prospective epidemiological study.

The details of the EI methodology are described by Genaidy et al. [31]. The structure of the EI indicator takes into account the main and interactive effects of risk factors (e.g., repetition, force, and posture). A case study is given in Ref. 31 to illustrate the use of the EI formula. It should be pointed out that the EI equation assumes equal weights for various factors. Thus a heavy-force/low-repetition operation can result in effects comparable to those of a light-force/high-repetition operation. However, it seems that one risk factor may be more important than another factor under various conditions. The EI formula in its current form does not account for these conditions.

In sum, the EI methodology is a general method that has also been adopted by other researchers in ergonomic task analysis. Variations of the EI methodology are the lifting strength rating [33–35] and job severity index [36] techniques that are discussed in the following section.

Table 6 Range-of-Motion Data (in degrees) for the Wrist, Elbow, Shoulder, Neck, Hip, Knee, and Ankle

Joint	Movement	Male		Female	
		Mean	SD	Mean	SD
Wrist	Flexion	90	12	79.7	15.1
	Extension	99	13	60.6	10.5
	Abduction	27	9	29.7	9.1
	Adduction	47	7	50.4	10.8
Elbow	Flexion	142	10	151.1	7.1
	Pronation	113	24	101.9	15.2
	Supination	77	22	88.9	17.1
Shoulder	Flexion	188	12	167.9	10.0
	Extension	61	14	41.5	10.0
	Abduction	134	17	—	—
	Adduction	48	9	—	—
Neck	Flexion	—	—	58.7	10.3
	Extension	—	—	89.3	9.9
	Lateral bending, right	—	—	50.5	7.6
	Lateral bending, left	—	—	47.2	8.0
	Rotation, right	—	—	83.1	10.2
	Rotation, left	—	—	78.5	11.2
Hip	Flexion	113	13	79.6	14.5
	Abduction	53	12	—	—
	Adduction	31	12	—	—
	Medial rotation	31	9	43.8	11.6
	Lateral rotation	30	9	55.8	9.5
Knee	Flexion (standing)	113	13	—	—
Ankle	Flexion	35	7	18.8	6.2
	Extension	38	12	49.7	8.6
	Abduction	23	7	37.6	10.8
	Adduction	24	9	30.0	9.5

Note: The hip medial and lateral rotations were measured with the subject in a sitting position for the male data.
Source: Compiled by Laubach [18] from Barter et al. [19] for the male data and from Harris and Harris [20] for the female data.

B. Ergonomic Task Analysis

1. Manual Materials Handling

As mentioned earlier, manual materials handling activities have received the attention of most investigators. This is attributed to their strong association with low back problems.

Static strength measurements have been used extensively by Chaffin and coworkers in industry to predict the future risk of low back injuries posed by infrequently performed manual handling tasks [33–35]. Chaffin and others devised a method to evaluate infrequent lifting tasks based on the lifting strength rating (LSR), defined as the ratio of the maximum weight handled on the job to the isometric strength exerted in the same load position. It was found that an LSR value of 0.2 or less results in a nominal risk of injury. When the LSR exceeded 0.5, the number of low back injuries increased significantly.

The job severity index (JSI) was developed by Ayoub and coworkers [36] for evaluating frequent lifting tasks. They developed a set of regression models designed to predict the

Table 7 Truncal Flexibility Measures (cm)

Flexibility measure	Age (yr)	Male Mean	Male SD	Female Mean	Female SD
Sit and Reach	20–29	2.1	8.8	2.9	8.1
	30–39	−0.5	9.0	2.4	7.9
	40–49	−3.5	9.2	1.5	8.7
	50–59	−7.4	9.3	−0.4	8.2
	60+	−9.0	10.3	−3.5	7.9
Side bending, right	20–29	23.4	3.9	22.8	3.8
	30–39	21.9	3.6	21.7	3.3
	40–49	19.8	3.6	19.1	3.4
	50–59	17.8	3.4	18.7	3.1
	60+	17.8	3.8	16.7	4.4
Side bending, left	20–29	23.4	4.0	23.2	3.8
	30–39	22.1	3.8	21.7	3.3
	40–49	20.1	3.8	19.3	3.5
	50–59	18.0	3.6	18.7	3.2
	60+	17.7	3.6	17.5	4.5

Note: Side bending was measured as the difference between the position of the fingertips in erect standing and in maximal side bending to the right and left sides. During measurements, the subject is instructed to keep the knees straight and both heels on the floor when bending to the side. Sit and reach is measured as the distance between the fingertips and the toes, with the knees straight and the ankles maintained at 90° by having the soles of the feet pressed perpendicular to the sitting surface.
Source: Ref. 21.

lifting capacity for various height ranges, namely, floor to knuckle, floor to shoulder, floor to reach, knuckle to shoulder, knuckle to reach, and shoulder to reach. These models were based on experimental data collected from a large industrial population using the psychophysical approach. Ayoub and coworkers found that if a frequent lifting job required a JSI value in excess of 1.5, the statistics of low back injuries increased at an alarming rate. A strong positive correlation between frequent lifting and back injury statistics was established.

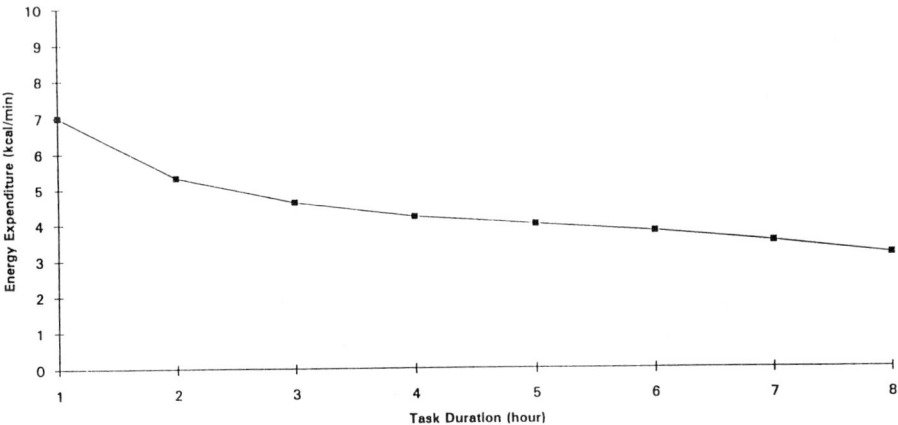

Figure 2 Energy expenditure as a function of task duration for continous lifting tasks.

Table 8 Repetition, Force, and Posture Guidelines for Upper Extremity Tasks

Risk factor	Joint	Level	Range
Repetition (times/day, assuming an 8-hr workday with two 15-min breaks and 30 min for lunch)	Fingers	Very low	0–3; 656
		Low	3,657–7,312
		Moderate	7,313–14,624
		High	14,625–21,936
		Very high	≥ 21,937
	Wrist	Very low	0–1,951
		Low	1,952–3,902
		Moderate	3,903–7,804
		High	7,805–11,706
		Very high	≥ 11,707
	Elbow, Shoulder, Neck	Very low	0–473
		Low	474–946
		Moderate	947–1,893
		High	1,894–2,838
		Very high	≥ 2,839
Force (% maximum force)	—	Very light	0%–1.6%
		Light	1.7%–3.2%
		Moderate	3.3%–6.4%
		Heavy	6.5%–9.6%
		Very heavy	≥ 9.7%
Body posture (angular deviation as percent of range of motion)	—	Very mild	0%–5%
		Mild	6%–10%
		Moderate	11%–20%
		Severe	21%–30%
		Very severe	≥ 31%

As pointed out earlier, Snook and Ciriello [25] developed a comprehensive database of manual materials handling capacity for lifting, lowering, pushing, pulling, and carrying activities. In an earlier study, Snook [37] found that a worker was exposed to a high risk of low back injury if he/she was handling a load that was equivalent to less than 75% of the working population.

2. Upper Extremity Work Tasks

Genaidy et al. [31] established a procedure for assessing upper extremity ergonomic deficiencies in the workplace. No epidemiological study was conducted to test the effectiveness of this procedure. However, the findings of Silverstein and coworkers [38] were used to perform preliminary testing of the methodology.

C. Job-Simulated Exercise Programs

The concept of repetition maximum was utilized by Asfour et al. [12] and Genaidy [13] as an instrument to evaluate the gains in dynamic strength acquired when individuals are conditioned to perform manual handling tasks. Endurance time was extensively used by Genaidy

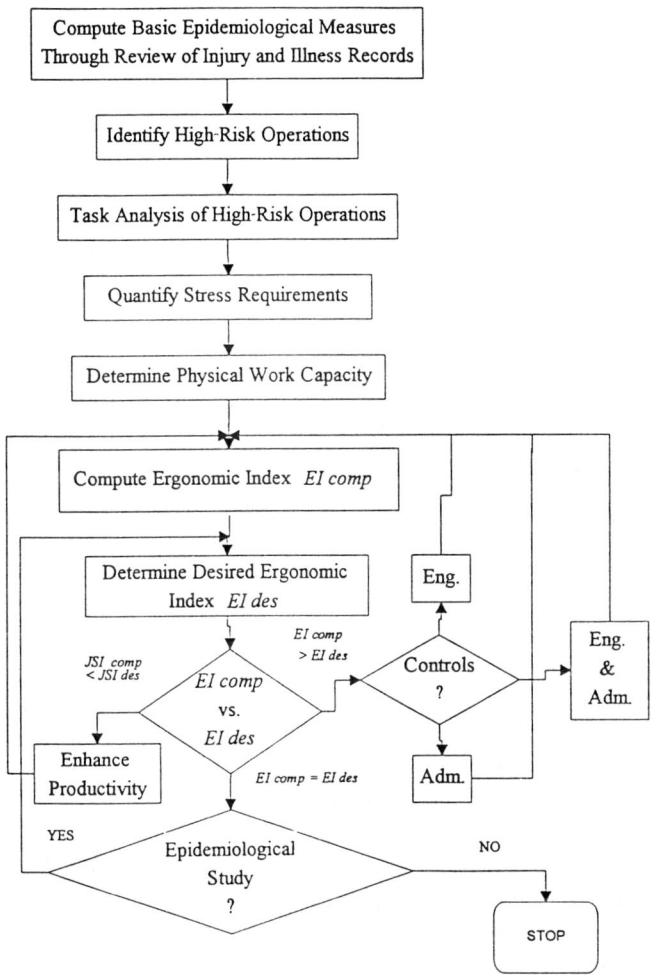

Figure 3 Use of physical work capacity within the context of ergonomic job design cycle.

and coworkers in evaluating the muscular endurance of individuals engaged in job-simulated exercise programs [13,14]. Joint flexibility was also examined in these studies.

D. Preemployment Placement and Screening and Essential Job Functions

Physical work capacity measures can provide a tool to assess the physical ability of individuals after they are hired. Employees can be advised about their ability to perform a certain job and the degree of injury risk associated with it. However, the physical work capacity cannot be used by employers to deny an employee a certain job if the employee elects to perform that job.

Physical work capacity values can be used to monitor injured employees during rehabilitation and those assigned to light duty jobs. Moreover, the physical capacity data can assist the human resource, medical, and safety departments in developing a data bank of the essential job functions.

V. CONCLUDING REMARKS AND FUTURE CONSIDERATIONS

The principles and applications of physical work capacity have been discussed in this chapter. It is evident from this review that most studies have concentrated on developing PWC limits for manual materials handling activities. These data were incorporated as part of ergonomic task analysis to identify trigger points for the control of occupational low back injuries. These efforts were largely augmented because it has been widely reported that musculoskeletal injuries resulting from manual materials handling, particularly those of the lower back, constitute the single most costly health disorder in the workplace. This is further supported by the first incidence of occupational (i.e., low back) injury reported in the history of mankind during the work process of building the pyramids in ancient Egypt.

Despite decades of extensive research in the field of manual materials handling, the cost of low back injuries is still rising at an alarming rate. So why aren't we able to significantly reduce the cost of such injuries and other forms of musculoskeletal disorders in the workplace?

Very recently, Leamon [39] provided a review on the validity of various criteria for the prevention of occupational low back injuries. He pointed out some of the limitations of prior studies and emphasized the importance of extending research findings to reality through independent replication and validation of research findings. Although he indicated "the need for a validation criterion for use in the workplace design in order to reduce low back pain disability or severity and the need for an international protocol to allow cross-study validation of present and future criteria," no recommendations were given that can benefit researchers and practitioners in their efforts to control injuries. Furthermore, Leamon did not examine the root cause of the onset of low back injuries although he claimed that "we simply do not know what causes low back disability."

Even though we do not possess the required research findings to fully understand the etiology of occupational musculoskeletal injuries (except in some cases), we should go back to the basics and attempt to understand the root cause of these disorders. I believe such attempts can be made and that they may significantly contribute to our understanding.

The root cause of musculoskeletal disorders can be discovered by looking closely at the role of humans in organizations, keeping in mind that the human engine is not a mechanical engine. Such an understanding will assist in developing a framework for the design of work in organizations. This is specially important in today's world, in which (1) customers take charge, (2) competition (locally and globally) heats up, and (3) government regulations intensify.

REFERENCES

1. E. R. Tichauer, The Biomechanical Basis of Ergonomics—Anatomy Applied to the Design of Work Situations, Wiley, New York, 1978.
2. P. O. Astrand and K. Rodahl, Textbook of Work Physiology: Physiological Bases of Exercise, McGraw-Hill, New York, 1986.
3. S. S. Asfour, A. M. Genaidy, and A. Mital, Physiological guidelines for the design of manual lifting and lowering tasks: the state of the art, Am. Ind. Hyg. Assoc. J. 49:150–160 (1988).
4. A. M. Genaidy, T. M. Khalil, S. S. Asfour, and R. C. Duncan, Human physiological capabilities for prolonged manual lifting tasks, IIE Trans. 22(3):270–280 (1990).
5. D. B. Chaffin and G. B. J. Anderson, Occupational Biomechanics, Wiley, New York, 1991.
6. National Institute for Occupational Safety and Health, *Work Practices Guide for Manual Lifting*, NIOSH Tech. Rep., DHHS, NIOSH, Cincinnati, OH, 1981.

7. A. M. Genaidy, S. M. Waly, T. M. Khalil, and J. A. Hidalgo, Compression tolerance limits of the lumbar spine, *Ergonomics* 36(4):415–434 (1993).
8. B. A. Silverstein, L. J. Fine, and T. J. Armstrong, Hand wrist cumulative trauma disorders in industry, *Bri. J. Ind. Med.* 43:779–784 (1986).
9. H. Hafez, *Manual Lifting Under Hot Environmental Conditions*, Ph.D. Thesis, Texas Tech Univ., Lubbock, TX, 1984, unpublished.
10. D. B. Chaffin, *Ergonomics Guide for the Assessment of Human Static Strength*, American Industrial Hygiene Association, Washington, DC, 1975.
11. E. Kamon and M. M. Ayoub, *Ergonomics Guides to Assessment of Physical Work Capacity*, American Industrial Hygiene Association, Washington, DC, 1976.
12. S. S. Asfour, M. M. Ayoub, and A. Mital, A., Effects of an endurance and strength training programme on the lifting capability of males, *Ergonomics* 27:435–442 (1984).
13. A. M. Genaidy, Truncal flexibility exercise effects on musculoskeletal capability for manual handling tasks, *Appl. Ergon.* 22(3):155–162 (1991).
14. L. Guo, A. Genaidy, J. Warm, W. Karwowski, and J. Hidalgo, Effects of job-simulated flexibility and strength-flexibility training protocols on maintenance employees engaged in manual handling operations, *Ergonomics* 35(9):1103–1117 (1992).
15. A. M. Genaidy, T. Gupta, and A. Al-Shedi, Improving human capabilities for combined manual materials handling through a short and intensive physical training program, *Am. Ind. Hyg. Assoc. J.* 51(11):810–814 (1990).
16. M. C. Batti'e, S. J. Bigos, L. D. Fisher, T. H. Hansson, M. E. Jones, and M. D. Wortley, Isometric lifting strength as a predictor of industrial back pain reports, *Spine* 14(8):851–856 (1989).
17. V. Mathiowetz, N. Kashman, G. Volland, K. Weber, and M. Dowe, Grip and pinch strength: normative data for adults, *Arch. Phys. Med. Rehab.* 66:69–74 (1985).
18. L. L. Laubach, Range of motion, in *Anthropometric Source Book*, Vol. I, *Anthropometry for Designers*, NASA Ref. Pub. 1024, Webbs Assoc., 1978, Chap. VI, pp. VI/1–VI/20.
19. J. T. Barter, I. Emanuel, and B. Truett, *A Statistical Evaluation of Joint Range of Motion Data*, WADC-TN-57-311, Wright Air Development Center, Wright-Patterson Air Force Base, OH, 1957.
20. M. L. Harris and C. W. Harris, A factor analytic study of flexibility, *National Convention of the American Association of Health, Physical Education and Recreation*, Research Section, St. Louis, MO, 1968.
21. M. C. Batti'e, S. J. Bigos, A. Sheehy, and M. D. Wortley, Spinal flexibility and individual factors that influence it, *Phys. Ther.* 67(5):653–658 (1987).
22. A. M. Genaidy and S. S. Asfour, Effects of frequency and load of lift on endurance time, *Ergonomics* 32(1):51–57 (1989).
23. S. S. Asfour, M. Tritar, and A. M. Genaidy, Endurance time and physiological responses for prolonged arm lifting tasks, *Ergonomics* 34:335–342 (1991)
24. S. S. Asfour, T. M. Khalil, A. M. Genaidy, M. Akcin, I. M. Jomoah, J. G. Koshy, and M. Tritar, *Ergonomic Injury Control in High Frequency Lifting Tasks*, Final Rep., NIOSH, Grants No. 1 R01, OH02591-01 and 5 R01 OH02591-02, 1991.
25. S. Snook and V. M. Ciriello, The design of manual handling tasks: revised tables of maximum acceptable weights and forces, *Ergonomics* 34:1197–1213 (1991).
26. M. M. Ayoub, N. J. Bethea, S. Deivanayagam, S. S. Asfour, G. M. Bakken, and D. Liles, *Determination and Modeling of Lifting Capacity*, Final Rep., DHHS (NIOSH) Grant No. 5-R01OH-00545-02, 1978.
27. A. Mital, Comprehensive maximum acceptable weight of lift database for regular 8-hr workshifts, *Ergonomics* 27:1127–1138 (1984).
28. A. Mital, Maximum weights of lift acceptable to male and female industrial workers for extended work shifts, *Ergonomics* 27:1115–1126 (1984).
29. T. R. Waters, V. Putz-Anderson, A. Garg, and L. J. Fine, Revised NIOSH equation for the design and evaluation of manual lifting tasks, *Ergonomics* 36:749–776 (1993).
30. J. A. Hidalgo, A comprehensive lifting model: the missing link, Unpublished Doctoral Dissertation, Univ. Cincinnati, Cincinnati, OH, 1993.

31. A. M. Genaidy, A. A. Al-Shedi, and R. L. Shell, Ergonomic risk assessment: preliminary guidelines for analysis of repetition, force and posture, *J. Hum. Ergol.* 22:45–55 (1993).
32. A. M. Genaidy, Physical ergonomics job design: an accident prevention approach to control upper extremity cumulative trauma disorders in manufacturing industry, *J. Occup. Accidents 13*:303–320 (1990).
33. D. B. Chaffin, G. D. Herrin, W. M. Keyserling, and J. A. Foulke, *Pre-employment Strength Testing*, NIOSH Tech. Rep., DHEW (NIOSH) Publ. No. 77-163, 1977.
34. D. B. Chaffin, G. D. Herrin, and W. M. Keyserling, Preemployment strength testing. An updated position, *J. Occup. Med. 20*(6):403–408 (1978).
35. W. M. Keyserling, G. D. Herrin, and D. B. Chaffin, Isometric strength testing as a means of controlling medical incidents on strenuous jobs, *J. Occup. Med. 22*(5):332–336 (1980).
36. D. H. Liles, S. Deivanayagam, M. M. Ayoub, and P. Mahajan, A job severity index for the evaluation and control of lifting injury, *Hum. Factors 26*(6):683–693 (1984).
37. S. Snook, The design of manual handling tasks, *Ergonomics 21*:963–985 (1978).
38. B. A. Silverstein, L. J. Fine, and T. J. Armstrong, Occupational risk factors and carpal tunnel syndrome, *Am. J. Ind. Med. 11*:343–358 (1987).
39. T. B. Leamon, Research to reality: a critical review of the validity of various criteria for the prevention of occupationally induced low back pain disability, *Ergonomics 37*:1959–1974 (1994).

10
Worker Participation
Approaches and Issues

Alexander L. Cohen
Occupational Human Factors, Cincinnati, Ohio

I. INTRODUCTION

This chapter summarizes the literature on participative approaches in addressing workplace problems, with mention made of their application to workplace safety and health issues in general and ergonomic problems in particular. The material reproduces with few modifications a section of a recent National Institute for Occupational Safety and Health (NIOSH) report [1] describing team techniques to define ergonomic problems in meatpacking jobs and the merits of this approach in proposing effective control measures. The text is used to set the stage for three case studies that are described in the report in elaborating on the processes involved in these types of interventions. The results of these studies are discussed briefly later in the chapter.

The term *participatory approach* as used in the work setting has a number of meanings. In this chapter, its essential meaning is *worker involvement*. Hence, references to teams, groups, and committees formed to deal with work-related issues (ergonomic hazards in this instance) are assumed to include front-line employees or their representatives. Other members of such bodies may be supervisory or managerial persons, staff from other departments whose duties pertain to matters at issue, and outside consultants. Lawler [2] characterizes employee participation as the movement of decision making, information, knowledge and skill, and rewards to the lower levels of an organization. References to these and other elements are apparent in the various forms and levels of worker participation noted in the chapter.

II. BACKGROUND

A. Rationale for Worker Participation

A review of the industrial psychology and organizational behavior management literature makes clear the benefits that can accrue from worker involvement in dealing with organizational issues, along with some important qualifiers [2-4]. The results are summarized in the following paragraphs.

1. Enhanced Worker Motivation and Job Satisfaction

An employee's work motivation and job satisfaction are increased not only by added pay but also by the opportunities to provide input into decisions affecting work methods, everyday job routines, and performance goals. Having control over one's own work is especially satisfying and enhances commitment and quality effort. Positive results, though, are conditioned by a number of factors such as the following.

1. The perception that an important work performance matter is at issue, not some trivial concern (e.g., the color of the hallways).
2. The perception that the work is interesting and challenging. Worker participation to address repetitive, simplistic, standard tasks in and of itself would not be a good candidate unless the concern was to consider job redesign or other changes.
3. Workforce educational level and knowledge that indicate capabilities for offering meaningful input. Today's workforce, being better educated than their forebearers, have greater expectations about job roles and greater need for self-esteem.

2. Added Problem-Solving Capabilities

Employee involvement in decisions affecting their work situations can capitalize on their unique and relevant experience. Indeed, the person doing the job often has the best knowledge of the problem elements and insights into ways to improve the work. Effectiveness here can depend upon whether the individuals have the problem-solving skills needed to identify valid solutions and the ability to argue effectively for their adoption. Another factor is whether the issue is a local one in which the group can make decisions and take action or one having broader implications that require higher level review and approval. If the latter is the case, delays in responsiveness to recommendations can create cynical attitudes about the participative process.

3. Greater Acceptance of Change

There is evidence that participation in decision making regarding a major organizational change can lead to a significant reduction in resistance to that change. Creating better understanding of the needs for change though improved communications and the involvement of those affected in structuring the change in desirable ways can do much to gain employee commitment to a successful implementation. Lacking these efforts, change can be perceived as threatening job security or as having other negative consequences.

4. Greater Knowledge of the Work and the Organization

Taking part in problem solving in workplace conditions and in decision making in work design with those in one's own work group and/or with others from different units or areas will invariably increase the employee's knowledge of his or her own job if not of the overall company operation. An important payoff from such interaction can be improved communications and coordination among the members and their respective departments. However, for this to occur requires employee training in communication skills and motivation to use the knowledge of how other units function.

B. Forms and Levels of Participation[1]

Employee participation in work organizations can take a variety of forms. Among the shaping factors are (1) the nature of the issues requiring consideration, (2) whether the matters

---

[1]The legality of management forming certain groups with employee participants to address productivity, quality, and safety matters has been questioned. The National Labor Relations Act forbids such actions, fearing domination of such groups by management. In response, some employers have gone to self-directed work teams, and others are keeping with the existing forms but including volunteer employees as members who represent themselves in such groups. The issue may be resolved through court tests or OSHA reform legislation, which may exempt well-intentioned groups such as those dealing with safety and health concerns. See LaBar [5] for further details on this subject.

are broad-based or specific to a local operation or group, (3) whether the needs for response or action are time-limited or necessitate continuing efforts, (4) the abilities of the group most affected, and (5) the organization's prevailing practices for joint labor-management or participative approaches in resolving workplace issues. The degree or level of involvement may also vary. At one extreme may be simple consultations with individual workers or groups to obtain their reactions to ideas of superiors who will make the final decision. At the other extreme may be obtaining worker ideas along with those from management and other affected parties in addressing issues, with decisions based on consensus. The fact-finding report from the Commission on the Future of Worker-Management Relations [6] outlines the variety and scope of employee participation and labor-management cooperation in U.S. workplaces. In this section, common forms of worker participation found in industry are described as are different levels of sharing in decision making and other factors reflecting the degree of actual worker involvement.

1. Quality Circles

Quality circles are generally defined as small groups of worker volunteers from the same work area who, with their supervisor, agree to meet regularly to identify, analyze, and solve quality and related problems in their areas of responsibility [2,7]. They usually consist of 8-10 members who meet once a week during work hours. The volunteers typically receive training in some form of problem-solving techniques as part of this activity.

The use of quality circles is attributed to Deming's introduction of data-based quality control techniques in Japan to rebuild their industry after World War II [7]. Although originally intended as a program for troubleshooting by engineers, the movement quickly evolved to include line workers in accord with Deming's view that quality must concern every employee rather than be limited to the engineers or the quality control department. The success of Japanese industry in capturing large market shares for their products in the early 1970s led American businesses to emulate their techniques. In 1986 it was reported that more than 40% of U.S. companies employing more than 500 workers were using quality circles [8].

As Krigsman and O'Brien [7] note, quality circles in Japan were focused on performance data and quality control issues. Worker involvement was based on the underlying idea that workers ought to be responsible for the quality of their work and are in the best position to troubleshoot it. In the United States, quality circles became more of a participatory management technique intended not only to yield increased productivity and product quality but also to enhance employee motivation and job satisfaction. Although experiences in the United States tended to support these various outcomes, the results were not always up to expectations [9, 10]. For example, Griffin [10], in his study of U.S. electronics plants, found that quality circles produced initial improvements in job satisfaction, organizational commitment, and performance measures but that over time and in the absence of other supportive measures, these indicators reverted back to their original levels. When asked about this end result, quality circle members in this study stated that they felt that management was no longer interested in their recommendations and that their supervisors asked fewer questions as to how the group was functioning and displayed less enthusiasm about evaluating suggestions. Without continued management support for this program, the early improvements could not be sustained. On this point, Lawler [2] and Griffin [10] view quality circles as a building block to other forms of worker participation that could ultimately create a more participative culture in an organization. Cascio [3] notes too that worker participation programs can die out eventually if the organization does not change in a manner consistent with the democratic values that characterize such practices.

Safety circles represent a variation on the quality circle form of worker involvement, the difference being that the thrust of the group effort is directed to identifying, analyzing, and solving safety and related health risk problems in their work area [11,12]. The National Safety Council [13] describes a step-by-step approach to establishing safety circles. Needs for management support and resources for implementing recommendations, decision-making authority to be invested in the group, training of members in safety subjects, and interpersonal relationships are duly noted.

2. Labor-Management Committees

Whereas quality circles are small in size, composed of volunteers from a single work area who are brought together to address problems specific to their job tasks, labor-management committees are more expansive, including elected or appointed members from different areas within an organization and charged with a broader agenda. Also, unlike quality circles whose members can actually implement solutions, most committees only recommend actions, and their recommendations are then forwarded to other parties for concurrence or coordination in determining how and when approved actions can be carried out.

Joint labor-management committees offer opportunities to identify areas of mutual concern and to engage in cooperative activities that can reduce the level of traditional adversarial behavior between the two parties [2,14]. Two areas, quality of working life (QWL) and occupational safety and health, have been the focus of much joint committee activity. QWL committees seek ways to improve working conditions so as to enhance worker job satisfaction and morale and therein increase company productivity. QWL efforts can encompass recommendations for making a more pleasant physical environment, furnishing educational opportunities during off-job hours and facilities for recreation. In some instances, collective bargaining agreements struck between unions and management have enabled QWL committees to also address certain aspects of job classification and work schedule issues. The reader is referred to Lawler [2] for more details and examples of QWL committee work.

Joint labor-management safety and health committees offer opportunities for cooperative problem solving with regard to hazard recognition and control concerns as well as for the recommendation of preventive measures [14]. The effectiveness of these groups is the topic of a later discussion.

The membership of joint labor-management QWL and safety and health committees includes representatives from affected groups. Worker participation may be attained through elected workers or local union leaders, with management represented by department heads or other key figures. The success of such groups in effecting actions depends upon their own decision-making authority or links to others who have that role. As already noted, the committees make recommendations whose implementation may take the form of establishing task forces, work teams whose job is to formulate and carry out specific plans. A by-product of the committee deliberations and follow-on actions by these groups is that information is shared widely in the organization and more channels are opened for communication. As a result, more employees can understand the business better and participate more effectively in problem-solving activities.

3. Work Teams

Work teams are referred to in the literature as "self-regulating" work groups in that they can make decisions in terms of setting production goals, selecting work methods and quality control procedures, and managing inventory [2]. In some cases, such groups may have even more autonomy, for example, determining pay rates and hiring/firing policies. Management maintains oversight of the group's practices and operations and has the right to challenge any of its decisions. Work teams include all of the employees working in a given area who, with a chosen lead worker or supervisor, are given the responsibility for producing an entire product or offering a complete service. Because of their broadened roles, work team members are cross-trained so that each can do the various tasks that fall within the domain of the team, and frequently workers rotate their work assignments. In addition to the extensive training that may be needed to perform these multiple job functions, work team members require instruction in interpersonal skills. As explained, these skills are necessary to ensure positive, effective interactions among the group members. Indeed, their varied responsibilities demand that work teams meet often to discuss and reach agreement on numerous matters. Experiences with work teams in mining and various manufacturing companies have demonstrated gains in rate and quality of output, reduced turnover, and improvements in overall work efficiency (summarized in Peters [15] and Lawler [2]). There are also cases where work teams in these establishments did not survive. This appears to be most evident in companies having a more traditional management approach.

4. Gain Sharing

Gain sharing acknowledges worker participation in efforts to improve company economic performance through increasing the sales value of production relative to labor costs [3]. In one such plan, a ratio of the two factors is set, based upon the past year's experience, that, if exceeded, will result in cost savings to be shared by the employees and management in accordance with some agreed-upon formula. Another plan sets a production/performance standard that, if met in fewer than the expected work hours, yields savings that will be distributed. The participative structure in each instance uses a formal suggestion system that invites workers to submit in writing their ideas for improving work efficiency. Department production and screening committees made up of worker and management representatives review these inputs and select those to be implemented.

Company experiences with gain sharing and other incentive plans as reported in the literature show a roughly 20% increase in productivity but at the same time much variability in these results [16]. In some cases the plans yielded a 75% increase in output, and in others a 5% decrease. Success seems to be a function of many factors such as whether the market can absorb the increased production, whether product costs are controllable by employees, whether there is top management commitment and supervisory support for the plan, and whether management is open in sharing financial results and giving other evidence of trustworthiness.

5. Levels of Participation

Worker participation can also be viewed along a number of different dimensions. Liker et al. [17], for example, offer models reflecting variations in two dimensions. One is the locus of decision making, whether at the management level with consultations sought from affected individuals or groups or delegated downward with little management involvement. The second dimension is the manner of employee input into such processes—whether each person in an affected group has direct involvement or is represented by a subgroup. Quality circle and work teams as described above would appear to fit the model in which all workers are involved and have authority to make and carry out decisions. In contrast, joint labor-management committees would be categorized as representative in makeup with authority limited to making recommendations, not actual decisions. By itself, the formal suggestion system inherent in gain sharing would offer opportunity for direct input but no decision-making power, this being assumed by other committees or retained by management.

As noted by Liker et al. [17], Lawler [2], and others, the success of worker participation efforts in solving workplace problems and enhancing productivity, worker motivation, and satisfaction is not dependent on any one form of involvement but on what is best suited to the issues to be addressed and the situational factors that are present. Also, certain forms may evolve into others as conditions change, which may be important to sustain or further the positive effects seen in such practices.

III. WORKER PARTICIPATION APPROACHES IN WORKPLACE HAZARD CONTROL

Evidence indicating the effectiveness of worker involvement in efforts to reduce work-related risks of injury and disease is the subject of this section. Such participation has taken various forms akin to those previously mentioned. Reports documenting the importance of these approaches in cause-and-effect terms as well as defining factors of major consequence to successful outcomes are not numerous. Indeed, field studies in this area do not allow for easy isolation or manipulation of these variables or for comparisons with adequate control or nontreatment conditions. Due caution is thus advised in either interpreting or generalizing results. In this section, worker involvement in general injury and disease control problems is first described, followed by a discussion of efforts directed to controlling ergonomic hazards. The literature reviewed in these cases is admittedly selective and to some extent summary in nature. Its purpose is to illustrate worker participation approaches as applied to these kinds of concerns, highlighting certain aspects of their implementation, and resultant findings.

A. Joint Labor-Management Safety and Health Committees

The most common institutionalized form of worker participation in workplace safety and health matters is though membership on joint labor-management committees set up for that purpose [14]. Collective bargaining agreements between unions and management, especially after the passage of the Occupational Safety and Health Act of 1970, contained provisions for the establishment of these committees that were attempts to resolve these issues on a local level. The Bureau of National Affairs [18] reported that in 1970, 31% of industrial contracts covering 1000 or more workers had such provisions; this proportion rose to 39% in 1975 and to 45% in 1983. Boden et al. [19], in a survey of manufacturing companies having 500 or more employees in one state (Massachusetts) in the early 1980s, found that 67% of the unionized establishments had a joint labor-management committee addressing safety and health issues and that 49% of non-union workplaces had similar groups with employee-management representations. A 1993 national poll by the National Safety Council [20] found that 66% of the respondent companies had joint committees. The survey acknowledged sampling and other limitations that led the authors to feel that this figure might be higher than the national average.

The more cogent question, however, is whether the existence of these committees has had a positive impact on worker safety and health. The literature suggests mixed findings. For example, Cooke and Gautschi [21] used data from the state of Maine for compensable injuries and OSHA citations in 113 manufacturing firms during the period 1970–1976. Controlling for the size of the production workforce, business cycle effects, and OSHA citation experience, they found that the presence of joint labor-management safety and health committees was associated with a small and statistically insignificant decrease in lost-time injuries over the period in question. Similarly, Boden et al. [19] found virtually no effect in a study of whether the existence of a joint safety and health committee was correlated with either the number of OSHA complaints or serious hazards as measured by citations for 127 Massachusetts manufacturing firms. More detailed study of a subsample of companies with these committees, however, showed these outcome measures to co-vary in inverse fashion with the number of the powers of the committee to act, its opportunities to access and review different types of data (hazard, injury, and medical reports), and perceptions of a strong management commitment to worker health and safety. The authors concluded that maintaining a joint health and safety committee as a formality yields little results on company safety and health experience and that its impact is a function of activity level and a company environment truly supportive of its efforts.

Reinforcing the point just made, California in 1984 [18] reported the benefits of organizing joint labor-management committees to conduct self-inspections of safety and health conditions at major construction sites in the state as part of a voluntary compliance program. For work at three sites that employed 200-2600 workers, the injury and illness incidence rate dropped far below those averaged for the construction industry as a whole or the individual employer's rate at other similar projects. At one site, the decrease was from 7.4 cases per 100 full-time workers per year at program start-up to 4.2 cases afterwards. Project managers attributed the improved safety performance to increased awareness of hazards by employers and employees, better communications between the parties, and a belief by the workers that they could influence safety on the job.

Evidence that the effectiveness of joint health and safety committees in reducing injury and enforcement rates and in enhancing problem solving in workplaces is dependent on added considerations is also seen in the survey results from two Canadian provinces (Ontario and Quebec [22]). In analyzing performance on such indicators between the sample sites surveyed relative to all other workplaces in their peer group, it was found that a *committee capacity* factor was an important determinant in attaining a positive outcome. Elements of committee capacity, based on factor analyses, were found to consist of opportunities to conduct a broad scope of activities, use of institutionalized procedures in decision making, policies supporting improved resources for member training in safety and health, and sharing of information. While showing the apparent benefits of legislation mandating the establishment of joint labor-management health and safety committees in Canadian businesses, the authors acknowledge that the committee capacity factor accounts for only a limited amount of variance and that workforce experience and stability may be even more important factors in the final analysis.

Joint labor-management committees by themselves do not appear to be a major determinant in studies contrasting organizational program practices in companies that have exemplary safety and health records with those of more poorly performing cohorts. While perhaps facilitating worker participation, other direct means for promoting worker inputs into the program seem to be more influential than a formal committee. For reasons stated by Boden et al. [19], committees can vary greatly in their activities and roles that can affect workplace safety and health. Most studies comparing program factors in companies with good versus poor safety performance lack details as to whether there are functional differences between the committees found in the contrasting samples and details of their relationship to other participative efforts that may be of consequence. A commonly expressed view about safety and health committees is that without their establishment workers would have little means for involvement in any safety and health activities [20].

Joint labor-management health and safety committees have also been formed nationally to support continuing education of their respective members and to sponsor research work to address pressing health and safety problems of mutual benefit.

B. Work Teams for Hazard Control

Case studies and other reports in the popular and technical literature illustrate how work teams, safety circles, or equivalent groups, each of small size and composed of worker members engaged in similar jobs and from the same area, have made positive contributions to hazard control efforts [12, 23–25]. Typical is a report by Edwards [12], who studied the impact of a quality circle (QC) technique on safety issues in a large surface mine. Setup elements included (1) forming a screening committee of department heads and a QC-trained facilitator to set ground rules for the plan, (2) composing QCs of five to eight persons from worker volunteers in four selected departments, and (3) giving QC members plus mine safety committee persons 8 h of training on subjects such as brainstorming, data collection, and group dynamics. Subsequent 1-h weekly meetings were held during which the QCs focused on problems that would be expected at most mines, i.e., tool shortages, poor communications, unavailability of parts or supplies, lack of support equipment, inadequate housekeeping, etc. The circle members chose a problem they wanted to solve, collected data for delineating its nature, and then offered possible solutions taking into account cost-effectiveness considerations. A number of recommendations were implemented that had a significant effect on both productivity and safety. With regard to the latter, QC members experienced 58% of the accident occurrences on their shift in the 6 months before the program and 31% in the following 6-month period. For departments with circles, the accident frequencies decreased by 18% in before/after comparisons over the same 6-month periods.

Some difficulties in organizing or maintaining work team efforts directed to hazard control have also been noted. For example, a county engineering department reported marked improvement in the safety performance of work crews in one section when a quality circle approach was adopted to elicit worker inputs into ways of making their operations safer and feedback and incentives were used to reinforce the program efforts [24]. Injury frequency dropped by 52% and their associated cost by 92% after the plan was instituted for work crews who previously had the worst safety record of the various sections in the department. However, expanding this program to another division within the engineering department proved problematic for a number of reasons. The job routines of these workers did not require a natural team effort, and workers enjoyed their independence in fulfilling their specific responsibilities. As a consequence, the team problem-solving effort was viewed more as a "gimmick" of management. The program was nevertheless implemented wherein team members began blaming each other and management for failure to achieve any positive results. As a remedy, and at the suggestion of the workers with their supervisors, the teams were redrawn to take account of mutual needs for working relationships and compatibility among the partners. This worker input into the program helped reduce the earlier resistance. An 18% drop in injuries was noted after the revamped teams were formed though costs remained unchanged.

Peters [15], in reviewing research on organizational and behavioral factors associated with mine safety, mentions a study assessing the benefits of a self-regulated work team as introduced in a Pennsylvania coal mine on an experimental basis. The miners received training to make them each capable of performing any job in their section and to teach them about mine safety laws and violations. Periodic meetings and feedback were used to motivate worker interest in safety, and the autonomous nature of the group made each miner responsible for maintaining safe working conditions. Supervisors had responsibility and authority for the safety of their work crews with lesser concerns for production. This group experienced fewer violations and shutdowns than others in the same mine. The work crews also put into place more safe work practices and were more proactive toward safety than they were before the intervention. Despite these positive findings, however, an effort to expand the program to other mine sections was voted down by the union. One reason for the rejection was the perception that the special treatment given to the experimental group created an elitist attitude among its members that was resented by the miners in the other sections. This effect was unintended, and efforts to overcome the negative fallout either were not taken or were not sufficient to correct the situation. Peters [15] notes that the intervention effort in the mine disappeared 4 years after it was first initiated.

C. Direct Worker Inputs in Hazard Control

Reports of direct worker inputs having been formally solicited into hazard control programs, as contrasted with using a team or committee approach, are not that common. One case study of this type, conducted by Lin and Cohen [26], is important in showing both the merits of worker involvement for this purpose and some of its limitations. The site for the work was a 500-bed hospital with 1800 full- and part-time employees where a worker hazard detection program was put into place on a trial basis. Employees were first surveyed to determine their current level of awareness of workplace hazards and the means to control them. This was followed by a campaign to motivate employee reports of hazards by locating forms at convenient places, requiring a prompt follow-up response by safety staff to all such submissions, and highlighting actions taken in newsletters and posters.

Comparisons were made of the hazard reporting rates of employees, the number of recorded staff injuries or illnesses of staff members, and the content of the hazardous conditions reported by the employees as related to their recorded injuries and illnesses during 12-month periods before and after the start of this worker-based reporting system. Results showed that the frequency of hazard reporting increased during the intervention period and the frequency of actual injuries and illness declined during the last 6 months of the trial, by the end of which most of the hazard control recommendations had been implemented. This finding suggested an increased safety consciousness among the workers and a consequent reduction in the number of job mishaps. In analyzing the content of injury and illness records with the hazard recognition reports, there were instances of the hazard reports far exceeding the recorded cases of related injuries, which in turn became a basis for prioritizing control needs. Indeed, in several instances during the trial period, accident risk factors identified in worker reports were not acted upon soon enough to prevent an injury that occurred soon afterwards.

On the other hand, there were also instances where some hazards resulting in a high percentage of injury cases went undetected by the workers. Needle puncture wounds and physical exertion back injury from patient lifting were particularly notable. Because these mishaps are inherent in job routines and procedural in nature, their risks appeared less obvious to the workers than those posed by fixed physical features in their work environment. This omission indicated the need for employee training in appreciating functional kinds of hazards as a means of improving their overall hazard recognition skills. The latter was one of the basic recommendations agreed to by management, who, being satisfied with the overall findings of the trial, decided to adopt this worker participation effort as a permanent hospital program.

D. Worker Participation in Ergonomic Problem Solving

Ergonomics addresses the interaction of job demands and worker capabilities, the aim being to design the work requirements and/or workplace conditions in ways that will optimize productivity and at the same time preserve the health and safety of the workforce. Although the subject is much broader in scope [27], the rising incidence of musculoskeletal disorders of the upper extremities and the unabated numbers of costly low back problems in U.S. industry have focused ergonomic concerns on these types of problems in the main. Much is already known about occupational risk factors for these kinds of disorders, the major ones being forceful exertions, awkward body postures, local contact stresses, and repetitive motions [28]. Some efforts at controlling these hazards through the redesign of tools, improvement of workstation layouts, and the use of less fatiguing work organization methods have been reported, and guidelines for carrying out these actions have also been publicized [29-31]. Worker involvements in such activities and aspects of their participation are described below.

The automobile and auto parts industry has been the primary site for participatory ergonomics programs in the United States as well as in other countries. Indeed, the tradition of assembly line work with numerous workers engaged in short-cycle tasks requiring repetitive turning and twisting actions with tools and/or frequent lifting or other forms of manual materials handling make it a natural candidate for ergonomic study and problem solving. Reports in the popular literature cite a number of cases where worker participation has been instrumental in achieving successful outcomes. LaBar [32], for example, describes how the introduction of quality circles in a U.S. tire manufacturing plant following a takeover by a Japanese corporation turned around sagging production levels and an increasing injury incidence rate. The quality circles, referred to as Employee Involvement Groups (EIGs), were set up in different departments and run in accordance with Japanese practices with a steering committee overseeing their activities. While addressing a variety of safety, production, and quality control topics, a sampling of improvements made or recommended by these groups indicated a focus on ergonomic problems and solutions. One improvement noted was replacing an 18 stitches per tire procedure with one requiring just two stitches, thus reducing problems of repetitive motions believed responsible for the excess cases of carpal tunnel syndrome and tendinitis found for workers engaged in this task. Another was installing hydraulic systems to lift and turn 115-pound tires for inspection as opposed to having workers lift them, and using similar powered systems to lift heavy sheets of rubber. The apparent benefits were reductions in the incidence and severity of back injury. Overall, these and other types of hazard control measures in the plant caused a fivefold reduction in the incidence rate of worker injury over a 4-year period after the introduction of the EIGs. Inquiries of senior level management and union persons who remained with the company after the takeover credit these and other positive changes to listening to workers' suggestions and getting workers more involved in company activities. Quality circle concepts were instrumental in accomplishing this purpose.

LaBar's [33,34] descriptions of ergonomics efforts in two other automobile assembly plants emphasize the need to train the workforce at all levels with respect to recognizing relevant risk factors and early symptoms, the importance of engineering controls, and the role of employees in identifying problem areas and developing solutions. Regarding the latter, mention is made of over 200 suggestions received in 1 year from employees at one plant for ergonomic improvements, many of which were implemented and some carried over to a new plant facility as well. However, the reports are not clear in defining whether there are recognized formal groups where workers interact with others in providing this input or whether the suggestions are made strictly on an individual basis. References to teams, committees, and task forces acknowledge persons from the medical, safety, and engineering departments who appear to spearhead the hazard control program with workers advised to report problems to them. Nevertheless, successes are noted. One plant [34] reported a 50% drop in the number of ergonomics-related injuries one year after the training program, and the second [33] a 27% reduction.

Unlike the reports cited above, which offer popularized accounts of worker participation efforts in ergonomic activities within the auto industry, Liker et al. [35] provide a detailed critical analysis of such experiences in two auto plants, one engaged in stamping auto parts and the other in machining and assembling auto chassis. The programs, as described, grew out of collaborations between the nation's largest automobile manufacturers and the auto workers union to study ergonomic issues in their work operations. For this purpose, it was agreed to engage outside parties to offer needed training and consultations. It is noted that university faculty and staff with specialties in this area played a large role in facilitating the development of programs within the two plants.

The study was undertaken to determine if a participatory ergonomics approach could yield benefits in reducing work-related injury, given downsizing and the need for the workforce to quickly adapt to new and different production technologies. At the time of the study, both plants were under threat of closing as a cost-saving measure and apparently were kept open only by management and labor efforts to come up with innovative plans that would keep them competitive. The two plants were each subdivided into two major areas with separate ergonomic groups to address their respective problems, propose solutions, and put them into place. An advisory committee was also established at each plant to provide direction for the overall effort and monitor progress. Three stages of ergonomic program development are described at each plant, referred to as laying the groundwork (stage 1), program development (stage 2), and maintenance (stage 3). The authors describe how differences in leadership style, makeup and motivation of the advisory committee and members of the ergonomics group, training in and use of job analytical methods, and experience in group decision making affected the processes in each of these stages and the resultant outcomes of the program. For example, leaders who were most trained in ergonomics but poor at facilitating group processes did little to engage the rest of their group members and thus lost their contributions. Others committed to ergonomics and participative management practices were most effective as judged by the satisfaction ratings of members attending meetings and observer ratings of ergonomic project reports and accomplishments at each meeting. Having connections to secure or lobby for outside resources was considered an added leader asset in that implementation of some of the approved changes required support from plant departments other than those involved in this program.

In another example, managers and engineer members of ergonomics groups who used their formal authority to assert their views in meetings were found to stifle the inputs of production level members who took a more backseat role. Attendance at regular meetings ultimately dropped off despite efforts to break this pattern of domination. While the few who remained active made recommendations that improved operations, their outputs paled in comparison to the number of workstation improvements made by other groups whose efforts took account of the ideas and views of all group members.

In still another example, the ergonomics group that achieved the most active involvement of its members showed more deliberateness in undertaking job analyses and in reaching a consensus on a problem-solving strategy than those groups in which the level of participation was less apparent. Though the former group's effort took more time, it yielded more in-depth changes per workstation and a greater number actually implemented than the latter groups' efforts. Further mention of the Liker et al [35] report will be made in a later section dealing with key factors in worker participation efforts to effect ergonomic improvements.

Aside from experiences in the automobile manufacturing industry, descriptions of ergonomic problem-solving activities in warehousing, textile manufacture, and shipping and mail delivery operations have appeared in which worker involvement has been emphasized [25,34]. Of these cases, only the warehousing example is described here because it offers the most detail and has other features deserving mention. Embodying a companywide program for gaining worker input into efforts aimed at enhancing product quality, operational efficiency, and workplace safety, a team formed of seven storekeepers who receive, stock, and then move raw materials from the warehouse to the production assembly line noted two problems posing potential hazards. One was that employees engaged in materials movement work were subject to undue numbers of injuries. Using a problem-solving process that included analyzing accident and medical reports, it was found that back injuries from lifting constituted the major hazard. Team brainstorming sessions plus the use of consultants in materials handling identified major vendor contributions to the problem. Specifically, it was found that vendors routinely exceeded both package weight and size specifications in their deliveries. Some cartons weighed twice the load limit, and others were so large that they had to be broken down to fit the tote boxes of the materials handling systems that were in use. These factors not only increased the risk of overexertion injuries but also required extra labor. Steps recommended by the storekeeper team to remedy this problem consisted of debiting vendors for any deliveries received that did not meet the packaging limits and tagging cartons in violation to alert workers to take added precautions in handling them. Both of these recommendations were accepted by management with estimates that back injuries could be cut by 50%, which, along with the net gain from the debit charged back to vendors for packaging violations, would result in substantial cost savings for this operation.

A second potential hazard noted by forklift operators in this warehouse was that their route of travel posed a risk of pedestrian accidents, especially to other workers who were engaged in product testing and other operations in the same area. During peak times many of these workers were present and stood in the aisles to do their jobs. Adding to the problem was the many blind alleys and intersections where approaching vehicles could not be seen by pedestrians until they were almost directly in front of them. Although there was not a single accident to cite, the forklift truck operators felt strongly that this was a problem that had to be addressed. They proceeded to log near-miss incidents, which occurred at a rate of at least one per day. They set a goal of reducing near-misses by 75% and through team brainstorming sessions drew up a list of solutions that were agreed to by consensus. Relocating product test stations, installing mirrors to aid viewing around corners, and redesigning pedestrian walkways were among the remedies offered. After implementing these and other solutions, near-miss observations were repeated and found that the goal set forth had been achieved. Through the reaction of one team member, the report acknowledges the team-building experience that took place during this problem-solving effort. Indications of the growth of interactive skills and increasing trust based upon ratings by team members taken over the course of team meetings are mentioned though no data are actually presented.

Mention was made earlier that this discussion of participatory approaches was taken from a NIOSH report [1] that described three case studies using this form of intervention in addressing ergonomic hazards in meat packing. The observations from these three studies viewed the processes of team building and follow-through activities in each case against a set of pointers or guidelines that are presented in Section IV. The results are summarized in Section V.

In sum, the aforementioned reports of employee involvement in solving workplace health and safety problems in general and ergonomic hazards in particular show the merits of such an approach. At the same time, conclusions and generalizations from these results require tempering. For example, because popular as well as scientific periodicals are more prone to publish work showing positive results, cases where worker participation efforts may have failed to produce successful outcomes go unreported. Also, most cases have not included controls for other influences that could be affecting results apart from worker participation per se. Increased management attention to worker groups and related affairs, irrespective of any efforts to solicit their inputs into work conditions, can produce positive effects on job performance (see the Hawthorne studies described in Schermerhorn et al. [4]). However, these and other criticisms notwithstanding, the cases speak for themselves in demonstrating worker contributions to positive hazard control accomplishments in the workplace.

IV. CRITICAL FACTORS AFFECTING RESULTS

In viewing the literature on worker participation as a whole, certain elements appear common to many of the documented reports on successful application of this approach to workplace issues or problems. The more prominent of these elements, reflecting organizational factors as well as methodology, are elaborated below. While there have not yet been any systematic efforts to study and assess the significance of these elements in facilitating both the process and outcomes of worker participation, some supportive evidence for their importance is noted based upon the cases reviewed earlier as well as other references to be cited. Most of the commentary focuses on a work team approach to worker participation with special attention to ergonomics-type problems.

A. Commitment and Responsiveness of Top Management and Supervisory Staff

Before even beginning discussion of a worker participation program, top management's commitment to the program is necessary, as is the support of supervisory personnel, union officials, and other worker leaders. Expressions of commitment can take various forms. The presence of such officials serving on committees that set the overall goals for the program and monitor progress is one mode of expression. Another is a policy that formally delegates authority downward, allowing more worker input into decisions on working conditions via their participation on teams or other working groups set up for that purpose. Still another is a positive response to recommendations from such groups and the supply of resources to implement acceptable solutions. Liker et al. [35], in analyzing the ergonomic program experiences at two auto plants, note that committees serving steering or oversight functions for lower level groups should not overreach their roles. They mention how one such committee undertook some job analyses and dictated suggestions for change that proved unfeasible. A top-down approach nullifies the whole concept of worker participation and was perceived in that way by the workers. It was later rectified. The support of middle-level supervisors for worker participation efforts can be problematic if they see their usual responsibilities being diluted. Many quality circle efforts stated in many U.S. plants, though showing some initial benefits, did not last, the suspicion being that the resistance of middle managers was one of the factors that led to the program demise. Supervisors who remain supportive see their role as that of mentoring workers on ways to improve their job performance and helping worker groups refine their suggestions and presentations to top management committees to secure approval.

B. Management and Worker Training

Organizational changes enabling front-line workers to have more input into decisions on workplace conditions and the manner of implementation necessitates additional training of both management and the affected workers that goes beyond the usual subjects of instruction. For workers, one major need is to improve their communication skills and their ability to interact with others in group projects. As Lawler [2] notes, quality circles and work teams, in particular, require numerous meetings during which positive interactions among the worker members and other parties can be critical to effective group action. Worker training in empowerment techniques now being offered in union-sponsored safety and health courses stresses these and other objectives in efforts to promote change for reducing the risks of injury and disease [36].

Managers at various levels may also need training in the listening and feedback skills necessary to work with groups of workers who are assuming responsibilities for decision making. Cascio [3] notes that both groups need to learn the basic interpersonal skills necessary to build respect for each other. On the technical side, and where emergent problems are at issue, special training for workers, managers, and supervisory staff may be warranted. Ergonomic hazards fall into this category, and most of the reports reviewed above mention some form of additional instruction given to both the workers and managers to facilitate efforts in defining ergonomic risk factors and ways to control them. Resources for covering assorted training needs must be a consideration in a worker participation program, including provisions for outside consultants if necessary.

Aside from the subject of training, increasing attention is being paid to the manner of instruction in the area of occupational safety and health [36,37]. Adult learning techniques stressing active forms of instruction through case study and demonstrations and targeting issues directly related to the trainees' experiences appear to have the most merit. Special needs of some who because of language problems or other deficiencies have trouble in comprehending material are also being met by the use of interpreters or visual aids.

C. Composition of Participants

As already noted, there is no single form of worker participation that meets all needs. The approach depends on the nature of the problem to be addressed, whether local to a group or having wide-range implications, the skills and abilities of those to be involved, and the penchant of the organization for joint labor-management or participative approaches in problem-solving ventures. By their very nature, ergonomic problems, though perhaps specific to a given job or operation, typically require a response that cuts across a number of organizational units. Indeed, hazard identification through job analyses, records of injuries, or symptom surveys and the development of control measures and their implementation can necessitate inputs from safety/hygiene, human resource, engineering, maintenance, and medical staffpersons plus ergonomics specialists. These people plus workers and management representatives are considered essential players in a meaningful program effort. In listing possible members of an ergonomics team, Vink et al. [38] also include members from purchasing units as the issues raised can have implications for procurement actions, e.g., added or revised specifications on new equipment orders.

Drawing front-line workers or their representatives for any work team approach to ergonomic problem solving from the real or suspected problem areas or operations is the natural choice. For reasons already stated, the intimate knowledge of the job scene and insights into problems possessed by these individuals can be tapped for decision making and can facilitate implementation. Emphasizing the importance of this kind of input, some recommend that workers themselves prioritize all proposed solutions in making final decisions or before a final review by experts [38]. Supervisors and specialist members of a work team must be careful not to dominate discussions or to allow their stature or expert knowledge to intimidate the workers, as either will limit their contribution to the group process. Consultants brought in to advise on a problem also present this risk. Rather than dictating solutions to those who know the job through everyday experience, consultants who make an effort to work with the group in formulating procedures for defining and solving problems are far more likely to produce successful outcomes. Benefits from these kinds of services can be experiences that build in-house resources for tackling future concerns.

While there is no "correct" size for a work team, the range appears to be 7–15 members for obtaining optimal results. Larger groups present difficulties in creating effective group interactions and cohesiveness, both considered critical to effective decision making [2]. A need for larger representation may be met by setting up parallel smaller groups and establishing a second-level steering or coordinating group to monitor the overall effort as necessary.

D. Information Sharing

Effective worker participation in problem solving requires access to information. In terms of addressing hazard control issues, records of accidents, injury data, and cost figures for proposed control measures need to be made available to the teams who are expected to come up with realistic recommendations for solving such problems in their work areas or operation. Knowledge of other departmental functions and business matters in general may also be essential if the problem at hand and its solution have broader implications. As already noted, ergonomic issues readily transcend the areas of immediate impact, which gives greater importance to communication and cooperation between the various organizational units and parties involved. Even more important is that management motivations for establishing or maintaining work teams be made clear to the participants and that the value of their activities be appropriately recognized and rewarded. Misinformation or misperceptions here can be damaging. The perception that management opts for suggestions from work teams that cut costs or improve productivity without demonstrating equal regard for those that benefit worker welfare can extinguish the program. Cascio [3] notes that in order for workers to be convinced that working harder and smarter will not cost them their jobs, they must be assured of job security.

E. Activities and Motivation

OSHA inspections, citations for violations, and work-related injury or illness statistics can prompt organizations to take actions for hazard control. Teams or groups formed for that purpose follow a common set of steps, typically holding discussions to define the problem, gathering and analyzing data to sort out key elements, and developing and agreeing on recommendations for control actions and plans for implementation. According to the reports of Liker et al. [35] and Lewis et al. [25], actions taken by groups that reflect deliberate discussions of ideas, more orderly forms of data collection, and the use of analytical techniques have the best chance of furnishing effective solutions to problems. But these points aside, what can drive the activity level of a work team and motivate its members to be responsive to its tasks or objectives? The psychology literature indicates that setting a goal and providing frequent feedback marking progress toward attainment of that goal are potent ways for bringing about behavioral actions toward the prescribed ends. Applying these ideas, a wealth of studies exist in the occupational safety and health literature that show the merits of goal setting and feedback in enhancing safety performance among worker groups who are at risk [37,39,40]. Similarly, several of the worker participation cases described in this chapter make mention of goal setting by the work team and evaluations to determine if and when the goal was met. Other factors are more subtle but nevertheless important. The commitment of the workers themselves and the team leader in the belief that their efforts are going to make a difference can be a driver. Liker et al. [35] note how the success of worker groups in the ergonomic study at the two auto plants was shaped by leaders who were totally committed to the process of group problem solving. Management's recognition and rewards for accomplishments of the work teams in solving problems can serve to reinforce these actions and further the team's efforts to tackle other issues. The literature notes too that worker participation programs are perceived positively by those members who participate directly; those not involved do not share that view.

F. Evaluation

References to feedback and goal attainment presume that some measurable indicators of team performance are being applied. The ergonomic cases in the auto plants reviewed above used observer and participant ratings of team meetings in terms of satisfaction with their accomplishments, number of work situations studied for problems, and recommendations made and/or actually implemented. These represent process-type measures. Continuation of the program also represents this type of measure although not expressly mentioned in these cases. Outcome indicators such as changes in frequency and severity data of work-related injury and illness before and after the formation of work teams for addressing ergonomic hazards have also been used but have limitations. For one thing, unless applied to large databases, i.e., employers with large labor forces, musculoskeletal disorders from ergonomic hazards remain statistically rare events and lack sufficient variability for meaningful evaluation. For another, the use of these measures can necessitate an extended time frame to determine whether the intervention has had any beneficial effects. Other influential factors apart from work team efforts may occur in this time period that can confound observations of this type. The cases cited in the general occupational safety and health literature have used surrogate indicators for assessing interventions such as near-misses for accident potential, extent or adherence to safe work practices, and/or the use of personal protective devices as evidence of reduced exposure and risk for more chronic disorders [37]. In this regard, data on the actual reduction of risk factors or levels of exposure to them could also serve to indicate the before and after benefits of ergonomic interventions stemming from work team efforts. In addition, surveys indicating fewer complaints or less fatigue or discomfort among workers following changes instituted by the work team could be taken as a positive sign of ergonomic job improvement. Of course, without baseline data or control groups to rule out intervening influences, there will be questions as to whether any of the aforementioned changes are truly due to the work team's actions. It is to be stressed that judgments of the efficacy of worker participation in team approaches to ergonomic hazard control or other endeavors will require data collection on measures that are valid reflections of this type of intervention.

Table 1 offers a series of pointers or guidelines for framing worker participation and general team-building programs that summarize the major thoughts of this section. Table 2 captures the main observations of the NIOSH case studies [1] using a work team approach to ergonomic problem solving at three meatpacking plants as viewed against these guidelines. Both similarities and differences are noted among these findings with regard to the guideline considerations. Most adhere to the pointers; some add important qualifiers or other considerations. An extensive analysis of these results and the lessons learned from the exercise is found in the NIOSH report [1].

V. EMERGENT ISSUES AND NEEDS

A. Political and Economic Factors of Consequence

Both political and economic factors continue to give increasing importance to worker input into decisions affecting company business matters and operations. OSHA reform legislation, the adoption of total quality management concepts, and the downsizing and restructuring of

Table 1 Important Pointers in Framing Worker Participation and Team-Building Approaches to Workplace Problem Solving as Suggested by the Current Literature

Issue	Pointer
Management commitment	1. Top management's commitment and support of worker participation approaches to company problem-solving needs is critical, as is the cooperation of lower level supervisors and union officials or recognized worker leaders. 2. Policy declarations on the importance of participative approaches in addressing workplace issues require follow-up management actions to prove credibility. Those having merit are worker memberships on existing or newly formed groups at various levels within the organization, including those that have authority to make decisions in local areas of operation; providing timely responses to worker generated proposals for problem solving and resources to implement. 3. Efforts will be needed to redefine the roles of midlevel supervisors as mentors to workers, to work with them in promoting ideas for work improvement and ways in which they can be implemented.
Training	1. Workers and mangement staff plus others who may be formed into a work team, task group, or committee will require added training to ensure effective joint actions. Workers will need training in communication skills and abilities to interact in group problem-solving tasks; managers, in listening and feedback skills. 2. Both workers and managers plus other participant members of a work team or task group should be given the necessary technical training to appreciate the targeted problems at issue. Resources for this and other add-on training should include provisions for outside consultants or experts as may be necessary. 3. Training practices should stress active forms of instruction focused on issues relevant to the trainees' experience. Special needs of those having language difficulties or other impediments to comprehension should be addressed.
Composition	1. No single form of worker participation can effectively fit all needs. Approaches depend upon the problem(s) to be addressed, whether limited to one group, area, or operation or having broader ramifications, the abilities of the workforce involved, and the climate of the organization in terms of using participative approaches in problem solving. 2. Teams formed to address workplace problems that cut across different units in an organization should include representatives from such groups in addition to impacted workers, management persons, and technical consultants as needed. Groups of 7–15 persons can afford ample interactions and cohesiveness in actions. 3. Precautions should be taken to prevent supervisors/managers, specialists, and consultant persons on a team from intimidating front-line worker members of a team or dominating discussion.
Information Sharing	1. Effective worker participation and team efforts to solve problems demand access to information germane to the issues in question. 2. As the team participants may represent different operations and be at different staff levels, the success of group efforts can hinge on sharing information.

(continued)

Table 1 Continued

Issue	Pointer
	3. Management must be up-front and honest in communicating their support for participative decision making and in acknowledging possible consequences of actions that may be proposed. Worker concerns for job security are certain to raise questions.
Activities and motivation	1. Team-building activities invariably include meetings to clarify aspects of the problem, doing data gathering and analyses to isolate causal or contributing factors, and developing remedial suggestions and planned efforts at implementation. Procedures reflecting orderly, systematic ways for dealing with each of these elements offer the best chances for success.
	2. Goal setting and frequent feedback to mark progress in a group's problem-solving efforts loom as key ways for motivating performance.
	3. Team leader commitments to the objectives of the group can facilitate accomplishments.
	4. Management's recognition and rewards for team success in problem-solving work can reinforce and sustain the continued interest of team members.
Evaluation	1. Team performance efforts need to be evaluated, and suitable process and/or outcome measures should be used for that purpose.
	2. Surrogate indicators may offer alternatives to more basic measures in cases where the latter data do not satisfy conditions for meaningful evaluations.

Source: Ref. 1.

businesses are particularly relevant to the topic of this report, and brief comments stressing the connection are noted below.

1. OSHA Reform Legislation

Discussions of OSHA reforms in Congress have considered a provision requiring companies with 11 or more workers to create joint management and employee safety committees [41]. The rationale is that forming such a group would enhance both the employers' and employees' commitment to addressing workplace hazards. By-products of this experience are also noted, such as greater workforce morale, increased worker responsibility for their own safety, and improved trust and cooperation between management and employees. A National Safety Council survey [20] found that responses from companies without such committees agreed with these views. At the same time, both these respondents and others who have existing worker-management safety committees indicated that the formation of such committees was not the only way to increase worker participation in safety and health matters. Other forms were surveys, group meetings, and the solicitation of individual suggestions. Perhaps the issue is not so much the form of worker involvement as the provision of appropriate and effective mechanisms to ensure worker input. OSHA's current guidelines [46] for meatpacking plants in establishing a program to deal with ergonomic hazards cites the need for employee involvement including membership in safety and health committees who could process information to target problem areas, analyze risk factors, and make recommendations for corrective action. Regardless of the outcome of the legislative process, the push for worker involvement in company safety and health program practices committees is apparent.

Table 2 Observations in Meatpacking Case Reports re Pointers in Worker Participation/Team Approaches to Ergonomic Problem Solving as Suggested by the Current Literature

Condition	Observations
Management commitment	*Case 1*: Formalized policy on ergonomics hazard control efforts involving worker participation. Plantwide committee formed to deal with such problems comprising department heads, worker representatives, and others instrumental in accomplishing goals. Made resources available to implement team-proposed solutions in a minimal time period. *Case 2*: Instituted program in 1986. Issued formal policy on worker participation in ergonomics problem solving. Designated an ergonomics program coordinator to oversee multiplant efforts who sat on the top decision-making group of the corporation. Ergonomics committees formed in each plant with representatives from management, worker groups, and others in position to effect proposed changes. *Case 3*: Offered resources to support team-building activities including overtime pay for workers to attend meetings. Ranking managers/directors sat on ergonomics teams with workers.
Training	*Case 1*: Provisions made for training team members in both team building and ergonomics problem solving, the latter including opportunities for practicing methods and techniques. General awareness training on ergonomic problems given to all plant employees. Company safety officer capable of handling efforts with university investigator assistance. *Case 2*: Specialty training on ergonomic issues given to team members. Awareness training on ergonomic hazards given to all employees, even office staff, as part of overall corporate policy. Corporate ergonomics coordinator assumed responsibility for all such training. *Case 3*: Formal training limited in time and focused completely on ergonomic issues. No in-house expertise; handled exclusively by outside university consultant.
Composition	*Case 1*: Team memberships ensured inputs from production workers engaged in the problem jobs, supervisory and engineering personnel, plus maintenance persons from the same department, or a combination who could facilitate data gathering, development, and implementation of proposals. Teams were 7–9 members in size and apparently small enough to be effective, considering overall results reported. Second-level, plantwide ergonomics committee representatives included the purchasing head, which is a recommended practice, and other members who provided close team support (e.g., nurse member supplied injury/medical data in defining problem jobs). *Case 2*: With two exceptions, departmental teams were formed similar to Case 1 as was a plantwide ergonomics committee at the intervention site. One difference was the presence of the corporate ergonomics coordinator, who served in an advisory capacity at both the team and plant committee level. The presence of the ergonomics coordinator at this site and other plants in the corporation suggested close oversight of all company ergonomics activities and possible limits on individual team/plant autonomy. *Case 3*: Teams as formed did include production workers assigned to the problem jobs plus supervisory staff and maintenance people from the areas of concern. Also members were top plant officials whose presence could have limited openness of discussion and inputs from production workers,

(continued)

Table 2 Continued

Condition	Observations
	although one top official was intentionally absent from many meetings so as to not to exert disproportional influence on the team. The teams experienced some turnover in production worker members and had to cope with language/literacy limitations of some participants. Reasonable efforts were made to deal with some of these problems.
Information sharing	*Case 1*: Individual teams received company information on CTD prevalence, workmens' compensation claims and costs, sick absence, and employee turnover to assist in defining problem jobs, though the means of access and/or mode of data presentation were not described. A more direct way for workers to track injuries was recommended. Opportunities to collect other data reflecting risk factors, interviewing workers as to complaints, were freely granted. Varied efforts made to publicize and keep all plant employees informed of team's activities, progress, and accomplishments.
	Case 2: It is intimated that teams shared data similar to that noted in Case 1 for the jobs that were preselected by management and the corporate ergonomics coordinator for study at the plant intervention site. Also the teams had access to ergonomics risk factor information and could collect other information that went into the decisions to focus on these jobs. Monthly and quarterly reports on the team's progress were circulated to other plants in the corporation.
	Case 3: Team members were provided injury statistics and workers' compensation data at the start of the project, but the teams did not review these records as the project progressed. Team activities were publicized in a quarterly newsletter distributed to all employees.
Activities and motivation	*Case 1*: Teams attempted to follow an orderly approach in defining and rank-ordering jobs through the use of injury/medical record data and risk factor evidence, then brainstorming and prioritizing ideas for improvement along with means for implementation. These experiences should build team member skills and lay a strong foundation for future efforts. Proposed solutions took account of ease of implementation, feasibility, and cost and opted primarily for engineering changes, a preferred approach. Those actually implemented proved to have positive effects but did not meet the expectations of some teams and the workforce as a whole. This resulted in feelings of dissatisfaction with the overall program. More realistic goal setting would seem indicated.
	Case 2: Procedures used customized forms, checklists for data gathering on risk factors, and decisions on solutions developed by the company. These gave order to team activities. Teams focused efforts on preselected problem jobs that were recognized as posing difficult problem-solving elements based on earlier attempts. Easier job targets could have provided the teams with some early success and positive motivation; the teams expressed disappointment that proposed changes would take some time to implement.
	Case 3: Two jobs for study were preselected by management and the outside investigator-consultant. A team was formed for each job. Team activities almost solely directed to brainstorming preselected jobs for solutions, which were then prioritized as to feasibility and cost factors. Approach jumps to solution without allowing for much team understanding of the problem. Although some improvements were made to the jobs, certain elements were intractable, making it difficult for the teams to sense success.

Table 2 Continued

Condition	Observations
Evaluation	*Case 1*: Data collection addressed both team-building and performance issues in ways that showed changes over time, including first indications of positive results of team-generated ergonomic improvements following implementation. Both subjective survey methods and traditional objective measures were included in the evaluation, with efforts made to tap not just team responses but the workforce as a whole and to analyze the results in terms of those whose jobs were affected and those not affected. *Case 2*: Data collection included self-report surveys of team members on how well meetings were run, productivity, representations, quality of leadership, and other team-building issues. Data also collected in symptom surveys to corroborate problems and risk factors and set a baseline for determining benefits of improvements along with the more traditional injury/medical data points. None of the proposed solutions could be implemented and/or evaluated within the study time frame. *Case 3*: Surveys of teams concentrated on aspects of member interaction and team effectiveness, as well as responses to the objectives of the program as a morale builder, some given at the beginning and end of the study period. Data analyzed by different representative groups to show differences in views between management/supervisory staff and production worker team members. Besides symptom surveys, a plan was included to collect measurements of hand/wrist motions and forces before and after some proposed job improvements so as to offer quantitative indications of the potential benefits of certain job changes in more immediate ways.

Source: Ref. 1.

2. Total Quality Management (TQM) Movement

Adding impetus to worker participation approaches in industrial management practices is the growing acceptance of total quality management (TQM) principles first introduced by Deming and others [42–44]. Empowering workers to solve problems, help improve processes, and foster ongoing teamwork to ensure quality efforts at each stage of producing a product or providing a service is a key element in the TQM plan. Others are provisions for education, retraining, and self-improvement of the workforce, leadership roles that support or enable workers to do a better job, and continual striving to improve company operations and productivity. Auditing of performance at all stages is implicit to attaining the goal of a total quality effort. Safety and health objectives can be readily folded into the TQM program wherein cases of work-related injury and illness are treated as defects in the quality of the work process. Signs of unsafe conditions, poor work practices, and risky worker behaviors are targets for joint worker-management actions aimed at their elimination. Millar [43] and others, in extolling the virtues of TQM with respect to occupational safety and health, report that companies who have adopted this style of management show both a reduction in work injuries and number of lost workdays and an increase in productivity.

3. Downsizing and Restructuring of Businesses

The need to remain competitive in global markets and to maintain profitability has caused many U.S. businesses to reduce their workforces and restructure their operations. As a streamlining, cost-saving move, layers of middle management or supervision have been removed in many cases, giving work units at lower levels more autonomy in directing operations including those concerned with workplace safety and health. Greater worker involvement is seen as a key to success in making this change. LaBar [45, p. 30] paraphrases the statements of one executive of a major U.S. corporation:

> We used to have supervisors watching people, and if something wasn't being done right, the supervisor would walk over and correct it. With fewer management people around, self-directed worker groups must assume responsibility for everything—productivity, quality, safety.

Added training for workers is considered crucial to getting workers involved in safety as well as other issues. It is recognized too that garnering worker involvement in these efforts can be complicated if layoffs are also occurring in their ranks, which can produce morale problems. Labor-management cooperation on ways to resolve this conflict will be necessary.

B. Other Needs

The political and economic factors just described make apparent the trend for workers to have greater input into defining and solving workplace problems, and the literature takes account of the merits of such an approach and factors of consequence. What remains is to expand the knowledge base of applications, given that forms of worker participation, problems at issue, and situational circumstances may all vary. Recognizing that ergonomic issues cut across so many different work situations, it is important that any efforts made to solve such problems through worker participation approaches be reported. The examples documented so far offer only selective views of the processes involved in the efforts, and thoughts about generalizability and robustness await further confirmation.

Moreover, the successes being reported are largely based on first or early observations and relate to problems that are most amenable to solution. Even here, and as shown in the NIOSH meatpacking cases [1], worker expectations about having their recommendations approved, implemented, and producing results often can exceed realistic time lines. This suggests the need for techniques to overcome frustrations and sustain worker participation and interest where circumstances slow the adoption of recommended actions for ergonomic improvement or where the problems under study are complicated and do not lend themselves to any immediate solutions. Perhaps redirecting efforts to other, more tractable kinds of workplace issues may be one tactic to maintain the motivation of those involved, given delays or difficulties in the ergonomics area. Longer term study of worker participation efforts directed to ergonomic problem solving or other hazard control matters would help clarify the dynamics of this situation.

As already mentioned, employee participation in company efforts to solve problems raises the motivation and productivity levels of those involved. This finding together with the fact that most ergonomic improvements in work operations also result in productivity gains ought to be a big selling point for promoting this type of intervention. How best to use this approach in overcoming the usual business concerns of added cost and "bottom line" impacts on competitive market position in companies having suspected ergonomic problems needs to be considered.

REFERENCES

1. C. C. Gjessing, T. F. Schoenborn, and A. Cohen, *Participatory Ergonomic Interventions in Meatpacking Plants*, DHHS (NIOSH) Publ. No. 94-124, National Institute for Occupational Safety and Health, Cincinnati, OH, 1994.
2. E. E. Lawler III, *High Involvement Management*, Jossey-Bass, San Francisco, CA, 1991.
3. W. F. Cascio, *Applied Psychology in Personnel Management*, Prentice-Hall, Englewood Cliffs, NJ, 1991, Chap. 6.
4. J. R. Schermerhorn, Jr., J. G. Hunt, and R. N. Osborn, *Managing Organizational Behavior*, Wiley, New York, 1985, Module A, pp. 5-8.
5. G. LaBar, Is your safety committee legal?, *Occup. Hazards* 11:35-38 (1993).
6. Commission on the Future of Worker-Management Relations, Fact Finding Report, U.S. Dept. of Labor and U.S. Dept of Commerce, Washington, DC, May 1994.
7. N. Krigsman and R. M. O'Brien, *Organization Behavior Management and Statistical Process Control*, Haworth, Binghamton, NY, 1987, pp. 67-82.
8. M. L. Marks, The question of quality circles, *Psychol. Today* 3:36-45 (1986).
9. K. I. Miller and P. R. Monge, Participation, satisfaction and productivity: a meta-analysis, *Acad. Manage. J.* 29(4): 727-753 (1986).
10. R. W. Griffin, Consequences of quality circles in an industrial setting: a longitudinal assessment, *Acad. Manage. J.* 31(2):338-358 (1988).
11. H. H. Cohen, Employee involvement: implications for improved safety management, *Prof. Safety* 6:30-35 (1983).
12. S. Edwards, Quality circles are safety circles, *Natl. Safety News*, June 1983, pp. 31-35.
13. National Safety Council, *Safety Circles,* Occup. Safety and Health Data Sheet 738 Rev. 93, Natl. Safety Council, Itasca, IL, 1993.
14. Office of Technology Assessment, *Preventing Illness and Injury in the Workplace*, OTA-H-256, U.S. Congress, Washington, DC, 1985, pp. 315-323.
15. R. H. Peters, *Review of Recent Research on Organizational and Behavioral Factors Associated with Mine Safety*, Bur. of Mines Inf. Circ. 9232, U.S. Dept. of Interior, Washington, DC, 1989.
16. R. A. Guzzo, R. D. Jette, and R. A. Katzell, The effects of psychologically based intervention programs on worker productivity, *Personnel Psychol.* 38:275-281 (1985).
17. J. K. Liker, M. Nagamachi, and Y. R. Lifshitz, A comparative analysis of participatory ergonomics programs in the U.S. and Japan manufacturing plants, *Int. J. Ind. Ergon.* 3:185-199 (1989).
18. Bureau of National Affairs, Lower injury rates, costs seen as a result of voluntary compliance, California/OSHA states, Jan. 5, 1984.
19. L. I. Boden, J. A. Hall, Levenstein, and L. Punnett, The impact of health and safety committees, *J. Occup. Med.* 26(11):829-834 (1984).
20. T. W. Planek and K. P. Kolosh, Survey shows support for safety and health committees, *Safety Health* 1:76-79 (1993). (Also summarized in *Survey of Employee Participation in Safety and Health*, An executive summary available from the National Safety Council, 1121 Spring Lake Drive, Itasca, IL.)
21. W. Cooke and F. Gautschi, OSHA, plant safety programs and injury reduction, *Ind. Relations* 20(3):245-257 (1981).
22. C. Touhy and M. Simard, *The Impact of Joint Health and Safety Committees in Ontario and Quebec*, Spec. Rep. to the Canadian Association of Administrators of Labour Law, Toronto, Canada, August 1992.
23. K. J. Saarela, An intervention program utilizing small groups: a comparative study, *J. Safety Res.* 21:149-156 (1990).
24. E. B. Lanier, Jr., Reducing injuries and costs through team safety, *Prof. Safety* 7:21-25 (1992).
25. H. B. Lewis, A. S. Imada, and M. M. Robertson, Xerox leadership through quality: merging human factors and safety through employee participation, *Proc. 32nd Annual Meeting*, Anaheim, CA, 1988, Vol 2, pp. 756-759.
26. L. Lin and H. H. Cohen, *Development and Evaluation of an Employee Hazard Reporting and Man-

agement Information System in a Hospital, Safety Sciences Contract Rep. No. 210-81-3102, Division of Safety Research, National Institute for Occupational Safety and Health, Morgantown, WV, 1983.
27. A. Cohen and F. Dukes-Dobos, Applied ergonomics, in *Patty's Industrial Hygiene & Toxicology*, 2nd ed., Vol. 3B, *Biological Responses*, Wiley Interscience, New York, 1985, pp. 375-430.
28. W. Keyserling, T. Armstrong, and L. Punnett, Ergonomic job analysis: a structured approach for identifying risk factors associated with overexertion injuries and disorders, *Appl. Occup. Environ. Hyg.* 6:353-363 (1991).
29. S. Ulin, T. J. Armstrong, and G. G. Herrin, Preferred tool shapes for various horizontal and vertical work locations, *Appl. Occup. Environ. Hyg. J.* 7:327-337 (1992).
30. T. R. Waters, V. Putz-Anderson, A. Garg, and L. J. Fine, Revised NIOSH equation for the design and evaluation of manual lifting tasks, *Ergonomics* 36:749-776 (1993).
31. E. Grandjean, *Ergonomics in Computerized Offices*, Taylor & Francis, London, England, 1987.
32. G. LaBar, Employee involvement yields improved safety record, *Occup. Hazards* 5:101-104 (1989).
33. G. LaBar, Ergonomics, the Mazda way, *Occup. Hazards* 4:43-46 (1990).
34. G. LaBar, Succeeding with ergonomics, *Occup. Hazards* 4:29-33 (1992).
35. J. K. Liker, B. S. Joseph, and S. S. Ulin, Participatory ergonomics in two U.S. automotive plants, in *Participatory Ergonomics*, K. Noro and A. Imada, Eds., Taylor & Francis, London, England, 1991, Chap. 6.
36. N. B. Wallerstein and M. Weinger, Health and safety education for worker empowerment, *Am. J. Ind. Med.* 22(5):619-635 (1992).
37. A. Cohen and M. J. Colligan, *Assessing Occupational Safety and Health Training: A Literature Review*, Div. Training and Manpower Development, Natl. Inst. for Occupational Safety and Health, Cincinnati, OH, 1993.
38. P. Vink, E. Lourijsen, E. Wortel, and J. Dul, Experiences in participatory ergonomics: results of a round-table session during the 11th IEA Congress, Paris, July 1991, *Ergonomics* 35(2):123-127 (1992).
39. J. S. Chhokar and J. A. Wallin, Improving safety through applied behavior analysis, *J. Safety Res.* 13: 141-151 (1984).
40. B. Sulzer-Azaroff, T. C. Harris, and K. B. McCann, Beyond training: occupational performance management techniques, in *Occupational Safety and Health Training—State of the Art Reviews*, M. J. Colligan, Ed., Hanley & Belfus, Philadelphia, PA, 1994, pp. 321-339.
41. M. P. Weinstock, OSHA reform: the push for worker involvement, *Occup. Hazards* 12: 37-39 (1991).
42. J. Roughton, Integrating a total quality management system into safety and health programs, *Prof. Safety* 6:32-37 (1993).
43. J. D. Millar, Valuing, empowering employees vital to quality health and safety management, *Occup. Health Safety* 9:100-101 (1993).
44. S. M. Mottzko, Variation, system improvement and safety management, *Prof. Safety* 8:17-20 (1989).
45. G. LaBar, Safety management in tight times, *Occup. Hazards* 6:27-30 (1993).
46. Occupational Safety and Health Administration, *Ergonomic Program Guidelines for Meatpacking Plants*, OSHA Rep. 3123, U.S. Department of Labor, Washington, D.C., 1993.

11
Job Analysis

Katharyn A. Grant
National Institute for Occupational Safety and Health, Cincinnati, Ohio

I. INTRODUCTION

Ergonomic job analysis is the methodology used by engineers and safety professionals to describe work activities for the purpose of comparing existing task demands to human capabilities. Modern job analysis is founded in the work of early industrial engineers such as Frederick Taylor and Frank and Lillian Gilbreth [1]. At its inception, job analysis was primarily viewed as a tool for improving efficiency and productivity. However, in recent years job analysis techniques have been applied increasingly to identify work conditions and job demands associated with the onset of fatigue, overexertion, injuries, and chronic musculoskeletal disorders.[1] There is increasing evidence that if job analysis techniques are used to identify these conditions, effective control methods can be introduced to eliminate or reduce the risk of injury to the worker [2].

II. BACKGROUND AND SIGNIFICANCE TO OCCUPATIONAL ERGONOMICS

Renewed interest in job analysis techniques for identifying ergonomic hazards has its roots in several sources. First, there is research to indicate that the benefits of well-designed jobs include improved efficiency, safety, and satisfaction for employees. Many industries faced with rising workers' compensation costs and disability insurance premiums for musculoskeletal disorders have found job analysis and ergonomic redesign effective for reducing losses associated with musculoskeletal injuries [3–5]. Further, as companies face pressures to enhance quality and remain competitive in a global marketplace, there is hope that ergonomic improvements will ultimately translate into greater productivity and profitability.

Second, the growing interest in ergonomics parallels the increasing impact of society's expectations about occupational health. Although ergonomics was not mentioned by name in the Occupational Safety and Health Act of 1970 (OSH Act), the OSH Act states that it is the general duty of all employers to provide their employees with a workplace free of recognized hazards, whether or not they are regulated by specific federal standards [6]. In recent years, the Occupational Safety and Health Administration (OSHA) has taken an active role in investigating ergonomic hazards under authority of the General Duty clause. Furthermore, recog-

---

[1]Fatigue is usually transient; if given sufficient recovery time, muscles can overcome fatigue without permanent damage. Overexertion injuries can be caused by a single event and include most sprains and strains and some back injuries. Chronic musculoskeletal disorders are usually caused by repeated trauma to the musculoskeletal system.

nition of the significance of ergonomic problems in the workplace has led OSHA to call for an "ergonomics standard" to define requirements for a comprehensive ergonomic safety and health management program for general industry [7]. Under the proposed model, employers will be required to implement a system for identifying risk factors associated with musculoskeletal injuries and determining if these factors have been eliminated or reduced to the extent feasible [8]. If this proposal is enacted, job analysis will become increasingly important as a component of this program.

III. FUNDAMENTALS OF JOB ANALYSIS

The steps involved in performing an ergonomic job analysis depend, in part, on the purpose and scope of the study. Some situations will demand a comprehensive survey of all jobs in a facility; in other circumstances, the evaluation may be limited to a specific group of workers in response to a specific complaint. Job analysis procedures are also used to evaluate the effectiveness of various control measures. In general, however, the fundamental elements of job analysis are as follows [9]:

Identifying potential hazards
Preparing for field study
Conducting the field study
Interpreting the results

A. Identifying Potential Hazards

The process of identifying hazards arising from poor work design begins with two activities: (1) reviewing records to identify jobs associated with a high rate of accidents and injuries and (2) becoming familiar with the processes and job activities that are performed in each work area.

1. Review Injury Data

The most obvious indication that a job poses excessive work demands is a high rate of injury or absenteeism among workers who perform the job. Excessive turnover can also indicate that a job poses difficulties for a large percentage of the workforce. Information about injuries, absenteeism, and turnover can come from two sources: (1) existing plant records (e.g., accident reports, dispensary logs, OSHA 200 reports, and worker compensation records) and (2) surveys of current employees (see Sec. IV for information on medical surveillance for ergonomic programs).

Based on surveillance information, jobs should be selected for analysis based on the number of injuries or complaints associated with the job, the severity of those injuries and complaints, and the number of workers affected by current job conditions [10]. If significant hazards are identified during the analysis and control methods are introduced, continued surveillance can later provide a measure of the effectiveness of these interventions.

2. Review Processes and Job Activities

Before the investigator can begin the process of collecting data needed for the analysis at the worksite, background information about the job processes and work activities is needed. Basic information should be obtained by reviewing process and job descriptions, interviewing supervisors and employees, and conducting a "walk-through" survey of the worksite. In some cases a checklist may be used to identify potential risk factors that may be present at the worksite and require further investigation. The information collected should answer questions such as

1. How many workers are employed in each job?
2. What are the characteristics of the workforce (e.g., gender, age, education level)?
3. What are the primary tasks involved in each job? Do workers perform the same tasks over and over throughout the workshift, or do they perform a large number of different tasks?
4. Is there an established work rate? If so, how is the work rate determined (e.g., line paced, time standards, etc.)?
5. Is there an opportunity for workers to rotate to other jobs?
6. How much do workers employed in this job earn? What type of pay system is used (e.g., hourly wage, piece rate)?
7. How many hours do employees work per week? Is work organized into shifts? How much overtime is worked per week?

Although these questions do not address the presence of specific hazards, the information gained from responses is often useful in identifying resources needed for the evaluation, developing an analysis strategy, and interpreting data that are gathered later in the evaluation.

B. Preparing to Conduct a Job Analysis

1. Gather Necessary Equipment

Most information needed to identify and evaluate ergonomic hazards is collected through observation and direct measurement. Equipment needs depend on the nature of the suspected hazards and the desired level of sophistication in the measurement; however, in most cases, the following items are useful for gathering data for job analysis [11]:

1. Cameras and film for recording workers' postures and motions during job activities
2. Tape measures and rulers for measuring workstation dimensions and reach distances
3. Force gauges and spring scales for measuring the force of exertions (e.g., the force needed to push or pull a hand cart) and the weight of tools or objects
4. Stopwatches for measuring the duration of work activities, breaks, etc.

Videotape can be invaluable as a tool for documenting biomechanical stressors in the workplace. Guidelines for recording work activities on videotape are provided in Table 1. After work activities are recorded on videotape, one or more analysts can review tapes using slow-motion or real-time playback to accurately measure task durations or detect subtle or rapid movements [12]. Analysts can also record job activities on video before and after changes are made to determine if changes were effective in reducing exposure to hazards. Finally, videotape can be useful as a visual aid for ergonomic training and demonstration purposes.

More sophisticated methods for analyzing motions and evaluating responses to work demands are available, although most are rarely used outside of the research laboratory (Table 2). The primary disadvantage of most of these measurement systems is that the equipment tends to be expensive and not well suited to industrial environments. In addition, many of these methods require extensive training and calibration, and data reduction and analysis can be time-consuming [13]. However, under the right conditions, some of these methods have been used effectively in industrial settings. For example, Armstrong et al. [14] successfully applied elec-

Table 1 Guidelines for Recording Work Activites on Videotape

- If the video camera has the ability to record the time and data on the videotape, use these features to document when each job was observed and filmed. Recording the time on videotape can be especially helpful if a detailed motion study will be performed at a later date (time should be recorded in seconds). Make sure the time and date are set properly before videotaping begins.
- If the video camera cannot record time directly on the film, it may be useful to position a clock or a stop watch in the field of view.
- At the beginning of each recording session, announce the name and location of the job being filmed so that it is recorded on the film's audio track. Restrict subsequent commentary to facts about the job or workstation.
- For best accuracy, try to remain unobtrusive; i.e., disturb the work process as little as possible while filming. Workers should not alter their work methods because of the videotaping process.
- If the job is repetitive or cyclic in nature, film at least 10–15 cycles of the primary job task. If several workers perform the same job, film at least two or three workers performing the job to capture differences in work method.
- If necessary, film the worker from several angles or positions to capture all relevant postures and the activity of both hands. Initially, the worker's whole body posture should be recorded (as well as the work surface or chair on which the worker is standing or sitting). Later, close-up shots of the hands should also be recorded if the work is manually intensive or extremely repetitive.
- If possible, film jobs in the order in which they appear in the process. For example, if several jobs on an assembly line are being evaluated, begin by recording the first job on the line, followed by the second, third, etc.
- Avoid making jerky or fast movements with the camera while recording. Mounting the camera on a tripod may be useful for filming work activities at a fixed workstation where the worker does not move around much.

tromyography (EMG) to examine forearm muscle forces during poultry-cutting tasks. Habes [15], Christensen [16], Aaras and Westgaard [17], Gomer et al. [18], and Milerad et al. [19] also used EMG to examine muscle load and fatigue among assembly line workers, postal workers, and dentists. Marras et al. [20] employed a triaxial electrogoniometer to document the three-dimensional trunk motion characteristics associated with over 400 industrial lifting jobs in 48 industries. Garg et al. [21] used portable devices for measuring heart rate and oxygen consumption to examine the physiological responses of grocery warehouse workers to their work tasks.

2. Identify Strategy

The primary objective of the field study should be to collect sufficient information to allow the analyst to completely describe the job as it is currently being performed (i.e., what the worker is doing and how it is being performed in time). Generally, this requires the analyst to observe the job during a "typical" work period under "normal" operating conditions. The number of workers employed in a job, the number of tasks, the type of rotation system employed, and the hours of work all have an impact on the sampling strategy employed in gathering data. If workers perform a well-established set of tasks that tend to be repeated in a fixed sequence, the time needed for data collection is usually minimized. However, if jobs do not involve repetitive cycles or if irregular elements that occur only once or twice a day (e.g., machine setup) are a concern, then it may be necessary to observe workers at randomly selected times over an extended period (i.e., several days) to gather information about all of the various activities they perform. If necessary, work-sampling procedures can be used to estimate the percentage of time workers spend performing various activities [1].

Table 2 Direct Measurement Systems for Characterizing Ergonomic Hazards

Equipment	Use	Applications	Limitations
Accelerometers	Describe acceleration of body segments	Tasks involving high speed motions	Six or more accelerometers required at each joint—can interfere with motion.
Electromyograph (EMG)	Assess muscle activation patterns, muscle force or effort, and fatigue	Tasks involving static muscle exertions	Applicable only to static work; requires calibration; signal subject to noise.
Finger force sensors	Assess finger forces, evaluate contact stresses	Manual tasks requiring force exertions with the hands and fingers	Sensors must be attached to tool or equipment surface or to fingertips (may affect tactile sensitivity).
Electrogoniometer	Describe posture and motion (velocity and acceleration)	Tasks requiring highly dynamic movements	Goniometer may affect motion pattern. Data must be telemetrized or logged to permit free movement.

(continued)

Table 2 Continued

Equipment	Use	Applications	Limitations
2-D and 3-D video motion analysis systems	Describe posture and motion (velocity and acceleration)	Lifting tasks; static tasks; tasks with highly dynamic movements	Requires good lighting; 3-D requires multiple cameras and adequate floor space.
Pendulum potentiometer	Describe trunk and shoulder posture (continuous)	Static tasks	Inertial effects can induce errors; provides no data about trunk loading. Signal may be subject to noise.
Heart rate monitor	Describe metabolic load and physical stress	Hot tasks; highly static or dynamic tasks, tasks with mental demands	Equipment often bulky; requires calibration.
Oxygen uptake monitor	Describe metabolic load/physical stress	Tasks requiring whole body exertion	
Wrist activity monitors	Describe work-rest patterns, movement frequency and intensity	Repetitive tasks; tasks with high metabolic demands; shift work	Additional work needed to link data to specific tasks; movement intensity data are limited.

C. Conducting the Field Study

1. Observe Work Processes

In almost all cases, the bulk of job-related information will be collected through direct observation of the work tasks at the worksite. For best accuracy, the work process should be disturbed as little as possible during the observation period. Because workers who perform the same job may use different methods due to differences in training, stature, or strength, several workers who perform the same job should be observed. The analyst should also attempt to observe the same task at different times during the work shift to determine if fatigue affects workers' performance or if the workload changes during the workday.

Formalized procedures for organizing and recording observational data have been developed and are described in Table 3. Because many of these systems are applicable to only one or two of the risk factors that may be present at a worksite, they should be implemented only after a preliminary evaluation indicates that specific problems (e.g., awkward upper extremity postures) are present. For most of these systems, there is also an important trade-off between the time required to analyze and record the data and the level of detail produced by the analysis. For example, although the Ovako Working Posture Analysis System (OWAS) [23] is easily learned and applied, the procedure provides only a rough description of posture. Systems developed by Priel [22], Corlett et al. [24], and Armstrong et al. [14] provide a more detailed description of posture but require more time for training and data analysis.

2. Interview Supervisors and Employees

Informal discussions with workers and supervisors can also provide useful information for the analysis. Workers who perform a job on a daily basis are often the best source of information about the specific elements that may pose a hazard. Supervisors can provide a "big picture" view of operations as well as insight into the feasibility of proposed changes. Interviews also give workers the opportunity to participate and to provide input to the job analysis process. Soliciting workers' input may increase their willingness to accept changes to their job if redesign is necessary.

3. Take Measurements

Observing the job also provides the analyst the opportunity to make measurements at the job site. A sketch or drawing of the workstation is useful for identifying the location of fixtures, equipment items, etc. Sketches can be labeled with dimensions indicative of work surface heights, reach distances, walking distances, and so on [11]. If workers handle tools or objects, the size, shape, and weight of these items should be recorded.

D. Interpreting the Results

Once data collection is completed, the first step in analyzing the results is to break each job into a series of tasks and subtasks. The goal of this analysis is to link excessive job demands to a specific aspect of the job or work environment. Once jobs are broken into tasks, the work methods, the workstation, and the tools and equipment required to perform each task can be examined to determine if one or more of these elements contributes to the biomechanical stress of the task [34]. Specifically, each task should be examined in terms of the following dimensions [10,35]:

Forces required to perform the task. Force requirements have a direct impact on the muscular effort that must be expended by the worker to perform a task. As muscular effort increases in response to heavy task loading, circulation to the muscle decreases,

Table 3 Observational Systems for Characterizing Ergonomic Hazards

System	Hazards described	Applications
Posturegram [22]	Static postures (head, trunk, upper and lower limbs)	Tasks involving static postures
Ovako Working Posture Analysis System (OWAS) [23]	Static postures (back, upper and lower limbs)	Tasks involving static postures (e.g., steel production, construction)
Posture targeting [24]	Static postures (head, trunk, upper and lower limbs)	
Postural Stability Diagram (PSD) [25]	Static exertions with hands and feet	Tasks requiring static exertions (e.g., push-pull forces)
AET [26]	Static postures, heavy muscular work, strenuous or repetitive exertions	"All purpose" system—tasks presenting both physical and mental loads
Upper extremity posture analysis system [14]	Awkward upper limb postures, manual force exertion	Highly repetitive manual activities (e.g., poultry processing)
ARBAN [27]	Postures (neck, shoulders, trunk, legs), muscle stress (static and dynamic), vibration and shock	Repetitive work; can be used to compare alternative work methods
Classification system for the trunk and shoulders [28]	Awkward postures (trunk, shoulders)	Repetitive nonseated jobs (e.g., automotive assembly)
VIRA [29]	Awkward postures (neck, shoulders)	Seated, repetitive arm work
Posture recording [30]	Awkward/static sitting postures	Seated work
Physical Work Stress Index (PWSI) [31]	Posture (whole body, hand), acceleration and/or vibration, thermal load	Jobs with significant physiological demands (e.g., lifting, digging)
Hand exertion classification system [32]	Repetitive or forceful manual exertions; awkward shoulder, arm, wrist, and hand postures; contact stresses; vibration	Repetitive manual activities (e.g., manufacturing and warehouse jobs)
Rapid Upper Limb Assessment (RULA) [33]	Awkward posture (neck, trunk, arm, leg, wrist), muscle use (static and dynamic), muscle force	Repetitive tasks (e.g., manufacturing, grocery VDT, microscope work)

causing more rapid muscle fatigue. In addition, high contact forces that create pressure over one area of the body can inhibit nerve function and blood flow. If possible, the force required to perform a task should be measured at the worksite and compared to the known capacities of the workforce (see below). Contact points between the body and work surfaces or tools should also be identified and described.

Postures assumed during the task. Postures determine which muscles are used in an activity and how forces are translated from the muscles to the part or tool being handled. In general, muscle strength about a joint is maximized near the midpoint of the range of motion. Therefore, performing a task in a neutral posture generally requires less effort than performing the same task in a more extreme posture. Furthermore, postures that are sustained through active muscle contraction consume energy and produce waste metabolic products that can cause fatigue and pain. The analyst should note situations where the tool design, workstation layout, or equipment design causes excessive reaches or sustained awkward shoulder, elbow, wrist, or trunk postures.

Frequency of muscle activation (repetition rate). Frequent repetition of the same work activities can exacerbate the effects of awkward work postures and forceful exertions. Tendons and muscles can often recover from the effects of stretching or forceful exertion if sufficient time is allotted between exertions. However, if tasks are repeated frequently, fatigue and muscle strain can accumulate, producing permanent tissue damage. There is evidence that even low-level exertions, if repeated at a very high rate, can cause muscle fatigue and injury if sufficient recovery time is denied. If a job requires workers to repeat a simple task over and over throughout the workday, repetition can be quantified by counting the number of similar motion patterns that occur within a specific time period or estimated subjectively using qualitative descriptions [34].

Duration of work and recovery periods. Work duration can have a substantial effect on the likelihood of both localized and general fatigue. Even at low levels of effort, biomechanical or physiological strain will accumulate over time; therefore, the provision of frequent rest breaks is encouraged to give tissues the opportunity to recover from exertions during work. Work duration and the number of breaks allotted to employees during the workshift can usually be determined through observation or interview.

Exposure to vibration and cold. Exposure to whole body and segmental vibration has been associated with increased rates of back pain and upper extremity musculoskeletal disorders, although the pathogenic mechanism is not yet well understood. Exposure is frequently associated with activities such as the operation of motor vehicles and the use of powered hand tools [11]. Cold temperatures can reduce the dexterity and sensibility of the hand, causing workers to apply more grip force to tool handles and objects than necessary. There is also evidence that cold tends to exacerbate the effects of segmental vibration. Because of these relationships, vibrating or cold objects and surfaces should be identified and the frequency and duration of exposure should be recorded.

In some cases, it will be obvious that task demands exceed the capabilities of the workers performing the job. For example, most analysts will recognize a job as being too repetitive if the worker has difficulty keeping up with the required pace or if the body is in constant motion. However, in other instances it may not be as clear what force or postural demands are appropriate for the population of workers performing a task. For example, a workforce of young males may be able to tolerate job demands that would be inappropriate for an older or predominantly female population. Currently, it is not known how much force or repetition can be endured safely by a majority of workers. However, it may be possible to estimate the percentage of workers for whom a job may be difficult. Information that can be used for this purpose is discussed in the following paragraphs.

1. Strength Data

Strength data for various populations and muscle groups have been published in a number of sources [35-37]. If the force requirements of a task are known, it may be possible to compare these requirements with existing strength data to estimate the percentage of the population for which the job may be difficult. Successfully applying strength data requires knowledge of the subject population upon which the data are based, the posture in which the measurements were made, whether the measurement was static or dynamic, and how long the effort was sustained [12]. Recently, computerized *biomechanical models* have been developed to predict the percentage of males and females capable of exerting static forces in certain postures. The advantage of these models is that most recognize that a worker's capacity for force exertion is rarely dependent on the strength of a single muscle group. Rather, the capacity for exerting force is dependent on the moment created at each joint by the external load and the muscle strength at that joint. The models compute the moment created at each joint by an exertion and compare the moments against static strength data to estimate the percent of the population capable of performing a specific exertion for each joint and muscle function. Currently, both two- and three-dimensional models are available for these analyses [38]. (Additional information about biomechanical models is contained in Chapter 4 of this book.)

2. Anthropometry

Anthropometric data on body size and range of joint motion can be used to assess the appropriateness of workplace, equipment, and product designs relative to workforce capacities for reach, grasp, and clearance [11]. Compilations of anthropometric data for various populations are available from numerous sources [39,40]. Anthropometric data have also been used to generate computerized man-models that can be incorporated into existing CAD programs for direct application in the design and evaluation of new and existing workstations [41]. For additional information on anthropometry, see Chapter 1.

3. Physiological Data

In activities such as repetitive lifting and load carrying, large muscle groups perform submaximal, dynamic contractions. During these activities, a worker's endurance is primarily limited by the capacity of the oxygen transport and utilization systems (maximum aerobic power) [42]. Data compiled on the maximum aerobic capacities of working populations and the energy demands of common industrial tasks can be found in a number of sources [12,42,43]. In general, maximum aerobic capacity declines with age, increases with physical fitness level, and is 13-30% lower for women compared to men [12,44]. Several researchers suggest that the maximum energy expenditure rate for an 8-hr workday should not exceed 33% of maximum aerobic power [12]. Limits of 5.2 kcal/min for average healthy young males or 3.5 kcal/min for populations containing women and older workers have been proposed based on this recommendation [45-47].

Because table values provide only a rough approximation of the metabolic costs of a given job, models to predict metabolic energy expenditure for simple tasks based on a combination of personal and task variables have also been developed [48-50]. It has been demonstrated that the energy expenditure rate of complex jobs can be predicted if the energy expenditure rates of the simple tasks that make up the job and the time duration of the job are known. Comparison of measured and model-predicted rates for 48 tasks indicated that models can account for up to 90.8% of the variation in measured metabolic rates [50].

4. Psychophysical Data

Difficulties in applying strength data to dynamic tasks involving more than one muscle group, and the recognition that motivational factors play an important role in determining an individual's capacity for physical work, have led to the use of *psychophysics* to develop guidelines for the evaluation and modification of repetitive work tasks [51–53]. Psychophysical limits are generally based on data derived from laboratory simulations of a specific task in which the participants are allowed to adjust their workload to a level subjectively defined as the maximum acceptable. Limiting workload in this manner should allow workers to perform work tasks without overexertion or excessive fatigue [54]. Although there are few data to indicate how psychophysically derived limits relate to the risk of injury during work, many researchers believe that use of these limits may be the most accurate method of determining whether a given task is acceptable [55]. Psychophysical limits for various lifting, pushing, and pulling tasks and other manual operations have been developed and are widely available for application [56–58].

5. Integrated Models

More sophisticated evaluation techniques that integrate biomechanical, physiological, and psychophysical considerations to assess the appropriateness of job tasks are also available. Probably the best known method for evaluating the demands of materials handling tasks is the *NIOSH lifting equation*. The NIOSH lifting equation was first published in 1981 to assist safety and health practitioners in evaluating sagittal plane lifting tasks [43]. The equation has recently been revised to reflect new research findings and provide methods for evaluating asymmetrical lifting tasks, lifts of objects with less than optimal hand–container couplings, and jobs with a larger range of work durations and lifting frequencies [47]. Using criteria from the fields of biomechanics, psychophysics, and work physiology, the equation defines a recommended weight limit (RWL) based on specific task parameters (e.g., the location of the load relative to the body and the floor, the distance the load is moved, the frequency of the lift). The RWL represents a load that nearly all healthy workers can lift over a substantial period of time without placing an excessive load on the back, causing excess fatigue, or otherwise increasing the risk of low back pain. The actual weight of lift can be compared to the RWL for a given task to derive an estimate of the risk presented by the task and to determine whether measures to reduce the risk of injury to workers are needed. (Additional information about the NIOSH lifting equation can be found in Chapter 31 of this handbook).

E. Developing Solutions

Ultimately, the results of a job analysis should suggest ways to eliminate, reduce, or control ergonomic hazards through modification of equipment, workstations, or work methods that contribute to excessive work demands. In all cases, the best ergonomic solutions are those in which safe work is a natural result of the job design and are independent of specific worker capabilities or work techniques. However, in situations where design changes are infeasible, it may be possible to limit exposure to ergonomic hazards by reducing production rates, providing additional rest breaks, periodically rotating workers to less stressful jobs, and increasing the number of employees assigned to specific work tasks. In some situations, control of ergonomic hazards may require a combination of engineering or administrative tactics. When possible, the proposed solution should be tested among a small group of workers to allow adjustments to be made before widespread changes are implemented. Finally, follow-up job analyses should be performed to ensure that the solution effectively reduces the hazard without imposing new demands on the worker. (Additional information about ergonomic control strategies is contained in Chapter 21.)

F. Documentation

The importance of documenting the results of an ergonomic job analysis cannot be underestimated. Benefits of careful documentation include the following [58]:

Documentation permits the analyst to track the progress of the analysis, i.e., to determine what has been done and what remains to be done.
Careful documentation provides a compendium of facts for use in justifying requests for resources and intervention.
Adequate documentation allows the analyst to keep track of successful solutions that may work again in later projects and to track mistakes that should be avoided.
Good documentation forms the foundation for evaluating the overall success of an ergonomics project. Data obtained before and after the introduction of control measures can be compared to determine whether the interventions were effective in reducing worker exposures to hazards, the occurrence of complaints or injuries, and the cost of medical insurance and workers' compensation.

The final report of a job analysis should describe why the study was done, the tools and equipment used in the study, the problems identified, and the solutions recommended to correct these problems. An outline for this type of report is shown in Table 4.

Table 4 Job Analysis Report Format

Introduction
    Purpose of report
Definition of Problem
    Area layout (including dimensions and photos)
    Worker population description
    Injury statistics
    Discomfort survey results
    Interview findings
Procedure
    Tools and equipment used
    Work sampling plan
Analysis and Discussion
    Problems identified
Conclusions
Recommendations
    Recommended solutions to problems identified (prioritized)
    Associated costs
    Associated benefits
References
Appendix
    Examples of questionnaires/survey forms used
    Analysis charts
    Computer output (if computer software was used in the analysis)
    Other supporting data and analyses

G. Job Analysis Example

The following example illustrates how the methods and information discussed in the previous sections were used to identify potential hazards associated with jobs in one department of a manufacturing facility that produces printed circuit boards.

1. Background

The production of printed circuit boards is a multistep process. One step in the process is a cleaning procedure to remove impurities from the surface of each panel before it is printed and etched. This procedure is performed by a computer-operated "preclean" machine. Panels are automatically loaded into the machine from a cart and are transported through the machine on a conveyor belt. An operator monitors the operation of the machine and performs required maintenance activities (e.g., adding water or chemicals to the machine).

2. Problems Identified

As part of a plantwide screening survey to identify early indications of musculoskeletal injuries, a body part discomfort survey (Figure 1) was administered to 12 machine operators in the preclean area. Despite the job's seemingly low physical demands, the survey indicated that 8 of the 12 workers experienced arm pain, 6 of 12 experienced back pain, and 9 of 12 reported moderate to severe foot pain.

A site visit was conducted to gather background information from operators working in the area. Interviews with employees and supervisors revealed that the machine operators' job is nonstereotypical; i.e., operators perform a variety of tasks at different times throughout the workday. Therefore, the job was observed again at three separate times during a single work shift for periods of 10–15 min each. In addition, operators demonstrated a task (adding pumice to the machine) that is performed only once or twice during the work shift. Measurements were made of machine dimensions, reach distances, and work heights. In

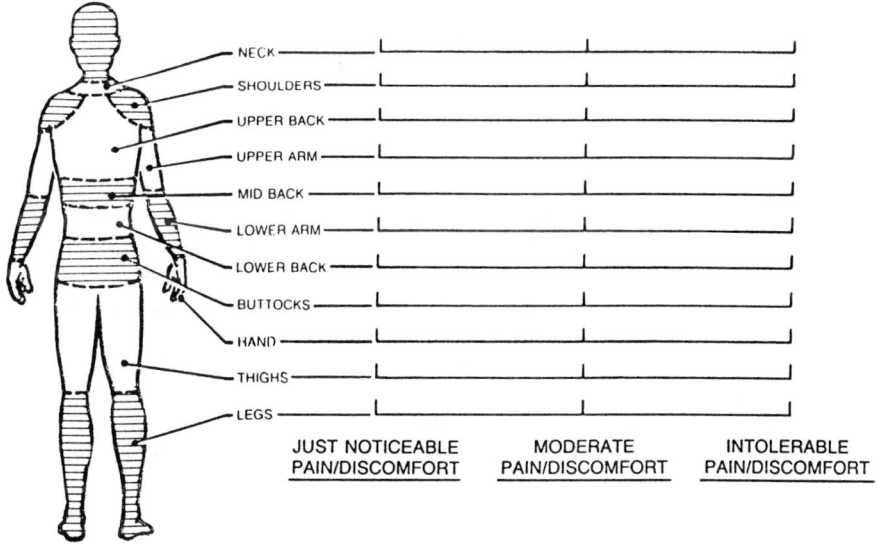

Figure 1 Body part discomfort survey used to collect data in the preclean area.

addition, the weights of objects handled by the operators (hoses, pumice bags) were obtained. These measurements were compared to recommended workstation dimensions based on anthropometric information and used to compute recommended weight limits (RWLs) using the 1991 NIOSH Lifting Equation.

From discussions with the operators and information obtained at the worksite, several ergonomic problems were identified. These are summarized in Table 5. The principal hazards included awkward lifts and excessive reaches. Based on an evaluation of the frequency with which each task was repeated, the workers' average yearly exposures to these awkward

Table 5 Problem Summary—Preclean Machine Operator

Preclean Operator Task(s)
1. Test chemicals in the preclean machine.
2. Add pumice or water to the machine.
3. Perform machine maintenance or housekeeping as needed.

Body Part Discomfort Data
  12 workers surveyed
    6 of 12   neck pain
    7 of 12   shoulder pain
    4 of 12   wrist pain
    8 of 12   back pain
    9 of 12   foot pain

1991 NIOSH Lifting Guide Data
Pumice Bag (lift from pallet)
  Object weight: 35 lb
  Horizontal distance (back of pallet): 32 in.
  Horizontal distance (front of pallet): 18 in.
  Final distance: 11 in.
  Origin of lift: 19 in.
  Distance traveled: 17 in.
  Frequency: 2 lifts/8 hr
  Asymmetric angle: 0°
  Coupling: poor

NIOSH RWL (back of pallet): 16.7 lb
  Lift Index: 2.10
NIOSH RWL (front of pallet): 21.7 lb
  Lift Index: 1.61
Ergonomic Costs
  35,000 lb lifted per year
  3900 extended reaches per year
  3900 trunk flexions per year
Machine Lid
  Object weight: 22 lb
  Horizontal distance: 24 in.
  Final distance: 10 in.
  Origin of lift: 44 in.
  Distance traveled: 16 in.
  Frequency: 2 lifts/8 hr
  Asymmetric angle: 30°
  Coupling: poor
NIOSH RWL: 14.4 lb
Lift Index: 1.53

Problems Identified
1. To add pumice to the preclean machine, the operator must lift pumice bags from a pallet to the top of the machine. The weight of the pumice bags exceeds the NIOSH RWL; therefore, the operator may be at risk for back injury. Additionally, bags often leak, creating dust in the area.
2. Spraying water inside the machine requires the operator to remove plastic covers from the top of the machine. Each lid weighs 22 lb, which exceeds the NIOSH RWL. Removing the lids from the machine also requires an extended reach of 24 in. with the trunk flexed. According to anthropometric tables in Eastman Kodak [40], most people cannot reach more than 18 in. in front of the body without bending or leaning.
3. Only one water hose is provided in the preclean area. Adding water to the preclean machine or washing the floor requires the operator to uncoil the hose and drag it up to 30 ft, placing stress on the operator's back and shoulders. An extended reach of 24 in. is required to access the machine with the hose.
4. The operator is occasionally required to activate controls and valves located on a panel above the preclean machine. An extended reach of 24 in. is required to access the panel. A large machine located next to the preclean machine hampers the operators' ability to obtain test samples.
5. Operators are required to stand on concrete surface or metal gratings throughout their work shift. Prolonged standing may cause foot/leg discomfort.

Job Analysis

motion patterns were estimated. Although tasks such as adding pumice and water to the preclean machine are performed only a few times each day, comparisons with NIOSH RWLs indicated that workers who performed these tasks were at risk for back injury.

3. Recommended Solutions

Methods of reducing or eliminating the number of awkward lifts and reaches were proposed to mitigate the hazards posed by this job. Recommendations included changing the way supplies (i.e., pumie and water) are provided to the operator and providing additional equipment (e.g., hoses and stools) in the preclean area (Table 6). As indicated, the recommended changes were anticipated to eliminate many of the workers' exposures to biomechanical hazards for musculoskeletal injury. Follow-up is necessary to determine if interventions are actually effective in reducing exposures to anticipated levels.

IV. CRITICAL REVIEW OF CURRENT STATUS

A. Use of Checklists

Increases in the demand for ergonomic expertise in industrial settings have led to the development of various checklists for use in conducting workplace ergonomic audits. Generally, checklist evaluations should not be substituted for a full-scale ergonomic evaluation of the workplace; most checklists are designed to help investigators document their efforts in a systematic fashion and target areas for further analysis (see Sec. III.A.2). They can also be used as a rapid screening tool to allow investigators with limited training to identify common ergonomic hazards in a short period of time.

The limitations of checklists are that they usually do not provide enough information to identify specific work attributes responsible for excessive stress nor do they provide insights as to how jobs can be redesigned to reduce biomechanical stress [59]. Furthermore, check-

Table 6 Proposed Solutions—Preclean Machine Operator

Ergonomic Savings
  35,000 lb lifted per year
  3900 extended reaches per year
  3900 trunk flexions per year

Problem Solutions
1. Instead of buying pumice in 35-lb bags, pumice should be purchased in drums, which can be stored next to the preclean machine. Providing pumice in drums instead of bags will prevent operators from having to lift the bags and should permit better dust control (leaky bags will be eliminated). A hoist would be used to lift the drums when pumice is needed. An in-line scale should be provided to allow operators to measure the amount of pumice required to maintain concentrations at proper levels.
2. The design of the machine covers should be modified to improve access to the inside of the machine. Lids should be hinged or should slide over or under one another so that lifts are not required.
3. Additional hoses with longer nozzles are needed. A longer nozzle would reduce the reach needed to spray water into the machine, and an additional hose (stored on a spring-loaded coil) would allow the operators to access the machine more easily (i.e., with less stress on the back and shoulders) from all directions.
4. The control panel over the preclean machine should be moved so that the operator can access the panel with a reach of less than 18 in. The machine next to the preclean machine should be relocated so that the operator can more easily access the valves needed for testing.
5. Provide a chair or stool in the preclean area to allow operators to sit occasionally. Providing antifatigue mats around the machine may also alleviate foot and leg discomfort.

lists do not substitute for an understanding of the risk factors for musculoskeletal disease. The danger in relying on checklists is that hazards may be overlooked if they are not specifically described by the checklist. For example, ergonomic hazards in an office environment are likely to be different from those in a manufacturing facility.

To ensure that the questions are appropriate to the workplace of interest, checklists should be customized and evaluated in the walk-through survey before the actual evaluation is conducted. Examples of various ergonomic checklists are found in the Appendix of this handbook.

B. Worker Assessment vs. Expert Assessment

Traditionally, the job analysis task has been left to experts with professional training in ergonomics or industrial engineering. Because many industrial facilities lack in-house ergonomic expertise, some companies have filled this void with consultants. A drawback to this approach is that consultants may lack an understanding of the intricacies that can impact the effectiveness of various control methods; also, their involvement usually ends before the intervention process is complete. In these situations, companies are frequently left with little or no mechanism to ensure that changes will be implemented or that interventions will be effective.

It is increasingly recognized that the workers who perform a particular job possess a large knowledge base that could be helpful in recognizing ergonomic problems and implementing effective changes. As a result, several companies have established "ergonomics committees" composed of representatives of management, labor, engineering, and safety to identify and resolve ergonomic problems in the workplace. A participatory approach to the control of ergonomic hazards offers at least two advantages. First, ergonomic committees frequently possess the variety of experience needed to solve problems with multiple causation. Second, allowing employees to participate in decision-making processes creates ownership and improves commitment to solutions [60].

According to Joseph, participative programs can be an effective mechanism for implementing ergonomic changes in the workplace [61]. However, the success of a participatory ergonomics program will depend on the commitment of management to the principles of prevention through ergonomic job design. Furthermore, training is necessary to ensure that employees are informed of ergonomic hazards to which they may be exposed and are able to participate actively in their own protection. Supervisors and managers should also be trained to recognize ergonomic hazards and to understand their medical consequences. Finally, engineers and maintenance personnel should be trained to prevent and correct ergonomic hazards through design and maintenance procedures.

V. FUTURE CONCERNS

Job analysis is only one element in an overall program to control and prevent musculoskeletal disorders in the workplace. To be useful, information derived from job analysis activities must lead to appropriate intervention strategies for eliminating ergonomic stresses or reducing stresses to acceptable levels. It is hoped that future research will provide additional evidence to demonstrate that a program that includes a systematic method of identifying ergonomic hazards can be effective in reducing the incidence and cost of musculoskeletal injury. Furthermore, it is anticipated that future developments will result in new data gathering techniques and the development of more accurate models to predict human capabilities in relationship to job demands.

REFERENCES

1. B. W. Niebel, *Motion and Time Study,* 8th ed., Irwin, Homewood, IL, 1989.
2. T. J. Armstrong, R. G. Radwin, D. J. Hansen, and K. W. Kennedy, Repetitive trauma disorders: job evaluation and design, *Hum. Factors* 28(3):325–336 (1986).
3. D. T. Ridyard, A successful applied ergonomics program for preventing occupational back injuries, in *Advances in Industrial Ergonomics and Safety*, Vol. II, B. Das, Ed., Taylor & Francis, Philadelphia, 1990, pp. 125–132.
4. F. McKenzie, J. Storment, P. Van Hook, and T. Armstrong, A program for control of repetitive trauma disorders associated with hand tool operations in a telecommunications manufacturing facility, *Am. Ind. Hyg. Assoc. J.* 46:674–678 (1985).
5. G. Lutz and T. Hansford, Cumulative trauma disorder controls: the ergonomics program at Ethicon, Inc., *J. Hand Surg.* 12A(2Pt2):863–866 (1987).
6. Occupational Safety and Health Act, 1970, PL 91–596.
7. Code of Federal Regulation, 29 CFR 1910, Ergonomic Safety and Health Management Program, 1991.
8. *Ergonomics Program Management Guidelines for Meatpacking Plants*, OSHA 3123, U.S. Dept. of Labor, Occupational Safety and Health Administration, Washington, DC, 1990.
9. A. D. Hosey, General principles in evaluating the occupational environment, in *The Industrial Environment—Its Evaluation and Control*, G. D. Clayton, Ed., Natl. Inst. for Occupational Safety and Health, Cincinnati, OH, 1973, p. 95.
10. S. H. Rodgers, Functional job analysis technique, in *Occupational Medicine: State of the Art Reviews*, J. S. Moore and A. Garg, Eds., Hanley & Belfus, Philadelphia, 1992, p. 680.
11. W. M. Keyserling, T. J. Armstrong, and L. Punnett, Ergonomic job analysis: a structured approach for identifying risk factors associated with overexertion injuries and disorders, *Appl. Occup. Environ. Hyg.* 6:353–363 (1991).
12. Eastman Kodak Co., *Ergonomic Design for People at Work,* Vol. 2, Van Nostrand Reinhold, New York, 1986.
13. L. Punnett and W. M. Keyserling, Exposure to ergonomic stressors in the garment industry: application and critique of job-site work analysis methods, *Ergonomics* 30(7):1099–1116 (1987).
14. T. J. Armstrong, J. A. Foulke, B. S. Joseph, and S. A. Goldstein, Investigation of cumulative trauma disorders in a poultry processing plant, *Am. Ind. Hyg. Assoc. J.* 43:103–116 (1982).
15. D. J. Habes, Use of EMG in a kinesiological study in industry, *Appl. Ergon.* 15(4):297–301 (1984).
16. H. Christensen, Muscle activity and fatigue in the shoulder muscles of assembly-plant employees, *Scand. J. Work Environ. Health* 12:582–587 (1986).
17. A. Aaras and R. H. Westgaard, Further studies of postural load and musculo-skeletal injuries of workers at an electro-mechanical assembly plant, *Appl. Ergon.* 18(3):211–219 (1987).
18. F. E. Gomer, L. D. Silverstein, W. K. Berg, and D. L. Lassiter, Changes in electromyographic activity associated with occupational stress and poor performance in the workplace, *Hum. Factors* 29(2):131–143 (1987).
19. E. Milerad, M. O. Ericson, R. Nisell, and A. Kilbom, An electromyographic study of dental work, *Ergonomics* 34(7):953–962 (1991).
20. W. S. Marras, S. A. Lavender, S. E. Leurgans, S. L. Rajulu, W. G. Allread, F. A. Fathallah, and S. A. Ferguson, Industrial quantification of occupationally related low back disorder risk factors, *Proc. Hum. Factors Society 36th Annual Meeting*, Human Factors Society, Santa Monica, CA, 1992.
21. A. Garg, G. Hagglund, and K. Mericle, A physiological evaluation of time standards for warehouse operations as set by traditional work measurement techniques, *IIE Trans.* 19:235–245 (1986).
22. V. Z. Priel, A numerical definition of posture, *Hum. Factors* 16:576–584 (1974).
23. O. Karhu, P. Kansi, and I. Kuorinka, Correcting working postures in industry: a practical method for analysis, *Appl. Ergon.* 8:199–210 (1977).
24. E. N. Corlett, S. J. Madeley, and I. Manenica, Posture targeting: a technique for recording working posture, *Ergonomics* 22(3):357–366 (1979).

25. D. W. Grieve, The Postural Stability Diagram (PSD): personal constraints on the static exertion of force, *Ergonomics* 22(10):1155–1164 (1979).
26. W. Rohmert and K. Landau, *Das Arbeitswissenschaftliche Erhebungsverfahren zur Tatigkeitsanalyse (AET)*, Verlag Hans Huber, Bern, 1979.
27. P. Holzmann, ABRAN—a new method for analysis of ergonomic effort, *Appl. Ergon.* 13(2):82–86 (1982).
28. W. M. Keyserling, Postural analysis of the trunk and shoulders in simulated real time, *Ergonomics* 29(4):569–583 (1986).
29. A. Kilbom and J. Persson, Work technique and its consequences for musculoskeletal disorders, *Ergonomics* 30(2):273–279 (1987).
30. H. J. C. Gil and E. Tunes, Posture recording: a model for sitting posture, *Appl. Ergon.* 20(1):53–56 (1989).
31. J. G. Chen, J. B. Peacock, and R. E. Schlegel, An observational technique for physical work stress analysis, *Int. J. Ind. Ergon.* 3(3):167–166 (1989).
32. D. Stetson, W. M. Keyserling, B. A. Silverstein, T. J. Armstrong, and J. A. Leonard, Observational analysis of the hand and wrist: a pilot study, *Appl. Occup. Environ. Hyg.* 6:927–937 (1991).
33. L. McAtamney and E. N. Corlett, RULA: a survey method for investigation of work-related upper limb disorders, *Appl. Ergon.* 24(2):91–99 (1993).
34. D. J. Habes and V. Putz-Anderson, The NIOSH program for evaluating biomechanical hazards in the workplace, *J. Safety Res.* 16:49–60 (1985).
35. S. S. Ulin and T. J. Armstrong, A strategy for evaluating occupational risk factors of musculoskeletal disorders, *J. Occup. Rehab.* 2(1):35–50 (1992).
36. E. Kamon and A. Goldfuss, In-plant evaluation of muscle strength of workers, *Am. Ind. Hyg. Assoc. J.* 39:801–807 (1978).
37. V. Mathiowetz, N. Kashman, G. Volland, K. Weber, M. Dowe, and S. Rogers, Grip and pinch strength: normative data for adults, *Arch. Phys. Med. Rehab.* 66:69–72 (1985).
38. D. B. Chaffin, Biomechanical modeling for simulation of 3D static human exertions, in *Computer Applications in Ergonomics, Occupational Safety and Health*, M. Mattila and W. Karwowski, Eds., Elsevier, Amsterdam, 1992, pp. 1–11.
39. NASA, *Anthropometric Source Book,* Vols. I–III (Ref. Pub. 1024), NASA Sci. Tech. Inf. Office, Yellow Springs, OH, 1978.
40. Eastman Kodak Co., *Ergonomic Design for People at Work*, Vol. 1, Van Nostrand Reinhold, New York, 1983.
41. S. S. Ulin, T.J. Armstrong, and R. G. Radwin, Use of computer aided drafting for analysis and control of posture in manual work, *Appl. Ergon.* 21:(2)143–151 (1990).
42. P. O. Astrand and K. Rodahl, *Textbook of Work Physiology: Physiological Bases of Exercise*, 3rd ed., McGraw-Hill, New York, 1986.
43. NIOSH, *Work Practices Guide for Manual Lifting,* NIOSH Tech. Rep. No. 81-122, U.S. Dept. of Health and Human Services, Natl. Inst. Occup. Safety and Health, Cincinnati, OH, 1981.
44. J. V. G. A. Durnin and R. Passmore, *Energy, Work and Leisure*, Heineman, London, 1967.
45. B. Bink, The physical working capacity in relation to working time and age, *Ergonomics* 5(1):25–28 (1962).
46. F. H. Bonjer, Actual energy expenditure in relation to the physical working capacity, *Ergonomics* 5(1):29–31 (1962).
47. T. R. Waters, V. Putz-Anderson, A. Garg, and L. Fine, Revised NIOSH equation for the design and evaluation of manual lifting tasks, *Ergonomics* 36(7):749–776 (1993).
48. B. Givoni and R. F. Goldman, Predicting metabolic energy cost, *J. Appl. Physiol.* 30:429–433 (1971).
49. W. H. VanderWalt and C. H. Wyndham, An equation for prediction of energy expenditure of walking and running, *J. Appl. Physiol.* 34:559–563 (1973).
50. A. Garg, D. B. Chaffin, and G. D. Herrin, Prediction of metabolic rates for manual materials handling jobs, *Am. Ind. Hyg. Assoc. J.* 39:661–674 (1978).

51. S. H. Snook and C. H. Irvine, Psychophysical studies of physiological fatigue criteria, *Hum. Factors* 11(3):291–300 (1969).
52. F. Gamberale, Perception of effort in manual materials handling, *Scand. J. Work Environ. Health* 16(Suppl. 1):59–66 (1990).
53. V. Putz-Anderson and T. L. Galinsky, Psychophysically determined work durations for limiting shoulder girdle fatigue from elevated manual work, *Int. J. Ind. Ergon.* 11(1):19–28 (1993).
54. S. H. Snook, The design of manual handling tasks, *Ergonomics* 21(12):963–985 (1978).
55. D. B. Chaffin and G. B. J. Andersson, *Occupational Biomechanics,* Wiley, New York, 1991.
56. S. H. Snook and V. M. Ciriello, The design of manual handling tasks: revised tables of maximum acceptable weights and forces, *Ergonomics* 21:1197–1213 (1991).
57. S. H. Snook, D. R. Vaillancourt, V. M. Ciriello, and B. S. Webster, Psychophysical studies of repetitive wrist flexion and extension. *Ergonomics* 38(7):1488–1507 (1995).
58. *The UAW-Ford Job Improvement Guide*, UAW-Ford Natl. Joint Committee on Health and Safety, 1988, p. 6-3.
59. W. M. Keyserling, D. S. Stetson, B. A. Silverstein, and M. L. Brouwer, A checklist for evaluating risk factors associated with upper extremity cumulative trauma disorders, *Ergonomics* 36(7):807–831 (1993).
60. A. S. Imada, Macroergonomic approaches for improving safety and health in flexible, self-organizing systems, in *The Ergonomics of Manual Work,* W. S. Marras, W. Karwowski, J. L. Smith, and L. Pacholski, Eds., Taylor & Francis, Philadelphia, 1993, pp. 477–480.
61. B. S. Joseph, A participative ergonomic control program in a U.S. automotive plant: evaluation and implications, Ph.D. Dissertation, Univ. Michigan, Ann Arbor, MI, 1985.

12
Workstation Evaluation and Design

David R. Clark
GMI Engineering & Management Institute, Flint, Michigan

I. INTRODUCTION

This chapter presents methods and recommendations for the evaluation of existing or proposed workstations and their design or redesign to minimize the risk of potential ergonomics problems. The reader is cautioned that the comprehensive evaluation and design of workstations often requires the application of wide varieties of knowledge available on ergonomics. Further, the purpose of this chapter is not to repeat that which is covered in much more detail elsewhere within this book, but to stress the importance of taking a systematic approach to the synergistic application of that knowledge. Although traditional examples of workstation design criteria are given, this chapter avoids the common practice of providing detailed checklist-oriented guidelines.

II. BACKGROUND AND SIGNIFICANCE TO OCCUPATIONAL ERGONOMICS

There can be no more fundamental aspect of occupational ergonomics than the concern with the design of the "local" space in which workers must spend considerable time and effort doing their jobs. In this respect, it is a very personal process in that workers feel a sense of ownership for the workstation, and the morale, health and safety, and productivity of workers can be significantly affected both positively and negatively by the success or failure of design efforts.

Ergonomics, or the "laws of work," by definition and tradition, has focused primarily on the physical aspects of work, such as force and energy requirements. That these are inexorably intertwined with the specific design of workstations hardly need be mentioned. The spatial and temporal arrangements of the task being done in a workstation defines the parameters necessary for biomechanical analysis and design of work–rest cycles, among other things. In some sense, this part of workstation design and evaluation can be thought of as "applied anthropometry."

But one must look beyond traditional ergonomics to traditional human factors engineering. While the former focused on the physical aspects of work, the latter examined the psychophysical interface between the worker and the task. Here one is concerned with the input-output loop of stimulus and response involving information processing and action initiation, i.e., the senses and motor control. Workers usually have to sense something about the dynamics of the process around them, make decisions, and take appropriate actions. The design of the workstation to facilitate this is critical.

Therefore, with respect to workstations, the traditional boundary between ergonomics and human factors engineering is blurred. It is interesting to note that the United States-based Human Factors Society changed its name to the Human Factors and Ergonomics Society in 1993 to reflect the fact that, to many practitioners, the terms are synonymous and are used interchangeably.

III. MODELS FOR WORKSTATION EVALUATION AND DESIGN

A. The Human–Machine Interface Model

To properly appreciate the importance of the design of the workstation, a "system" must be defined that includes not only the "hardware" but also the worker and the interface between the two. Figure 1 illustrates the components of a model of the human–machine interface system.

Human. Humans can play a multitude of roles in systems, which can be arranged dimensionally (Figure 2). In the dimension of *control*, or the degree to which humans determine which way the process will operate, their role can vary from one extreme, that of monitor or supervisor, to the other extreme, that of controller, meaning that no action takes place in the process without intimate human involvement. In many systems, the control role will be somewhat less extreme, most commonly as an initiator, humans being able to choose the course and timing of the system's next action based on its current state and the knowledge and experience of the worker. Traditional human factors engineering is most likely to be applied in the analysis of the dimension of control.

A second dimension of the human role in systems is that of *physical work*, or the degree to which the worker is required to exert force or expend energy for the process to continue properly. At one extreme, the human may not be required to provide any significant force or energy, such as in the case of activating "zero-force" and/or "zero-displacement" controls. It should be noted here that studies have shown that mental processes such as information processing and decision making do not significantly increase energy expenditure. At the other extreme of the physical work dimension would be found those cases in which the worker provides all the force and energy necessary for the task, such as in hammering a nail. More likely is the scenario of a combination of subtasks involving both extremes, mixed with those in which the worker provides some of the force or energy necessary to get the process moving, such as through the use of mechanical aids, servomechanisms, or simply letting gravity take over. It is in this dimension that traditional ergonomics is applied.

Machine. The concept of the machine, as used here, is quite broad and includes not only the hardware of the process itself, including tools, but also that of the product, includ-

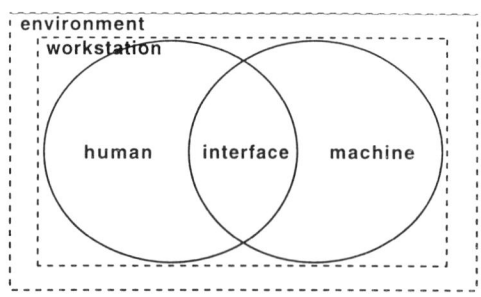

Figure 1 The human–machine interface model of the workstation system.

Workstation Evaluation and Design

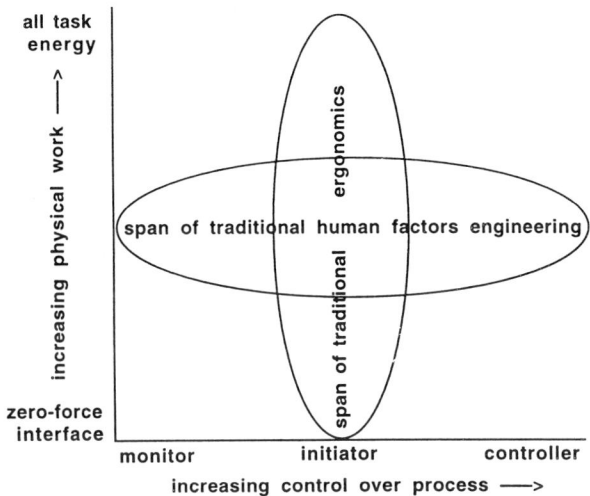

Figure 2 The human role in the workstation system.

ing raw materials, finished work, and by-products. The information connected with both of these that is necessary for the system to operate is also part of the machine. In addition, the sources of energy other than the human are included.

Interface. The interface between human and machine can be viewed as either the physical characteristics of the connection between the two or the process of establishing, maintaining, and terminating such a connection. A detailed task analysis is very helpful in identifying the important elements of the interface.

The physical characteristics include location, size, shape, texture, color, force, and movement. The connection process characteristics include regularity and foreseeability (i.e., operation vs. service vs. supervisory tasks), frequency (including repetitiveness), duration, and competing tasks (i.e., primary vs. secondary tasks).

Environment. Although often difficult or impossible to control due to both technical and nontechnical issues, the environment in which the process is operating can play an important role in the success of the system. Environmental issues include those of engineering concern such as noise and vibration, temperature and humidity, and light and other radiation. But in addition there are sometimes subtle factors at work, such as workplace esthetics, privacy, communication, sense of community; management goals, strategies, policy, work rules, and performance measures; and worker goals, motivations, and morale.

B. The Input–Process–Output Model

Another modeling approach is to consider the "what" and "how" of the tasks to be accomplished. The basic components of such a model are (1) the inputs to the process, (2) the process itself, and (3) the outputs of the process. The inputs and outputs will often simply be outputs and inputs of other processes. Figure 3 shows the components of this model.

Input. System inputs consist of (1) information, (2) raw material, and (3) energy. Raw material is to be transformed. Information is either to be transformed or provides the dynamic "instructions" as to how transformations are to take place. The non-human energy powers these transformations, either totally or in concert with human efforts.

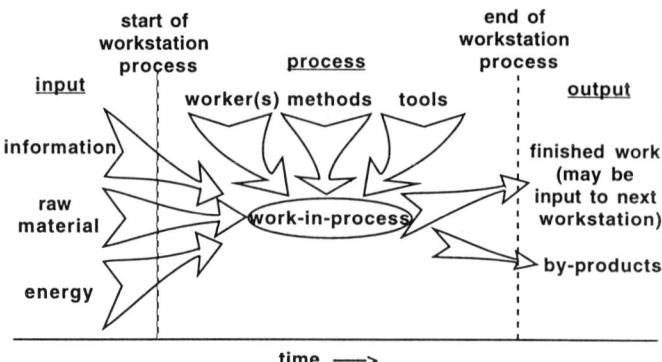

Figure 3 The input–output–process model of the workstation system.

Process. Processing components consist of (1) workers, (2) methods, (3) tools, and (4) work-in-process. Workers control transformation, albeit at different levels as discussed in the preceding section. Methods are the static "instructions" as to how transformations are to take place. Tools are the "hardware" used in the transformation and can include machines, jigs, fixtures, measurement devices, and manual and powered hand tools. Work-in-process is any altered state of process inputs prior to becoming process outputs. This is important in that if the process is interrupted the state of work-in-process can be critical to the methods of resuming the process (e.g., retaining vs. discarding partially completed work).

Output. System outputs consist of (1) finished work, and (2) by-products. The distinction between these two outputs is often in the eye of the beholder. A narrow view of a flow-through "supersystem" of many connected systems might view all outputs not required as subsequent inputs as by-products. A broader view must take into account that the overall enterprise is responsible for the proper disposal of by-products, and thus they are simply inputs to auxiliary systems charged with that function and subject to their own process definitions.

C. The Object-Oriented Process Model

One could also consider the ergonomics attributes of the requirements of the work as found in the objects and processes associated with the workstation. A table of some such attributes is shown in Table 1. Note that for each category—information, objects, and energy—a "flow" exists. For example, the flow of energy through the workstation entails the potential sequence: enable, control, apply, disable, dissipate.

Thus, for each requirement of the work, the designer must consider all the appropriate ergonomics factors, both individually and in combination with others, to reduce the possibility of undue strain.

D. The Ergonomics Stressors Model

Another way of assessing the human's role in the system is to look at the sources of stress, i.e., the causes of strain. These can be divided into three categories: physical, psychophysical, and psychological. Not surprisingly, many of the factors already noted fall into these categories.

Table 1 Ergonomics Attributes of Workstation Requirements

Workstation requirements			Ergonomics attributes	
Do . . .	With . . . , or to what . . .	When . . . , or how . . .	Physical	Sensory
Input	Information	Interruption-driven (push) Process-driven (pull)	Access, connection	Signal-to-noise ratio Parsimony
	Objects	Locate	Location, reach	Amplitude, direction
		Select	Separability	Texture, frequency
		Grasp, control	Size, orientation, handhold, hand clearance	
		Move	Mass, path, velocity	
	Energy	Enable	Location, action	Identification, assessment
Process	Information	Detect, understand, decide		Distraction, confusion, capacity, error-proofing
	Objects	Handle, regrasp	Location, assembly order, fit	Visibility
		Modify	Force, energy	
		Combine	Cycle time, duration, repetitiveness	
	Energy	Control, apply	Location, action	Identification
Output	Information	Communicate, archive		
	Objects	Move	Mass, path	
		Locate		
		Release	Destination	
	Energy	Disable, dissipate	Location, action	Identification

1. Physical Stressors

Physical sources of stress include force, movement, and repetition.

Force. Force can be exerted statically or dynamically. In static force exertion, where the force generation is often localized, isolated muscle groups may be working very hard while others are not working at all. In addition, when muscles maintain contraction under static exertion, the blood vessels within the muscles are constricted, leading to reduced blood flow, which in turn reduces the amount of oxygen being delivered to the muscle and reduces the ability of the circulatory system to remove the by-products of static muscle work, notably lactic acid. This leads to generally lower forces and shorter fatigue times than would be present for dynamic work.

In the case of dynamic work, the often rhythmic body movements, such as in walking, cause elements of paired muscle groups to activate in an alternating fashion, reducing the amount of time any one muscle group is activated in comparison with static work. The associated pattern of blood vessel constriction and relaxation, in concert with the one-way "check valves" in the venous system, actually assists the heart in pumping blood back to the heart and lungs.

Movement. Moving a mass as a result of exerting a force is the definition of work. In the case of static and many rhythmic dynamic force exertions, no net external work is performed because the distance moved is zero, although the internal work can be considerable. Also of concern is the necessity for the worker to effect whole body movement, such as is the case when reach capability is exceeded. The long-term concern would be energy expenditure, especially over the course of a workday. In tasks involving significant worker movement, it can often be shown that most of the energy expended goes to just moving the worker around.

Repetition. Whenever the worker must repeat the same force exertions or movements, especially within short time frames, the repetitiousness of the events can become of primary concern. In fact, even very small forces or movements can be significantly stressful if they are repeated often enough. The types of strains that the body manifests in these cases can be broadly described as *cumulative trauma disorders* or *repetitive motion disease* or some other similar term and include such notable specific strains as carpal tunnel syndrome, tendinitis, and epicondylitis.

2. Psychophysical Stressors

Psychophysical sources of stress include components of the "input–output" or "stimulus–response" loop: displays and other stimuli information processing and decision making, and controls.

Displays and Other Stimuli. Stimuli can be thought of as useful plants and weeds. Useful stimuli, like plants, must be sufficient for attention getting (here I am!), discriminability (this is what I am), and appropriateness (you need me), while not being crowded out by "weedy," distracting or masking stimuli. Parsimony, or the KISS (Keep It Simple yet Sufficient) principle, applies here but is often complicated by the lack of adequate control over the surrounding environment.

Information Processing and Decision Making. Any useful stimulus received by the worker must be processed, even if only minimally, and a decision reached as to the appropriate response. The factors that affect this process include the sufficiency of the stimulus from the worker's perspective and the availability of a corresponding, objective decision rule. Worker education, training, and experience, as well as the use of visual controls in the workstation can greatly affect this part of the process.

Controls. One of the possible responses of a worker in the stimulus–response loop is to operate a control. A control is defined as a discrete mechanism provided in the workstation designed to cause the system to operate in a specific way. As in stimuli, controls must have sufficiency, including attention getting (can I find it?), discriminability (do I know what it is for and how it operates?), and appropriateness (will it do what I need?).

3. Psychological Stressors

Psychological sources of stress include non-occupational vs. occupational stressors, motivation and reward systems, and the environment.

Non-Occupational vs. Occupational Stressors. In a perfect world, workers would arrive at their workstations without the "baggage" of what is happening to them outside of work. This is obviously not the case, and while these outside stresses cannot be controlled, they can be recognized, appreciated, and, often, accommodated. In a robust workstation design, reasonable variance in worker performance, whatever its cause, should not lead to catastrophic results. If it could and the worker was aware of the possibility, it would result in even more stress on the worker.

Motivation and Reward Systems. Most workers want to do a good job. However, work and workstations are often designed in such a way that the worker has no awareness of the quality of the work being done (due to the lack of a standard of performance, a measurement system, or feedback) and/or no control over the quality of that work (due to the inability to remedy or stop output of poor quality work).

Environment. Outside the immediate workstation environment, over which control may be substantial, there is most likely a larger, more uncontrollable environment that contains a "meta" version of the same types of stressors as those discussed in the previous paragraphs. The extent to which they can affect individual workers affects the overall success of workstations and therefore the efforts that must go into their analysis and design.

E. The Product-Process Model

Another view regarding workstations is that they are just a reflection of the characteristics of the product design and the basic processes that are necessary to produce that product (or service). It is important to isolate those characteristics that are truly fundamental and cannot be changed without undue effort from those that can be changed to accommodate more ergonomically designed tasks. Table 2 defines and gives examples of such characteristics.

F. The Life-Cycle Task Model

Comprehensive workstation evaluation and design should address the ergonomics of associated tasks throughout the life cycle of the workstation. The tasks of the worker include not only those that occur regularly and consistently, but also those that do not but are nonetheless foreseeable, such as dealing with breakdowns in the process due to bad materials, tools, controls, and information. The span of time over which such tasks can be considered can be from conception to disposal, i.e., the entire workstation life cycle. A list of life-cycle tasks is shown in Table 3. The tasks that the workstation analyst or designer should focus on are those shown in boldface—operation, service, quality assurance, and repair.

IV. CRITERIA FOR WORKSTATION EVALUATION AND DESIGN

A. Criteria Development

Inevitably, the practitioner desires fairly specific criteria for the evaluation and design of a workstation. This is always the most problematic part of these discussions, because the wide range of applications and conditions makes for almost limitless combinations. Much information must be gathered before a proper analysis can be performed, which, when coupled with general engineering methods, will often lead to the selection of fairly specific design crite-

Table 2 Types of Product and Process Characteristics

Type	Fundamental	Nominal	Discretionary
Definition	Cannot be changed without an overall change in design	Can be changed with localized change in design	Can be changed without any significant change in design
Product example	Snap-fit vs. screw-fastened cover	Self-drilling vs. self-tapping screws	Hex- vs. Torx-head screw
Process example	Batch build vs. one-piece flow	Manual vs. automatic part unloading	Dedicated staffing vs. job rotation

Table 3 Tasks Associated with the Life Cycle of a Workstation

Timing	Regular tasks	Irregular tasks Product-oriented	Process-oriented
More predictable or schedulable	**Operation** **Service** including job setting and scheduled maintenance **Quality assurance**	Conception Design Prototyping Validation Pilot building Validation	Conception Design Construction Validation Distribution Installation **Repair**
Less or not predictable or schedulable			Shutdown Mothballing Demolition Transportation from site Recycling Disposal

ria. To organize the following discussion of information gathering, the model shown in Figure 4 will be used, in a somewhat outside-in manner.

1. Environment

The presumption with regard to the environment is that it is not subject to redesign. If this is not the case, then the analysis/design process must move up hierarchically to include other workstations affected by change. Understanding the environment can affect workstation design decisions because the environment can act on (1) the motivation and (2) the capability of workers to perform their tasks. Some of the questions that should be answered to gain this understanding are:

Does the worker feel a sense of isolation or sense of community, with respect to both the workforce and the overall process of which the workstation is but a part?
Is the worker afforded privacy as well as lines of communication when needed?
Is there workstation access to and from the general environment so as to promote the timely flow of needed people, information, equipment, and material?
Are the worker's senses (e.g., vision, hearing) overwhelmed so as to prevent or degrade the proper transfer of necessary information to and from the worker not only about that worker's workstation but also about adjacent workstations and the environment itself?

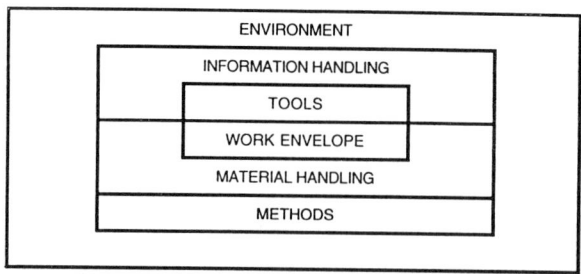

Figure 4 Applied workstation evaluation and design model.

2. Methods

A well-designed workstation includes prescribed methods, where possible, for how the work is to be performed. Questions to be answered are:

What general skills, specific skills, and training are or will be necessary for the worker prior to deployment?

Will periodic training or skills assessment/enhancement will be necessary after deployment?

What information (e.g., visual controls) will be readily accessible to the worker within the workstation regarding the methods to be used?

How much latitude will the worker have in following the method? This is a double-edged sword, because a totally inflexible method may render the worker incapable of reacting to irregular or unforeseen events, whereas too much flexibility may result in no method at all.

Does the design of the product and the process provide sufficient controls as to method? For example, do two parts go together in a single, unambiguous fashion?

What is the worker's span of control over his/her work?

How much variety is provided? What is the frequency of fundamental motions (i.e., bending, twisting, gripping, etc.)?

3. Information Handling

Information is any stimulus in the environment that provides the worker with necessary data about how to perform his or her task successfully. This includes static information about products and processes as well as dynamic information about their current state. Questions to be answered include:

What information must be obtained by the worker? Where is this information obtained? Does the information exhibit persistence, or is it of momentary quality? What is the predictability of the location, frequency, and form of this information? What is the proper response if information is not available when expected or needed?

What information must be provided by the worker? To whom or what does this information flow? What is the predictability of the location, frequency, and form of this information flow? What is the proper response if information flow is held up?

What decision-making processes must the worker perform to convert incoming information and existing knowledge into outgoing information and action? What is the duration of these processes?

What is the likelihood that environmental "noise" can affect any of the above?

4. Materials Handling

Materials include any objects that flow into, are consumed within, or flow out of the workstation. Although most materials are incorporated into the product, there are others that must be considered, such as consumable tools (e.g., sandpaper) and fuels (e.g., batteries for a test gauge). Questions to be answered are:

What materials must be obtained by the worker? What is the location, frequency, and characteristics of these materials? How predictable are they? What is the proper response if material is not available when expected or needed?

What materials are produced by the worker? To whom or what do these materials flow? What is the predictability of the location, frequency, and characteristics of this materials flow? What is the proper response if materials flow is held up?

What materials handling processes must the worker perform to convert incoming materials into outgoing materials? What is the duration of these processes?
What is the proper response if any material qualities adversely affect the workstation tasks? What mechanical aids are necessary?

5. Tools

Tools are defined broadly to include both portable and fixed tools, equipment, and machinery. They are further characterized as being more or less assigned to a workstation (or a related group of workstations) and are usually found within or in the vicinity of the workstation. Questions include:

What tools, equipment, or machines must be obtained by the worker? What are their location, frequency, and characteristics? How predictable are these? What is the proper response if the tools are not available or suitably operational when expected or needed?
What are the prescribed methods for using tools, equipment, or machines? What is their duration?
What is the proper response if any material qualities adversely affect the workstation tasks? What are the energy sources? What controls are provided? Are controls unambiguous?

6. Work Envelope

It is not accidental that the work envelope is the last of the workstation model components to be discussed. It is often the most visible aspect of workstation design, sometimes at the expense of the other components. Further, it is impossible to lay out a work envelope if complete information is not available on all the tasks and the information, materials, and tools to be used. The questions to be asked at this point are actually quite simple:

Is there enough room for the worker to comfortably fit and move into, within, and out of the workstation, before, during, and after the task?
Can the worker "reach" without undue effort all of the necessary materials, information, tools, and controls in the workstation? (*Reach* here means both physically and with the senses, e.g., seeing and hearing.)

In spite of the simplicity of the questions, the answers require a basic understanding of the variability of the size of workers (i.e., anthropometry) and the use of anthropometric data. There are three basic choices when applying anthropometric data in the design of a workplace, each of which has its applications: design for extremes, design for the average, and provide for adjustability.

Design for Extremes. The extremes are either the largest or smallest individual who may be expected to use a workplace.

Design to fit. Dimensions for the largest individual, the 95th or 99th percentile, are used to determine clearance dimensions such as the minimum height for overhead conveyors or the top of a doorway that will permit the majority of the population to walk beneath without hitting their heads. The large individual also determines the necessary width for aisles, the size of access openings for machine repair, and the necessary clearance between a chair seat and the bottom of a table or workbench.

Design to reach. Dimensions for the smallest person, the 5th or 1st percentile, are used to determine reach dimensions, such as the maximum height of shelves or controls, the location of parts bins in the workplace, or the height of nonadjustable chair seats that will permit the small person's feet to touch the floor.

Design for the Average. Designing for average body dimensions is usually a mistake. An average dimension will lead to discomfort in a large part of the workforce. The smaller person in the workforce will not be able to reach a part or control placed at the reach distance of the average person, and the larger person in the workforce may not fit in the workplace or may have to assume unusual postures to perform the task. A conveyor line set to the reach height of the average person will have the smaller individual reaching up to work, causing arm and shoulder fatigue, while the larger operators will be bending down, causing neck and back fatigue.

Provide for Adjustability. The previous two design criteria assume that the workplace components (conveyors, workbenches, parts bins, etc.) are fixed and cannot be adjusted. The alternative to these approaches is to provide adjustability in the workplace. The advantage of this approach is that a large proportion of the workforce can be accommodated. The disadvantages include higher design and manufacturing costs, time of adjustment, and greater likelihood of the adjustable component breaking.

Where adjustment is a practical approach, the proportion of the population to be accommodated must be determined. Typically the range of adjustment will include the middle 90% or 95% of the population. For example, a standing workplace for light assembly work should be at approximately elbow height. Obtaining the required percentile dimensions from a typical anthropometric table, accommodating the 5th percentile female through the 95 percentile male, might require the following range of adjustment:

Female 5th percentile = 96.8 cm
Male 95th percentile = 118.1 cm
Range of adjustment = 21.3 cm

Only the smallest 5% of the female workforce and the largest 5% of the male workforce will not be able to adjust this workplace properly.

To accommodate a different percentage of the population might require using the appropriate Z value from a normal probability table and the dimension's mean and standard deviation from a typical anthropometric table. For example, accommodating 0.5th percentile female to the 99.5th percentile male:

For 0.5th–99.5th percentile: $Z = \pm 2.576$
Female: 50th percentile = 102.6 cm, s = 3.6 cm
    0.5th percentile = 102.6 − (2.576)(3.6) = 93.5 cm
Male: 50th percentile = 110.5 cm, s = 4.6 cm
    99.5th percentile = 110.5 + (2.576)(4.6) = 122.2 cm
Range of adjustment = 28.7 cm

As can be seen, the range of adjustment would have to be increased by 7.4 cm to accommodate this additional part of the workforce. Often this is not practical or may be too costly; therefore, even in adjustable workplaces, the proportion of those to be accommodated must be reduced.

Alternatives to Adjustment. Accommodating individuals in fixed height workstations can often be accomplished through appropriate placement of workers with respect to the tasks. The assembly of large items, such as car bodies or engines, usually involves component installations at a variety of work heights. By matching operators to tasks with inherently more appropriate work heights rather than by random assignment, a greater percentage of operators can be accommodated.

Where fixed work heights must be used, yet all work is performed at approximately the same height, such as work with presses or welders, designing the work height to fit the large individual may be the practical alternative. To fit smaller individuals, adjustable platforms can be moved into the workplace when needed.

Job rotation can also be used to minimize an individual's exposure to uncomfortable postures caused by improper work heights.

B. Examples of Specific Design Criteria

1. Sitting Workstations

Advantages. Energy expenditure is approximately 20% less for sitting than for standing. A sitting workstation provides a high degree of body stability and reduces fatigue when work periods exceed 2 hr. Workers can use feet for control actions.

Indications and Contraindications. A seated workplace can be used under the following conditions:

1. All items needed to perform the task can be easily placed within the seated workspace.
2. Frequent reaches above the work surface are not required.
3. Tasks requiring fine manipulative hand movements are predominant.
4. The handling of parts or application of force does not exceed 45 N.
5. There is no interference with leg clearance.

Criteria. The commonsense criterion for physical access is the ability to reach, which in turn defines the *reach envelope*, which consists of that spatial volume that can be reached by the worker while remaining seated (i.e., static) in a "normal" chair without undue stretching, bending, or twisting.

Work surface height. In any workplace, work height is often a trade-off between (1) a close view of the work and the need for precise movements and (2) the freedom to perform gross movement and the ability to generate force. The normal work height for seated tasks is about 5 cm below elbow height. This clearance provides a reasonably close view of the work while permitting movement uninterfered with by obstructions on the work surface. This clearance can then be increased for more movement and force capability, and decreased, even to where work height is above elbow height, for precise physical and visual tasks. In this last case, it may be necessary to provide elbow support for prolonged tasks.

It is important to carefully consider what constitutes the work surface. Often, it is not the physical desk or benchtop but the location of the hands while working, such as locations on machines, fixtures, or parts. In the case of computer keyboards, the home row of the keyboard is the work surface. It is recommended that the home row be at elbow height to facilitate the relatively precise movements required.

To select which height dimension within the workstation to set first, one should identify which dimension is the least flexible from such possible dimensions as the heights of points of operation, the desk or benchtop, the elbow, or the seating. The subsequent determination of the initial height of that dimension will then drive the process of finding most other height dimensions. For instance, using the work surface height as the initial dimension will most likely require accommodating workers through adjustable seat and foot rest dimensions.

Reach dimensions. Normally, workers will have to reach away from the normal point of operation to complete their task, perhaps to obtain material or tools, or to operate controls. Reaches should be kept within certain regions depending upon the frequency of the reach, as illustrated later in Figure 7. The majority of work should be performed within 15 cm of the work surface. Although both should be kept to a minimum, reaches below the level of the work surface are preferable to those above the work surface; reaches above shoulder level and behind the body should be kept to a minimum. Repetitive work should be performed with the shoulders and upper arms relaxed and the elbows bent at approximately 90°. Padded forearm rests should be provided for precision work.

Clearances. Clearance below the work surface for legs and toes is necessary. Sufficient thigh clearance (25–30 cm) should be provided between the seat pan and the underside of the work surface. Sufficient clearance also permits the worker to vary posture to reduce fatigue.

Seat dimensions. The seat area should be large enough to permit movement to relieve pressure points. The seat pan and back rest should be padded and covered with a permeable material that will allow absorption of sweat. The back support should adjust both horizontally and vertically. Figure 5 gives typical dimensions for a seated workplace.

2. Standing Workstations

   Advantages. When operators are required to stand for long periods of time, fatigue can become an important factor in their job performance. To minimize the effects of fatigue, it is important that the task be designed to eliminate excessive reaches, stooping, bending or twisting motions, and unnatural head positions because of the visual requirements of the task.

   Indications and Contraindications. A standing workplace is needed under the following work conditions:

1. Mobility or walking is required.
2. Handling of objects weighing more than 45 N is frequently required.
3. Extended reaches are often required.
4. Large downward forces must be exerted.
5. The workplace has insufficient leg clearance for a seated operator.

   Criteria. The standing reach envelope is a bit larger and extends the sitting reach envelope to include that which can be reached by more effective bending forward and twisting than is available while seated, and by a nominal repositioning of the feet that does not result in a significant movement away from "home" position of the workstation.

   Work surface height. Normal work height for light assembly tasks is just below elbow height. For more precise work, the work surface should be raised above elbow height and padded rests provided for the forearms. For heavy work where large downward or upward forces are required, the work surface should be 10–20 cm below elbow height. Remember,

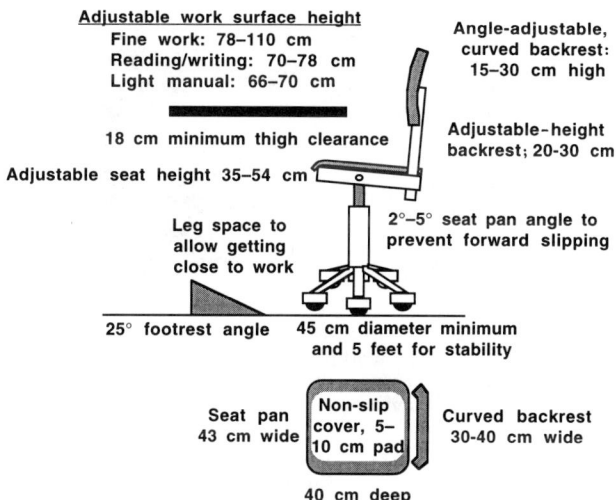

Figure 5 Seated workplace dimensions.

work height is where work is being done and is not necessarily the height of the work surface.

Reach distances. Allow for the shortest arms when reaching up or out. Keep frequent work within forearm reach and below mid-chest height. Avoid reaches behind the body.

Clearances. Allow at least 200 cm of overhead clearance. Access into and out of the workplace should allow for the largest individual. Clothing and personal protective equipment require additional clearance. Allow toe space and knee clearance at workbenches and conveyors. Provide sufficient room for the operator to move about and change postures while working.

Task characteristics. Avoid the use of foot pedals in standing workplaces. Avoid placing displays or other visuals above eye height. In general, the primary visual task should be within 10° downward of the direct line of sight and within 15° right and left. Avoid or minimize work that requires fine manipulations or precision adjustment.

Other considerations. Providing floor padding or shoe inserts may improve worker comfort when continuous standing is required. Provide a stool or seat that the operator may use during breaks or temporary work stoppages. Figures 6 and 7 give typical dimensions for a standing workplace.

3. Sit/Stand Workstations

Advantages. A sit/stand workplace is often preferred from both a physiological and an orthopedic point of view and is recommended when the following work conditions are present:

1. Multiple tasks are performed, with some best performed while seated and others while standing.
2. Reach requirements occasionally extend more than 40 cm forward and/or more than 15 cm above the work surface.

Indications and Contraindications. The minimum height for the work surface during standing tasks is 100 cm. However, this height may present a hazard from excessive chair height during seated work. A 90 cm work surface with a platform that folds down or slides

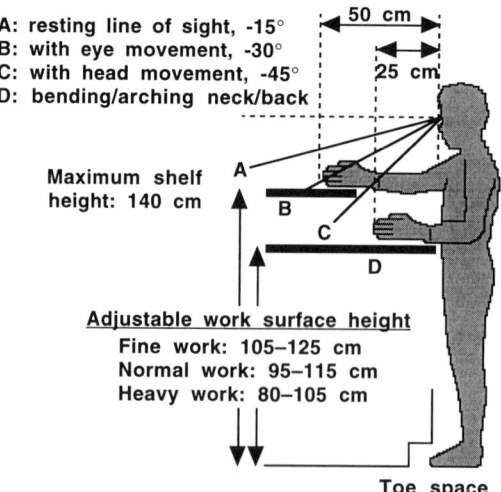

Figure 6 Dimensions for standing work.

Workstation Evaluation and Design

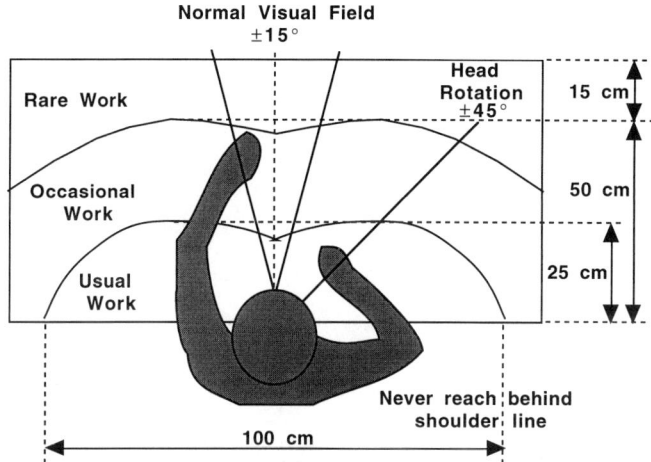

Figure 7 Reach dimensions for standing or seated work.

onto the work surface for standing work can eliminate this problem. A work surface that adjusts between 90 and 105 cm would provide the best range of heights for both sitting and standing.

C. Examples of Workstation Evaluation and Design

This section presents two examples of applying the principles covered in this chapter. The first example is an evaluation of an industrial assembly workstation, and the second is the design of an office workstation. The steps in the evaluation/design process include: (1) definition of tasks, (2) application of workstation models and ergonomic principles, (3) layout, (4) input by worker(s) and modification, (5) prototyping, (6) refinement, and (7) utilizing feedback. The reader should note that all models and criteria presented in this chapter are not necessarily applicable or useful for every evaluation or design exercise.

1. Industrial Example of Evaluation

This example is of an existing workstation that is part of a transfer line where radios are installed in automotive instrument panels. The nominal work height is governed by the transfer line and is 80 cm. Lateral space in the workstation is limited to 1.5 m. The overall cycle time for installation is 60 sec. A layout of the existing workstation is shown at the top of Figure 8.

Task Definition. The existing sequence of tasks is as follows:

1. Read manifest and determine radio option for next installation.
2. Get radio from stock.
3. Place in test fixture.
4. Connect test cable (contains power and signal).
5. Perform specified test sequence (Power on, Select station, Volume increase/decrease, Power off).
6. Remove radio from fixture (if necessary, discard failures and replace from pretested buffer stock).
7. Place radio in IP opening from bottom.

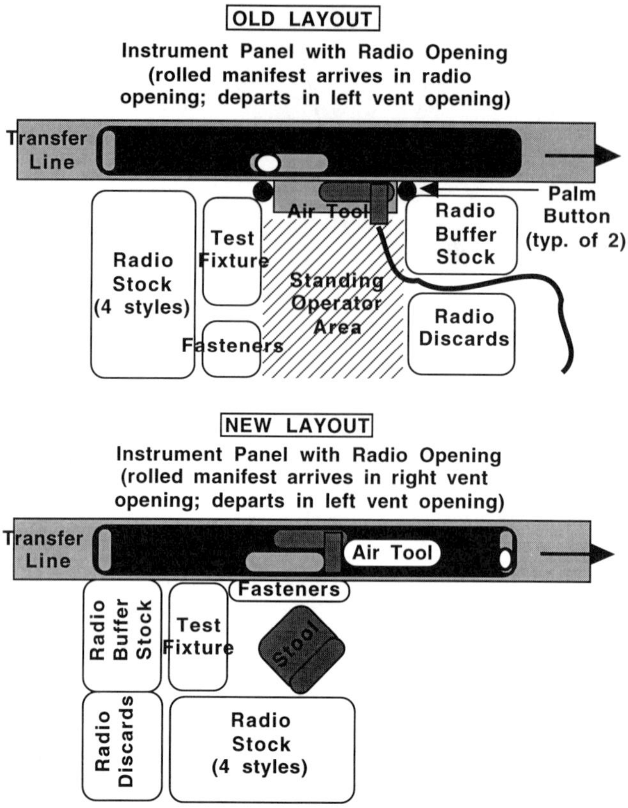

Figure 8 Old and proposed new layouts for industrial workstation example.

8. Secure radio from top with two fasteners using pistol-grip air tool.
9. Index transfer line using palm buttons.

Application of Workstation Models and Ergonomic Principles. Table 4 presents a summary of possible analyses based on the models and principles shown in Tables 1–3.

Layout. Figure 8 (bottom) shows proposed layout of a new workstation arrangement incorporating some of the improvements suggested in Table 4.

Worker Input and Modification. Workers were asked for their assessment of the proposed workstation. The first questions were, not surprisingly, What's wrong with the way it is? and Who requested this? Ultimately, however, several good suggestions were received, including improving visibility between this workstation and the two adjacent workstations to allow workers to anticipate things that might affect their job. As a result, modifications were proposed to materials storage rack heights.

Prototyping, Refinement, and Feedback. The project is now set to proceed to appropriate prototyping to prove out spatial and temporal relationships and equipment selection. At least some small refinements will be realized during this process. After implementation, continuous improvement would be fostered through the use of feedback form the workers and others.

2. Office Example of Design

This example is of a proposed new workstation that will process requests for data entry to and retrieval from a computer database. Approximately 9 m² of floor space is available, but there are presently no serious constraints on the equipment that can be used.

Task Definition. The existing sequence to tasks is as follows:

1. Batch enter data from hand-completed forms (approximately 100 letter-size forms per day, each of about 80 total characters in 10 fields; average input time about 2 min/form; total time about 200 min.
2. File forms after entry.
3. Receive individual customer requests (in person, by mail, or by phone) for retrieval or change of existing data (in-person and phone requests handled while you wait; mail requests printed and mailed by next day; expectation is 10 in-person, 25 mail, and 25 phone requests per day; average service time about 4 min each; total time about 240 min).

Application of Workstation Models and Ergonomic Principles. Table 5 presents a summary of possible analyses based on the models and principles shown in Tables 1–3.

Layout (New). Figure 9 shows a proposed layout of a workstation arrangement incorporating some of the improvements suggested in Table 5.

Worker Input and Modification. In this case, no workers were directly available, so reliance will be placed on subsequent prototyping, refinement, and feedback processes.

Prototyping, Refinement, and Feedback. Proceed as in first example, with greater emphasis on feedback because there was no worker input during the initial stages of the design process.

V. CRITICAL REVIEW OF CURRENT STATUS

A. Independent vs. Nonindependent Design Criteria

As should be evident from a review of this book, ergonomics is a collection of a fairly large number of areas of concern. Taken separately, most of these areas have been systematically researched, and meaningful guidelines have been developed. However, the design of something as potentially open-ended as a workstation involves the simultaneous application of these multiple guidelines. As I see it, the primary problem in the effective design of workstations is the independent (albeit understandable) application of these nonindependent guidelines. No one would be foolish enough to suggest that one could apply anthropometry without considering biomechanics, or make decisions on task allocation between workers and machines independent of a knowledge of the effects of repetition on cumulative trauma disorders.

Unfortunately for the practitioner, the number of combinations and permutations of ergonomics stressors that can be present in the real world of work design is such that research has not had time to examine them all. When attempts are made to reduce complex multidimensional phenomena to simplistic guidelines, much can be lost in their general applicability. For instance, the NIOSH lifting guidelines sought to reduce the admittedly complex analysis of manual lifting tasks to a function of four (1981 version) or six (1991 version) easily measured variables. But the fine print in the "instruction manual" cautions about applying the guidelines in situations that are quite common, including highly dynamic lifting. It can be said without fear of contradiction that the use of these guidelines has often strayed beyond their limits.

Table 4 Summary of Analyses of Industrial Workstation Example

No.	Do . . .		With . . . , or to what . . .	When . . . , or how . . .
A1	IO	Wait	IP	From pervious palm buttons
A2	IO	Get	Manifest	From IP
A3	II	Read	Option	From manifest
A4	IO	Select	Radio	From workstation stock
A5	PO	Put	Radio	To test fixture
A6	PI	Control	Test protocol	At loaded test fixture
A7	II	Get	Test result	During test protocol
A8	PO	Get	Radio	From test fixture
A9	OO	Put	Defective Radio	From test result negative
A10	IO	Get	Pretested Radio	From workstation stock
A11	PO	Put	Radio	Into IP
A12	IO	Get	Fasteners	From stock
A13	PO	Get, use, put	Air tool use	Radio in IP, fasteners in hand
A14	II	Sense	Installation complete	End of air tool use
A15	OI	Press	Palm buttons	End of installation complete
A16	OO	Wait	IP w/radio	From palm buttons

What the practitioner must do, therefore, in the evaluation and design of workstations is to be thorough in the identification of ergonomics issues and in the analysis of each such issue. Common sense must then be applied.

VI. FUTURE CONCERNS

A. Two-Sided Standards vs. One-Sided Guidelines

In the application of guidelines that have been developed for engineers and workstation designers, a pattern has been observed that should give pause. It is natural for the users of guidelines to want definitive, objective criteria to divide the domain of designs into two all-inclusive, mutually exclusive regions, "OK" and "not OK" (i.e., two-sided standards). Real-world phenomena do not often succumb to such categorization, at least for the layperson using a relatively easy to measure dimension. The reality is that the distribution of workstations

Product and process characteristics			Ergonomics attributes	Possible remediation /improvements
Fundamental	Nominal	Discretionary		
Frequency	Concurrent tasks		Frequency, work–rest	Job rotation, task reallocation
Timing	Location		Visual task, reach	Locate to leading end of IP to reduce waiting time/reach
Location	Coding		Visual task, discrimination	KISS coding, if necessary
Number of types		Quantity, location, identification	Reach, force, posture, discrimination	Multiple, single-part presentation in optimal location; visual controls
Fixture design		Location	Reach, force, posture	Optimal location out of way of conveyor
Procedure			Control design	Eliminate through 100% pretest
Criteria		Sensing	Display design, noise	Environmental noise control
Fixture design		Location	Reach, force, posture	Optimal location out of way of conveyor
Not usable		Location	Reach, force, posture	Gravity discard
Number of types		Quantity, location, identification	Reach, force, posture	Multiple, single-part presentation in optimal location
Location		Orientation and holding	Reach, force, posture	Rotating fixture on conveyor; design to hold radio in place
Type		Quantity, location	Reach, force, posture	Optimal location; auto feed to tool
	Energy source	Type, location	Reach, force, posture, noise, vibration	Counterbalanced, torque-limited in-line with vibration dampening
			Decision making	Objective decision criteria
Signal	Type of control	Location	Reach, force, posture	One-hand no/low force control with light curtain
Frequency	Concurrent tasks		Frequency, work–rest	Job rotation, task reallocation

within the two regions along a given dimension generally overlaps as illustrated in Figure 10. As can readily be seen, at any value of the dimension there are finite possibilities that a workstation may be in either region. Thus, when the question is asked, Is the workstation OK at this value?, the answer is, "It depends." This can be very disconcerting to one trying to evaluate a workstation.

The response of those who prepare guidelines has generally been to pick a value at which the odds of a making a wrong "OK" decision are held to a reasonable level, accepting an increased chance of a wrong "not OK" decision (i.e., controlling α risk at the expense of β risk). As an example, a guideline might state that clamp forces should not exceed 0.3 N or that energy expenditure should not exceed 3.5 kcal/min. Thus associated decision rules become: If forces are less then 0.3 N, or energy expenditure is less than 3.5 kcal/min, the task is OK; else it depends.

What it depends on are values of other dimensions associated with the task; this is the very topic of the preceding section on the multidimensionality of workstation evaluation. For

Table 5 Summary of Analyses of Office Workstation Example

No.		Do . . .	With . . . , or to what . . .	When . . . , or how . . .
B1	IO	Get	Batch	1/day at start of day
B2	IO	Get	Form	~100/batch as need
B3	II	Sequence read	Character	~80/form
B4	OI	Select, press	Key stroke	1/character
B5	II	Read	Character	1/key stroke
B6	PI	Select, press	Correct	Bad character
B7	OO	Put	Form	Successful
B8	OO	Put	File	Forms
B9	II	Communicate	Request	10 in-person/day
B10	II	Communicate	Request	25 phone/day
B11	II	Get	Request	25 mail/day
B12	PI	Code	Request	As needed
B13	OI	Select, press	Key stroke	As needed per request
B14	II	Read	Character	1/key stroke
B15	PI	Select, press	Correct	Bad character
B16	OI	Communicate	Verbal response	As needed per in-person and phone request
B17	PI	Select, press	Print response	As needed per mail request
B18	OI	Get, put	Printed response	As needed

Workstation Evaluation and Design

Product and process characteristics			Ergonomics attributes	Possible remediation/ improvements
Fundamental	Nominal	Discretionary		
Batching > 1 day	Multiple batches per day	Time of day, location	Location, workload	Delivery to workstation
Form design		Location	Location, repetition, visual task, discrimination	Adjustable multiple-document holder
Field design		Location	Location, visual task, discrimination	Discrete information design (check-off boxes, etc.)
Discrete entries	Coding	Location	Location, posture, repetition	Equipment selection (including adjustability)
	Context checking	Location	Location, posture, visual task	Adjustable VDT with antiglare screen and/or appropriate task lighting
	Auto correct		Repetition	Discrete information input (menu selection, check-off boxes, etc.)
Form design		Location	Location, repetition	Multiple-document holder
Filing > 1 day	Multiple filing per day	Time of day, location	Location, posture, repetition	Active filing locations at reasonable distances and at heights minimizing reach
Option	Timing	Location, request, drop-off	Location, repetition, energy expenditure, auditory task	Workstation integrated into request counter, including height to eliminate sit/stand/sit
Option	Timing	Location, voice mail	Location, repetition, distraction, auditory task	Telephone headset; noise control applied to environment (equipment selection, separation)
Option	Receipt timing	Processing timing	Location, repetition, visual task	Delivery to workstation
Timing	Visual controls		Identification, matching, and coding	Discrete information design
Discrete entries	Coding	Location	Location, posture repetition	Equipment selection (including adjustability)
	Context checking	Location	Location, posture, visual task	Adjustable VDT with antiglare screen and/or appropriate task lighting
	Auto correct		Repetition	Discrete information input (menu selection, check-off boxes, etc.)
Option	Timing	Location, revisit, recall, or mail response	Location, repetition, energy expenditure, auditory task	Workstation integrated into request counter, including height to eliminate sit/stand/sit
Form design, timing		Location	Repetition	Discrete information input (menu selection, check-off boxes, etc.), batching output
Option	Timing	Location	Locations, postures, repetition	Printer at workstation; mail pick-up at workstation

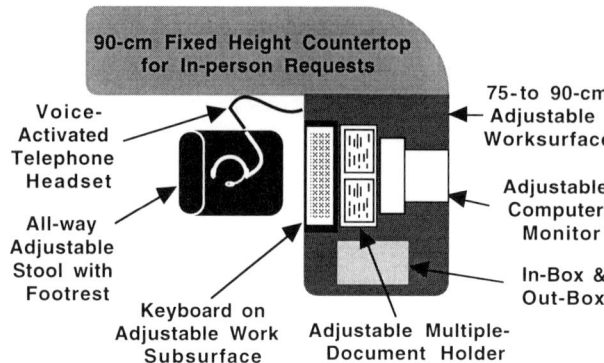

Figure 9 Proposed layout for office workstation example.

instance, 1 N of force is acceptable in a preferred posture using a power grip; or 5 kcal/min is acceptable for a younger, male worker, This, then, is the one-sided guideline; that is, the decision is clear on only one side of the question.

This is the way it should work. However, it often results that practitioners start using the one-sided "OK–depends" guideline as a two-sided "OK–not OK" standard. This occurs for two reasons: (1) a lack of sufficient knowledge or training of the users on the basis for and the interpretation of the guidelines being used or (2) a lack of clear organizational policy as to the proper use of guidelines or the failure to follow such a policy. Thus, I see an unfortunate scenario occuring in many organizations that practice ergonomics, at least in part, in this fashion:

The inevitable: using one-sided standards as two-sided
The only good news: declaring "unsafe" jobs as unsafe
The bad news: declaring some "safe" jobs "unsafe," causing resources to be invested to solve problems that do not exist while other "unsafe" jobs go uncorrected
The worse news: institutionalizing the misuse of standards to the point where it becomes impossible to correct the misuse

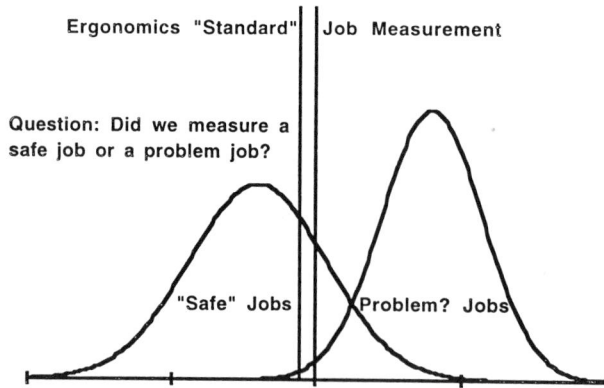

Figure 10 One-sided versus two-sided ergonomics standards.

The worst news: giving ergonomics an undeserved reputation for being too easy on workers and compromising productivity, almost guaranteeing its ultimate failure to be an effective tool, both for worker health and safety and product and process design

REFERENCES

1. D. C. Alexander and B. M. Pulat, *Industrial Ergonomics: A Practitioner's Guide*, Industrial Engineering and Management Press, Atlanta, GA, 1985.
2. R. W. Bailey, *Human Performance Engineering: A Guide for System Designers*, Prentice-Hall, Englewood Cliffs, NJ, 1982.
3. J. H. Burgess, *Designing for Humans: The Human Factor in Engineering*, Petrocelli Books, Princeton, NJ, 1986.
4. Eastman Kodak Company, *Ergonomic Design for People at Work*, Vol. I, Van Nostrand Reinhold, New York, 1983.
5. Eastman Kodak Company, *Ergonomic Design for People at Work*, Vol. II, Van Nostrand Reinhold, New York, 1986.
6. E. Grandjean, *Fitting the Task to the Man*, 4th ed., Taylor & Francis, London, 1988.
7. Human Factors Society, American National Standard for Human Factors Engineering of Visual Display Terminal Workstations, ANSI/HFS 100-1988, Santa Monica, CA, 1988.
8. S. Konz, *Work Design: Industrial Ergonomics,* 3rd ed., Publishing Horizons, Worthington, OH, 1990.
9. National Safety Council, *Making the Job Easier: An Ergonomics Idea Book*, Chicago, 1988.
10. D. J. Oborne, *Ergonomics at Work*, 2nd ed., Wiley, New York, 1987.
11. S. Pheasant, *Bodyspace: Anthropometry, Ergonomics and Design*, Taylor & Francis, London, 1986.
12. M. S. Sanders and E. J. McCormick, *Human Factors in Engineering and Design*, 7th ed., McGraw-Hill, New York, 1993.
13. W. E. Woodson, *Human Factors Design Handbook*, McGraw-Hill, New York, 1981.

13
Tool Evaluation and Design

Andris Freivalds

The Pennsylvania State University, University Park, Pennsylvania

I. INTRODUCTION

Tools are as old as the human race itself. The hands and feet could be considered tools given to the human by nature. However, tools as we know them were developed as extensions of the hands and feet to amplify the range, strength, and effectiveness of these limbs. Thus, the early human, by picking up a stone, could make the fist heavier and harder to produce a more effective blow. Similarly, by using a stick, a longer and stronger arm was created.

The exact time when humans began to use and to make tools is not known. Leaky [1], during his excavations in Africa, uncovered evidence that more than a million years ago the prehistoric human was already a toolmaker using stones for chipping and bones for leatherwork. Similarly, Napier [2,3] indicated that with changing tasks, such as converting from the power grip to the precision grip, there was a similar change in the anatomy of the hand as well as the development of tools. An important milestone occurred when stone tools were provided with handles some 35,000 years ago. The addition of the handle increased the range and speed of action and increased the kinetic energy for striking tasks [4]. A still later change in tool development occurred with the chnage in tasks from food gathering to food production. New tools were required and accordingly developed. Surprisingly, many of these tools, with minor improvements and refinements, are still in use today. The reasons for such stagnation could be twofold: either the tool reached an optimal form very quickly with no room for improvement, or there was no impetus for further improvement. The latter is the resigned view that since a tool has been used by so many people for so many yeras, no further improvement is possible. The former view is obviously not true, because Lehmann [5] noted the existence of over 12,000 different styles of shovels in Germany in the 1930s, all essentially used for the same task. Indeed, the last great change in tool development occurred with the start of the Industrial Revolution when tasks changed from food production to the manufacture of goods.

The parallel development of tools with changing technology has given rise to another problem. The current technology explosion has proceeded too quickly to permit the gradual development of tools appropriate for the new industrial tasks. The instant demands for new and specialized tools to match the needs of technology has, in many cases, bypassed the testing needed to fit these tools to their human users. This has resulted in a variety of hand tool-generated work stressors, trauma, and chronic problems, reducing productivity, disabling individuals, and increasing medical costs for industry.

II. BACKGROUND AND SIGNIFICANCE TO OCCUPATIONAL ERGONOMICS

Poor design and excessive use of hand tools are associated with increased incidence of both acute and subacute cumulative trauma of the hand, wrist, and forearm [6,7]. Acute trauma includes burns, cuts, lacerations, abrasions, fractures, strains, sprains, dislocations, and even amputations caused by the upper extremity being caught, cut, or burned by the tool.

Cumulative trauma disorders (CTDs) are injuries to the musculoskeletal system that develop gradually as a result of repeated microtrauma. Because of the slow onset and relatively mild nature of the trauma, the condition is often ignored until the symptoms become chronic and more severe injury occurs. These problems are a collection of a variety of problems including repetitive motion disorders, carpal tunnel syndrome, tendinitis, ganglionitis, tenosynovitis, and bursistis, with these terms sometimes being used interchangeably. There are four major work-related factors that seem to lead to the development of CTD: (1) use of excessive force during normal motions, (2) awkward or extreme joint motions, (3) high amounts of repetition of the same movement, and (4) the lack of sufficient rest for the traumatized joint to recover. The most common symptoms associated with CTD are pain, restriction of joint movement, and soft tissue swelling. In the early stages there may be few visible signs, but if the nerves are affected, sensory responses and motor control may be impaired. If left untreated, CTD can result in permanent disability.

Tenosynovitis is a disease of the tendon sheaths due to overuse or unaccustomed use of improperly designed tools. It is often experienced by trainees. Improperly designed tools causing ulnar deviation coupled with supination of the wrist may increase the occurrence of tenosynovitis. Repetitive motions and impact shocks may further aggravate the condition.

Carpal tunnel syndrome is a disorder of the hand caused by injury of the median nerve inside the wrist. The median nerve may be injured from extreme flexion and extension of the wrist, which causes inflammation and thickening of the tendon sheaths, which in turn compress the median nerve. Symptoms include impaired or lost nervous function in the first three and a half digits manifesting as numbness, tingling, pain, and loss of dexterity. Again, proper tool design is very important to avoid these extreme wrist positions [6].

Extreme radial deviations of the wrist are conducive to pressures between the head of the radius and the adjoining part of the humerus resulting in tennis elbow. Similarly, simultaneous dorsiflexion of the wrist with concurrent full pronation is equally stressful on the elbow. Tichauer and Gage [8] used the tennis elbow as a good example of the basic principle that work strain may affect a site distant from the site of work stress.

Trigger finger results from a work situation in which the dital phalanx of the index finger must be bent and flexed against resistance before more proximal phalanges are flexed. Excessive isometric forces impress a groove on the tendon, or the tendon enlarges due to inflammation and then snaps back into position when it moves within the sheath, usually with an audible click [8].

White finger results from excessive vibration from power tools inducing the constriction of arterioles within the digits. The resulting lack of blood flow appears as a blanching of the skin, with a corresponding loss of motor control. Other such CTDs are discussed by Meagher [9] and Stetson et al. [10].

Short-term fatigue and discomfort have also been considered risk factors and have been shown to be related to handle and work orientation in hammering [11,12] and to tool shape and work height in work with screwdrivers [13,14]. Also, it has been shown that poor design of the grip of a tool leads to exertion of higher grip forces [15,16] and to extreme wrist deviations [8] and therefore to more fatigue.

The cost of CTDs in U.S. industry, although not all due to improper tool design, is quite high. Data from the National Safety Council [17] suggest that 15–20% of workers in key industries (meatpacking, poultry processing, auto assembly, and garment manufacturing) are at potential risk for CTD and that in 1991 some 223,600 cases or 61% of all occupational injuries were associated with repetitive actions. The worst industry was manufacturing, while the worst occupational title was butchering with 222 CTD claims per 100,000 workers [18]. With such high rates and average costs of $30,000 per case, NIOSH, in its Year 2000 Objectives, has targeted the reduction of CTD incidence from 82 to 60 cases per 100,000 overall workers and from 285 to 150 in certain manufacturing industries [19].

The proper selection, evaluation, and use of hand tools is a major ergonomic concern. The following review discusses the basic principles involved in tool design, provides a general evaluation process for tools, and discusses the attributes desirable for specific tools.

III. TOOL EVALUATION AND DESIGN

A. Principles of Tool Design

1. General Principles

An efficient tool has to fulfill some basic requirements [4]:

1. It must perform effectively the function for which it is intended. Thus, an axe should convert a maximum amount of its kinetic energy into useful chopping work, separate wood fibers cleanly, and be easily withdrawn.
2. It must be properly proportional to the body dimensions of the operator to maximize efficiency of human involvement.
3. It must be designed to match the strength and work capacity of the operator. Thus, allowances have to be made for the gender, age, training, and physical fitness of the operator.
4. It should not cause undue fatigue; that is, it should not demand unusual postures or practices that will require more energy expenditure than necessary.
5. It must provide sensory feedback in the form of pressure, some shock, texture, temperature, etc. to the user.
6. Its capital and maintenance costs should be reasonable.

2. Anatomical Concerns

Anatomy of the Hand and Types of Grip. To better understand the design principles of hand tools, it is necessary to have a brief description of the anatomy and functioning of the human hand and some of the diseases that can result from its misuse. The human hand is a complex structure of bones, arteries, nerves, ligaments, and tendons (Figure 1). The fingers are controlled by the extensor carpi and flexor carpi muscles in the forearm. The muscles are connected to the fingers by tendons that pass through a channel in the wrist formed by the bones of the back of the hand on one side and the transverse carpal ligament on the other. Through this channel, called the carpal tunnel, pass also various arteries and nerves. The bones of the wrist connect to two long bones in the forearm, the ulna and the radius. The radius connects to the thumb side of the wrist, and the ulna connects to the little finger side of the wrist. The orientation of the wrist joint allows movement in only two planes, each at 90° to the other (Figure 2). The first gives rise to palmar flexion and dorsiflexion (or extension). The second movement plane gives ulnar and radial deviation.

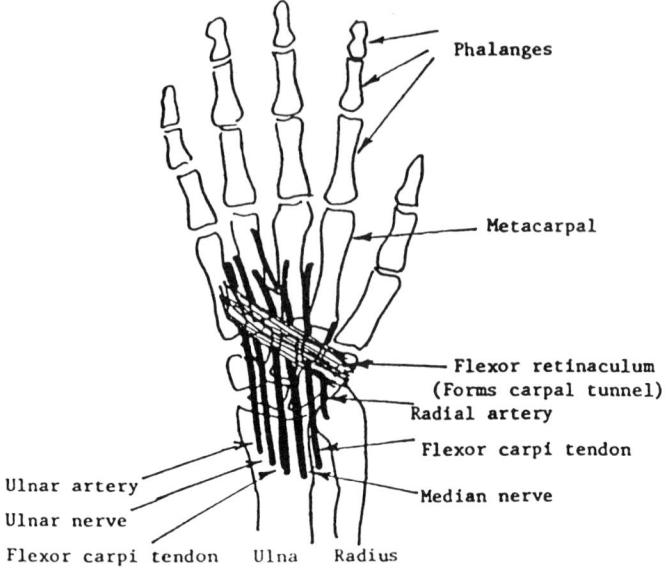

Figure 1 Anatomy of the hand.

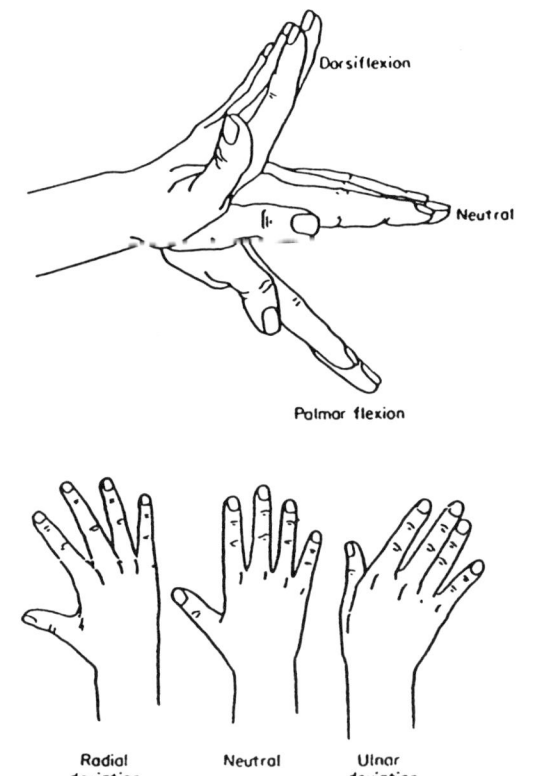

Figure 2 Types of wrist movement.

The manual dexterity prduced by the hand can be defined in terms of a power grip and a precision grip. In a power grip, the tool, whose axis is more or less perpendicular to the forearm, is held in a clamp formed by the partly flexed fingers and the palm, with opposing pressure being applied by the thumb (Figure 3). There are three subcategories of the power grip differentiated by the line of action of force: (1) force parallel to the forearm, as in sawing; (2) force at an angle to the forearm, as in hammering; and (3) torque about the forearm, as when using a screwdriver. As the name implies, the power grip is used for power or for holding heavy objects.

In a precision grip, the tool is pinched between the flexor aspects of the finger and the opposing thumb. The relative position of the thumb and fingers determines how much force is to be applied and provides a sensory surface for receiving feedback necessary to give the precision needed. There are two types of precision grip: (1) internal, in which the shaft of the tool (e.g., knife) passes under the thumb and is thus internal to the hand; and (2) external, in which the shaft (e.g., pencil) passes over the thumb and is thus external to the hand. The precision grip is used for control. Other grips are just variations of the power or precision grip and include the hook grip, for holding a box or handle, a two-point pinch, and a lateral pinch (Figure 3).

Static Muscle Loading. When tools are used in situations in which the arms must be elevated or when they have to be held for extended periods, muscles of the shoulders, arms, and hands may be loaded statically, resulting in fatigue, reduced work capacity, and soreness. Abduction of the shoulder with corresponding elevation of the elbow will occur if work has to be done with a pistol-grip tool on a horizontal workplace. An in-line or straight tool reduces the need to raise the arm and also allows for a neutral wrist posture [20].

Prolonged work with arms extended can produce soreness in the forearm for assembly tasks done with force. By rearranging the workplace so as to keep the elbows at 90°, most of the problem can be eliminated (Figure 4). Similarly, continuous holding of an activation switch can result in fatigue of the fingers and reduced flexibility.

Awkward Wrist Position. As the wrist is moved from its neutral position there is loss of grip strength. Starting from a neutral wrist position, pronation decreases grip strength by 12%, flexion/extension by 25%, and radial/ulnar deviation by 15% [22]. The percent of maximum grip strength available can be quantified by

$$\text{Grip} = 95.7 + 4.3 \text{ PS} + 3.8 \text{ FE} - 25.2 \text{ FE}^2 - 16.8 \text{ RU}^2 \tag{1}$$

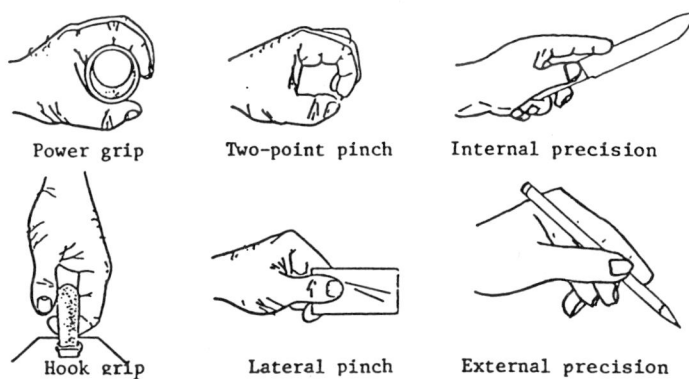

Figure 3 Types of grip.

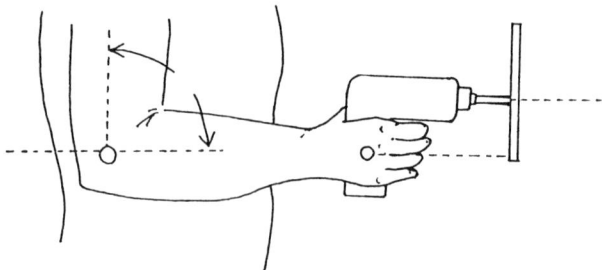

Figure 4 Optimum working posture with elbow bent at 90°.

where
PS = 1 if the wrist is fully pronated or supinated and 0 if in a neutral position
FE = 1 if the wrist is fully flexed or extended and 0 if in a neutral position
RU = 1 if the wrist is fully in radial or ulnar deviate and 0 if in a neutral position

Furthermore, awkward hand positions may result in soreness of the wrist, loss of grip, and, if sustained for extended periods of time, the occurrence of carpal tunnel syndrome. To reduce this problem, the workplace or tools should be redesigned to allow for a straight wrist—lowering work surface and edges of containers, tilting jigs toward the user (Figure 5), using a pistol grip on knives (Figure 6) [23], and using a pistol handle on powered tools for vertical surfaces and in-line handles for horizontal surfaces (Figure 7). Similarly, the tool handle should reflect the axis of grasp and should be oriented so that the eventual tool axis is in line with the index finger, e.g., Tichauer's [21] "bent" plier handles (Figure 8).

Tissue Compression. Often, in the operation of hand tools, considerable force is applied by the hand. Such actions can concentrate considerable compressive force on the palm of the hand or the fingers, resulting in ischemia, obstruction of blood flow to the tissues, and

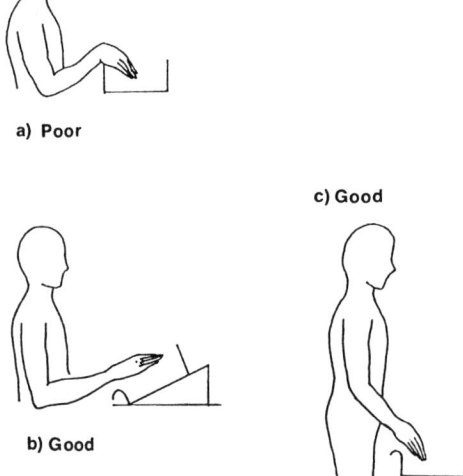

Figure 5 Proper orientation (b, c) of jigs and containers.

Tool Evaluation and Design

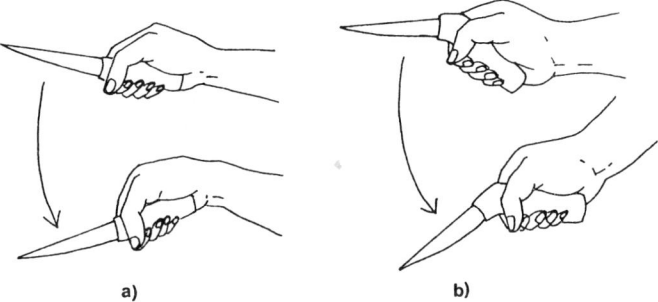

Figure 6 Pistol grip (b) knife. (Adapted from Ref. 23.)

eventual numbness and tingling of the fingers. Handles should be designed to have large contact surfaces to distribute the force over a larger area (Figure 9) or to direct it to less sensitive areas such as the tissue between the thumb and index finger. Similarly, finger grooves or recesses in tool handles should be avoided. Because hands vary considerably in size, the grooves will accommodate only a fraction of the population.

Gender. Female grip strength typically ranges from 50 to 67% of male strength [24–26]; that is, the average male can be expected to exert approximately 500 N, whereas the average female can be expected to exert approximately 250 N. An interesting survey by Ducharme [27] examined how tools and equipment that were physically inadequate for female workers hampered their performance. The worst offenders were crimpers, wire strippers, and soldering irons. Females have a twofold disadvantage—an average lower strength and an

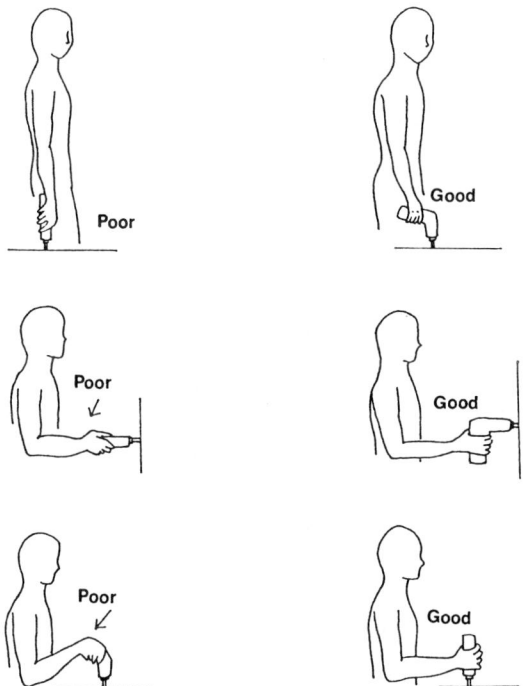

Figure 7 Proper orientation of power tools in the workplace.

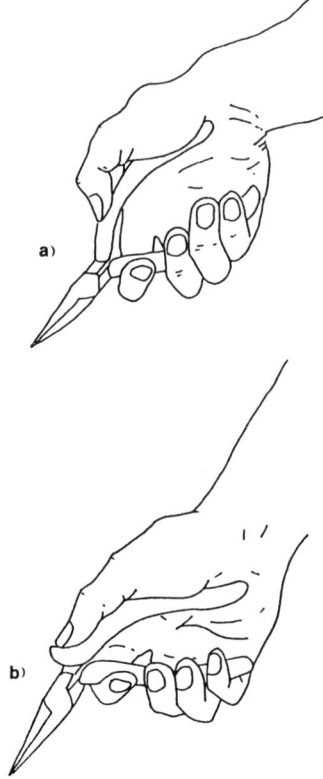

Figure 8 (a) Traditional and (b) redesigned pliers. (Adapted from Ref. 21.)

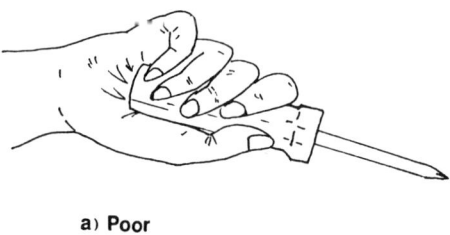

a) Poor

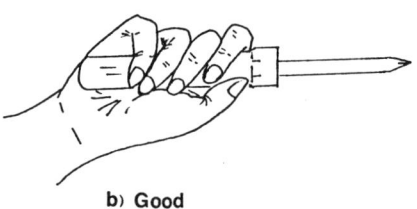

b) Good

Figure 9 Avoiding tissue compression in tool design.

average smaller grip span. Ducharme concluded that women could be integrated more quickly and safely into the work if tools were designed to accommodate smaller dimensions.

On the other hand, Pheasant and O'Neill [28] challenged Ducharme's assertions on the basis of their findings that optimal performance for both males and females occurred at similar conditions. Males had sufficient strength to overcome the deficiencies in tool design, which posed much greater problems for females.

Handedness. Alternating hands permits reduction of local muscle fatigue. However, in many situations this is not possible as the tool use is one-handed. Furthermore, if the tool is designated for the user's preferred hand—which for 90% of the population is the right hand—then 10% are left out [29]. Laveson and Meyer [30] gave several good examples of right-handed tools that cannot be used by a left-handed person; these included a power drill with side handle on the left side only, a circular saw, and a serrated knife leveled on one side only.

A few studies have compared task performance using dominant and nondominant hands. Shock [31] indicated that the nonpreferred hand grip strength is 80% of the preferred hand grip strength. Miller and Freivalds [32] found that right-handed males show a 12% strength decrement in the left hand, whereas right-handed females show a 7% strength decrement. Surprisingly, both left-handed males and females had nearly equal strengths in both hands. He concluded that left-handed subjects were forced to adapt to a right-handed world. Using time study ratings, Konz and Warraich [33] found decrements ranging from 9% for an electric drill to 48% for manual scissors for ratings using the nonpreferred hand as opposed to the preferred hand.

Posture. A series of studies were performed by Mital and colleagues to examine various tool and operator factors on torque capability [34–37]. In general, unless the posture is extreme, e.g., standing versus lying down, torque exertion capability was not affected substantially. The height at which torque was applied had no influence on peak torque exertion capability. On the other hand, torque exertion capability decreased linearly with increasing reach distance. Another interesting requirement for proper tool use is the volume or space envelope generated during operation of the tool. Comprehensive data on a variety of tools were collected by Baker et al. [38].

Repetitive Finger Action. If the index finger is used excessively for operating triggers, symptoms of trigger finger develop. Thus, trigger forces should be kept low, preferably below 10 N [20], to reduce the load on the index finger. Two- or three-finger-operated controls are preferable (Figure 10); finger strip controls or a power grip bar are even better. For a two-handled tool, a spring-loaded return saves the fingers from having to return the tool to its starting position [20]. In addition, the high number of repetitions must be reduced. Although critical levels of repetitions are not known, attempts have been made to identify the maximal number of exertions per hour or shift that can be tolerated. Most of these have concerned wrist movements but could reasonably apply to the fingers as well. Obolenskaja and Goljanitzki [39] observed high rates of tenosynovitis in teapackers with 7000–12,000 hand movements per day. Luopajarvi et al. [40] found high rates of muscle-tendon disorders in assembly line packers with over 25,000 movements per day. NIOSH [41] found similar problems in workers exceeding 10,000 motions per day.

3. Single Handles

Theory of Cylindrical Grip. One theory of gripping forces has been described by Pheasant and O'Neill [28] and by Grieve and Pheasant [42]. The hand gripping a cylindrical handle forms a closed system of forces in which portions of the digits and palm are used, in opposition to each other, to exert compressive forces on the handle (Figure 11). The strength

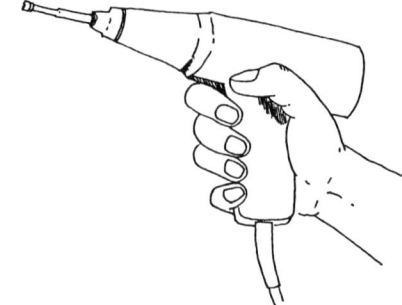

Figure 10 Three-finger trigger for power tools.

of the grip (G) may be defined as the sum of all components of forces exerted normal to the surface of the handle.

When exerting a turning action on the handle, the maximum torque, as given at the moment of hand slippage, is given by

$$T = SD \tag{2}$$

where T = torque (Nm)
S = total shear/frictional force (N)
D = handle diameter (m)

and where S can be defined by

$$S = \mu G \text{ (N)} \tag{3}$$

where μ = coefficient of friction
G = grip (N)

Thus, torque is directly dependent upon handle diameter because handle diameter determines the leverage of shear forces. This was confirmed experimentally by Pheasant and O'Neill [28]. For thrusting motions in the direction of the long axis of the handle, the diameter is not involved, and determination of maximum force is more complicated.

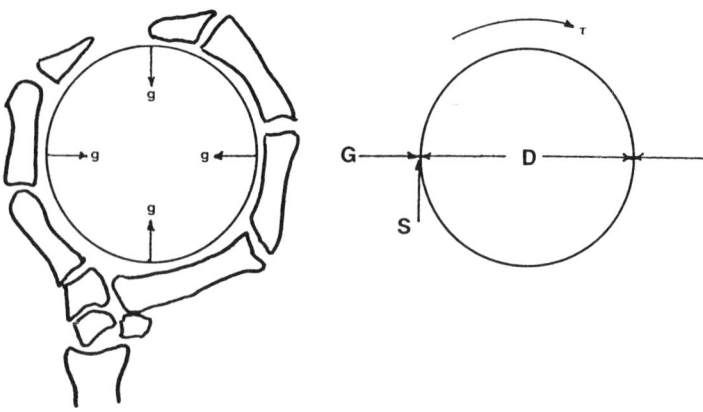

Figure 11 Theory of the cylindrical grip. (Adapted from Ref. 28.)

For handles larger than the grip span diameter, the gripped area no longer increases in proportion to the diameter. An analysis of such conditions was performed by Replogle [43], who concluded that for handles up to twice the grip span diameter, the relative ungripped area of the handle increases, reducing the effective gripped area. Torque then can be expressed by

$$T = 3d^2(4 - d)/(d + 2)^2 \tag{4}$$

where d is the ratio of handle diameter to grip span diameter. For larger handle diameters, the expression for torque becomes much more complicated [43].

Handle Diameter. Power grips around a cylindrical object should surround more than half the circumference of the cylinder, but with the fingers and thumb barely touching [44]. For a power grip on a screwdriver. Rubarth [45] recommended a diameter of 40 mm. Basing their recommendations on empirical judgments of stair rails, Hall and Bernett [46] suggested 32 mm. Based on minimum EMG activity, Ayoub and Lo Presti [47] found a 51-mm handle diameter to be best. However, based on the maximum number of work cycles completed before fatigue and the ratio of grip force to EMG activity, they suggested a 38-mm diameter. Pheasant and O'Neill [28] found that muscle strength deteriorates when handles greater than 50 mm in diameter are used. Rigby [48], for heavy loads and full encirclement of the hand, recommended 38 mm. For handles on boxes, Drury [49] found diameters of 31–38 mm to be best in terms of least reduction in grip strength. Using various handles of noncircular cross section, Cochran and Riley [50,51] found the greatest thrust forces in handles of 41.4 mm equivalent circular diameter (based on their 130-mm circumference) for both males and females. For manipulation, however, the smallest handles of 22 mm were found to be the best [52]. Replogle [43], in validating his grip model, found maximum torques with handle diameters of 50 mm. Eastman Kodak [20], based on company experience, recommends 30–40 mm with an optimum of 40 mm for power grips and 8–16 mm with an optimum of 12 mm for precision grips. Thus, one can summarize that handle diameters should be in the range of 31–50 mm, with the upper end of the range best for maximum torque and the lower end best for dexterity and speed.

Handle Length. The length of the handle has been studied to a lesser extent. For cut-out handles, there should be enough space to admit all four fingers. Hand breadth across the metacarpals ranges from 71 mm for a 5th percentile female to 97 mm for a 95th percentile male [53]. Thus, 100 mm may be a reasonable minimum, but 125 mm may be more comfortable [24]. Eastman Kodak [20] recommended 120 mm. If the grip is enclosed or gloves are used, even larger openings are recommended. For an external precision grip, the tool shaft must be long enough to be supported at the base of the first finger or thumb. A minimum value of 100 mm is suggested [24]. For an internal precision grip, the tool should extend past the palm, but no so far as to hit the wrist (Figure 9) [24]. It is interesting to note that screwdriver torque was experimentally found to be proportional to the handle grip length [54].

Handle Shape. As early as 1928, Rubarth [45] investigated handle shape and concluded that, for a power grip, one should design for maximum surface contact so as to minimize unit pressure of the hand. Thus, a tool with a circular cross section was found to give the greatest torque. Pheasant and O'Neill [28] concluded that the precise shape of the handle was irrelevant and recommended simple knurled cylinders. Evaluation of handle shape on grip fatigue in manual lifting (which is a different action than for tool use) did not indicate any significant differences in shapes [55]. Maximum pull force, though, was obtained with a triangular cross sectoin, apex down. For thrusting forces, the circular cross section was found to be worst and a triangular best [50]. However, for a rolling type of manipulation, the triangular shape was slowest [52]. A more comprehensive study indicated that no one shape may be perfect and that optimal shape may be more dependent on the type of task and motions involved than was initially thought [51]. A rectangular shape of width/height ratios from 1:1.25 to 1:1.5 appeared to be a good compromise. A further advantage of a rectangular cross section is that the tool does not roll when placed on a table [24]. It should also be noted that handles should not have the shape of a true cylinder except for a hook grip. For screwdriver-type tools, the handle end is rounded to prevent undue pressure at the palm, and for hammer-type tools the handle may have some flattening curvature to indicate the end of the handle.

In a departure from circular, cylindrically shaped handles, Bullinger and Solf [56] proposed a more radical design using a hexagonal cross section, shaped as two truncated cones jointed at the largest ends. Such a shape fits the contours of the palm and thumb best in both precision and power grips and yielded highest torques in comparison with more conventional handles. A similar dual truncated conical shape was also developed for a file handle. In this case, the heavily rounded square cross section was found to be markedly superior to more conventional shapes.

A final note on shape is that T-handles yield much better performance than straight screwdriver handles. Pheasant and O'Neill [28] reported as much as a 50% increase in torque. Optimum handle diameter was found to be 25 mm, and optimum angle was 60°, i.e., a slanted T [57]. The slant allows the wrist to remain straight and thus generate larger forces.

Grip Surface, Texture, and Materials. For centuries, wood was the material of choice for tool handles. Wood was readily available and easily worked. It has good resistance to shock, poor thermal and electrical conductivity, and good frictional qualities even when wet. Because wooden handles can break and stain with grease and oil, there has been a shift to plastic and even metal. However, metal should be covered with rubber or leather to reduce shock and electrical conductivity and increase friction [58]. Such compressible materials also dampen vibration and allow a better distribution of pressure, reducing the feeling of fatigue and hand tenderness [59]. The grip material, however, should not be so soft that sharp objects, such as metal chips, can become embedded in the grip and make it difficult to use. Grip surface area should be maximized to ensure a pressure distribution over as large an area as possible. Excessive localized pressure sometimes causes pain that forces workers to interrupt their work. Pressure/pain thresholds of around 500 kPa for females and 700 kPa for males have been found, with the thenar and os pisiforme areas being most sensitive [60]. During maximal power grips, these values are greatly exceeded.

The frictional characteristics of the tool surface vary with the pressure exerted by the hand, the smoothness and porosity of the surface, and the type of contamination [61,62]. Sweat increases the coefficient of friction, whereas oil and fat reduce it. Adhesive tape and suede provide good friction when moisture is present [61]. The type of surface pattern as defined by the ratio of ridge area to groove area shows some interesting characteristics. When the hand is clean or sweaty, the maximum frictions are obtained with high ratios (i.e., maximizing the hand–surface contact area), whereas when the hand is contaminated, maximum frictions are obtained with low ratios (i.e., maximizing the capacity to channel away contaminants) [62].

Angulation of Handle. As discussed previously, deviations of the wrist from the neutral position under repetitive load can lead to a variety of cumulative trauma disorders as well as decreased performance. Therefore, angulation of tool handles such as those of power tools may be necessary to maintain a straight wrist. The handle should reflect the axis of grasp, i.e., about 78° from the horizontal, and should be oriented so that the eventual tool axis is in line with the index finger [58]. This principle has been applied to various tools such as pliers and soldering irons, as mentioned previously.

An interesting extension of this concept has been promoted as Bennett's handle [63]. Bennett developed this concept based on the angle formed by the index finger and the life line

Tool Evaluation and Design

under the thumb. This angle of 19°, used for his handles, is claimed to maintain a straight wrist, generate increased strength and control, and reduce stress, shock, and fatigue [64]. Bennett's claims initially were supported by anecdotal evidence of improved performance [63]. Since then Konz and his colleagues [65–67] have conducted a variety of tests to evaluate the effectiveness of Bennett's handle on a hammer in comparison with a standard hammer. In the second study a variety of angled handles were evaluated, and subjects rated a 10° bend as being most preferred. In the third study, performance in driving nails was evaluated using various bent hammers. No performance difference was found, but the 10° bend was again rated significantly higher. In the final study, using a semantic-differential questionnaire, Konz [67] concluded that although no significant performance effects were found, subjects preferred a slight (5–10°) bend rather than the 19° of Bennett's handle. An independent study by Knowlton and Gilbert [68] used cinematography to evaluate curved and conventional claw hammers. Bilateral grip strength was measured before and after the task of nail driving. The curved hammer produced a smaller strength decrement and caused less ulnar deviation than the conventional hammer. Thus, a bent handle does give some benefits.

4. Grip Span for Two-Handled Tools

Grip strength and the resulting stress on finger flexor tendons vary with the size of the object being grasped. A maximum grip strength is achieved at about 45–50 mm on a dynamometer with parallel sides [25] and at about 75–80 mm on a dynamometer with handles angled inward [26]. At distances different from the optimum, percent grip strength decreases (Figure 12) as defined by

$$\text{Grip} = 100 - 0.11S - 10.2S^2 \tag{5}$$

where S is the given grip span minus optimum grip span in centimeters.

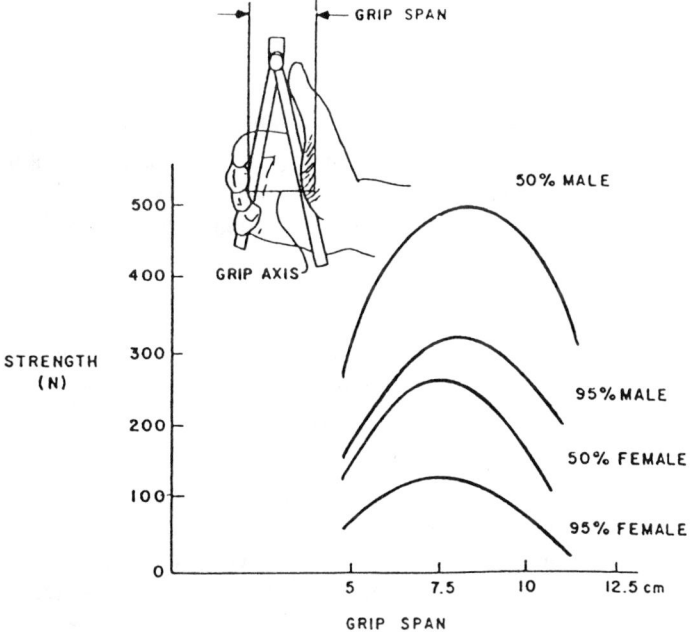

Figure 12 Grip strength as a function of grip span. (Reproduced with permission from Ref. 26.)

Because of the large variation in individual strength capacities, and to accommodate 95% of the population, maximal grip requirements should be limited to less than 90 N [69].

5. Weight

The weight of the hand tool will determine how long it can be held or used and how precisely it can be manipulated. For tools held in one hand with the elbow at 90° for extended periods of time, Greenberg and Chaffin [69] recommended a load of no more than 2.3 kg. A similar value was suggested by Eastman Kodak [20]. For precision operations, tool weights greater than 0.4 kg are not recommended unless a counterbalanced system is used. Heavy tools, used to absorb impact or vibration, should be mounted on a truck to reduce effort for the operator [20]. In addition, the tool should be well balanced, with its center of gravity as close as possible to the center of gravity of the hand (unless the purpose of the tool is to transfer force as in a hammer). Thus, the hand or arm muscles do not need to oppose any torque developed by an unbalanced tool.

6. Gloves

Gloves are often uesd with hand tools for safety and comfort. Safety gloves are seldom bulky, but gloves worn in subfreezing climates can be very heavy and interfere with grasping ability. The wearing of woolen or leather gloves may add 5 mm to the hand thickness and 8 mm to the hand breadth at the thumb, whereas heavy mittens add 25 and 40 mm, respectively [70]. More important, gloves reduce grip strength by 10-20% [71-73] and manul dexterity performance times by 12-64% [74]. Neoprene gloves slowed performance times by 12.5% over barehanded performance, terry cloth by 36%, leather by 45%, and polyvinyl chloride (PVC) by 64%. In some cases, by protecting the hand, gloves could improve operational speed [75,76]. On the other hand, gloves consistently reduced torque production [77]. Thus, there is a trade-off to be considered between increasing injury and reduced performance without gloves and reduced performance with gloves. Perhaps the tool should be redesigned even more so as to compensate for the glove effects.

7. Vibration

Vibration is a separate and very complex problem with powered hand tools. Vibration can induce white finger syndrome, the primary symptom of which is a reduction in blood flow to the fingers and hand due to vasoconstriction of the blood vessels, which lead to a loss of sensory feedback and decreased performance. In addition, vibration may contribute to the development of carpal tunnel syndrome, especially in jobs with a combination of forceful and repetitive exertions [78].

It is generally recommended that vibrations in the critical range of 40-130 Hz [79] or a slightly larger range of 2-200 Hz [80] be avoided. The exposure to vibration can be reduced through a reduction in the driving force and the use of vibration-damping materials [81].

8. Rhythm

The operation of hand tools involves repetition of a particular pattern of motion. A skilled operator requires a basic motor pattern that will be most economical in terms of energy expenditure and is thus one attribute of skill. Once this pattern is established, it is continued with very consistent velocity and acceleration through kinesthetic and aural feedback. Optimum rhythms have been observed by Drillis [4] as follows: filing, 78 strokes/min; chiseling, 60 strokes/min; shoveling, 14-17 strokes/min; and cranking, 35 rev/min.

9. Other Aspects of Tool Design

Tools should not have protruding sharp edges or corners. Two-handled tools should have stop limits to limit closure of the tools and prevent pinching of the fingers. Locking tools should not engage until the tool closes to the point where the fingers cannot be inserted. Tool surfaces should have matte surfaces to reduce glare [58].

B. Attributes of Common Industrial Hand Tools

1. Shovels

Shovels are used to lift, move, and toss loose dirt or sand or other material. The blade is fastened to the shaft through a socket, which, if stamped from a flat sheet, is generally rolled over to form a crimp known as a frog. The shaft may either taper to an end or have a handle. The handle traditionally has been of a T form but more lately is of a D form. Long shafts are generally 1.2–1.7 m long, whereas the short D handle is about 0.7 m long. However, for an unconstrained posture and task, long-handled shovels are 18% more efficient than short-handled shovels [82]. The angle of the shaft with respect to the horizontal, called the lift, provides the tool with added leverage. Based on a trade-off between smaller low back compressive forces and lower energy cost, Freivalds [82] found the optimum lift angle to be approximately 32°. Shovel weight should be as small as possible, especially if one considers that this is unproductive weight that has to be lifted along with the load moved. Weights should be below 1.5 kg [83,84]. Blade size depends very much on the density of the material being handled; the less dense the material, the larger the blade size. For a given material (foundry sand), Freivalds and Kim [84] found that a blade size/weight ratio of 0.0676 m^2/kg is optimum; that is, for a 1.5-kg shovel the optimum blade size is 0.1 m^2 with a load of 4.4 kg.

Optimum shoveling rate is in the range of 18–21 scoops/min [5,83,85]. The optimum shovel load ranges from 5 to 11 kg, depending on the decision criterion to be used [86,87]. For high rates of shoveling (18–20 scoops/min), the lower end of the load range (5–7 kg) may be more appropriate (which follows the principle of reducing static loading), whereas for lower rates (6–8 scoops/min), the higher end of the load range (8–11 kg) may be acceptable (which follows the principles of increasing efficiency with larger loads) [88]. Shoveling throw height is a trade-off between increased efficiency for higher heights and the cost of an increase in energy expenditure. Because shoveling performance stays reasonably constant up to a height of 1.3 m [89], an acceptable throw height may be as much as 1–1.3 m. The same conflicting criteria apply to throw distance. However, with shoveling performance remaining fairly constant up to a distance of 1.2 m [89], that would be an acceptable distance.

2. Hammers

Hammers are striking tools designed to transmit a force to an object by direct contact and thereby change its shape or drive it forward. The tool's efficiency in doing this may be defined as the ratio of the energy used in striking to the energy available in the stroke. This efficiency is maximized by placing the tool mass center as close as possible to the center of action, i.e., increasing the mass of the tool head relative to that of the handle. Another aim is to transform as much as possible of the kinetic energy of the hammer into deforming an object's shape. Thus, the mass of the hammer should be small relative to the mass of the forging and anvil. On the other hand, in driving a nail the intent is to transform the kinetic energy of the hammer into the kinetic energy of the nail. Then the mass of the hammer should be great in relation to the mass of the nail. The overall mechanical efficiency for hammering a 6-in. nail into a wooden block can be as high as 57% [4]. However, there is a limit to the weight that can be placed in the head of the hammer. Increasing the head weight decreases angular velocity and ultimately the total kinetic energy in addition to increasing physiological energy costs [90,91].

3. Saws

The action of heavy sawing requires a power grip with repetitive flexion and extension at the elbow, whereas the action of light sawing involves a precision grip with manipulation of the wrist. For the former, pistol grips are used, whereas for the latter a cylindrical screwdriver type of handle provides the best precision grip. Gläser [92] found that for forestry work, a two-handed action provided more force and better performance but at a higher energy cost. The most efficient was the kneeling posture because less torso support was needed and less energy was expended. Typically, western saws cut as they are pushed through the wood, whereas Japanese saws cut as they are pulled through the wood [93]. Although sawing times were not significantly different, energy expenditure is significantly lower for Japanese saws [93].

4. Pliers

Pliers and related tools—wire strippers, pincers, and nippers—are tools with a head in the form of jaws that can have a variety of configurations, i.e., joints may be simple or complex. Although sometimes the handles are straight, more typically they are curved outward to conform roughly to the shape of the grasp. The grasp, depending on use, can be of the precision or power type. In their simple form, pliers are a very common tool and, if used casually for short periods of time, will give reasonable performance with little fatigue. However, the relationship of the handles to the head forces the wrist into ulnar deviation (Figure 8a), a posture that cannot be held repeatedly or for prolonged periods of time without fatigue or the occurrence of cumulative trauma disorder. A further problem is that such a deviation reduces the range of wrist rotation by 50%, thus reducing productivity [94]. By bending the handles of the pliers (Figure 8b) instead of the wrist, Tichauer [94] was able to reduce the stress on the operator's wrists and reduce overall injury rates by a factor of 6 at Western Electric Co. A similar design (with similar success) was implemented for diagnoal cutting pliers at Eli Lilly and Co. [95].

Other factors to be considered in the design of pliers were detailed by Lindstrom [96]: a working grip width of 90 mm for men and 80 mm for women and a handle length of 110 mm for men and 100 mm for women. For repeated or continuous operation, the required working strength should not exceed 33–50% of the individual's maximum strength. To minimize the applied pressure to the soft tissue of the palm (less than 200–400 kPa), the handles should be enlarged and flattened. Thus, indentation of the handles for the fingers is undesirable. Encasing the basic metal handles in a rubber or plastic sheath provides insulation and improves the tactile feel.

5. Screwdrivers

The handles of screwdrivers (and similar tools—files, chisels, etc.) can be used with either a precision grip for stabilization or a power grip for torque. The handle must also be capable of being approached equally effectively from all directions. Other crucial factors are the size, shape, and texture of the handle. In terms of size, applied torque increases with an increase in the diameter of the handle [28]. Differences in the precise shape of handles appear not to be significant as long as the hand does not slip around the handle [51]. Thus, knurled cylin-

Tool Evaluation and Design

ders allow for significantly greater torque production than smooth cylinders. Further details can be found in Sec. III.A.3.

6. Knives

Although a very old tool, the knife has recently appeared in the literature as a possible cause of the increase in cumulative trauma disorders suffered by food processors [23,97]. For poultry processing, Armstrong et al. [23] suggested a pistol-type grip (Figure 6) to allow the operator to hold the blade and the forearm horizontal so as to eliminate ulnar deviation and wrist flexion. A circular or elliptical handle with a large circumference (99 mm) and a strap was recommended to allow the hand to relax between exertions without losing a grip on the knife. Similarly, for the fish canning industry, Karlqvist [97] fitted some knives with a pistol grip handle and others with larger diameter handles for better balance and movement. Cochran and Riley [15] explained the injuries due to long hours of static loading on the forearm flexors resulting in fatigue. In addition, body fluids cause slippery handles, with the operator's hand slipping from the handle over to the blade. They recommended tangs (barriers on the handle perpendicular to the blade) of a minimum length of 1.5 cm to prevent injury.

C. Attributes of Common Industrial Power Tools

1. Power Drills

In a power drill or other power tool, the major function of the operator is to hold, stabilize, and monitor the tool against a workpiece, while the tool performs the main effort of the job. Although the operator may at times need to shift or orient the tool, the main function of the operator is to effectively grasp and hold the tool. A drill comprises a head, body, and handle, with all three, ideally, being in line. The line of action is from the line of the extended index finger so that in the ideal drill the head is off-center with respect to the central axis of the body. Handle configuration is important, with the choices being pistol grip, in-line, or right angle. As a rule of thumb, in-line and right angle grips are best for tightening downward on a horizontal surface, while pistol grips are best for tightening on a vertical surface (Figure 7), with the aim being to obtain a standing posture with a straight back, upper arms hanging down, and a straight wrist (Figure 4). For the pistol grip, this results in the handle being at an angle of approximately 78° to the horizontal [58].

Another factor of importance is the center of gravity. If it is too far forward in the body of the tool, a turning moment is created, which must be overcome by the muscle of the hand and forearm, creating muscular effort additional to that required for holding, positioning, and pushing the drill into the workpiece [69]. Greenberg and Chaffin recommend placing the primary handle directly under the center of gravity so that the body juts out behind the handle as well as in front. For heavy drills, a secondary supportive handle may be needed, either to the side or preferably below the tool, such that the supporting arm can be tucked in against the body rather than being abducted.

2. Nutrunners

Nutrunners, especially common in the automobile industry, are used to tighten nuts, screws, and other fasteners. They come in a variety of handle configurations, torque outputs, shutoff mechanisms, speeds, weights, and spindle diameters and are commercially available from a variety of sources. Torque levels range from 0.1 to 5000 N-m and, for pneumatic tools, are generally lumped into approximately 22 power levels (M1.6–M45) depending on motor size and the gearing required to drive the tool. The torque is transferred from the motor to the spindle through a variety of mechanisms in such a way that the power (often air) can be quickly shut off once the nut or other fastener is tight. The simplest and cheapest mechanism is a direct drive, which is under the operator's control but, because of the long time needed to release the trigger once the nut is tightened, transfers a very large reaction torque to the operator's arm. Mechanical friction clutches will allow the spindle to slip, reducing some of this reaction torque. A better mechanism for reducing the reaction torque is the airflow shutoff, which automatically senses when to cut off the air supply as the nut is tightened. A still faster mechanism is an automatic mechanical clutch shutoff [98]. The most recent mechanisms include the hydraulic pulse system where the rotational energy from the motor is transferred over a pulse unit containing an oil cushion (filtering off the high-frequency pulses as well as noise) and a similar electrical pulse system, both of which, to a large extent, reduce the reaction torque [99].

Variation of torque delivered to the nut depends on a variety of conditions such as the properties of the tool, the operator of the tool, and properties of the joint [i.e., the combination of the fastener and material being fastened (ranging from soft, with the materials having elastic properties such as body panels, to hard, when two stiff surfaces, such as pulleys on a crankshaft, are brought together)], and stability of the air supply. The torque experienced by the user (the reaction torque) depends on these factors plus the torque shutoff system [100]. In addition, other potential problems include noise from the pneumatic mechanism, which reaches levels as high as 95 dB(A), vibration levels exceeding 132 dB(V), and dust or oil fumes from the exhaust air [101].

Reaction torque, the excess torque produced by the nutrunner and transferred to the operator's hand and/or arm once the fastener is tightened and before the power source can be decreased, is believed to contribute to the development of cumulative trauma disorders [100]. A study by Freivalds and Eklund [98] examined relevant power tool parameters for their effect on operator stress levels as measured by electromyography, wrist angle, and subjective ratings of perceived discomfort. In general, using electric tools at lower than normal rpm levels or underpowering pneumatic tools resulted in larger reaction torques and more stressful ratings. Pulse-type tools produced the lowest reaction torques and were rated least stressful. It was hypothesized that the short pulses "chop up" or allow the inertia of the tool to resist the reaction torque. Also, subjective ratings correlated significantly with peak reaction torques, indicating that the operators were good judges of stress levels and that subjective ratings could be valid measures of operator stress.

Reaction torque bars should be provided if the torque exceeds 6 N-m for in-line tools used in a downward action, 12 N-m for pistol grip tools used in a horizontal mode, and 50 N-m for right-angled tools used in a downward or upward motion [102,103].

D. Tool Evaluation Checklist

The previous information can be summarized in the form on an evaluative checklist (Table 1). If the tool does not conform to the recommendations and desired features, it shoud be replaced or redesigned.

IV. CRITICAL REVIEW OF THE CURRENT STATUS

Currently, the most important issue regarding tool design is the reduction in the potential for the development of cumulative trauma disorders. Until about 10 years ago, tool use was little

Tool Evaluation and Design

Table 1 Tool Evaluation Checklist

	Yes	No
Basic Principles		
1. Does the tool perform the desired function effectively?	☐	☐
2. Does the tool match the size and strength of the operator?	☐	☐
3. Can the tool be used without undue fatigue?	☐	☐
4. Does the tool provide sensory feedback?	☐	☐
5. Are the tool capital and maintenance costs reasonable?	☐	☐
Anatomical Concerns		
1. If force is required, can the tool be grasped in a power grip (i.e., handshake)?	☐	☐
2. Can the tool be used without shoulder abduction?	☐	☐
3. Can the tool be used with a 90° elbow angle (i.e., forearms horizontal)?	☐	☐
4. Can the tool be used with the wrist straight?	☐	☐
5. Does the tool handle have large contact surfaces to distribute forces?	☐	☐
6. Can the tool be used comfortably by a 5th percentile female operator?	☐	☐
7. Can the tool be used in either hand?	☐	☐
Handles and Grips		
1. For power uses, is the tool grip 30–50 mm in diameter?	☐	☐
1a. Can the handle be grasped with the thumb and fingers slightly overlapped?	☐	☐
2. For precision tasks, is the tool grip 8–16 mm in diameter?	☐	☐
3. Is the grip cross section circular?	☐	☐
4. Is the grip length at least 100 mm (125 mm if gloves are worn)?	☐	☐
5. Is the grip surface finely textured and slightly compressible?	☐	☐
6. Is the handle nonconductive and stain-free?	☐	☐
7. For power uses, does the tool have a pistol grip angled at 78°?	☐	☐
8. Can a two-handled tool be operated with less than 90 N grip force?	☐	☐
9. Is the span of the tool handles between 70 and 80 mm?	☐	☐
Power Tool Considerations		
1. Are trigger activation forces less than 10 N?	☐	☐
2. For repetitive use, is a finger strip trigger present?	☐	☐
3. Are less than 10,000 triggering actions required per shift?	☐	☐
4. Is a reaction bar provided for torques exceeding		
6 N-m for in-line tools	☐	☐
12 N-m for pistol-grip tools	☐	☐
50 N-m for right-angled tools?	☐	☐
5. Does the tool create less than 85 dBA for a full day of noise exposure?	☐	☐
6. Does the tool vibrate?	☐	☐
6a. Are the vibrations outside the 2–200 Hz range?	☐	☐
Miscellaneous and General Considerations		
1. For general use, is the weight of the tool less than 2.3 kg?	☐	☐
2. For precision tasks, is the weight of the tool less than 0.4 kg?	☐	☐
3. For extended use, is the tool suspended?	☐	☐
4. Is the tool balanced (i.e., center of gravity on the grip axis)?	☐	☐
5. Can the tool be used without gloves?	☐	☐
6. Does the tool have stops to limit closure and prevent pinching?	☐	☐
7. Does the tool have smooth and rounded edges?	☐	☐

changed from the days of the Industrial Revolution. The operator used tools manually in the manufacture of goods. The operations required considerable force, which was somewhat leveraged by the appropriate tool. Because of the manual nature of the tasks and the forces involved, the operations were fairly slow. With the advent of automation, the excessive force levels were eliminated and many task elements relegated to the human could be eliminated. The operator performs a smaller part of the original task, which now can be speeded up because the machine does most of the work. Unfortunately, the elements still left to the human operator become more limited in scope and thus more repetitious in nature. This incomplete and unergonomic automation has led to an upsurge in CTD cases, especially if the repetition is combined with excessive wrist deviations and forceful exertions. The threshold for injury based on frequency of repetition is not known. A couple of studies have indicated repetition rates ranging from 10,000 [41] to 25,000 [40] per shift. This is a critical number that undoubtedly depends also on the posture and force exerted and needs to be fine-tuned. In fact, the exact trade-offs between the three factors are not known but are very important to quantifying the exposure and threshold levels for CTD.

Tied in with frequency is the trade-off with productivity. Any reduction in frequency of tool use will have a direct result in decreasing productivity. One alternative to maintaining constant productivity is to rotate operators for a critically repetitive task. Then again, it is necessary to know threshold levels of frequency and to know how much rest, at what intervals, must be provided to the tool operator to ensure recovery from the trauma induced from repetitive tool use. Also, it is important to know whether performing a greater variety of movements than those in the injurious task will allow the body to recover or only delay recovery. These are all issues that have not been fully addressed.

Another issue not fully resolved is the trade-off between manual and power tool use. Most researchers [24,102,103], on the basis of information on the force capacity and greater fatigability of humans compared to machines, have advocated the use of power tools. Unfortunately, power tools, whether powered electrically or pneumatically, produce some vibration. Vibration damping typically requires either an increase in the inertial mass, at the cost of increasing the weight of the tool and increasing the fatigue of the user, or vibration-absorbing systems that introduce a bit of "slop" in the hand/handle interface that absorbs the vibrations but at the cost of reducing control of the tool. Power tools also have a tendency to produce reaction torques, which can be reduced by using pulse-type tools [98] but at the cost of increasing vibration, or by using reaction bars, again at the cost of limiting the control or maneuverability of the tool. These are issues that need to be clarified further.

V. FUTURE CONCERNS

Recent research indicates that power grip capabilities can be increased through a better understanding of the pressure distribution of the hand during tool use [60,104] or by improving the frictional characteristics of the tool handle surface [62]. Perhaps the development of new polymers for application to tool handles can improve the efficiency of tool use. Also, new ways of measuring the hand/handle interface, such as the "data glove" of Yun [104], can provide more accurate information on this topic.

Most current work addresses the power grip for tools. However, most power requirements are being fulfilled by machines, leaving the human operator to perform more precise tasks that currently cannot be easily replicated by the machine. Unfortunately, there is very little information on precision or pinch grips and the precision aspects of tools. Questions on grip design and force exertion capabilities for precision grips and on occupational injury risk during work with high demands on precision need to be studied further.

Epidemiological considerations are also important in substantiating proper ergonomic designs. Unfortunately, at present, there are few good studies that support good ergonomic tool design or clearly indicate the deficiencies in such designs. More morbidity data for both hand and powered tools are needed.

A final but very important consideration is the adaptation of tools for a more diverse population. For example, with the aging of the worker population and the passage of the Americans with Disabilities Act, it is imperative that tools also be usable by individuals with a wide range of capabilities. This is both a challenge and an opportunity for ergonomists and tool designers to put their skills to effective use.

REFERENCES

1. L. S. B. Leaky, Finding the world's earliest man, *Natl. Geogr.* 118:420–435 (1960).
2. J. Napier, The evolution of the hand, *Sci. Am., 207*:56–62 (1962).
3. J. Napier, Early man and his environment, *Discovery 24*:12–18 (1963).
4. R. J. Drillis, Folk norms and biomechanics, *Hum. Factors 5*:427–441 (1963).
5. G. Lehmann, *Praktische Arbeitsphysiologie,* Thieme Verlag, Stuttgart, Germany, 1953, pp. 182–197.
6. T. J. Armstrong, *An Ergonomic Guide to Carpal Tunnel Syndrome*, American Industrial Hygiene Association, Akron, OH, 1983.
7. F. Aghazadeh and A. Mital, Injuries due to handtools, *Appl. Ergon.* 18:273–278 (1987).
8. E. R. Tichauer and H. Gage, Ergonomic principles basic to hand tool design, *Am. Ind. Hyg. Assoc. J. 38*:622–634 (1977).
9. S. W. Meagher, Hand tools: cumulative trauma disorders caused by improper use of design elements, in *Trends in Ergonomics/Human Factors*, W. Karwowski, Ed., Elsevier, Amsterdam, 1986, pp. 581–587.
10. D. S. Stetson, T. J. Amstrong, L. J. Fine, B. A. Silverstein, and K. Tannen. A survey of chronic upper extremity disorders in an automobile upholstery plant, in *Trends in Ergonomics/Human Factors*, W. Karwowski, Ed., Elsevier, Amsterdam, 1986, pp. 623–630.
11. R. W. Schoenmarklin and W. S. Marras, Effects of handle angle and work orientation on hammering, Part I. Wrist motion and hammering performance, *Hum. Factors 30*:397–411 (1989).
12. R. W. Schoenmarklin and W. S. Marras, Effects of handle angle and work orientation on hammering, Part II. Muscle fatigue and subjective ratings of body discomfort, *Hum. Factors 30*:413–420 (1989).
13. S. S. Ulin and T. J. Armstrong, Effect of tool shape and work location on perceived exertion for work on horizontal surfaces, *Proc. Int. Ergon. Assoc.,* 1991, pp. 1125–1127.
14. S. S. Ulin, C. M. Ways, T. J. Armstrong, and S. H. Snook, Perceived exertion and discomfort versus work height with a pistol-shaped screwdriver, *Am. Ind. Hyg. Assoc. J.* 51:588–594 (1990).
15. D. J. Cochran and M. W. Riley, The effects of handle shape and size on exerted forces, *Hum. Factors 28*:253–265 (1986).
16. Å. Kilbom and J. Ekholm, *Handgreppsstyrka,* MUSIC study, Stockholm, Sweden, 1991.
17. National Safety Council, *Accident Facts*, Chicago, IL, 1993.
18. V. Putz-Anderson, *Cumulative Trauma Disorders*, Taylor & Francis, London, 1988.
19. NIOSH, *Occupational Safety and Health, Year 2000 Objectives*, National Institute for Occupational Safety and Health, Centers for Disease Control, Atlanta, GA, 1989.
20. Eastman Kodak Co., *Ergonomic Design for People at Work*, Lifetime Learning, Belmont, CA, 1983, pp. 140–159.

21. E. R. Tichauer, Some aspects of stress on forearm and hand in industry, *J. Occup. Med.* 8:63–71 (1966).
22. R. Terrell and J. Purswell, The influence of forearm and wrist orientation on static grip strength as a design criterion for hand tools, *Proc. Hum. Factors Soc. 20th Annual Meeting*, Santa Monica, CA, 1976, pp. 28–32.
23. T. J. Armstrong, J. A. Foulke, B. S. Joseph, and S. A. Goldstein, Investigation of cumulative trauma disorders in a poultry processing plant, *Am. Ind. Hyg. Assoc. J.* 43:103–116 (1982).
24. S. Konz, *Work Design*, Publishing Horizons, Worthington, OH, 1990, pp. 237–258.
25. S. T. Pheasant and S. J. Scriven, Sex differences in strength—some implications for the design of handtools, in *Proceedings of the Ergonomics Society*, K. Coombes, Ed., Taylor & Francis, London, 1983, pp. 9–13.
26. D. B. Chaffin and G. Andersson, *Occupational Biomechanics*, Wiley, New York, 1984, pp. 355–368.
27. R. E. Ducharme, Problem tools for women, *Ind. Eng.*:46–50 (Sept. 1975).
28. S. T. Pheasant and D. O'Neill, Performance in gripping and turning—a study in hand/handle effectiveness, *Appl. Ergon.* 6:205–208 (1975).
29. S. Konz, Design of handtools, *Proc. Hum. Factors Soc. 18th Annual Meeting*, Santa Monica, CA, 1974, pp. 292–300.
30. J. K. Laveson and R. P. Meyer, Left out "lefties" in design, *Proc. Hum. Factors Soc. 20th Annual Meeting*, 1976, pp. 122–125.
31. N. Shock, The physiology of aging, *Sci. Am.* 206:100–110 (1962).
32. G. Miller and A. Freivalds, Gender and handedness in grip strength, *Proc. Hum. Factors Soc. 31st Annual Meeting*, Santa Monica, CA, 1987, pp. 906–909.
33. S. Konz and M. Warraich, Performance differences between the preferred and non-preferred hand when using various tools, in *Ergonomics International '85*, I. D. Brown, R. Goldsmith, K. Coombes, and M. A. Sinclair, Eds., Taylor & Francis, London, 1985, pp. 451–453.
34. A. Mital, Effects of tool and operator factors on volitional torque exertion capabilities of individuals, in *Ergonomics International '85*, I. D. Brown, R. Goldsmith, K. Coombes, and M. A. Sinclair, Eds., Taylor & Francis, London, 1985, pp. 262–264.
35. A. Mital, N. Sanghavi, and T. Huston, A study of factors defining the operator–hand tool system at the work place, *Int. J. Prod. Res.* 23:297–314 (1985).
36. A. Mital, Effects of body posture and common hand tools on peak torque exertion capabilities, *Appl. Ergon.* 17:87–96 (1986).
37. A. Mital and N. Sanghavi, Comparison of maximum volitional torque exertion capabilities of males and females using common hand tools, *Hum. Factors* 27:283–294 (1986).
38. P. T. Baker, J. M. McKendry, and G. Grant, Volumetric requirements for hand tool usage, *Hum. Factors* 2:156–162 (1960).
39. A. J. Obolenskaja and I. Goljanitzki, Die seröse Tendovainitis in der Klinik und im Experiment, *Deut. Z. Chirurg.* 201:388–399 (1927).
40. T. Luopajarvi, I. Kuorinka, M. Virolainen, and M. Holmberg, Prevalence of tenosynovitis and other injuries of the upper extremities in repetitive work, *Scand. J. Work, Environ. Health* 5(Suppl. 3):48–55 (1979).
41. NIOSH, *Health Hazard Evaluation—Eagle Convex Glass Co.,* HETA-89-137-2005, Natl. Inst. Occupational Safety and Health, Cincinnati, OH, 1989.
42. D. Grieve and S. Pheasant, Biomechanics, in *The Body at Work,* W. T. Singleton, Ed., Cambridge Univ. Press, Cambridge, UK, 1982, pp. 142–150.
43. J. O. Replogle, Hand torque strength with cylindrical handles, *Proc. Hum. Factors Soc. 27th Annual Meeting*, Santa Monica, CA, 1983, pp. 412–416.
44. B. Jonsson, T. Lewin, P. Tomsic, G. Garde, and P. Forssblad, *Handen som Arbetsredskap,* Arbetarskyddsstyrelsen, Stockholm, Sweden, 1977.
45. B. Rubarth, Untersuchung zur Festgestaltung von Handheften für Schraubenzieher und ähnliche Werkzeuge, *Ind. Psychotech.* 5:129–142 (1928).
46. N. B. Hall and E. M. Bernett, Empirical assessment of handrail diameters, *J. Appl. Psychol.* 40:381–382 (1956).

47. M. Ayoub and P. LoPresti, The determination of an optimum size cylindrical handle by use of electromyography, *Ergonomics 14*:509–518 (1971).
48. L. V. Rigby, Why do people drop things? *Qual. Prog.* 6:16–19 (1973).
49. C. G. Drury, Handles for manual materials handling, *Appl. Ergon.* 11:35–42 (1980).
50. D. J. Cochran and M. W. Riley, An evaluation of handle shapes and sizes, *Proc. Hum. Factors Soc. 26th Annual Meeting*, Santa Monica, CA, 1982, pp. 408–412.
51. D. J. Cochran and M. W. Riley, An evaluation of knife handle guarding, *Hum. Factors 28*:295–301 (1986).
52. D. J. Cochran and M. W. Riley, An examination of the speed of manipulation of various sizes and shapes of handles, *Proc. Hum. Factors Soc. 27th Annual Meeting*, Santa Monica, CA, 1983, pp. 432–436.
53. J. Garrett, The adult human hand: some anthropometric and biomechanical considerations, *Hum. Factors 13*:117–131 (1971).
54. R. Magill and S. Konz, An evaluation of seven industrial screwdrivers, in *Trends in Ergonomics/Human Factors*, W. Karwowski, Ed., Elsevier, Amsterdam, 1986, pp. 597–604.
55. W. L. Scheller, The effect of handle shape on grip fatigue in manual lifting, *Proc. Hum. Factors Soc. 27th Annual Meeting*, Santa Monica, CA, 1983, pp. 417–421.
56. H. J. Bullinger and J. J. Solf, *Ergononomische Arbeitsmittel-gestaltung, II. Handgeführte Werkzeuge-Fallstudien,* Bundesanstalt für Arbeitsschutz und Unfallforschung, Dortmund, Germany, 1979.
57. C. Saran, Biomechanical evaluation of T-handles for a pronation supination task, *J. Occup. Med. 15*:712–716 (1973).
58. T. M. Fraser, *Ergonomic Principles in the Design of Hand Tools*, International Labor Office, Geneva, Switzerland, 1980.
59. G. L. Fellows and A. Freivalds, Ergonomics evaluation of a foam rubber grip for tool handles, *Appl. Ergon.* 22:225–230 (1991).
60. C. Fransson-Hall and Å. Kilbom, Sensitivity of the hand to surface pressure, *Appl. Ergon.* 24:181–189 (1993).
61. B. Buchholz, L. J. Frederick, and T. J. Armstrong, An investigation of human palmar skin friction and the effects of materials, pinch force and moisture, *Ergonomics 31*:317–325 (1988).
62. O. Bobjer, S. E. Johansson, and S. Piguet, Friction between hand and handle. Effects of oil and lard on textured and non-textured surfaces; perception of discomfort, *Appl. Ergon.* 24:190–202 (1993).
63. J. T. Emanual, S. J. Mills, and J. F. Bennett, In search of a better handle, *Proc. Symp. Hum. Factors Ind. Design in Consumer Products*, Tufts Univ., Meford, MA, 1980, pp. 34–40.
64. Bennett Ergonomic Labs, *Why Bennett's Biocurve*, Minneapolis, MN, 1983.
65. R. Krohn and S. Konz, Bent hammer handles, *Proc. Hum. Factors Soc. 26th Annual Meeting*, 1982, pp. 413–417.
66. S. Konz and B. Streets, Bent hammer handles performance and preferences, *Proc. Hum. Factors Soc. 28th Annual Meeting*, 1984, pp. 438–440.
67. S. Konz, Bent hammer handles, *Hum. Factors, 27*:317–323 (1986).
68. R. G. Knowlton and J. C. Gilbert, Ulnar deviation and short term strength reductions as affected by a curve handled ripping hammer and a conventional claw hammer, *Ergonomics 26*:173–179 (1983).
69. L. Greenberg and D. B. Chaffin, *Workers and Their Tools*, Pendell Press, Midland, MI, 1976.
70. A. Damon, H. W. Stoudt, and R. A. McFarland, *The Human Body in Equipment Design,* Harvard Univ. Press, Cambridge, MA, 1966.
71. H. Hertzberg, Engineering anthropometry, in *Human Engineering Guide to Equipment Design,* H. Van Cott and R. Kincaid, Eds., U.S. Govt. Printing Office, Washington, DC, 1973, pp. 467–584.
72. D. J. Cochran, T. J. Albin, M. W. Riley, and R. R. Bishu, Analysis of grasp force degradation with commercially available gloves, *Proc. Hum. Factors Soc. 30th Annual Meeting*, Santa Monica, CA, 1986, pp. 852–855.
73. L. R. Sudhakar, R. W. Schoenmarklin, S. A. Lavender, and W. S. Marras, The effects of gloves

on grip strength and muscle activity, *Proc. Hum. Factors Soc. 32nd Annual Meeting*, Santa Monica, CA, 1988, pp. 647-650.
74. B. Weidman, *Effect of Safety Gloves on Simulated Work Tasks*, AD 738981, Natl. Techn. Inf. Service, Springfield, VA, 1970.
75. J. V. Bradley, Effects of gloves on control operation time, *Hum. Factors 11*:13-20 (1969).
76. J. V. Bradley, Glove characteristics influencing control manipulability, *Hum. Factors 11*:13-20 (1969).
77. A. D. Swain, G. G. Shelton, and L. V. Rigby, Maximum torque for small knobs operated with and without gloves, *Ergonomics 13*:201-208 (1970).
78. B. A. Silverstein, L. J. Fine, and T. J. Armstrong, Occupational factors and carpal tunnel syndrome, *Am. J. Ind. Med. 11*:343-358 (1987).
79. D. E. Wasserman and D. W. Badger, *Vibration and the Worker's Health and Safety*, NIOSH TR-77, U.S. Govt. Printing Office, Washington, DC, 1973.
80. R. Lundstrom and R. S. Johansson, Acute impairment of the sensitivity of skin mechanoreceptive units caused by vibration exposure of the hand, *Ergonomics 29*:687-698 (1986).
81. E. R. Andersson, Design and testing of a vibration attenuating handle, *Int. J. Ind. Ergon.* 6:119-125 (1990).
82. A. Freivalds, The ergonomics of shoveling and shovel design—an experimental study, *Ergonomics 29*:19-30 (1986).
83. E. A. Müller and K. Karrasch, Die grösste Dauerleistung beim Schaufeln, *Int. Z. Angew. Physiol. Einschl. Arbeitsphysiol. 16*:318-324 (1956).
84. A. Freivalds and Y. J. Kim, Blade size and weight effects in shovel design, *Appl. Ergon. 21*:39-42 (1990).
85. C. H. Wyndham, J. F. Morrison, C. G. Williams, R. Heyns, E. Margo, A. M. Brown, and J. Astrup, The relationship between energy expenditure and performance index in the task of shovelling sand, *Ergonomics 9*:371-378 (1966).
86. G. Kommerell, Die Schaufelarbeit in gebückter Haltung, *Arbeitsphysiologie 1*:278-295 (1929).
87. G. Dressel, K. Karrasch, and H. Sptizer, Arbetisphysiologische Untersuchungen beim Schaufeln, Steinetragen un Schubkarreschieben, *Zentralbl. Arbeistwiss. Soz. Betriebsprax.* 3:33-48 (1985).
88. A. Freivalds, The ergonomics of tools, *Int. Rev. Ergon. 1*:43-75 (1987).
89. A. G. Stevenson and R. L. Brown, An investigation on the motion study of digging and the energy expenditure involved, with the object of increasing efficiency of output and economising energy, *J. Roy. Army Med. Corps 40*:39-45, 99-111, 340-349, 423-434 (1923).
90. C. J. Widule, V. Foley, and F. Demo, Dynamics of the axe swing, *Ergonomics 21*:925-930 (1978).
91. D. L. Corrigan, V. Foley, and C. J. Widule, Axe use efficiency—a work theory explanation of an historical trend, *Ergonomics 24*:103-109 (1981).
92. H. Gläser, Beiträge zur Form der Waldsäge und zur Technik des Sägens, Ph.D. Dissertation, Everswalde, Germany, 1933.
93. A. S. Bleed, P. Bleed, D. J. Cochran, and M. W. Riley, A performance comparison of Japanese and American hand saws, *Proc. Hum. Factors Soc. 26th Annual Meeting*, Santa Monica, CA, 1982, pp. 403-407.
94. E. R. Tichauer, Biomechanics sustains occupational safety and health, *Ind. Eng.* February 1976, pp. 46-56.
95. T. A. Yoder, R. L. Lucas, and C. D. Botzum. The marriage of human factors and safety in industry, *Hum. Factors 15*:197-205 (1973).
96. F. E. Lindstrom, *Modern Pliers*, Bahco Verktyg, Enköping, Sweden, 1973.
97. L. Karlqvist, Cutting operations at canning bench—a case study of handtool design, *Proceedings of the 1984 International Conference on Occupational Ergonomics*, Human Factors Association of Canada, Rexdale, ON, 1984, pp. 452-456.
98. A. Freivalds and J. Eklund, Reaction torques and operator stress while using powered nutrunners, *Appl. Ergon. 24*:158-164 (1993).
99. Atlas Copco, *Industrial Power Tools* (Catalog), Atlas Copco Tools AB, Box 81510, Stockholm, Sweden, 1990.

100. R. G. Radwin, E. VanBergeijk, and T. J. Armstrong, Muscle response to pneumatic hand tool torque reaction forces, *Ergonomics 32*:655–673 (1989).
101. M. Vuori, M. Rauko, and S. Herranen, *Evaluating and Choosing Pneumatic Screwdrivers and Nutrunners*, The Finnish Work Environment Fund, Helsinki, Finland, 1989.
102. A. Mital and Å. Kilbom, Design, selection and use of hand tools to alleviate trauma of the upper extremities: Part I. Guidelines for the practitioner, *Int. J. Ind. Ergon. 10*:1–5 (1992).
103. A. Mital and Å. Kilbom, Design, selection and use of hand tools to alleviate trauma of the upper extremities: Part II. The scientific basis for the guide, *Int. J. Ind. Ergon. 10*:7–21 (1992).
104. M. H. Yun, A hand posture measurement system for evaluating manual tool tasks, *Proc. Hum. Factors Ergon. Soc. 37th Annual Meeting*, Santa Monica, CA, 1993, pp. 754–758.

14
Manual Materials Handling

Thomas R. Waters and Vern Putz-Anderson
National Institute for Occupational Safety and Health, Cincinnati, Ohio

I. INTRODUCTION

The intent of this chapter is to (1) provide basic information to identify and quantify the risk factors associated with manual materials handling (MMH) and (2) to discuss typical methods used to reduce or eliminate these factors so that work-related musculoskeletal injuries (MSIs) can be prevented. The chapter was designed to meet the needs of those individuals who are responsible for providing technical occupational ergonomics support but lack formal training in ergonomics. The chapter includes sections describing

Who is at risk
What risk factors are associated with work-related injuries
Methods of identifying and analyzing hazards
Methods of preventing work-related injuries

II. HEALTH RISK OF MMH

A. Adverse Health Effects

Despite the spread of mechanization in industry, MSIs attributed to MMH are still a major cause of lost work time, increased costs, and human suffering in the American workforce. These MSIs include a variety of injuries or disorders of the wrist, arm, shoulder, neck, and back as well as the lower extremities.

Problems with the low back are especially problematic, and low back disorders resulting in low back pain (LBP) continue to be one of the leading occupational health and safety issues facing preventive medicine. Despite efforts at control, including programs directed at both workers and jobs, work-related back injuries still account for a significant proportion of human suffering and economic cost to this nation.

The scope of the LBP problem was summarized in a report published in 1982 by the Bureau of Labor Statistics (BLS) [1]. According to the BLS report, nearly 20% of all injuries and illnesses in the workplace and nearly 25% of the annual workers' compensation payments are attributable to back injuries. A more recent report by the National Safety Council [2] indicated that overexertion is the most common cause of occupational injury, accounting for 31% of all injuries. The back, moreover, is the body part most frequently injured (22% of 1.7 million injuries) and the most costly to workers' compensation systems.

Data analyzed by Park, obtained from the National Health Interview Survey—Occupational Health Supplement (NHIS-OHS) [3] revealed that nearly half of the estimated 22 million cases of back pain reported by workers over a 12-month period in 1988 were work-related. These injuries accounted for more than 543 million lost working days, and about 12% of the respondents stopped working or changed jobs because of back pain.

B. Who Is At Risk

Approximately 30% of the American workforce is routinely engaged in jobs that expose the worker to the physical hazards associated with manual materials handling. This estimate is based on occupational exposure data collected in 1981–1983 in the National Occupational Exposure Survey (NOES) conducted by the National Institute for Occupational Safety and Health (NIOSH) [4]. The survey was designed to be representative of virtually all the nonagricultural, nonmining, and nongovernmental businesses covered under the Occupational Safety and Health Act of 1970. The NOES data were analyzed to determine the probabilities of being exposed to various physical hazards and agents such as frequent bending and/or unaided lifting, hand/wrist manipulations, arm movement, noise, and vibration. The analysis indicated that lifting activities were one of the most frequently identified ergonomic exposure hazards. Hand/wrist manipulations and arm transport motions were also included as frequently identified hazards by the NIOSH surveyors.

The industries with the greatest estimated total employees exposed to the hazards of lifting per Standard Industrial Classification (SIC) were health services, special trade contracting, general building contracting, food and kindred products, and trucking and warehousing.

The NOES report also summarized the composition of the workforce with respect to gender (see Table II-6, NOES Vol. III [4]). Four of the five industries with the largest number of workers exposed to the hazards of lifting (i.e., special trade contracting, general building contracting, food and kindred products, and trucking and warehousing) employed significantly more men than women (i.e., between 2 and 10 times as many male workers as females). One industry with a large number of exposed female workers, however, is the health services industry, which employs approximately four times as many female workers as males.

The National Center for Health Statistics (NCHS) recently published data from the National Health Interview Survey (NHIS), which included an Occupational Health Supplement (OHS). The survey was designed to provide an estimate of exposures to various physical activities that are thought to increase the risk of adverse health effects [3]. The report stated that, of the currently employed population, 16% reported spending 4 or more hours daily in repeated strenuous physical activity, and 30% reported spending 4 or more hours daily in repeated bending, twisting, or reaching. Presumably, much of this activity was spent performing work-related MMH activities, such as lifting, carrying, pushing, or pulling, in awkward postures.

C. Summary

It is clear that a tremendous number of workers are routinely exposed to the physical hazards of MMH, and many of them will develop one or more serious work-related musculoskeletal disorders during their working lifetime. Nevertheless, it should be recognized that these hazards can be identified and controlled so that work-related MSIs can be prevented.

III. RISK FACTORS

There are a variety of manual materials handling activities that increase a worker's risk of developing a musculoskeletal injury, including jobs that involve a significant amount of manual lifting, pushing, pulling, or carrying and jobs requiring awkward postures, prolonged sitting, or exposure to cyclic loading (whole body vibration). In addition to these frequent patterns of behavior, there are a variety of personal and environmental factors that may increase a worker's risk of developing an MSI. Table 1 contains a list of the more noteworthy work-related risk factors for developing MSI.

Personal risk factors are conditions or characteristics of the worker that may affect the probability that an overexertion injury may occur. Personal risk factors include attributes such as age, level of physical conditioning, strength, and medical history.

Environmental risk factors are conditions or characteristics of the external surroundings that may affect the probability that an overexertion injury may occur. Environmental risks include factors such as temperature, lighting, noise, vibration, and friction at the floor.

Job-related risk factors are conditions or characteristics of the MMH job that may affect the probability that an overexertion injury may occur. Job-related risk factors include attributes such as the weight of the load being moved, the location of the load relative to the worker when it is being moved, the size and shape of the object moved, and the frequency of handling. These factors may be the most important for prevention because they directly affect the magnitude of physical hazard to the worker.

Work-related MSIs are typically attributed to a direct trauma, a single exertion ("overexertion"), or multiple exertions ("repetitive trauma"). It is not always possible, however, to determine the specific cause of the injury, and the pathophysiology of many types of MSIs is poorly understood. Moreover, it is important to note that personal characteristics such as age, physical conditioning, and concomitant diseases can modify the way the body responds to stressful exertions. Therefore, an injury could occur at different loading levels for different workers, and even for an individual worker a load may be tolerable one day and excessive on another day due to fluctuations in muscular strength and aerobic fitness [5].

Table 1 Risk Factors Associated with Manual Material Handling Injuries

Personal factors	Environmental factors	Job-related factors
Gender	Humidity	Location of load relative to the worker
Anthropometry (body weight and height)	Light	Distance object is moved
	Noise	
Physical fitness and training	Vibration	Frequency and duration of handling activity
	Foot traction	
Lumbar mobility		Bending and twisting
Strength		Weight of object or force required to move object
Medical history		
Years of employment		Stability of the load
Smoking		Postural requirements
Psychosocial factors		
Anatomical abnormality		

Knowledge about job-related, environmental, and personal risk factors is far from complete. Although a large volume of research literature exists in which multiple factors contributing to LBP have been studied, many are conflicting, and it is difficult to compare these studies because the methods used and the variables studied are rarely similar [6]. For example, the populations studied, sampling and data collection methods, and task and time scales often vary across studies [7]. In addition, studies are confounded by the high prevalence of LBP in the general population, the wide range of disorders, nonspecific symptoms, and frequent association with nonoccupational factors.

A need exists for long-term prospective and controlled studies of large samples of workers to identify and quantify workplace factors that contribute to LBP. Prospective studies are needed so that causal relationships between factors involved in the development of LBP can be differentiated from factors resulting from LBP [8]. Also, relationships between LBP and risk factors would be more apparent if precise and accurate outcome measures were available [9].

A considerable volume of literature exists suggesting that there is a significant relationship between psychosocial factors, such as job control and satisfaction, and risk of MSI [10]. A report by Bongers and de Winters [11] contains a detailed review of the literature up to 1992 on psychosocial factors and musculoskeletal injuries. Bongers and de Winters concluded that monotonous work, high perceived workload, and time pressure are related to MSI symptoms and that low control of work demands and lack of social support of colleagues are positively associated with musculoskeletal disease. These conclusions are based on the review of studies such as those by Linton [12] and Bigos et al. [13]. They also suggest that the role psychosocial factors play in affecting the risk of injury may be as important as the actual physical demands of the job [13]. The data indicate that as the perceived demands of the job increase and the workers' control over those demands and social support decrease, the rate of injury increases. It is probable that these psychosocial factors lead to an increase in the physical effects on the worker such as increased muscle tension, lower endurance, or modified body mechanics. More research is needed to develop a comprehensive control strategy that considers these psychosocial factors.

In summary, there is ample evidence that manual materials handling can result in serious injury and that these injuries are attributable to numerous risk factors that must be considered in the design of a safe workplace. The interpretation of the research linking MSIs and manual handling is problematic, however, because of the high prevalence of certain injuries such as low back pain in the general population and their frequent association with nonoccupational factors. In addition, the relationship is further obscured by the wide range of disorders, the nonspecific nature of the condition, and the general lack of objective data relating different risk factors to overexertion injury. What is generally recognized, however, is that musculoskeletal injuries are a function of a complex set of variables that include aspects of job design, work environment, and personal factors. How all of these factors are combined to precisely determine the overall risk associated with a specific manual materials handling job remains to be resolved.

IV. HAZARD IDENTIFICATION/JOB ANALYSIS

A variety of analytic tools are available for the ergonomic evaluation of manual materials handling tasks, especially manual lifting tasks. These tools range in complexity from simple checklists that are designed to provide a general indication of the physical stress associated with a particular manual materials handling job to complicated computer models that provide detailed information about specific risk factors. Although not exhaustive, Table 2 summarizes the advantages and disadvantages of a variety of ergonomic assessment tools. As described below, these tools provide objective information about the physical demands of manual handling tasks so that risks can be reliably assessed and an effective prevention strategy can be developed.

In general, these tools are based on scientific studies that provide a relationship between sources of physical stress and risk of musculoskeletal injury, particularly when those stressors exceed the physical capacity of the workers [14–16]. Any assessment of physical stress or human capacity is complicated by the influence of a variety of factors including work performance, motivation, expectation, and fatigue tolerance.

Table 2 Ergonomic Assessment Tools

Assessment tool	Advantages	Disadvantages
Checklists	Simple to use Best suited for use as a preliminary assessment tool Applicable to a wide range of MMH jobs	Does not provide detailed information about the specific risk factors Does not quantify the extent of the risk factors
Biomechanical models	Provide detailed estimates of mechanical forces on musculoskeletal components Can identify specific body structures exposed to high physical stress	Not applicable for estimating effects of repetitive activities Difficult to verify accuracy of estimates Rely on a number of simplifying assumptions
Psychophysical tables	Provide population estimates of worker capacities that integrate biomechanical and physiological stressors Applicable to a wide range of MMH activities	Reflects more about what a worker will "accept" than about what is "safe" May over- or underestimate demands for infrequent or highly repetitive activities
Physiological models	Provides detailed estimates of physiological demands for repetitive work as a function of duration Applicable to a wide range of MMH activities	Not applicable for estimating effects of infrequent activities Lack of strong link between physiological fatigue and risk of injury
Integrated assessment models	Simple to use Use the most appropriate criterion for the specified task	Require a significant number of assumptions Limited range of application
Videotape assessment	Economical method of measuring postural kinematics Can be used to analyze a large number of samples	Labor-intensive analysis generally required Limited to two-dimensional analysis for field measurements
Exposure monitors	Provide direct measures of posture and kinematics during MMH activities Applicable to a wide range of MMH activities	Require the worker to wear a device on the body Lack of data linking monitor output and risk of injury

According to the Occupational Safety and Health Administration (OSHA) [17], the objectives of the worksite analysis are to identify existing physical hazards and conditions, operations that create hazards, and areas where hazards may develop. Worksite analysis not only includes a systematic method to identify those work tasks or operations requiring hazard control but may also include a comprehensive review and analysis of injury and illness records to find any evidence of musculoskeletal injuries or to identify any trends in injury patterns relating to particular departments or jobs. This information is necessary to identify those jobs that are believed to possess a significant risk of injury and require a detailed quantitative analysis of physical hazards.

A. Checklists

A checklist may be the first choice for a rapid ergonomic assessment of a particular workplace. Checklists are designed to provide a general evaluation of the extent of hazard associated with a manual materials handling task or job. A checklist usually consists of a series of questions that are indicative of physical hazards. For example questions about frequent bending, heavy lifting, awkward or constrained postures, poor couplings at the hands or feet, and hazardous environmental conditions. Some checklists use a yes/no format, whereas others use a numerical rating format. Checklists are easy to use, but they lack specificity and they are imprecise. An example of a manual handling checklist is presented in Figure 1. For more information on checklists, see Grandjean [18], Eastman Kodak [19], Alexander and Pulat [20], or the National Occupational Health and Safety Commission [21].

B. Biomechanical Models

Biomechanical assessment involves the systematic application of engineering concepts to the functioning human body to predict the distribution of internal musculoskeletal forces resulting from the interaction with externally applied forces of the task. Biomechanical modeling provides a method for predicting the pattern and magnitude of these internal forces during manual handling. These predicted forces can then be compared to predetermined tissue tolerance limits to assess the biomechanical stress associated with specific loading conditions. When a worker is performing a manual lifting task, for example, the internal reaction forces that are needed to provide equilibrium between the body segments and the external forces are supplied by muscle contractions, tendons, and ligaments at the body joints (i.e., the human body acts as a lever system). Specifically, the external forces at the hands and the body segments create rotational moments or torques at various body joints, especially the low back. The skeletal muscles exert forces that result in moments about the joints so as to counteract the moments due to external load and body segment weights. Since the moment arms of the muscles (and ligaments) are much smaller than the moment arms of the external forces and body segment weights, small external forces can produce large muscle, tendon, ligament, and joint reaction forces. On the other hand, the muscles can produce large motions with small degrees of shortening. The concept of muscles loading skeletal structures is extremely important in the biomechanics of MSIs, because handling light loads in certain postures can create large mechanical loads on the muscles, ligaments, and joint surfaces.

The complexity of mathematical formulation and ease of use of biomechanical models varies, depending on which factors are included. Some important factors to consider when using a biomechanical model are

NIOSH HAZARD EVALUATION CHECKLIST FOR LIFTING, CARRYING, PUSHING, OR PULLING		
Risk Factors	YES	NO
1. General		
1.1 Does the load handled exceed 50 lbs.?	[ ]	[ ]
1.2 Is the object difficult to bring close to the body because of it's size, bulk, or shape?	[ ]	[ ]
1.3 Is the load hard to handle because it lacks handles or cutouts for handles, or does it have slippery surfaces or sharp edges?	[ ]	[ ]
1.4 Is the footing unsafe? For example, are the floors slippery, inclined, or uneven?	[ ]	[ ]
1.5 Does the task require fast movement, such as throwing, swinging, or rapid walking?	[ ]	[ ]
1.6 Does the task require stressful body postures, such as stooping to the floor, twisting, reaching overhead, or excessive lateral bending?	[ ]	[ ]
1.7 Is most of the load handled by only one hand, arm, or shoulder?	[ ]	[ ]
1.8 Does the task require working in environmental hazards, such as extreme temperatures, noise, vibration, lighting, or airborne contaminants?	[ ]	[ ]
1.9 Does the task require working in a confined area?	[ ]	[ ]
2. Specific		
2.1 Does lifting frequency exceed 5 lifts per minute?	[ ]	[ ]
2.2 Does the vertical lifting distance exceed 3 feet?	[ ]	[ ]
2.3 Do carries last longer than 1 minute?	[ ]	[ ]
2.4 Do tasks which require large sustained pushing or pulling forces exceed 30 seconds duration?	[ ]	[ ]
2.5 Do extended reach static holding tasks exceed 1 minute?	[ ]	[ ]
Comment: "Yes" responses are indicative of conditions that pose a risk of developing low back pain. The larger the percentage of "yes" responses, the greater the possible risk.		

Figure 1 Manual materials handling checklist.

1. The mechanical nature of the model (static or dynamic)
2. Dimensionality of the model (two- or three-dimensional)
3. Accuracy of the representation (e.g., single or multiple muscles, intra-abdominal pressure, muscular cocontraction, and active and passive elements)
4. Complexity of the input needed to use the model (e.g., musculoskeletal geometry, mechanical parameters, physiological measures of muscle function)

Some of the more complex biomechanical models of the low back, for example, integrate three-dimensional mechanical data and electromyographic data from the back muscles as input to estimate internal loading on the spine [22].

Although biomechanical models are typically used to provide guidance for the design of infrequent stressful activities with high levels of exertion, recent studies have provided some guidance for the use of biomechanical models in assessing repetitive lifting tasks. In studies of spinal compression tolerance, for example, Brinckmann et al. [23] showed that repeated compression loading of the spinal motion segments causes them to fail at lower forces than those required in a single loading cycle.

For more information on the use of biomechanical models to assess MMH tasks, refer to Chaffin and Andersson [24], National Research Council [25], or Garg [26].

C. Psychophysical Tables

Psychophysics is a branch of psychology that examines the relationship between the perception of human sensations and physical stimuli. Stevens [27] and Snook and Ciriello [28] suggest that a worker's actual level of physical stress can be assessed by her or his subjective judgment or perception of the physical stress. In other words, the worker's perception of workload is used to assess the combined effects of physiological and biomechanical stress created by various manual materials handling factors.

Typical psychophysical studies of MMH capacity consist of measurement of acceptable levels of workload for various task conditions and for various workers. The results are usually provided in tables of acceptable weights of lift or carry or acceptable forces of pushing and pulling [28]. These psychophysical measures are contained in databases that provide acceptable workloads for various segments of the population. A portion of a psychophysical database for maximum acceptable sustained push force is shown in Table 3. Acording to the database, the push force acceptable to 50% of the female industrial subjects for a 15.2-m push (50 ft) at a height of 89 cm (35 in.) and a rate of 1 push every 5 min was 14 kg (31 lb).

Most psychophysical research involving MMH activities has emphasized lifting tasks, although the use of psychophysical techniques is not restricted to lifting. Psychophysics also is applicable to lowering, pushing, pulling, holding, and carrying activities. The use of psychophysical data to assess the physical demands of manual handling is most appropriate for repetitive activities that are performed more often than once a minute. For more information on psychophysical databases, see Snook [16], Snook and Ciriello [28], or Ayoub and Mital [29].

Other psychophysical assessment methods have been developed to assess various MMH activities. For example, self-report measures such as rating of perceived exertion (RPE) [30] and body part discomfort (BPD) [31] have been used to assess a variety of lifting jobs. These assessment measures provide useful information about the worker's perception of the physical demands of the job. Moreover, RPE and BPD compare favorably with measures of physical demand.

Databases containing whole body and segmental strength measures have also been developed for the design of manual materials handling tasks. These include isometric, isokinetic, and isoinertial strength databases for whole body activities such as lifting and various databases for the arms, legs, and back. For more information on strength measurement, detailed summaries of these databases are available in Chaffin and Andersson [24] and Ayoub and Mital [29].

Table 3 Portion of a Psychophysical Table for Maximum Acceptable Push Force

Height from floor to hands (cm)	Percent of industrial females	Maximum acceptable forces of sustained push (kg) for 15.2-m push						
		One push every						
		6 sec	12 sec	1 min	2 min	5 min	30 min	8 hr
89	90	5	6	6	7	7	8	10
	75	7	8	9	10	11	11	14
	50	9	11	13	13	14	15	19
	25	12	14	16	16	18	19	24
	10	14	17	19	19	21	23	28

Source: Adapted from Ref. 28.

It should be noted that psychophysics relies on self-report from subjects, and consequently the perceived "acceptable" limit may differ from the "safe" limit. Also, the psychophysical approach is not valid for some task conditions, for which other assessment measures may be more appropriate [8].

D. Physiological Models

One goal in designing a manual materials handling task is to avoid the accumulation of physical fatigue that may contribute to an MSI. This fatigue can affect specific muscles or groups of muscles, or it can affect the whole body by reducing the aerobic capacities available to sustain work. The physiological factors that affect the suitability of a manual handling task at the local muscle effort level include (1) the duration of force exertion and (2) the frequency of repetition per minute. Local muscular fatigue will develop if a heavy effort is sustained for a long period. With heavy loads, the muscles will need a substantially longer recovery period to return to their previous state. Small changes in workplace layout or handling heights, however, can often solve a local muscle fatigue problem through a reduction in holding duration. In addition, local muscle fatigue associated with maintaining awkward postures or with constant bending can reduce the capacity of the muscles needed for lifting and therefore can increase the potential for an MSI injury to occur.

Although fatigue in one or more muscles at the local level may limit the acceptable workloads for manual materials handling tasks that are performed for short, intensive periods during a work shift, it is the energy expenditure demands of the muscles, as measured by oxygen consumption, that have the most profound effect on what a worker is able to do. Energy expenditure demands depend on the extent of muscular exertion, frequency of activity, and duration of continuous work. A worker's limit for physiological fatigue is often affected by a combination of discomfort in local muscle groups and more centralized (systemic) fatigue associated with oxygen demand and cardiovascular strain [19,32].

To assess the cardiovascular demands of manual materials handling tasks, physiological parameters such as heart rate (HR), oxygen (O_2) consumption, and ventilatory rate may be used. Electromyographic (EMG) assessments and blood lactate provide a relative measure of the instantaneous level of physiological status and muscular fatigue. As described below, these measures can be used to help prevent MSI by predicting the limits of fatigue for repetitive handling tasks.

Physiological models provide a method for estimating the cardiovascular demands associated with a specific manual materials handling activity. One such model developed by Garg [33] allows the analyst to estimate the energy expenditure demands associated with a complex manual materials handling job. First, the job is separated into distinct elements or subtasks such as standing and bending, walking, carrying, vertical lifting or lowering, and horizontal arm movement, for which individual energy expenditure values can be predicted by the model. Next, the amount of time engaged in each of the subtasks is determined, and a subtotal for the energy expenditure requirements for each subtask is computed by taking the product of the incremental expenditures and the amount of time spent performing that activity. Finally, the total energy expenditure of the job, which determines the overall physiological demands of the job, is estimated by taking the sum of the subtotals of the subtasks.

Direct measures of oxygen consumption may provide the best estimate of physiological demand. For this type of measurement, it is assumed that the consumption of 1 liter of oxygen per minute is approximately equal to an energy expenditure of 5 kilocarlories (kcal) per minute. As mentioned above, heart rate is also useful in predicting physiological demand but is less reliable than direct oxygen consumption measures due to individual differences in the relationship between heart rate and energy expenditure for different people. To simplify measurement, portable monitors are available for acquiring heart rate and oxygen consumption data during manual materials handling tasks to accurately determine physiological demands. For information on assessing physiological demands, see Astrand and Rodahl [34] or Eastman Kodak [19].

E. Integrated Assessment Models

An integrated assessment model involves a unique approach that considers all three of the primary stress measures—biomechanics, physiology, and physchophysics. The integrated approach provides a method of estimating the relative magnitude of physical demand for a specific manual materials handling task, which is based on the most appropriate stress measure for that unique set of task factors. The result of the assessment is typically represented as a weight or force limit or as an index of relative severity. An integrated model considers the combined effects of the various task factors and uses the most appropriate stress measure to estimate the magnitude of hazard associated with each task factor. For example, the model will predominantly rely on the biomechanical assessment for an infrequent heavy lift but will rely more heavily on the physiological assessment for a highly repetitive lifting task.

Examples of integrated assessment models include Ayoub's job severity index (JSI) [35] and the National Institute for Occupational Safety and Health (NIOSH) recommended weight limit (RWL) equation [8]. Details on the revised NIOSH lifting equation are presented elsewhere in this book (see Chapter 31.)

F. Videotape Assessment

Most ergonomic assessments include the use of videotape analysis, where a video camera is used to record the work activity for later analysis. Videotape recordings make it easier to stop or freeze the action so that various measures of body posture or workplace layout may be evaluated. Videotape analysis may consist of general observation by an analyst, which results in subjective estimates of physical hazards, or more detailed video assessment may be used to objectively quantify the extent of the hazard.

Complex computerized video analysis systems are available that are capable of automatically capturing individual frames from videotape recordings of workers performing manual materials handling activities. These video frames can then be analyzed to provide more detailed assessments of spatial or dynamic biomechanical hazards that may not be apparent from the observational approach. These systems may include automatic digitizing capabilities for marking body joint locations and/or automatic scoring for assessing joint or body segment positions. Automated analysis systems are uniquely suited for certain types of assessments in which large amounts of videotape must be analyzed to determine the extent of the physical hazards.

Operationally, videotape analysis is generally easy to use, and the output data are presented in a form that is easy to understand and apply. Video analysis is limited, however, in its capability to analyze activities that occur outside of the camera's focal plane (the plane parallel to the face of the camera lens). For example, when a body segment or group of segments move outside the camera's focal plane, the joint angles and positions measured from the digitized frame are distorted. The amount of distortion depends on the degree of displacement of the joint or segment from the focal plane. In an attempt to compensate for this problem, some systems have incorporated distortion correction algorithms. Correcting distortion problems is problematic, however, and requires a high degree of technical training.

Manual Materials Handling

Guidelines for videotape job analysis that have been developed by NIOSH researchers are provided elsewhere in this book (see Chapter 10.) For more information on videotape analysis and motion analysis, see Chaffin and Andersson [24] or Eastman Kodak [19].

G. Exposure Monitors

Monitoring devices have been developed to measure various aspects of physical activity, including position, velocity, and acceleration of movement. Some monitors can even measure three-dimensional joint angles in real time. These systems consist of mechanical sensors that are attached to various parts of the worker's body such as the wrist, back, or knees. The mechanical sensors convert angular displacement (rotation) into voltage changes that can be displayed in real time and/or saved to a computer for later analysis. The position measures acquired from the sensors can then be differentiated to obtain rotational velocities and acceleration components. These movement characteristics can be used to estimate the extent of risk associated with a particular task and help to identify potential ergonomic solutions. Examples of positional monitoring equipment include potentiometer-based wrist motion monitors and back motion monitors (illustrated in Figure 2) such as those developed at The Ohio State University [36,37] and strain gauge–based strip goniometers such as those developed by Penny and Giles (Santa Monica, California).

Another device that has been used to assess the extent of exposure to repetitive movement is an accelerometer-based motion-recording system or activity monitor. An activity monitor consists of one or more accelerometers mounted within a small aluminum container attached to a velcro strap that is attached to a worker's wrist, leg, or trunk. The accelerom-

Figure 2 Lumbar motion monitor.

eters are sensitive to the movements of the body and are capable of counting and recording rapid movements inherent in a specific task or activity. These measures are important because "highly dynamic" movements that occur over an extended period of time, such as an 8-h work shift, are believed to increase workers' risk of musculoskeletal injury. The data acquired from the activity monitor are typically plotted as a series of temporal histogram plots showing the extent of dynamic movement as a function of time as illustrated in Figure 3. The greater the total dynamic activity, the greater the height and density of the sequential histogram bars and the greater the potential for injury [38].

It is important to note that the output from exposure monitors alone cannot provide all the information that is needed to assess the overall extent of physical demand associated with a manual materials handling task. The weight of the load handled and its position, velocity, and acceleration relative to the body during the task are also important in assessing the demands of the task. This approach may be best suited for repetitive or high-speed manual materials handling tasks where the internal forces on the body may be affected more by extreme postures or rapid movements rather than the weight or position of the external load.

V. PREVENTION AND CONTROL

In the past, attempts to prevent overexertion injuries associated with MMH focused on adopting arbitrary weight limits for lifting loads, hiring strong workers, or using training procedures that emphasized "correct" (but not necessarily safe) lifting techniques. None of these approaches, however, has proven to be effective in significantly reducing overexertion injuries

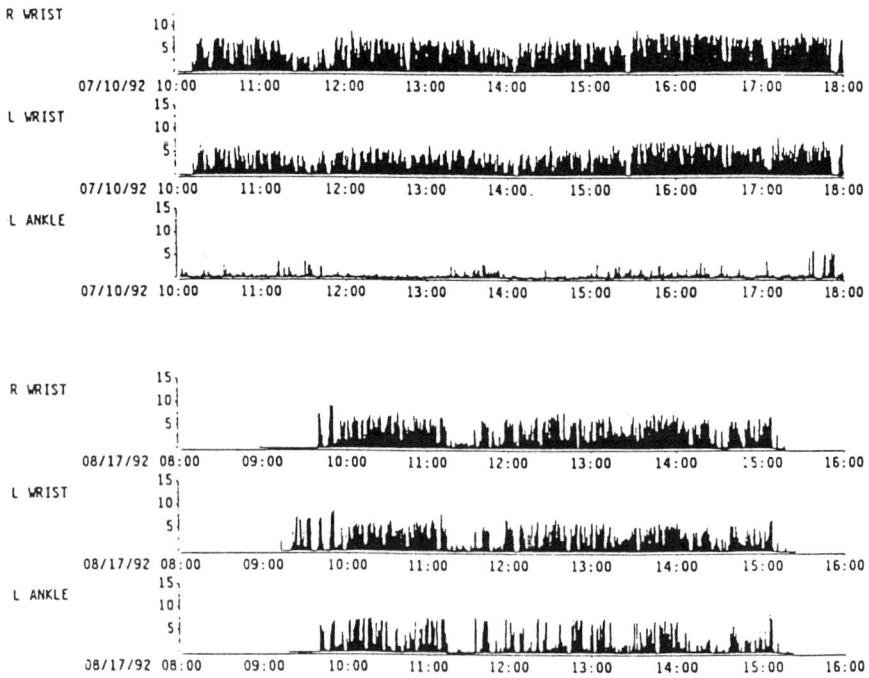

Figure 3 Temporal histogram plots from "activity monitor" show the extent of dynamic movement as a function of time.

[39,40]. More recently, industry has recognized the risks associated with MMH and, to reduce costs and increase productivity, companies have begun to implement ergonomic programs or practices aimed at preventing these injuries.

In many cases, these ergonomic programs rely on guidelines provided by the federal government. For example, the Occupational Safety and Health Administration (OSHA) published a management guideline for implementing an ergonomics program for "red meat" packing plants. SIC code 2011 [17]. The guideline was designed to cover both cumulative trauma disorders (CTDs) of the upper extremities and low back disorders. The "red meat" packing guideline may serve as the model for development of a broader ergonomics guideline or standard that would apply to many industries.

The OSHA "red meat" packing guideline describes the basic requirements of an effective ergonomics program. According to OSHA, an effective ergonomics program requires

1. A *commitment by the employer* to provide visible involvement of top management, so that all employees, from management to line workers, fully understand that management has a serious commitment to the program
2. A *written program* that establishes clear goals and objectives that are communicated and understood by all members of the organization
3. A commitment to provide for and *encourage employee involvement* in the identification and control of ergonomic problems
4. A regular program of *review and evaluation* to evaluate the implementation of the program and to monitor the progress of the program accomplishments

According to OHSA, hazard prevention and control is an essential element of an ergonomics program.

A. Hazard Prevention and Control

Once hazards have been identified through systematic worksite analysis, design measures must be implemented to prevent or control these hazards. According to the OSHA guideline [17], ergonomic hazards are prevented primarily by effective design of the workstation, tools, and job. To be effective, an employer's program should use appropriate engineering and administrative controls to correct or control ergonomic hazards. These controls should include design or modification of workstations, tools, or work methods to eliminate excessive exertion and awkward postures and modification of work schedules and rates to reduce the duration, frequency, and severity of exposures to ergonomic stressors.

Attempts at ergonomic control of MMH problems have included both worker-directed and workplace-directed programs. Worker-directed approaches primarily deal with attempts to maintain a match between the worker's capacity and the demands of the job. This is usually attempted through worker screening, increases in the physical capacity of the worker, or protection of the worker from the physical hazard in some way. Consequently, these approaches are developed on a worker-to-worker and job-to-job basis. More important, this approach requires detailed information about the capacity of each worker as well as the demands of the job. Moreover, worker-directed approaches fail to directly reduce the extent of the physical hazard associated with the task or job.

In comparison, workplace-directed approaches rely on changes in the design of the work to eliminate or minimize the effects of the MMH problem. Consequently, these approaches are more generalized and less dependent on having detailed knowledge of the capacity of the specific worker performing the task or job. Workplace-directed approaches include elimination or manual materials handling from the process through automation, reduction of the amount of physical exertion required to perform the MMH activity by using mechanical aids to assist the worker, or modification of the layout of the workplace to reduce the physical hazard.

B. Workplace-Directed Approaches

1. Automation

Workplace automation should be a first approach when the work requires high physical demands, is highly repetitive, or is performed in a hazardous environment. Automation may consist of one or more machines or machine systems such as conveyors, automated handling lines, automated storage and retrieval systems, and robots. This approach is best suited for the design of new work processes or activities or for the design of highly stressful tasks. Since automation may require large capital expenditures, this approach could be prohibitively costly for small companies with few workers.

2. Mechanical Aids

In cases where the physical demands are high and automation is not practical, mechanical aids should be used to reduce the extent of physical demands of a task or job. Mechanical handling aids include machines or simple devices such as hand trucks, cranes, hoists, lift tables, powered mobile equipment and lift trucks, overhead handling and lifting equipment, and vacuum lift devices that provide a mechanical advantage during the MMH task. NIOSH has developed a list of information sources pertaining to the materials handling industry that provides a brief overview of the materials handling equipment available in the United States [41]. The list also provides contact telephone numbers and addresses for more detailed information needs.

3. Job Modification Through Ergonomic Design

The use of job modification through ergonomic design has been a basic tenet of occupational safety and health practice for achieving hazard control in preference to other methods such as worker selection and testing, training in safe work practices, or the use of personal protective equipment, which are less reliable and often less effective. Tasks and tools are modified by using ergonomic principles to reduce the effects of biomechanical stress. For MMH jobs, ergonomic design or redesign may be accomplished by modifying the job layout and procedures to reduce bending, twisting, horizontal extensions, heavy lifting, forceful exertions, and repetitive motions.

The ergonomic approach is largely based on the assumption that work activities involving less weight, less repetition, less time in awkward postures, and less applied force are less likely to cause injuries and disorders. The ergonomic approach is desirable because it seeks to eliminate potential sources of problems within the job itself. Ergonomics also seeks to make safe work practices a natural result of the tool and worksite design.

There are at least four advantages to adopting an ergonomic design/redesign strategy.

First, an ergonomic approach does not depend on specific worker capabilities such as strength and conditioning.
Second, human biological factors and their variations are accounted for in ergonomic approaches because there is consideration of design data accommodating large segments of the population.
Third, compared to worker-directed approaches, which may be temporary, ergonomic intervention is relatively permanent because the workplace hazard is eliminated.

Fourth, to the extent that sources of biomechanical stress at the worksite are eliminated or significantly reduced, the difficult issues involving potential worker discrimination, lifestyle modifications, or attempts at changing behavioral patterns of workers (training and education issues) will be of lesser practical significance.

Although this approach is preferred over worker-directed approaches, it should be noted that no single job modification will always be sufficient; numerous adjustments may be required to properly redesign a job.

C. Worker-Directed Approaches

1. Training and Education

According to OSHA [17], the purpose of training and education is to "ensure that employees are sufficiently informed about the ergonomic hazards to which they may be exposed and thus are able to participate actively in their own protection." Training should include general instruction on the types of injuries that may occur, what risk factors may contribute to these injuries, how to recognize and report symptoms, and how to prevent these injuries. Training should also include job-specific instruction on the proper use of essential manual materials handling equipment and techniques for the specific task or job. The training program should include the following individuals:

All exposed workers
Supervisors
Managers
Engineers and maintenance personnel
Health care providers

The term *training* has also been used to describe two distinctly different approaches to injury prevention and control: (1) instructional training in "safe" material handling and (2) fitness training (e.g., conditioning, strengthening, or work hardening). The basic premise of instructional training in safe materials handling is that people can safely handle greater loads when they perform the task correctly than if they perform the task incorrectly. Fitness training, on the other hand, is based on the premise that people can safely handle greater loads when their strength or aerobic capacity is increased.

Instructional Training. Although the use of instructional training is fundamentally sound and should be encouraged, there are some potential problems in its application. First, care must be taken to ensure that appropriate, scientifically based work practices are being taught. This is not always the case. In some training courses, for example, workers are commonly taught to lift with a straight back and bent legs (i.e., squat posture) rather than bending over to pick up the load (i.e., stoop posture). There is evidence, however, that this recommendation does not always provide the "safest" lifting style [40]. From a biomechanical perspective, for example, the horizontal distance of the load from the worker may be more important in determining spinal loading forces than the amount of forward bending. For this reason, NIOSH does not recommend a single, correct lifting style for all manual lifting tasks, but rather suggests that a free-style lift, in which workers choose whatever style they prefer, may be appropriate in most instances [42]. Second, the instructional approach relies on the worker's ability to comply with a set of recommended practices that may be forgotten or changed from time to time.

Regardless of the potential problems associated with manual materials handling training, all workers who perform MMH activities should receive basic instructional training in the recognition of hazardous tasks and should have a through knowledge of what to do when a hazardous task is identified. Furthermore, the instructional training should provide information to workers on how they can become involved in the process of preventing and controlling injuries on the job.

Fitness Training. Unlike instructional training, the basic premise of fitness training is that a worker's risk of injury will decrease if his strength or fitness increases. Although this seems like a reasonable assumption, it is not easy to determine the relationship between an individual's strength and his risk of MSI. Certainly a worker's capacity to perform heavy work might be increased, but there is controversy about the relationship between worker strength and risk of injury [43]. For example, stronger workers may not actually be able to tolerate stressful lifting demands any better than weak workers. This issue is discussed in more detail in the next section. Finally, it is not known how the soft tissues of the body respond to increased loads associated with stronger muscles.

Although both types of training programs have been used to prevent MMH injuries, the effectiveness of training in preventing or controlling injuries is unclear at the present time. Therefore, training programs should be used primarily as a supplement to workplace-directed approaches. Additional information on training programs is provided by Ayoub and Mital [29] and Eastman Kodak [19].

2. Employee Screening

Some ergonomic experts advocate the use of screening methodologies, which rely on the assessment of one or more physical characteristics of the worker, to select specific workers for certain MMH jobs. In general, screening approaches are designed to (1) identify workers with a high risk of overexertion injury or (2) screen workers according to some preselected set of strength or endurance criteria in an attempt to match the capacity of the worker to the demands of the job.

Risk Assessment Screening. Attempts to identify workers with a high risk of overexertion injury have included such activities as spinal radiographs (X-rays), psychological testing, and medical examinations, which are designed to provide an objective basis for excluding certain individuals from stressful MMH jobs. None of these methodologies, however, has been shown to be reliable in predicting an individual's risk of overexertion injury [39]. It is now widely accepted that the medical risks from radiation associated with radiography far outweigh any potential benefit derived from routine spinal X-ray screening [42]. Although no psychological tests have been found to quantify a worker's risk of overexertion injury, psychological testing may provide an indication of how a worker might respond to a severe injury. Similarly, other than identifying workers with previous low back injury, medical examinations also have failed to reliably identify workers who may have an above-average risk of overexertion injury.

Physical Capacity Screening. Another type of screening approach that has been used to select workers for MMH tasks includes individual testing of physical characteristics such as strength, aerobic capacity, or functional capability. The underlying basis for using tests such as these to screen workers for MMH jobs is the belief that the risk of injury is dependent on the relationship between the capacity of the worker and the demands of the job. When the physical demands of the job exceed the capacity of the worker, then the worker is at risk of developing a musculoskeletal injury (MSI). Thus, it has been proposed that workers should be matched to jobs according to the demands of the job.

A number of studies have been conducted to develop databases of maximum strength capacities (i.e., population averages) that could be used to rate an individual's physical capacity. These studies, however, disagree as to which of the three principal testing methods—

isometric (static), isokinetic (constant velocity), or isoinertial (constant acceleration)—is most useful for determining strength capacity guidelines.

Maximum isometric lifting strength (MLS) has been studied and reported extensively in the research literature, has a well-established testing procedure [44], and has been reported in field tests to predict risk of injury [14,45]. Extensive measures of MILS have been made for various work postures and activities. In one study, for example, the isometric strength of 1239 workers in rubber, aluminum, and electronic component industries was measured [46]. In another study, the standardized isometric strengths (i.e., arm lift, torso lift, and leg lift) of 2178 aircraft manufacturing workers were measured [43], and the employees were followed for more than 4 years to document back pain complaints. The investigators concluded that worker height, weight, age, and gender are poor predictors of standardized isometric strength (a finding that agrees with those of other studies [46]). In addition they concluded that standardized isometric strength is a poor predictor of reported back pain (a finding that does not agree with some other studies) (cf [47]). More data are needed to clarify this area of controversy. If isometric strength is a poor predictor of risk of injury, then care should be taken in using the results of isometric strength tests to place workers in stressful MMH jobs.

Some researchers argue that traditional isometric lifting strength measurements, by which thousands of workers have been tested, are limited in assessing what workers can do under dynamic task conditions. These researchers suggest that dynamic strength testing is more appropriate than static strength testing for determining strength capacity [48], because dynamic tests replicate actual job requirements better than static tests. Other researchers claim that isokinetic lifting strength measurements probably have no greater inferential power to predict risk of injury or job performance than any other form of testing [49]. On the other hand, Kroemer [50] suggested that the isoinertial strength testing method is the most appropriate for MMH because it most closely matches actual lifting conditions. Isoinertial methods, however, have not been generally validated regarding their ability to predict risk of injury [51].

In terms of measurement issues, guidelines have been proposed for assessing dynamic muscle performance. Marras et al. [52] recently published an ergonomics guide for assessing dynamic measures of low back performance. This guide provides information on elements of dynamic performance, techniques to assess dynamic performance, and relationships between testing techniques and internal forces. Little information is available, however, for other MMH tasks.

3. Personal Protective Equipment (PPE)

Personal protective equipment (PPE) is defined as devices or items used by workers to protect themselves from recognized hazards such as heat, cold, vibration, and other physical hazards. PPE includes gloves, shoes, arm guards, and protective clothing. Musculoskeletal stressors should be considered when selecting PPE, because the use of protective devices could contribute to extreme postures or excessive force. For example, the use of gloves may cause workers to unnecessarily use greater grip forces than needed to perform the job, which may increase their risk of MSI.

There is little published evidence that certain devices that have been advertised as PPE, such as braces, splints, back belts, and other similar devices, provide any realistic protection from injury for healthy workers performing manual materials handling activities. For this reason, OSHA does not consider them to be PPE [17]. In fact, there is some evidence that prolonged use of back belts by healthy workers may actually increase the risk of low back injury above what it would have been if the workers had not used the back belt [53].

Another approach that has been used to protect workers from the risk factors associated with manual materials handling include a variety of devices designed to force workers to use "good body mechanics." For example, a back inclinometer alarm, which sounds an alarm when the worker's back exceeds a certain flexion angle, was developed to prevent excessive back flexion. These devices, however, have not been shown to reduce the incidence or severity of manual materials handling injuries and cannot replace careful ergonomic job design.

D. Summary

In summary, although there is controversy about what the most appropriate approach should be for preventing MMH-related MSIs, it is clear that solutions do exist. Ideally, workplace-directed approaches should be the first choice, followed by worker-directed approaches when necessary.

VI. CRITICAL REVIEW OF CURRENT STATUS

Few would argue that certain manual materials handling activities such as excessive lifting, pushing, pulling, and carrying continue to represent a serious hazard for musculoskeletal injury for many workers. Historically, however these injuries were simply considered one of the costs of doing business. This is no longer considered an acceptable viewpoint, especially when the staggering costs in human suffering as well as the financial costs are considered. Perhaps such costs have increased to the point where injuries can no longer be tolerated and action will be taken to eliminate hazards in the workplace. We must come to the realization that these injuries can and must be prevented.

In this chapter, we have attempted to describe the manual materials handling problem and provide an overview of analytical tools useful in preventing musculoskeletal disorders. The analytical tools are needed to identify physical hazards that may result in overexertion injury and are effective in reducing the potential for risk of work-related overexertion injury.

A discussion of medical management of musculoskeletal injuries is presented elsewhere in this book (see Chapters 22 and 23). Special attention, however, must be given to the medical management of low back pain (LBP), which is the leading health disorder associated with MMH. From a medical perspective, LBP represents one of the more challenging conditions to diagnose and treat. This was confirmed by a panel of experts convened by the Agency for Health Care Policy and Research (AHCPR), who indicated that due to a lack of consensus, there is a significant variation in assessment and treatment of LBP, which results in inappropriate or at least less than optimal care for many patients with low back problems. To address this concern, the AHCPR panel has been working on a comprehensive report designed to help physicians and their patients in making decisions about appropriate care for acute low back problems. The report, which summarizes results from LBP studies, was released in 1995. The document is entitled "Clinical Practice Guidelines for Acute Low Back Problems." To obtain a copy of the report, contact the AHCPR at 1-800-358-9295.

VII. FUTURE CONCERNS

Although much is known about the assessment of the physical hazards of manual materials handling, difficult questions remain that require further research. For example,

What level of accuracy is attainable in measuring a worker's maximum safe capacity? This is an important question, because worker selection approaches, which are generally believed to be less effective in reducing MSIs than ergonomic approaches, rely on the assumption that it is possible to accurately determine a worker's safe capacity. Also, how much does a worker's capacity vary from year to year, month to month, day to day, and hour to hour?

If a worker's maximum safe capacity cannot be accurately determined, what are appropriate "margins of safety" for specific MMH jobs? What criteria should be used for determining the acceptable margin of safety? What are the implications of designing jobs for large segments of the population?

How do performance standard or incentive programs affect workers' risk of injury? If performance standards must be used, what criteria should be used to ensure that they are safe?

What is the accuracy of existing assessment tools for the design and evaluation of MMH tasks (e.g., biomechanical, psychophysical, physiological, and integrated models)? Also, what are the limitations of the assessment tools?

There are also numerous questions remaining concerning methods for preventing manual materials handling injuries. For example,

What criteria should be used to decide when and how much automation is needed in a job?

What is the effectiveness of equipment designed to reduce the stress of manual materials handling problems?

Little has been done to encourage technology transfer between industries. What can be done to encourage the sharing of solutions to similar MMH problems between industries? Databases are needed to provide access to practical, economical solutions to MMH problems for industries in a competitive environment.

The workforce is aging, and more women are employed than ever before. How will workforce changes affect our approach to prevention?

How will be the Americans with Disabilities Act affect future research efforts to develop hazard control solutions?

These questions illustrate the complexity of many of the issues that safety and health experts and industrial leaders must resolve to reduce work-related injuries. As demonstrated in the past, no effort will be successful without the full cooperation of management, organized labor, government, and the workers themselves. The solution must be a team effort with a total commitment to identification and elimination of hazardous materials handling tasks.

Due to space limitations, it has not been possible to discuss all of the important issues related to MMH. We have not, for example, discussed issues such as slip, trip, or fall hazards or certain MMH activities such as one-handed lifting. Information on these issues, however is provided in Refs. 54–57.

REFERENCES

1. Bureau of Labor Statistics, *Back Injuries Associated with Lifting,* Dept. of Labor, Bureau of Labor Statistics, Bull. 2144, 1982.
2. National Safety Council, *Accident Facts,* 1990 ed., Natl. Safety Council, Chicago, 1990.
3. C. H. Park, D. K. Wagener, D. M. Winn, and J. P. Pierce, *Health Conditions Among the Currently Employed: United States, 1988,* Dept. of Health and Human Services, Natl. Center for Health Statistics; *Vital Health Stat. 10*(186), 1993.
4. NIOSH, *The National Occupational Exposure Survey,* Publ. No. 89-103, Dept. of Health and Human Services, Natl. Inst. Occup. Safety Health, Cincinnati, OH, 1989.
5. M. H. Pope, J. W. Frymoyer, and G. Andersson, *Occupational Low Back Pain,* Praeger, New York, 1984.

6. R. C. Jensen, Disabling back injuries among nursing personnel: research needs and justification, *Res. Nursing Health 10*:29–38 (1987).
7. J. D. G. Troup and F. C. Edwards, *Manual Handling and Lifting: An Information and Literature Review with Special Reference to the Back*, Her Majesty's Stationery Office, London, 1985.
8. T. R. Waters, V. Putz-Anderson, A. Garg, and L. J. Fine, Revised NIOSH equation for the design and evaluation of manual lifting tasks, *Ergonomics 36*(7):749–776 (1993).
9. J. W. Frymoyer, M. H. Pope, J. H. Clements, D. G. Wilder, B. McPherson, T. Ashikaga, and B. Vermont, Risk factors in low back pain, *J. Bone Joint Surg. 65A*(2):213–218 (1983).
10. S. Bigos, D. M. Spengler, N. A. Martin, J. Zeh, L. Fisher and A. Nachemson, Back injuries in industry: a retrospective study, III. Employee-related factors, *Spine 11*:252–256 (1986).
11. P. M. Bongers and C. R. deWinters, *Psychosocial Factors and Musculoskeletal Disease: A Review of the Literature*, Nederlands Inst. Praventive-Gezondheidsonderzoek (NIPG-TNO), NIPG Pub. 92.029, 1992.
12. S. J. Linton, Risk factors for neck and back pain in a working population in Sweden, *Work Stress 4*:41–49 (1990).
13. S. J. Bigos, D. M. Battie, D. M. Spengler, L. D. Fisher, W. E. Fordyce, T. H. Hansson, A. L. Nachemson, and M. D. Wortley, A prospective study of work perceptions and psychosocial factors affecting the report of back pain, *Spine 16*(1):1–6 (1991).
14. D. B. Chaffin and K. S. Park, A longitudinal study of low-back pain as associated with occupational weight lifting factors, *Am. Ind. Hyg. Assoc. J. 34*:513–525 (1973).
15. J. W. Frymoyer, M. H. Pope, M. C. Costanza, J. C. Rosen, J. E. Goggin and D. G. Wilder, Epidemiologic studies of low back pain, *Spine 5*:419–423 (1980).
16. S. H. Snook, The design of manual handling tasks, *Ergonomics 21*:963–985 (1978).
17. OSHA, *Ergonomics Program Management Guidelines for Meatpacking Plants*, OSHA Document No. 3123, Occupational Safety and Health Administration, Dept. Labor, Washington, DC, 1990.
18. E. Grandjean, *Fitting the Task to the Man*, Taylor & Francis, London, 1982.
19. Eastman Kodak Company, Ergonomics Group, *Ergonomic Design for People at Work*, Vol. 2 Van Nostrand Reinhold, New York, 1986.
20. D. C. Alexander and B. M. Pulat, *Industrial Ergonomics: A Practitioner's Guide*, Ind. Eng. Manage. Press, Inst. Industrial Engineers, Norcross, GA, 1985.
21. National Occupational Health and Safety Commission, *Safe Manual Handling: Discussion Paper and Draft Code of Practice*, Australian Gov. Pub. Service, Canberra, 1986.
22. W. S. Marras and C. M. Sommerich, A three dimensional model of loads on the lumbar spine. Part I, Model structure, *Hum. Factors 33*:123–137 (1991).
23. P. Brinckmann, M. Biggemann, and D. Hilweg, Fatigue fracture of human lumbar vertebrae, *Clin. Biomech. Suppl.* 1 (1988).
24. D. B. Chaffin and G. B. J. Andersson, *Occupational Biomechanics*, Wiley, New York, 1991.
25. National Research Council, *Ergonomic Models of Anthropometry, Human Biomechanics, and Operator-Equipment Interfaces, Proceedings of a Workshop*, K. H. E. Kroemer, S. T. Snook, S. K. Meadows, and S. Deutsch, Eds., 1986. Available from the National Research Council, Washington, DC.
26. A. Garg, *The Biomechanical Basis for Manual Lifting Guidelines,* Natl. Tech. Inf. Service, Rep. No. 91-222-711, 1991.
27. S. S. Stevens, The psychophysics of sensory function, *Am. Sci. 48*: 226–253 (1960).
28. S. H. Snook and V. M. Ciriello, The design of manual handling tasks: revised tables of maximum acceptable weights and forces, *Ergonomics 34*:1197–1213 (1991).
29. M. M. Ayoub and A. Mital, *Manual Materials Handling,* Taylor & Francis, London, 1989.
30. G. Borg, Psychophysical scaling with applications in physical work and the perception of exertion, *Scand. J. Work Environ. Health 16*(Suppl. 1):55–58 (1990).
31. E. N. Corlett and R. P. Bishop, A technique for assessing postural discomfort, *Ergonomics 19*:175–182 (1976).
32. S. H. Rodgers, J. W. Yates, and A. Garg, *The Physiological Basis for Manual Lifting Guidelines,* Natl. Tech. Inf. Service, Rep. No. 91-227-330, 1991.
33. A. Garg, A metabolic rate prediction model for manual materials handling jobs, Ph.D. Dissertation, Univ. Michigan, 1976.

34. P. O. Astrand and K. Rodahl, *Textbook of Work Physiology*, 3rd ed., McGraw-Hill, New York, 1986.
35. M. M. Ayoub, N. J. Bethea, S. Deivanayagam, S. S. Asfour, G. M. Bakken and D. Liles, *Determination and Modeling of Lifting Capacity, Final Report,* Dept. Health and Human Services, Natl. Inst. Occup. Safety Health, Grant No, 5-RO1-OH-00545002, 1978.
36. W. S. Marras and F. Fattalah, Accuracy of a three dimensional lumbar motion monitor for recording dynamic trunk motion characteristics, *Int. J. Ind. Ergon.* 9(1):75–87 (1992).
37. W. S. Marras and R. W. Schoenmarklin, Wrist motions in industry, *Ergonomics* 36(4):341–351 (1993).
38. K. A. Grant, T. L. Galinsky, and P. W. Johnson, Use of the actigraph for objective quantification of hand/wrist activity in repetitive work, *Proc. 37th Annual Human Factors Meeting,* Seattle, WA, 1993.
39. S. H. Snook, R. A. Capanelli, and J. W. Hart, A study of three preventive approaches to low back injury, *J. Occup. Med.* 20(7):478–481 (1978).
40. A. Garg, What basis exists for training workers in "correct" lifting technique?, in *The Ergonomics of Manual Work*, W. S. Marras, Karwowski, Smith, and Pacholski, Eds., Taylor & Francis, London, 1993.
41. T. M. Bernard, *Material Handling Information Sources,* Unpublished NIOSH Report, Div. Phys. Sci. Eng., Cincinnati, OH, 1991.
42. NIOSH, *Work Practices Guide for Manual Lifting*, NIOSH Tech. Rep. No. 81-122, U. S. Dept. Health Hum. Services, Natl. Inst. Occup. Safety Health, Cincinnati, OH, 1981.
43. M. C. Batti'e, S. J. Bigos, L. D. Fisher, T. H. Hansson, M. E. Jones, and M. D. Wortley, Isometric lifting strength as a predictor of industrial back pain reports, *Spine* 14(8):851–856 (1989).
44. D. B. Chaffin, Ergonomics guide for the assessment of human static strength, *Am. Ind. Hyg. Assoc. J.* 36:505–511 (1975).
45. G. D. Herein, M. Garret, and C. K. Anderson, Prediction of overexertion injuries using biomechanical and psychophysical modes, *Am. Ind. Hyg. Assoc. J.* 47(6):322–330 (1986).
46. W. M. Keyserling, G. D. Herein, and D. B. Chaffin, An analysis of selected work muscle strength, *Proc. 22nd Annual Meeting Hum. Factors Soc.,* Detroit, MI, 1978.
47. W. M. Keyserling, G. D. Herein, and D. B. Chaffin, Isometric strength testing as a means of controlling medical incidents on strenous jobs, *J. Occup. Med.* 22(5):332–336 (1980).
48. K. H. E. Kroemer, Testing individual capability to lift material: repeatability of a dynamic test compared with static testing, *J. Safety Res.* 3(6):4–7 (1985).
49. J. M. Rothstein, R. L. Lamb, and T. P. Mayhew, Clinical uses of isokinetic measurements, *Phys. Ther.* 67(12):1840–1844 (1987).
50. K. H. E. Kroemer, An isoinertial technique to assess individual lifting capacity, *Hum. Factors* 25(5):493–506 (1983).
51. K. H. E. Kroemer, Matching individuals to the job can reduce manual labor injuries, *Occup. Safety Health News Dig.* 3(6):4–7 (1987).
52. W. S. Marras, J. D. McGlothlin, D. R. McIntyre, M. Nordin, and K. H. E. Kroemer, *Dynamic Measures of Low Back Performance: An Ergonomics Guide,* Am. Ind. Hyg. Assoc., Fairfax, VA, 1993.
53. C. R. Redell, J. J. Congleton, R. D. Huchingson, and J. F. Montgomery, An evaluation of a weightlifting belt and back injury prevention training class for airline baggage handlers, *Appl. Ergon.* 23(5):319–329 (1992).
54. M. M. Ayoub and A. Mital *Manual Materials Handling*, Taylor & Francis, London, 1989.
55. D. B. Chaffin and G. B. J. Andersson, *Occupational Biomechanics*, Wiley, New York, 1991.
56. Eastman Kodak Company, Ergonomics Group, *Ergonomic Design for People at Work,* Vol. 2 (Van Nostrand Reinhold, New York, 1986.
57. M. H. Pope, G. B. J. Andersson, J. W. Frymoyer, and D. B. Chaffin, Eds., *Occupational Low Back Pain: Assessment, Treatment, and Prevention*, Mosby Year Book, St. Louis, MO, 1991.

15
Manual Materials Assist Devices

Jeffrey C. Woldstad* and Roderick J. Reasor†
Virginia Polytechnic and State University, Blacksburg, Virginia

I. INTRODUCTION

Many manual materials handling tasks pose a significant musculoskeletal risk to workers as they are currently performed. The analytical methods discussed in earlier chapters of this book provide the means to identify these potentially harmful tasks and to prioritize them for intervention efforts. Unfortunately, ergonomic efforts to reduce the risk associated with manual materials handling tasks usually require fundamental changes to the work area and the component work tasks. These changes are often expensive, also, if not done carefully, they can create new engineering and ergonomic problems.

This chapter addresses one of the most common and least expensive solutions proposed to address harmful manual materials handling problems, that being the use of a manual materials assist device. For the purposes of this chapter, manual materials assist devices are defined as devices that can be used to assist the worker in performing some component of a manual materials handling task. This includes work components related to lifting, lowering, carrying, and holding a load. We restrict our discussion to devices intended to assist workers in performing these work activities as opposed to automation intended to replace human workers.

The intent of this chapter is to review both the ergonomic and manufacturing issues associated with the selection of a manual materials assist device. To accomplish this, we first review some of the many different types of devices that can be classified as manual materials assist devices using the above definition. Then we discuss the ergonomic issues that should be considered when selecting and installing a manual materials assist device.

II. BACKGROUND AND SIGNIFICANCE TO OCCUPATIONAL ERGONOMICS

The use of manual materials assist devices is recommended by NIOSH [1] and others [2–5] as an engineering solution to potentially harmful manual materials handling tasks. Practitioners new to occupational ergonomics often mistakenly assume that installing such a device immediately solves the ergonomics and biomechanics problems associated with a work task. Although the installation of these devices is often successful, just as often devices are disliked by workers, underutilized, and eventually discarded in favor of the old manual methods. In

---

Present affiliations
*Texas Tech University, Lubbock, Texas
†Eastman Chemical Company, Kingsport, Tennessee

our experience, it is rare to go into a large U.S. industrial facility and not see one or more manual materials assist devices lying unused in a corner of the plant.

To ensure that manual materials assist devices are used to their fullest potential, it is necessary for the ergonomist to go beyond just suggesting that such a device be employed. We must become actively involved in the design, selection, and installation of these systems. This chapter is an initial attempt at providing some basis for this process.

III. TYPES OF MANUAL MATERIALS ASSIST DEVICES

A variety of equipment is available to assist operators with the manual movement of materials in the workplace. This equipment is made by many companies in the United States and throughout the world. Although our discussion will not be able to adequately reflect this diversity, three general categories of equipment will be briefly reviewed. These are industrial manipulators, lift tables, and cart and trolley systems.

A. Industrial Manipulators

Industrial manipulators are often employed to assist operators in lifting, supporting, and moving heavy, awkward, or fragile loads [6]. The manipulator grasps the load and then uses pneumatic, hydraulic, or vacuum force to overcome the effects of gravity. An operator is then able to move the load to a desired location while maintaining control of the load throughout its movement. Specialized manipulators can permit the operator to roll the load over, tilt the load, and pour or dump the load.

There are a large variety of industrial manipulators available on the market (see Figures 1–4 for some examples). These devices are typically capable of handling loads up to 1700 lb within a work area of 8–20 ft. Of the many different designs, three basic types of manipulators are often identified: triaxial manipulators, balancing hoists, and articulated jib cranes [7]. Each of these is briefly discussed below.

1. Triaxial Manipulators

A triaxial manipulator is used when objects vary in size, configuration, or weight and when the pick-up and put-down locations vary. A triaxial manipulator has structural members arranged in a parallelogram that permit loads to be moved vertically, horizontally, or diagonally. They can be fixed overhead as shown in Figure 1 or mounted to an overhead trolley system. Besides supporting the load, often these devices are effective in extending an operator's reach capability and permit more precise load positioning.

There are several factors that must be considered when selecting a triaxial manipulator for a given application [8]. These include:

Headroom requirements. Typically 12 ft of headroom is needed unless a low-profile manipulator is used.
Reach and lifting capacity. Lifting capacity will vary with the design of the device and should be closely tailored to the application. Reach is a function of lifting capacity and is defined as the total horizontal travel.
Reach-in capability. The manipulator may need the ability to be extended beyond the normal work envelope for special applications.
Side-loading capability. The ability to insert loads from the side can be limited; if required, the capability of a given unit must be determined.

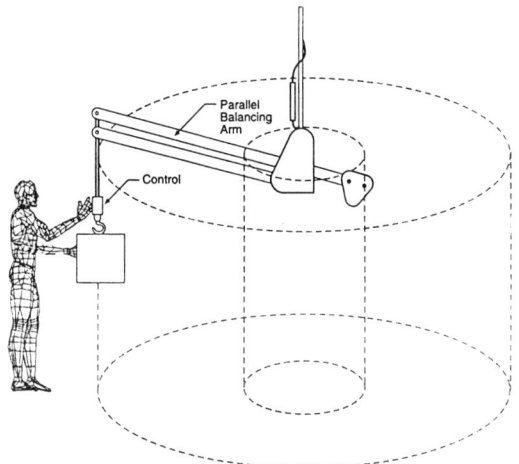

Figure 1 Triaxial manipulator. (From Ref. 8)

2. Balancing Hoists

Balancing hoists float a load (overcome the effects of gravity) much like a triaxial manipulator, but they have no reach-in capability. Their primary function is load positioning by an operator. They are typically mounted on an overhead trolley running on a jib or bridge crane and can handle loads of up to 1400 lb. Figure 2 illustrates a common trolley-suspended hand hoist. Figure 3 shows a simple tool and fixture suspension (where the tool remains on the hook all the time).

3. Articulated Jib Cranes

A jib crane has a rotating boom attached to a vertical column. The vertical column can be mounted from an overhead trolley, on a floor pedestal, or from a wall bracket. Jib cranes typically use hoists for lifting and lowering a load. An articulated jib crane has an additional pivoting arm that provides additional reach capabilities as shown in Figure 4. Articulating jib cranes have a working area of up to 40 ft in diameter. They are especially useful when overhead space is limited, when loads are heavy or awkward, or when smooth, lateral movement along a conveyor is essential [8].

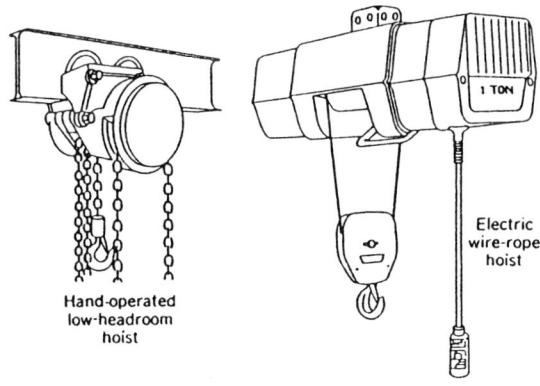

Figure 2 Trolley-suspended hoists. (From Ref. 25.)

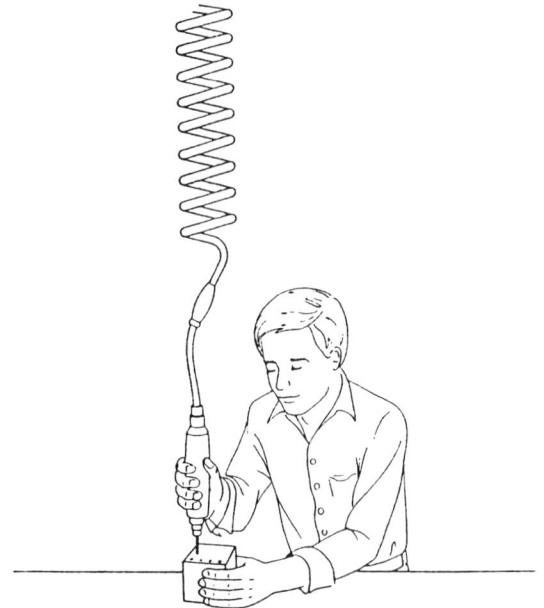

Figure 3 Balancing hoist—tool and fixture suspension. (From Ref. 25.)

4. End-of-Arm Tooling

Industrial manipulators are used to lift a wide variety of loads. These include sacks of flour, TV tubes, computers, carpeting, heavy steel coils, furniture, and more. Thus, the end-of-arm tooling is almost always custom designed for the application. Common end-of-arm tooling includes vacuum cups, clamps, magnets, C-hooks, scissor grabs, core grips, and grabs. Figure 5 illustrates the variety of end-of-arm tooling commonly used.

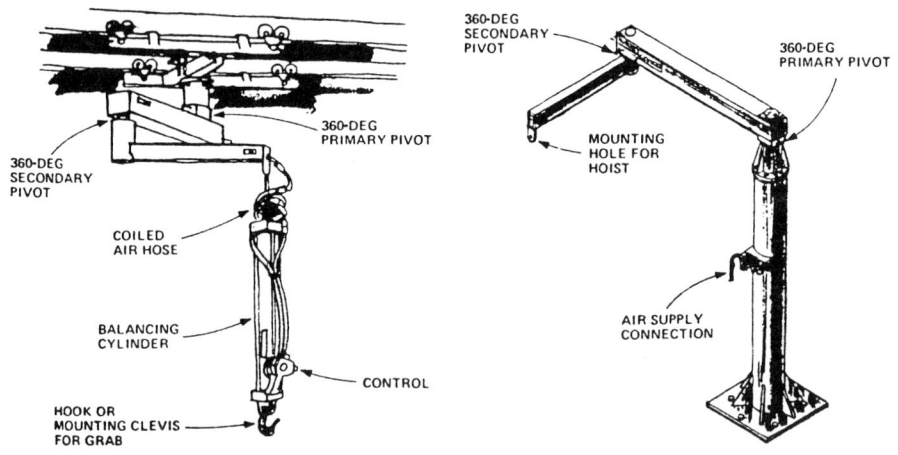

Figure 4 Articulating jib cranes. (From Ref. 8.)

Manual Materials Assist Devices

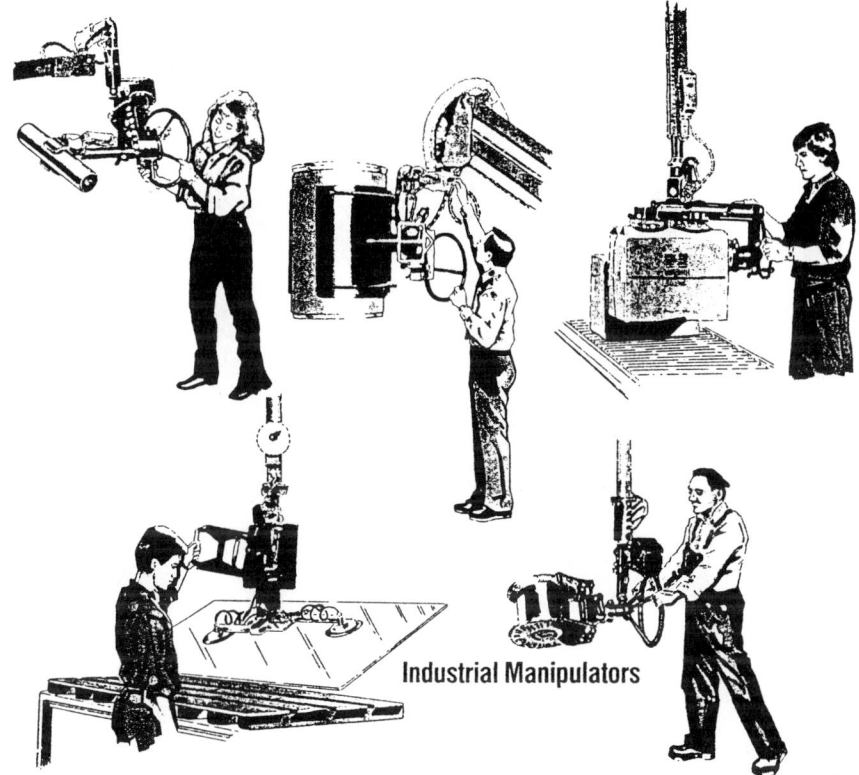

Figure 5 End-of-arm tooling for industrial manipulators. (From Ref. 6.)

B. Lift Tables

Lift tables provide a versatile means to lift, lower, or hold material. They are often an effective means of reducing the horizontal and vertical distances associated with lifting tasks (see Chapter 4, Sec. II and Chapter 3, Sec. VI). Lift tables use a mechanical, hydraulic, or pneumatic mechanism to move a platform up or down (see Figure 6). The lifting mechanism most commonly operates in a scissor-like manner, but it may act directly on the platform. Lift tables are available in a wide variety of sizes (table tops from 3 to 240 sq ft) and capacities (200–100,000 lb). Table tops can be simple flat surfaces, or they can be designed to tilt, rotate, or contain some special-purpose superstructure that supports the handling of tools, coils, and special equipment (such as wheels, rollers, or ball transfers).

Lift tables have been used in a variety of applications. These fall into five major areas: stacking and unstacking, lift and tilt, work positioning, upenders and downenders, and dock lifts. Each of these application areas is briefly discussed below.

Stacking and unstacking.   Product is brought to the work height and then stacked or unstacked by hand. Stacking is easier if the top surface of the table or stacked pile is slightly below the work height. Conversely, unstacking is easier if the stack height is slightly above the work height.

Lift and tilt.   Primarily used for dumping or feeding small, individual materials in bulk or mass.

Work positioning.   Work is positioned at the ideal height for a given operator.

Figure 6 Lift table. (From Ref. 26.)

Upenders and downenders. Change the vertical or horizontal axis of a product or process.
Dock lifts. Equalize the height difference between a loading/unloading dock and a shipping/receiving container (e.g., truck or railcar).

C. Industrial Hand Trucks

Industrial hand trucks are widely used to assist workers in transporting or carrying the load from one location to another. They represent a versatile method of performing intermittent movements over a variable path. Figure 7 shows examples of several types of industrial trucks.

Hand trucks are also known as floor trucks. Hand trucks are used in warehousing, manufacturing, shipping, and distribution. They are most commonly used when the travel distance is relatively short, the volume and frequency of moves are low, physical limitations (narrow aisles, doors, etc.) prohibit the use of other types of equipment, or low initial cost

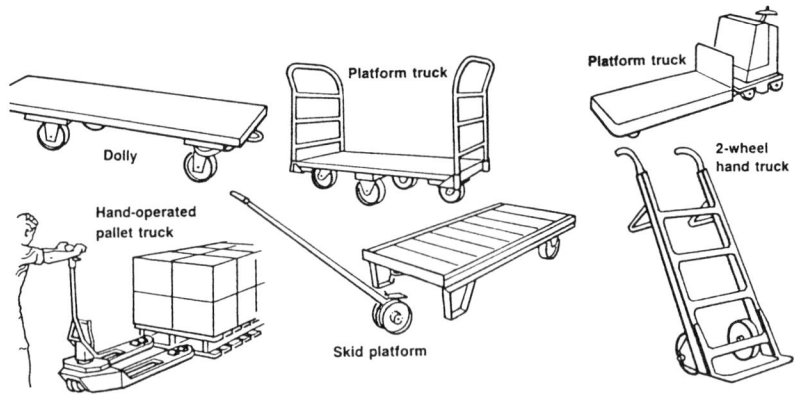

Figure 7 Industrial hand trucks. (Adapted from Ref. 27.)

and maintenance requirements are constraining considerations. Hand trucks can be classified as two-wheel hand trucks, multiple-wheel floor trucks, and hand lift trucks.

A two-wheel hand truck is the most common and most widely used type of hand truck. It basically operates as a lever with wheels and is used to move materials that are too heavy or too awkward to be carried. Typical capacity is up to 1000 lb. Examples include barrel trucks, bag trucks, beverage trucks, appliance trucks, and a wide variety of special-purpose trucks (cylinder trucks, cotton trucks, grain trucks, etc.).

Multiple-wheel floor trucks typically consist of a rectangular, load-carrying platform with four or six wheels. They are used to transport up to 4000 lb. These platforms may have attached superstructures ranging from four corner stakes to completely enclosed containers. The platforms can be interconnected and used in towline operations. Examples include dollies, platform trucks, semi-live skid platforms, tilt- and nontilt type trucks, and wagon-type trucks.

Hand lift trucks contain a wheeled load carrier and a lifting device. The load carrier (a platform or forks) can be rolled under a unit load (pallet or skid). The lifting device can be hydraulically or mechanically operated and is typically used to raise the unit load only 2–3 in. above the floor for transport. Hand lift trucks are commonly used to handle 2500–5000 lb.

IV. ERGONOMICS ISSUES IN DESIGN, SELECTION, AND INSTALLATION

This section describes some of the ergonomic issues relevant to the design, selection, and installation of manual materials assist devices. The discussion is directed primarily at different types of industrial manipulators and hand trucks. Many of the issues relevant to the installation of lift tables have been covered in previous chapters of this text (see Chapter 2).

A. Horizontal Pushing and Pulling

Most manual materials assist devices function to support the load vertically during lift, carry, position, and lower tasks. By supporting the load in this manner, they eliminate the static force required to counteract the effects of gravity on the load. However, to manipulate these devices, operators are usually required to supply the horizontal push and pull forces needed to move the load from one position to another. Unfortunately, the magnitude of these push and pull forces can become sizable depending upon the configuration of the assist device and the work performed. Peak push and pull forces ranging up to 500 N and 80% of the subject's maximum force capability have been recorded in experiments with manual materials assist devices [9,10].

Similar to other whole body strength exertions, the ability of operators to generate horizontal push and pull forces is highly dependent on factors such as individual strength capability and body posture. In attempting to establish limits for push and pull activities, several methods have been used, including strength evaluations [11, 12], intra-abdominal pressure [13–15], biomechanical models [16,17], and psychophysical methods [18,19]. For those engineers interested in establishing push and pull force limits, the article by Snook and Ciriello [19] provides a relatively complete table of recommended maximum push and pull force limits based on a number of psychophysical evaluations.

B. Floor Surface Coefficient of Friction

The definition of coefficient of friction (COF) and some requirements for various tasks have been addressed earlier in this text (see Chapter 5). For pushing and pulling activities, it is clear that the COF should exceed the ratio of the horizontal push or pull force and the operator's body weight [20]. As an example, if we have an operator who weighs 170 lb and the operator is required to produce 100 lb of horizontal push force, then a minimum COF of 100/170 = 0.58 would be reasonable.

It might seem that this value would be an absolute minimum requirement if 100 lb of force is to be produced. However, high push and pull forces can be generated with relatively low COF values if the handle used for pushing and pulling is high enough [21]. With high handle heights, operators are able to generate a force vector with both an upward component and a horizontal component. The vertical force component will serve to increase the force normal to the floor surface and, as a result, the frictional force. Using this method, operators are able to generate much higher initial push or pull forces than anticipated; unfortunately, the same procedure does not work very well in stopping the device.

In an investigation of the required COF for push and pull tasks, Redfern and Andres [22] found that increased cart resistance and faster pushing speeds produced a statistically significant increase in the required COF. However, they were unable to find any statistically significant effect of handle height or exertion direction (push or pull). In addition, they note that the required COF was extremely variable between the different subjects tested. Clearly, more experimental evaluations are required in this area before concrete COF guidelines can be established for pushing and pulling tasks.

In summary, push and pull tasks are clearly tasks that demand a high COF between the shoe and the floor surface. When installing manual materials assist devices, ergonomists should evaluate the slipperiness of the floor surface and consider the installation of one of the many high-traction flooring surfaces available on the market. Such precautions are especially important if high push and pull forces will be required, if pushing or pulling will be done along an inclined surface, or if floor contaminants are present in the work area (see Chapter 5).

C. System Mass and Inertia

Several recent studies have shown that the force generated in using a manual materials assist device and the operator's perception of task difficulty are substantially increased by the inertial mass of the load being moved [9,10]. This is a particular concern with manual materials assist devices because of the large loads that are sometimes transported with the systems and the mass of the devices themselves.

The inertial mass of a device can be thought of as the characteristic of the device that resists motion or acceleration when a force is applied. Technically, mass is the quantitative measure of inertia; however, throughout this chapter we use the terms mass, inertia, and inertial mass interchangeably. The larger the mass of the device, the more force that will be required to produce a constant acceleration. This relationship is quantified in Newton's second law of motion:

$$F = m \times a \qquad (1)$$

where F is the applied force, m is the inertial mass of the system, and a is acceleration [23]. In practical terms, this means that to keep push and pull forces within reasonable bounds, operators must accelerate and decelerate an assist device with a large inertial mass much more slowly than a similar device with a small inertial mass. In many cases, operators simply refuse to control their motions to the extent required, and high acceleration and deceleration forces result. In still other cases, time standards make a reduction in movement time impossible.

In an investigation of these issues, Woldstad and Chaffin [10] recently conducted an experiment that measured the peak hand forces associated with the movement of a small overhead balancing hoist. Four inertial loads were considered, ranging from 81.7 to 217.9 kg. The experimental task had seven inexperienced operators push the load from a standstill and position it over a specified target. Peak acceleration and deceleration forces for this experiment are shown for the 2.74-m condition in Figure 8. Note that the forces for both acceleration and deceleration were much higher for the larger inertial loads. Movement times, shown in Figure 9, indicated that subjects did slow down for the higher inertia trials but not to the extent required to maintain moderate acceleration and deceleration forces. In fact, a careful review of the average forces demonstrated in Figure 8 will show that at the highest inertial loads the deceleration forces approached maximum recommended levels for push and pull activities [19].

A similar set of experiments were recently conducted by Resnick [9]. The first two of these experiments used a triaxial manipulator assist device and measured peak hand force during acceleration and deceleration for symmetric and asymmetric positioning tasks. In addition to hand force, the experiment measured the subjects' perceived level of exertion using the Borg CR-10 scale. The results of manipulating the inertial mass of the system were mixed for these experiments. For sagittally symmetric manipulations, increasing the load manipulated (loads considered were 0, 22.5, 45, and 67.5 kg) significantly increased both the peak push force and the peak pull force but did not affect the CR-10 rating. For asymmetric manipulations, changes to the load had no significant effect on any of the three measures. The third experiment reported by Resnick considered three different types of assist devices: a triaxial manipulator, a hoist with a fixed pivot, and a hoist on an overhead rail. A 10-kg and a 30-kg load were used in this experiment. The results showed significant increases in both push and pull forces for the heavier load. This was especially true for the hoist with the fixed pivot, where average push forces increased by approximately 220% and average pull forces increased by approximately 200%.

Although the results reported above are from only a small set of investigations, they coincide well with the experience of many operators and engineers. Operators often complain of fatigue when moving systems with large inertial masses, especially if the load must be moved rapidly or over a large distance. Similarly, performance times are often much longer

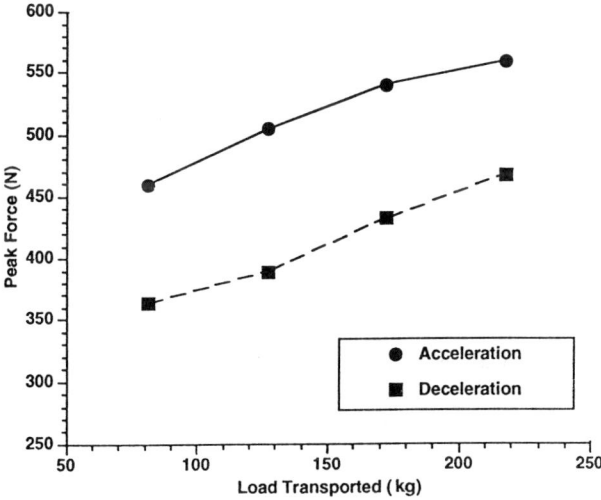

Figure 8 Peak acceleration and deceleration forces (in newtons) for hoist movements using different loads. (See Ref. 10.)

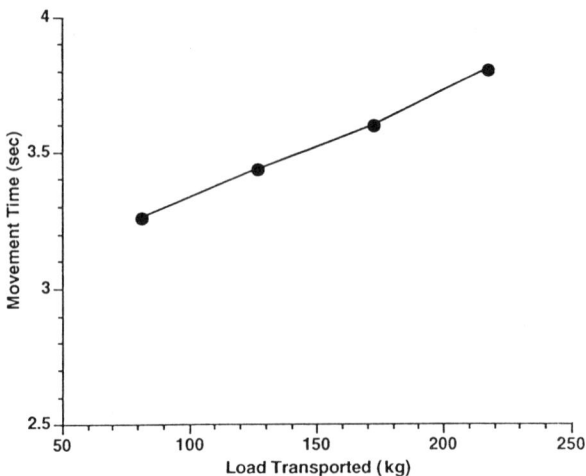

Figure 9 Movement times (in seconds) for hoist movements using various loads. (See Ref. 10.)

for large mass systems. To avoid these problems, every attempt should be made to reduce the inertial mass of manual materials assist devices. This includes avoiding devices that use counterweights to compensate for the load and selecting a device that is specifically designed for the size of load to be handled. In addition, both operators and industrial engineers should recognize that large inertial loads have the potential for creating large acceleration and deceleration forces. Operators need to be trained to accelerate and decelerate large loads slowly and smoothly, even though this may require longer movement and cycle times. At the same time, industrial engineers may need to consider the increase in time required to safely manipulate large loads when rating performance and developing time standards.

D. System Friction or Resistance to Movement

Frictional forces result when objects come into contact with each other. They are the forces that resist a sliding of one object over another. In manual materials assist devices, there are frictional forces that resist motion at any point where two components come into contact. This includes joints and axles where rotations occur, and also areas where wheels in the device come into contact with an internal or external surface. Frictional forces are often separated into a static frictional force component and a dynamic frictional force component. For manual materials assist devices, the static frictional force will resist any attempt to put the assist device in motion from a stopped or resting position, and the dynamic frictional force will resist the motion of the device as it is moved. For a number of reasons, the static frictional component is usually much larger than the dynamic component (see Chapter 5).

To evaluate the effects of frictional forces on manual materials assist devices, the movement of these devices must be considered in separate phases or components. Consider that the typical manual movement of an assist device consists of an acceleration phase during which the operator accelerates the device to walking speed, a transport phase during which the operator provides only the amount of force necessary to maintain a walking velocity, a deceleration phase during which the operator provides the required force to stop the assist device at the intended position or target, and a final adjustment phase during which the operator provides any final positioning of the load. As we argued in the previous section, increasing the inertial mass of the assist device will increase the force used for all of these different movement components, except perhaps the middle transport component. The effect of frictional forces, however, will be dramatically different for different phases of the movement.

For the acceleration phase of the movement, the static frictional force will resist initial movement of the device, and the dynamic frictional force will resist the attempt by the operator to accelerate the load. If the operator has a desired acceleration profile, then frictional forces will increase the force used during acceleration. For the transport phase of the movement, the dynamic frictional forces will determine the force that must be used by the operator. Because we usually assume that the operator is not accelerating or decelerating the load during the phase, the only force that is required to maintain the desired velocity is the force needed to oppose the frictional forces.

Although the dynamic frictional forces will increase the effort required of the operator to accelerate and transport the manual materials assist device, they will decrease the force required to decelerate the system. For the deceleration phase of the movement, the frictional forces act in the same direction as the forces the operator must supply. The higher the frictional force, the lower the deceleration force. The final positioning phase of the movement is usually composed of several small accelerations and decelerations of the load. Similar to the accelerations and decelerations described, increased resistance or frictional force will increase the force associated with the accelerations and decrease the force associated with deceleration. As a general rule, increased frictional forces will increase the effort associated with the final positioning component. This is of particular concern if the work task requires precise positioning of the load.

Woldstad and Chaffin [10] and Resnick [9] (as described previously in Section IV.C) have tested several of these assumptions in the laboratory by measuring the force profiles created for subjects using different manual materials assist devices while manipulating the frictional forces associated with the system. In the first experiment [10], a small overhead balancing hoist with a rigid connection between the overhead trolley and the load was used. The results of this experiment showed that increasing the system friction (in this case the friction of the rail system) from 24.2 to 48.4 N had no significant effect on the force used for accelerating the load but did significantly reduce the force used to decelerate the load. The time taken for these movements under the higher friction condition was not affected for short distance movements (0.9 m) but was significantly longer for longer distance movements (2.7 m).

Resnick [9] used a triaxial manipulator assist device and controlled system friction by using disk brakes at the joints of the articulated arm. The frictional resistance levels used were 0 and 25N. For sagittally symmetric movements, the results showed significantly higher forces for both acceleration and deceleration and significantly higher subjective exertion ratings (using the Borg CR-10) with the high-friction condition. For sagittally asymmetric movements, the results showed significantly higher acceleration forces and subjective exertion ratings (using the Borg CR-10) but no significant difference in deceleration force. A curious result of these experiments was that, in contrast to the results reported by Woldstad and Chaffin [10], the increase in frictional forces increased or did not affect the deceleration. However, in evaluating the force profiles created by subjects in this experiment, Resnick did not separate the deceleration component of the movement from the final positioning component. Further, he notes that peak deceleration forces occurred during push components of the positioning task as opposed to during the decelerations.

In summary, it is clear that larger frictional forces within manual materials assist devices will increase the push and pull forces generated by workers for the greater part of the work cycle. The exception to this will be during deceleration of the load. Although frictional forces in general should be minimized to decrease the hand forces required, there may be situations where the deceleration of the load is of such concern that system friction is desirable, as when extremely large inertial loads are being transported. In these situations, operators can accelerate the load to a level where they have difficulty controlling and stopping the system within the work area. Increased frictional forces in these situations will decrease the operator's ability to accelerate the system and increase their ability to decelerate it. However, in this same situation, the addition of some type of brake to the device would likely achieve the same level of control while not increasing the effort associated with acceleration.

Although high levels of system friction are clearly not desirable for a majority of applications, the effects of low levels of friction are not yet clear. Evidence from investigations using purely inertial loads suggests that the addition of a small amount of system friction may improve operators' ability to control manual materials assist devices [24]. It is hoped that future investigations will provide us with a better understanding of these issues.

E. Physical Layout and Linkage

The large variety of manual materials assist device designs presents a variety of problems in the physical placement of these devices within the workplace. For many devices, the system resistance and inertial mass are often different for different directions of movement. As an example, consider the simple overhead cable hoist depicted in Figure 10. For this device, large movements parallel to the dual overhead rails require that the entire cross-bridge assembly

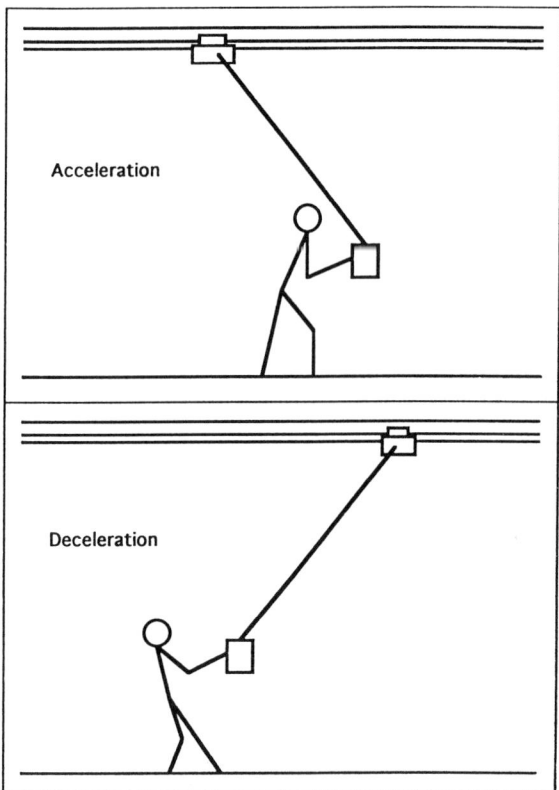

Figure 10 Position of the overhead carriage of a balancing hoist during acceleration and deceleration.

be moved. Because the load is connected to the cross-bridge assembly via a chain or cable as opposed to a rigid connection, the cross-bridge assembly will lag behind the operator as shown in the figure. When the operator stops, the overhead assembly will pass over the top of the operator and produce a delayed jerk as it reaches the end of its travel envelope (see Figure 10). This jerk can be substantial if the mass of the cross-bridge assembly is large. For this same device, the dynamics of the movement are much different if the movement is perpendicular to the dual rails. In this situation, only the lift mechanism must be moved and not the entire cross-bridge assembly.

Similar problems exist with many triaxial manipulators. Typical manipulations of these devices often combine rotational and linear translations of components within the device. Device components often have different inertial mass and frictional resistance characteristics in different movement directions, resulting in a dynamic response of the device that is highly dependent on the movement trajectory. In addition, many linkage systems prevent straight-line paths between different points within the work envelope. To help alleviate many of these trajectory problems, some designers have increased the degrees of freedom or number of moving joints in the system. Although this provides more flexibility in the number of possible trajectories of the device, it often creates a problem similar to that described above for overhead hoist systems, where components of the device continue to move after the load has stopped moving. In addition, operators are often not able to predict the response of the device for different force inputs when it has many degrees of freedom thus creating a safety problem.

There have been no quantitative investigations of the linkage and arrangement problems described above. In many cases, such research will be difficult due to the variety of devices on the market. However, on the basis of a number of years of working with these devices, the following practical recommendations are proposed:

1. Use the device with the smallest number of moving components that will still do the job. Every attempt should be made to use the simplest possible design if human operators are to control the device.
2. Attempt to provide as rigid a link as possible between major mass components of the device and the operator's hands. Large moving mass components controlled through multiple-link arm assemblies, cables, or chains should be avoided.
3. When installing a device, identify the trajectories within the work area that will be associated with the work tasks that will use the device. Orient the device in a manner that allows linear movement and minimizes the inertial mass along these trajectories.
4. Carefully consider work height and control design for these devices. More detail on work height considerations can be found in Chapter 2, and more on control design in Chapter 3.
5. Do not be constrained by previous work methods when redesigning a work area to incorporate a manual materials assist device. These devices present new problems, but they also present new opportunities. In most cases, assist devices are able to manipulate much larger loads than human workers. Designing work activities to take advantage of these differences often can result in substantial improvements in productivity.

V. CRITICAL REVIEW OF CURRENT STATUS

This chapter has attempted to provide a broad overview of some of the issues that should be considered in the design, installation, and use of manual materials assist devices. Experimental research in this area has only just begun. In reviewing the literature, we were able to locate only two sets of experimental investigations that involved the actual human manipulation of manual materials assist devices, and both of these were conducted in the same laboratory. Clearly, more evaluations are necessary before many of the issues discussed in this chapter can be resolved. However, while this work is proceeding, our present lack of information should not inhibit ergonomists from taking an active role in the engineering of these devices. The use of manual materials assist devices has increased dramatically in the last 10 years, and this trend is likely to continue into the foreseeable future. Attention to the effects of these devices on their human operators is critical to successfully integrating this technology into the workplaces of the future.

VI. FUTURE CONCERNS

A great many areas need to be investigated with respect to manual materials assist devices. We believe the following to be the most critical:

1. A better understanding of the biomechanics of pushing activities is needed. We have learned a great deal about lifting tasks in the last decade, but very little research has been directed at understanding pushing and pulling activities. Most manual materials assist devices replace lifting work activities with pushing and pulling activities.
2. More information is needed on how people physically control complex machinery. We have only a very limited understanding of how people use their musculoskeletal system to control and direct complicated mechanical systems. Manual materials assist devices are currently being developed with redundant degrees of freedom, braking systems, and a wide variety of levels of automation. At present, we are unable to predict whether such changes will improve or degrade operator performance.
3. Finally, the practical experiences of ergonomists and engineers working with manual materials assist devices need to be more effectively communicated. Because of the lack of established theory in this area, anecdotal and case study reports are extremely valuable. Unfortunately, few if any reports of this type are currently available.

REFERENCES

1. National Institute of Occupational Safety and Health, *A Work Practice Guide for Manual Lifting*, Tech. Rep. No. 81-122, U.S. Dept. of Health and Human Services, Washington, DC, 1981.
2. D. B. Chaffin and G. Andersson, *Occupational Biomechanics*, 2nd ed., Wiley, New York, 1991.
3. Eastman Kodak Company, *Ergonomic Design for People at Work*, Vols. 1 and 2, Van Nostrand Reinhold, New York, 1986.
4. S. A. Konz, *Work Design: Industrial Ergonomics*, 3rd ed., Publishing Horizons, Worthington, OH, 1990.
5. K. Kroemer, H. Kroemer, and K. Kroemer-Elbert, *Ergonomics: How to Design for Ease and Efficiency*, Prentice-Hall, Englewood Cliffs, NJ, 1994.
6. D. B. Torok, Industrial manipulators lend a hand to your material handling applications, *Mater. Handl. Eng.* 48(1): (1991).
7. G. Schwind, Manipulators: power and reach to the worker, *Mater. Handl. Eng.* 48(7): (1993).
8. G. A. Castleberry, Industrial manipulators, in *Material Handling Handbook,* 2nd ed., R. A. Kulwiec, Ed., Wiley, New York, 1985.
9. M. L. Resnick, *Biomechanics, kinematics, psychophysics and motor control in the application of material handling devices (MHDs),* Ph.D. Dissertation, Univ. Michigan, Ann Arbor, MI, 1993.

10. J. C. Woldstad and D. B. Chaffin, Dynamic push and pull forces while using a manual materials handling assist device, *IIE Trans.*, in press.
11. M. M. Ayoub and J. W. McDaniel, Effects of operator stance on pushing and pulling tasks, *AIIE Trans.* 6(3): 185–195 (1974).
12. K. H. E. Kroemer, Horizontal push and pull forces: applied when standing in working positions on various surfaces, *Appl. Ergon.* 5(2): 94–102 (1974).
13. P. R. Davis and D. A. Stubbs, Safe levels of manual forces for young males, 1, *Appl. Ergon.* 8(3): 141–150 (1977).
14. P. R. Davis and D. A. Stubbs, Safe levels of manual forces for young males, 2, *Appl. Ergon.* 8(3): 219–228 (1977).
15. P. R. Davis and D. A. Stubbs, Safe levels of manual forces for young males, 3, *Appl. Ergon.* 8(4): 33–37 (1978).
16. D. B. Chaffin, G. D. Herring, W. M. Keyserling, and A. Garg, A method for evaluating the biomechanical stresses resulting from manual materials handling jobs, *AIHA J. 38*: 662–675 (1977).
17. J. B. Martin and D. B. Chaffin, Biomechanical computerized simulation of human strength in sagittal-plane activities, *AIIE Trans.* 4(1): 19–28 (1972).
18. S. H. Snook, The design of manual handling tasks, *Ergonomics 21*: 963–985 (1978).
19. S. H. Snook and V. M. Ciriello, The design of manual handling tasks: revised tables of maximum acceptable weights and forces, *Ergonomics 34*(9): 1197–1213 (1991).
20. W. F. Fox, *Body Weight and Coefficient of Friction as Determiners of Pushing Capability,* Human Eng. Spec. Stud. Ser. No 17, Lockheed Co., Marrietta, GA, 1967.
21. D. W. Greive, Slipping due to manual exertion, *Ergonomics 26*(1): 61–72 (1983).
22. M. S. Redfern and R. O. Andres, The analysis of dynamic pushing and pulling: required coefficients of friction, *Proc. Int. Conf. Occup. Ergon.,* Toronto, 1984.
23. N. Özkaya and M. Nordin, *Fundamentals of Biomechanics: Equilibrium, Motion, and Deformation,* Van Nostrand Reinhold, New York, 1991.
24. R. M. Pagulayan, Determining the human ability to judge inertia during a dynamic pushing task, M. S. Thesis, Virginia Polytechnic Inst. State Univ., Blacksburg, VA, 1994.
25. S. A. Konz, *Facility Design: Manufacturing Engineering,* 2nd ed., Publishing Horizons, Scottsdale, AZ, 1994.
26. D. J. Quinn, Lift tables, *Material Handling Handbook,* 2nd ed., R. A. Kulwiec, Ed., Wiley, New York, 1985.
27. R. A. Kulwiec, *Advanced Material Handling*, Material Handling Institute, Charlotte, NC, 1983.

16
Evaluating Physical Qualifications of Workers and Jobs

Richard J. Wickstrom
Disability Control, Inc., Cincinnati, Ohio

I. INTRODUCTION

Safety and productivity are impacted by the extent to which work stresses the capacities or limits of the body's biomechanical, physiological, or psychological systems. When there is insufficient physical stress on our bodies, fitness tends to decline, and productivity may suffer. On the other hand, too much physical stress may cause us to be injured or break down over time. Investigation of the lifting strength ratio [1–4] and job severity index [5–7] have revealed significant correlations between injury and physical capacities of workers.

One of the problems with assessing the physical qualifications of workers for jobs is that our selection of equipment and methods is rarely based on a clear rationale about the class of task or skill represented. This chapter presents a taxonomy for evaluating the physical demands of jobs, provides examples of objective and relevant methods for assessing workers, and discusses future concerns related to fitness-for-duty screening.

II. BACKGROUND AND SIGNIFICANCE TO OCCUPATIONAL ERGONOMICS

A common task-descriptive language (or taxonomy) is needed to conceptualize and classify the factors or qualifications associated with the kinds of tasks that people perform. Table 1 includes a summary of important scientific/theoretical and applied/practical benefits of having a common task-descriptive language, as reported by Fleishman and Quaintance [8].

III. CRITICAL REVIEW OF CURRENT STATUS

The most widely used taxonomy for matching the physical qualifications of workers with jobs is the U.S. Department of Labor's (DOL) taxonomy of 20 physical demand factors [9,10]. Except for *Strength*, the physical demand factors are classified according to frequency [9,10].

Constantly (C): Present two-thirds or more of the time.
Frequently (F): Present up to two-thirds of the time.
Occasionally (O): Present up to one-third of the time.
Not present (N): Activity or condition does not exist.

Table 1 Benefits of a Common Task-Descriptive Language

Scientific/Theoretical Benefits
1. Conducting literature reviews
2. Establishing better bases for conducting and reporting research studies to facilitate their comparison
3. Standardizing laboratory methods for studying human performance
4. Generalizing research to new tasks
5. Exposing gaps in knowledge
6. Assisting in theory development

Applied/Practical Benefits
1. Job definition and job analysis
2. Human-machine system design
3. Personnel selection, placement, and human resource planning
4. Training
5. Performance measurement and enhancement
6. Development of retrieval systems and databases

The DOL Strength physical demand factor is classified as sedentary, light, medium, heavy, and very heavy, based on the amount of standing, walking, sitting, lifting, carrying, pushing, and pulling required [9,10].

Sedentary: Exerting up to 10 lb of force occasionally or a negligible amount of force frequently to lift, carry, push, pull, or otherwise move objects, including the human body. Sedentary work involves sitting most of the time but may involve walking or standing for brief periods of time. A job is sedentary if walking and standing are required only occasionally and all other sedentary criteria are met.

Light: Exerting up to 20 lb of force occasionally, up to 10 lb of force frequently, or a negligible amount of force constantly to move objects. Physical demand requirements are in excess of those for sedentary work. Even though the weight lifted may be only a negligible amount, a job should be rated as light work (1) when it requires walking or standing to a significant degree, (2) when it requires sitting most of the time but entails pushing or pulling of arm or leg controls, or (3) when it requires working at a production rate pace entailing the constant pushing or pulling of materials, even though the weight of those materials is negligible. Note: The constant stress and strain of maintaining a production rate pace, especially in the industrial setting, is physically demanding of a worker even though the amount of force exerted is negligible.

Medium: Exerting 20–50 lb of force occasionally, or 10–25 lb of force frequently, or greater than negligible up to 10 lb of force constantly to move objects. Physical demand requirements are in excess of those for light work.

Heavy: Exerting 50–100 lb of force occasionally, or 25–50 lb of force frequently, or 10–20 lb of force constantly to move objects. Physical demand requirements are in excess of those for medium work.

Very heavy: Exerting in excess of 100 lb of force occasionally, or in excess of 50 lb of force frequently, or in excess of 20 lb of force constantly to move objects.

The distribution of levels of physical demand factors 2–20 across the 12,741 job titles of the 1991 *Dictionary of Occupational Titles* [10] is summarized in Table 2.

In the categories that follow, operational definitions are presented along with examples of methodologies for assessing workers on each factor.

Table 2 Physical Demand Factor Distribution

DOL factor		Constant	Frequent	Occasional	Never
2	Climbing	0.0%	3.1%	10.2%	86.7%
3	Balancing	0.1%	1.8%	5.3%	92.8%
4	Stooping	0.1%	11.5%	23.5%	64.9%
5	Kneeling	0.0%	3.7%	10.7%	85.6%
6	Crouching	0.0%	6.1%	15.6%	78.3%
7	Crawling	0.0%	0.5%	2.5%	97.0%
8	Reaching	10.6%	81.0%	7.5%	0.9%
9	Handling	11.1%	81.1%	7.0%	0.8%
10	Fingering	4.6%	50.5%	28.9%	4.6%
11	Feeling	0.4%	4.8%	10.7%	84.7%
12	Talking	0.9%	25.5%	9.5%	64.1%
13	Hearing	1.0%	26.4%	12.5%	60.1%
14	Tasting/smelling	0.0%	0.3%	0.6%	99.2%
15	Near acuity	5.9%	66.9%	13.5%	13.7%
16	Far acuity	0.4%	6.2%	5.3%	88.1%
17	Depth perception	1.2%	30.7%	14.7%	53.4%
18	Accommodation	1.7%	16.6%	20.1%	61.7%
19	Color vision	0.7%	10.3%	26.9%	62.1%
20	Field of vision	0.4%	3.8%	3.2%	92.6%

Source: ref. 10.

A. Lifting

Lifting is the ability to exert force upon materials from a stationary position (includes raising, lowering, holding, pushing, or pulling). This relates to Physical Demand Factor 1 (Strength) as defined by the U.S. Department of Labor [9,10] and ICIDH Codes D 48 (Lifting disability) as defined by the World Health Organization [11].

When placing a worker in a specific job, one must know more than only the maximum weight of the load being handled. A worker may be unable to lift 10 lb overhead due to a shoulder injury yet still be well matched for a job that requires lifting up to 50 lb at waist level or below. A worker with a back or shoulder injury can often be accommodated by being placed in a job where heavier items are lifted near waist level. The *Applications Manual for the Revised NIOSH Lifting Equation* [12] provides an excellent method for characterizing and evaluating lifting tasks based on the maximum and average force exerted, the location of the hands and angle of asymmetry at the origin and destination of the lift, the frequency and duration of lifting, and coupling to the object.

Before constructing a fitness-for-duty assessment for lifting, it is useful to summarize the lifting demands of the job in terms of the frequency and force exerted at a lower level (below knuckle height), at midrange (near waist level), and at an upper level (above shoulder height):

	Occasionally (≤32 reps/day)	Frequently (≤200 reps/day)	Continuously (>200 reps/day)
Upper level:	____lb	____lb	____lb
Midrange:	____lb	____lb	____lb
Lower level:	____lb	____lb	____lb

Figure 1 PAT Lower Lifting Strength Test.

The JAMAR Physical Agility Tester (PAT) and the Jackson Strength Evaluation System are examples of systems that may be used to assess lifting capacity [13–15]. When occasional lifting at a lower level is the critical job demand, lifting strength may be measured dynamically with the PAT Lower Lifting Strength Test as shown in Figure 1 [13] or measured statically with the Jackson Back Lift Test as shown in Figure 2 [15,16]. When occasional lifting at a midrange or upper level is the critical job demand, lifting strength may be measured dynamically with PAT Upper Lifting Strength Test as shown in Figure 3 [13] or measured statically with the Jackson Arm Lift Test as shown in Figure 4 [14,15]. When frequent or continuous lifting is the critical job demand, lifting endurance may be measured with the PAT Lower Lift Endurance Test or the PAT Upper Lift Endurance Test [13].

Standardized tests of cardiorespiratory fitness may also be used to screen workers for physiologically demanding jobs [16,17]. Submaximal step tests are often selected for this purpose because they have been shown to be inexpensive, safe, and simple to administer and tend to be most similar to lifting demands. To avoid excessive cardiorespiratory fatigue, NIOSH recommends energy expenditure limits of 50%, 40%, and 33% of maximum aerobic capacity for tasks lasting up to 1 hr, 1–2 hr, and 2–8 hr, respectively [18].

B. Carrying

Carrying is the ability to hold materials in the hands or arms or on the shoulders while walking. This relates to Physical Demand Factor 1 (Strength) as defined by the U.S. Department

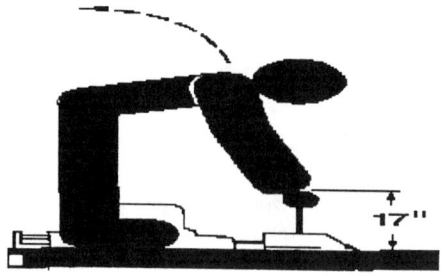

Figure 2 Jackson Back Lift Test.

Figure 3 PAT Upper Lifting Strength Test.

of Labor [9] and ICIDH Code D 48 (Lifting disability) as defined by the World Health Organization [11]. This may be measured in terms of frequency (percent of the day), maximum and average weight carried (pounds), average and maximum distance carried (yards), and number of repetitions. When carrying strength is the critical job demand, it may be measured dynamically with the PAT Carrying Strength Test as shown in Figures 5a and 5b [13], or estimated from results on the Jackson Arm Lift Test as shown in Figure 4 [14,15]. Carrying tasks that occur over longer distances are not routinely assessed, because it is usually possible to reduce these physical demands by using a cart or other mechanical device for transporting.

C. Transporting

Transporting is the ability to push or pull materials while walking. This relates to Physical Demand Factor 1 (Strength) as defined by the U.S. Department of Labor [9] and ICIDH Code D 48 (Lifting disability) as defined by the World Health Organization [11]. This may be measured in terms of frequency (percent of the day), maximum and average force exerted

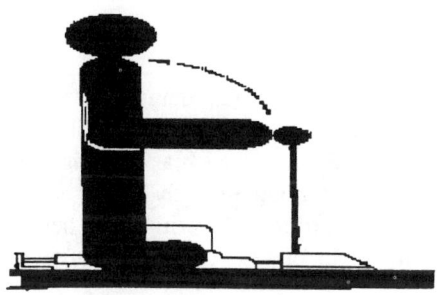

Figure 4 Jackson Arm Lift Test.

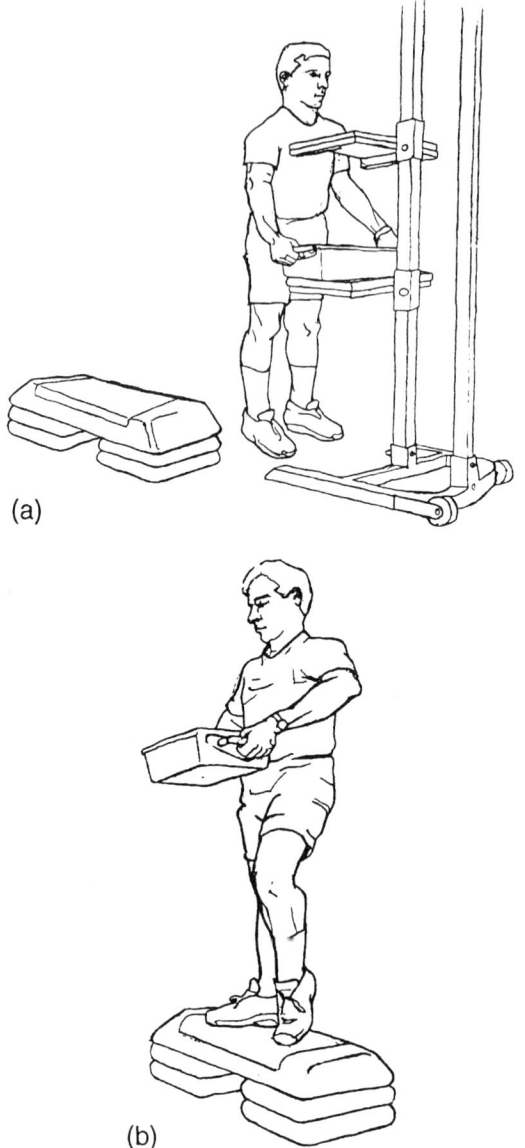

Figure 5 PAT Carrying Strength Test.

during transporting (pounds), average and maximum distance of transporting (yards), and number of repetitions. When transporting strength is the critical job demand, it may be measured dynamically by a work simulation task such as pushing a weighted sled (not shown) or estimated from results on the PAT Carrying Strength Test as shown in Figures 5a and 5b [13]. When frequent or continuous transporting is the critical job demand, then transporting endurance is usually estimated from results on tests of cardiorespiratory fitness, such as a step test [19–22].

D. Sitting

Sitting is the ability to remain in a sitting position. This relates to Physical Demand Factor 1 (Strength) as defined by the U.S. Department of Labor [9] and ICIDH Code D 71.0 (Disability in sustaining positions) as defined by the World Health Organization [11]. This may be measured in terms of frequency (percent of the day) and maximum and average time spent sitting for any given function.

E. Standing

Standing is the ability to remain standing without moving about. This relates to Physical Demand Factor 1 (Strength) as defined by the U.S. Department of Labor [9] and ICIDH Code D 71.0 (Disability in sustaining positions) as defined by the World Health Organization [11]. This may be measured in terms of frequency (percent of the day) and maximum and average time spent standing for any given function.

F. Walking

Walking is the ability to move about on foot. This relates to Physical Demand Factor 1 (Strength) as defined by the U.S. Department of Labor [9] and ICIDH Codes D 40 (Walking disability), D 41 (Traversing disability), D 42 (Climbing stairs disability), and D 44 (Running disability) as defined by the World Health Organization [11]. This may be measured in terms of frequency (percent of the day), maximum and average distance ambulated, and maximum and average speed of ambulation. When walking is done occasionally or frequently, the PAT Balance Test as shown in Figure 6 [13] may be used to measure fitness for walking. When walking is required continuously, fitness for walking may be measured by a step test [19–22], self-paced walking test [23], 12-min run test [24], or treadmill test. Although a treadmill test is the most accurate method [25,26], this mode of testing is less common for field applications due to expense and lack of portability.

G. Climbing

Climbing is the ability to ascend or descend ladders, stairs, scaffolding, ramps, poles, and the like using feet and legs or hands and arms. This relates to Physical Demand Factor 2 (Climbing) as defined by the U.S. Department of Labor [9] and ICIDH Code D 42 (Climbing stairs disability) as defined by the World Health Organization [11]. This may be measured in terms of frequency (percent of the day); height, steepness, and type of structure climbed; and maximum and average duration of climbing. When climbing is required occasionally, the PAT Balance Test as shown in Figures 6a and 6b [13] and other measures such as the JAMAR Grip Strength Test [24,25] may be used to estimate fitness for climbing. When climbing is required frequently, standardized tests of cardiorespiratory fitness such as a treadmill or step test may be used to evaluate fitness for climbing.

H. Balancing

Balancing is the ability to maintain equilibrium when standing in place or moving on varying surfaces. This relates to Physical Demand Factor 3 (Balancing) as defined by the U.S. Department of Labor [9,10] and ICIDH Codes D 58 (Postural disability) and D 41 (Traversing

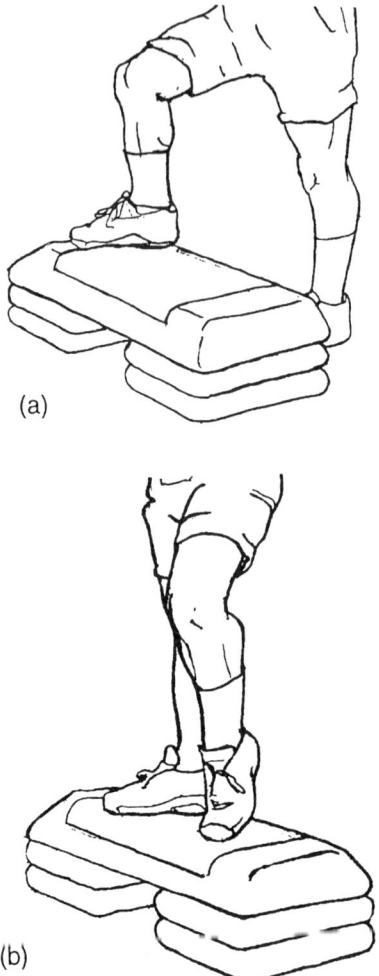

Figure 6 PAT Balance Test.

disability) as defined by the World Health Organization [11]. This may be measured in terms of frequency (percent of the day), type or condition of surface, activities during which balance must be maintained, and maximum and average duration of balancing. The PAT Balance Test as shown in Figures 6a and 6b [13] may be used along with more static balance tasks such as walking forward, backward, and/or sideways along a beam or line to estimate fitness for balancing.

I. Stooping

Stooping is the ability to bend at the waist. This relates to Physical Demand Factor 4 (Stooping) as defined by the U.S. Department of Labor [9] and ICIDH Code D 56 (Crouching disability) as defined by the World Health Organization [11]. This may be measured in terms of frequency (percent of the day) and maximum and average duration of stooping. When

Figure 7 PAT Manipulation—Stooping.

stooping is required occasionally, this can be measured with the PAT Forward Reach Test or a modification to the PAT Lower Manipulation Test as shown in Figure 7, or estimated based on a test of lower lifting strength test such as the PAT Lower Lifting Strength Test as shown in Figure 1 or the Jackson Back Lift Test as shown in Figure 2. When stooping is required frequently, fitness for stooping may be measured with the PAT Lower Lifting Endurance Test [13].

J. Kneeling

Kneeling is the ability to support body weight on one or both knees. This relates to Physical Demand Factor 5 (Kneeling) as defined by the U.S. Department of Labor [9] and ICIDH Code D 55 (Kneeling disability) as defined by the World Health Organization [11]. This may be measured in terms of frequency (percent of the day) and maximum and average duration of kneeling. In many instances, it is up to the worker to decide what method of working to employ at a lower lever, e.g., whether to kneel, crouch, or stoop. When kneeling is required occasionally, the PAT Lower Manipulation Test may be administered with a kneeling method (Figure 8).

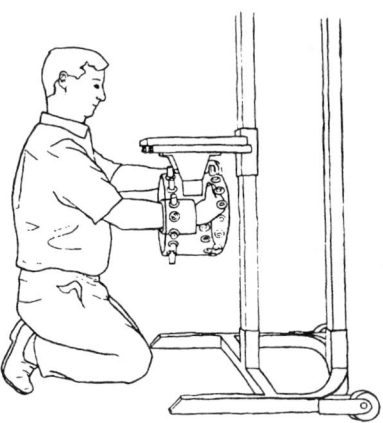

Figure 8 PAT Manipulation—Kneeling.

K. Crouching

Crouching is the ability to bend at the knees and waist. This relates to Physical Demand Factor 6 (Crouching) as defined by the U.S. Department of Labor [9] and ICIDH Code D 56 (Crouching disability) as defined by the World Health Organization [11]. This may be measured in terms of frequency (percent of the day) and maximum and average duration of crouching. When crouching is required occasionally, the PAT Lower Manipulation Test as shown in Figure 9 may be used to measure fitness for crouching. When crouching is required frequently, the PAT Lower Lifting Endurance Test should also be done.

L. Crawling

Crawling is the ability to move about on the hands and knees or hands and feet. This relates to Physical Demand Factor 7 (Crawling) as defined by the U.S. Department of Labor [9] and ICIDH Code D 57 (Other body movement disability) as defined by the World Health Organization [11]. This may be measured in terms of frequency (percent of the day), distance, and maximum and average duration of crawling. When crawling is required occasionally, the PAT Forward Manipulation Test (Figure 9) and the PAT Balance Test (Figures 6a and 6b) may be administered. When crawling is required frequently, the PAT Lower Manipulation Test (Figure 9) and the PAT Upper Manipulation Test (Figure 10) should be done as well.

M. Foot Use

Foot use is the ability to push, treadle, kick, or otherwise manipulate items with one or both feet. This relates to Physical Demand Factor 1 (Strength) as defined by the U.S. Department of Labor [9] and ICIDH Code D 67 (Foot control disability) as defined by the World Health Organization [11]. This may be measured in terms of frequency (percent of the day), type of foot movement or coordination required, and maximum and average duration and speed of foot use. It is important to note whether the dominant foot, other foot, both feet, or either foot may be used for the task. Foot use ability is usually estimated from other tests such as the PAT Balance Test (Figure 6) and the PAT Lower Manipulation Test (Figure 9).

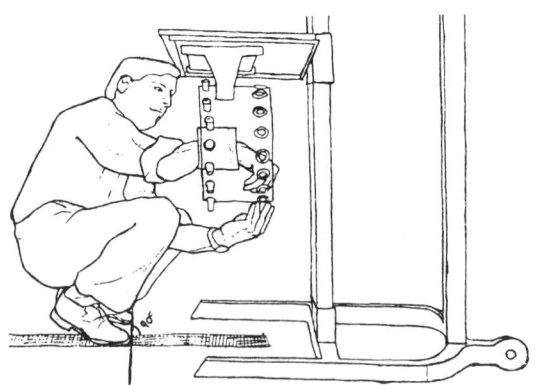

Figure 9 PAT Lower Manipulation Test.

Evaluating Physical Qualifications

Figure 10 PAT Upper Manipulation Test.

N. Reaching

Reaching is the ability to extend the hand(s) and arm(s) in any direction. This relates to Physical Demand Factor 8 (Reaching) as defined by the U.S. Department of Labor [9] and ICIDH Code D 53 (Reaching disability) as defined by the World Health Organization [11]. This may be measured in terms of frequency (percent of the day) and the maximum and average duration for reaching in the mid and upper range. It is important to note whether the dominant arm, other arm, both arms, or either arm may be used for reaching. Reaching may be evaluated by the PAT Forward Manipulation Test (Figure 11) and PAT Upper Manipulation Test (Figure 10).

Figure 11 PAT Forward Manipulation Test.

O. Handling

Handling is the ability to seize, hold, grasp, turn, or otherwise manipulate items with one or both hands. The fingers are involved only to the extent that they are an extension of the hand, such as to turn a switch or shift automobile gears. This relates to Physical Demand Factor 9 (Handling) as defined by the U.S. Department of Labor [9] and ICIDH Codes D 63 (Gripping disability), D 61.4 (Handling objects), D 64 (Holding disability), D 65 (Handedness disability), and D 66 (Other manual activity disability) as defined by the World Health Organization [11]. Handling may be measured in terms of frequency (percent of the day), type of movement or coordination required, speed, and maximum and average duration and speed of hand/wrist use. The evaluator should note whether the dominant hand, other hand, both hands, or either hand may be used for handling. Handling may be evaluated by the PAT Forward Manipulation Test (Figure 11), the PAT Upper Manipulation Test (Figure 10), and the JAMAR Grip Strength Test (not shown).

P. Fingering

Fingering is the ability to pick, pinch, or otherwise manipulate small items between the fingers of one or both hands. This relates to Physical Demand Factor 10 (Fingering) as defined by the U.S. Department of Labor [9] and ICIDH Code D 62 (Fingering disability) as defined by the World Health Organization [11]. This may be measured in terms of frequency (percent of the day), type of movement required, speed, and maximum and average duration and speed of finger use. The evaluator should note whether the dominant hand, other hand, both hands, or either hand may be used for fingering. Fingering may be evaluated by the Purdue Pegboard Test [29], the B & L Pinch Strength Test [30], the PAT Forward Manipulation Test as shown in Figure 11 [13], and the PAT Upper Manipulation Test as shown in Figure 10 [13].

Q. Feeling

Feeling is the ability to perceive attributes of objects, such as size, shape, temperature, or texture, by touching with skin, particularly that of the fingertips. This relates to Physical Demand Factor 11 (Feeling) as defined by the U.S. Department of Labor [9]. This may be measured in terms of frequency (percent of the day) and sensitivity required. Sensation may be directly measured by the Semmes-Weinstein Monofilament Test [31] or indirectly assessed during testing of finger dexterity with the Purdue Pegboard Test [29].

R. Talking

Talking is the ability to communicate orally in a clear and understandable manner. This relates to Physical Demand Factor 12 (Talking) as defined by the U.S. Department of Labor [9] and ICIDH Code D 21 (Disability in talking) as defined by the World Health Organization [11]. This may be measured in terms of frequency (percent of the day) and clarity or loudness required. Talking is usually assessed indirectly during the interview process.

S. Hearing

Hearing is the ability to detect or distinguish between sounds that vary in pitch or loudness. This relates to Physical Demand Factor 13 (Hearing) as defined by the U.S. Department of

Labor [9] and ICIDH Codes D 23 (Disability in listening to speech) and D 24 (Other listening disability) as defined by the World Health Organization [11]. This may be measured in terms of frequency (percent of the day) and amplitude of hearing required over a broad range of frequencies. Federal Motor Carrier Regulations [34] require that truck drivers be able to first perceive a forced whispered voice in the better ear at not less than 5 ft with or without the use of a hearing aid or, if tested by use of an audiometric device, to not have an average hearing loss greater than 40 dB at 500 Hz, 1000 Hz, and 2000 Hz frequencies with or without a hearing aid when the audiometric device is calibrated to American National Standard (formerly ASA Standard) Z24.5—1951. While this level of screening may be done with an inexpensive and hand-held device such as the Oto-Screen 1 by Handtronix, a more elaborate and expensive setup is required to meet the rigid specifications by OSHA [35] for initial and periodic audiograms on workers exposed to 90 dB (A) for 8 hr exposure. OSHA hearing exams must be done in a suitable quiet environment, the testing apparatus (audiometer, headphones, test room) must be calibrated, a wider range of frequencies must be evaluated (500, 1000, 2000, 3000, 4000, and 6000 Hz), the procedure must be standardized, and the person responsible must be properly trained in test administration, interpretation of results, and provision of ear protection.

T. Tasting

Tasting is the ability to detect or distinguish between flavors that vary in quality or intensity, using the tongue. This relates to Physical Demand Factor 14 (Tasting/Smelling) as defined by the U.S. Department of Labor [9]. This may be measured in terms of frequency (percent of the day). Given that this factor is not often present in most jobs, it needs to be evaluated in only a few occupations. Taste may be grossly evaluated by having the subject close the eyes and identify a variety of substances by taste (e.g., salt, sugar).

U. Smelling

Smelling is the ability to detect or distinguish between odors that vary in quality of intensity, using the nose. This relates to Physical Demand Factor 14 (Tasting/Smelling) as defined by the U.S. Department of Labor [9]. This may be measured in terms of frequency (percent of the day) and degree of smell required. Given that this factor is not often present in most jobs, it needs to be evaluated in only a few occupations. Smell may be grossly evaluated by having the subject close the eyes, occlude one nostril, and identify a variety of substances by smell (e.g., tobacco, coffee, soap, vanilla).

V. Near Vision

Near vision is the ability to see close environmental surroundings. This relates to Physical Demand Factor 15 (Near acuity) as defined by the U.S. Department of Labor [9] and ICIDH Code D 26 (Disability in detailed visual tasks) as defined by the World Health Organization [11]. This may be measured in terms of frequency (percent of the day) and degree of near vision required. Near vision needs to be specified by aptitude as well as by frequency so that it is clear what degree is needed to work. Levels of aptitude for near vision may be evaluated in Snellen equivalents using the Professional Vision Tester (PVT) Acuity Both Eyes "Near," Acuity Right Eye "Near," and Acuity Left Eye "Near" Tests [36].

W. Far Vision

Far vision is the ability to see distant environmental surroundings. This relates to Physical Demand Factor 16 (Far acuity) as defined by the U.S. Department of Labor [9] and ICIDH Code D 25 (Disability in gross visual tasks) as defined by the World Health Organization [11]. This may be measured in terms of frequency (percent of the day) and degree of far vision required. Federal Motor Carrier Safety Regulations [34] require truck drivers to have far vision acuity of at least 20/40 (Snellen) in each eye without corrective lenses or visual acuity separately corrected to 20/40 (Snellen) or better with corrective lenses and far vision acuity of at least 20/40 (Snellen) using both eyes with or without corrective lenses. Levels of aptitude for far vision may be evaluated by Snellen equivalents with the PVT Acuity Both Eyes "Far," Acuity Right Eye "Far," and Acuity Left Eye "Far" Tests [36].

X. Depth Perception

Depth perception is the ability to distinguish the distance of objects or the relative distance between objects. This relates to Physical Demand Factor 17 (Depth perception) as defined by the U.S. Department of Labor [9] and ICIDH Code D 25 (Disability in gross visual tasks) as defined by the World Health Organization [11]. This may be measured in terms of frequency (percent of the day) and degree of depth perception required. Levels of aptitude for depth perception may be evaluated by Shepard–Fry percentages with the PVT Stereo Depth Perception "Far" Test [36].

Y. Color Discrimination

Color discrimination is the ability to detect or distinguish between colors that vary in purity (saturation) or brightness (brilliance). This relates to Physical Demand Factor 19 (Color vision), as defined by the U.S. Department of Labor [9] and ICIDH Code D27.1 (Disability in color recognition) as defined by the World Health Organization [11]. This may be measured in terms of frequency (percent of the day) and degree of color vision required. Federal Motor Carrier Safety Regulations [34] require truck drivers to be able to recognize the colors of traffic signals and devices showing standard red, green, and amber. Levels of color vision may be evaluated with the PVT Color Discrimination Test [36].

Z. Peripheral Vision

Peripheral vision is the ability to see objects or movements toward the edges of the visual fields. This relates to Physical Demand Factor 20 (Field of vision) as defined by the U.S. Department of Labor [9] and ICIDH Code D 26 (Disability in gross visual tasks) as defined by the World Health Organization [11]. This may be measured in terms of frequency (percent of the day) and degree of peripheral vision required. The Department of Transportation requires truck drivers to have a field of vision of at least 70° in the horizontal meridian in each eye. Levels of peripheral vision may be evaluated with the PVT Peripheral Test—Right and Peripheral Test—Left [36]. The lights flash at 85°, 70°, and 55° temporally and approximately 35° nasally, so a possible total of 120° for each eye can be attained (highest temporal reading plus highest nasal reading).

IV. FUTURE CONCERNS

A. Impact of Civil Rights Legislation

Methods of assessing a worker's fitness for duty have been influenced by civil rights legislation and court decisions related to employment practices. Title VII of the Civil Rights Act of 1964 prohibits employers from discriminating against applicants and workers on the basis of race, color, religion, sex, or national origin. Section 503 of the Rehabilitation Act of 1973 requires employers with government contracts of $2500 or more to take affirmative action to employ and advance qualified handicapped individuals. Title I of the Americans with Disabilities Act (ADA) of 1990 couples many of the principles of the Rehabilitation Act of 1973 with the remedies and procedures set forth in the Civil Rights Act of 1964 for other types of discrimination [38]. ADA prohibits covered entities with 15 or more employees from discriminating against a qualified individual with a disability where that individual, with or without reasonable accommodation, can perform the essential job functions.

The Americans with Disabilities Act (ADA) includes a number of provisions that impact methods for assessing physical qualifications:

1. *Before making a job offer*, medical examinations and inquiries are not permitted. However, an employer may give a physical agility test to determine physical qualifications, provided the test is given to all similarly situated applicants or employees.
2. *After making a conditional job offer and before the individual starts to work*, unrestricted examinations and medical inquiries are permitted; however, the employer may not reject a person with a disability unless the reason is job-related and consistent with business necessity. Post-offer examinations must be given to all entering employees within the same job category.
3. *After employment*, any examination or medical inquiry required of employees must be job-related and consistent with business necessity, except voluntary examinations by employee health programs and examinations that are mandated by federal laws.

The Equal Employment Opportunity Commission's Uniform Guidelines on Employee Selection Procedures [38] state that selection procedures having adverse impact constitute discrimination unless validated. Records concerning impact are to be maintained by sex, race, and ethnic groups, and a selection rate for any race, sex, or ethnic group that has less than four-fifths of the rate for the group with the highest rate will generally be regarded as evidence of adverse impact but must be statistically significant.

Employers may use selection procedures that are not fully supported if (1) substantial evidence of validity exists and (2) a study is in progress to provide additional evidence of validity within a reasonable time. Procedures for conducting acceptable types of validity studies are detailed in other publications.

B. NIOSH Criteria

This emphasis of federal regulations on more objective and job-relevant evaluations of workers is consistent with NIOSH's criteria for evaluating employee screening programs [39]:

Is the test safe to administer? (Safety)
Does the test give reliable, quantitative values? (Reliability)
Is the test related to specific job requirements? (Validity)
Is the test practical? (Practicality)
Does the test predict risk of future injury or illness? (Utility)

1. Safety

Safety becomes a concern during strength or endurance screening when there is potential for an overly motivated or deconditioned subject to sustain an overexertion injury. This may occur when

1. The applicant's performance on a preemployment physical agility test is used to determine whether she or he will be offered the job.
2. The applicant must pass a post-offer evaluation to determine his or her fitness for duty in a specific job.
3. A person recovering from an injury or illness is evaluated to determine treatment progress.
4. A person is evaluated to determine his or her or ability to return to work or the extent of disability.

Because of potential for injury, Caldwell et al. [40] cautioned against providing a "specific goal, stated or direct feedback given, as to how great the exertion is during the test." This is a common problem with pass/fail work samples such as lifting 100 lb of rice from floor to table height. When employment is contingent on completing this work sample without some form of warm-up or gradual progression, there is significant potential for overexertion.

One commonly employed loss control strategy has been to require medical screening prior to administering a physical agility test [40]. Under the ADA [27], the employer has two options:

1. To describe the activity and ask the attending physician if the applicant can safely perform the test.
2. To administer the physical agility test after making a conditional job offer and obtaining any necessary medical information permitted under ADA.

It is important to note that "blanket" exclusions such as having a history of a back injury may not be sufficient justification to keep someone from participating in a physical agility test. An effective loss control strategy by itself or in conjunction with medical screening is to require completion of less stressful or warm-up activities before attempting maximum tests of strength or endurance. For example, if a person is not able to crouch briefly to complete the PAT Lower Reach Test [13], it would not be advisable to proceed with a maximum isometric leg lift test.

Although isometric strength testing of healthy subjects has been documented to be safe in a number of studies [2, 40,41], others have reported incidents of back discomfort and injuries during isometric strength testing [42]. Hansson et al. [43] performed a biomechanical evaluation of compression loading of four subjects during various isometric strength testing positions and reported high compression loads on L3 for the Torso Lift and Leg Lift, ranging from 5000 to 11,000 N. This level of compression loading has been shown to cause structural failures of the vertebral endplates in vitro. The arm lift and isometric tests of the trunk flexors and extensors were found to cause considerably less loading. Safety becomes even more critical when evaluating the work functional abilities of workers following an injury or illness.

For reasons of safety, the American Heart Association's Committee on Exercise recommends that patients with known or suspected heart disease not be tested without a qualified physician at the site to provide life-saving emergency care. Symptom-limited graded exercise testing is appropriate for high-risk persons and sedentary persons more than 45 years of age who intend to participate in moderate- to high-intensity activities [44].

2. Reliability

Reliability must be present for a test to be considered objective. A common measure of reliability is the coefficient of variation, which is the standard deviation divided by the mean. Chaffin et al. [2] report that during isometric strength testing it should be possible to achieve a coefficient of variation of less than 15%, with 5% often achievable in a well-controlled laboratory. Khalil et al. [45] reported similar reliability for testing injured workers in a pain center setting. Comparable reliability has been reported by other researchers for tests of dynamic lifting capacity with healthy and injured workers [46-51].

3. Content Validity

Content validity must be demonstrated for all physical qualification standards, tests, or selection criteria in accordance with EEOC's Uniform Guidelines [38] and provisions of the Americans with Disabilities Act [37]. Under ADA, a "job-related" standard or selection criterion may evaluate or measure all functions of a job, and employers may continue to select and hire people who can perform all of these functions. It is only when an individual's disability prevents or impedes performance of marginal job functions that the ADA requires the employer to evaluate this individual's qualifications solely on his or her ability to perform the essential functions of the job with or without accommodation.

Content validity is important because one physical attribute may not correlate with another; for example, grip strength may not predict other strength factors [41] such as the ability to lift from a lower level. Chaffin et al. [2] reported a high unexplained variance (27%) when anthropometric and specific strength scores were used to predict physical performance. Rosecrance et al. [52] reported only low to moderate correlations between isometric strength tests and dynamic lifting capacity in patients with low back pain. These results suggested that dynamic lift tests are a better simulation of the task being assessed and may be more appropriate for a back-injured population.

EEOC's Uniform Guidelines [38] state that content validity should be demonstrated in the following manner:

1. A job analysis should identify important work behaviors and associated tasks for successful performance.
2. Abilities being measured must be operationally defined.
3. User must show that the selection procedure measures and is a representative sample of abilities used to perform critical work behaviors. This means that the selection procedure should closely approximate an observable work behavior or its product should closely approximate an observable work product in a manner and setting that closely approximate the work situation. When the content differs substantially from the actual job, other measures of validity are necessary.
4. Statistical measures of reliability should be obtained.
5. Cutoff scores should normally be set to be reasonable and consistent with normal expectations of acceptable proficiency within the workforce.
6. Ranking based on content validity studies must show by job analysis or other method that a higher score is likely to result in better performance.

4. Practicality

Practicality dictates the need for minimal hardware expense, versatility to evaluate physical qualifications for different job conditions, minimum administration time, minimum space requirements, and minimum instruction time. For example, the most reliable and valid test of cardiorespiratory fitness is obtained by direct measurement of Vo_{2max} using established and standardized procedures. Test-retest reliability under these conditions has been reported to be $r = 0.95$ or higher with a standard error of 1 mL/(kg-min) or 0.3 MET [44]. Jette et al. [20] report several drawbacks to this approach: "It is time-consuming and expensive; it is difficult to motivate a person to exert himself maximally; and such exertion is potentially dangerous to a person's health, particularly the older sedentary individual and the person with cardiac disease." A safer and more practical method for evaluating cardiorespiratory function would be to estimate Vo_{2max} from measurement of heart rate and work at submaximal loads. The Canadian Aerobic Step Test and other submaximal step testing methods have demonstrated test-retest reliability in the range $r = 0.70$–0.90, with a standard error of 5–10% or 5–6 mL/(kg-min) or 1.8 METs [44]. Submaximal testing is appropriate for the purpose of nondiagnostic functional capacity evaluations of low-risk subjects younger than 45 years of age. The cost of equipment and administration is considerably less, and a physician does not need to be present, provided the health care personnel directing the tests are trained to the satisfaction of the responsible physician in cardiorespiratory resuscitation and emergency cardiac care.

5. Utility

Utility measures the degree to which the selection procedure is predictive of or significantly correlated with important elements of job performance. This is probably the most difficult criterion to meet, because it is costly and burdensome to conduct epidemiological studies to gather injury and illness data to support the use of a selection and placement procedure for both those who are well-matched and those who are not. For example, the use of X-rays for routine placement screening is considered by many to be unwarranted, considering the radiation hazard, costs, and lack of reliable predictive value [53].

C. Conclusion

The taxonomy presented in this chapter may be used to identify the critical physical demands of jobs. Developing a "functional" profile of job demands is a critical first step in the job modification and selection process. Objective tests of an individual's ability to perform the critical demands of the job help to bridge the gap in our understanding of the relationship between medical impairment and job demands. To avoid legal pitfalls, it is important that our selection of equipment and methods for functional screening be based on a clear rationale about the class of task or skill represented. This is still a relatively new area of investigation, and there have been only a limited number of studies to suggest that any one method is more accurate than another.

REFERENCES

1. D. B. Chaffin, Human strength capability and low back pain, *J. Occup. Med.* 16:248–254 (1974).
2. D. B. Chaffin, G. B. Herrin, and W. M. Kesserling, *Pre-employment Strength Testing in Selecting Workers for Materials Handling Jobs*, NIOSH Tech. Rep. 77-163, 1977.
3. D. B. Chaffin and K. S. Park, A longitudinal study of low-back pain as associated with occupational weight lifting factors, *Am. Ind. Hyg. Assoc. J.* 34:513–535 (1973).
4. W. M. Keyserling, G. D. Herrin, and D. B. Chaffin, Isometric strength testing as a means of controlling medical incidents on strenuous jobs, *J. Occup. Med.* 22(5):332–336 (1980).
5. M. M. Ayoub, N. J. Bethea, S. Deivanayagam, S. S. Asfour, G. M. Bakken, and D. Liles, Determination and modeling of lifting capacity, DHHS (NIOSH) Grant No. 5-R01-OH-00545-02, 1978.

6. D. H. Liles, S. Deivanayagam, M. M. Ayoub, and P. Mahajan, A job severity index for the evaluation and control of lifting injury, *Hum. Factors 26*(6):683–693 (1984).
7. M. M. Ayoub, R. Dryden, Joe McDaniel, R. Knipfer, and D. Dixon, Predicting lifting capacity, *Am. Ind. Hyg. Assoc. J. 40*(12):1075–1084 (1979).
8. E. A. Fleishman and M. K. Quaintance, *Taxonomies of Human Performance*, Academic, Orlando, FL, 1984.
9. U.S. Department of Labor, *The Revised Handbook for Analyzing Jobs*, U.S. Govt. Printing Office, Washington, DC, 1991.
10. U.S. Department of Labor, *Dictionary of Occupational Titles*, 4th ed., U.S. Govt. Printing Office, Washington, DC, 1991.
11. World Health Organization, *International Classification of Impairments, Disabilities, and Handicaps*, Geneva, 1980.
12. T. R. Waters, V. Putz-Anderson, and A. Garg, *Applications Manual for the Revised NIOSH Lifting Equation*, DHHS (NIOSH) Publ. No. 94-110, 1993.
13. Lafayette Instrument Company, *Instruction Manual*, JAMAR Physical Agility Tester (PAT), Model No. 32450, Lafayette, IN.
14. Lafayette Instrument Company, *Instruction Manual*, Jackson Strength Evaluation System, Model No. 32450, Lafayette, IN.
15. A. S. Jackson, *Preemployment Isometric Strength Testing Methods—Medical and Ergonomic Values and Issues*, Technical report prepared for Lafayette Instrument Company, Lafayette, IN, 1990.
16. G. E. Caple, Energy expenditure modeling in the return-to-work decision process, *Appl. Ind. Hyg. 3*:348–352 (1988).
17. C. K. Anderson and M. J. Catterall, The impact of physical agility testing on incidence rate, severity rate, and productivity, in *Trends in Ergonomics/Human Factors*, Vol. 4, S. S. Asfour, ed., North-Holland/Elsevier, Amsterdam, 1987, pp. 577–584.
18. T. R. Waters, V. Putz-Anderson, A. Garg, and L. J. Fine, Revised NIOSH equation for the design and evaluation of manual lifting tasks, *Ergonomics 36*:749–776 (1993).
19. R. J. Shephard, The current status of the Canadian Home Fitness Test, *Br. J. Sports Med. 14*(2–3):114–125 (1980).
20. M. Jette, J. Campbell, J. Mongeon, and R. Routhier, The Canadian Home Fitness Test as a predictor of aerobic capacity, *CMA J. 114*:680–682 (1976).
21. Canadian Society for Exercise Physiology, *Canadian Standardized Test of Fitness (CSTF) Operations Manual*, 3rd ed., 1986.
22. S. F. Siconolfi, C. E. Garber, T. M. Lasater, and R. A. Carleton, A simple valid step test for estimating maximal oxygen uptake in epidemiologic studies, *Am. J. Epidemiol. 121*(3):382–390 (1985).
23. E. J. Bassey, P. H. Fentem, I. C. MacDonald, and P. M. Scriven, Self-paced walking as a method for exercise testing in elderly and young men, *Clin. Sci. Mol. Med. 51*:609–612 (1976).
24. K. H. Cooper, *New Aerobics*, Bantam Books, New York, 1970.
25. N. L. Jones, *Clinical Exercise Testing*, 3rd ed., W. B. Saunders, Philadelphia, 1988.
26. American College of Sports Medicine, *Guidelines for Exercise Testing and Prescription* 4th ed., Lea & Febiger, Philadelphia, 1991.
27. V. Mathiowetz, N. Kashman, G. Volland, K. Weber, M. Dowe, and S. Rogers, Grip and pinch strength: normative data for adults, *Arch. Phys. Med. Rehab. 66*:69–74 (1985).
28. JAMAR Hand Dynamometer, Model No. J00105, Lafayette Instrument Co., Lafayette, IN.
29. J. Tiffin, *Purdue Pegboard Examiner Manual*, Model No. 32020, Lafayette Instrument Co., Lafayette, IN.
30. B & L Pinch Gauge, Model No. 78005, Lafayette Instrument Co., Lafayette, IN.
31. Semmes-Weinstein Monofilaments, Model No. 16009, Lafayette Instrument Co., Lafayette, IN.
32. Instruction manual, Oto-Screen I, Handtronix, Inc., Salt Lake City, UT.
33. L. S. Alford, Handheld screening audiometers: reliability factors of a new screening tool, *Hearing Instrum. 42*(10):49 (1991).
34. U.S. Department of Transportation, Federal Motor Carrier Safety Regulations, Part 391, 1990.
35. *Code of Federal Regulations*, Title 29, Sec. 1910.95 (General Industry).

36. Stereo Optical Company, *Reference and Instruction Manual*, "Optec 2000" Vision Tester Industrial Model, Chicago, IL
37. Equal Employment Opportunity Commission, *A Technical Assistance Manual on the Employment Provisions (Title I) of the Americans with Disabilities Act*, 1990.
38. Equal Employment Opportunity Commission, *Uniform Guidelines on Employee Selection Procedures*, Sec. 60-3, 43 FR 38295 (Aug. 25):126–153 (1978).
39. NIOSH, *Work Practices Guide for Manual Lifting*, NIOSH Tech. Rep. No. 81-122, U.S. Dept. of Health and Human Services, Natl. Inst. Occupational Safety and Health, Cincinnati, OH, 1981.
40. L. S. Caldwell, D. B. Chaffin, F. N. Dukes-Dobos, K. H. E. Kroemer, L. L. Laubach, S. H. Snook, and D. E. Wasserman, A proposed standard procedure for static muscle strength testing, *Am. Ind. Hyg. Assoc. J. 35*:201–205 (1974).
41. L. L. Laubach, Comparative muscular strength of men and women: review of the literature, *Aerosp. Med. 47*:534–542 (1976).
42. M. C. Battie, S. J. Bigos, L. D. Fisher, T. H. Hansson, M. E. Jones, and M. D. Wortley, Isometric lifting strength as a predictor of industrial back pain reports, *Spine 14*(6):851–856 (1989).
43. T. H. Hansson, S. J. Bigos, M. K. Wortley, and D. M. Spengler, The load on the lumbar spine during isometric strength testing, Spine 9(7):720–724 (1984).
44. M. L. Pollock and J. H. Wilmore, *Exercise in Health and Disease: Evaluation and Prescription for Prevention and Rehabilitation*, 2nd ed., W. B. Saunders, Philadelphia, 1990.
45. T. M. Khalil, M. L. Goldberg, S. S. Asfour, E. A. Moty, R. S. Rosomoff, and H. L. Rosomoff, Acceptable maximum effort (AME): a psychophysical measure of strength in back patients, *Spine 12*(4):372–376 (1987).
46. W. Karwowski and J. W. Yate, Reliability of the psychophysical approach to manual lifting of liquids by females, *Ergonomics 29*(2):237–248 (1986).
47. K. H. E. Kroemer, Testing individual capability to lift material: repeatability of a dynamic test compared to static testing, *J. Safety Res. 16*:1 (1985).
48. T. G. Mayer, D. Barnes, N. D. Kishino, G. Nichols, R. J. Gatchel, H. Mayer, and V. Mooney, Progressive isoinertial lifting evaluation: I. A standardized protocol and normative database, *Spine 13*(9):993–997 (1988).
49. S. H. Snook and V. M. Ciriello, The design of manual materials handling tasks: revised tables of maximal acceptable weights and forces, *Ergonomics 34*:1197–1213 (1991).
50. J. Alpert, L. Matheson, W. Beam, and V. Mooney, The reliability and validity of two new tests of maximum lifting capacity, *J. Occup. Rehab. 1*(1):13–29 (1991).
51. A. B. Griffin and J. D. G. Troup, Tests of lifting and handling capacity: their repeatability and relationship to back symptoms, *Ergonomics 27*(3):305–320 (1984).
52. J. C. Rosecrance, T. M. Cook, and N. S. Golden, A comparison of isometric strength and dynamic lifting capacity in men with work-related low back injuries, *J. Occup. Rehab. 1*(3):197–205 (1991).
53. M. L. Rowe, Are routine spine films on workers in industry cost- or risk-benefit effective?, *J. Occup. Med. 24*(1):41–43 (1982).

17
Office Ergonomics

Mary Brophy
School of Public Health, State University of New York, Albany, New York

Christin Grant
Center for Ergonomics, The University of Michigan, Ann Arbor, Michigan

I. INTRODUCTION

Computer use has increased dramatically over the past decade, with over 50 million personal computers now in use throughout the United States. The use of video display terminals (VDTs) in the workplace is widespread. Most office workers use computers for at least some of the tasks that they perform, and many use VDTs for the majority of time that they spend at work.

Video display terminal workstations are different from traditional office workstations in several important ways. Keying at a VDT can be faster and more continuous than on a typewriter because VDTs do not have a carriage return, nor do they require each sheet of paper to be manually inserted. New lighting and vision concerns are introduced by the use of the display screen. Many VDT jobs do not provide the opportunity to shift one's body position or perform tasks away from the VDT station. Increased psychological stress has been associated with the use of VDTs. The additional stress may be due to the introduction of new technology and technical procedures, or to conditions created by VDT technology, such as monitoring and reduced interpersonal interaction.

The changes brought about by the development of VDT technology may have contributed to the increase in cumulative trauma disorders (CTDs) and vision problems associated with VDT use. Office workers in the United States have experienced an increase in CTDs since 1986, as reported by the United States Bureau of Labor Statistics [1]. Additional factors, such as increased awareness on the part of office workers and physicians as well as better recording of CTDs may also have contributed to this increased incidence [2]. The effect of increased publicity about CTDs, legislation aimed at regulating VDTs, and fines imposed on employers by the United States Occupational Safety and Health Administration (OSHA) on the reported incidence of CTDs is difficult to measure.

Despite all the publicity about CTDs of the upper extremity, an office worker is 40 times more likely to experience back pain than pain in the upper extremity [3]. Although the incidence of back pain has not increased, both the severity, as indicated by the number of lost work days, and the cost of back injuries have increased [4,5].

Eyestrain, however, is the most frequent physical symptom suffered by VDT users [6,7]. The ocular symptoms reported by VDT users are usually neither permanent not disabling, and an accurate estimate of their incidence is not available. Many optometrists now feel that a significant percentage of the patients they evaluate have sought medical help for symptoms that are associated with VDT use [8].

It is important to remember in any discussion of CTD, back pain, and visual difficulties that good medical management and aggressive problem surveillance programs need to be implemented along with appropriate ergonomic interventions [9]. Furthermore, a professional health care provider should be consulted by workers who are consistently experiencing physical symptoms associated with the use of VDTs.

II. BACKGROUND AND SIGNIFICANCE

Although CTDs among office workers have been described in the medical literature since the end of the nineteenth century, the last quarter of the twentieth century has seen an increase in both the incidence of CTDs and the perception that they are associated with activities performed in the office. Both of these factors have resulted in increased awards from workers' compensation. The increased cost of workers' compensation has been one of the driving forces behind the efforts to reduce CTDs in the workplace.

It is important to remember that several personal predisposing factors have been described in the literature as well. These factors include various diseases, such as diabetes and thyroid disease, as well as traumatic injury to the wrist [10]. Despite the diverse etiology of CTDs, a substantial body of knowledge about the occupational risk factors associated with their development in the workplace has developed over the past 25 to 30 years. Many of these concepts developed from the work done by Dr. Thomas J. Armstrong and his associates at the University of Michigan in the meat-processing plants of the midwestern United States. In this chapter these concepts are applied specifically to the office environment. In addition, some of the literature pertaining to low back injury and eyestrain is reviewed and summarized.

A. Cumulative Trauma Disorders

The definition of cumulative trauma disorder involves three concepts. The injury itself occurs in a "soft tissue" of the body, usually a muscle, tendon, or nerve. The symptoms of injury involve pain, inflammation, and decreased function of the body part affected. The symptoms may be caused, precipitated, or aggravated by repeated motions or sustained exertions of the particular part of the body exhibiting the symptoms [11].

The risk factors associated with the development of CTDs in office workers are:

Repetition: repetitive work without adequate alternative activity to allow for physiological recovery
Sustained or awkward posture: prolonged and/or non-neutral position of any joint
Forceful exertion: any activity requiring excessive strength or accelerated motion
Contact stress: pressure on soft tissues caused by external surfaces
Psychosocial stress: organizational or interpersonal factors resulting in increased actual or perceived stress

1. Repetition

Another name for cumulative trauma disorders is repetitive strain injuries, which emphasizes the association between repeated exertions and the increased incidence of CTDs [12–14]. Keying is the most obvious example of repetitive movements associated with the use of VDTs. When keying at a rate of about 70 words per minute, a VDT operators performs approximately

2100 exertions per hour with each finger, assuming an average of 5 letters per word and an equal number of key strokes for each of the 10 fingers. One of the earliest recommended limits for human tendons is 1500 to 2000 repetitions per hour [15].

One of the best ways to reduce repetition for VDT operators is to incorporate other tasks, such as phoning or making copies, into the daily routine. The introduction of frequent breaks is another important way. Some studies have reported that muscle fatigue as well as perceived discomfort are reduced when frequent breaks are taken [16-20]. Increased performance has also been correlated with the use of frequent breaks [19]. Frequent short breaks, sometimes referred to as "minibreaks," appear to be more helpful than the traditional 15-minute break every 2 hours. Research on data entry work suggests that minibreaks of one minute or less, several times each hour, are more effective in reducing muscle fatigue and perceived discomfort than longer, less frequent breaks [21]. Another study reports that frequently taking two-second breaks reduces fatigue and increases endurance [19].

The effect of performing exercise for the upper extremity during breaks is controversial. Some reports indicate that active exercise may provide more benefits than passive breaks, massage, or resisted exercise [16,20]. Scientists at the National Institute for Occupational Safety and Health (NIOSH) have evaluated 126 different exercises suggested for VDT operators [22]. One of the problems identified by their study is that some of these suggested exercises involve the same kind of musculoskeletal stress produced by VDT use. In addition, the local tissue concentration of the inflammatory neurohormone substance P can be increased with certain exercises. Increased levels of substance P have been associated with the development of upper extremity CTDs [23]. Research has also reported a decrease in the incidence of upper extremity CTDs with increased aerobic exercise and conditioning [24].

2. Sustained or Awkward Postures

Insufficient recovery time is one of the contributing factors to muscle fatigue [25-28]. In sustained static exertions the activity is continuous. Some examples of activities that require sustained static exertion are prolonged sitting, holding the hands above the keyboard, prolonged gripping of a computer mouse or a phone handset, and maintaining the shoulders in a set position while keying.

Several factors contribute to sustained static exertions, including the configuration of the workstation and the postural habits developed by the worker. The adverse effect of sustained static exertion can often be reduced when the worker learns to vary body position and relax tensed muscles. A "pocket" electromyography (EMG) monitor, which senses muscle contraction, can provide feedback to remind an individual to relax the hands or shoulders.

Workplace modification can also help reduce the adverse effects of sustained static exertions. The use of a well-designed telephone headset can be beneficial to workers who spend considerable amounts of time on the phone. It is important that the headset properly fit the individual using it. One cause for continuously elevating or hunching the shoulders is a keyboard that is too high for the user. When the keyboard can be adjusted downwards, it is easier for the VDT operator to relax the shoulder muscles and thereby reduce sustained static exertion in the neck and shoulder region.

When a mouse is used continuously, sustained static exertion can occur. Under these circumstances it is worthwhile to consider using an alternative pointing device as well, such as a trackball. One drawback, however, is that it is not easy for all users to switch back and forth between different devices. Another concern associated with use of a mouse is reaching to grip the mouse when it is not positioned properly. The use of a mouse tray to keep the mouse pad properly positioned close to the user may be helpful.

Several postures have been associated with an increased incidence of CTDs. These nonneutral postures include bending the wrists upwards (extension) or downwards (flexion) during keying, bending the wrist sideways away from the thumb (ulnar deviation), reaching (arm flexion) for an improperly positioned mouse, and raising the shoulders and abducting the arm to position the hands over a keyboard that is too high. Holding a phone headset to the ear, either with the shoulder or with the hand, for extended periods of every work shift can place stress on the shoulders and neck [29–31].

Different physiological effects are produced by different awkward postures. Muscle or tendon fatigue can result in soreness and discomfort. Flexion or extension of the wrists can result in increased pressure in the carpal tunnel [32–34]. Under extreme conditions the pressure in the carpal tunnel is increased 20 to 30 times the normal pressure.

One common cause of flexion or extension of the wrists when keying is improper adjustment of the keyboard height. If the keyboard is too high, VDT operators will often flex the wrist to compensate. Adjustability of the keyboard height and angle so that the VDT operator can key with the wrists in the neutral or straight position often resolves this problem.

Another device that may be helpful in positioning the hands during keying is the wrist rest. The wrist should be approximately the same thickness as the front of the keyboard. If the wrist rest is too thick it can result in the VDT operator flexing the wrist in order to reach the keys, or it can cause pressure in the wrist when resting. The wrist rest should not be used when the operator is actively keying.

Several nontraditional keyboards have recently become available. These keyboards may be split in the middle so that VDT operators may position their hands and forearms from the side instead of directly in front of them, or they may be split in the middle like an A-frame so that operators can position their hands in an inclined position instead of flat on the table or work surface.

Ulnar deviation may occur when an operator bends the wrist laterally away from the thumb to reach keys at the side of the keyboard, for example, the number pad, which is frequently located at the right side of the keyboard. In addition, the tendency to deviate the wrist is increased when the keyboard is positioned too high for the operator. The operator can reduce this risk factor by moving the arm to position the hands over the keys rather than bending the wrist to achieve this goal. Workplace modification such as introducing modified split keyboards can achieve a similar effect.

3. Forceful Exertions

Although evaluation of industrial workers has indicated that excessive force is associated with an increased incidence of CTDs [35–37], muscular exertion is usually not associated with office work. A VDT operator when keying may use more force than is necessary [38]. When this happens, the extra energy is transferred to the tendons and soft tissue of the hand and forearm. Use of excessive force while keying can be a result of habit, internal mood state of the VDT user, or deadlines and other job-induced stress. For example, monitoring of performance without fully involving the employee in the monitoring process can lead to increased stress, which is then translated to increasingly forceful exertions while keying.

Although the use of excessive force may be behavioral, it appears that different keyboards require different amounts of force to depress the keys [38]. A keyboard with a light touch reduces the required amount of force, but this translates into less stress on the upper extremity only if the VDT operator makes the behavioral modifications that result in the application of less forceful exertions to depress the keys.

Other examples of forceful exertion in office work can arise from the use of manual staplers and the physical arrangement of the workstation. The forceful exertion needed for manual staplers can be eliminated by replacing them with electric staplers. When a VDT operator must lift a heavy manual from a high shelf or retrieve a phone book from under the work surface, forceful exertions are necessary. These forceful exertions are exacerbated by the awkward postures, such as a flexed shoulder or back, necessary to complete the action. Workplace modification can correct these types of problems. Objects used often need to be placed within easy reach; objects used less frequently may be placed nearby, but in a location that requires the VDT operator to get out of the chair to retrieve them. This type of workstation layout not only reduces the incidence of forceful exertions in awkward postures, but also provides the opportunity for the operator to take a minibreak during the workday.

Another approach to reducing forceful exertions is worker awareness and training programs. These programs provide awareness of the excessive force associated with keying by showing workers videotapes of themselves.

4. Localized Mechanical Stress

Localized mechanical stress, or contact stress, is caused by the continuous pressure of a sharp or hard surface against the soft tissue of an upper or lower limb. For example, injury of the ulnar nerve in a data entry clerk has been described as the result of the desk edge pressing against the soft tissues of the upper extremity [39]. When the back of the thigh presses against the hard edge of the front of a chair seat, contact stress occurs in the lower limb, which can result in decreased blood flow and nerve compression. Modifying or replacing furniture to eliminate the hard edges reduces the adverse effects of contact stress.

5. Psychosocial Stress

During the past decade, the role of psychosocial stress has become increasingly recognized as a significant factor in the development of CTDs. Several reports have linked psychosocial stress not only to an increased incidence of low back pain but also to increased loss of work time associated with low back pain [6,16,18,31,40–43]. The physiological mechanisms associated with the development of pain include increased muscle tension [16] and metabolic changes, which alter biochemical processes at the cellular level [44]. In addition to physiological changes, stress often induces behavioral changes such as the amount of force used during keying, increased tension in shoulder muscles and/or back muscles, and absenteeism, as well as an increased tendency to report problems and seek professional assistance [45].

Stress in office work of VDT operators has two components: the first is associated with introducing new technology inherent in the use of VDTs; the second is associated with the job demands and job position. The stress contributed by new technology is often transient. Electronic monitoring, however, is a technology-related stress that may not be transient. Electronic monitoring has been used in jobs as diverse as truck drivers, nurses, and telephone operators [6,46].

Stress in VDT operators may be related more to the total job and organizational structure than to the VDTs themselves. Some research has reported that job level is a better indicator of stress than VDT use [6]. For example, those with better jobs are more likely to be able to set their own priorities. Stress has been linked to jobs that include rigid work procedures, lack of social support, monotony, and insecurity [47,48]. Many individuals in these jobs also express dissatisfaction with their position.

Education is an essential part of any ergonomics program for VDT operators at all levels. Education provides the individual with awareness and, thereby, some measure of control over conditions in the workplace. Effectiveness of ergonomic programs has been linked with the availability of adjustable workstations, responsiveness of management, and development of team spirit. When individuals have some measure of control over the work environment, as well as over the organization of their work, the stress level often decreases [6,49,50].

The risk factors described above interact with each other, often synergistically. For example, when a job requires both force and repetition, the risk of developing a CTD is greater than the sum of each risk factor evaluated independently [51].

B. Back Pain Risks in Video Display Terminal Work

Most VDT work is done while sitting. Sitting is generally viewed as preferable to standing to perform VDT work because it uses less energy than standing (it is easier for the cardiovascular system to return blood from the lower limb when a person is sitting) and the lower limb does not have to bear the entire weight of the body. Sitting, however, especially for extended periods of time, introduces specific physiological problems such as a flattening of the lower back (eliminating the lumbar curve or lumbar lordosis) and decreased muscle tone in some muscle groups, for example, the abdominals. Studies have shown that when the lumbar curve is flat there is an increase in pressure in the intervertebral discs in the lumbar region [51–56]. Improperly designed chairs, or chairs that do not properly fit the individual, may exert pressure on the backs of the legs and buttocks and provide poor body support.

Behavior may add further to the stress of sitting. Some studies indicate that sitting, even in healthy postures, for extended periods of time stresses the spine and increases the risk of pain in the low back [57–61]. This theory, however, is not supported by all researchers [62–65]. Slouching and twisting can also contribute to the risk of low back pain [66]. Lifting, or twisting and lifting, while sitting can result in high compressive force on the lumbar discs [66], for example, in office workers who reach for phone books or heavy manuals while performing work at a VDT station.

Another area of concern among VDT users is neck and shoulder pain or discomfort. Often times VDT operators bend the neck in various positions to see the screen or read the material they are copying. Craning the neck, looking up at a screen, bending the neck forward, and twisting the neck to the side have been associated with the development of neck and shoulder disorders [30,67–70].

In addition to physiological risk factors, there are some psychological risk factors. In an office with poor organization and unsupportive management, the incidence of neck and shoulder pain was twice the incidence in offices with a healthier psychological atmosphere [48]. Job dissatisfaction has been correlated with an increased incidence of back injury [71]. The single best predictor of future back injury, however, is previous back injury.

1. Posture and the Work Environment

Application of ergonomics principles to office workstations requires both behavior modification and workstation modification. Although workstation modification initially may demand more expenditure of time and money, behavior modification is ultimately more difficult because it requires changing people's habits. One habit that is particularly helpful to instill is to change position often. This can involve major changes, such as getting up and performing to a task other than keying, or it can involve relatively minor postural changes such as leaning forward, crossing the legs, or simply stretching in the chair. There is no one perfect posture; any posture maintained for extended periods of time can produce physiological stress [72].

Workstation modification should be based on observed risk factors such as awkward postures. For example, a document holder can minimize the amount of neck flexion and twisting necessary to glance back and forth from the computer screen and the paper copy. Correct placement of the monitor eliminates neck extension. Neck extension, craning the neck backwards to look at an upwards angle, creates stress in the neck and shoulders and can lead to neck pain. Arrangement of the work area can reduce occasional stress on the low back if heavy manuals are stored between hip and shoulder height. Placement of often-used items in easy reach of the VDT operator can reduce twisting and reaching.

Several studies have indicated a relationship between lack of aerobic exercise or conditioning and the development of cumulative trauma disorders [18]. In addition, general body conditioning and aerobic exercise have been suggested as a way to counteract the adverse effects of prolonged sitting [16]. The development and implementation of exercise, or wellness, programs have been described in the literature [73,74].

2. Seating

The single most important piece of office furniture is arguably an adjustable chair. When people use VDT workstations for the majority of their time at work, it is essential that the chair be adjustable in height, provide lumbar support, and have an adjustable seat pan and backrest, as well as rounded edges to reduce contact stress. Armrests are desirable and should be adjustable.

Adjustability of the seat height and depth are two important features of an ergonomically designed chair. Workstations that are used by more than one person *must* have a chair with an adjustable seat height, especially if the individuals will be using the workstation for any length of time. Seat height adjustability also allows an individual to adjust to different work demands and posture.

Because of the increased compressive force on the lumbar discs when the lumbar curve is flattened during sitting, it is important to provide lumbar support in the sitting position. This support restores at least some of the lumbar curve. Many chairs have a lumbar support built into the seat back. For this support to be effective, it must be at the appropriate level, that is, at the level of the L5 disc. This is approximately at, or slightly below, belt level. A rolled towel or contoured pillow can be used in chairs without a lumbar support. The thickness of the roll should not be more than 2 in. (5 cm). Research indicates that back rolls that are too thick have resulted in increased muscle activity in the lower back [75].

Another way to decrease the compressive force on the lumbar discs in the sitting position is to open the angle between the thighs and the torso. This can be done by tilting the seat pan forward or by reclining the backrest. Reclining the backrest at least 10 degrees rotates the pelvis and helps to, at least partially, restore the lumbar curve to the low back and transfer some of the torso weight to the backrest [76–78].

Another way to transfer some of the body weight from the spine is to use armrests [56,79]. These must be used at the proper height, which is approximately elbow height when the arms are hanging freely at the side of the chest. Chairs with armrests that are not adjustable may cause more problems than they solve. When armrests are too low, the person using the chair tends to slouch or bend to make the body fit the chair. If the armrests are too high, they may force the chair user to maintain abducted arms, which can contribute to neck and shoulder pain.

In addition to addressing back concerns, a well-designed office chair minimizes leg discomfort associated with compression of the back of the thigh and postural immobility [31,80,81] by providing rounded edges on the seat front as well as adequate cushioning over a firm support. Some research indicates that the seat height contributes to leg discomfort [82]. For example, if the seat height is too high, it will compress the back of the thigh.

All the best features in a chair will not contribute to worker health, comfort, and productivity if the chair does not fit the individual and is not adjusted properly. It is important to emphasize that not only should the chair be adjusted to each individual, but that each individual may need to make small adjustments periodically during the workday.

C. Eyestrain Risks in Video Display Terminal Work

Despite the number of complaints from VDT operators concerning visual discomfort, VDT use has not been scientifically confirmed as the cause of these complaints [83–85]. It is generally accepted, however, that eye discomfort, blurred vision, and headaches can be caused, at least temporarily, by improper viewing or focal distance, inappropriate lighting, glare, and screen characteristics such as contrast, flicker, and jitter.

The concerns about viewing distance can be divided into three categories: (1) looking at objects that are too close, (2) looking at close objects for long periods of time, and (3) looking back and forth between objects that are at different focal distances. Focusing the eye requires muscle activity in the muscles that control the eyeball as well as in the muscles that control the lens curvature. When a person reaches the age of about 40, the lens becomes less pliable and its shape does not change as easily or as readily. Therefore, external lenses (i.e., glasses) are required to help the eye accommodate to various viewing distances.

When objects are too close, both the ocular muscles and the lenticular muscles must perform work, which is the extra effort required for near focusing. Both monitors and documents that are too close to the eyes require extra effort to see. In general, a viewing distance of 50–75 cm (20 to 30 in.) is recommended [86], although the exact distance is often a matter of personal preference and capability. In the traditional VDT workstation, placement of the keyboard a foot or more in front of the monitor can be achieved with an adjustable keyboard tray or stand. Use of a keyboard drawer for this purpose is not recommended unless it is adjustable.

It is important to place the monitor not only at the correct distance, but also at an appropriate level for viewing. Monitors that are placed in a position that requires neck extension, that is, looking up, have been associated with reports of increased discomfort and pain in the neck and shoulders [87,88]. Placing the top of the monitor at or below the line of sight provides the VDT operator with more options for slight postural changes [89].

The laptop computer introduces the problem of a keyboard and monitor that are connected. When the keyboard is in an acceptable position, the screen may be too close. In these cases, it is important to look into the distance periodically to relax the muscles used for close viewing. If the laptop is used in an office setting, it may help to use a detached, full-sized monitor.

The focusing mechanisms of the eye becomes fatigued when there are too few changes in focusing distance. A variety of temporary visual disorders, including blurred vision and possibly nearsightedness, have been associated with constantly focusing on near objects [90,91]. Video display terminal operators can reduce eye fatigue by occasionally looking at more distant objects for a few seconds. Young people can relax their eyes by looking at objects only about 75 cm (30 in.) away. As people get older, the distance necessary to relax the eyes increases, and a distance of about two m (80 in.) is recommended for most workers [92,93]. For people with bifocals or reading glasses, the need for occasionally looking at distant objects disappears, however, because the corrective lenses are doing all of the focusing.

Looking back and forth between two nearby objects that are different distances from the eyes can also be tiring because of the effort to continually refocus. Therefore, the document and the monitor should be about the same distance from the eyes. When both are 50 cm (20 in.) away, a 10-cm (4-in.) difference between them is acceptable. When both are 75 cm (31 in.) away, a 20-cm (8-in.) difference between them is acceptable* [94].

The appropriate amount of light in a particular workstation depends on the worker's age as well as the quality of the printed material. As people get older, they require more light to perform any task. Between 40 and 50 years of age, there is a 35% increase in the required brightness. By age 65 most people need twice the amount of light they required at age 40 [95].

Video display terminal environments have introduced new challenges for providing appropriate lighting in offices. Traditional office lighting recommendations are for about 1000 lux (100 footcandles). There have been reports of increased eye complaints when people use VDTs with this level of illumination [96]. In general, light levels of 200–500 lux (20–50 footcandles) have been recommended for offices where VDTs are used [86,97]. Some research indicates that VDT workers prefer light levels closer to 100 lux (10 footcandles) [98]. Task lighting is recommended in the instances where low illumination levels are provided. Furthermore, it must be recognized that there are personal preferences for lighting levels. The best way to achieve satisfactory results is to allow people to choose, within reasonable limits, the lighting environment that works best for them.

There are two types of glare. Direct glare results from light shining directly into the eyes. Reflected, or indirect, glare results from light bouncing off shiny or reflective surfaces. Direct glare can be detected by shielding the eyes with a hand to obtain immediate relief [99]. Shielding or changing the spatial relationship between the light source and the glare is the easiest way to reduce or eliminate direct glare.

Reflected or indirect glare can appear as hot spots or washout on the monitor screen. Several remedies are available. A good-quality filter appropriate for the monitor can reduce indirect glare. In addition, moving either the light source or the orientation of the monitor, changing the angle of the monitor, or shielding the face of the monitor from the light source can reduce or eliminate reflected glare.

Reflected glare can also occur on documents, particularly those with shiny paper or ink. Changing the angle of the document or shielding the document from the light source often eliminates the problem.

It is important to avoid the introduction of musculoskeletal problems when trying to reduce glare. Changing the position or angle of the monitor may force the VDT operator to maintain an awkward neck posture to read the screen. Furthermore, it may limit the changes in posture the VDT operator may assume and still be able to read the screen. Either of these situations introduces a risk factor discussed in Section II.A.

---

*To calculate more precisely whether differences in focusing distances are more likely to require the eyes to refocus, use this formula: $100/x - 100/y$ should not exceed 0.5, where x is the distance to the farther object (in cm) and y is the difference to the nearer object. The acceptable difference decreases under the dim lighting conditions.

Screen characteristics, such as contrast, flicker, and jitter, can affect the operator's ability to view the screen. These characteristics can usually be adjusted on the monitor itself. Contrast between the screen and the document and nearby walls is also important. Looking back and forth between a dark screen and bright paper document is particularly uncomfortable [100–102]. The document, or background, should be no more than 10 times brighter than the VDT screen, and many experts recommend that it should be no more than 3 times brighter than the VDT screen [89,90]. To achieve the low ratio levels, the ambient illumination should be in the range of 200–500 lux (20–50 footcandle).

In response to the vision stress experienced by some VDT operators, the use of computer glasses has become more prevalent. When a vision concern becomes apparent, the VDT operator should be encouraged to visit a vision professional. Most reading glasses and bifocals are made for an optimal distance of approximately 40 cm (16 in.). The recommended distance for a VDT monitor is between 50 and 75 cm (20 and 30 in.). Instead of having the monitor and document brought closer together, computer glasses can be prescribed for a distance of approximately 50–75 cm (20–30 in.).

III. CRITICAL REVIEW OF CURRENT STATUS

This critical review of the literature indicates that there is a scientific basis for considering cumulative trauma disorders as diagnosable pathological entities. As with all sickness, there is a psychological component. This relationship between the psychological and the physiological needs further study, especially in light of the current research on neurotransmitters and pathway facilitation throughout the central nervous system.
Several aspects of the relationship between the office environment and the development of cumulative trauma require further research, including the role of medical surveillance, break length and scheduling, exercise and stretching routines, management style, and the psychosocial infrastructure.

Several important concerns to VDT operators, including indoor air quality, electromagnetic radiation, and personal predisposing characteristics, are addressed elsewhere. The causes associated with the development of CDTs, back pain, and vision problems are interactive, and it is important to consider their relationship in any office ergonomic evaluation. Because of the multifactorial nature of office ergonomics, remediation needs to include several approaches. It is also important to remember that ergonomics is a process—a way of looking at the world—and that as office conditions change, the scope of office ergonomics will change.

IV. FUTURE CONCERNS

It is always difficult—if not impossible—to predict the future. Based on current trends, however, two scenarios suggest themselves. The first is that computer/office technology constantly changes—and will continue to change rapidly. People performing office work will need to enjoy keeping up with new developments and learning new technologies. This requires full participation in developing the uses to which each new technology is applied.

The second scenario is a radical change from offices as we know them today. Computer technology provides the ability for people who are not in the same physical location to communicate almost as effectively as those who are. When commuting time and hassles are considered, telecommunications may be more effective and less time consuming than actually travelling to the same physical location. This trend will bring about the demise of the large office currently in existence. With that demise, some of the psychosocial and management problems will change dramatically. The physiological concerns, however, will remain more or less consistent with the upper extremity and the low back stressors described in this chapter.

REFERENCES

1. Bureau of Labor Statistics, BLS Report on Survey of Occupational Injuries and Illnesses in 1990 (press release USDC-91-600), and unpublished BLS analyses for finance, insurance, and real estate workers, U.S. Department of Labor, Washington, DC, 1991.
2. J. Franklin, quoted in Labor department: Half of all job illnesses are RSIs. *VDT News 8* (1): 7-8 (1991).
3. A. Pramer, S. Furner, and D. P. Rice, *Musculoskeletal Conditions in the United States*, American Academy of Orthopaedic Surgeons, Park Ridge, IL, 1992.
4. B. S. Webster and S. Snook, The cost of compensable low back pain, *J. Appl. Occup. Med. 1*: 13-15 (1990).
5. B. Webster and S. Snook, July-August 1992 (private conversations). Liberty Mutual Insurance Co., Hopkinton, MA.
6. M. J. Smith, B. G. F. Cohan, and L. W. Stammerjohn, An investigation of health complaints and job stress in video display operations, *Human Factors 23*:387-400 (1981).
7. S. J. Dain, A. K. McCarthy, and T. Chan-Ling, Symptoms in VDU operators, *Am. J. Optom. Physiol. Optics 65*:162-167 (1988).
8. J. E. Sheedy, *Vision Problems at Video Display Terminals: A Survey of Optometrists*, University of California, Berkeley, 1991.
9. *Ergonomics Program Management Guidelines for Meatpacking Plants*, U.S. Department of Labor, OSHA, Washington, DC, 1990.
10. K. H. E. Kroemer, Avoiding cumulative trauma disorders in shops and offices. *Am. Ind. Hyg. Assoc. J. 53*:596-604 (1992).
11. T. J. Armstrong, *Introduction to Occupational Disorders of the Upper Extremities*, Center for Ergonomics, Ann Arbor, MI, 1991.
12. T. Luopajarvi, I. Kuorinka, M. Virolainen, and M. Holmberg, Prevalence of tenosynovitis and other injuries of the upper extremities in repetitive work, *Scan. J. of Work Environ. Health 5*:48-55 (1979).
13. L. Cannon, E. Bernacki, and S. Walter, Personal and occupational factors associated with carpal tunnel syndrome, *J. Occup. Med. 23*:255-259 (1981).
14. T. J. Armstrong, Ergonomics and cumulative trauma disorders of the hand and wrist, in *The Worker and Performer in a Hand Rehabilitation Setting*, W. B. Saunders, Philadelphia, 1988.
15. A. Hammer, Tenosynovitis, *Med. Rec*: 353-355 (1934).
16. K. Maeda, W. Hunting, and E. Grandjean, Localized fatigue in accounting machine operators, *J. Occup. Med. 22*:810-816 (1980).
17. R. A. Henning, S. L. Sauter, G. Salvendy, and E. F. King, Microbreak length, performance and stress in a data entry task, *Ergonomics 328*:855-864 (1989).
18. R. A. Green and C. A. Briggs, Anthropometric dimensions and overuse injury among Australian keyboard operators, *J. Occup. Med. 31*:747-750 (1989).
19. M. Hagberg and G. Sundelin, Discomfort and load on the upper trapezius muscle when operating a word processor, *Ergonomics 32*:1-9 (1986).
20. D. Ferguson and J. Duncan, A trial of physiotherapy for symptoms in keyboard operating, *Austra. J. Physiotherapy 22*:61-72 (1976).
21. *Potential Health Hazards of VDTs*, National Institute for Occupational Safety and Health, Research Report No. 81-129, U.S. Government Printing Office, Washington, DC, 1981.

22. K. Lee, N. Swanson, S. Sauter, R. Wickstrom, A. Waiker, and K. Mangum, A review of physical exercises recommended for VDT operators, *Appl. Ergonomics 23*:387–408 (1992).
23. S. Blair, Office ergonomics and VDTs, American Industrial Hygiene Conference and Exposition, Anaheim, CA, May 21–27, 1994.
24. P. Nathan, R. C. Keniston, L. D. Myers, and K. D. Meadows, Obesity as a risk factor for slowing of sensory conduction of the median nerve in industry, *J. Occup. Med. 34*:379–383.
25. A. C. Guyton, *Physiology of the Human Body*, Saunders, New York, 1989, pp. 650–658.
26. G. B. Sjogaard, K. Kiens, K. Jorgensen, and B. Saltin, Intramuscular pressure, EMG, and blood flow during low-level prolonged static contraction in man, *Acta Physiol. Scand. 128*:475–484.
27. M. W. Lee, A stochastic model of muscle fatigue in frequent strenuous work cycles, Ph.D. dissertation, University of Michigan, Ann Arbor, MI, 1979.
28. G. Sjorgaard, G. Savard, and C. Joel, Muscle blood flow during isometric activity and its relation to muscle fatigue, *Eur. J. Appl. Physiol. 57*:327–335 (1988).
29. D. Ferguson and J. Duncan, Keyboard design and operating posture, *Ergonomics 17*:731–744 (1974).
30. M. A. Life and S. T. Pheasant, An integrated approach to the study of posture in keyboard operation, *Appl. Ergonomics 15*:83–90 (1984).
31. S. L. Sauter, L. M. Schleifer, and S. J. Knutson, Work, posture, workstation design and musculoskeletal discomfort in a VDT data entry task, *Human Factors 33*:151–167 (1991).
32. R. Gelberman, P. Hergenmoeder, A. Hargans, G. Lundborg, and W. Akeson, The carpal tunnel syndrome: A study of carpal canal pressures, *J. of Bone Joint Surg. 63A*:380–383 (1981).
33. T. J. Armstrong, R. A. Werner, J. P. Waring, and J. A. Foulke, Intra-carpal canal pressure in selected hand tasks, Proceedings of the International Ergonomics Association, San Francisco, Sept. 1991.
34. D. Rempel, T. Bloom, R. Tal, A. Hargens, and L. Gordon, A method of measuring intracarpal pressure and elementary hand maneuvers, *Arbete och Halsa*, Stockholm, May 12–14, 1992, pp. 249–251.
35. T. J. Armstrong, Cumulative trauma disorders and office work (course material). The University of Michigan, Ann Arbor, April, 1991.
36. W. M. Keyersling, T. J. Armstrong, and L. Punnett, Ergonomic job analysis: A structured approach for identifying risk factors associated with overexertion injuries and disorders, *Appl. Occup. Environ. Hyg. 6*:333–363 (1991).
37. T. J. Armstrong and D. Chaffin, Carpal tunnel syndrome and selected personal attributes, *J. Occup. Med. 21*:481–486 (1979).
38. T. J. Armstrong, J. Foulke, B. Martin, and D. Rempel, An investigation of finger forces in alphanumeric keyboard work, Proceedings from the International Ergonomics Association, San Francisco, Sept., 1991.
39. P. Todano, A safety/prevention program for VDT operators: One company's approach, *J. Hand Ther. Apr.–Jun*:64–71 (1990).
40. R. Arndt, Working posture and musculoskeletal problems of video display terminal operators: Review and reappraisal, *Am. Ind. Hyg. Assoc. J. 44*:437–446 (1983).
41. M. S. Oxenburgh, S. A. Rowe, and D. B. Douglas, Repetition strain injury in keyboard operators, *J. Occup. Health and Safety 1*:106–112 (1985).
42. B. McPhee, Deficiencies in the ergonomic design of keyboard work and upper limb and neck disorders in operators, *J. Human Ergol. 11*:31–36 (1982).
43. M. Waersted, R. A. Bjorlund, and R. H. Westgaard, Shoulder muscle tension induced ny two VDU-based tasks of different complexity, *Ergonomics 34*:137–150 (1991).
44. D. M. Kiser and S. H. Rodgers, The physiological basis of work, in *Ergonomic Design for People at Work*, Van Nostrand Reinhold, New York, 1986, pp. 17–90.
45. E. A. Scalet, *VDT Health and Safety*, Ergosyst, Lawrence, KS, 1987.
46. S. Danann and T. Moavero, *Stories of Mistrust and Manipulation: The Electronic Monitoring of the American Workforce*, 9 to 5 Working Women Education Fund, Cleveland, 1990.

47. E. Grandjean, W. Hunting, and K. Nishiyama, Preferred VDT Workstation settings, body posture and physical impairments, *Appl. Ergonomics 15*:99–104 (1984).
48. S. J. Linton and K. Kamwendo, Risk factors in the psychosocial work environment for neck and shoulder pain in secretaries, *J. Occup. Med. 31*:609–613 (1989).
49. A. M. Rossignol, E. P. Morse, V. M. Summers, and L. D. Pagnatto, Video display terminal use and reported health symptoms among Massachusetts clerical workers, *J. Occup. Med. 29*:112–118 (1987).
50. "If people think they have even a modest personal control over their destinies, they will persist at tasks and they will do better at them." Quote from T. J. Peters, *In Search of Excellence: Lessons From America's Best-run Companies*, Harper and Row, New York, 1982.
51. B. Silverstein, L. Fine, and T. J. Armstrong, Occupational factors and carpal tunnel syndrome, *Am. J. Ind. Med. 1*:343–358 (1987).
52. A. Nachemson and J. M. Morris, In vivo measurements of intradiscal pressure, *J. Bone Joint Surg. 46*:1077 (1964).
53. H. Okushima, Study on hydrodynamic pressure of lumbar intervertebral disc, *Arch. Jap. Chir. 39*:45 (1970).
54. A. Nachemson and G. Elfstrom, Intravital dynamic pressure measurements in lumbar discs, *Scand. J. Rehab. Med. (Suppl.) 1*:1–40 (1970).
55. L. L. Tzivian, V. H. Rayhinstein, V. F. Motov, and F. F. Ovseychik, Results of clinical study of pressure within the intervertebral lumbar discs, *Otrop. Traumatol. Protez. 6*:31 (1971).
56. G. B. Andersson and R. Ortengren, Lumbar disc pressure and myoelectric back muscle activity during sitting, *Scand. J. Rehab. Med. 3*:104–121 (1974).
57. L. Hult, Cervical, corsal and lumbar spinal syndromes, *Acta Orthop. Scand. Suppl. 17*: (1954).
58. J. L. Lawrence, Rheumatism in coal miners, Part III: Occupational factors, *Br. J. Ind. Med. 12*:249–261 (1955).
59. K. H. E. Kroemer and J. C. Robinette, Ergonomics in the design of office furniture, *Industr. Med. Surg. 38*:115 (1969).
60. R. E. Patridge and J. A. Anderson, Back pain in industrial workers, Proceedings of the International Rheumatology Congress, Abstract 284, Prague, Czechosolovakia, 1969.
61. A. Magora, Investigation of the relation between low back pain and occupation. 3. Physical requirements: Sitting, standing and weight lifting, *Industr. Med Surg. 415*: (1972).
62. W. Braun, Ursachen des Lumbalen Banscheiberverfalls, *Die Wirbelsaule in Forschung and Praxis 43*: (1969).
63. C. G. Westrin, Low back sick-listing: A nosological and medical insurance investigation, *Scand. J. Soc. Med. Suppl. 7*: (1973).
64. M. Bergquist-Ullman and U. Larsson, Acute back pain in industry, *Acta. Orthop. Scand. Suppl. 170*: (1977).
65. H. O. Svensson and G. B. J. Andersson, Low back pain in 40–47 year old men: Work history and work environment factors, *Spine 8*:272–276 (1983).
66. G. B. Andersson and R. Ortengren, Myoelectric back muscle activity during sitting, *Scand. J. Rehab. Med. 3*:73–90 (1974).
67. D. B. Chaffin, Localized muscle fatigue: Definition and measurement, *J. Occup. Med. 15*:346 (1973).
68. S. Kumar and W. G. S. Scaife, A precision task, posture and strain, *J. Safety Research 11*:28–36 (1979).
69. K. Schuldt, J. Ekholm, K. Harms-Ringdahl, G. Nemeth, and U. P. Arborelius, Effects of changes in sitting and work posture on static neck and shoulder activity, *Ergonomics 29*:1525–1537 (1986).
70. W. Hunting, T. Lauble, and E. Grandjean, Postural and visual loads at VDT workplaces. Part 1: Constrained postures, *Ergonomics 24*:917–931 (1981).
71. S. J. Bigos, M. C. Battie, D. M. Spengler, L. D. Fisher, W. E. Fordyce, T. H. Hansson, A. L. Nachemson, and M. D. Wortley, A prospective study of work perceptions and psychosocial factors affecting the report of back injury, *Spine 16*:1–6 (1991).

72. J. Kramer, *Biomechanische Veranderungen in Lumbalen Bewegungsegment*, Hippokrates, Stuttgart, 1973.
73. L. A. Hebert, *Worksmart: The Industrial Athlete*, IMPACC, Bangor, ME, 1989.
74. D. W. Apt, *Back Injury Prevention Handbook*, CRC Press, Boca Raton, FL, 1991.
75. D. Zacharkow, *Posture: Sitting, Standing, Chair Design and Exercise*, Charles C Thomas, Springfield, IL, 1988, pp. 338-342.
76. J. J. Keegan, Alterations of the lumbar curve related to posture and seating, *J. of Bone-Joint Surg. 35*:567-589 (1953).
77. T. Bendix, *Adjustment of the Seated Workplace: With special Reference to Heights and Inclinations of Seat and Table*, Laegeforeningens Forlag, Copenhagen, 1987, pp. 1-23.
78. H. Schoberth, *Sitzhaltung, Sitzchaden, Sitzmobel*, Springer, Berlin, 1962.
79. D. B. Chaffin and G. Andersson, *Occupational Biomechanics*, John Wiley and Sons, New York, 1984.
80. M. Pottier, A. Dubreuil, and H. Monod, The effects of sitting posture on the volume of the foot, *Ergonomics 12*:753-758 (1969).
81. J. Winkel and K. Jorgensen, Evaluation of foot swelling and lower limb temperatures in relation to leg activity during long-term seated office work, *Ergonomics 6*:217 (1963).
82. U. Burandt and E. Grandjean, Sitting habits of office employees, *Ergonomics 6*:217 (1963).
83. M. J. Dainoff, Occupational stress factors in VDT operation: A review of empirical research, *Behavioral and Information Technology 1*:141-176 (1982).
84. T. Laubli and E. Grandjean, The magic of control groups in VDT field studies, *Ergonomics and Health in Modern Offices*, Taylor and Francis, London, 1984.
85. M. G. Helander, P. A. Billingsley, and J. M. Schurik, An evaluation of human factors on VDTs in the workplace, *Human Factors Review 55*:129 (1984).
86. *Human Factors Engineering of Visual Display Terminal Workstations*, American National Standards Institute and the Human Factors and Ergonomics Society (ANSI/HFS 100), 1988.
87. E. Grandjean, W. Hunting, and M. Pidermann, VDT workstation design: Preferred settings and their effects, *Human Factors 25*:161-175 (1983).
88. D. R. Ankrum and K. J. Nemeth, Posture, comfort and monitor placement, *Ergonomics in Design (April)*:7-9 (1995).
89. D. R. Ankrum, E. E. Hansen, and K. J. Nemeth, The vertical horopter and viewing distance at computer workstations, *Symbiosis of Human and Artifact*, Y. Anzai, K. Ogawa, and H. Mori, Eds., Elsevier Science, New York, 1995.
90. T. Grosvenor, A review and a suggested classification system for myopia on the basis of age-related prevalence and age of onset, *Am. J. Optometry and Physiol. Optics 64*:545-554 (1987).
91. W. R. Baldwin, Clinical research and procedures in refraction, *Synopsis of the Refractive State of the Eye: A Symposium*, M. J. Hirsch, Ed., Burgess, Minneapolis, 1967.
92. E. Grandjean, *Ergonomics of Computerized Offices*, Taylor and Francis, London, 1987, pp. 20-21.
93. J. E. Sheedy, personal conversation, School of Optometry, University of California, Berkeley, CA.
94. J. E. Sheedy, personal communication, The University of California, Berkeley, CA.
95. G. C. Fortuin, *Visual Power and Visibility*, Philips Research Report, Eindhoven, 1957, p. 95.
96. J. Nemecek and E. Grandjean, Das Grossraumbure in Arbeitsphysiologischer Sicht, *Industrielle Organisation 40*:233-243 (1971).
97. *Illuminating Engineering Society: IES Lighting Handbook*, IES, New York, 1981.
98. G. van der Heiden, U. Brauninger, and E. Grandjean, Ergonomic studies on computer-aided design, *Ergonomics and Health in Modern Offices*, Taylor and Francis, London, 1984.
99. J. E. Sheedy, Video display terminals: Solving the environmental problems, *Problems in Optometry 2*:17-31 (1990).
100. S. K. Guth, Light and comfort, *Industrial Med. and Surg. 27*:570-574 (1958).

101. S. K. Guth, The science of seeing: A search for criteria, *Am. J. of Optometry and Physiol. Optics* *58*:870–885 (1981).
102. M. Luckiesh and S. K. Guth, Brightness in visual field at borderline between comfort and discomfort, *Illumination Engineering* *44*:650–670 (1949).

18
The Human Factors Aspects of Shiftwork

Debra K. Dekker
COMSIS Corporation, Silver Spring, Maryland

Donald I. Tepas
University of Connecticut, Storrs, Connecticut

Michael J. Colligan
National Institute for Occupational Safety and Health, Cincinnati, Ohio

I. INTRODUCTION

The usual concept of the normal workday is an 8-hour period occurring during the daytime; all other schedules are often viewed as abnormal, unusual, or unnatural [1]. The challenges an individual faces when working outside "normal" hours are complex and multidimensional [2]. Yet, extended and around-the-clock services are rapidly becoming a fact of life for many industries and workers. Work schedules encompassing hours other than daytime are becoming more, not less, prevalent for a greater number of workers. The notion of a "nine-to-five" work schedule may well be a thing of the past.

The term *shiftwork* is often applied to schedules that include hours of work other than daytime, but there is no universally accepted definition of the term [3]. The most comprehensive estimate of the prevalence of shiftwork in the United States is based on the 1991 Current Population Survey, a household sample survey conducted monthly by the Bureau of the Census. The data indicated that of the 80.5 million full-time wage and salary employees, about 18%, or 14.5 million people, had work schedules that differed from daytime hours. Of the 18%, evening schedules were most prevalent at 5.1%, night shifts and "employer-arranged irregular schedules" followed, each at 3.7%, and rotating shifts were reported for 3.4% of total workers [4].

The reasons for around-the-clock hours of work are varied. Economic factors such as recouping a large capital investment on equipment, production cycle time, a high demand for products and services, or lower utility costs during non-peak usage hours [5] may force 24-h operations. Shiftwork is also prevalent in various transportation-related occupations including trucking, railroad, airlines, and shipping. The business concept of just-in-time manufacturing requires a continual movement of product and raw materials. With an increasingly mobile society, airline flights at all hours of the day are extremely convenient for transporting both passengers and freight.

As a society, we are demanding that more services be available on a 24-h basis. Constant medical, police, and fire protection are a requirement in our contemporary society. We realize the merit of grocery and drug stores, gas stations, and convenience stores that are open 24 h a day. During our relaxation periods, we enjoy having restaurants, entertainment, and recreational activities available at any time of the day or night. Nevertheless, in order to accommodate us, someone else must be at work to provide the services we desire!

For the individual, shiftwork can have biological, psychological, or social effects with both short- and long-term consequences. It is not possible within the confines of this chapter to provide an in-depth discussion of every variable that might affect a shiftworker. Those who wish to explore a particular issue in more detail should consult other reviews [3]. The purpose of this chapter is to highlight some of these issues and to alert the reader to things to think about when implementing or evaluating work schedules.

II. BACKGROUND AND SIGNIFICANCE TO OCCUPATIONAL ERGONOMICS

Ergonomics is defined as "the analysis of problems of people in their various working conditions within their real-life situations" [6]. This may require investigations into the individual as well as into the individual's living and working situations. All of these elements can relate to and influence each other. As will be discussed throughout this chapter, shiftwork does affect workers' health and well-being and hence is a topic for ergonomic study. Nonwork activities such as the timing and placement of sleep or domestic responsibilities can also affect work performance. In certain instances, any of these influences upon job performance can pose an immediate threat to the health and safety of the individual worker.

In other extreme instances, the actions (or inaction) of the worker can have disastrous consequences for the greater society at large. For example, the near meltdown at Three-Mile Island occurred during the night shift when workers were presumably tired. Poor judgment related to sleep loss and shiftwork was also cited as a factor in the space shuttle *Challenger* accident [7]. Although shiftwork appears inevitable, interventions are possible to help the worker cope with unusual work hours. Work schedules can be designed to enhance the work environment for the individual worker.

III. WORK SCHEDULING CONSIDERATIONS

A. Types of Shift Schedules

The conventional approach to shiftwork is to divide the day equally into three 8-hr intervals spanning the daytime, afternoon–evening, and nighttime hours. Such a simplistic scheme implies little variety in work hours. But quite the opposite is true with hundreds, if not thousands, of work schedules of unknown origin and rationale in U.S. industry [1]; among U.S. firefighters, for example, over 150 different schedules have been documented [8]. Some of the complexity is due to inconsistent use of terminology. The following are some definitions to assist in describing and communicating work schedule information [1].

The fundamental definition is that of *shift*: the time of day a worker is required to be at the workplace. By this definition, all workers who are scheduled to be at a workplace on a regular basis, including day workers, are shiftworkers. *Work system* describes all the work schedules in the given workplace if formal methods ensure that work requirements are met. When a work system includes schedules for most operations on all seven days of a week, it is said to have *continuous work weeks; discontinuous work weeks* regularly exclude most Sunday and/or Saturday work. Workers who work at the same time of day and no more than one shift per day have *permanent work hours*. If the time of day that an individual works changes according to a planned schedule, the person is said to have *rotating hours*. Some work systems require operations where the duration and time of day when the shift begins are not defined in advance because of unscheduled events; this work schedule is an *irregular shift*.

The *first shift*, also called the *day shift*, in most cases consists of at least seven consecutive hours somewhere between 0600 and 1600 hours (6 a.m. and 4 p.m.). The second or afternoon-evening shift, usually falls such that at least seven consecutive hours are between 1500 and 0100 hours (3 p.m. and 1 a.m.). The third or night *shift* usually falls so that at least seven consecutive hours are between 2200 and 0700 hours (10 p.m. and 7 a.m.). However, these shift designations are not universally accepted; shift designations in U.S. workplaces can vary from one location to another, so references to any work schedule should be as specific as possible.

B. Brief Review of Chronobiology

Human beings are biological systems. Many of our physiological processes display a rhythmic fluctuation on a regular basis. The periodicity or time taken to complete one cycle varies greatly for different processes. Some cycles can be measured in seconds; an example is the electrical activity of the brain as measured by an electroencephalogram (EEG). Other cycles have a periodicity measured in days; the female menstrual cycle has a period of around 28 days.

Biological rhythms with a period of about 24 hr are very common [9]; they are called circadian rhythms, originating from the Latin circa (about) and dies (day). If left to run on their own accord in an environment without time constrains, circadian rhythms tend to have a cycle length of around 25 hr [10]. Body functions such as temperature, sleep/wakefulness, secretion of certain hormones, and the functioning of organ systems including the cardiovascular, pulmonary, renal, and immune systems display circadian rhythms. Certain psychological and mental functions also display a daily rhythmicity [11]. Figure 1 illustrates the circadian rhythms of oral temperature and self-rated alertness over a 48-hr period.

Two sets of components help maintain a daily rhythmicity; endogenous and exogenous oscillators [9,10]. The endogenous oscillators are internal to the body. Some researchers have suggested that one area in the brain, the suprachiasmatic nucleus, is the site of this internal biological clock [12]. However, the notion of only one internal clock has been questioned.

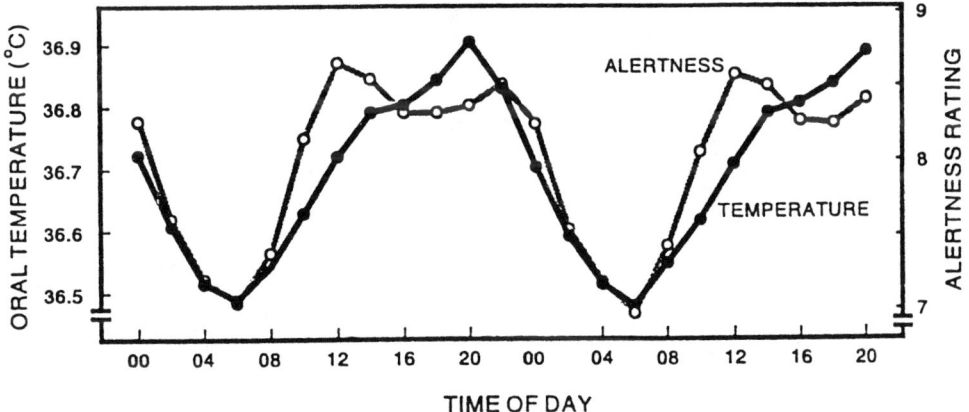

Figure 1 The circadian rhythm of oral temperature (solid circles) and self-rated alertness (open circles). Two cycles of the 12 points defining each rhythm have been plotted. (Reprinted by permission from Ref. 29.).

Exogenous components of circadian rhythms are related to environmental and social cues outside of the body—the cycle of light and dark, social interaction with other people, and the timing of meals are some examples [9]. Exogenous cues act as entrainment agents or zeitgebers (German, time-giver) to synchronize the body's clock so it coincides with the 24-hr solar day [9,10]. Problems are thought to arise when an individual attempts to do activities at a time of day that is discrepant with the body clock activity, for example, a shiftworker on the night shift attempting to maintain alertness and wakefulness when the body clock indicates that it is time to be sleeping. This may result in feelings of fatigue, lethargy, or insomnia-like symptoms.

C. Sleep

One of the most robust findings of shiftwork research is a reduction in sleep length associated with night shiftwork [13–18]. Afternoon-evening shiftworkers have the longest sleep length, and day workers sleep somewhere in between the two; Figure 2 is an illustration of sleep lengths for permanent day, afternoon, and night shiftworkers [17]. People who work on rotating shifts sleep even less on the night shift than permanent night workers. This sleep loss is cumulative, so that by the end of the week the night shiftworker has lost the equivalent of at least one night's sleep. The sleep quality of night workers may also be different from that of day workers [19].

The reason for shorter sleep lengths among night shiftworkers is twofold. First, night workers are attempting to take their main sleep period during the day when the body wants to be awake. Second, humans are also social creatures. In a day-oriented society, there are pressures to conform to the daily behavior of everyone else. Thus, many night shiftworkers may consciously elect to cut short their sleep time in order to spend more time interacting with family, friends, and society [20,21].

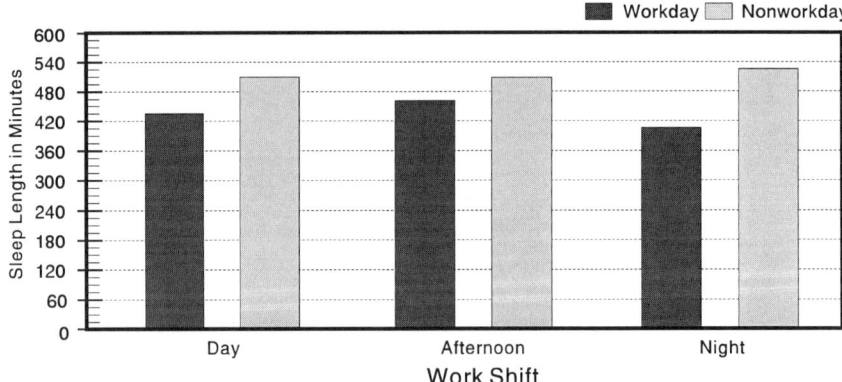

Figure 2 Mean workday and nonworkday sleep length in minutes for a sample of 1262 permanent discontinuous day, afternoon-evening, and night workers. A statistically significant difference was found between workday and nonworkday sleep length. For workday sleep, statistically significant differences were found among the various combinations of shifts. Nonworkday comparisons did not result in differences between any of the shift combinations. (Data adapted and reprinted by permission from Ref. 17.)

A shorter sleep length for a single night may not be detrimental or have a noticeable effect upon performance the next day. In fact, acute sleep deprivation is quite easy to make up during the next sleep period [22]. However, cutting sleep periods short for an extended period of time, such as weeks or months, leads to a condition of chronic sleep deprivation (CSD). The effects of CSD are not as easy to recover from, if one can recover at all. A common but mistaken notion is that one can always catch up on sleep during nonworkdays. If this were true, short sleepers including night workers should sleep longer than day workers on their days off. Research shows, though, that everyone sleeps about the same length of time on days off [17]; this is illustrated in Figure 2.

Chronic sleep deprivation may have long-term effects on a worker's health. It may also affect a worker's job performance. Researchers have demonstrated a correlation between sleep differences and performance decrements [23]. Frequently, CSD can result in lapses of attention in which the individual actually falls asleep for a very brief period of time. This condition is known as a microsleep [22]. Depending on the job, brief inattention may have negative and quite possibly disastrous consequences.

D. Work Performance Effects

Cognitive activity, performance, and subjective feelings like alertness (see Figure 1) also show circadian rhythms and fluctuate with the time of day [24–27]. However, the optimal time of day for performance varies as a function of task demands. A parallel relationship between performance efficiency on a variety of perceptual-motor tasks and the circadian rhythm of body temperature have been demonstrated [28]. Also, reaction time [15] and vigilance may suffer during and as a result of night work. Similar patterns of time of day variation do not appear to be the case for more cognitive tasks. Figure 3 illustrates differences in performance for a memory-based task depending on the cognitive load of the task. Differences in peak performance time among different tasks further suggest that there may be more than one biological clock within the body [29–31].

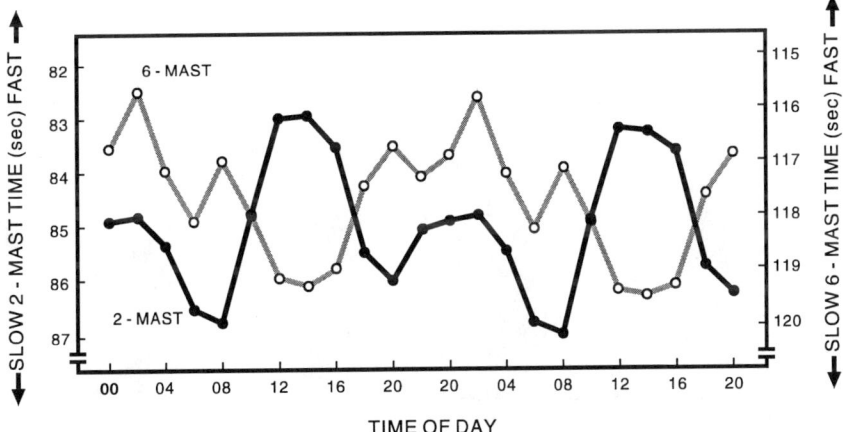

Figure 3 The circadian rhythms of low (solid circles) and high (open circles) memory load performance. The MAST (Memory and Search Task) is a task in which a certain number of items are held in short-term memory while being searched for within a larger list. The more items to be held, the higher the cognitive load. (Reprinted by permission from Ref. 29.)

IV. LIFESTYLE ISSUES FACING THE EMPLOYER AND SHIFTWORKER

A. Gender Differences

As of 1991, in the United States, about 18% of all full-time employed women and 25% of all employed men aged 16 and over worked other than a regular day shift. Women with preschool or school-aged children were over 50% more likely to work fixed nights than those with children over 14 years old [4]. Many European countries follow conventions proposed by the International Labor Organization (ILO) that either totally ban or severely restrict women from working during nighttime hours. A number of leading countries including the United States have never ratified these conventions, objecting that such restrictions lead to unfair employment practices. Amid this debate, the issue remains of whether there are differences between males and females in their job performance as well as in their reactions when working during nondaytime hours.

Gender difference studies of shiftworkers have investigated a wide variety of topics such as child care issues, domestic responsibilities, and time for family and leisure interests [32,33]. Yet with respect to the most robust finding in shiftwork research, sleep length, few studies have been done addressing gender differences. In a study of three samples totaling 2370 shiftworkers, Dekker and Tepas [34] found that women who work night shifts sleep even less than men on the same shift and report a greater amount of sleep difficulties, i.e., falling and staying asleep (Figure 4). Using same-sex day workers as a baseline, however, results indicated a greater percentage of sleep time loss for male night workers relative to their day shift counterparts. A difference score between day and night workers suggests that males may have a harder time than females adjusting to night work.

One factor that may affect sleep length is having a second job, and domestic responsibilities may be just like having a second job. In the previous study [34], over 75% of the females indicated that they had the responsibility for doing the housework whereas less than 20% of the males responded in a similar fashion. A shorter sleep length for females between 18 and 49 years of age was replicated in a recent study [35]. In the 50–59 age group, however, results indicated that males and females sleep about equal lengths of time. This is not surprising because the children of those around the age of 50 are usually leaving the home or

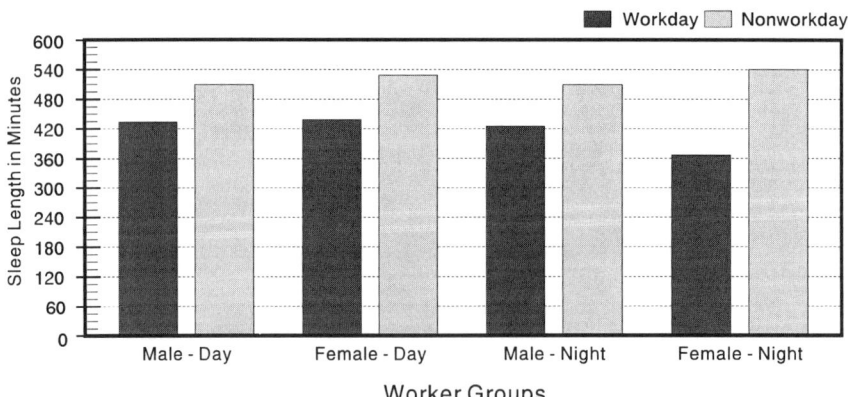

Figure 4 Mean sleep length in minutes for four groups of 101 workers each who were matched on the basis of age and job tenure. These groups were defined as males and females on the day and night shifts. As shown in this figure, females on the night shift sleep the least amount of time on workdays. These results were replicated in two other samples of workers. (Data adapted from Ref. 34.)

are at least of an age where they are less demanding of their mother's attention for immediate care needs. Domestic activities are quite likely to compete for time available for sleep during nonwork hours. Thus, it is not only shiftwork but also nonwork factors that affect sleep behavior.

B. Aging Workers

The question of whether older workers should work at night is a difficult one to answer for many reasons. First, there is considerable debate as to what constitutes "old." Second, there have been few longitudinal studies of the long-term effects of shiftwork, so the effects of working at unusual hours for an extended period of time are not completely understood [30,36]. Finally, many biological and behavioral processes that could be affected by shiftwork are also affected by the aging process itself, making it difficult to draw singular cause-and-effect conclusions.

Although primarily cross-sectional, a few studies that are both timely and commendable have addressed aging issues with respect to shiftwork. For example, using sleep length as a marker variable, Tepas et al. [35] reported that male workers on the afternoon-evening shift show a 50-min decrease in average sleep length from the 18-29 age group to the 40-49 age group; comparable results were found for both male and female night shiftworkers. As expected, age had a minimal effect on the sleep length of male and female day shiftworkers. Other research [18] has demonstrated a similar decrease in sleep quality and sleep length in workers after about the age of 45.

Many workplace practices in the United States allow workers to use seniority status to choose their shiftwork. Usually, senior workers bid out of night work to a more desirable daytime job. The exception to this practice may be the worker to remain on night shift if the plant management pays a high wage differential for night work. Then the earning potential for a worker may far outweigh any reason to leave the night shift. Although the wage differential is relatively small in the United States, averaging 5-10% more for night work, shift differentials in European countries are sometimes as high as 25%.

Studies of former shiftworkers who left shiftwork support the notion that people often leave night work because of a constitutional weakness and/or because shiftwork has a direct effect on their health. Workers who remain are referred to as a "survivor population." This implies that workers who cannot tolerate shiftwork will self-select out of a schedule they cannot tolerate, thereby eliminating the need to be concerned with issues related to shiftwork itself or tenure on a particular shift. Notions like the survivor hypothesis and other motivating factors such as high shift premiums may mask aging issues. Carefully controlled studies of aging and shiftwork are needed because the population in the United States is aging at a rapid rate.

C. Social Considerations of Shiftwork

Although workers may complain about difficulties while working the night shift, people may choose to work the shifts they do for numerous reasons. For many dual-income couples with small children, it may be desirable or even necessary to stagger work hours between the two parents. One parent can then assume child care while the other parent works, with parenting responsibilities shared between the two. For single mothers, an advantage of night work may be easier child care accommodations when the children are sleeping, and the mothers can sleep during the daytime when the children are at school [37]. Other reasons people choose night shiftwork is the time it makes available to participate in daytime activities including taking classes, having a second job for income, or the ease of making doctor, dentist, and other business and personal appointments [38].

At other times, advantages of night work may not outweigh the disadvantages. Further, many people do not have a choice of shift assignment. By far the most frequent complaint among night shiftworkers is the lack of nonwork time available for family and friends, placing a strain on other family members as well [1, 7, 38]. Workers on an afternoon-evening shift also experience social disruption, particularly if their children are school-aged. In this case, the worker is at work during the evening hours when the children are home. In the morning, little if any time is available to spend with the children before they must leave for school. People on afternoon shifts may not see their children at all until the weekend.

The time for socialization with non-family members is also compromised. Most night workers report that their friends do not work the same schedule as they do [14]. In order to interact with these individuals, night workers may elect to cut short their sleep time [20]. In households with children, workers incur an even greater sleep debt than workers without children because parents will delay their sleep until parenting responsibilities are completed [38]. Further, the night worker must sleep during the day at a time when there are external noises from other family members or from the outside, which may also shorten the length of the sleep period.

D. Eating Behavior

An obvious problem with respect to dietary concerns facing night workers is the timing of meals [7, 39]. Usually night workers are asleep during one major daytime meal, either breakfast or, more likely, the noontime meal. When night workers are on the job, eating a main meal during nightime hours may not seem a natural thing to do. In addition, night workers may have limited access to food. Quite often, plant cafeterias are closed during the evening and night hours, with a limited amount of food choices from vending machines as the only alternative. Further, the timing of the ingestion of certain substances such as caffeine may have significant carryover effects to later times of the day or even to later days [40]. In a study of locomotive engineers on irregular work schedules [41], the effects of caffeine upon mood and sleep latency carried over from workdays to nonworkdays.

Many night workers are dissatisfied with their eating habits [39]. Some researchers have suggested that the diet and eating patterns of shiftworkers are out of phase with digestion-related circadian functions. On a long-term basis, these altered eating habits may eventually lead to medical problems. Although there have been some reports of a higher incidence of gastritis and ulcers among night shiftworkers [13, 16], other researchers cannot replicate those results. The relationships between night shiftwork, gastrointestinal complaints, and diet are not fully understood. Complaints by night workers of ulcers and gastritis may be due to a combination of stress, bad food, or other interactions rather than being solely caused by night work per se.

E. Worker Training Programs

Many work schedules can be improved by being redesigned. But if a work schedule cannot be redesigned or modified to worker needs, training is beneficial [22]. Training programs not only alert the worker to potential problems associated with unusual schedules but can also suggest ways to cope with these challenges. For a training program to be most effective, care must be taken in its design and implementation. Too often an encyclopedia of information is thrown at the shiftworker with little attempt at application [42]. A better program is one where objectives are clearly stated and practical solutions are offered [1].

Although many educational programs are available on the market, to date no objective study with suitable experimental controls has been done to assess whether in fact these programs do promote adjustment to shiftwork [22]. Obviously, many techniques to help the worker cope with night work involve lifestyle changes. Such alterations to daily habits are very hard to accomplish and maintain on a long-term basis. What is needed are long-term studies including multiple follow-ups of the training's effectiveness [7].

V. CURRENT STATUS OF SHIFTWORK RESEARCH

This section briefly discusses some of the current knowledge regarding shiftwork practices. A complete and exhaustive review is not possible or practical within the confines of a chapter. The topics selected for inclusion here represent some of the major concerns in regard to work schedules.

A. Permanent Versus Rotating Shifts

One concern when implementing a work system is how to schedule the shifts for 24-hr coverage. There is considerable debate over permanent versus rotating work shifts. Rotating shifts distribute the less desirable work times more evenly among the workforce [38]. However, especially among night workers, a permanent work schedule may not be a permanent lifestyle schedule because nearly all night workers revert to a daytime routine on their non-workdays; permanent night workers are in essence rotating their sleeping and awake hours on a weekly basis [17]. The distinction between permanent and rotating shiftwork is not as absolute as the terms imply.

If rotating shifts are warranted, the next two decisions to be made concern the rate and the order of rotation through the shift schedule. It is generally recommended to rotate in a forward or clockwise direction [1, 7, 30, 43]. That is, regardless of where in the cycle one begins, shifts should move in the order of day, afternoon-evening, and night. The issue is not as clear-cut for rate of rotation. Arguments can be made for either a fast or a slow rotation [44].

In the United States, the two most common rates of rotation are either on a weekly ("fast") or a monthly ("slow") basis. In Europe, a fast rotation is defined as changing work shifts every 2 or 3 days, and a weekly rotation is considered slow. The purpose of a fast rotation is to get the worker through the dyssynchronous phase (i.e., night work) as quickly as possible. The few shifts of night work are not a problem because the individual's body clock continues to run on a daytime schedule. Two disadvantages of rapid rotations are very low nighttime alertness and poor daytime sleep because of the day orientation of the rhythm [7].

On the other hand, the argument for slow rotation is that when an individual is allowed to work for an extended period of time on one shift, the body clock has time to readjust to the new work schedule. Nonetheless, it usually takes about 21 consecutive days [7] on the same schedule for the body temperature circadian rhythm to adjust; other circadian rhythms may adjust at different rates. The problem for slow rotators is similar to that of permanent night shift workers—rarely does anyone work for 21 consecutive days without time off interspersed. On nonworkdays, people revert to a day-oriented routine that throws off the adjustment of the biological rhythms.

This debate on rotation rate has been addressed in a series of review articles [45–47]. Folkard [47] proposed a "best compromise" shift system. According to this advice, the individual designing and implementing a work system should analyze and identify the goals of the overall system and design a work schedule to achieve those goals. If safety is of utmost importance, such conditions might favor a permanent night shift. On the other hand, if social factors are a priority, then rapid rotation (less than three consecutive night shifts) may be more favorable.

B. Should Workers Nap?

As night shift workers tend to have shorter sleep lengths, one way to increase the amount of sleep per 24-hr period is to take naps throughout the day. Depending on their placement, naps can serve various purposes. For example, a *prophylactic* nap is taken prior to beginning an extended work period or unusual work hour to help alleviate feelings of fatigue that may be encountered during the following work period. A *recuperative* nap is taken after the work period to compensate for previous sleep loss. *Maintenance* naps occur on the job during the work shift. Although maintenance napping is not a systematic practice in the United States, it is quite common in Japan [7].

The effectiveness of napping is not yet fully understood. Napping may be beneficial in the short term to maintain alertness, bridge low points in arousal, and provide better recovery than rest without sleep when a worker is changing from day or evening to the night shift. Conversely, napping can hinder adaptation to night work by providing an excuse for sacrificing regular sleep. A study of sleep inertia, the brief time period upon awakening when an individual is not fully functioning, suggests that the time period for sleep inertia may vary depending on when an individual awakens from various stages of sleep [48]. This could be detrimental in a situation where split second decisions are required. Naps can also hinder adaptation of the circadian rhythm because sleep can be an anchor point or zeitgeber for circadian rhythms [7].

C. Do Workers Ever Adjust to Night Shiftwork?

The answer to this is a fairly definitive no [1, 2, 7, 15, 22, 30], although there are large individual differences in tolerance to shiftwork. The problems of reverting to daytime schedules during nonwork periods for night shift workers is a large obstacle to complete adjustment. In both a cross-sectional and a longitudinal study [2], results found no differences in reported workday sleep length for experienced versus inexperienced shiftworkers. Such results suggest that people do not fully adjust to shiftwork.

D. How Should One Select, Implement, or Assess a Work System?

At present, there is no single best work schedule system for all industries, and no silver bullet to solve shiftwork problems for workers in every setting. Rather, we need to evaluate both work schedule problems and solutions on a site-by-site basis—different strokes for different plants [14]. Although customized solutions may not be profitable for some consultants if they are not able to take advantage of generic routines, a customized solution considers the specific needs of the client. Multiple solutions may be appropriate for any industry, and an individualized approach can evaluate all practical interventions and choose the one that shows the best solution to the problem.

The scheduling of work for the operations of any particular plant must take into consideration both the industry's profitability and the labor force's ability to perform in a particular environment. To accomplish this, representatives from both labor and management should be present and involved in determining or assessing any work schedule. The workers who will actually be practicing these schedules should be included among the labor representatives. A cooperative attitude will ensure that work schedules are accepted and adhered to.

Finally, the work system must be continually reevaluated [1]. When a new schedule is implemented, a follow-up study should be scheduled for at least 6 months to 1 year after the initial implementation. At that time, any necessary adjustments in the work schedule system should be made. It is well known among behavioral researchers that any intervention probably will show an initial improvement in behavioral response. This is called the *Hawthorne effect*, first demonstrated in a series of studies at a plant in Cicero, Illinois [49]. Hopefully, the behavior change will be in the direction the practitioner desires! What is more important, however, is to sustain the change for a long period of time. Long-term results are not as easy as they sound; most of the recommendations for coping with shiftwork require lifestyle changes on the part of the worker's and the worker's family.

VI. FUTURE CONCERNS

A. What Do We Still Need to Know About the Effects of Shiftwork?

Although the efforts of researchers to date have been highly commendable, there still remain gaps in our practical knowledge of the effects of subjecting workers to non-daytime work schedules. The following list of research needs is by no means exhaustive or comprehensive. However, it is a list of immediate needs and, it is hoped, ones that for all practical purposes can be implemented promptly.

First, we need more real-life performance studies of on-the-job behavior at all hours of the day and night. Most of the present performance studies are conducted in controlled environments using laboratory-based measures and, in many cases, with college-aged people as experimental subjects. Although this type of research is informative, the degree to which the results generalize to a population of workers is not known. Further, it is extremely difficult, and perhaps impossible, to reproduce in the laboratory the motivation, teamwork, practice, and realistic consequence of work in a factory. Studies that do use real shiftworkers and give them a standard laboratory task to perform before, after, or on top of their real work do not really overcome these problems: the task is still an artificial task rather than motivated and consequential work [46]. Field studies should also include longitudinal designs so that the long-term effects of shiftwork on both health and performance can be adequately assessed [38].

Before we can do controlled field studies, we must first operationally define variables of interest. For example, "fatigue" is a popular notion among lay persons and some researchers. However, an operational definition of fatigue along with measurement methods has yet to be proposed. Subjective measures of mood and alertness [50–52] attempt to quantify subjective feelings; often constructs like fatigue are inferred from these responses. Although these kinds of scales have been validated, one must recognize that they are subjective, state-dependent measurements and are prone to recall bias. Physiological measurements such as eye blink rate or EEG monitors [12] are promising lines of research; validation and practical applications of these methods are yet to be seen.

The prevalence and extent of shiftwork in the United States is not well documented. One reason for this is a lack of standard definitions of terms so that all involved are communicating on the same level. In addition, a complete database of work schedules is not available. It has been recommended that the Bureau of Labor Statistics maintain such a registry of work schedules. We would highly endorse such a practice.

Another recommendation is that "hours of service laws" in all modes of transportation be reviewed immediately. These laws [3] attempt to regulate the number of hours worked for a defined period of time, usually on a weekly basis. Rest periods are prescribed by both the total hours worked during the entire weekly period and minimal off-duty time per 24 hr. For the railroad industry, rules regarding work schedules are documented in the *U.S. Code of Federal Regulations* 49 CFR 228; similar regulations are found in 49 CFR 395 for motor carrier operators. However, some of these laws are based on antiquated practices, if they are based on any practices at all! It is recommended that the laws be assessed and revised given the current state of knowledge regarding unusual work hours. Government, management, and labor support of research in this area is required.

Finally, ethical and public policy considerations of the effects of shiftwork must be addressed. Although disruptions caused by work schedules may seem to be of little immediate concern to the employer, ignoring the potential mishaps that can occur when workers are not performing at peak is unethical practice. For example, one small lapse of attention in either the transportation or nuclear power industries can seriously impair not only the health and well-being of the individual worker but also public safety. Shiftwork will continue to increase in the United States and in the global market. Problems and challenges for both workers and management will arise. These issues must be recognized and dealt with at once.

REFERENCES

1. D. I. Tepas and T. H. Monk, Work schedules, in *Handbook of Human Factors*, G. Salvendy, Ed., Wiley, New York, 1987, pp. 819–843.
2. D. I. Tepas, Adaptation to shiftwork: fact or fallacy? *J. Hum. Ergol.* 11(Suppl.):1–12 (1982).
3. Office of Technology Assessment, *Biological Rhythms: Implications for the Worker*, Rep. No. PB92-117589, 1991, Washington, DC.
4. U. S. Department of Labor, Bureau of Labor Statistics, Washington, DC, 1993.
5. D. L. Tasto, M. J. Colligan, E. W. Skjet, and S. J. Polly, *Health Consequences of Shiftwork*, DHEW Pub. (NIOSH), U.S. Govt. Printing Office, Washington, DC, 1978, pp. 78–154.
6. W. Rohmert, Physiological and psychological work load measurement and analysis, in *Handbook of Human Factors*, G. Salvendy, Ed., Wiley, New York, 1987, pp. 402–428.
7. R. R. Rosa, M. H. Bonnet, R. R. Bootzin, C. I. Eastman, T. H. Monk, P. E. Penn, D. I. Tepas, and J. K. Walsh, Intervention factors for promoting adjustment to night work and shiftwork, *Occup. Med. State Art Rev.* 5(2):391–415 (1990).
8. D. L. Tasto and M. J. Colligan, *Shift Work Practices in the United States*, U. S. DHEW Pub. No. (NIOSH) 77-148, Washington, DC, 1977.
9. D. S. Minors and J. M. Waterhouse, Introduction to circadian rhythms, in *Hours of Work*, S. Folkard and T. H. Monk, Eds., Wiley, New York, 1985, pp. 1–14.
10. T. H. Monk, Research methods of chronobiology, in *Biological Rhythms, Sleep and Performance*, W. B. Webb, Ed., Wiley, New York, 1982, pp. 27–57.
11. T. H. Monk and D. E. Embrey, A field study of circadian rhythms in actual and interpolated task performance, in *Night and Shift Work: Biological and Social Aspects*, A. Reinberg, N. Vieux, and P. Andlauer, Eds., Pergamon, Oxford, England, 1981, pp. 473–480.
12. M. Moore-Ede, *The Twenty-Four Hour Society*, Addison-Wesley, Reading, MA, 1993.
13. M. J. Colligan and D. I. Tepas, The stress of hours of work, *Am. Ind. Hyg. Assoc. J.*, 47:686–695 (1986).
14. D. I. Tepas, D. R. Armstrong, M. L. Carlson, J. C. Duchon, A. Gersten, and D. V. Lezotte, Changing industry to continuous operations: different strokes for different plants: *Behav. Res. Methods, Instum. Comput.* 17(6):670–676 (1985).

15. A. J. Tilley, R. T. Wilkinson, P. S. G. Warren, B. Watson, and M. Drud, The sleep and performance of shift workers, *Hum. Factors* 24(6):629–641 (1982).
16. M. Agervold, Shiftwork—a critical review, *Scand. J. Psychol.* 17:171–188 (1976).
17. D. I. Tepas and A. B. Carvalhais, Sleep patterns of shift workers, *Occup. Med. State Art Rev.* 5(2):199–208 (1990).
18. T. Akerstedt and L. Torsvall, Shift work: shift-dependent well-being and individual differences, *Ergonomics* 24(4):265–273 (1981).
19. J. K. Walsh, D. I. Tepas and P. D. Moss, The EEG sleep of night and rotating shift workers, in *The Twenty-Four Hour Workday: Proceeding on Variations in Work-Sleep Schedules*, L. C. Johnson, D. I. Tepas, W. P. Colquhoun, and M. J. Colligan, Eds., DHHS/NIOSH Pub. No. 81-127, U.S. Govt. Printing Office, Washington, DC, 1981, pp. 371–381.
20. R. P. Mahan, A. B. Carvalhais, and S. E. Queen, Sleep reduction in night-shift workers: is it sleep deprivation or a sleep disturbance disorder?, *Percept. Motor Skills* 70:723–730 (1990).
21. T. H. Monk, Social factors can outweight biological ones in determining night shift safety, *Hum. Factors* 31(6):721–724 (1989).
22. D. I. Tepas and R. P. Mahan, The many meanings of sleep, *Work & Stress* 3(1):92–102 (1989).
23. D. I. Tepas, J. K. Walsh, P. D. Moss, and D. Armstrong, Polysomnographic correlates of shift worker performance in the laboratory, in *Night and Shift Work: Biological and Social Aspects*, A. Reinberg, N. Vieux, and P. Anlauer, Eds., Pergamon, Oxford, England, 1981, pp. 179–186.
24. P. Colquhoun, Biological rhythms and performance, in *Biological Rhythms and Performance*, W. B. Webb, Ed., Wiley, New York, 1982, pp. 59–86.
25. S. Folkard and T. H. Monk, Circadian rhythms in human memory, *Br. J. Psychol.* 71:295–307 (1980).
26. T. H. Monk, P. Knauth, S. Folkard, and J. Rutenfranz, Memory based performance measures in studies of shiftwork, *Ergonomics* 21(10):819–826 (1978).
27. S. Folkard, Shiftwork and performance, in *Biological Rhythms, Sleep and Shift Work*, L. C. Johnson, D. I. Tepas, W. P. Colquhoun, and M. J. Colligan, Eds., DHHS/NIOSH Pub. No. 81-127, U.S. Govt. Printing Office, Washington, DC, 1981, pp. 283–305.
28. S. Folkard, and T. H. Monk, Shiftwork and performance, *Hum. Factors* 21(4):483–492 (1979).
29. T. H. Monk and D. E. Embrey, A field study of circadian rhythms in actual and interpolated task performance, in *Night and Shift Work: Biological and Social Aspects*, A. Reinberg, N. Vieux, and P. Andlauer, Eds., Pergamon, Oxford, England, 1981, pp. 473–480.
30. T. H. Monk and D. I. Tepas, Shift work, in *Job Stress and Blue Collar Work*, C. L. Cooper and M. J. Smith, Eds., Wiley, New York, 1985, pp. 65–84.
31. T. H. Monk, E. D. Weitzman, J. E. Fookson, M. L. Moline, R. E. Kronauer, and P. H. Gander, Task variables determine which biological clock controls circadian rhythms in human performance, *Nature* 304:543–545 (1983).
32. H. B. Presser, Young American parents as shift workers: their distinctive sociodemographic characteristics, in *Contemporary Advances in Shiftwork Research*, A. Oginski, J. Pokorski, and J. Rutenfranz, Eds., Medical Academy, Krakow, 1987, pp. 171–180.
33. G. L. Staines and J. H. Pleck, *The Impact of Work Schedules on the Family*, Univ. Press, Michigan, Ann Arbor, MI, 1983.
34. D. K. Dekker and D. I. Tepas, Gender differences in permanent shift worker sleep behavior, in *Shiftwork: Health, Sleep and Performance. Studies in Industrial and Organizational Psychology*, Vol. 10, G. Costa, G. Cesana, K. Kogi, and A. Wedderburn, Eds., Peter Lang, Frankfurt, Germany, 1990, pp. 77–82.
35. D. I. Tepas, J. C. Duchon, and A. H. Gersten, Shiftwork and the older worker, *Exp. Aging Res.* 19, 295–320.
36. T. Akerstedt, Sleepiness as a consequence of shift work, *Sleep* 11(1):17–34 (1988).
37. H. B. Presser, The growing service economy: implications for the employment of women at night, in *Shiftwork: Health, Sleep and Performance. Studies in Industrial and Organizational Psychology*, Vol. 10, G. Costa, G. Cesana, K. Kogi, and A. Wedderburn, Eds., Peter Lang, Frankfurt, Germany, 1990, pp. 131–136.

38. M. J. Colligan and R. R. Rosa, Shiftwork effects on social and family life, *Occup. Med. State Art Rev.* 5(2):315–322 (1990).
39. D. I. Tepas, Do eating and drinking habits interact with work schedule variables?, *Work & Stress,* 4(3):203–211 (1990).
40. M. J. Muehlbach, P. K. Schweitzer, M. L. Stuckey, and J. K. Walsh, The effect of caffeine on continuous performance at night, *Sleep Res.* 20:464 (1991).
41. D. K. Dekker, M. J. Paley, S. M. Popkin, and D. I. Tepas, Locomotive engineers and their spouses: coffee consumption, mood and sleep reports, *Ergonomics* 36(1–3):233–238 (1993).
42. D. I. Tepas, Educational programmes for shift workers, their families and prospective shift workers, *Ergonomics* 36(1–3):199–209 (1993).
43. C. A. Czeisler, M. C. Moore-Ede, and R. M. Coleman, Rotating shift work schedules that disrupt sleep are improved by applying circadian principles, *Science* 217:460–463 (1982).
44. T. H. Monk, Advantages and disadvantages of rapidly rotating shift schedules—a circadian viewpoint, *Hum. Factors* 28(5):553–557 (1986).
45. R. T. Wilkinson, How fast should the night shift rotate?, *Ergonomics* 35(12):1425–1446 (1992).
46. A. A. I. Wedderburn, How fast should the night shift rotate? A rejoinder, *Ergonomics* 35(12):1447–1451 (1992).
47. S. Folkard, Is there a "best compromise" shift system? *Ergonomics* 35(12):1453–1463 (1992).
48. P. Naitoh, T. Kelly, and H. Babkoff, *Sleep Inertia: Is There a Worst Time to Wake Up?*, Naval Medical Res. Rep. No. 91-45, 1993, Washington, DC.
49. F. J. Roethlisberger and W. J. Dixon, *Management and the Worker*, Harvard Univ. Press, Cambridge, MA, 1939.
50. E. Hoddes, V. Zarcone, H. Smythe, R. Phillips, and W. C. Dement, Quantification of sleepiness: a new approach, *Psychophysiology* 10(4): 431–436 (1973).
51. L. C. Johnson and P. Naitoh, *The Operational Consequences of Sleep Deprivation and Sleep Deficit*, AGARD Rep. No. AG-193, NATO, London, 1974.
52. D. M. McNair, M. Lorr, and L. F. Droppleman, *The Profile of Mood States*, Educational Testing Service, San Diego, CA, 1971.

19

ErgoMOST: An Engineer's Tool for Measuring Ergonomic Stress

Kjell B. Zandin, David L. Gardner, Edward J. Gill, and James R. Wilk
H. B. Maynard and Company, Inc., Pittsburgh, Pennsylvania

I. INTRODUCTION

In recent years, industry has become increasingly concerned about cumulative trauma disorders (CTDs) and other effects of ergonomic stress in the workplace. To address this concern, many different approaches have been taken to implement ergonomic programs. One approach, enacted by larger companies, is to engage the services of professional ergonomists to help determine causes of CTDs and implement solutions to these concerns. This approach satisfies many requirements of an ergonomic program but tends to be very costly.

A second approach is for companies to empower their manufacturing engineers, who have the responsibility for designing workplaces, methods, and tooling, to investigate and address the ergonomic problems that arise in the company. To accomplish this, companies have employed professional ergonomists to instruct these engineers in the identification of ergonomic stressors that occur in manufacturing operations. This approach provides engineers with a tool to evaluate ergonomic stress of operations. However, when the instruction is completed, the technique is found to be applicable to only a small range of operations, due to the extended application time required to apply the technique.

Although a company may initiate one or the other of these two approaches, the expense and time required makes these solutions slow to implement.

The flaws of these approaches create the need for a tool that provides the accuracy of a professional ergonomist and application time sufficient to cover a large number of operations while satisfying the ergonomic stress evaluation requirement of a comprehensive ergonomics program. To achieve this goal, a software application called ErgoMOST has been developed that provides a direct link between ergonomic analysis and a computer-based work measurement system [1]. This software program allows the engineer to perform ergonomic analyses on predefined methods that evaluate ergonomic stress and pinpoint potential ergonomic problems of an operation. The engineer can then use the software to simulate method improvements and job rotation to find the situations that minimize the operator's exposure to ergonomic stress.

II. BACKGROUND AND SIGNIFICANCE TO OCCUPATIONAL ERGONOMICS

Ergonomic analysis requires expertise in ergonomics and the principles of work as well as a significant degree of judgment and interpretation of the observations.

Many ergonomists incorporate task analysis, either formally or informally, when reviewing jobs and making ergonomic recommendations. Gathering and documenting the detailed steps of a method in order to facilitate ergonomic evaluation is a large part of this approach. In *Occupational Biomechanics,* Chaffin and Anderson [2] cite the potential for linking ergonomic analysis with elemental task analysis such as that used for time study or, on a more detailed level, MTM (methods-time measurement) analysis.

ErgoMOST is a tool that links ergonomic analysis with task analysis. This link reduces the application time due to the fact that the task analysis is completed by using the MOST (Maynard Operation Sequence Technique) work measurement technique. MOST is a predetermined motion time system developed in the early 1970s that combines a variety of engineering techniques into a single analytical tool.

This chapter discusses how ErgoMOST can be incorporated into a comprehensive ergonomics program. A case study is presented to demonstrate how ErgoMOST is used to detect ergonomic problems and to stimulate and evaluate possible alternatives to an operation, thus providing the information necessary to determine the method that minimizes ergonomic stress.

III. OVERALL ERGONOMIC PLAN

ErgoMOST is an important part of a comprehensive ergonomics program. The system provides the engineer the means by which to evaluate the ergonomic aspect of operations that have a method defined in the MOST work measurement system. Ergonomic evaluation is one part of a comprehensive ergonomics program that should include the following elements:

Employee and workplace audit
Ergonomic stress evaluation
Ergonomic design of products, workplaces, methods
Environmental evaluation and improvement
Ergonomic program organization
Education and training program
Fitness and rehabilitation program
Reporting, feedback, and follow-up

All of these elements are essential to the successful implementation of ergonomic principles into the everyday operation of a facility. The following is a description of each element of a successful ergonomics program.

A. Employee and Workplace Audit

The employee and workplace audit employs questionnaires to document employees' observations. This input is combined with observations of the engineer of methods and workplace design. The employee feedback and workplace design observations are compared to injury histories to determine relationships. Finally, a list of operations that should receive immediate attention with respect to improvements is created.

B. Ergonomic Stress Evaluation

ErgoMOST is used to evaluate the ergonomic stress of the operations that were listed in the audit. The system considers the ergonomic factors of force, posture, repetition, grip, and vibration, making it possible to identify the ergonomic stress that a joint incurs when performing a series of motions. Ergonomic problems are then identified and pinpointed at the method step level. An ergonomic stress index (ESI) is established for each joint for the entire job to indicate whether the operation presents a low, medium, or high risk for each of the factors (force, posture, repetition, grip, and vibration). An ESI is a 1–5 rating that classifies the risk for each body member. This ESI will be discussed in a later section.

C. Ergonomic Design of Products, Workplaces, and Jobs

Based on the results of the audit and the ergonomic stress evaluation, the products, workplaces, and jobs may need to be redesigned or controlled to meet human requirements. This activity will add the ergonomic component to a methods engineered process for productivity improvement. A combined effort to improve job ergonomics and economics can result in synergistic effects and attractive cost reductions.

To the greatest possible extent, the objective of this phase is to adapt the physical and organizational work conditions to better fit the human physiology. Ideal operator positions (sitting and standing) can be defined and applied in the design and modification of the workplace. ErgoMOST can be used for evaluating these method improvements and the effects of job rotation.

D. Ergonomics Program Organization

The commitment and support of top management is critical to the successful implementation of an ergonomics program. Because the program will affect all employees in an organization, it is essential to establish a project organization to reflect this fact.

A steering committee sets the direction, reviews the project, and makes necessary decisions. An ergonomics coordinator is responsible for the projects and activities relating to ergonomic evaluations and improvements. The coordinator will report to the steering committee, and involvement by the safety and medical departments is also required.

E. Education and Training Program

To increase awareness and understanding of ergonomics in the workplace, an educational program for all levels of the organization is designed and implemented. It is important to make training sessions and training material simple and understandable for all employees. The relationship between job design and employee health is reviewed, as are methods to reduce the risk of injury. The employees are encouraged to discuss problems and improvement ideas pertaining to their own workplaces and jobs with management and/or the ergonomics coordinator.

F. Fitness and Rehabilitation Program

Through consultations with medical personnel, a fitness program is developed to improve individuals' physical capacity, psychological awareness, and motivation to participate in the program. The medical personnel also prescribe rehabilitation procedures for those individuals who have experienced a CTD or any other injury.

G. Reporting Feedback and Follow-Up

The ergonomics program is a continuous improvement program linked to other industrial engineering programs such as productivity improvements based on methods engineering. Therefore, to assess the progress and results of ergonomic efforts and to meet OSHA requirements, a reporting and feedback system is developed to include employee input and injury monitoring.

IV. ERGOMOST AND THE ERGONOMIC PLAN

ErgoMOST is used to evaluate and quantify the ergonomic stress of operations, which is one component of the comprehensive ergonomics program. ErgoMOST attaches ergonomic analyses to work measured with the MOST system. The basic task analysis is an easy-to-follow, step-by-step method that is part of a MOST analysis.

Given the details of a task analysis with time values for each step in MOST, the engineer now needs to apply ergonomic analysis to each of the steps. The engineer requires some additional training to gain an understanding of postures and their potential effects on the worker. Posture analysis is best explained by Armstrong et al. [xx].

Using posture classifications determined from research, ErgoMOST allows the engineer to assess the posture of various body members (wrists, elbows, shoulders, back, neck, and knees) for each step of a process. The engineer classifies the posture for each joint for each method step. The system also allows the engineer to classify the grip that the operator is using in each method step as well as the vibration stress to which the operator is exposed.

The method, with the accompanying posture, grip, and vibration analysis, is then supplemented with information regarding the forces employed at each step of the process. The engineer easily determines the forces by weighing the various articles handled, by measuring the forces using a hand-held dynamometer, or by estimating the force requirement.

The ErgoMOST system then uses the input provided by the engineer to quantify the ergonomic stress of an operation by assigning an ESI for each joint in terms of force, posture, and repetition. The system also assigns ESI values for the grip stress and vibration. The vibration ESI is determined by taking both the severity and duration of exposure in the process into account.

An ergonomic stress index (ESI) is a 1–5 rating system. These ratings indicate potential risk for each joint in the following manner:

1–2	Low risk
3	Medium risk
4–5	High risk

A rating of 1 or 2 represents low ergonomic risk, indicating that the method can be implemented. A rating of 3 represents medium ergonomic risk, indicating that the method should be monitored to ensure that operators are not experiencing symptoms of CTDs. A rating of 4 or 5 represents high ergonomic risk, indicating that controls must be implemented to reduce the ergonomic stress to which the operator is exposed.

Measurement of ergonomic stress should not be considered an end but rather a means for improving operations. Because absolute limits are not known and the variations of dealing with the human body are so diverse, ErgoMOST is recommended for use in a comparative sense. Method steps measuring the highest ergonomic stress levels are targeted for analysis to determine whether they can be improved.

After the initial ErgoMOST analysis has been completed, it can be used in a variety of ways to pinpoint the ergonomic stressors contributing to the ESI ratings and then to simulate method improvements. The following capabilities make the system a useful tool for the engineer to assess ergonomic stress in an operation or group of operations.

The ErgoMOST system pinpoints problems in an operation. This is beneficial in identifying and documenting specific ergonomic problems of an operation in terms of force, posture, repetition, grip, and vibration. Documentation of problems and improvements is essential to obtaining a complete evaluation of ergonomic stress.

It simulates method improvement. Using ErgoMOST, an engineer can simulate an ergonomic improvement to determine if the improvement will help alleviate the problem. A simulation of a new or revised operation allows the engineer to identify potential ergonomic stress before the operation is put into effect.

It analyzes groups of processes. Since many ergonomic stressors are alleviated by job rotation, various combinations of tasks can be simulated and analyzed to assign a rating for a group of processes to determine the effects of leveling the peak stresses. ErgoMOST allows the engineer to specify exact quantities produced against each standard in a single shift.

It documents problems and solutions. ErgoMOST provides standard documentation of ergonomic problems and the actions taken to reduce them, thus showing the efforts made to reduce ergonomic problems in the facility.

The ErgoMOST technique has been applied in a wide variety of manufacturing, distribution, agricultural, food processing, and other industries. ErgoMOST can be applied in any industry in which ergonomic risks are likely to be a problem. It is important to note that operations that appear to be inefficient and at a high degree of ergonomic risk usually produce the highest ESI ratings. Even more interesting is the fact that basic method improvements usually result in a reduction of the ergonomic risk of the operation. Specific examples of method improvements and their effects on ergonomics are shown in the case study in Section VI, but the following are a few examples of basic methods engineering principles:

Workplace heights. Workplace heights should be adjustable to accommodate the varying heights of operators.

Material placement. Simple arrangement of materials with consideration given to wrist, elbow, and shoulder postures can result in greatly reduced stress. For example, placement of material in a container with an angled opening enables an operator to remove objects from the container or replace them with minimal wrist deviation.

Forces. Updating equipment to use modern actuating devices such as proximity switches rather than palm buttons can reduce forces that pose a risk of injury to the operator.

Job rotation. ErgoMOST measurements have shown that combinations of jobs can be devised that level the risk of certain high-stress operations because the operation is distributed among several employees rather than concentrated on one.

Tooling application and design. Obvious posture improvements can be seen and measured when a poorly designed tool is replaced with the proper tool or a specifically designed tool that meets the requirements of its application.

ErgoMOST was designed to provide the engineer with a technique that can be applied on a rather large scale to perform ergonomic analyses of operations and to ensure that these analyses translate into improvements implemented on the floor to provide safe and efficient processes for workers to perform.

VI. ERGOMOST CASE STUDY: LIGHT ASSEMBLY OPERATIONS

A. Situation

Acme Company had a number of reported cumulative trauma disorders in their light assembly department and was faced with the task of lowering the staggering compensation costs that were being incurred within the facility. Acme wanted to eliminate the ergonomic stressors from their operations without reducing productivity levels. To combat the problems in the assembly department, Acme, a MOST computer system user, set up a comprehensive ergonomic plan that included the use of ErgoMOST as a tool for evaluating ergonomic stress. The goal of the program was to eliminate existing ergonomic problems as well as to prevent the implementation of new operations that have serious ergonomic stressors present.

Acme followed the prescribed ergonomic plan discussed earlier. The following paragraphs present the results of implementing this plan.

B. Employee and Workplace Audit

The first step in alleviating the ergonomic stressors was to identify the jobs that were responsible for them. The assembly department was studied and the following observations were made:

1. The assembly department contained many operations that had highly repetitive work associated with them.
2. Some of the operations had workstations that were not designed properly for the work performed there or for the operator performing the job.
3. Job rotation was not being used as a tool to give an operator a break from these highly repetitive jobs.
4. Employees complained of discomfort from performing these light assembly operations. This discomfort was often experienced during the nighttime hours.
5. Injury history indicated that jobs 10 and 11 had the highest incidence of injury per 1000 hours worked.
6. Some operations consisted of 100% manual assembly whereas others had a process time embedded in the operation that allowed the operator to rest.
7. The operations were responsible for some posture problems associated with the introduction and dispersion of finished assemblies.
8. Some operators experienced awkward postures due to their orientation to the line.

Employees were also interviewed to obtain their input concerning any discomfort or other problems they might have in performing the operations.

After the conclusion of the Employee and Workplace Audit, a list of operations was made to determine the order in which operations would be studied within a facility or department. This list was constructed by investigating injury history and studying the employee feedback and workplace observations that were collected during the audit.

The approach at this initial stage was reactive because the company was addressing preexisting problems rather than designing efficient, safe methods at the beginning. This reactive approach to attacking ergonomic problems was imperative because CTDs were being reported. After issues concerning existing operations were resolved, a proactive approach could be taken.

A proactive approach to ergonomic stress should be the goal of every company. When a proactive approach is taken, ergonomic problems in methods and workplaces are designed out of the job or controlled before the process is introduced on the floor. To determine the suitability of a method when a new process is introduced, sound methods engineering must be used to ensure the accuracy and reliability of any analysis of a chosen method.

Because ErgoMOST assigns ergonomic ratings within a predetermined motion time system, processes can be simulated before they are performed. If a potential problem is perceived, then method improvements and job rotation can be simulated to assess various alternatives. This will ensure that the best possible method is implemented on the floor only after a proper cost/benefit analysis has been performed.

C. Ergonomic Stress Evaluation

ErgoMOST was used to evaluate the ergonomic stresses of the operations in the assembly department. The ErgoMOST ratings for the operations in the assembly department are displayed in Figure 1. It can be seen that operation 11 contained the most posture problems, with an unacceptable ESI=5 for the right shoulder. Further investigation into the method determined that the problems occurred in the introduction of materials to the operator and the extreme body motions involved in placing the finished assemblies aside. Following an improvement of the method, the ESI was reduced to 3, which is acceptable.

Operation 10 was identified to be most repetitive in nature, with an unacceptable ESI=5 for both the left elbow and the left shoulder. To alleviate the problems in this operation, operators were rotated through this station; this remedy resulted in an acceptable ESI=3.

The ErgoMOST ratings provided the engineer with quantitative ratings of problems that exist in light assembly. The ESI ratings and the Employee and Workplace Audit provided the engineer with ideas for reducing the ergonomic stress to which the operators were exposed. From this point, ErgoMOST was used to simulate implementation of engineering controls such as refining the method and administrative controls such as job rotation to reduce the high posture and repetition ratings.

D. Ergonomic Design of Products, Workplaces, and Jobs

From the evaluation, further simulations of possible method improvements and job rotation were explored. Further investigation into the ErgoMOST ratings targeted some potential method improvements through the reorientation of the operator to the assembly line to facilitate smoother transition of the materials to and from the line. An ErgoMOST analysis performed on the improved operation showed a reduction in the awkward postures associated with the operation as evidenced by the lower posture ratings for the operation. However, the operation still had some ESI ratings of 3 or greater, indicating that more had to be done to make the operation safer. Figure 2 shows the ratings of the current method versus the improved method.

Job rotation was explored as a means of alleviating the repetitive ergonomic stress that was present in operation 10. Production quantities were entered into the ErgoMOST system to simulate a rotation within a group of four jobs to determine if there would be a reduction in the overall ratings, with all operators being exposed to approximately the same levels of ergonomic stress. Simulation of various job rotations indicated that the lowest ratings would be achieved by combining operations 8 and 11 and combining operations 9 and 10. Figure 3 shows the ratings of these combinations.

This rotation would allow each operator to perform one operation that is machine-aided and one operation that is strictly manual assembly, thereby eliminating the extremely high

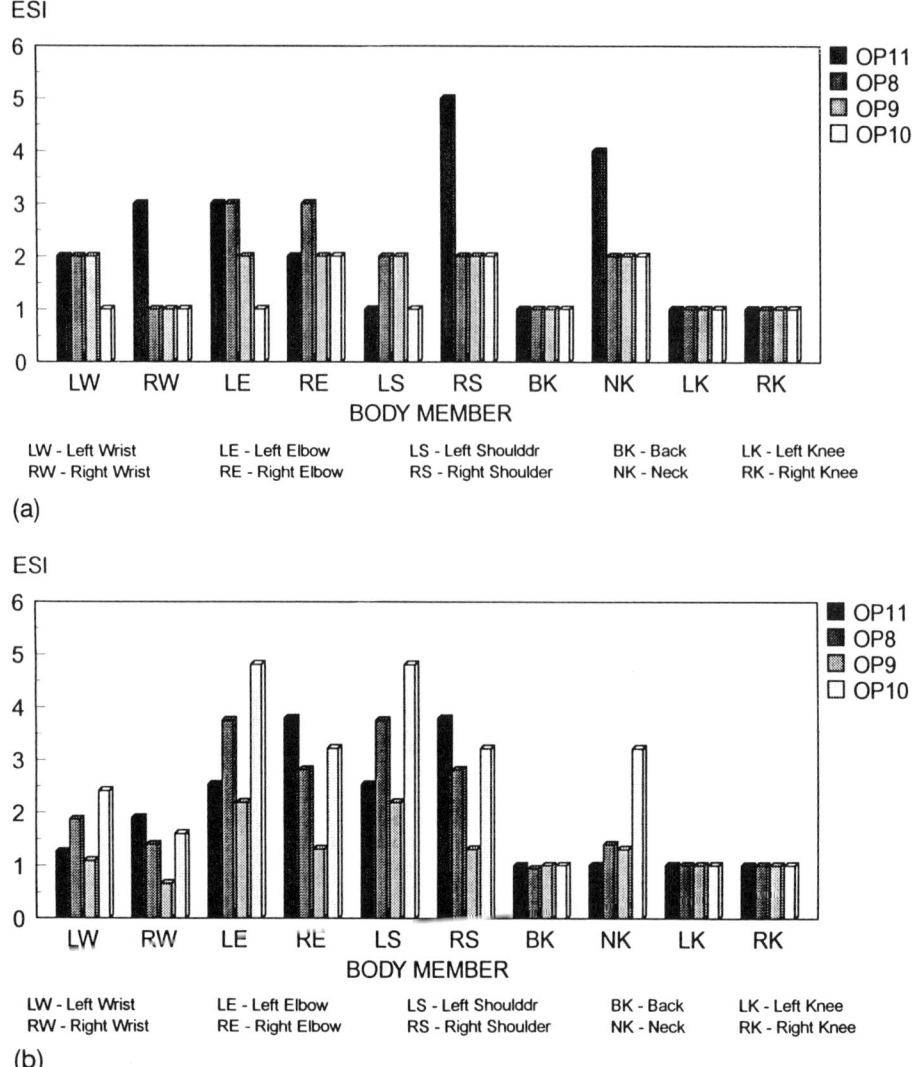

Figure 1 Ergonomic stress indices for light assembly. (a) posture ESI; (b) repetition ESI.

ratings that were present in some of the jobs that did not contain process times. Conversely, it increased some of the ratings relative to those of the operations that did contain a process time. Implementation of the engineering and administrative controls discussed above "smoothed" the ratings for all operators and eliminated all the extremely high ratings that might contribute to workplace injuries.

E. Ergonomic Program Organization

During the audit, a steering committee was organized. This committee consisted of representatives from management, medical personnel, and members of the workforce. The commit-

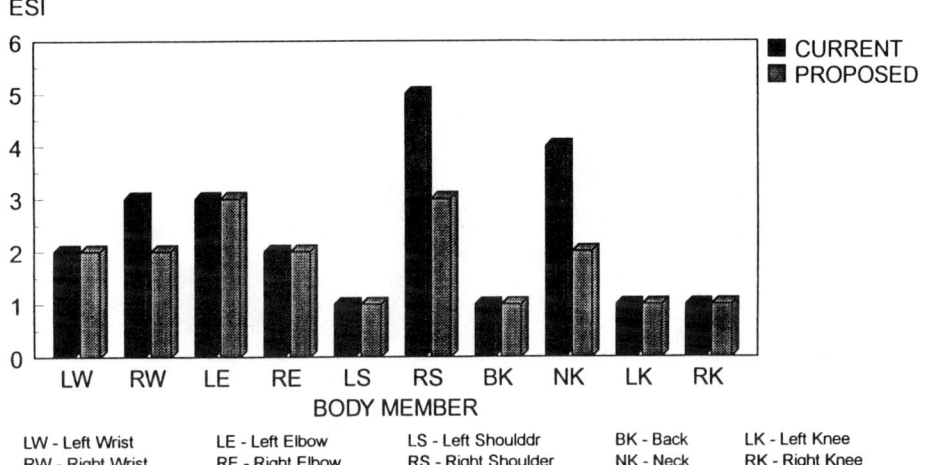

Figure 2 Posture ESI values for current and revised methods.

tee reviewed and monitored the implementation of the ergonomic program. The proposed job rotation was presented to the steering committee and approved.

F. Education and Training Program

An education and training plan was devised by the steering committee. It provided general ergonomics training for all employees, including operators, line supervisors, maintenance personnel, and managers, so that they could recognize situations in which someone might be exposed to excessive ergonomic stress.

A decision was also made to require all new employees and workers entering new jobs to be fully trained in the use of all tools and machines that are required to perform the operation. They were also trained to identify symptoms of CTDs so that they could report any possible problems. Line supervisors were trained in proper work practices and also in the detection of early symptoms of CTDs.

G. Fitness and Rehabilitation

The medical department devised a plan to offer employees the opportunity for exercise and rehabilitation. This included stretch breaks during the shift as well as fitness programs during off-hours.

H. Reporting, Feedback, and Follow-Up

According to employee feedback, the operators performing operations 10 and 11 thought that the improvements were adequate because no discomfort was experienced from performing the operations. The operators performing operations 8 and 9 felt that more effort was needed to complete the assembly operation but that no noticeable discomfort was experienced.

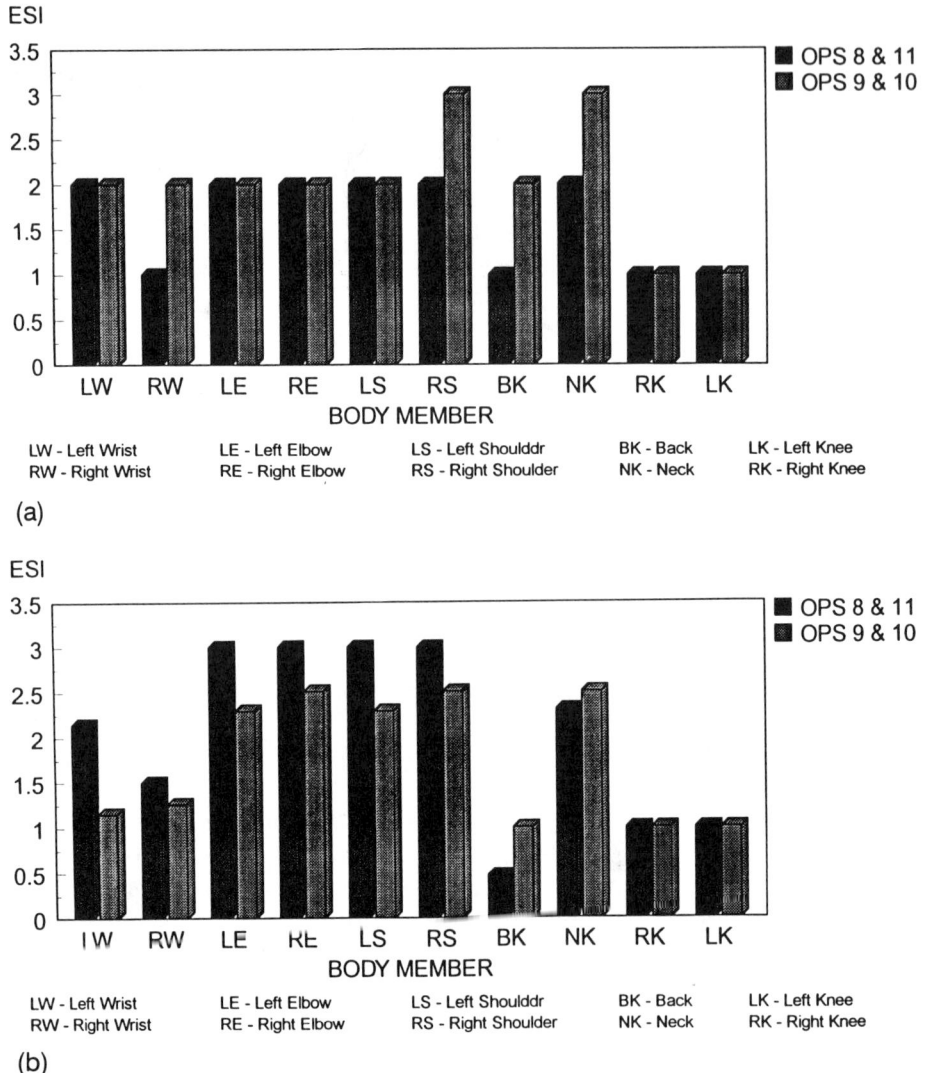

Figure 3 Ergonomic stress indices for the proposed job rotation. (a) Posture ESI; (b) repetition ESI.

Acme continues to monitor employee feedback and injuries that arise to determine if the controls implemented to limit ergonomic stress are successful. If injuries continue, the operation will have to be reevaluated.

I. Conclusion

Job rotation and method improvements were the solutions in this particular situation. Methods improvements should always be made and implemented as a means to alleviate ergonomic stress. Job rotation is also an effective way of spreading ergonomic stress exposure among operators as long as the operations in the rotation vary sufficiently in work content or require different motions in their performance. The appropriate combination of solutions must be determined to minimize the potential for workplace-related injuries.

VIII. CRITICAL REVIEW OF CURRENT STATUS

Until a few years ago, ergonomics was mainly a subject of university research. With the exception of a few industries, U.S. companies did not pay much attention to ergonomic stress. However, the situation was somewhat different in other countries. In Europe, for instance, ergonomics has been a matter of concern since the 1950s. With the increase in work-related injuries in the United States and the increasingly aggressive auditing activities of the Occupational Safety and Health Administration (OSHA), ergonomics has become a subject that must concern everyone who performs or manages work.

Although ergonomics may not have been the focal point, workers have often gained ergonomic benefits as a result of productivity improvements. Frequently, improvements in workplaces, tools, and methods made with the aim of improving productivity results in simultaneous, although serendipitous, reduction in ergonomic stress. Therefore, it was logical to expand the MOST Work Measurement System to include the capability to perform ergonomic analyses in addition to methods analyses and improvements. With the addition of ErgoMOST, this system has become a more comprehensive analytical tool for improvement, affecting both productivity and safety.

ErgoMOST is primarily an analytical tool to be used for improving workplace designs, methods, and tools. These improvements result in a reduction or elimination of pain and discomfort. It must be keep in mind, however, that since the science of ergonomics has not been fully developed and the limits for ergonomic stress not firmly established, ErgoMOST is limited to currently available research data. ErgoMOST considers posture, force, repetition, grip, and vibration based on limits that have been established by the most recent research. As more research is conducted, published, and accepted, these limits will be adjusted and adapted to the new findings. ErgoMOST is a tool under constant development and refinement.

There are some aspects of ergonomic analysis that cannot be incorporated into a computer application because of the level of human judgment required, the limited scope of current research, or the limitations of the computer itself. ErgoMOST is a tool for evaluating workplace design and simulating the effect of workplace and method changes. Although ErgoMOST can be used to make relative evaluations, it cannot be used to define absolute stress limits for human work. ErgoMOST can quickly identify where high stress occurs and where the stress point is, but it cannot predict injury or the long-term effect on ergonomic stress. ErgoMOST can identify the body member that is under stress in a quantifiable way, but it cannot determine stress as a function of human body characteristics. ErgoMOST can be used to develop awareness with respect to ergonomic stress evaluations, but it will not suggest steps to reduce stress. This is a task for the engineer to perform. As stated above, however, ErgoMOST can help the engineer to quickly identify where stress occurs, enabling him or her to concentrate on the solution rather than the problem.

IX. FUTURE CONCERNS

Because of the present interest in the relatively young science of ergonomics and the attention it has received during recent years, we predict that knowledge of work-related stress will increase substantially in the next few years. We will most likely witness increases in the development of ergonomic science in college and universities as more students specialize in ergonomics and perhaps make it a career. It is important, though, that the research performed in universities be related to practical situations in industry. In order to advance the science of ergonomics as well as to better assist industry, the research and tests conducted must resemble actual work activities.

Because of the concern about ergonomic stress during recent years, new experiences resulting from increased application of tools such as ErgoMOST, along with training and awareness, will produce new approaches based on superior knowledge. Therefore, it is essential that research and industry work together to improve models and data, allowing the science of ergonomics to mature quickly.

Thanks to mechanization and a focus on productivity improvements, workplaces have become more "humanized." There are, however, many jobs in industry that are causing ergonomic stress at an unfortunately increasing rate. Therefore, it is important that ergonomic factors be considered at the product design stage so that they can be taken into account in the design of workplaces, tools, and equipment. It is equally important to consider the method of performing the job. It should be natural to include ergonomic factors in the design of both the workplace and the method to ensure that the risk of injury to the worker is minimized.

The fact that an operation has been performed in the same workplace for many years does not ensure its safety. Therefore, it is important to be able to quickly identify problems through ergonomic audits. In addition to the audit, a company should institute a training program for all employees to increase their understanding and awareness of ergonomic conditions and problems. By making employees aware of ergonomic stress, the incentive is created for everyone to make improvements and thereby reduce the ergonomic stress.

As for the future of ErgoMOST, it will be refined as new research results and data become available. Its capabilities will increase, more specific data will be built into the system, and stress limits will be recalibrated. New factors and additional body members will be added to the model. Also, environmental factors will be included. The system will become easier and faster to apply.

More specifically, the new features in ErgoMOST will include

Energy expenditure
Graphics displays using split screens
Standard posture database
Prioritization of problem areas
Suggestion of improvements for simulations
Composite ergonomic stress index (ESI) for the entire body. (Now the ESI is calculated for each joint.)

It is up to researchers to provide the data, industry to apply ergonomics through an ergonomics program, and companies supplying ergonomic tools to use their application experience to further develop and refine their tools.

ACKNOWLEDGMENTS

ErgoMOST was initially developed in cooperation with the Occupational Biomechanics Laboratory, a department of mechanical, industrial, and nuclear engineering at the University of Cincinnati. Input was also obtained from other universities and research projects.

MOST Work Measurement System and ErgoMOST were developed by H. B. Maynard and Company, Pittsburgh, Pennsylvania, with the purpose of enhancing productivity and safety in the workplace.

REFERENCES

1. K. B. Zandin, *MOST Work Measurement Systems*, 2nd ed., Marcel Dekker, New York, 1990.
2. D. B. Chaffin and G. Anderson, *Occupational Biomechanics*, Wiley, New York, 1984.
3. E. Grandjean, *Fitting the Task to the Man. A Textbook of Occupational Ergonomics*, 4th ed., Taylor & Francis, London, 1988.
4. J. Sanders, ErgoMOST as a tool for productivity improvement, *Proc. 1993 MOST User's Conference*, Pittsburgh, 1993, pp. 161–163.
5. U.S. Department of Health and Human Services, *Work Practices Guide for Manual Lifting*, Public Health Service, Centers for Disease Control, and National Institute for Occupational Safety and Health, DHHS (NIOSH) Pub. No. 81-122, Cincinnati, OH, 1981.
6. W. K. Hodson, *Maynard's Industrial Engineering Handbook*, 4th ed., McGraw-Hill, New York, 1992.
7. U.S. Department of Labor, Occupational Safety and Health Administration, *Ergonomics Program Management Guidelines for Meatpacking Plants*, OSHA Pub. No. 3108, Washington, DC, 1990.
8. T. J. Armstrong, J. A. Foulke, B. S. Joseph, and S. A. Goldstein, Disorders in a poultry processing plant, *Am. Ind. Hyg. Assoc. J.* (1982).
9. W. M. Keyserling, T. J. Armstrong, and L. Punett, Ergonomic job analysis: a structured approval for identifying risk factors associated with overexertion injuries and disorders, *Appl. Occup. Environ. Hyg.* (1991).
10. W. M. Keyserling, Ergonomic consideraterions in job analysis, *Proc. 1992 MOST User's Conference*, Pittsburgh, 1992, pp. 3–24.

20
The Design and Evaluation of a Musculoskeletal and Work History Questionnaire

Grace Kawas Lemasters
University of Cincinnati Medical School, Cincinnati, Ohio

Margaret R. Atterbury
Greater Cincinnati Occupational Health Center, Cincinnati, Ohio

I. INTRODUCTION

The cornerstone of many studies investigating occupationally related disorders is the development of a quality survey instrument. Accuracy of the conclusions, however, depends on how carefully a questionnaire has been constructed and evaluated. Before developing a survey instrument, the investigators must become thoroughly acquainted with the topic under investigation, knowledgeable about the disease process, and familiar with the targeted population. In ergonomic field studies, a questionnaire is a useful tool for assessing the presence of symptoms of work-related musculoskeletal disorders (WMDs), for characterizing job factors associated with the development of these disorders, and for assessing symptoms and associated outcomes before and after an ergonomic intervention.

II. BACKGROUND AND SIGNIFICANCE TO OCCUPATIONAL ERGONOMICS

The design and complexity of a musculoskeletal questionnaire depends on its intended use, method of administration, and time availability. The significance of this chapter to occupational ergonomics is that it describes the process of designing and evaluating a symptom and work history questionnaire associated with musculoskeletal disorders. The questionnaire described here was used to estimate the prevalence of WMDs among over 500 currently employed union carpenters and as part of an ergonomic intervention study among journeymen carpenters. In addition, this instrument is being used in a longitudinal study of apprentice carpenters. Apprentices are being followed prospectively to determine the incidence of WMDs.

This instrument was pretested, pilot tested, and evaluated for its reliability and validity. In Secs. III–X, we describe the steps in its development and results of the evaluation. The final questionnaire is presented in the Appendix.

III. REVIEW OF OTHER INSTRUMENTS

The first step in questionnaire construction begins before the first question is written. In addition to a literature review of relevant musculoskeletal disorders, a review of available instruments is critical. First we reviewed prior studies of musculoskeletal disorders [1–4].

Next, several questionnaires used in NIOSH Health Hazard Evaluation (HHE) studies of WMDs were reviewed. These included the *Los Angeles Times*, *Newsday*, Shoprite, Perdue, and Morrell questionnaires [5-9]. These instruments were administered to office workers, newspaper writers and editors, grocery clerks, and meatpackers. Therefore, they needed to be tailored to our study population of carpenters. Additionally, changes were made to adapt those that were self-administered to one that was to be administered over the telephone by trained interviewers who were unfamiliar with the fields of medicine and ergonomics.

Several of the questionnaires reviewed began with the question "In the past year have had pain, aching, stiffness, burning, numbness, or tingling in the area shown in this diagram?," and the respondent was referred to a body diagram. Because our interview was to be administered via telephone, a multicolored body part diagram was mailed to the carpenter's homes. However, realizing that the respondent might not have the diagram in hand at the time of interview, the questionnaire was designed to function independently of the body figure. In fact, only half the respondents actually had the diagram present when contacted. Therefore, each body part was specified by name, with some degree of misclassification inevitable. For example, the respondent might be confused about where the shoulder begins and the neck ends. Since we were concerned about misclassification, the upper and lower back were not differentiated but were referred to as the "back."

Another difference in our questionnaire was that questions about the quality of symptoms were asked separately; i.e., the respondent could answer just "burning" or "pain" or "numbness or tingling" or any combination of these symptoms. This separation of symptoms was done to determine the prevalence of specific types of disorders. For instance, symptoms of nerve-related disorders were likely to include numbness and tingling.

In asking about frequency and duration of symptoms, the format used was similar to one found in the NIOSH Health Hazard Evaluation (HHE) instruments, but the responses were simplified by reducing the number of categories from seven to four. This simplification was necessary because the questions and possible responses were asked over the telephone, and it was thought that subjects might not remember such a long list. Additionally, it was thought that subjects might recall symptoms over the past year as occurring either daily, weekly, or monthly rather than every 2 or 3 weeks or once every 2 months.

Another modification was made to expand the section on injuries. Injuries were hypothesized to be a major cause of musculoskeletal symptomatology among carpenters, in contrast to what might be expected among office workers, news reporters, editors, meatpackers, or those in manufacturing. In the questionnaires reviewed, only one injury question was included for the purpose of excluding that respondent's symptoms as a WMD. Since many carpenters have on-the-job injuries, additional questions were included to further characterize types and causes of injuries.

We composed an entirely original work history section. Characterizing job tasks and musculoskeletal stressors in carpentry and construction was complex because of the wide variety of tasks performed by carpenters in contrast to manufacturing where workers repeat the same task over and over. Input from retired and active union carpenters was invaluable in designing this section; in particular, they assisted in constructing a comprehensive list of specialties.

In summary, the review of these questionnaires helped by building upon the successes and avoiding the mistakes of others. Thus, the instrument review, though time-consuming, was infinitely more useful than having to start from the beginning.

IV. KNOWING THE TARGET POPULATION

In the initial phase of questionnaire construction an in-depth understanding of the study group is needed. It is important to know the general knowledge base of the respondents. Discussions with individuals in the occupation under investigation help in understanding their health concerns, work activities, knowledge base, and literacy level. Input solicited from several carpenters and visits to construction sites were crucial to gaining an understanding of the nature of the work tasks and possible health problems associated with these tasks.

Site visits to construction areas helped identify the body areas at greatest risk of damage. A body parts diagram and symptom check adapted from previous studies [10, 11] was given to a group of carpenters to complete at the end of their work shift. Reviewing the results of this initial walk-through was critical in designing the larger in-depth instrument. It not only helped define the body regions with the most discomfort for carpenters, but also familiarized the study team with the variety of activities performed by carpenters. Specifically, from these data we identified that the hip area and the upper extremities may be important target areas of damage. It was learned, for example, that carpenters wear tool belts weighing from 6 to 15 lb and that much of their work requires the use of power tools with arms extended overhead. Interviews with carpenters and visits to work sites facilitated the process of gaining knowledge about the study population, their work, and their potential clinical problems.

V. DEFINING WMDS

The term *work-related musculoskeletal disorders* (WMDs) is used in reference to conditions that are also called cumulative trauma disorders (CTDs), repetitive strain injuries (RSIs), or overuse syndromes. All these terms refer to conditions that involve muscles, tendons, and/or nerves and usually manifested as pain, discomfort, and/or tingling in a body area. Many studies of these conditions have investigated symptoms in the neck and upper extremities [3,12,13]. Some WMDs such as carpal tunnel syndrome (CTS) and de Quervain's tenosynovitis are relatively specific clinical diagnoses. Others have less specific signs and symptoms (e.g., myofascial pain, tension neck syndrome, lumbar strain) and may encompass a broader anatomical area and/or lack specific criteria for diagnosis. The term WMD is preferred because of the multifactorial nature of these problems [13]. Most of these disorders have been associated with occupational factors (e.g., force, repetition, awkward or static postures, vibration) and/or nonoccupational factors such as sports activities, hobbies, underlying disease, and personal factors. These additional risk factors also need to be addressed in the questionnaire.

We defined an individual as having a WMD if all the following criteria were satisfied:

1. Onset of symptoms after starting work as a carpenter
2. Severity reported as mild, moderate, severe, or the worse discomfort ever
3. Symptoms occurring at least once per week (frequency) or lasting at least one week (duration)
4. No history of injury to that body part

NIOSH used a similar case definition in their HHE studies of WMDs [5–8].

Defining WMDs by questionnaire is useful for surveillance purposes because a range of severity of symptoms can be assessed and localized to a particular body area in a specified population. Although numerous studies have been carried out using symptoms and standardized physical exams to identify WMDs [1,3,5,14], these have not been validated as diagnostic for specific clinical disorders.

VI. QUESTION CONSTRUCTION

There are two general approaches to question development. The first is unstructured and open-ended. The unstructured question allows more flexibility; the question content, sequence, and wording are under the control of the interviewer, and the answers of the respondents often direct the interview. The second approach, the standardized closed-ended instrument, has structured questions with a fixed format. The latter approach was selected for the carpenter study in order to (1) increase efficiency in terms of time, (2) eliminate difficult coding problems and a need for interpretation by judges, and (3) increase the ease of administration by a nonmedical interviewer group.

The advantage of asking standardized questions is that it provides greater uniformity of measurement responses and the investigator can obtain the responses that the question was intended to address. For example, when asking questions concerning an injury, we wanted to know about specific types of injuries and therefore provided a list: "a fall, an object hitting or falling on you [as is common in the carpentry trade], an auto accident, sports accident, and other sudden injury" (Part II, neck, Q12). This question was followed by one asking for information on the type of injury: "a broken bone, whiplash, ruptured disc, burn or other." These fixed, generally closed-ended responses reduced the need for coding and provided information that helped in determining if an individual symptom was related to a WMD or an injury at the workplace or elsewhere. In contrast, an open-ended question such as "What type of injury did you have?" requires the review and coding of every response.

Several types of standardized questions were used in this instrument. A combination strict-choice and item checklist was used to gather prior medical history (Part II, Q8). The respondent was asked to answer yes or no (strict choice) to each disease condition provided in a checklist of medical conditions. The disease conditions that were listed are risk factors that may be associated with WMDs. The checklist is probably the most direct approach for providing a memory jog for prior medical conditions possibly related to a WMD.

The medical symptom questionnaire portion used a rating scale to determine the degree of pain that the individual was experiencing. The rating scale defined pain ranging from mild, moderate, and severe to the worst discomfort ever experienced. Originally it was felt that only pain at the moderate or greater level should be considered in our case definition for a WMD. As our experience with the carpenters grew, it became apparent that this group was relatively stoic, accustomed to a high degree of physical labor and to considerable pain as a part of their job. Hence, the decision was made in the analysis stage to include those reporting mild pain in the case definition.

In reviewing other instruments we occasionally found a "double-barreled" question, one that asks for a single response to a combination of questions. An example of a double-barrelled question is: "In the past year, did you have severe pain lasting longer than a week?" That question should be divided and asked as: "How long did the symptom last?" (Part II, neck, Q4) and "How severe is the pain?" (Q5).

One common problem in question construction concerns the response set. A response set is a tendency of a subject to answer in a certain way, such as "never" or "always," irrespective of the content of the question. We were particularly concerned about a potential response set problem in the work history section (Part IV, Work History, Q1). The first question asked, "From when you started as a carpenter in 19— have you done any of the following jobs?" The goal was to learn about each specialty (flooring, welding/burning, form work, drywall, scaffolding, framing, piledriving, finish, ceiling, millwright, fixtures, supervision, and office work) that the carpenters had ever performed. The follow-up question asked, "How often have you done this job in the past year?," with the categories ranging from "never" to "all the time." If both questions were asked sequentially for each specialty, the concern was that the participant would grow tired of answering the "How often" question and would say "no" to the subsequent specialties in order to end the interview. Therefore, we decided to first ask a strict-choice response of "yes" or "no" to each specialty area for part A. Then we returned to each of those having an affirmative response and asked the second question (B). Though this may not have completely avoided a response set, it helped to minimize the problem.

In summary, there are some general tenets of question construction that should be followed:

1. Questions should be asked in an unbiased manner with no leading.
2. Every question should be clear, unambiguous, and understandable for both the respondent and the interviewer. Avoid using unfamiliar medical items and ambiguous words such as always, may, large, many, important, some, most, much, near, several, generally, occasionally, and often.
3. *Every* word in a question is important; *each* should be examined for possible elimination. Questions should not exceed 20 words in length.
4. Ask easy questions first and save the difficult and sensitive ones for last.

In formulating questions, the investigator needs to think backwards from the analysis and forward from the hypothesis. Before writing the first question, it is of utmost importance to know specifically what information is needed in order to answer the overall research question(s).

VII. QUESTIONNAIRE COMPONENTS

In order to assess prevalence or incidence as well as characterize other issues pertaining to WMDs, the questionnaire needed to contain the following sections: (1) demographics and medical history, (2) musculoskeletal symptoms and injuries, and (3) a work history.

A. Demographics and Medical History

The demographics and medical history section included factors such as age, sex, height, weight, and hand dominance, which may play a role in the development of musculoskeletal disorders. The question concerning income was saved until last. Sports activities and hobbies were probed as possible risk factors for musculoskeletal disorders. The medical history (Part II, Q8) establishes the presence of self-reported physician-diagnosed medical conditions related to the development of WMDs. These may have been either other risk factors (e.g., diabetes mellitus related to the development of CTS) [15] or conditions that may have caused symptoms similar to those of WMDs (e.g., cervical spine disease) [16]. Additionally, prior diagnoses of CTS, arthritis, tendinitis, or other conditions that could be work-related were specified. These items specifically referred to "physician-diagnosed" conditions in order to eliminate inaccurate self-diagnosis that may occur for such conditions as rheumatoid arthritis.

Diabetes mellitus, systemic lupus erythematosus ("lupus"), rheumatoid arthritis (RA), and hypothyroidism are systemic illnesses that are associated with the development of carpal tunnel syndrome (CTS) [15,17]. Gout, lupus, and rheumatoid arthritis are inflammatory disorders that also cause musculoskeletal disease, especially arthritis and/or tendinitis. A herniated or "ruptured" cervical disc with nerve root compression may cause symptoms (paresthesias and weakness in the hand) similar to CTS, though usually neck and arm pain are present. Cervical nerve root compression also has been associated with coexistent CTS [15,18]. CTS is the most common nerve entrapment syndrome [16,17] that has been associated with repetitive motion and force at work [1,19]. Thoracic outlet syndrome (TOS) which is much less common, may result in symptoms of numbness and tingling in the hand, usually in the ulnar distribution. TOS is often difficult to diagnose and is usually a result of anatomical conditions [20] but may be aggravated by work postures [21]. Raynaud's phenomenon, which is paresthesias and white or blue discoloration of the fingers on exposure to cold temperatures, may be an isolated syndrome, part of a systemic illness or a result of vibration exposure at work [22].

A smoking history was included in the questionnaire because of its association with increased risk of low back pain and intervertebral disc disease [23]. Likewise, pregnancy [15] and the use of oral contraceptives [24] were included because these factors have been associated with CTS. Finally, questions on alcoholism and diabetes mellitus were included because these are common causes of diffuse peripheral neuropathies that may mimic nerve-related WMDs such as CTS.

B. Musculoskeletal Symptoms

We were most interested in assessing the presence of persistent musculoskeletal symptoms, so the interview began with a screening question (see Part II, Q12): "Within the past 12 months have you experienced any recurring symptoms such as pain, aching, numbness or any other symptom in your . . ."; specific body areas were then listed. An affirmative response generated further questions characterizing musculoskeletal symptoms in the identified area. The same questions were asked for each body area in order to establish onset, quality, frequency, duration, and intensity of symptoms and a history of injury to that body part. These questions addressed musculoskeletal symptom characteristics needed in establishing whether or not a respondent met the case definition for a WMD.

Questions were added to assess a respondent's opinion as to whether or not a musculoskeletal problem was temporally associated with and/or caused by carpentry work. Severity of symptoms and resulting disability were ascertained by questions on whether medical attention was sought, number of workdays lost, and job task modification, and on whether or not the symptoms awoke the respondent from sleep.

C. Injuries

Injuries were hypothesized to be a major cause of musculoskeletal symptomatology among carpenters. Consequently, the questionnaire had a group of questions characterizing injuries to individual body areas. These questions were asked only if the patients had had symptoms in the past 12 months, and subsequently assessed only injuries with recent recurring symptoms or with symptoms in that area due to another cause (e.g., a coexistent WMD).

Injury was defined as a sudden event such as a fall or an object hitting a worker. Our intent was to separate symptoms due to sudden injuries at work from those due to WMDs.

We did not want to exclude strains (e.g., low back) that may have been a result of cumulative job-related musculoskeletal stresses but were manifested following a single event. Therefore, we chose not to include "strain" as a sudden injury category but had "other" with an open-ended option.

D. Work History

When designing the work history, it is crucial that investigators have a predefined plan by which to model exposures. Will surrogate measures of exposure such as duration, latency, and job tasks be used, or will intensity measures also be made? Because of the variety of tasks performed by carpenters and the job movement from one specialty to another, it was decided that duration of employment in the carpentry trade was to be the primary measure of exposure. Also of interest was knowing if one specialty (e.g., finish, ceiling, form) was more ergonomically hazardous than other specialties. For example, it was hypothesized that those who did ceiling work might report more shoulder disorders. Therefore, as mentioned earlier, the carpenters were queried about how much time they had spent in each specialty in the last month, in last year, and in their entire career. Because this study also included an intervention phase, other general questions were asked. These had to do with what tools they wanted to change and how they might institute changes.

Workplace psychosocial stress also has received attention as a risk factor for musculoskeletal disorders [25]. In a study by Leino [26], musculoskeletal disorders as measured by rheumatic-type symptoms were associated with occupational stress in both men and women and for all occupational class groups. These findings are in support of the demand-control model proposed by Karasek [27]. This model is based on the combined joint effects of task demands and individual task control as predictors of stress-related illness. This theory hypothesizes that a worker is at greatest risk for a stress-related illness when the demands of the job are high and the worker's control over the tasks is low. Hence, as part of the work histories, questions were included that focused on job demands and job control; these questions were drawn from another instrument shown to have good reliability [28]. The first set of psychosocial questions asked about the level of influence the individual had over the "amount of work, availability of materials, policies and procedures, pace of your work, quality of the work, and hours or schedule" (Part IV, Work History, Q7). Demands of the job were addressed with questions on how fast and hard they had to work, how physically exhausted they felt, and the extent of their work load.

In summary, the work history section served several purposes. Exposure models were developed that addressed duration of time in the trade and the development of a WMD and identified the relationship between a WMD and work in particular specialties. In addition, questions were formulated that assisted in devising possible interventions in either tool design or application. Finally, the work history included psychosocial elements related to job function that might also be risk factors for a WMD.

VIII. PRETEST AND PILOT TEST

Pretesting an instrument is a means of evaluating its effectiveness before using it in the final study. The pretest is usually conducted on a small group in an interactive manner. Its goal is to determine if the questionnaire is clear, is understandable, and contains important elements as viewed from the subject's perspective. A pretest was performed on this instrument with the help of carpenters at a local union meeting. In teams of two investigators and one subject, we administered the questionnaire to five volunteers. The pretest identified questions that were inappropriate, delineated those that were misleading, and identified questions that should be added.

With a "final" draft in hand, the instrument was then pilot tested. A pilot study is a small-scale test using the exact procedures planned for the larger study. For the pilot study, carpenters belonging to a union in another city were recruited. They were sent a letter and a body part diagram and were interviewed by telephone using the same interview team planned for the larger study. Thus, the pilot study not only helped to refine the instrument but also assisted in interviewer training.

Though much was gleaned from the pilot study, one major finding is worth discussing. Originally it was planned to ask carpenters about musculoskeletal symptoms by systematically proceeding through body regions one by one. During the pilot study, it was learned that once the carpenters became aware of how many additional questions there were when an affirmative response was given for one body area, (approximately 15), they became more and more reluctant to report symptoms. There was a decreasing frequency of affirmative responses progressing from the neck to the ankles. To prevent this from occurring in the larger study, a screening question (Part II, Q12) that asked about recurring musculoskeletal symptoms for all eight body areas was added. Then for each affirmative response, the extra unit of questions were asked (for an example, see neck questions 1–15).

Without the pilot study and the addition of the screening questions, the conclusions of the study could have been biased simply due to the way the instrument was formatted. The pilot study also revealed that on average the questionnaire could be administered in 20 minutes.

IX. RELIABILITY AND VALIDITY OF THE INSTRUMENT

The reliability of a questionnaire is the extent to which results obtained by a procedure, in this case telephone interviews, can be replicated [29]. A questionnaire may lack reliability for several reasons: The factor under investigation lacks stability, there is divergence among interviewers in the way the instrument is administered, or question construct is faulty.

Instrument reliability was assessed using the test-retest methodology. Portions of the questionnaire were administered to approximately 9% (N-49) of a subgroup randomly selected from the original population. The reinterview (retest) occurred between 1 and 3 weeks after completion of the original interview. A detailed report of this reliability assessment is given in Ref. 30. The areas addressed in the reliability assessment included recall of reported symptomatology, prior medical conditions, prior injuries, and work histories. The observed agreements ranged from 75.5% to 95.7%. Not surprisingly, prior medical history (94.7%), injury to body region (88.4%), and job specialty performed (87.9%) showed the highest agreement with kappa (κ) estimates ranging from 0.70 to 0.77. A kappa estimate of 1.0 indicates perfect reproducibility, and one of 0.0 indicates no agreement.

The observed agreement for the musculoskeletal symptoms was 75.5% ($\kappa = 0.46$). Thirty-seven of the 49 reported exactly the same symptoms as they did in the original interview; they agreed that they did ($N = 26$) or did not ($N = 11$) have a musculoskeletal symptom at a particular body region. Eight changed from yes to having a symptom to no on the retest, and four did the reverse. Certainly, some of this change is due to instability of the response. In contrast to items like prior medical diagnoses and prior injuries that are established past events, current symptomatology is in a dynamic state with changes expected over time. Levine et al. [31] showed a higher agreement of report of symptoms on a test-retest evaluation with a Pearson correlation coefficient of 0.91. In their study, however, the retest was done the day after the original interview, maximizing the stability of a response by minimizing the elapsed time between interviews. Overall, our questionnaire was found to have acceptable reliability.

The validity of the instrument also was evaluated. Validity is the extent to which an instrument measures what it purports to measure. There are three basic types of validity: content, construct, and criterion. All three forms of validity assessment were undertaken.

The least powerful, although certainly not less important, is content validity. Content validity implies that the instrument incorporates the domain of the phenomena under study, i.e., work-related musculoskeletal disorders. Since our instrument was adapted from those developed by other experts in the field of WMD, content validity was, in part, gained by association. A panel of experts also was convened to further evaluate the content validity of this instrument.

Construct validity is defined as the extent to which the measurement corresponds to theoretical constructs. One test of construct validity is to determine the instrument's predictive ability. In this study, duration of employment as a carpenter was associated with a significant increase in the relative risk of having a WMD. As an example of construct validity, the prevalence study demonstrated that, compared to the baseline group having less than 10 years in the trade, those with 20 years in the trade as a carpenter showed a significant increase in shoulder disorders, relative risk = 2.8 (95% confidence interval 1.5, 5.3) [32]. Another example of construct validity is found in a study that compared median sensory distal amplitude and latency using an ergonomic checklist to define the exposed and unexposed groups [33]. Those who were identified on the checklist as gripping greater than 6 lb were shown to have significantly smaller amplitudes and longer latencies. Thereby, the checklist was able to define a theoretical construct of an ergonomic hazard.

The third form of validity measure is criterion. Criterion validity is the degree to which the measures correlate with an external criterion commonly referred to as a gold standard. Finding a criterion for evaluation of symptoms of musculoskeletal disorders is problematic because there is no gold standard in diagnosing these conditions. Clinical diagnoses require the presence of a combination of symptoms, physical signs, often results of electrodiagnostic or other laboratory testing, and sometimes response to treatment. Lack of money and time limit the use of a complete clinical evaluation when evaluating a cohort of workers. When determining the prevalence of WMDs among a large group, WMD cases may be defined by symptom criteria alone or with physical findings.

Symptoms are by definition subjective. Since symptoms are thought to be the earliest clinical manifestation of a musculoskeletal disorder, establishing the presence and severity of symptoms is critical to evaluating the prevalence of WMDs. A common feature among WMDs is that they manifest as discomfort in various anatomical locations and result in a broad range of disabilities. Measures of symptoms and disability using a self-administered questionnaire have been found to be reproducible and consistent in assessment of CTS [31].

Physical examination outcomes have limited use as a gold standard for several reasons. First, most physical examination maneuvers have not been tested for sensitivity and specificity for specific disorders when the examiners are blinded to symptoms. Second, most examination techniques used in the evaluation of musculoskeletal disorders require a subjective rating of discomfort on the part of the examinee. Finally, there may be intra- and interexaminer variation in examination techniques.

Nerve conduction outcomes do provide an objective measure of median nerve function and are often referred to as a gold standard in the diagnosis [20] of CTS. Nerve conduction studies (NCSs) can quantitate median nerve function, measuring the latency, or the time required for a nerve impulse to travel a set distance, and the amplitude of sensory and motor nerves. These electrodiagnostic tests are used to establish whether or not a median nerve abnormality, or mononeuropathy, at the wrist (i.e., CTS) exists. Even NCSs have limited use as a gold standard in diagnosing CTS. There is some variation in electrodiagnostic criteria for diagnosing a median mononeuropathy. Further, these tests were developed to be part of the entire clinical evaluation, as an extension of the physical examination, to further assess nerve function. Electromyographers use patients' clinical history, physical exam findings, and NCS results in making diagnoses. And finally with respect to CTS, this diagnosis requires the presence of particular symptoms as well as physical findings and/or NCS findings of a median mononeuropathy.

Thus far there have been relatively few published studies that have examined the validity of a broad musculoskeletal symptom questionnaire [34,35]. A few studies have addressed the validity of a variety of surveillance case definitions for carpal tunnel syndrome by comparing symptoms and/or physical findings [36,37] with nerve conduction results.

In our study, 100 carpenters who were either asymptomatic or met the questionnaire WMD case definition for hand/wrist, shoulder, or knee underwent a standardized musculoskeletal examination performed by two physician examiners. This examination included Phalen's test and Tinel's tests, which are commonly used in the diagnosis of CTS. All 25 hand/wrist cases and 29 asymptomatic carpenters underwent nerve conduction studies performed by a single electromyographer blinded to case status.

Thus far we have analyzed the data with respect to hand/wrist symptoms, physical exam signs of CTS, i.e., Phalen's and Tinel's tests, and electrodiagnostic findings. With respect to positive physical findings of Phalen's or Tinel's tests among hand/wrist cases, we found fair agreement over that which might occur by chance. Forty-eight percent of cases had a positive response to one of these tests. Overall agreement, or percent of cases with positive findings and noncases with negative findings, was 70% [38].

Hand/wrist cases were found to have mean NCS values representing a median mononeuropathy at the wrist (including prolonged mean median sensory and motor distal latencies and diminished median sensory and motor amplitudes) compared to noncases. Approximately 80% of the hand/wrist cases met the electrodiagnostic criteria for a median mononeuropathy at the wrist [39]. There was greater agreement between hand/wrist case status and a median mononeuropathy than between hand/wrist case status and physical findings.

In summary, given the above limitations, there were more physical examination and NCS findings of CTS among the questionnaire-defined hand/wrist cases than among noncases. NCS results in particular indicate that the questionnaire case definition, based on symptoms, was useful in estimating the prevalence of CTS among carpenters.

X. ANALYSIS STRATEGY

The data analysis plan should be developed prior to questionnaire administration. Without an a priori plan, the investigator may find that an important risk factor needed for the analysis has been overlooked. There is usually minimal recourse for recovery of missing information after interviewing has been completed. The symptom survey described in this chapter was conducted using direct computer entry during the telephone interview. Hence, the data cleanup step was shortened by eliminating the need to prepare the data for keypunch, have it keypunched, usually with 100% verification (double entry), and then check for errors associated with the keypunch process.

At the end of our questionnaire the interviewers were asked to evaluate the respondent's recall abilities and level of cooperation. This evaluation is, of course, the interviewer's subjective interpretation. The interviewers were asked (1) "How well did the respondent seem to remember his/her health information?"; "How well did the respondent seem to remember his/her work history?"; "How cooperative was the respondent?" Respondents who were scored as "not very well" or "not at all cooperative" on all three questions were excluded. This evaluation resulted in the exclusion of only eight individuals but was important to ensure higher quality data.

We were then ready to begin data analysis and to proceed logically from the simple to the complex. The first analytic step was the generation of simple descriptive statistics—means, medians, proportions, standard deviation, and ranges. These statistics serve not only to provide a quick assessment of data quality (i.e., ensuring that the minimum and maximum observed values for each variable are within acceptable ranges), but also to provide descriptive characteristics of the study population.

The next phase of analysis was to compare exposure (the independent variable) to the outcome WMDs (the dependent variable). Based solely on questionnaire data, two "surrogate" measures of exposure to ergonomic risk, duration of time in the trade and primary carpenter specialty, were used. Number of years in the trade was divided into three categories, less than 10, greater than 10 but less than 20, and greater than 20. These categories were chosen because they divided the study sample into three approximately equal groups. Those meeting the case definition were compared with those who did not. Mantel Haenzel and relative risk estimates were calculated comparing the two longer duration categories to the baseline of less than 10 years. An "unexposed" noncarpentry group would have been preferred as a baseline comparison group, however. Using the "low" tenure group probably underestimates the true risk. The final analysis step was to use a multivariate approach. For this study, a multivariate logistic regression analysis was conducted using duration of time in the trade as a continuous variable while adjusting for other risk factors, e.g., chronic medical conditions and age. In conclusion, the analytic procedures were more complex that was presented in this brief summary.

XI. SUMMARY

This chapter discusses some of the principles in questionnaire development and the methods used to construct questions. The NCS findings indicated that the questionnaire case definition based on symptoms was useful in estimating the prevalence of carpal tunnel syndrome among carpenters. Though a questionnaire is not a substitute for clinical evaluation in the diagnosis and treatment of individuals, it was shown to be effective in estimating the prevalence of musculoskeletal disorders, at least in this population.

APPENDIX: MUSCULOSKELETAL SYMPTOM AND WORK HISTORY QUESTIONNAIRE

Version 1 © GCOHC, 1992

DATE: (of completing questionnaire) ☐☐ ☐☐ 19 ☐☐   1-80
Month Day Year Card

Interviewer Identification: ☐☐☐ (Interviewer enter your three initials)

Carpenter Identification: ☐☐☐

START TIME (Use Military Time) __ __ : __ __

"Hello, Mr/Ms _____, this is _____ calling for the Greater Cincinnati Occupational Health Center. Did you recently receive a letter from us about our health study on carpenters?"

(If no) May I read you the letter?   (READ LETTER)

(If yes) Then you know we are trying to find out some things about the strains and pains that carpenters get because of their work."

(THEN READ FOLLOWING STATEMENT)
"Also, this interview is completely voluntary. If you need to stop at any time just let me know and we can call you back to finish when it's convenient for you. Would this be a convenient time for you?"

> Have you been retired for more than a year? (If no continue)
>
> (If yes) Well today we are talking only to people who are currently working as a carpenter but thank you for your time.

"Do you happen to have the diagram of the body we sent you?"

    1   yes
    2   no

1. What is your date of birth? (DK/REFUSED = 99)
   ☐☐ ☐☐ ☐☐
   (month) (day) (year)

2. What is your height in stocking feet? (DK/REFUSED = 9,99)
   ☐ Feet ☐☐ Inches

3. What is your weight without shoes? (DK/REFUSED = 999)
   ☐☐☐ Pounds

4. Are you: 1 Right-handed?
              2 Left-handed?
              3 Both-handed?
              4 Refused

5. Which hand do you use most at work?

              1 Right
              2 Left
              3 Both Equally
              4 Refused

6. Are you: 1. Male or 2. Female?
   (Circle appropriate response)

7. What is your ethnic background? (read list)
   1___Black, not of Hispanic origin
   2___Hispanic
   3___White, not of Hispanic origin
   4___Other
   5___DK/REFUSED

Musculoskeletal and Work History Questionnaire

PART II

"Now we would like to get some information on your medical background."

8. "Have you ever been told by a physician that you had any of the following?"

   a. diabetes   1 yes   2 no
   b. gout   1 yes   2 no
   c. hypothyroidism or   1 yes   2 no
      underactive thyroid
   d. lupus   1 yes   2 no
   e. ruptured disc in the neck   1 yes   2 no
   f. ruptured disc in the back   1 yes   2 no
   g. rheumatoid arthritis   1 yes   2 no
   h. carpal tunnel syndrome   1 yes   2 no
      side: 1 left   2 right   3 both
   i. thoracic outlet syndrome   1 yes   2 no
      side: 1 left   2 right   3 both
   j. Raynauds syndrome   1 yes   2 no

9. "Have you ever been told by a **PHYSICIAN** that you had any ther type of arthritis, tendinitis or other joint problem?
   1 yes
   2 no   (GO TO QUESTION 10)

For each yes circle body location and write the year condition began.

IF YES: Did this problem occur in your?	(Read Each)		Did this occur in the (left right middle) part of your body?			What year was the condition diagnosed by a doctor?
1 neck?	1 yes	2 no	1 left	2 right	3 middle	19__ __
2 shoulder?	1 yes	2 no	1 left	2 right	3 both	19__ __
3 elbow/forearm?	1 yes	2 no	1 left	2 right	3 both	19__ __
4 hand/wrist?	1 yes	2 no	1 left	2 right	3 both	19__ __
5 back?	1 yes	2 no	1 left	2 right	3 middle	19__ __
6 hip/thigh?	1 yes	2 no	1 left	2 right	3 both	19__ __
7 knee?	1 yes	2 no	1 left	2 right	3 both	19__ __
8 ankle?	1 yes	2 no	1 left	2 right	3 both	19__ __

10a. "Have you ever smoked cigarettes regularly (at least 1 pack per week)?"
   1 never smoked
   2 ex smoker
   3 current smoker

10b. "On average, how many alcoholic beverages do you have a week?"

   ____ ____ ____

IF FEMALE
11. "Are you currently:"
    a. pregnant   1 yes   2 no
    b. using birth control pills   1 yes   2 no

12. "Within the past 12 months have you experienced any recurring symptoms such as pain, aching, numbness, or any other symptom in your:"

    1   neck?
    2   shoulder?
    3   elbow/forearm?
    4   hand/wrist?
    5   back?
    6   hip/thigh?
    7   knee?
    8   ankle?

CIRCLE ALL THAT APPLY AND ASK THE SERIES OF QUESTIONS FOR EACH BODY PART CIRCLED. IF NO SYMPTOMS SKIP TO WORK HISTORY SECTION.

Now I would like to ask you a few questions about your NECK.

1. "In the **PAST YEAR** have you experienced any of the following in your **NECK**": (CIRCLE ALL THAT APPLY)
    1. pain?
    2. aching?
    3. burning?
    4. numbness or tingling?
    5. none of the above? **(IF NONE GO TO Q. 12)**

2. "Please tell me the month and year you first had this **NECK** problem(s)."

    ___ ___ /19 ___ ___
    month   year

3. "How often have you had this **NECK** problem in the **PAST YEAR?**"
    1. daily
    2. about once a week
    3. about once a month
    4. generally less than once a month

4. "In the **PAST YEAR** when you had this **NECK** problem, on average how long did it usually last?"
    1. less than a week
    2. between a week and 1 month
    3. longer than a month but not constant
    4. constant

5. "Would you call this discomfort mild, moderate, severe, or just about the worst discomfort you have ever suffered in your life?"
    1. mild discomfort?
    2. moderate discomfort?
    3. severe discomfort?
    4. worst discomfort ever?

6. "Did you become aware of this problem while employed as a carpenter?"
    1. yes
    2. no

7. "Do you feel this problem was caused by your work as a carpenter?"
    1. yes
    2. no

8. "Have you seen a doctor, nurse, or other health care provider for the problem you've had with your **NECK** ?"
    1. Yes If yes, what was the diagnosis?_____

                            _____ [ ][ ][ ]code
    2. No

9. "In the past year how many days of work did you miss because of this problem?"
    ___ ___ ___ days

Musculoskeletal and Work History Questionnaire

10. "Did you have to change your job task(s) because of this problem?"
    1 yes
    2 no

11. "Does this problem you have with your **NECK** wake you from sleep at night?"
    1 yes
    2 no

INJURIES

12. "Did you ever have an injury to your **NECK** as a result of:" (RECORD MOST SEVERE)
    1 a fall?
    2 an object hitting you or falling on you?
    3 an auto accident?
    4 a sports activity?
    5 other sudden injury? (PLEASE SPECIFY)
    _____
    _____
    _____ [ ][ ][ ]code
    6 none of the above (IF NONE GO TO NEXT BODY PART WITH A YES ON Q 12 Page. 3)

13. "When did this injury occur?"
    ___ ___ /19___ ___
    month   year
    (IF MORE THAN ONE INJURY TAKE DATE OF THE MOST SEVERE)

14. "What type of injury was this?"   (CIRCLE MOST SEVERE)
    1 broken bone?
    2 whiplash?
    3 ruptured disc?
    4 burn?
    5 other?_____ [ ][ ][ ] CODE

15. "Do you think your current **NECK** problem is a result of this injury?"
    1 yes
    2 no

Now I would like to ask you a few questions about your SHOULDER.

1. "In the **PAST YEAR** have you experienced any of the following in your **SHOULDER**": (CIRCLE ALL THAT APPLY)
    1 pain?
    2 aching?
    3 burning?
    4 numbness or tingling?
    5 none of the above?   **(IF NONE GO TO Q 13)**

2. "In which **SHOULDER** does (did) this occur?"
    1 left
    2 right
    3 both

3. "Please tell me the month and year you first had this **SHOULDER** problem(s)."
    ___ ___ /19___ ___
    month   year

4. "How often have you had this **SHOULDER** problem in the **PAST YEAR?**"
    1 daily
    2 about once a week
    3 about once a month
    4 generally less than once a month

5. "In the **PAST YEAR** when you had this **SHOULDER** problem, on average how long did it usually last?"
    1 less than a week
    2 between a week and 1 month
    3 longer than a month but not constant
    4 constant

6. "Would you call this discomfort mild, moderate, severe, or just about the worst discomfort you have ever suffered in your life?"
    1   mild discomfort?
    2   moderate discomfort?
    3   severe discomfort?
    4   worst discomfort ever?

7. "Did you become aware of this problem while employed as a carpenter?"
    1   yes
    2   no

8. "Do you feel this problem was caused by your work as a carpenter?"
    1   yes
    2   no

9. "Have you seen a doctor, nurse, or other health care provider for the problem you've had with your <u>SHOULDER</u>?"
    1   Yes   If yes, what was the diagnosis?_____
    _____[ ][ ][ ]code
    2   No

10. "In the past year how many days of work did you miss because of this problem?"
    ___ ___ ___ days

11. "Did you have to change your job task(s) because of this problem?"
    1   yes
    2   no

12. "Does this problem you have with your <u>SHOULDER</u> wake you from sleep at night?"
    1   yes
    2   no

INJURIES

13. "Did you ever have an injury to this <u>SHOULDER</u> as a result of:"   (RECORD MOST SEVERE)
    1   a fall?
    2   an object hitting you or falling on you?
    3   an auto accident?
    4   a sports activity?
    5   other sudden injury? (PLEASE SPECIFY)
    _____ _____
    _____
    _____[ ][ ][ ]code
    6   none of the above   (IF NONE GO TO NEXT BODY PART SECTION WITH A YES ON Q 12 Page 3)

14. "When did this injury occur?"
    ___ ___ /19___ ___
    month   year
    (IF MORE THAN ONE INJURY TAKE DATE OF THE MOST SEVERE)

15. "What type of injury was this?"   (CIRCLE MOST SEVERE)
    1   broken bone?
    2   dislocation?
    3   severe bruise?
    4   cut?
    5   burn?
    6   other?_____ [ ][ ][ ] CODE

16. "In which <u>SHOULDER</u> did this injury occur?"
    1   right
    2   left

17. "Do you think your current <u>SHOULDER</u> problem is a result of this injury?"
    1   yes
    2   no

Musculoskeletal and Work History Questionnaire

Now I would like to ask you a few questions about your ELBOW/FOREARM.

1. "In the **PAST YEAR** have you experienced any of the following in your **ELBOW/FOREARM**": (CIRCLE ALL THAT APPLY)
    1. pain?
    2. aching?
    3. burning?
    4. numbness or tingling?
    5. none of the above? **(IF NONE GO TO Q 13)**

2. "In which **ELBOW/FOREARM** does (did) this occur?"
    1. left
    2. right
    3. both

3. "Please tell me the month and year you first had this **ELBOW/FOREARM** problem(s)."
    ___ ___ /19___ ___
    month   year

    FREQUENCY
4. "How often have you had this **ELBOW/FOREARM** problem in the **PAST YEAR?**"
    1. daily
    2. about once a week
    3. about once a month
    4. generally less than once per month

    DURATION
5. "In the **PAST YEAR** when you had this **ELBOW/FOREARM** problem, on average how long did it usually last?"
    1. less than a week
    2. between a week and 1 month
    3. longer than a month but not constant
    4. constant

6. "Would you call this discomfort mild, moderate, severe, or just about the worst discomfort you have ever suffered in your life?"
    1. mild discomfort?
    2. moderate discomfort?
    3. severe discomfort?
    4. worst discomfort ever?

7. "Did you become aware of this problem while employed as a carpenter?"
    1. yes
    2. no

8. "Do you feel this problem was caused by your work as a carpenter?"
    1. yes
    2. no

9. "Have you seen a doctor, nurse, or other health care provider for the problem you've had with your **ELBOW/FOREARM** ?"
    1. Yes   If yes, what was the diagnosis?_____
    _____ [ ][ ][ ]code
    2. No

10. "How many days of work did you miss because of this problem?"
    ___ ___ ___ days

11. "Did you have to change your job task(s) because of this problem?"
    1. yes
    2. no

12. "Does this problem you have with your **ELBOW/FOREARM** wake you from sleep at night?"
    1. yes
    2. no

INJURIES

13. "Did you ever have an injury to your <u>ELBOW/FOREARM</u> as a result of:" (RECORD MOST SEVERE)
    1. a fall?
    2. an object hitting you or falling on you?
    3. an auto accident?
    4. a sports activity?
    5. other sudden injury? (PLEASE SPECIFY)

    _____

    _____

    _____ [ ][ ][ ]code
    6. none of the above (IF NONE GO TO NEXT BODY PART SECTION WITH A YES ON Q 12 Page 3)

14. "When did this injury occur?"
    __ __ /19__ __
    month   year
    (IF MORE THAN ONE INJURY TAKE DATE OF THE MOST SEVERE)

15. "What type of injury was this?"   (CIRCLE MOST SEVERE)
    1. broken bone?
    2. dislocation?
    3. severe bruise?
    4. cut?
    5. burn?
    6. other?_____ [ ][ ][ ] CODE

16. "To which <u>ELBOW/FOREARM</u> did this injury occur?"
    1. right
    2. left
    3. both

17. "Do you think your current <u>ELBOW/FOREARM</u> problem is a result of this injury?"
    1. yes
    2. no

Now I would like to ask you a few questions about your <u>HANDS and WRISTS.</u>

1. "In the <u>PAST YEAR</u> have you experienced any of the following in your <u>HAND/WRIST</u>": (CIRCLE ALL THAT APPLY)
    1. pain?
    2. aching?
    3. burning?
    4. numbness or tingling in the cold?
    5. numbness or tingling any other times?
    6. fingers turning white or blue?
    7. none of the above? (IF NONE GO TO Q 13)

2. "In which <u>HAND/WRIST</u> does (did) this occur?"
    1. left
    2. right
    3. both

3. "Please tell me the month and year you first had this <u>HAND/WRIST</u> problem(s)."
    __ __ /19__ __
    month   year

4. "How often have you had this <u>HAND/WRIST</u> problem in the <u>PAST YEAR?</u>"
    1. daily
    2. about once a week
    3. about once a month
    4. generally less than once a month

5. "In the <u>PAST YEAR</u> when you had this <u>HAND/WRIST</u> problem, on average how long did it usually last?"
    1. less than a week
    2. between a week and 1 month
    3. longer than a month but not constant
    4. constant

Musculoskeletal and Work History Questionnaire

6. "Would you call this discomfort mild, moderate, severe, or just about the worst discomfort you have ever suffered in your life?"
   1. mild discomfort?
   2. moderate discomfort?
   3. severe discomfort?
   4. worst discomfort ever?

7. "Did you become aware of this problem while employed as a carpenter?"
   1. yes
   2. no

8. "Do you feel this problem was caused by your work as a carpenter?"
   1. yes
   2. no

9. "Have you seen a doctor, nurse, or other health care provider for the problem you've had with your **HAND/WRIST**?"
   1. Yes If yes, what was the diagnosis?_____
      _____ [ ][ ][ ]code
   2. No

10. "How many days of work did you miss because of this problem?"
    ___ ___ ___ days

11. "Did you have to change your job task(s) because of this problem?"
    1. yes
    2. no

12. "Does this problem you have with your **HAND/WRIST** wake you from sleep at night?"
    1. yes
    2. no

INJURIES

13. "Did you ever have an injury to your **HAND/WRIST** as a result of:" (RECORD MOST SEVERE)
    1. a fall?
    2. an object hitting you or falling on you?
    3. an auto accident?
    4. a sports activity?
    5. other sudden injury? (PLEASE SPECIFY)
       _____
       _____
       _____ [ ][ ][ ]code
    6. none of the above (IF NONE GO TO NEXT BODY PART WITH A YES ON Q12 P.3)

14. "When did this injury occur?"
    ___ ___ /19___ ___
    month   year
    (IF MORE THAN ONE INJURY TAKE DATE OF THE MOST SEVERE)

15. "What type of injury was this?"   (CIRCLE MOST SEVERE)
    1. broken bone?
    2. dislocation?
    3. severe bruise?
    4. cut?
    5. burn?
    6. other?_____ [ ][ ][ ] CODE

16. "In which **HAND/WRIST** did this injury occur?"
    1. right
    2. left
    3. both

17. "Do you think your current **HAND/WRIST** problem is a result of this injury?"
    1. yes
    2. no

"Now I would like to ask you a few questions about your BACK."

1. "In the **PAST YEAR** have you experienced any of the following in your **BACK**": (CIRCLE ALL THAT APPLY)
    1. pain?
    2. aching?
    3. burning?
    4. none of the above? **(IF NONE GO TO Q 13)**

2. a. "In which part of your **BACK** does (did) this occur?"
    1. left?
    2. right?
    3. middle?
    4. both sides?
    5. whole back?

   b. "Did this occur in the:"
    1. upper back (above the waist)?
    2. lower back?
    3. whole back?

3. "Please tell me the month and year you first had this **BACK** problem(s)"
   ___ ___ /19___ ___
   month   year

4. "How often have you had this **BACK** problem in the **PAST YEAR?**"
    1. daily
    2. about once a week
    3. about once a month
    4. generally less than once a month

5. "In the **PAST YEAR** when you had this **BACK** problem, on average how long did it usually last?"
    1. less than a week
    2. between a week and 1 month
    3. longer than a month but not constant
    4. constant

6. "Would you call this discomfort mild, moderate, severe, or just about the worst discomfort you have ever suffered in your life?"
    1. mild discomfort?
    2. moderate discomfort?
    3. severe discomfort?
    4. worst discomfort ever?

7. "Did you become aware of this problem while employed as a carpenter?"
    1. yes
    2. no

8. "Do you feel this problem was caused by your work as a carpenter?"
    1. yes
    2. no

9. "Have you seen a doctor, nurse, or other health care provider for the problem you've had with your **BACK**?"
    1. Yes   If yes, what was the diagnosis?_____
       _____[ ][ ][ ]code
    2. No

10. "How many days of work did you miss because of this problem?"
    ___ ___ ___ days

11. "Did you have to change your job task(s) because of this problem?"
    1. yes
    2. no

12. "Does this problem you have with your **BACK** wake you from sleep at night?"
    1. yes
    2. no

Musculoskeletal and Work History Questionnaire

INJURIES

13. "Did you ever have an injury to your **BACK** as a result of:" (RECORD MOST SEVERE)
    1. a fall?
    2. an object hitting you or falling on you?
    3. an auto accident?
    4. a sports activity?
    5. other sudden injury? (PLEASE SPECIFY)
    _____
    _____ [ ][ ][ ]code
    6. none of the above (IF NONE GO TO NEXT BODY PART WITH A YES ON Q12 P.3)

14. "When did this injury occur?"
    ___ ___ /19___ ___
    month   year
    (IF MORE THAN ONE INJURY TAKE DATE OF THE MOST SEVERE)

15. "What type of injury was this?"   (CIRCLE MOST SEVERE)
    1. broken bone?
    2. ruptured/slipped disc?
    3. severe bruise?
    4. cut?
    5. burn?
    6. other?_____ [ ][ ][ ] CODE

16. "Do you think your current **BACK** problem is a result of this injury?"
    1. yes
    2. no

Now I would like to ask you a few questions about your HIP/THIGH.

1. "In the **PAST YEAR** have you experienced any of the following in your **HIP/THIGH**": (CIRCLE ALL THAT APPLY)
    1. pain?
    2. aching?
    3. burning?
    4. numbness or tingling?
    5. none of the above? (IF NONE GO TO Q 13)

2. "In which **HIP/THIGH** does (did) this occur?"
    1. left
    2. right
    3. both

3. "Please tell me the month and year you first had this **HIP/THIGH** problem(s)."
    ___ ___ /19___ ___
    month   year

4. "How often have you had this **HIP/THIGH** problem in the **PAST YEAR?** "
    1. daily
    2. about once a week
    3. about once a month
    4. generally less than once a month

5. "In the **PAST YEAR** when you had this **HIP/THIGH** problem, on average how long did it usually last?"
    1. less than a week
    2. between a week and 1 month
    3. longer than a month but not constant
    4. constant

6. "Would you call this discomfort mild, moderate, severe, or just about the worst discomfort you have ever suffered in your life?"
    1. mild discomfort?
    2. moderate discomfort?
    3. severe discomfort?
    4. worst discomfort ever?

7. "Did you become aware of this problem while employed as a carpenter?"
    1. yes
    2. no

8. "Do you feel this problem was caused by your work as a carpenter?"
    1 yes
    2 no

9. "Have you seen a doctor, nurse, or other health care provider for the problem you've had with your HIP/THIGH?"
    1 Yes   If yes, what was the diagnosis?_____
    _____ [ ][ ][ ]code
    2 No

10. "How many days of work did you miss because of this problem?"
    ___ ___ ___ days

11. "Did you have to change your job task(s) because of this problem?"
    1 yes
    2 no

12. "Does this problem you have with your HIP/THIGH wake you from sleep at night?"
    1 yes
    2 no

INJURIES

13. "Did you ever have an injury to your HIP/THIGH as a result of:" (RECORD MOST SEVERE)
    1 a fall?
    2 an object hitting you or falling on you?
    3 an auto accident?
    4 a sports activity?
    5 other sudden injury? (PLEASE SPECIFY)
    _____
    _____
    _____ [ ][ ][ ]code
    6 none of the above   (IF NONE GO TO NEXT BODY PART WITH A YES ON Q12 P3)

14. "When did this injury occur?"
    ___ ___ /19___ ___
    month   year
    (IF MORE THAN ONE INJURY TAKE DATE OF THE MOST SEVERE)

15. "What type of injury was this?"   (CIRCLE MOST SEVERE)
    1 broken bone?
    2 dislocation?
    3 severe bruise?
    4 cut?
    5 burn?
    6 other?_____ [ ][ ][ ] CODE

16. "Do you think your current HIP/THIGH problem is a result of this injury?"
    1 yes
    2 no

"Now I would like to ask you a few questions about your KNEE"

1. "In the PAST YEAR have you experienced any of the following in your KNEE":
    (CIRCLE ALL THAT APPLY)
    1 pain?
    2 aching?
    3 swelling?
    4 locking?
    5 giving out or giving way of your knee?
    6 none of the above?   (IF NONE GO TO Q 13)

2. "In which KNEE does (did) this occur?"
    1 left
    2 right
    3 both

3. "Please tell me the month and year you first had this KNEE problem(s)."
    ___ ___ /19___ ___
    month   year

Musculoskeletal and Work History Questionnaire

4. "How often have you had this **KNEE** problem in the **PAST YEAR?**"
    1. daily
    2. about once a week
    3. about once a month
    4. generally less than once a month

5. "In the **PAST YEAR** when you had this **KNEE** problem, on average how long did it usually last?"
    1. less than a week
    2. between a week and 1 month
    3. longer than a month but not constant
    4. constant

6. "Would you call this discomfort mild, moderate, severe, or just about the worst discomfort you have ever suffered in your life?"
    1. mild discomfort?
    2. moderate discomfort?
    3. severe discomfort?
    4. worst discomfort ever?

7. "Did you become aware of this problem while employed as a carpenter?"
    1. yes
    2. no

8. "Do you feel this problem was caused by your work as a carpenter?"
    1. yes
    2. no

9. "Have you seen a doctor, nurse, or other health care provider for the problem you've had with your **KNEE** ?"
    1. Yes   If yes, what was the diagnosis? _____
    _____ [ ][ ][ ]code
    2. No

10. "How many days of work did you miss because of this problem?"
    ___ ___ ___ days

11. "Did you have to change your job task(s) because of this problem?"
    1. yes
    2. no

12. "Does this problem you have with your **KNEE** wake you from sleep at night?"
    1. yes
    2. no

INJURIES

13. "Did you ever have an injury to your **KNEE** as a result of:" (RECORD MOST SEVERE)
    1. a fall?
    2. an object hitting you or falling on you?
    3. an auto accident?
    4. a sports activity?
    5. other sudden injury? (PLEASE SPECIFY)
    _____
    _____
    _____ [ ][ ][ ]code
    6. none   (IF NONE GO TO NEXT BODY PART WITH A YES ON Q12 P.3)

14. "When did this injury occur?"
    ___ ___ /19___ ___
    month   year
    (IF MORE THAN ONE INJURY TAKE DATE OF THE MOST SEVERE)

15. "What type of injury was this?"   (CIRCLE MOST SEVERE)
    1. broken bone?
    2. severe bruise?
    3. torn ligament?
    4. torn meniscus?
    5. dislocated kneecap?
    6. other? _____ [ ][ ][ ] CODE

16. "In which **KNEE** did this injury occur?"
    1. right
    2. left
    3. both

17. "Do you think your current **KNEE** problems are a result of this injury?"
    1. yes
    2. no

"Now I would like to ask you a few questions about your ANKLE"

1. "In the **PAST YEAR** have you experienced any of the following in your **ANKLE**": (CIRCLE ALL THAT APPLY)
    1. pain?
    2. aching?
    3. swelling?
    4. none of the above? **(IF NONE GO TO Q 13)**

2. "In which **ANKLE** does (did) this occur?"
    1. left
    2. right
    3. both

3. "Please tell me the month and year you first had this **ANKLE** problem(s)."
    ___ ___ /19___ ___
    month   year

4. "How often have you had this **ANKLE** problem in the **PAST YEAR**?"
    1. daily
    2. about once a week
    3. about once a month
    4. generally less than once a month

5. "In the **PAST YEAR** when you had this **ANKLE** problem, on average how long did it usually last?"
    1. less than a week
    2. between a week and 1 month
    3. longer than a month but not constant
    4. constant

6. "Would you call this discomfort mild, moderate, severe, or just about the worst discomfort you have ever suffered in your life?"
    1. mild discomfort?
    2. moderate discomfort?
    3. severe discomfort?
    4. worst discomfort ever?

7. "Did you become aware of this problem while employed as a carpenter?"
    1. yes
    2. no

8. "Do you feel this problem was caused by your work as a carpenter?"
    1. yes
    2. no

9. "Have you seen a doctor, nurse, or other health care provider for the problem you've had with your **ANKLE** ?"
    1. Yes If yes, what was the diagnosis?_____
       _____ [ ][ ][ ]code
    2. No

10. "How many days of work did you miss because of this problem?"
    ___ ___ ___ days

11. "Did you have to change your job task(s) because of this problem?"
    1. yes
    2. no

12. "Does this problem you have with your **ANKLE** wake you from sleep at night?"
    1. yes
    2. no

Musculoskeletal and Work History Questionnaire

INJURIES

13. "Did you ever have an injury to your **ANKLE** as a result of:" (RECORD MOST SEVERE)
    1. a fall?
    2. an object hitting you or falling on you?
    3. an auto accident?
    4. a sports activity?
    5. other sudden injury? (PLEASE SPECIFY)
    _____
    _____
    _____ [ ][ ][ ]code
    6. none   (IF NONE GO TO PART III)

14. "When did this injury occur?"
    __ __ /19__ __
    month   year
    (IF MORE THAN ONE INJURY TAKE DATE OF THE MOST SEVERE)

15. "What type of injury was this?"   (CIRCLE MOST SEVERE)
    1. broken bone?
    2. severe bruise?
    3. torn ligament or sprain?
    4. dislocated ankle?
    5. other?_____ [ ][ ][ ] CODE

16. "In which **ANKLE** did this injury occur?"
    1. right
    2. left
    3. both

17. "Do you think your current **ANKLE** problems are a result of this injury?"
    1. yes
    2. no

PART III

1. "Of the problems you have just described, which ONE do you consider to be the most serious or troublesome?"
    1. neck
    2. back
    3. shoulder
    4. elbow/forearm
    5. hand/wrist
    6. hip/thigh
    7. knee
    8. ankle

2. "Have you ever filed a Worker's Compensation claim for problems in any of the following body parts?"
    1. neck?   (CIRCLE ALL THAT APPLY)
    2. back?
    3. shoulder?
    4. elbow/forearm?
    5. hand/wrist?
    6. hip/thigh?
    7. knee?
    8. ankle?
    9. none of the above?

PART IV

Now we would like to ask you some questions about your carpentry work.

1. "When were you first employed as a carpenter?"   __ __MO   19__ __

2. "Over the past five years, on average, how many weeks per year were you employed as a carpenter?"   __ __ weeks

    (IF LAID OFF FOR A LONG TIME ASK: WHEN YOU WERE WORKING AS A CARPENTER ON AVERAGE HOW MANY WEEKS WERE YOU EMPLOYED?)

3. A. "Since you began working as a carpenter did you leave the trade or have you been unemployed for six months or longer?"
        1   yes
        2   no   (IF NO GO TO Q 4)

    B. (If yes) "What is the total amount of time you were not working as a carpenter?"
        __ __ . __ years/months

4. A. "Are you currently working in the carpentry industry?"
        1   yes
        2   no (IF NO GO TO PART D)

    B. "What is your current primary job activity?"   (CIRCLE APPROPRIATE NUMBER)

        1   Flooring?
        2   Welding/Burning?
        3   Form work?
        4   Drywall?
        5   Scaffolding?
        6   Framing?
        7   Piledriving?
        8   Finish?
        9   Ceiling?
        10 Millwright?
        11 Fixtures?
        12 Supervision with no actual carpentry work?
        13 Office work?
        14 Something else (specify)?
        _____

        _____ [ ][ ][ ]

    C. "What is the most physically demanding task on your current job?" (try for one answer)
        1   lifting?
        2   carrying?
        3   climbing?
        4   other (please specifiy)
        _____

        5   DK / Refused

    D. "When was the last time you worked as a carpenter?"
        __ __/19__ __
        mo   year

WORK HISTORY:

1. "We would like to get some information about the jobs you have done over the years.

    FROM WHEN YOU STARTED AS A CARPENTER IN 19__ __, have you done any of the following jobs?"

    (read whole job list then ask weeks for each yes)

Musculoskeletal and Work History Questionnaire

How often have you done this job in the **PAST YEAR**?

		(1) never	(2) seldom	(3) sometimes	(4) most of the time	(5) all of the time
Flooring	1 Yes 2 No	___	___	___	___	___
Welding/Burning	1 Yes 2 No	___	___	___	___	___
Form work	1 Yes 2 No	___	___	___	___	___
Drywall	1 Yes 2 No	___	___	___	___	___
Scaffolding	1 Yes 2 No	___	___	___	___	___
Framing	1 Yes 2 No	___	___	___	___	___
Piledriving	1 Yes 2 No	___	___	___	___	___
Finish	1 Yes 2 No	___	___	___	___	___
Ceiling	1 Yes 2 No	___	___	___	___	___
Mill wright	1 Yes 2 No	___	___	___	___	___
Fixtures	1 Yes 2 No	___	___	___	___	___
Supervision with no actual carpentry work	1 Yes 2 No	___	___	___	___	___
Office work	1 Yes 2 No	___	___	___	___	___
other (specify)						
_____	1 Yes 2 No	___	___	___	___	___
_____	1 Yes 2 No	___	___	___	___	___
_____	1 Yes 2 No	___	___	___	___	___
_____	1 Yes 2 No	___	___	___	___	___

2. "In **JUST** the past one month which one of the just mentioned jobs have you done the most?"
   1. Flooring?
   2. Welding/Burning?
   3. Form work?
   4. Drywall?
   5. Scaffolding?
   6. Framing?
   7. Piledriving?
   8. Finish?
   9. Ceiling?
   10. Millwright?
   11. Fixtures?
   12. Supervision with no actual carpentry work?
   13. Office work?
   14. other(specify)

   _____
   _____ [ ][ ][ ]

3. "In your entire career as a carpenter what specialty/job have you done the most?"
   (NAME TOP 3 IN RANK ORDER)

1	Flooring
2	Welding/Burning
3	Form work
4	Drywall
5	Scaffolding
6	Framing
7	Piledriving
8	Finish
9	Ceiling
10	Millwright
11	Fixtures
12	Supervision with no actual carpentry work
13	Office work
14	other(specify)

   _____

   _____ [ ][ ][ ]

   TOP THREE
   1_____
   2_____
   3_____

4. "On average, during a typical year, how many weeks did you work overtime, that is more than 5 days or more than 40 hours?"
   ____weeks

5. "If you could change any of the tools or equipment to make your job easier on your body, which tools would you change?"
   1. _____
   2. _____
   3. _____
   4. _____
   5. _____

6. "How would you change these tools?" **(PROBE AND CLARIFY)**

   _____
   _____
   _____
   _____

7. "The next series of questions asks HOW MUCH INFLUENCE you felt you have had in your work situation over the past twelve months. By influence, we mean the degree to which YOU CONTROL what is done by others and have freedom to determine what you do yourself.

   Please answer the following questions with a number on a scale from 1 to 5, with 1 being little to no control, and 5 being very much control."

   Scale: none or very little   1
           a little   2
           a moderate amount 3
           much   4
           very much   5

   "OVER THE PAST TWELVE MONTHS:"   **(CIRCLE CORRECT NUMBER)**

   a. How much influence do you have over the <u>AMOUNT OF WORK</u> you do?   1 2 3 4 5

   b. How much influence do you have over the <u>AVAILABILITY OF MATERIALS</u> and <u>EQUIPMENT</u> you need to do your work?   1 2 3 4 5

   c. How much do you influence the <u>POLICIES AND PROCEDURES</u> in your work group?   1 2 3 4 5

Musculoskeletal and Work History Questionnaire

d. How much influence do you have over THE PACE OF YOUR WORK, that is, how FAST or SLOW you work?   1 2 3 4 5

e. How much influence do you have over THE QUALITY OF THE WORK that you do?   1 2 3 4 5

f. How much influence do you have over THE HOURS OR SCHEDULE that you work?   1 2 3 4 5

8. "The next series of questions asks HOW OFTEN certain things happen at your job, **over the past twelve months.**

   Please answer the following questions with a number on a scale from 1 to 5, with 1 being rarely, and 5 being very often."

   Scale: rarely 1
   occasionally . . 2
   sometimes . . . 3
   often 4
   very often . . . 5

   "OVER THE PAST TWELVE MONTHS:"

   a. How often does your job require you to work VERY FAST?   1 2 3 4 5
   b. How often does your job require you to work VERY HARD?   1 2 3 4 5
   c. How often are you PHYSICALLY exhausted at the end of the work day?   1 2 3 4 5
   d. How often does your job leave you little time to get things done at work?   1 2 3 4 5
   e. How often is there a great deal to be done at work?   1 2 3 4 5

9. "All in all, HOW SATISFIED are you with being a carpenter?"

   1 Very satisfied ?
   2 Somewhat satisfied ?
   3 Not too satisfied ?
   4 Not at all satisfied ?

10. "If you could change any aspect of your job, what would you change?"

   _____
   _____
   _____
   _____

11. "Do you do any of the following sports activities or hobbies three or more hours per week?"

   racket sports . . . . 1 . yes   2 no
   musical instrument 1 . yes   2 no
   ball sport . . . . 1 . yes   2 no
   fish   . . . . 1 . yes   2 no
   knit/sew . . . . 1 . yes   2 no
   hunt/shoot . . . . . . 1 . yes   2 no

12. "Is your family income about:"

    1 less than $15,000 per year?
    2 between $15,000 and $25,000?
    3 between $25,000 and $50,000?
    4 over $50,000?

This concludes our interview. Do you have any additional questions or comments? We will be sending you a summary of our findings. Thank you very much for your participation.

VERIFY NAME/ADDRESS/PHONE AND RECORD CORRECT INFORMATION

STOP TIME _ _:_ _

INTERVIEWER:

How well did the respondent seem to remember his/her health information?

1 very well
2 fairly well, some problems
3 not very well

How well did the respondent seem to remember his/her work history?

1 very well
2 fairly well, some problems
3 not very well

How cooperative was the respondent?

1 very cooperative
2 fairly cooperative
3 not at all cooperative

REFERENCES

1. B. Silverstein, L. Fine, and T. Armstrong, Occupational factors and carpal tunnel syndrome, *Am. J. Ind. Med. 11*:343–358 (1987).
2. I. Kuorkina, B. Jonsson, A. Kilbom, H. Vinterberg, F .Biering-Sorinsen, G. Andersson, and K. Jorgensen, Standardised Nordic questionnaires for the analysis of musculoskeletal symptoms, *Appl. Ergon. 18*(3):233–237 (1987).
3. L. Punnett, J. Robins, D. Wegman, and W. M. Keyserling, Soft tissue disorders in the upper limbs of female garment workers, *Scand. J. Work Environ. Health 11*:417–425 (1985).
4. E. Holmström, Musculoskeletal disorders in construction workers, Dept. of Physical Therapy, Lund Univ. Lund, Sweden, 1992.
5. NIOSH, Health Hazard Evaluation Report: Los Angeles *Times*, Los Angeles, California. U.S. Dept of Health and Human Services, Public Health Service, Centers for Disease Control, Natl. Inst. Occupational Safety and Health, Cincinnati, OH, NIOSH Rep. No. HETA 90-132277, 1993.
6. NIOSH, Health Hazard Evaluation Report: *Newsday*, Inc., Melville, New York. U.S. Dept. of

Health and Human Services, Public Health Service, Center for Disease Control, Natl. Inst. Occupational Safety and Health, Cincinnati, OH, NIOSH Rep. No. HETA 89-250-2046, 1990.
7. NIOSH, Health Hazard Evaluation Report: Shoprite Supermarkets, New Jersey-New York, U.S. Dept. of Health and Human Services, Public Health Service, Centers for Disease Control, Natl. Inst. Occupational Safety and Health, Cincinnati, OH, NIOSH Rep. No. HETA 88-344-2092, 1990.
8. NIOSH, Health Hazard Evaluation Report: Perdue Farms, Inc. Lewiston, North Carolina and Robertsonville, North Carolina, U.S. Dept. of Health and Human Services, Public Health Service, Center for Disease Control, Natl. Inst. Occupational Safety and Health, Cincinnati, OH, NIOSH, Rep. No. HETA 89-307-2009, 1990.
9. NIOSH, Health Hazard Evaluation Report: John Morell & Co., Sioux Falls, South Dakota, U.S. Dept. of Health and Human Services, Public Health Service, Centers for Disease Control, Natl. Inst. Occupational Safety and Health, Cincinnati, OH, NIOSH Rep. No. HETA 88-180-1958, 1990.
10. E. Corlett and R. Bishop, A technique for assessing postural discomfort, *Ergonomics* 19:175-182 (1976).
11. S. Sauter and L. Schliefer, Work posture, work station design and musculoskeletal discomfort in a VDT data entry task, *Hum. Factors* 33(2):151-167 (1991).
12. M. Hagberg and D. H. Wegman, Prevalence rates and odds ratios of shoulder-neck diseases in different occupational groups, *Br. J. Ind. Med.* 44:602-610 (1987).
13. T. J. Armstrong, B. Buckle, L. Fine, M. Hagberg, B. Jonsson, A. Kilborn, I. Kuorinka, B. Silverstein, G. Sjogaard, and E. Viikara-Juntur, A conceptual model for work related neck and upper limb disorders, *Scand. J. Work Environ. Health* 19:73-84 (1993).
14. S. R. Stock, Workplace ergonomic factors and the development of musculoskeletal disorders of the neck and upper limbs: a meta-analysis, *Am. J. Ind. Med.* 19:87-107 (1991).
15. B. Dorwart, Carpal tunnel syndrome: a review, *Semin. Arthritis Rheum.* 14(2):134-140 (1984).
16. M. Bleeker, Medical surveillance for carpal tunnel syndrome in workers, *J. Hand Surg.* 12A(2 Pt 2):845-848 (1987).
17. R. T. Katz, Nerve entrapments: an update, *Orthopedics* 12(8):1097-1197 (1989).
18. A.R.M. Upton and A. J. McComas, The double crush in nerve entrapment syndromes, *Lancet* August 18: 359-360 (1973).
19. B. Silverstein, L. Fine, and T. Armstrong, Carpal tunnel syndrome: causes and a preventive strategy, *Semin. Occup. Med.* 1(3):213-221 (1986).
20. D. Dawson, Entrapment neuropathies of the upper extremities, *N. Engl. J. Med.* 329(27):2013-2018 (1993).
21. P. Waris, Occupational cervicobrachial syndromes, *Scand. J. Work Environ. Health* 6(Suppl. 3):3-14 (1980).
22. D. Rempel, R. Harrison, and S. Barnhart, Work related cumulative trauma disorders of the upper extremity, *J. Am. Med. Assoc.* 267(6):838-842 (1992).
23. J. L. Kelsey, A. L. Golden, and D. J. Mundt, Low back pain/prolapsed lumbar intervertebral disc, *Rheum. Dis. Clin. N. Am.* 16(3):699-716 (1990).
24. M. Sabour and H. Fadel, The carpal tunnel syndrome: a new complication ascribed to the pill, *Am. J. Obstet. Gynecol.* 107:1265-1267 (1970).
25. K. O. Anderson, L. A. Bradley, L. D. Young, L. K. McDaniel, and C. M. Wise, Rheumatoid arthritis: review of psychological factors related to etiology, effects and treatment, *Psychol. Bull.* 98:358-387 (1986).
26. P. Leino, Symptoms of stress predict musculoskeletal disorders, *J. Epidemiol. Community Health* 43:293-300 (1989).
27. R. Karasek, Control in the workplace and its health-related aspects, in *Job Control and Worker Health*, S. Sauter, J. Hurrell, and C. Cooper, Eds.), Wiley, New York, 19xx, p. 130.
28. J. Hurrell and M. McLaney, Exposure to job stress—a new psychometric instrument, *Scand. J. Work Environ. Health* 4(1):27-28 (1988).
29. J. M. Last, *A Dictionary of Epidemiology,* 2nd ed., Oxford Univ. Press, New York, 1988.

30. A. Booth-Jones, G. Lemasters, P. Succop, A. Bhattacharya, M. Atterbury, H. Applegate, and R. Stinson, A test-retest reliability study of work-related musculoskeletal disorder questionnaire, *Am. Ind. Hyg. Conf. Expo.,* Anaheim, CA, 1994.
31. D. W. Levine, B. P. Simmons, M. J. Koris, L. H. Daltroy, G. G. Hohl, A. H. Fossel, and J. N. Katz, A self-admininstered questionnaire for the assessment of severity of symptoms and functional status in carpal tunnel syndrome, *J. Bone Joint Surg.* 75-A(11):1585–1592 (1993).
32. G. Lemasters, M. Atterbury, H. Applegate, A. Bhattacharya, R. Stinson, Y. Li, H. Pierson, A. Booth-Jones, and C. Forrester, Prevalence of work related musculoskeletal disease in carpenters, *Am. Ind. Hyg. Conf. Expo.,* Anaheim, CA, 1994.
33. D. S. Stetson, B. A. Silverstein, W. M. Keyserling, R. A. Wolfe, and J. W. Albers, Median sensory distal amplitude and latency: comparisons between nonexposed and managerial professional employees and industrial workers, *Am. J. Ind. Med.* 34:175–189 (1993).
34. E. Holmström and M. Ulrich, Low back pain—correspondence between questionnaire interview and clinical examination, *Scand. J. Rehab. Med.* 23:119–125 (1991).
35. K. Ohlsson, R. Attewell, B. Johnsson, A. Ahlm, and S. Skerfving, An assessment of neck and upper back extremity disorders by questionnaire and clinical examination, *Ergonomics* 37(5):891–897 (1994).
36. J. N. Katz, M. Larson, A. Fossel, and M. Liang, Validation of a surveillance case definition of carpal tunnel syndrome, *Am. J. Public Health* 81(2):189–193 (1991).
37. A. Franzblau, R. Werner, J. Valle, and E. Johnston, Workplace surveillance for carpal tunnel syndrome: a comparison of methods, *J. Occup. Rehab.* 3(1):1–4 (1993).
38. M. Atterbury, G. Lemasters, J. Limke, Y. Li, H. Applegate, R. Stinson, H. Pierson, and C. Forrester, Physical examination and nerve conduction results in carpenters with hand wrist symptoms, *Am. Ind. Hyg. Conf. Exp.,* Anaheim, CA, 1994.
39. M. Atterbury, J. Limke, G. Lemasters, Y. Li, H. Applegate, R. Stinson, H. Pierson, and C. Forrester, Nerve conduction studies in carpenters with hand/wrist symptoms, *Am. Occup. Health Conf.,* Chicago, IL, 1994.

21
Fall Prevention in Industry Using Slip Resistance Testing

Mark S. Redfern
University of Pittsburgh, Pittsburgh, Pennsylvania

Timothy P. Rhoades
Applied Safety and Ergonomics, Inc., Ann Arbor, Michigan

I. INTRODUCTION

Occupational slips and falls are a major ergonomic problem across most industries. Occupational falls are estimated to cause about 13% of all work-related deaths in the United States and between 10 and 20% of all accidental injuries [11]. Only overexertion and musculoskeletal injuries and being struck by an object rank higher. Falls have been a particularly serious problem in the construction industry [2–4]. In a survey of worker injuries involving disability, the National Safety Council [1] found that the construction industry had the highest percentage at 22%. Other industries, such as service, transportation, agriculture, and mining, are also highly susceptible, with 16–19% of injuries involving disabilities due to falls.

Causes of falls in the workplace are multidimensional and complex, involving environmental and human factors. Environmental factors include characteristics of work surfaces, elevation, and changes in elevation. The dominant environmental factor in falls is the slip resistance characteristic of the floor [5]. Changes in frictional characteristics of the floor surface can cause a loss of traction resulting in a fall.

Injury can also result from a slip without a fall. Often a slip occurs but stability is recovered; yet injury can still occur from striking an object or from a muscular strain. Anderson and Langerhof [6] found that 37% of non-falling accidents were reported to involve a foot slip in a particular manufacturing environment. Low back injuries can also be initiated by a foot slip [7,8]. Injuries of this type are usually not reported to be related to a slip; thus, the number of low back injuries caused by slips is underestimated.

II. BACKGROUND AND SIGNIFICANCE TO OCCUPATIONAL ERGONOMICS

Due to the high frequency of slips as a cause of injury in the workplace, slip prevention is critical if musculoskeletal injuries are to be reduced. Slip prevention must address both human movement characteristics and environmental factors. This chapter focuses on methods used to prevent slips in the workplace. The topic is addressed from an ergonomic perspective that evaluates the relationship between the worker and the work environment. Engineers have primarily sought solutions to reduce slip hazards by their design of the workplace, but environmental designs need to be developed with the human movement requirements in mind. The biomechanics of slips and falls are presented to describe how slips occur. This information then forms the basis for a discussion of preventive ergonomic measures.

III. BIOMECHANICS OF WALKING AND SLIPS

Slips that occur during walking and carrying loads on a level surface are related to forces generated at the feet and frictional properties of the shoe/floor interface. A slip occurs when the shear forces at the shoe/floor interface exceed the frictional capabilities of that interface. Since the shear forces are highest at the heel contact and push-off phases of walking, these are the points where slips most often occur. Heel contact is the critical phase where slips can result in falls. During heel contact, body weight is being transferred to the lead foot. Should a slip occur at heel contact, the new base of support is not able to accept the body weight. The result is a fall.

The ratio of shear to normal foot force components has been used to understand the biomechanics of slips. This value, described by Strandberg and Lanshammer [16] as the instantaneous "friction used," may also be thought of as the "required" coefficient of friction (RCOF) for those phases of gait where heel movement is not expected [36]. These RCOF data can be used to assess frictional requirements of the shoe/floor interface. The RCOF during a normal step is biphasic with maximum resistance to forward movement just after heel contact and maximum resistance to rearward movement just prior to toe-off (Figure 1). Heel movement will occur if the "friction use" values at heel contact exceed the coefficient of friction (COF) of the shoe/floor interface. The peak of these values during the heel contact phase has been used to predict slip potentials for various gait activities [9–11]. When proper allowances are made for normal heel strike and toe-off displacement, the remaining "friction use" data represent the RCOF for the movement analyzed. It is believed that the peak RCOF value represents the maximum frictional requirement (in terms of shoe/floor COF) during the task.

Since heel contact is most often the critical point in the gait cycle for slips that result in falls, the dynamics of heel contact are important biomechanical factors to consider in understanding the mechanisms of a fall. Even during normal walking, heel contact is a dynamic event. Small slips of the heel commonly occur at heel contact but do not result in falls or loss of stability. These "microslips," as Perkins [12] calls them, are usually less than 2 cm [13,14]. Slips lead to falls when the length of the slip is greater than about 10 cm, termed a "slide" by Leamon and Lee [15]. Slips with recovery occur for heel movements of about 2–10 cm.

Heel dynamics of individuals who were carrying loads were investigated by Redfern et al. [14]. This study was undertaken to determine precisely how the foot moves during the heel contact phase. Motion of the heel during heel contact was measured while individuals were carrying boxes of varying weights at differing speeds. Table 1 shows the results of this study. Microslips occurred, and their length was affected by cadence or walking speed. The heel rapidly decelerates just prior to heel contact (Figure 2). The horizontal velocity decreases from a pre-heel contact maximum of about 450 cm/s to between 14 and 24 cm/s at heel contact. The exact heel contact velocity was dependent on cadence (see Table 1). During a microslip condition, the heel comes to a complete stop about 100 ms after impact. Shoe angle to the floor also changes rapidly. A maximum of almost 25° is established just prior to heel contact, and the transition to foot flat occurs about 100 ms after contact. The shoe angle at heel contact is about 20° but is changing rapidly at that time. These results are similar to those

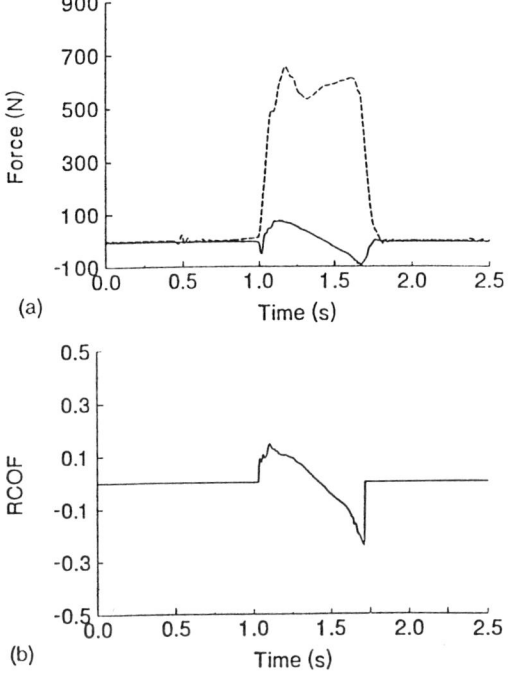

Figure 1 Foot force reactions taken by a force plate during walking at a natural cadence. (a) Vertical (dashed) and anteroposterior (solid) forces during a step. (b) The estimated required coefficient of friction (RCOF) for the same step.

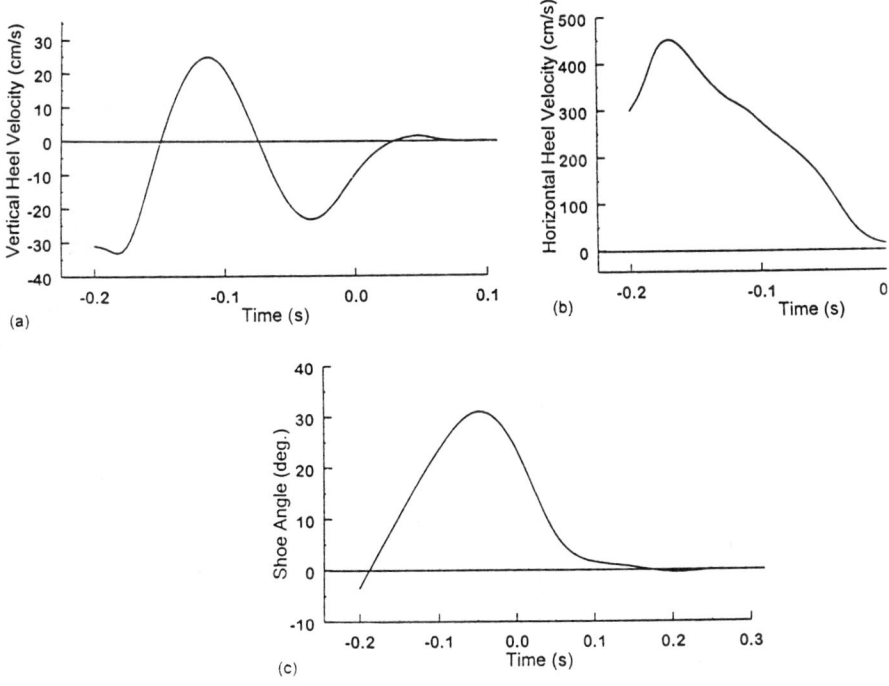

Figure 2 Dynamics of the foot during heel contact. Heel contact is at $t = 0$ s. (a) Vertical heel velocity; (b) horizontal anteroposterior heel velocity; (c) shoe sole angle to the floor.

Table 1 Characteristics of Heel Contact Dynamics During Gait

Steps/min	Results (by cadence)				Strandberg and Lanshammar [16] 90,100,110,120
	70	90	110	Avg.	
Microslip (cm)	1.13 (0.34)	1.34 (0.39)	1.56 (0.55)	1.34 (0.47)	1.2 (mini) (0.4)
Angle at heel strike (deg)	22.9 (3.5)	24.2 (3.3)	25.3 (4.2)	24.2 (3.8)	22.0 (grips) (5.3)
Velocity (Y) at heel strike (cm/s)	13.9 (9.9)	19.0 (15.1)	24.1 (20.8)	19.1 (16.4)	28.2 (grips)
Deceleration (Y) before heel/strike (cm/s^2)	1577.3 (502.0)	2931.3 (426.4)	3831.2 (517.6)	2780.5 (1056.1)	
Velocity (Y) 100 ms before heel strike (cm/s)	171.7 (53.1)	312.2 (40.6)	407.3 (55.9)	297.1 (110.3)	
Time of foot flat (s)	0.122 (0.027)	0.115 (0.045)	0.102 (0.044)	0.113 (0.040)	

found by Strandberg and Lanshammar [16] during different gait experiments in terms of microslips, velocities, and shoe angles and indicate the highly dynamic nature of walking and carrying. This dynamic quality of walking and carrying has direct implications for the relevance of slip resistance testing, which will be discussed in the following section.

During an actual slip and fall while walking, the foot rarely comes to a stop but instead continues to move after heel contact [13, 17]. Under these conditions, a foot slip at heel contact can be considered dynamic and without a static component. Once a slip is initiated and continues beyond about 10 cm, the leading foot accelerates and moves out in front of the body. This causes an increase in the ratio between horizontal and vertical forces that lessens the chance that a slip can be stopped. Since the lead foot is the base of support for the anticipated step, a slip causes a loss of that base of support. Substantial compensatory actions (movements of the arms and other leg) are used to attempt to reestablish a base of support, often with limited success. In this sliding situation the center of mass of the body is in front of the trailing foot during the period when the slip of the leading foot occurs, normally resulting in a loss of any base of support and a fall.

IV. SLIP RESISTANCE TESTING

A. Slip Resistance Measurement

Slip and fall prevention has focused on measurement of the "tractive" or slip resistance properties of flooring and shoes. Most slip resistance tests attempt to evaluate the coefficient of friction (COF) of the shoe/floor interface under various relevant conditions such as dry, wet, oily, etc. These measurements are then used to rate the slip potential of an industrial environment.

The two basic measures used for rating slip resistance of the shoe/floor interface are the static and dynamic coefficients of friction. A static coefficient of friction (SCOF) is most often

Fall Prevention in Industry

used and can be defined as the shear force required to initiate sliding of the shoe material over the floor divided by the vertical force on that same material. Numerous devices have been developed to measure the SCOF of floor surfaces using various shoe materials. Some of these devices have been tested for usability and reliability [18,19]. Two basic types of SCOF measurement devices are the drag sled tester and the articulated strut tester (Figure 3). Probably the most widely used field device is the sled-type tester. In this test, a weight (usually 10 lb) is applied to a sample of shoe sole material that is placed on the floor. A force gauge is then used to apply a force in the horizontal direction until the sled begins to move. The horizontal force is recorded and divided by the vertical force to calculate the SCOF. In contrast, articulated strut systems have a weight mounted at the top of an articulated arm located above the shoe sole material. The articulated arm moves laterally, altering the ratio of normal to shear forces on the shoe sole material. The angle of the articulated arm when the sole material slips is noted and directly related to the SCOF.

The choice of a SCOF tester is difficult for the safety practitioner because SCOF measurement values can vary substantially across devices. In other words, different SCOF testers will give different SCOF values under the same conditions. As a result, there is considerable controversy regarding which device or devices provide a valid measure. As testers are expensive to purchase and no one tester has been established as best for field use, many practitioners choose to conduct inexpensive, manually controlled drag-sled SCOF tests using home-

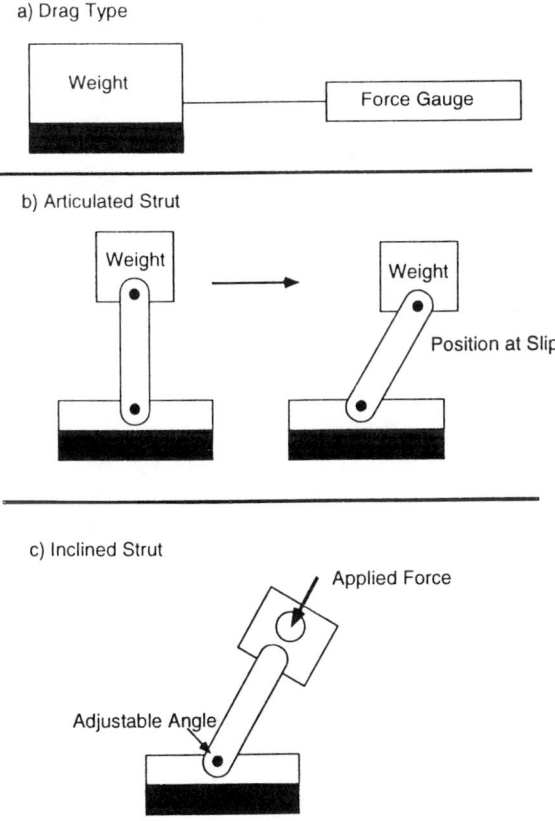

Figure 3 Diagrams of three types of static COF testing devices. (a) Drag tester; (b) articulated strut; (c) inclined strut.

made test fixtures. This measurement method has several shortcomings, but the practitioners' reluctance to invest in equipment that may not stand the test of time is understandable.

Among the most widely known static testers are the horizontal pull slipmeter, the James machine, and the NBS-Brungraber tester. The horizontal pull slipmeter (HPS) measures SCOF in a manner like the generic drag-sled test described earlier, except that the HPS uses a motor to apply horizontal force in a controlled manner and can also provide a dynamic COF measurement in addition to the SCOF measure. The James machine, an articulated strut design developed in the early 1940s, is the most established SCOF tester. The James machine is particularly important because it is used to test floor finishes used nationwide in both industrial and commercial settings. The SCOF acceptance criteria of 0.5 that seems to be gaining consensus is most closely tied to the James machine. Unfortunately, the James machine is too large to permit field testing. To accommodate field testing, the NBS-Brungraber tester (also called the Mark I) was developed. This device successfully incorporates the articulated strict design concept of the James machine into a portable device and has had increasing acceptance as a field measurement device.

A major limitation of the horizontal pull slipmeter, James machine, and NBS-Brungraber tester is that they each suffer from an adhesion-related problem that can often result in readings that are too high under wet conditions. In particular, with each of these testers there is a period of time prior to application of a horizontal force when only a vertical load is applied to the shoe sole sample. During this period of time adhesive forces can develop at the shoe/floor interface under wet conditions. Thus, when the horizontal force is applied to move the shoe sole material it must overcome adhesive forces that are not found during heel strike, and a falsely high SCOF reading is possible. To address this problem one could restrict testing to dry conditions, as is done with the James machine, or resort to dynamic coefficient of friction (DCOF) measures. Yet a third possibility is to measure SCOF in such a way that horizontal and vertical forces are applied at the same time. This approach is the design principle behind two inclined strut testers recently developed. These devices, the Brungraber Mark II and the Ergodyne, each feature an adjustable inclined strut that defines the vector of travel for the shoe sample as it contacts the floor material (see Figure 3c). With a vertical vector, no slip resistance is required. As the vector moves toward horizontal, the ratio of horizontal to vertical forces increases until the shoe sample finally slides along the floor surface. The SCOF is defined by the shallowest angle at which the sole material will not slip.

Although new, the inclined strut tester may overcome the problem of adhesion found with other SCOF testers. There is debate, however, about whether SCOF is the appropriate measure of slip resistance under all conditions [13]. This debate stems from studies of gait and actual slips such as those discussed in the previous section. On the basis of these studies, many researchers believe dynamic COF measures better reflect the actual frictional properties of the shoe/floor interface for slips during walking or load carrying.

A dynamic COF (DCOF) is simply the shear force required to sustain movement of the shoe material divided by the vertical force. Some DCOF devices that have been used in evaluating the shoe/floor interface are the horizontal pull slipmeter [20], the SATRA frictional tester [17, 22], the PSRT [23], the Finnish slip tester [24] and the Tortus [18]. The horizontal pull slipmeter and the Tortus are portable devices that can be used in the field. The SATRA, Gronqvist, and PSRT testers are laboratory-based machines. In general, dynamic measurements can vary greatly from static COFs, with static measures being higher than dynamic measures. Some tests, however, have shown dynamic COF actually higher than the static measures for certain shoe/floor interfaces under specific contaminant conditions.

Although no thorough comparisons among dynamic devices have been made, there do appear to be differences. These differences are due to a number of factors or device characteristics. Variables such as the angle of the shoe sole material to the floor, sliding velocity, and vertical force can all have an effect on the resulting COF. Redfern et al. [25] investigated the effects of velocity and vertical force on DCOF recordings using a sled-type device [26]. It was found that for this type of device, pulling velocity has a significant influence on the resulting DCOF. Vertical force was statistically significant but accounted for only small changes in DCOF. Although this work provided important information about the effects of vertical force and velocity, the sled device used was limited in its ability to incorporate other important parameters. Other parameters, such as angle of the shoe, contact time before testing, and the contact application point of the shoe have been found to affect test results [22,24] and also to be relevant to the biomechanics of slips and falls [16].

Laboratory devices that incorporate some of these parameters have recently emerged [22, 24, 27]. These devices move a shoe over a floor surface at a velocity typical of heel contact. Figure 4 shows a diagram of the device used by Redfern and Bidanda [23], called the programmable slip resistance tester (PSRT). This device allows slip resistance tests to be performed under varying conditions that better reflect the interactions of the shoe and floor during human gait. The effects of shoe materials, floors, shoe angle, heel velocity, vertical force, and contaminants on DCOF recordings using the PSRT were investigated. Test parameters that were found to be significant were the shoe velocity and vertical force as they interacted with the contaminants and shoe materials used. In general, higher test velocities produced lower DCOF measures; however, this was not always true. Thus, it is important in using the PSRT to choose the proper velocities, shoe angles, and vertical forces based upon knowledge of the basic biomechanics of tasks to be investigated.

Once the DCOF test parameters have been chosen, the effects of floors, shoes, and contaminants can be evaluated. Redfern and Bidanda [23] showed the effects of shoe material, floor type, and contaminant on DCOF measures (Figure 5). Floor effects are mixed and depend upon the shoe. A harder shoe sole [in this case polyvinyl chloride (PVC)] performed best on the rougher concrete surface under contaminated conditions. The softer rubber-soled shoes appeared to be better on smoother surfaces for either dry or wet conditions. Interac-

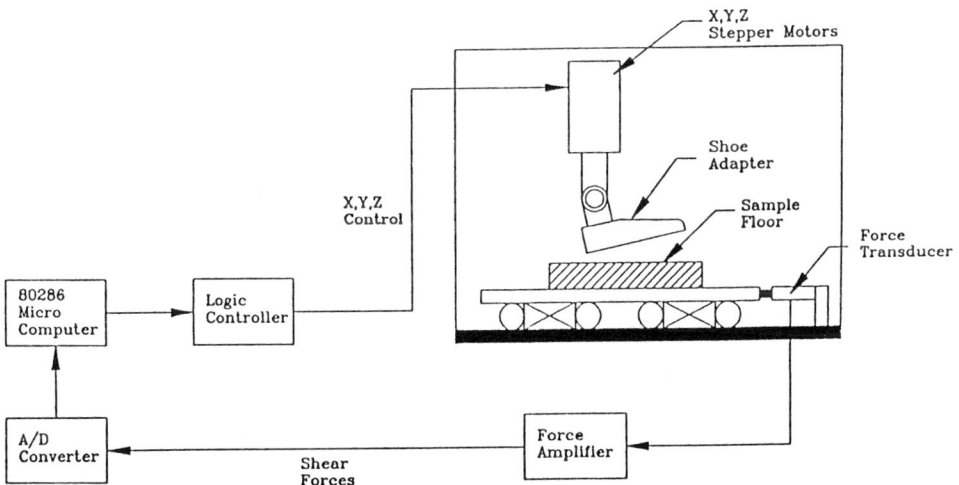

Figure 4 Diagram of the programmable slip resistance tester. The components of the system are (a) microcomputer; (b) logic controller; (c) analog-to-digital converter; (d) X, Y, Z stepper motors; (e) shoe adapter; (f) floor base; (g) force transducer; and (h) force amplifier. (Taken from Bidanda and Redfern [27].)

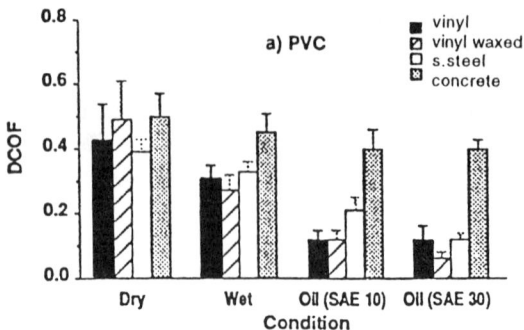
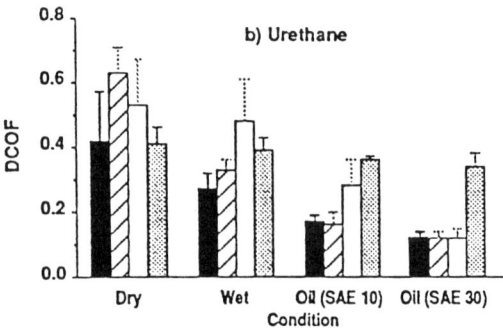
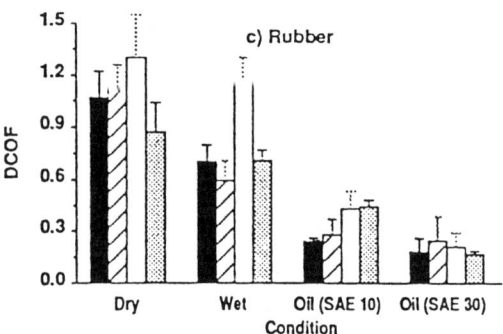

Figure 5 DCOF measurements using the PSRT under containment conditions for three shoe sole materials on four types of floors. (Taken from Redfern and Bidanda [23].)

tions of shoes, floors, and contaminants thus form complex results that can be compared using these test methods.

B. Slip Resistance Guidelines

Attempts have been made to establish safety classifications of shoe/floor conditions for both SCOFs and DCOFs. The first limits were developed for the SCOF in the 1940s using the

James machine. A SCOF criterion of 0.5 as measured by the James machine was established at that time by Underwriter's Laboratory (UL) using a leather shoe material on floor polishes under dry conditions. Since that time, a SCOF level of 0.5 has become increasingly accepted as "safe" for a variety of testers and conditions, with values less than 0.5 being increasingly more hazardous. Two scales for SCOF safety levels are from the American Society for Testing and Materials (ASTM) [28,29] and Rosen [21] (Table 2). In 1975, the ASTM suggested gradual limits for SCOF, with 0.5 being the minimum standard and SCOF of less than 0.3 hazardous [28]. Rosen [21] proposed a SCOF rating level with slightly different gradations, but with 0.5 as the key value to define safe walkways. Note that these SCOF classification schemes and others have been based upon consensus processes rather than scientific evidence [30]. As previously discussed, a major problem with a poorly defined 0.5 criterion is that COF measurements can vary substantially between devices, particularly under wet, oily, or other contaminant conditions [20].

Dynamic COF safety limits have also been explored. In the 1940s, Sigler developed a pendulum-type DCOF tester, now known as the Sigler pendulum tester, and suggested a DCOF minimum of 0.4 using this machine [31,32]. This guideline of 0.4 again seems to be promoted more from consensus and experience than from scientific investigation. The British Standard Institute (BSI) later developed a DCOF rating scale with a minimum DCOF level of 0.4 for slip prevention [33,34] (see Table 2).

More recently, experiments that investigated the required COFs during gait and other tasks were performed [10, 12] and related to slip resistance criteria. Perkins [12] showed that the peak RCOF at heel contact is about 0.28 for normal gait. Thus, the minimum required COF to prevent any slipping is suggested to be 0.28. While this level is regarded by some as a limit for static friction, it is probably better regarded as a limit for dynamic friction [16]. Gronqvist et al. [24] developed a DCOF classification based upon the studies of Strandberg and Lanshammar [16] that agrees with Perkin's [12] observations (see Table 2). This classification was focused on DCOF measures made with the tester that Gronqvist et al. [24] developed. They propose a minimum DCOF level of 0.20 to be "slip resistant." This guideline is much lower than traditional SCOF criteria and lower than previous DCOF limits suggested by BSI [33] and Sigler [31]. Redfern and Bidanda [23] stated in their study of slip resistance using a similar device that the Gronqvist [24] guidelines may be low. Redfern and Bidanda [23] found that DCOF levels for dry floors (believed to be safe) were all greater than 0.3. For oily floors, however, DCOF values were between 0.2 and 0.3 and are probably hazardous based on experience in industry. One possible explanation for this discrepancy is the fact that the Gronqvist scale [24] was based on normal gait. However, in industrial environments that are a wide variety of tasks and floor surface conditions that may increase frictional requirements to prevent slips. These tasks include walking on ramps, turning, and pushing/pulling. For example, frictional requirements of the surface on ramps can more than double those of level surfaces [9]. Also, recent data suggest that turning may also increase frictional requirements by as much as 50% [35]. Thus, slip resistance guidelines suggesting frictional requirements need to consider biomechanical requirements of the tasks.

C. Slip Resistance Application

Despite the potential problems, slip resistance testing can be an effective method for evaluating both the current workplace and possible redesign of it to reduce slip and fall potentials. The approach incorporates basic ergonomic principles of evaluation and design. Jobs are documented and evaluated in terms of worker tasks and the environment. Second, any constraints in the work area are noted. These constraints may be structural requirements of floor-

Table 2 Proposed Coefficient of Friction Recommendations for Safe Walkways

ASTM [28]		Rosen [21]		BSI [33]		Gronqvist et al. [24]	
SCOF	Description	SCOF	Description	DCOF	Description	DCOF	Description
1.00	Very good	>0.60	Very safe	>0.75	Very good	>0.30	Very slip resistant
0.80	Good	0.50–0.59	Relatively safe	0.40–0.75	Good	0.20–0.29	Slip resistant
0.50	Standard for nonhazardous walkway	0.40–0.49	Dangerous	0.40	Minimum	0.15–0.19	Unsure
0.40	Poor	0.35–0.39	Very dangerous	0.40–0.20	Poor	0.05–0.14	Slippery
0.30	Hazardous	<0.34	Unusually dangerous	<0.20	Hazardous	<0.05	Very slippery

ing, sanitation and maintenance conditions, and durability of the materials. Once this information has been collected, slip resistance tests can be used to establish a relative slip potential of the environment as it currently stands and any possible alternative designs.

One example of an industrial application of slip resistance testing is given by the evaluation and redesign of a balcony area within a manufacturing facility. There was a history of serious accidents due to slips and falls from the maintenance balconies in this area. It was determined that the balconies needed to be redesigned. First, the environment and worker tasks were recorded. The current flooring was a diamond plate steel floor. Contaminants were found to consist of water and hydraulic fluids. Worker tasks included routine maintenance of equipment and occasional emergency repair due to mechanical failure of the manufacturing equipment. During these tasks, it was not uncommon for hydraulic fluid to be present, particularly for the emergency repairs. The required worker tasks pertinent to slips on the balconies involved walking, leaning on the rails, pulling, and twisting. The amount of time spent on the balconies was relatively short. Structural constraints of the balconies imposed a need for solid steel floors. Based on this information, a number of candidate floor coverings were identified. These floors were then evaluated by using slip resistance tests in the laboratory under conditions found at the work site. These slip resistance measurements indicated that two floors had significantly higher slip resistance than other candidate floors and that there was no difference in slip resistance between the two. One floor was chosen on the basis of a significant difference in cost and availability. Since the installation of this new flooring on the balconies, fall injuries have been dramatically reduced and the same design has been used in other plants.

Another application of the effective use of slip resistance tests was in the footwear selection and even the design of shoe sole materials for a given environment. Although the development of a specific shoe design is usually not warranted, this may be a viable option for extreme environments where flooring cannot be changed. In one industry, a number of slips and falls off equipment were noted. The constraints were that the surface of the equipment where the workers walked could not be changed. A decision was made to focus on the shoe materials and attempt to develop a shoe that could reduce slips in these areas. A shoe manufacturer was contacted, and a design procedure was implemented. There were other constraints on the shoe materials that were conveyed to the shoe manufacturer, and a number of shoe sole materials were chosen as candidates based upon these criteria. Shoe soles were designed with different materials and different tread designs to be tested by using dynamic and static slip resistance tests. The contaminant conditions of greatest concern were wet and wet with detergent. Environmental temperature was also a concern and was included in the tests. All shoe soles were tested on the given floor with contaminants. The results of the slip resistance tests were compared across materials and treads and indicated that certain materials performed better under wet conditions and others under wet with detergent conditions. One tread pattern was generally superior, although the increase in slip resistance was small. Temperature had a large effect on some materials and less on others. After comparing all shoe soles and treads over all conditions, one sole configuration was chosen. The shoe was designed and worn in field tests with very positive results to date.

This example of designing specific shoes for industry is not common, but it has great potential for reducing slips and falls. In some ways it is analogous to the way in which tool manufacturers design hand tools to prevent hand and wrist disorders. Shoe manufacturers with the capability for flexible manufacturing could be used to develop specific shoes for difficult slip and fall areas that have specific contaminant and task requirements.

A warning with respect to the use of slip resistance tests in the design of shoes and floors for industry: Slip resistance is only one ergonomic factor to consider. High slip resistance work surfaces are not always appropriate, because very high slip resistance can lead to other problems such as trips or stumbles. Other ergonomic shoe/flooring considerations are fatigue and chronic musculoskeletal disorders of the lower extremities. Standing and walking on an extremely high friction surface can cause lower extremity and even whole body fatigue. Also, knee and hip pain can result from twisting and turning on surfaces with high frictional properties. In the balcony design example, workers were on these surfaces infrequently and falls were the overwhelming concern. The same floors would not be appropriate in an area where workers stand for long periods of time. Thus, the appropriate flooring depends on the ergonomic factors of the workplace.

Another concern in floor designs to reduce falls is that many falls occur at a point of transition in flooring characteristics. Transitions from high to low slip resistance can result in a slip. When going from low slip resistance to high, a trip can occur. Gait patterns appropriate for one frictional surface may not be appropriate for another. Gait appropriate for low to moderate slip resistant areas can lead to midswing trips on a high-friction surface, particularly if the floor surface is raised or rough. Transitions from a high-friction surface to a slippery surface can lead to heel slips if a person's gait is not changed. In designing the flooring system, transitions should be minimized, particularly in high-traffic areas, and clearly marked when they are unavoidable.

V. CRITICAL REVIEW OF CURRENT STATUS

Occupational slips and falls are an often overlooked important ergonomic problem. Ergonomic assessment of the workplace should include evaluation and interventions to reduce falls. Current slip measurement methods can be used to evaluate slip potentials through coefficient of friction testing. Guidelines have been proposed and used that give levels of COF values to determine slip and fall potentials. However, there is not full agreement on what level of slip resistance is "safe." There is also an ongoing debate in the ergonomics community on which COF measure, static or dynamic, should be used. A further disagreement exists regarding which device should be used within SCOF and DCOF testing procedures. Although these controversies exist, slip testing can and is used effectively for evaluating workplace slip potential and determining ergonomic interventions. Comparisons of alternative solutions using slip evaluations can quantify potential improvements to the workplace.

VI. FUTURE CONCERNS

More research is required to understand the relationship between slip resistance measurements and actual slip and fall accidents. These studies should involve biomechanical research under various slippery conditions correlated to results with slip resistance measurements. This will lead to an understanding of the effectiveness of these measures in evaluating foot slip potential. This could also lead to improved COF measurement systems or possibly some agreement on which current measurement systems are most appropriate for various conditions and work environments.

VII. CONCLUSIONS

Occupational falls are a continuing ergonomic problem that cuts across all industrial sectors. The causes of these falls are complex and involve environmental and human movement factors. Slip prevention needs to address both factors and be performed within the scope of other ergonomic concerns (i.e., fatigue, stability during manual tasks, etc.) to be effective. Foot forces generated during various tasks can vary greatly and must be considered in environmental design of work surfaces. The primary environmental concern for slip prevention is the frictional property of the shoe/floor/contaminant interfaces. Dynamic and static COF measures are currently used to evaluate slip resistance of the shoe/floor interface. There is no consensus in the ergonomics community on which measure is better; however, there is an increasing belief that dynamic testing may be more relevant to tasks that require walking and carrying. Criteria to establish safe levels of SCOF and DCOF are also not agreed upon throughout the field. Despite these shortcomings, slip resistance tests are useful in determining the relative slip resistance of the shoe/floor interface under various conditions. Slip resistance tests are particularly useful in evaluating current conditions and comparing alternative solutions. These tests should be used as one component of an ergonomic evaluation and design of the shoe/floor system in the workplace. Other factors such as worker task requirements, fatigue from long-term standing, inadequate lighting, and tripping hazards also need to be incorporated into ergonomic evaluations of workplaces.

REFERENCES

1. National Safety Council, *Accident Facts*, 1991 edition, NSF, Chicago, 1991.
2. R. G. Snyder, *Occupational Falls*, UM-HSRI-77-51, National Institute for Occupational Safety and Health, Cincinnati, OH, 1977.
3. P. Britt, Construction safety: wave goodby to work-site falls, *Safety Health 148*(3):54–57 (1993).
4. Construction Safety Association of Ontario (CSAO), *Human Factors Engineering Report on Mounting and Dismounting Construction Equipment*, September 1980.
5. *Federal Register 55*:13341–13360 (1990).
6. R. Andersson and E. Lagerlof, Accident data in new tech Swedish information system on occupational injuries, *Ergonomics 26*(1):33–42 (1983).
7. D. P. Manning and H. S. Shannon, Slipping accidents causing low-back pain in a gearbox factory, *Spine 6*(1):70–72 (1981).
8. D. G. Troup, J. W. Martin, and D. C. E. F. Lloyd, Back pain in industry, a prospective survey, *Spine 6*(1):61–69 (1981).
9. E. J. McVay and M. S. Redfern, Rampway safety: foot forces as a function of rampway angle, *J. American Industrial Hygiene Association*, 55(7):626–634 (1994).
10. M. S. Redfern and R. O. Andres, The analysis of dynamic pushing and pulling: required coefficients of friction, *Proc. Int. Conf. Occup. Ergon.*, 1984, pp. 569–571.
11. F. L. Buczek, P. R. Cavanagh, B. T. Iulakowski, and P. Pradhan, Slip resistance needs of the mobility disabled during level and grade walking, in *Slips, Stumbles and Falls: Pedestrian Footwear and Surfaces*, B. E. Gray, Ed., ASTM STP 1103: ASTM, Philadelphia, PA, 1990, pp. 39–54.
12. P. J. Perkins, Measurement of slip between the shoe and ground during walking, in *Walkway Surfaces: Measurement of Slip Resistance*, ASTM STP 649, Ed. C. Anderson and J. Senne, ASTM, Philadelphia, 1978, pp. 71–87.
13. L. Strandberg, On accident analysis and slip-resistance measurement, *Ergonomics 26*(1): 11–32 (1983).
14. M. S. Redfern, M. A. Holbein, D. Gottesman, and D. B. Chaffin, Kinematics of heelstrike during walking and load carrying: implications for slip testing, *Ergonomics*. In Press.
15. T. B. Leamon and K. W. Lee, Microslip length and the perception of slipping, *Proc. 23rd Int. Congr. Occup. Health,* 1990, p. 17.
16. L. Strandberg and H. Lanshammar, The dynamics of slipping accidents, *J. Occup. Accidents 3*: 153–162 (1981).

17. P. J. Perkins and M. P. Wilson, Slip resistance testing of shoes—new developments, *Ergonomics* 26(1):73–82 (1983).
18. R. O. Andres and D. B. Chaffin, Ergonomic analysis of slip-resistance measurement devices. *Ergonomics* 28(7):1065–1079 (1985).
19. B. T. Kulakowski, F. L. Buczek, P. R. Cavanagh, and P. Pradham, Evaluation of performance of three slip resistance testers, *J. Testing Eval.* 17(4):234–240 (1989).
20. W. English, Improved tribology on walking surfaces, in *Slips, Stumbles and Falls: Pedestrian Footwear and Surfaces*, B. E. Gray, Ed., ASTM STP 1103, ASTM, Philadelphia, PA, 1990, pp. 73–81.
21. S. I. Rosen, *The Slip and Fall Handbook,* Hanrow, Columbia, MD, 1983.
22. M. P. Wilson, Development of SATRA slip test and tread pattern design guidelines, in *Slips, Stumbles and Falls: Pedestrian Footwear and Surfaces*, B. E. Gray, Ed., ASTM STP 1103, ASTM, Philadelphia, PA, 1990, pp. 113–123.
23. M. S. Redfern and B. Bidanda, Slip resistance of the shoe-floor interface under biomechanically relevant conditions, *Ergonomics* 37(3):511–524 (1994).
24. R. Gronqvist, J. Roine, E. Jarvinen, and E. Korhonen, An apparatus and a method for determining the slip resistance of shoes and floors by simulation of human foot motions, *Ergonomics* 32(8):979–995 (1989).
25. M. S. Redfern, A. Marcotte, and D. B. Chaffin, The effects of velocity and applied vertical force on the dynamic coefficient of friction: a dynamic slip resistance study, *Ergonomics.* In Press.
26. M. S. Redfern, A. Marcotte, and D. B. Chaffin, A dynamic coefficient of friction measurement device for shoe/floor interface testing, *J. Safety Res.* 21:61–65 (1990).
27. B. Bidanda, and M. Redfern, Development of a microcomputer baesd slip resistance tester. *Comput. Ind. Eng. J.*, In Press.
28. American Society for Testing and Materials (ASTM), *Standard Method of Test for Static Coefficient of Friction of Polish-Coated Floor Surfaces as Measured by the James Mahcine,* ASTM Designation D 2047–75, ASTM, Philadelphia, 1975.
29. American Society for Testing and Materials, (ASTM), *Standard Test Method for Static Coefficient of Friction of Shoe Sole and Heel Materials as Measured by the James Machine,* reprinted from the *Annual Book of ASTM Standards,* ASTM, Philadelphia, 1977.
30. J. M. Miller, Slippery work surfaces: towards a performance definition and quantitative coefficient of friction criteria, *J. Safety Res.* 14(4): (1983).
31. P. A. Sigler, *Relative Slipperiness of Floor and Deck Surfaces*, U.S. Bur. of Standards-Building Materials and Structures, Rep. BMS 100, 1943.
32. P. A. Sigler, M. N. Geib, and T. H. Boone, Measurement of slipperiness of walkway surfaces, *J. Res. Natl. Bur. Stand.* 40:339–346 (1948).
33. British Standards Institution (BSI) *British Standard Code of Practice for Stairs,* BS5395, British Standards Institution, London, 1977.
34. P. E. Ballance, J. Morgan, and D. Senior, Operational experience with a portable friction testing device in university buildings, *Ergonomics* 28(7):1043–1054 (1985).
35. M. S. Redfern and E. McDonald, The effects of turning on frictional forces at the feet, Technical Report, University of Pittsburgh, 1995.
36. T. P. Rhoades and J. M. Miller, Measurement and comparison of "required" versus "available" slip resistance, *Proceedings of the 21st Annual Meeting of the Human Factors Association of Canada*, Edmonton, Alberta, Canada, Sept., 14–16 (1988).

22
Record-Based ("Passive") Surveillance for Cumulative Trauma Disorders

Shiro Tanaka
National Institute for Occupational Safety and Health, Cincinnati, Ohio

I. INTRODUCTION

Surveillance means different things to different people. A formal definition of surveillance for cumulative trauma disorders (CTDs) may be stated as "the ongoing and systematic collection, analysis and interpretation of exposure and health data necessary for the planning, implementation, and evaluation of programs for prevention and control of CTDs" (From Ref. 1, with slight modification).

Surveillance is a noun derived from the verb "survey," which means "to take a general or comprehensive view of a situation" and "to view in detail in order to ascertain a condition, value, etc." Further, the word *surveillance* implies the attitude of preparedness to make appropriate responses to the surveyed situation. Therefore, in a very broad sense, surveillance is a basic survival function of any living organism. Even a small creature performs surveillance of its surroundings and responds by flight from an attacker or a dash toward food. In our current context, business organizations, both small companies and large corporations, must maintain surveillance of production, customer's needs, market trends, and employee health status to survive and maintain a healthy existence.

II. TYPES OF SURVEILLANCE FOR CTDS AND DEFINITIONS

Since there are many types of surveillance, it is worthwhile to clarify the position of this chapter in the overall surveillance scheme. In the fields of public health and epidemiology, the term *active surveillance* has been traditionally used to indicate the activities of data collection (case finding) by reaching out to, and searching through, patients' records at doctors' offices and hospitals. In contrast, *passive surveillance* has been used to mean the data collection activity of waiting for and receiving the disease reports at the public health office, usually at the Department of Health of a state or city [2].

In recent years, however, a new meaning of active and passive surveillance has been added within the discipline of CTD epidemiology [3]. In this instance, the term active surveillance is used to denote the activities of generating the data, finding cases by administering (musculoskeletal) health questionnaires and/or conducting physical examinations among workers. In contrast, the term passive surveillance is used in CTD epidemiology to denote reviewing and analyzing preexisting records such as OSHA 200 logs and workers' compensation claims, which are usually kept at the employer's office. As a result, the terms may refer to entirely different surveillance activities.

In view of the increasing public health importance of CTDs and increasing interactions between the disciplines of public health and ergonomics, it is highly desirable to avoid confusion by clarifying these terms and definitions. Therefore, I have devised more descriptive terms:

Record-based surveillance: the use of available records such as OSHA 200 logs
Data-generating surveillance: Activities such as questionnaire administration and/or health examinations

To place our topic in perspective, various types of surveillance programs that can be performed for various purposes are listed in Table 1. In this chapter, most of the discussions will focus on record-based surveillance [B.1(a), Table 1] for an in-plant ergonomic program [B.2(a)]. At the same time, readers are reminded that the entire ergonomic program in the plant, from reviews of records to the assessment of effectiveness of ergonomic intervention (feedback), can also be included in a broad sense of surveillance, as shown in Figure 1.

Depicted in the upper part of the flowchart are the surveillance activities for CTDs in a narrower sense. These begin with record-based surveillance, with a systematic review of OSHA 200 logs, workers' compensation claims, sickness and absenteeism records, and other health care records. This analysis will assist identification of high CTD risk jobs.

The results of such an analysis should be compared with the results of various other assessment activities, which are presented in the middle of the flowchart. Such activities would include examination of the awareness of ergonomic issues among management and labor, assessment of the type and degree of exposure (e.g., ergonomic walk-through), and data-generating surveillance such as musculoskeletal health questionnaires and physical examinations. These results usually corroborate one another, leading to the identification of the problem areas and jobs in the plant as illustrated later in Case Report 2.

Shown in the bottom third of the flowchart is the intervention stage, in which various ergonomics intervention measures are implemented to abate the problem. When a follow-up assessment is performed after a sufficient time period, it can be expected that the data from assessment activities will show a decreasing trend of CTDs (provided that the cause of the problem was properly identified and the intervention was correctly targeted and implemented).

In a very broad sense, all of the stages depicted in Figure 1 may be included in a comprehensive surveillance scheme.

Table 1 Types and Definitions of Surveillance for CTDs

A. Exposure variables (measured in ergonomics surveys). Repetition of task, force, posture, duration of work/rest periods, vibration, workstation design, etc. (discussed elsewhere in this book)
B. Outcome variables. Health/morbidity status of workers
   1. Source of surveillance data
      (a) Record-based surveillance (passive surveillance[a])—the analysis of existing records and/or data
      (b) Data-generating surveillance (active surveillance[a])—the generation of data by use of questionnaires, medical examinations, etc. (discussed in Chapter 23)
   2. Scope of surveillance
      (a) In-plant surveillance—private industrial effort
      (b) Public health surveillance—governmental effort

[a]In this chapter, the author has avoided using the terms "active" or "passive" surveillance because they may be confusing or misleading to readers coming from different disciplines.

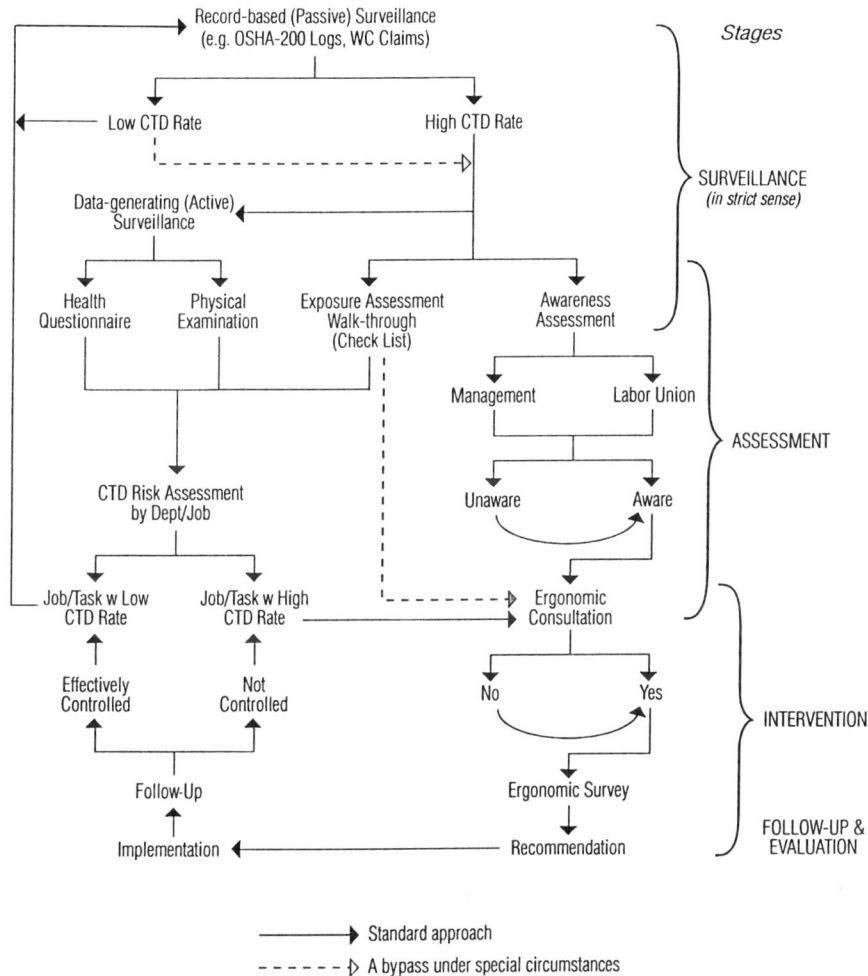

Figure 1 A flowchart of comprehensive CTD surveillance in industry (Solid arrow: standard approach. Dashed arrow: a bypass under special circumstances.)

III. BACKGROUND AND SIGNIFICANCE TO OCCUPATIONAL ERGONOMICS

Surveillance activities should provide the backbone of any ergonomics program. However, for people who are not familiar with ergonomics, it is not necessarily easy to understand and conceptualize the CTD surveillance functions in a plantwide ergonomics program. People who have been well trained in industrial hygiene, epidemiology, or occupational medicine may find that their learned principles and techniques are not quite suitable for dealing with CTDs in the workplace. The reasons for this may be illustrated by comparing surveillance methods for lead exposure and its health effects with those for CTDs (Table 2).

Lead has been known for many centuries to be toxic to humans. In the past several decades, much has been elucidated about the biochemical and toxicological details of lead poisoning. As a result, we now know how much exposure to lead can cause what toxic effects, and standards have been developed to control the exposure by setting the maximum limit for atmospheric lead as well as the blood lead level. Also, several biochemical indicators of

Table 2 Comparison of Exposure/Effect Assessments for Lead Poisoning and Cumulative Trauma Disorders

	Lead poisoning	CTDs
Exposure agent	Lead, lead compounds	Physical stress (force, repetition, posture, duration/lack of rest)
Assessment methodology	Established and specific (lead in air or blood)	Being developed but not established (e.g., ergonomic checklist)
Quantification	Precise	Difficult (except for repetition and duration)
Nonoccupational exposure	Identifiable and distinguishable	Identifiable but not easily distinguishable
Outcome/effect indicators	Nerve damage; kidney damage	Local fatigue, pain, discomfort (e.g., health questionnaire)
Diagnosis	If suspected, usually simple and definitive	Fairly simple for presumptive diagnosis; tests such as EMG, NCV are costly[a]
Technology for prevention	Known and available	Still much unknown or under development

[a]EMG = electromyography; NCV = nerve conduction velocity.

lead exposure and absorption have been identified and used as exposure-monitoring tools. Today, a very effective surveillance for occupational lead exposure and absorption can be conducted by using these refined methodologies to monitor lead levels in atmospheric and biological samples [4].

In contrast, the state of our current monitoring capability in medical surveillance for CTDs does not have the advantage of sensitive and accurate measures. This situation is analogous to that of several decades back in history when we had to wait until the manifestation of frank symptoms of lead poisoning such as lead colic and radial nerve paralysis or signs such as "lead lines" on the gum or in skeletal radiographs. Now we have sensitive tools to monitor exposure to lead. In contrast, this is not the case for CTDs. For example, despite the fact that carpal tunnel syndrome (CTS) is probably the most studied CTD and criteria for its surveillance have been developed [5], medical evaluation of CTS is still largely subjective. Although the measurement of nerve conduction velocity (NCV) is available as an objective method to test dysfunction of the median nerve [6], its high cost (due to the need for skilled technicians and professional interpretations) is rather prohibitive for the routine use of NCV measurement as an industrial screening tool.

Furthermore, the technology for the ergonomic assessment of exposure factors is still in its developmental stage, although some noteworthy advances have been made in recent years [7]. There have been a few documented success stories in which epidemiological and ergonomic investigations have led to some definitive intervention strategies [8]. These cases typically involved specific and obvious physical stresses such as the use of the knee kicker by carpet layers to stretch carpet [9,10]. However, to date, most epidemiological attempts to simply establish certain occupations or jobs as the cause of certain CTDs (e.g., CTS) have been unsuccessful [11]. Such failures are not surprising when one realizes that the causative agents are not the occupations or jobs per se but rather the physical stresses demanded by the job and how the job or task is performed. In other words, almost any manual job can lead to CTS or tendinitis depending on how the task is performed, while even meat packing jobs that have been known for their severe musculoskeletal stresses [12] could be performed, theoretically, without incurring a CTD if appropriate ergonomic practices were put into effect.

Therefore, we have to come to the realization that these difficulties are inherent in the medical and ergonomic surveillance of CTDs. Nonetheless, we must use currently available methodologies to perform the needed quantitative assessment and hope for technological advancements in coming years. It is not difficult for an epidemiological study to identify occupations or industries with an elevated risk of CTDs. However, epidemiological techniques alone cannot determine what exposure factors in the work are really causing the problem and how these can be controlled. Thus, cooperative and coordinated efforts of both ergonomists and epidemiologists are much needed to achieve this goal [13].

IV. RECORD-BASED SURVEILLANCE

In-plant record-based surveillance ("passive" surveillance) for CTDs involves reviewing and analyzing existing records or data systems that are normally kept by the employer [14]. Typically, available records would include OSHA 200 logs and workers' compensation (WC) claims, which will be the main subjects of discussion in this chapter. Record keeping of OSHA 200 logs is required by law for the purpose of surveillance, and the details are described in official publications [15]. In contrast, the use of WC claims for this purpose is for convenience, because the WC system was not created for surveillance purposes. Some companies keep so-called sickness and accident (S&A) records from which data for CTDs can be extracted. Medical records may be kept at the company's health unit or at the health care provider's office. Also, the health insurance records of employees have been used for CTD surveillance [16]. More detailed medical information is usually available from these health care record systems than from OSHA 200 logs or WC claims. However, routine extraction of necessary data from health insurance records is not always easy or simple unless the data are computerized and suitably coded. The insurance carriers are usually reluctant to release the data owing to the confidentiality issue and the proprietary nature of information.

For the purpose of basic CTD surveillance at the place of employment, periodic (preferably monthly) review of OSHA 200 logs and WC claims is recommended and should be effective as long as the information is recorded honestly and without bias or interference. Equally important are the keen sense and ability of the person in charge of surveillance to recognize workers' physical complaints or remarks, to detect abnormal trends, and to respond appropriately (see Sec. VII, Case Report 1).

The quality and usefulness of a record system for the purpose of CTD surveillance will depend on various factors. To calculate incidence (i.e., number of new cases) rates, the record should include, at a minimum, date of occurrence, personal identifier, department, job title, part of the body affected, and preferably the diagnosis. Conditions reportable in column 7-f of OSHA 200 logs are defined as "disorders associated with repeated trauma (DART)" with examples such as carpal tunnel syndrome, synovitis, tenosynovitis, bursitis, Raynaud's phenomenon, and noise-induced hearing loss [15]. Also, the denominator data, such as the number of workers by department and hours worked, should be available for calculation of incidence rates. The record should be easily accessible to and retrievable by authorized personnel for the purpose of conducting surveillance. Computerized data processing would make the analysis fast and simple.

An incidence rate (IR) per 100 full-time employees of illnesses may be computed by using the formulas [15]

$$IR = \frac{\text{Number of illnesses} \times 200{,}000}{\text{Employee hours worked}}$$

and

200,000 hr worked per year = 100 full-time employees (FTE)

In this calculation, only new cases are counted for a given time period. If one worker experienced the same CTD more than once during the reporting period, the occurrences are counted separately as long as there was a period of complete recovery in between. The severity or duration of the disorder is not considered in incidence rate.

The incidence rate should be computed for the entire company or plant, for each department, and for each section if within a large department. It may sometimes happen that only a small number of workers are performing highly repetitive manual work within a large company or department. In such a situation, the incidence rate for the company or department may not be high, and the CTD hazard for the small group of exposed workers can be overlooked (a dilution effect). Calculation by job/task groups, combined with an ergonomic evaluation, should be able to point out the problem areas (see Sec. VII, Case Report 2).

The calculation of the total hours worked can be obtained from computerized payroll data in most companies. If the number of hours worked is not computerized, it must be hand calculated using a calculator. For the purpose of obtaining a rough estimate, the average or usual number of workers may be used instead of hours worked, as long as the number of employees remained fairly stable during the year. (For example, if a lathe operator quit after working 6 months and was replaced by another lathe operator who worked the remaining 6 months of the year, they can be counted as 1 person-year.) However, it should be kept in mind that this rough estimation tends to underestimate the incidence rate if each worker did not work a full 2000 hr a year due to part-time work, vacation, illness, layoff, etc.

For official reporting purposes, annual incidence rates are required by OSHA. However, for the purpose of maintaining an effective in-plant ergonomics surveillance program, it is not recommended to wait for 12 months if one wants to detect an upsurge of CTDs at the earliest possible stage. Therefore, in addition to being in compliance with the OSHA requirement, examination of the current data and trend should be performed at least on a quarterly, and preferably on a monthly, basis.

In addition to the incidence rate, which considers the number of new cases in a set time period, the prevalence (rate) is sometimes used. The prevalence measures the frequency of all current cases of a disease, both new and old (continuing from a previous period), at a given point in time (point prevalence) or for a prescribed period of time (period prevalence). Thus,

$$\text{Point prevalence} = \frac{\text{number of new and old cases at a given point in time}}{\text{number of workers at the same point in time}}$$

$$\text{Period prevalence} = \frac{\text{number of new and old cases during a given time period}}{\text{number of workers at the mid-interval in the same time period}}$$

Period prevalence is of limited usefulness [17], as it does not distinguish new cases from old ones nor does it count repeated episodes of the same disease of the same person occurring in the time period.

For the purpose of estimating the overall seriousness of CTDs by departments or disorder and for setting the intervention priority, the severity index may be useful. The severity index (SI) may be calculated for all CTD cases or by diagnosis, or by part of the body affected, using the following formula:

$$SI = \frac{\text{Total number of workdays lost due to the defined disorder(s)}}{\text{Total number of workers or hours worked in a time period}}$$

However, it must be kept in mind that the severity index may be influenced by such factors as the type (conservative or invasive) of medical management [18], sickness benefit, and opportunity for transfer to less stressful jobs. It may also be skewed by unusually long illnesses experienced by a small number of employees.

V. ADVANTAGES AND DISADVANTAGES OF RECORD-BASED SURVEILLANCE

Major advantages of record-based surveillance using OSHA 200 logs or workers' compensation claims are its low cost and easy accessibility. Since the employer is required to collect and maintain the record, the reviewer's main task is to tabulate and analyze the data. This effort is small compared to the data-generating surveillance, which involves administration of a health questionnaire and/or physical examinations, and subsequent data analysis [3].

A major shortcoming of record-based surveillance is said to be its underreporting. Fine and others [3] estimated that record-based surveillance detected only one-seventh of potential CTD cases that were uncovered by data-generating surveillance. There are several reasons for this underestimation. First, the mere presence of pain or other symptoms is neither reportable in OSHA 200 logs nor sufficient for filing a WC claim. Second, some employers may be reluctant to list all reportable cases in the OSHA 200 logs. There were well-publicized cases of deliberate underreporting by some meat packing companies in the mid-1980s. After OSHA started imposing large fines for such neglect or concealment, the number of reported cases started to increase. Third, employees may hesitate to report their symptoms or illnesses to their supervisor, particularly at a time when jobs are scarce.

In contrast, musculoskeletal questionnaires or physical examinations of data-generating surveillance (described in more detail in Chapter 23) typically elicit reports of symptoms and signs in various body parts in a confidential manner. Pains and discomfort above certain levels and lasting more than a certain number of days are counted as positive cases. Therefore, it is not surprising that data-generating surveillance (particularly for research purposes) can detect many times more cases than record-based surveillance.

Worker's compensation (WC) claims are filed under specific rules and regulations that vary from state to state. Therefore, record-based surveillance using these databases may be subject to an equal or higher degree of underestimation compared to the review of OSHA 200 logs. On the other hand, WC claims usually contain a wealth of information related to each CTD claim, including the body part affected and diagnosis [19]. Under special circumstances such as for research, health insurance records can be obtained and analyzed for detection of potential CTDs [16]. However, widespread or routine use of health insurance records as a surveillance tool may not always be feasible or practical, particularly for small companies.

The primary objective of conducting an in-plant CTD surveillance is early detection and intervention. For this purpose, the concept of Sentinel Health Event—Occupational (SHE-O), which was first proposed by Rutstein and others [20], can be applied. For investigation of various occupational diseases, including CTDs, a single case could trigger more focused examinations of the health status of coworkers and the work area where the index case was detected [21]. Even if record-based surveillance may detect only a small portion of CTD cases, it can be said that one case detected by record-based surveillance may lead to several times as many unreported cases.

VI. CRITICAL REVIEW OF CURRENT STATUS AND FUTURE CONCERNS

As presented in the above, record-based surveillance for in-plant monitoring of CTDs can be conducted fairly rapidly and easily once the system is in place and managed by a knowledgeable and responsible individual. With the support of responsive management, this person can play an important role for timely intervention and control of CTD problems in the plant. At the plant level, disorders associated with repetitive trauma (DART) conditions reportable in column 7-f of OSHA 200 logs seem to be specific enough to include the part of the body in the data.

Nationwide surveillance for CTDs is very important for making effective policy decisions for prevention but difficult to conduct for various reasons. As discussed in the beginning of this chapter, record-based CTD surveillance is plagued by the ill-defined nature of the disorders and the difficulty of exposure assessment. The current OSHA-required record-keeping system (which is the basis of the BLS' annual report) has been criticized for its tendency to underestimate the incidence of CTDs [3]. Also, when the data were compiled by BLS to prepare the annual report, all of the DART conditions were lumped together. This process reduced its usefulness as a nationwide surveillance tool, although DART accounted for 56% of the total cases of occupational illnesses reported by private industry in 1990.

Since 1992, BLS has been using a redesigned occupational injury and illness surveillance program to collect more detailed information such as the demographics of the affected workers and the cirumstances of the incident for lost workday cases. This new method has generated information that is more useful toward prevention of CTDs, which was not available under the old reporting system [22].

Also, at the time of this writing, OSHA's effort to propose an ergonomic protection standard for general industries is stalled in a political process. Detailed methods of surveillance for exposure to musculoskeletal stress and health effects are described in the proposed draft [23]. Whether or not such a standard is eventually promulgated, the surveillance methods described in the OSHA draft are very useful for implementation of an in-plant ergonomics program on a voluntary basis. I am also confident that the basic aspects of record-based CTD surveillance described in this chapter would be applicable regardless of the type of standard or guidelines that might be finalized by OSHA.

VII. CASE REPORTS

Two cases have been selected to illustrate some of the points made in this chapter.

A. Case Report 1

This case is based on a paper by Luopajärvi et al. [24] reporting on a food (unspecified) production factory where approximately 200 female workers performed packing tasks on assembly lines in the 1970s. As shown in Table 3, the number of cases of occupational hand

Table 3 A Surveillance Record of a Food Packing Company (Case Report 1)

Year	Number of cases	Lost workdays	Hindsight remarks
1972	1	42	Something happened here, but no intervention?
1973	12	1117	
1974	16	1446	Many of these cases could have been prevented if problem had been identified and intervention begun in 1973.
1975	46	3670	
1976	51	5288	
1977	20	840	
1978	5	201	
1979	1	24	
1980	0	0	

Source: Data from Ref. 24 by permission.

disease in 1972 was only 1 with 42 lost workdays. In 1973, this jumped to 12 cases with 1117 lost days, and over the following years the numbers continued to increase. In 1976, the situation had become so serious that a project was started to deal with the problem with seven working groups directed by a multiprofessional leadership. The epidemic was eventually brought under control by 1979. However, during the 6-year span 1973–1978, a total of 150 cases and 12,562 lost workdays due to hand disorders were recorded.

Although the intent of the paper was to report on the effectiveness of various intervention measures to contain the epidemic, it seems to present a very interesting case study for surveillance. First, there was at least a basic surveillance system in this plant to record occupational illnesses and lost workdays. Through 1972, the packaging was probably done by a slow, old-fashioned method. Although it was not described in the paper, something happened during the year 1973. A likely scenario might be that a new company policy was implemented to increase the production rate by way of increased quota, possibly accompanied by the introduction of partial automation. Workers had to adjust the speed of manual work to that of the machine, but obviously it was far beyond their physical capacity.

The sudden increase of morbidity must have caught the eye of plant management in 1973. Strangely, however, nothing was done to deal with the surge of new cases until 1976. By hindsight, if the intervention effort had been initiated in 1973, they could possibly have prevented up to 130 cases and 11,000 lost workdays. This case amply illustrates that the failure of early detection of the problem by surveillance and resultant lack of timely intervention allowed the problem to continue and even increase in size for several years.

The moral of this case would be that an increased rate of CTDs must be examined to determine the reason for the increase and dealt with promptly to prevent an epidemic, which can be very costly.

B. Case Report 2

This case is based on my own experience at the National Institute for Occupational Safety and Health (NIOSH). In a joint surveillance project by NIOSH and Ohio Bureau of Workers' Compensation (BWC), WC claims were analyzed for "inflammation or irritation of joints, tendons, or muscles" resulting from "overexertion occurring over a protracted time period" for various parts of the body [19]. As a result, we were able to identify companies with a high incidence rate of such cases for the hand/wrist including carpal tunnel syndrome. (Before 1985, Ohio BWC did not include a separate code for CTS.) Subsequent telephone con-

Table 4 Comparison of Crude Incidence Rates of Carpal Tunnel Syndrome (CTS) and Related Disorders by Various Surveillance Methods at a Hose Manufacturing Plant (Case Report 2)[a]

Dept.	Number of workers	WC claims	From OSHA 200 logs		From questionnaire		
			CTS or like	Sprains/ strains	Neck/ Arm	Hand wrist	Nocturnal hand pain
A	67	2.4	1.5	13.4	31%	31%	32%
B–J	388	0.2	0.1	0.9	22%	17%	6%
Office	156	0	0	0	10%	3%	2%
Total	611	0.4	0.2	2.0	20%	15%	8%

[a]Rates are per 100 employees per year or percent as indicated.
Source: Condensed from tables in Ref. 25.)

tacts with seven of these companies all confirmed the existence of, and the management's concern with, work-related CTS among their employees. One company, a manufacturer of garden, automotive, and industrial hoses, cooperated with NIOSH in an ergonomic and epidemiological investigation of their CTD problem [25].

The investigation consisted of the review of WC claims and OSHA 200 logs, administration of a questionnaire, and an ergonomic walk-through assessment and analysis of videotapes. Table 4 shows a summary of crude incidence rate of CTS (and related disorders) by department.

It can be noted in Table 4 that positive responses by questionnaire (data-generating surveillance) were far more frequent than those detected by WC claims or OSHA 200 logs (record-based surveillance). This is consistent with the previous discussion on the varying degree of detection by different surveillance methods. However, regardless of surveillance methodology, the interdepartmental comparisons showed a consistently higher rate for a specific department (Department A). A later ergonomic survey revealed that the CTDs in this department were indeed caused by very forceful and repetitive manual work, which required frequent bending and twisting of the wrist, elbow, and shoulder.

It is interesting to note that this company was initially selected because of a very high overall incidence rate of WC claims for CTDs. However, it was later found that the high rate was spurious and was due to the erroneously small number of employees (denominator) listed in an industrial directory for the state. The company would not have been selected if the number of employees had been listed correctly. Nevertheless, upon further inquiry and site visit, it was revealed that Department A had a very high incidence rate of CTDs compared to other departments.

This case illustrates that for the purpose of in-plant surveillance for CTDs, WC claims and OSHA 200 logs can be used effectively to detect high-risk departments or jobs. At the same time, however, it was learned that for the purpose of public health surveillance, an overall low or moderate rate of CTDs determined by a simple calculation may be misleading, because the problem areas may be identified only after a detailed in-plant investigation.

REFERENCES

1. D. N. Klaucke, J. W. Buehler, S. B. Thacker R. G. Parrish, F. L. Trowbridge, R. L. Berkelman, and the Surveillance Coordination Group, Guidelines for evaluating surveillance systems, *Morbidity and Mortallity Weekly Report (MMWR) 37*(S-5):1–17 (1988).

2. W. A. Orenstein and R. H. Bernier, Surveillance for the control of vaccine-preventable disease, in *Public Health Surveillance,* H. Halperin, E. L. Baker, and R. R. Monson, Eds., Van Nostrand Reinhold, New York, 1992, pp. 80-82.
3. L. J. Fine, B. A. Silverstein, T. J. Armstrong, C. A. Anderson, and D. S. Sugano, Detection of cumulative trauma disorders of the upper extremity in the workplace, *J. Occup. Med. 28*:674-678 (1986).
4. P. J. Seligman, W. E. Halperin, R. J. Mullan, and T. M. Frazier, Occupational lead poisoning in Ohio: surveillance using workers' compensation data, *Am. J. Public Health 76*:1299-1302 (1986).
5. Anon., Occupational disease surveillance: carpal tunnel syndrome, *MMWR 38*:485-488 (1989).
6. J. N. Katz, M. G. Larson, A. Sabra, C. Krarup, C. R. Stirrat, R. Sethi, H. M. Eaton, A. H. Fossel, and M. H. Liang, The carpal tunnel syndrome—diagnostic utility of the history and physical examination findings, *Ann. Intern. Med. 112*:321-327 (1990).
7. W. S. Marras and R. W. Schoenmarklin, Wrist motions in industry, *Ergonomics 36*:341-351 (1993).
8. S. Tanaka, S. T. Lee, W. E. Halperin, M. J. Thun, and A. B. Smith, Reducing knee morbidity among carpetlayers, *Am. J. Public Health 79*:334-335 (1989).
9. M. J. Thun, S. Tanaka, A. B. Smith, W. E. Halperin, S. T. Lee, M. E. Luggen, and E. V. Hess, Morbidity from repetitive knee trauma in carpet and floor layers, *Br. J. Ind. Med. 44*:611-620 (1987).
10. A. Bhattacharya, M. Mueller, and V. Putz-Anderson, Traumatogenic factors affecting the knees of carpet installers, *Appl. Ergon. 16*:243-250 (1985).
11. S. R. Stock, Workplace ergonomic factors and the development of musculoskeletal disorders of the neck and upper limbs—a meta analysis, *Am. J. Ind. Med. 19*:87-107 (1991).
12. V. R. Masear, J. M. Hayes, and A. G. Hyde, An industrial cause of carpal tunnel syndrome, *J. Hand Surg. 11A*:222-227 (1986).
13. S. Tanaka and J. D. McGlothlin, A conceptual quantitative model for prevention of work-related carpal tunnel syndrome (CTS), *Int. J. Ind. Ergon. 11*:181-193 (1993).
14. M. Burke, *Applied Ergonomics Handbook,* Lewis, Boca Raton, FL, 1992.
15. Bureau of Labor Statistics (BLS), *Recordkeeping Guidelines for Occupational Injuries and Illnesses,* U.S. Dept. of Labor, Washington, DC, 1986.
16. R. M. Park, N. A. Nelson, M. A. Silverstein, and F. E. Mirer, Use of medical insurance claims for surveillance of occupational disease—an analysis of cumulative trauma in the auto industry, *J. Occup. Med. 34*:731-737 (1992).
17. B. MacMahon and T. F. Pugh, *Epidemiology: Principles and Methods,* Little, Brown, Boston, 1970.
18. N. M. Hadler, Arm pain in the workplace—a small area analysis, *J. Occup. Med. 34*:113-119 (1992).
19. S. Tanaka, P. Seligman, W. Halperin, M. Thun, C. L. Timbrook, and J. J. Wasil, Use of workers' compensation claims data for surveillance of cumulative trauma disorders, *J. Occup. Med. 30*:488-492 (1988).
20. D. D. Rutstein, R. J. Mullan, T. M. Frazier, W. E. Halperin, J. M. Melius, and J. P. Sestito, Sentinel health events (occupational): a basis for physician recognition and public health surveillance, *Am. J. Public Health 73*:1054-1062 (1983).
21. R. J. Mullan and L. I. Murthy, Occupational sentinel health events—an up-dated list for physician recognition and public health surveillance, *Am. J. Ind. Med. 19*:775-799 (1991).
22. T. Luopajärvi, I. Kuorinka, and R. Kukkonen, The effects of ergonomic measures on the health of the neck and upper extremities of assembly-line packers—a four year follow-up study, *Proc. 8th Congr. Int. Ergon. Assoc.* Tokyo, Japan, 1982, pp. 160-161.
23. S. Tanaka and D. Habes, Health Hazard Evaluation Report No. 87-428-2063, Anchor Swan Division, Harvard Industries, Inc., Bucyrus, OH; National Institute for Occupational Safety and Health (NIOSH), Cincinnati, OH, 1990.
24. Bureau of Labor Statistics: Survey of Occupational Injuries and Illnesses, 1993. BLS Form-9300 NO4, U.S. Department of Labor, Washington, D.C.

25. Occupational Safety and Health Administration: OSHA Draft Proposed Ergonomic Protection Standard: Summaries, Explanations, Regulatory Text, Appendices A and B. *Occupational Safety and Health Reporter,* 24(42) Special Supplement, pp. S1–S248; March 20, 1995; The Bureau of National Affairs, Inc., Washington, DC.

23
Active Surveillance of Work-Related Musculoskeletal Disorders
An Essential Component in Ergonomic Programs

Norka Saldaña
Johnson & Johnson Shared Services, Caguas, Puerto Rico

I. INTRODUCTION

Identification of work-related musculoskeletal problems in industry at an early stage allows for early control, a safer environment, and a healthier workforce. OSHA defined ergonomics surveillance as "the ongoing, systematic collection, assessment and interpretation of health incidence and exposure data in the process of describing and monitoring the circumstances which may be related to ergonomic hazards or the presence thereof" [1]. Thus ergonomic surveillance techniques are used (1) to detect, monitor, and control patterns of health and disease among employees, (2) to identify work-related risk factors and related ergonomic hazards that may cause, precipitate, or aggravate a condition, and (3) to evaluate the effectiveness of the ergonomic interventions in reducing the incidence or prevalence of musculoskeletal disorders.

Ergonomic surveillance attempts to anticipate work-related health problems before they occur. When ergonomic hazards cannot be eliminated completely, then ergonomic surveillance serves as a control mechanism aimed at identifying a disorder in its early stages before it becomes a more serious problem. To accomplish this, ergonomic surveillance should address the following questions:

1. What types of health problems are the employees experiencing?
2. What is causing these problems?
3. How can these health problems be eliminated?

Work-related musculoskeletal disorders (WMDs) accounted for 60% of all occupational illnesses among U.S. workers in 1991 [2]. Consequently, ergonomic surveillance concentrates mostly on the identification of WMDs referred to as WMD surveillance. The underlying assumption for WMD surveillance is that if musculoskeletal discomfort symptoms persistently recur from job exposure and there is no intervention and no provision for sufficient recovery time, then a musculoskeletal disorder may develop.

This chapter concentrates on WMD surveillance. There are two basic methods for collecting health data that can help determine past or present occurrences of musculoskeletal problems. These two methods are (1) passive surveillance and (2) active surveillance. The words *passive* and *active* relate to the time of occurrence of the events being studied relative to the researcher's place in time. Passive surveillance looks at past cases, whereas active surveillance anticipates new cases. Passive surveillance is discussed briefly here (for more information, refer to Chapter 22). Then two approaches for active surveillance are presented and discussed in terms of the effectiveness of each approach for the surveillance of WMDs.

II. BACKGROUND AND SIGNIFICANCE TO OCCUPATIONAL ERGONOMICS

Surveillance is from the French for keeping a watch over. It was first used by French police to describe the way in which criminals or potential criminals were watched [3]. In the medical arena, the term is applied to the observation of infectious diseases. Thus surveillance involves the planned follow-up of individuals at risk of developing a disease or individuals with a disease in an early stage who need close observation as the condition develops. In general, the objective of medical surveillance is to prevent diseases from claiming new victims.

Although medical surveillance programs are commonly established in a variety of industrial settings, the quality and effectiveness of such programs vary widely [4]. These programs incorporate medical examinations as well as laboratory tests and worksite exposure data. The purpose of occupational surveillance in large part is to detect adverse health conditions and to determine whether employees are at risk from continued exposure [5]. Ergonomic surveillance of upper extremity WMDs is most commonly used to identify the jobs with high rates of disorders so that an effective control program may be developed. However, identification of these disorders is very difficult due to many complex etiologic factors, long latency, effects of aging, and lack of standardized diagnostic criteria [6].

A. Definition of Work-Related Musculoskeletal Disorders

Work-related musculoskeletal disorders are conditions that affect the soft tissues. Other commonly used terms include ergonomic disorders (EDs), cumulative trauma disorders (CTDs), and repetitive strain injuries (RSIs). WMDs are a class of musculoskeletal disorders that include damage to the tendons, tendon sheaths, and synovial lubrication of tendon sheaths, and to the related bones, muscles, and nerves of the hands, wrists, elbows, shoulders, neck, and back. These disorders develop gradually over periods of weeks, months, or even years due to repeated exertions and movements of the body. These musculoskeletal disorders belong to a collection of health problems that are more prevalent among the working class than among the general population. The more frequently occurring WMDs in the workplace include carpal tunnel syndrome, epicondylitis (tennis elbow), tendinitis, tenosynovitis, synovitis, stenosing tenosynovitis of the finger, DeQuervain's disease, and low back pain. Refer to Putz-Anderson [7] for more information on WMDs.

Early symptoms of musculoskeletal disorders are referred to as unpleasant sensations or discomfort associated with fatigue, perceived exertion, and poor posture [8]. Aches, numbness, and burning are some symptoms associated with discomfort. At the beginning these discomfort symptoms may be transient and occur mostly at night. As the disorder develops, they become more persistent and painful. Musculoskeletal pain arises from injury, irritation, or inflammation and may be considered a condition affecting the soft tissues [9]. Thus, discomfort symptoms may be predictive of musculoskeletal problems.

B. Significance of Work-Related Musculoskeletal Disorder Surveillance

Work-related musculoskeletal disorder constitute a major source of employee disability and lost wages. The National Institute for Occupational Safety and Health (NIOSH) has identified WMDs as one of the ten leading occupational health problems of workers [10]. Officers of NIOSH have stated that over 5 million workers (some 4% of the American workforce) suffer WMDs each year, with predictions that the figure will reach 50% of the workforce by the year 2000. Compensation and disability claims for some of the most severe cases of WMDs that required extensive treatment and surgery ranged from $30,000 to $60,000 [11]. Carpal tunnel syndrome strikes 23,000 workers per year, costing about $3500 in benefits and rehabilitation per person and about $30,000 per person if the injury requires surgery [12]. Thus, active surveillance of WMDs should constitute an essential component in an ergonomic program used to control WMDs and reduce human suffering, lost workdays and wages, and compensation claims. As a result, the active surveillance of WMDs can have a significant impact on the overall safety and health of employees and on the overall safety and health costs to employers.

III. PASSIVE SURVEILLANCE OF WORK-RELATED MUSCULOSKELETAL DISORDERS

Passive surveillance is characterized by the collection of past data from available records. Thus passive surveillance involves looking first at the effects and then at the causes. Among the available records used for passive surveillance of WMDs are OSHA 200 logs, worker's compensation and insurance records, plant medical records, safety and accident reports, and payroll records. Records are examined to identify past cases of WMDs. This information can be used to determine which departments or jobs pose a risk to the workers. Thus, the emphasis of a retrospective approach is on the health effects of poor ergonomic design. The health information collected from a records review may provide an indication of jobs that may need immediate attention or modification.

There are various techniques for analyzing the data on medical and safety records to identify priority jobs. Three popular statistics used to monitor WMD experience are frequency of cases, incidence rate, and severity rate.

The frequency of cases is the number of cases reported during a specific time period. This statistic is useful only when it specifies the group (e.g., a department, a specific job, a particular operation, or the whole plant) in which the count is made and the type of cases (e.g., carpal tunnel syndrome, all shoulder problems, all ergonomics-related problems) counted. When the numbers of employees per group differ substantially, the frequency count is not enough to provide an indication of the priority jobs. In these cases, the frequency must be related to the number of people in each group before groups are compared. The computation of the incidence rate makes this adjustment and gives an indication of the number of new cases among a specific number of workers or for a given number of worker hours. Finally, the severity rate can be determined based on such factors as the number of days lost, the medical costs, and additional operational costs.

To determine which rates are high, baseline rates may be established. Frequently used baseline rates are the plantwide incidence and severity rates for work-related musculoskeletal injuries and illnesses or those for nonproduction personnel. Using these baselines, the analyst may determine the number of cases that could have been avoided in a specific group in relation to the baseline group.

The major benefits of reviewing available records for passive surveillance include (1) low cost, (2) the possibility of using a systematic approach to obtain and code the data, (3) the possibility of performing historical analyses, and (4) the possibility of providing a basis for comparing the health effects due to the lack of an adequate ergonomic design among departments, operations, and areas of the plant. Moreover, it provides a means for measuring and evaluating the impact of ergonomic interventions.

Reviews of available records at a worksite have been used for ergonomic surveillance to identify jobs with a high incidence of musculoskeletal problems. For example, workers' compensation claims data were found to be useful in locating high-risk operations causing cumulative trauma disorders [13]. Plant medical records have also been used in retrospective studies to demonstrate associations between musculoskeletal disorders and occupational risk factors [14] and to demonstrate the effects of ergonomic intervention on the development of WMDs [15].

One major disadvantage of the passive surveillance approach is that records are not complete in the sense that minor injuries and discomfort are often ignored or underreported in health records. The nature of WMDs is such that initial symptoms can be somewhat vague and discomfort symptoms may develop gradually over a period of weeks, months, or years. Thus, recording a condition in its early stage depends on employees taking the first step to report symptoms. As a result, the analysis of the injury and illness incidence rates does not adequately reflect the relative risks of the operation [16–18]. To overcome this problem, an alternative model of injury analysis was developed to predict subsequent major injury using work history and medical records [19].

Another disadvantage of analyzing health records retrospectively for passive surveillance is that the rates of recordable cases are not updated in a timely manner. Thus, it is not always possible to determine the present health status of the jobs. To determine whether the health status of a job was "under control," an injury-monitoring test was developed based on updating the incidence and severity rates in a timely manner [20].

Furthermore, the possibility of doing a retrospective study for passive surveillance depends on the availability and reliability of records on both the risk factor and the outcome factor. Striking differences in the plantwide incidence of WMDs were found depending on which source was examined (OSHA logs, workers' compensation claims, medical absences, and plant medical cases) [24]. This result may also be related to the lack of a standardized method for reporting WMDs or inconsistent diagnostic criteria. For example, a case of carpal tunnel syndrome may be reported as an industrial injury rather than as an illness [25].

Although passive surveillance is relatively low-cost, collecting information from records is very time-consuming and the data are subject to misclassification at the time they are recorded as well as when they are retrieved from the record. Furthermore, records used for passive surveillance rarely have exposure information such as the tools used, the employee's department and job, the length of time on the same job, overtime hours worked, previous jobs, or additional jobs. Thus, it is necessary to pursue this exposure information by other means.

In general, the main disadvantage of passive surveillance is that even when records are complete, consistent, and available for review, they only reflect past occurrences of health problems and may not give an accurate picture of the present situation.

IV. ACTIVE SURVEILLANCE OF WORK-RELATED MUSCULOSKELETAL DISORDERS

A. Traditional Pencil-and-Paper Surveys

Active surveillance involves looking first at causes and then at effects by actively seeking and collecting relevant information for a specific purpose and in a standardized manner. In an ergonomic program, the prevention of WMDs is a main goal. Active surveillance may offer the means to identify musculoskeletal problems early and thus prevent more serious disorders. Active surveillance involves the design of a data collection system and the collection of the necessary information from the workforce. One system commonly used in active surveillance for WMDs uses discomfort symptoms questionnaires or pencil-and-paper surveys, also referred to as "shop floor surveillance" [23], that are completed by all employees in a particular problem area or department or in the entire plant. This traditional survey method allows for active investigation of health problems by obtaining information directly from the employees about their current discomfort symptoms.

Questionnaires used for active surveillance may include the following information:

1. Location of discomfort
2. Onset of discomfort
3. Severity of discomfort
4. Frequency of discomfort episodes
5. Duration of discomfort episodes
6. Employee perception of what aggravates discomfort
7. Job-related information (e.g., shift, job title, time on the job)

OSHA [24] has included in the *Ergonomics Program Management Guidelines for Meatpacking Plants* an example of a symptoms survey for an ergonomics program developed by Silverstein [25].

Some questions used for WMD surveillance use anatomical discomfort charts depicting the anterior view, the posterior view, or both views of the human body to identify the location of discomfort. The location of discomfort is indicated by shading the affected areas. This information may be useful for identifying the elements of the job that may be causing discomfort in specific body areas.

Also, some questionnaires used for WMD surveillance may request an indication of the severity of discomfort in the affected areas. This information may be useful to (1) identify employees who need referral for medical supervision or, where needed, timely initiation of treatment, (2) identify employees particularly vulnerable to a certain health hazard, (3) evaluate the effectiveness of medical interventions and ergonomic interventions on the severity of discomfort, and (4) identify trends in discomfort severity with respect to time. The severity of discomfort is usually assessed in the surveys by means of rating scales.

Pain intensity levels have been assessed with a variety of rating scales [26]. Rating scales are particularly useful because they can be used quickly and easily with minimal instruction. There are two popular types of scales: visual analog scales and numerical or verbal scales. Visual analog scales are straight lines with ends defined by the extreme limits of the sensation or response to be measured. Verbal or numerical scales are straight lines with descriptive terms placed at intervals along the line. Numerous studies have shown that pain or discomfort can be mapped and rated using various scales [27–36].

Several investigators have adapted a method developed by Corlett and Bishop [37] that combines a rating scale and a pictorial presentation of body diagrams to assess the relationship between discomfort and postural loading at the joints [30,32,38], to assess the relationship between discomfort and working postures [8,39,40], and to assess the incidence and severity of musculoskeletal disorders in the workforce [41,42].

B. Real-Time Computerized Surveys

1. Computerized Data Collection Systems

Automated health data collection from patients is not a new idea. The first reported computerized questionnaire was developed in 1966 [44] to elicit information on allergy symptoms from patients. This initial computerized interviewing system consisted of a small general-purpose digital computer with a keyboard with four special function keys. Questions were displayed to patients on a small cathode ray oscilloscope. The patient would respond to questions by pressing one of the four special keys. This first computerized questionnaire established three important points: (1) that computers could obtain information from patients, (2) that computers could do this in a manner that was well received by patients, and (3) that patients could interact with a computer "interviewer" without any prior training or knowledge of computing.

Other investigators have developed and evaluated computerized questionnaires for health data collection [45-49]. Among the computerized questionnaires for health data collection reviewed, there were two that differed from the others in that they made use of a graphic display terminal and an electronic light pen to select a rectangular response area on the screen. These are the Automated Medical History (AMH) [49] and a computer-based interview system for patients with back pain [50]. These two systems incorporated advanced computer technologies such as graphics and a penlike input device to facilitate the interaction between the computer and the user.

Questions in general are routinely used to collect data that are usually graded by computers. Traditional computerized questionnaires have required patients to input information directly to the computer, thus reducing the time spent coding and entering the data for grading. However, these computerized questionnaires required patients to interact with the computer only once, to report health information. Examples where patients routinely enter health data directly into a computer have not been reported so far. The availability of personal computers now makes this possible.

2. The Discomfort Assessment System

The Discomfort Assessment System (DAS) is a computerized system developed for the collection of perceived musculoskeletal discomfort [51,52]. This system allows for real-time active surveillance of musculoskeletal discomfort data. It consists of a computerized questionnaire operated directly by the worker. The system has a graphical interface that makes extensive use of diagrams, windows, buttons, and menus to display information on the screen. There is no keyboard in this system.

The interface simply consists of a series of screens and a light pen. There are screens that request the patient identification number, job title, shift number, and length of time in the job, screens that request the location of discomfort, and screens that request the severity of discomfort in each area affected. When items on the screen (buttons, menu alternatives, body areas, etc.) are selected with the light pen they become highlighted.

DAS displays the anterior and posterior views of the human body with various body areas delineated to facilitate the recognition and selection of affected areas. The posterior view of the human body was adapted from the Nordic Questionnaire [53]. When a user selects a body area by pointing with the light pen, the area becomes highlighted.

The severity of discomfort is indicated by pointing at a desired level on numerical a scale with verbal anchors (a modification of Borg's scale [27]) that looks like a thermometer. The scale is vertical, and the numbers range from 0 to 10 with the words "nothing at all" [28] and "worst imaginable" [36] used as anchors at 0 and 10, respectively. When a level on the scale is selected, the "mercury" fills up to the selected level. Figure 1 shows the screen that requests the severity of discomfort displaying some information entered during an interaction with the system.

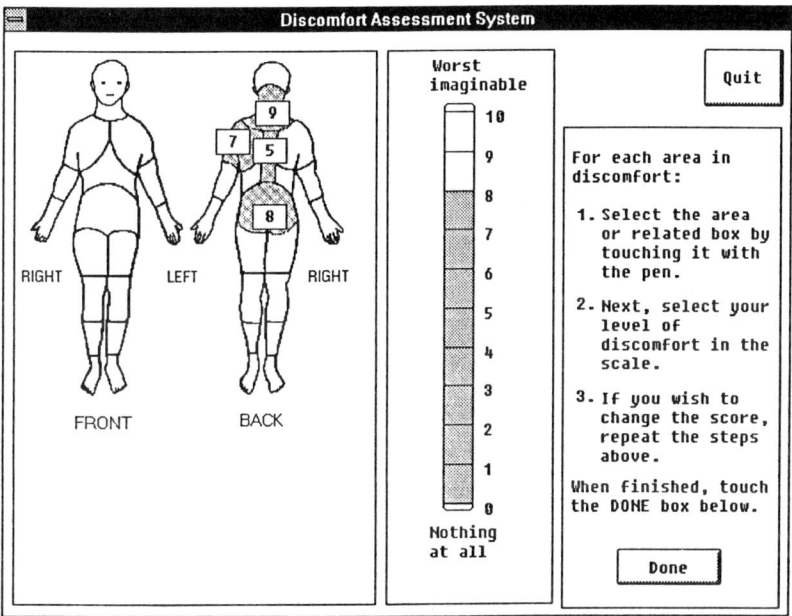

Figure 1 Example of a discomfort severity screen captured during an interaction with DAS.

3. Case Study

A musculoskeletal discomfort survey in Michigan demonstrated the potential capability of DAS for collecting discomfort symptoms information directly from the workforce [54]. In one study, musculoskeletal discomfort data were collected in a 5-day study of 11 rural mail carriers using DAS. Participants were asked to submit five reports, one at the beginning of the day and each of the remaining four at the end of a specific task during their workday. It was observed that different tasks required different muscle groups. The purpose of requesting a discomfort report after completing specific subtasks was to investigate the location of discomforts resulting from these and to determine the patterns of discomfort that occurred with time.

The data revealed that more than 50% of the participants were experiencing musculoskeletal discomfort in one or more anatomical areas (neck, shoulders, elbows, hands, back, and legs). This result was consistent with the high incidence rate of musculoskeletal disorders among rural mail carriers. The discomfort data revealed differences in the frequency of discomfort reports between body areas depending on the task preceding the report; this indicated that reported discomfort was task-related and not random. This result showed that data from a discomfort survey may be useful for identifying job hazards that may be stressing a particular body area in a significant portion of the employees. Finally, discomfort data collected from this case study were used to compare the patterns of discomfort severity reported by a healthy individual, an individual with carpal tunnel syndrome, and an individual who had undergone surgery for carpal tunnel syndrome. The results of the data analysis revealed significant patterns in discomfort severity among these three individuals. Thus active surveillance may also provide a means of evaluating individual cases and monitoring their progress.

V. CRITICAL REVIEW OF CURRENT STATUS

Complete elimination of ergonomic hazards in the workplace may not be a feasible alternative in some jobs. Thus ergonomic programs will operate as control mechanisms to identify musculoskeletal problems early enough to prevent more serious problems from developing. The effectiveness of an early detection system in identifying and controlling WMDs depends to a large extent on the completeness and timeliness of reporting and recording a person's symptoms data. At present, the best technique available to control WMDs uses active surveillance [23]. Yet active surveillance is in an early stage of development.

A. Advantages and Disadvantages of Traditional Pencil-and-Paper Surveys

Active surveillance allows the investigator to plan and control the collection of data and therefore be aware of problems that may result during data collection. The major benefit of active surveillance of the workforce is that it may be possible to obtain a better picture of the current health situation. This is so because active surveillance reaches out to employees and elicits information from them. Traditional survey data collection methods are advantageous because they can be administered to a large population and can be used to assess a large variety of health-related factors. Furthermore, reporting biases related to the definitions of different musculoskeletal disorders and requirements of specific records may be reduced.

Discomfort symptoms information obtained from the surveys can be used to identify a job or a department that needs further ergonomic investigation. The frequency and severity of discomfort symptoms can also be used as a baseline to evaluate the effectiveness of ergonomic interventions in reducing the discomfort symptoms. Furthermore, measurement of discomfort from time to time yields discomfort series (time series) that can be investigated for predictability [43].

The success of an active surveillance program also depends on the time taken to respond to the results of the survey. If feedback or interventions are delayed, then some conditions may become severe and employees may lose interest in reporting their discomfort symptoms. Furthermore, surveys rely on employees' willingness to report their health conditions to coworkers or supervisors. The level of awareness of employees about the objectives of the surveillance program and their understanding of the impact of their discomfort symptoms reports are extremely important for the success of active surveillance.

Traditional survey data collection methods such as pencil-and-paper surveys or questionnaires have several limitations associated with active surveillance. The administration of a discomfort symptoms survey may be costly because it requires trained personnel for the collection, interpretation, and coding of the data. On the other hand, if personnel involved in active surveillance are not properly trained, the effectiveness of the program will be reduced. Furthermore, surveys also rely on the employees' recall or recognition of their health conditions. However, surveys are conducted sporadically and, because in the early stages of WMDs many of the symptoms are transient and occur mostly at night, the employee may not recall or recognize the significance of symptoms during the survey.

Nevertheless, a study [21] showed that a pencil-and-paper questionnaire used in addition to physical examination data was as sensitive as an excellent plant medical department in detecting upper extremity musculoskeletal disorders. Other investigators [25,55,56] have also used pencil-and-paper questionnaires to assess employees' musculoskeletal discomforts.

Furthermore, when the purpose of the surveillance program is to detect early cases and some "false positive" are allowed, then pencil-and-paper questionnaires administered periodically to a random sample of employees to collect information about potential confounders as well as symptoms and exposure histories would provide sufficient information for an early warning system [57]. However, health surveillance information gathered at prescheduled intervals by means of questionnaires may not present a realistic picture of the number of cases because workers are exposed to job stresses that may change on a day-to-day basis, thus incorporating reporting biases [58].

B. Advantages and Disadvantages of Real-Time Computerized Surveys

The availability of personal computers and the rapid advances in computer technologies make the use of computers possible now for developing real-time computerized surveys like DAS. Automated surveys are an alternative to traditional methods involving pencil-and-paper surveys. Computerized surveys have the same advantages as the traditional surveys for active surveillance of WMDs. However, real-time computerized surveys like DAS may gather day-to-day information, reducing reporting biases and presenting a more realistic picture of musculoskeletal problems. Real-time computerized surveys may overcome some of the limitations of traditional pencil-and-paper surveys. For example, they gather information in a timely manner, thus allowing for the immediate investigation of the patterns of health and disease among groups of employees or within one individual.

Furthermore, real-time computerized surveys eliminate the need for trained personnel to administer the questionnaire and interpret and code the data. Thus, in that sense they may be less costly. Because the data are automatically coded, they are readily available for grading, which accelerates the intervention process.

Nevertheless, such surveys also have several shortcomings. First, they require computer hardware. This full implementation of a real-time computerized survey system in a large facility may be costly. Second, similar to traditional pencil-and-paper surveys, such a survey relies on a person's self-assessment of discomfort and willingness to report discomfort symptoms, in this case to a computer instead of to a coworker or supervisor. Third, computerized surveys need further testing in the field for validation.

Recent advances in technology have brought forth a competitive market of computer hardware and software. Thus, the implementation of a real-time computerized survey system for active surveillance of WMDs may be economically feasible for many companies. Furthermore, results obtained in 1966 [44] demonstrated that people were capable of interacting with a computer system without any prior training or knowledge of computers. With the proliferation of computers over the intervening years, that conclusion is even more valid today.

VI. FUTURE CONCERNS

The effectiveness of an early detection system to identify and control WMDs depends to a large extent on the completeness and timeliness of reporting and recording symptoms data. Real-time computerized systems that take advantage of new computer technologies should be developed for the collection of data in active surveillance programs. A real-time active surveillance system of WMDs should allow an employee to report discomfort symptoms at any time. That is to say, employees should have access to a computer terminal to input a report any time they experience discomfort. Furthermore, the discomfort reports database should automatically update itself after each report, thus providing real-time analysis of the data.

Because real-time computerized surveys depend on the employee's self-assessment of discomfort and willingness to report, employees should receive education on WMD prevention, with emphasis on early recognition of symptoms as well as on the importance of an early report to aid in control of the problem.

Furthermore, new methodologies should be developed for the analysis of the data collected that will indicate whether the patterns of discomfort symptoms are "in control" and provide warnings when the health status is "out of control" [51].

REFERENCES

1. U.S. Department of Labor, Occupational Safety and Health Administration, Ergonomic Safety and Health Management; Proposed Rule 57 FR 34192-34200, 192.
2. U.S. Department of Labor, Bureau of Labor Statistics, *BLS Report on the Survey of Occupational Injuries and Illnesses in 1991*, USDL Pub. No. 92-731, 1991.
3. R. E. Yodaiken, Surveillance, monitoring, and regulatory concerns, *J. Occup. Med. 28*(8):569-571 (1986).
4. D. K. Parkinson and M. J. Grennan, Establishment of medical surveillance in industry: problems and procedures, *J. Occup. Med. 28*(8):772-777 (1986).
5. B. W. Mintz, Medical surveillance of employees under the Occupational Safety and Health Administration, *J. Occup. Med. 28*(10):913-919 (1986).
6. J. D. Millar, Summary of proposed national strategies for the prevention of leading work-related diseases and injuries, Part 1, *Am. J. Ind. Med. 13*:223-240 (1988).
7. V. Putz-Anderson, *Cumulative Trauma Disorders: A Manual for Musculoskeletal Diseases of the Upper Limbs*, Taylor & Francis, New York, 1988.
8. I. Kuorinka, Subjective discomfort in a simulated repetitive task, *Ergonomics 26*:1089-1101 (1983).
9. R. Cailliet, *Soft Tissue Pain and Disability*, F. A. Davis, Philadelphia, 1988.
10. U.S. Department of Health and Human Services, Centers for Disease Control, National Institute for Occupational Safety and Health, DHHS (NIOSH) Pub. No. 89-129, 1989.
11. R. Hiltz, Fighting work-related injuries, *Natl. Underwriter 89*(13):15 (1985).
12. The Bureau of National Affairs, Inc., *Occupational Safety and Health Reporter*, BNA Pub. No. 5-31-89, 1989, p. 2165.
13. S. Tanaka, P. Seligman, W. Halperin, M. Thun, C. L. Timbrook, and J. J. Wasil, Use of worker's compensation claims data for surveillance of cumulative trauma disorders, *J. Occup. Med. 30*(6):448-492 (1988).
14. R. H. Westgaard and A. Aaras, Postural muscle strain as a causal factor in the development of musculo-skeletal illnesses, *Appl. Ergon. 15*:162-174 (1984).
15. R. H. Westgaard and A. Aaras, The effect of improved workplace design on the development of musculo-skeletal illnesses, *Appl. Ergon. 16*:91-97 (1985).
16. B. A. Silverstein, L. J. Fine, and T. J. Armstrong, Occupational factors and carpal tunnel syndrome, *Am. J. Med. 11*:343-358 (1987).
17. T. J. Armstrong, L. J. Fine, S. A. Goldstein, Y. R. Lifshitz, and B. A. Silverstein, Ergonomic considerations in hand and wrist tendinitis, *J. Hand Surg. 12A*(5; Part 2):830-837 (1987).
18. W. Margolis and J. F. Kraus, The prevalence of carpal tunnel syndrome symptoms in female supermarket checkers, *J. Occup. Med. 29*(12):953-956 (1987).
19. M. K. Chung, S. H. Wu, and G. D. Herrin, The use of a mixed Weibull model in occupational injury analysis, *J. Occup. Accid. 7*:239-250 (1986).
20. A. H. Boyd and G. D. Herrin, Monitoring industrial injuries: a case study, *J. Occup. Med. 30*(1):43-48 (1988).
21. L. J. Fine, B. A. Silverstein, T. J. Armstrong, C. A. Anderson, and D. S. Sugano, Detection of cumulative trauma disorders of upper extremities in the workplace, *J. Occup. Med. 28*(2):674-678 (1986).
22. D. S. Louis, Cumulative trauma disorders, *J. Hand Surg. 12A*(5; Part 2):823-825 (1987).
23. B. A. Silverstein, Shop floor surveillance to identify and control work-related musculoskeletal disor-

ders, Presented at the Occupational Disorders of the Upper Extremities Course, Mar. 29–30, 1990, Ann Arbor, MI.
24. U.S. Department of Labor, Occupational Safety and Health Administration, *Ergonomic Program Management Guidelines for Meatpacking Plants,* OSHA 3123, 1991 (reprinted).
25. B. A. Silverstein, The prevalence of upper extremity cumulative trauma disorders in industry, Doctoral Dissertation, Univ. Michigan, Ann Arbor, 1985.
26. C. R. Chapman, K. L. Casey, R. Dubner, L. M. Foley, R. H. Gracely, and A. E. Reading, Pain measurement: an overview, *Pain* 22:1–31 (1985).
27. G. Borg, A category scale with ratio properties for intermodal and interindividual comparisons, in *Psychophysical Judgment and the Process of Perception,* H. G. Geisler and P. Petzold, Eds., VEB Deutscher Verlag der Wissenschaften, Berlin, 1982, pp. 25–34.
28. F. Gaston-Johansson, Pain assessment: differences in quality and intensity of the words pain, ache and hurt, *Pain* 20:69–76 (1984).
29. K. Harms-Ringdahl, H. Brodin, L. Eklund, and G. Borg, Discomfort and pain from loaded passive joint structures, *Scand. J. Rehab. Med.* 15:205–211 (1983).
30. K. Harms-Rindgahl, On assessment of shoulder exercise and load-elicited pain in the cervical spine, *Scand. J. Rehab. Med. Suppl.* 14:1–40 (1986).
31. K. Harms-Ringdahl, A. M. Carlsson, J. Ekholm, A. Raustorp, T. Svensson, and H. Toresson, Pain assessment with different intensity scales in resopnse to loading of joint structures, *Pain* 27:401–411 (1986).
32. K. Harms-Ringdahl and J. Ekholm, Intensity and character of pain and muscular activity levels elicited by maintained extreme flexion position of the lower cervical-upper-thoracic spine, *Scand. J. Rehab. Med.* 18:117–126 (1986).
33. R. Melzack, The McGill pain questionnaire: major properties and scoring methods, *Pain* 1:277–299 (1975).
34. D. D. Price, P. A. McGrath, A. Raffi, and B. Buckingham, The validation of visual analogue scales as ratio scale measures for chronic and experimental pain, *Pain* 17:45–56 (1983).
35. J. Scott and E. C. Huskisson, Graphic representation of pain, *Pain* 2:175–184 (1976).
36. R. A. Seymour, J. M. Simpson, J. Charlton, and M. E. Phillips, An evaluation of length and end-phrase of visual analogue scales in dental pain, *Pain* 21:177–185 (1985).
37. E. N. Corlett and R. P. Bishop, A technique for assessing postural discomfort, *Ergonomics* 19:175–182 (1976).
38. M. Boussenna, E. N. Corlett and S. T. Pheasant, The relation between discomfort and postural loading at the joints, *Ergonomics* 25:315–322 (1982).
39. V. Bhatnager, C. G. Drury, and S. G. Schiro, Posture, postural discomfort, and performance, *Hum. Factors* 27:189–199 (1985).
40. S. F. Wiker, Effects of relative hand location upon human movement time and fatigue, Doctoral Dissertation, Univ. Michigan, Ann Arbor, 1986.
41. P. W. Buckle, D. A. Stubbs, and D. Baty, Musculo-skeletal disorders (and discomfort) and associated work factors, in *The Ergonomics of Working Postures* N. Corlett, J. Wilson, and I. Manenica, Eds., Taylor & Francis, London, 1984, pp. 19–30.
42. B. A. Silverstein and L. J. Fine, Evaluation of upper extremity and low back cumulative trauma disorders: a screening manual, Tech. Rep., Univ. Michigan School of Public Health, Environmental and Industrial Health Department, Ann Arbor, 1984.
43. G. Affleck, H. Tennen, S. Urrows, and P. Higgins, Individual differences in the day-to-day experience of chronic pain: a prospective daily study of rheumatoid arthritis patients, *Health Psychol.* 10:419–426 (1991).
44. W. V. Slack, G. P. Hicks, C. E. Reed, and L. J. Van Cura, A computer based medical history system, *N. Engl. J. Med.* 274(4):194–198 (1966).
45. W. I. Card, M. Nicholson, G. P. Crean, G. Watkinson, C. R. Evans, J. W. Wilson, and D. Russell, A comparison of doctor and computer interrogation of patients, *Int. J. Biomed. Comput.* 5:175–187 (1974).
46. J. Diaz, O. M. Z. Miranda, A. Faundes, and J. A. Pinotti, Preliminary experiment with computerized anamnesis in gynecology and reproductive health, *Int. J. Gyn. Obstet.* 24:285–290 (1986).

47. W. E. Hammond and W. W. Stead, The evolution of computerized medical information system, *Proc. Tenth Annual Symp. Comput. App. Med. Care,* Computer Society of the IEEE, Washington, DC, 1986.
48. R. W. Lucas, W. I. Card, R. P. Knill-Jones, G. Watkinson, and G. P. Crean, Computer interrogation of patients, *Br. Med. J.* 2:623–625 (1976).
49. J. G. Mayne, W. Weksel, and P. N. Sholtz, Toward automating the medical history, *Mayo Clinic Proc.* 43:1–25 (1968).
50. A. M. C. Thomas, J. T. C. Fairbank, P. B. Pynsent, and D. J. Baker, A computer-based interview system for patients with back pain, *Spine* 14:844–846 (1989).
51. N. Saldaña, Design and evaluation of a computer system operated by the workforce for the collection of perceived musculoskeletal discomfort: a tool for surveillance, Doctoral Dissertation, Univ. Michigan, Ann Arbor, 1991.
52. N. Saldaña, DAS: a graphical computer tool for the collection of musculoskeletal discomfort information from the workforce, Comput. Ind. Eng. 23:215–218 (1992).
53. I. Kuorinka, B. Jonsson, A. Kilbom, H. Vinterberg, F. Biering-Sorensen, G. Andersson, and K. Jorgensen, Standardised Nordic questionnaires for the analysis of musculoskeletal symptoms, *Appl. Ergon.* 18(3):233–237 (1987).
54. N. Saldaña, G. D. Herrin, T. Armstrong, and A. Franzblau, A computerized method for assessment of musculoskeletal discomfort in the workforce: a tool for surveillance, *Ergonomics* 37:1097–1112 (1994).
55. B. A. Silverstein, T. J. Armstrong, A. Longmate, and D. Woody, Can in-plant exercise control musculoskeletal symptoms?, *J. Occup. Med.* 30(12):922–927 (1988).
56. S. L. Sauter and L. M. Schleifer, Work posture, workstation design, and musculoskeletal discomfort in a VDT data entry task, *Hum. Factors* 33(2):151–167 (1991).
57. B. A. Silverstein, Patterns of cumulative trauma disorders in industry, Presented at the Engineering Summer Conferences: Occupational Ergonomics Course, June 15–19, 1987, Ann Arbor, MI.
58. J. D. McGlothlin, An ergonomic program to control work-related cumulative trauma disoders of the upper extremities, Doctoral Dissertation, Univ. Michigan, Ann Arbor, 1988, p. 119.

24

Development and Implementation of an Ergonomics Process in the Automotive Industry

Reactive and Proactive Processes

Bradley S. Joseph and Glenn Jimmerson
Ford Motor Company, Dearborn, Michigan

I. INTRODUCTION

Ergonomics examines the interaction between the worker and the work environment including such factors as machinery, the workstation, and climate. If the match between worker and work environment is poor, the worker's ability to perform the job will be severely compromised. Over a short period of time, this poor match may lead to fatigue and worker discomfort. If conditions persist, physical injury and disability may result.

II. BACKGROUND AND SIGNIFICANCE TO OCCUPATIONAL ERGONOMICS

Large industrial firms are struggling to incorporate occupational ergonomics into their operating practices. Pressure from outside (pending legislation from the federal government) and inside (aging workforce, increasing costs of medical care, pressure to become more efficient through productivity gains) are forcing manufacturing companies to use ergonomics information. To accomplish this task, manufacturing must develop a model to begin incorporating ergonomics into general operating procedures. Kvalseth's (1980) [1] research indicated that several factors were important in implementing an ergonomics program. They include: management's perception for the need of ergonomics, management knowledge of the benefits or ergonomics, and the degree of cooperation between the workers and management. These factors show how important it is to "walk" a plant before attempting to implement the program. This chapter helps develop a plan that can be used for a general ergonomics process. It has two components—reactive and proactive processes.

To begin to develop this model, it is important to develop a clear and concise working definition of ergonomics. Ergonomics literally means the study of work. Since this definition can encompass a wide range of issues, manufacturing companies need to narrow the scope of occupational ergonomics. In the case of Ford Motor Company, occupational ergonomics means fitting jobs to people.

It is important to note that even though the process described in this chapter was used in a large manufacturing company, the concept and lesson can be used in most manufacturing industries large or small.

III. DESCRIPTION OF REACTIVE AND PROACTIVE ERGONOMICS PROCESSES AT FORD MOTOR COMPANY

A. History of Ergonomics at Ford Motor Company

Practical ergonomics has always been a consideration at Ford Motor Company. In the past, ergonomics analysis concentrated on the interaction between vehicle design and customer satisfaction. Designers incorporate ergonomics principles when determining reach distances to critical controls, display and display layout, seat dimension, and interior volume (leg room, shoulder room, etc.) Then in the late 1970s and early 1980s, it became apparent that these same principles should be applied when designing a vehicle for ease of manufacture and assembly. Industrial engineering began to focus more on human capability rather than on just the time required to complete a task, and ergonomics became one of several issues. Today, ergonomic principles are being considered for serviceability issues.

Several years ago the need emerged for a more focused effort in ergonomics of manufacturing. Special programs were undertaken in several divisions of Ford Motor Company. For example, in 1982 Body and Assembly Operations (B&AO) began a long-term agreement with The University of Michigan Center for Ergonomics. Several goals were outlined, including the following.

Establish a university organization to facilitate and develop ergonomics in the manufacturing and assembly process
Develop an education program to disseminate ergonomics principles to engineers, managers, supervisors, and suppliers
Coordinate research efforts to improve engineering specifications and guidelines for use in the design and evaluation of existing and future body and assembly operations
Provide consultation services as requested by B&AO personnel.

This activity ran continually for six years (1982–1988) and involved a number of faculty members, research assistants, and graduate students. Even though the direct benefit of this relationship was difficult to measure, the relationship had a positive impact for both Ford Motor Company and the university. Below is a list of outcomes from the project:

Eighty technical reports and scientific papers
Three videotapes—ranging from an overview of low back pain and upper extremity cumulative trauma disorders to in-plant studies
A 250-page ergonomics technical manual
A guide to developing and implementing in-plant ergonomics programs (The guide and manual are incorporated into the UAW/Ford Ergonomics Process.)
Fifteen different training courses attended by over 500 Ford personnel from 25 B&AO plants in the last year.

It is particularly important to note that even though over 500 persons within the company were trained, ergonomics did not become an integral part of the operating culture. Even though there were instances when engineers applied ergonomics principles, in most cases, they did not apply them as a general practice. There are many explanations for this, but the simplest and most correct is time—the engineers had many things to do to complete their jobs and less and less time to do them. Adding ergonomics evaluation was something they did not want or recognize as a priority issue.

During the course of this university/Ford relationship, it became apparent that if ergonomics was to become truly part of the operating culture at Ford, the company would have to attack it from within. Several pilot projects with The University of Michigan, inside and outside Body and Assembly Division, were begun to study better ways to implement a program that would be owned by Ford Motor Company. Most of these projects were jointly sponsored with the United Auto Workers Union. These projects include

1983—The Milan and Saline Plastics Plant. A joint UAW–Ford in-plant ergonomics program was initiated in cooperation with The University of Michigan [1].

1984—The Van Dyke Chassis Plant. Research efforts concentrated on organizational issues related to a joint labor-management ergonomics team.

These and other programs at Ford demonstrated the benefits resulting from the use of ergonomics principles on the plant floor. They have yielded real job improvements as well as valuable scientific information. In 1987, UAW-Ford contract negotiations resulted in an agreement to expend significant effort in ergonomics. In essence, the collective bargaining agreement states that "the Company believes that applying ergonomics principles to the work environment, if done properly, can lead to reduced injuries, improved quality, and greater productivity." To achieve these goals, The Union and the Company intended to increase awareness of ergonomics principles and practices by jointly developing and implementing and/ or sponsoring education and training programs." More important, the agreement states that "each assembly and manufacturing unit of 125 or more employees may establish a Local Ergonomics Committee" [1].

At that point, the company began developing a formal action plan in ergonomics. The program involves a two-prong approach—reactive and proactive. The first prong was the development of the joint UAW-Ford Ergonomics Program. It is a program that reacts to existing concerns by reviewing problems jobs and redesigning existing equipment at the plant. The backbone of the program is the local ergonomics committee (LEC).

B. Development of the Reactive Program

Late in 1987, the National Joint Committee on Health and Safety initiated discussions on the best way to implement a reactive ergonomics process at Ford. Members of this committee identified three major requirements for success:

1. Ergonomics would be available to all levels of the plant.
2. Ergonomics would coexist with existing programs
3. Ergonomics would be developed with a participative approach.

These discussions yielded a process that was jointly developed by the UAW and Ford with the cooperation of The University of Michigan and with the services of Maritz Communications Company. This process is discussed in detail in the book *The UAW–Ford Ergonomics Process*, which consists of two volumes—*The UAW-Ford Implementation Guide* [2] and *The UAW-Ford Job Improvement Guide* [3]. The process has three steps: process start-up, job improvement cycle, and long-term development (Figure 1). The *Implementation Guide* is an instruction manual that outlines the methodology to implement an ergonomics process in an industrial facility. It assumes that cooperation exists between labor and management and that a joint labor-management team will have the authority to identify, evaluate, develop, and implement ergonomics changes. The *Job Improvement Guide* is a manual to show the LECs

Figure 1 The UAW-Ford ergonomics process.

the detailed methodology to evaluate ergonomics issues and develop and implement solutions. Following is a discussion of the process.

1. Process Start-Up

Process start-up is the first stage of the Ergonomics Process. It has five substeps (see Figure 2). The first three are securing leadership commitment, selecting local ergonomics committee members, and training. Securing leadership commitment is to be accomplished locally at each plant. The National Joint Committee on Health and Safety supports this task effort through the development of an Ergonomics Process Leadership Commitment videotape and training for health and safety engineers and union representatives [4]. The video was distributed during an Ergonomics Process orientation seminar at the December 1988 UAW-Ford health and safety conference.

Members of the LEC received at least four days of professional training. In April and May 1989, 73 LECs (over 600 participants) went through a four-day jointly developed introductory training course. The course had two goals:

1. To teach LEC members how to use the guide and execute the steps of this cycle
2. To teach LEC members an introductory level of technical ergonomics information to enable them to recognize ergonomics problems and develop solutions

2. Job Improvement Cycle

Figure 3 shows the substeps of the job improvement cycle. It is a six-step approach to practical ergonomics that includes methods for identifying priority jobs to fix, evaluating job stressors, developing and implementing job improvements, and documenting and following up on individual projects. Since this cycle is so important, a *Job Improvement Guide* [3] and videotape were developed that detailed these steps [5].

3. Long-Term Development

No successful plant ergonomics program can survive without a long-term strategy and vision. Much thought has already been given to the long-term development of the UAW-Ford Ergonomics Process. Three areas of critical need have been identified:

1. *A need for specialized training.* As the LEC becomes more and more involved in the process, members will require advanced and specialized training. More and more ex-

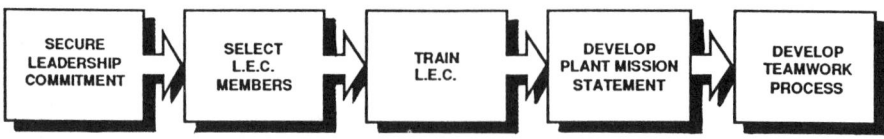

Figure 2 Process start-up.

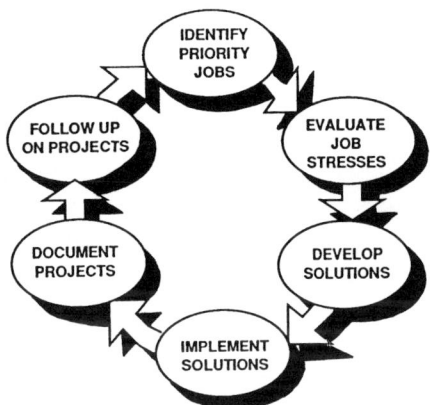

Figure 3 The job improvement cycle.

pertise will be required from plant and division engineers. Consequently, a core of engineers will have to be trained in the UAW-Ford Ergonomics Process and in advanced basic ergonomics principles.

2. *A need to communicate ergonomics.* A main reason for bringing together the ergonomics committee is the multidisciplinary nature of the science. Successful implementation of ergonomics principles is both reactive (changes to existing jobs) and proactive (ergonomically designing new jobs). Successful activities in identifying and correcting poorly designed jobs must be communicated to the engineers and others who are responsible for developing new processes in the plant, so the same "mistakes" will not be repeated.

3. *A need to review constantly and improve the process.* As time goes on, many Ford sites will recognize the benefits of ergonomics. They will want to increase their plant's level of activity. The LEC can only perform so many functions at one time.

The current ergonomics efforts are reactive in nature; that is, the LECs will mainly be reviewing existing problems and redesigning existing equipment at the plant. The information generated by a LEC is isolated within its own plant. A more effective and cost-efficient approach is for the company to be proactive—for ergonomics principles to be applied during product research, design, and plant layout in order to prevent ergonomic stressors. In this way, problems will be designed out of the product and process before reaching the operation phase. These efforts will help establish Ford Motor Company as the continued leader in ergonomics applications and research. However, it is essential to the success of these efforts that all sectors of product development, including research, design, and engineering, become an integral part of the ergonomics process.

C. Development of the Proactive Program

The proactive ergonomics process begins with an analysis of the way in which new equipment is engineered and installed in a facility. Figure 4 shows a summary of a typical process required for the installation and maintenance of industrial equipment. The process consists of a series of specialized functions performed by separate organizations or units. In general, the flow of information is from the corporation to the divisions to the plants. Units at the divisional level typically perform process study and design functions or phases; units at the plant perform implementation, operation, and maintenance phases. Often, these units are separated geographically and organizationally (e.g., division and plant), making it difficult to coordinate efforts.

The phases differentiate periods of time in which groups of units have to complete a task before going to the next phase. First, new processes are studied (study phase) and designed (design phase) at division engineering. Little plant input is solicited during this phase. Next, division and plant engineering together install and debug machinery (implementation phase). This procedure involves a complex series of actions whereby process machinery plans are sent to selected vendors, the plans are interpreted, and the machinery is built to specifications, delivered, and installed in the plant using resources from the plant, vendor, and division. Unless the plant is willing to bear costly delays and excessive expenditures, few changes can be made on machinery, for ergonomic or other reasons, between the time the vendor builds the machinery and the time he delivers it. Also, the vendor has to sign off on machine specifications and is under contract to build it to standards agreed upon by the plant and division. Any changes to the design must be negotiated before the vendor can implement them. Consequently, the plant may wait until the machines are delivered and operating under those specifications (known as final sign-off) before changes are made. After debugging, normal operation and maintenance proceeds (operation phase). Often there is a need for process improvements or other redesign to update equipment (redesign phase). Depending on the cost, the plant usually controls these activities. However, due to limitations on cash and human resources, lost production from shutting down the machines, and other plant priorities, this activity is often limited and takes a considerable amount of time to complete [4].

There are several points in time when the ergonomics information can be applied. The most important points are during process design, process building, and process installation. Significant impact on design can occur during process design and build. Less can occur during implementation and operation. Using the existing organizational process, all organizational groups participating in ergonomic efforts have a significant impact in solving current problems and in preventing future problems. For this to happen, information from the plant local ergonomics committees must be communicated to the proper divisional and corporate functions to ensure that the lessons learned through the LECs are incorporated into new plant and job designs (Figure 4).

This process is referred to as Design for Ergonomics (DFE). The DFE process was developed to look like a product warranty system in which the customer who uses the equipment feeds information back to the designers. In this system, the customer is the worker on the production line responsible for assembling or manufacturing the new product. This worker has several means of communicating back to the process designers about things that are right and things that are wrong with the process. This information is stored in a data bin or database and becomes institutional knowledge for the engineers the next time they are assigned to design a new process. When a new industrial process is going to be developed, the designers and engineers apply lessons learned from previous processes to the new process before specifications are finalized.

The type of knowledge in the data bin depends upon the source of information. Recall that the local ergonomics committees are primarily responsible for identifying ergonomics issues and developing and implementing solutions in the workplace. The outcome from this process is a data file indicating where ergonomics issues occurred. These issues are recorded in the Concern Log, and the details are recorded in a problem-solving record. At Ford this record is referred to as the 8-D form or eight-discipline problem-solving form.

1. Implementing the Design for Ergonomics Process—Process Start-Up

Implementing the Design for Ergonomics process is similar to implementing the reactive process. First leadership commitment must be secured. This usually occurs through a meet-

Ergonomics in the Automobile Industry 507

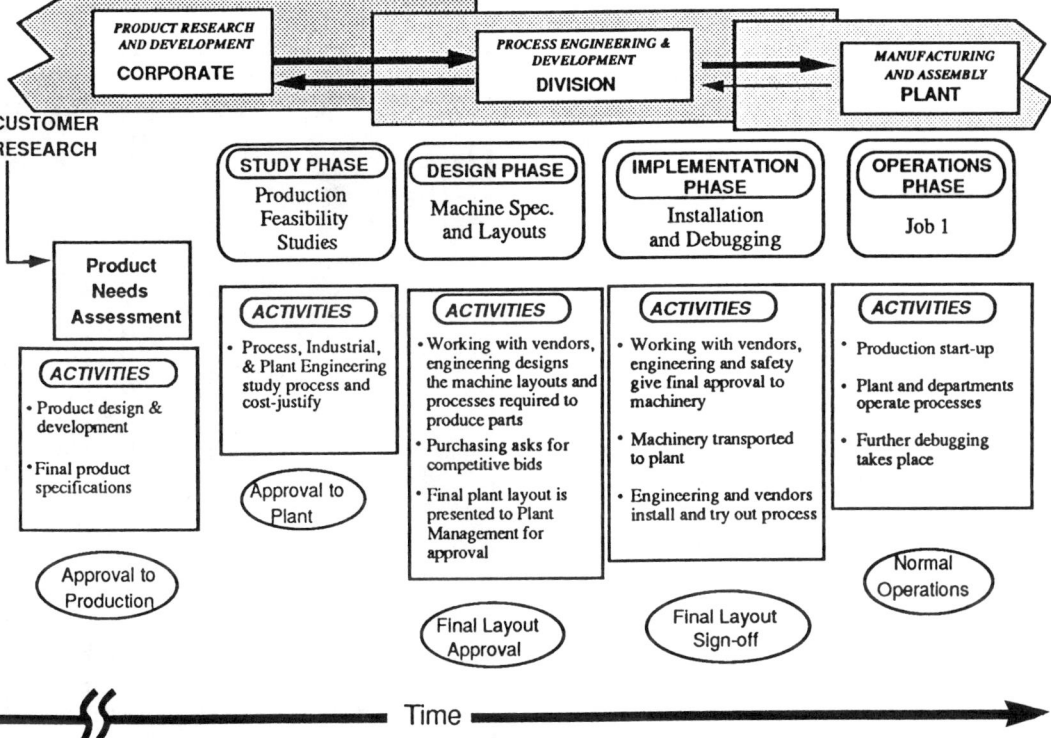

Figure 4 Product organizational and design process.

ing with the project or program manager. During this meeting, the manager must be convinced of the necessity for ergonomics and the need for the engineers on the project to participate.

Second, the Design for Ergonomics (DFE) team is selected. This team usually resembles a simultaneous engineering team consisting of facility, process, manufacturing, and industrial engineers; equipment vendors; product engineers; hourly representatives; and employee relations personnel. It is easy to see the reason for the engineering and vendors to participate. However, it is often difficult to explain the employee relations role in the process. Employee relations' primary role is to tell the group about the human side of the production process. It represents the voice of the workers (the voice of the customer) and may include injury/illness data, employee complaints, and absenteeism data.

Third, the DFE team must be trained. Recently a new Design for Ergonomics training program was developed. During the training, teams are brought together about two years before the job is to be completed. They are trained in basic methods of ergonomics analysis and methods to integrate ergonomics into their timing plans. Finally, the team forms a mission statement.

2. Implementing the Design for Ergonomics Process—Design Improvement Cycle

The design improvement cycle is similar to the job improvement cycle except that the process is evaluated proactively during the study, design, and implementation phases. To accomplish this, the team must have a process to identify and evaluate and develop solutions to its problems. Figure 5 shows a detailed flowchart of the design improvement cycle.

The chart shows a large cycle beginning at the top left-hand corner and continuing counterclockwise. The design improvement cycle begins with product development. As product design is released, engineers begin to develop a process to build the product. These processes

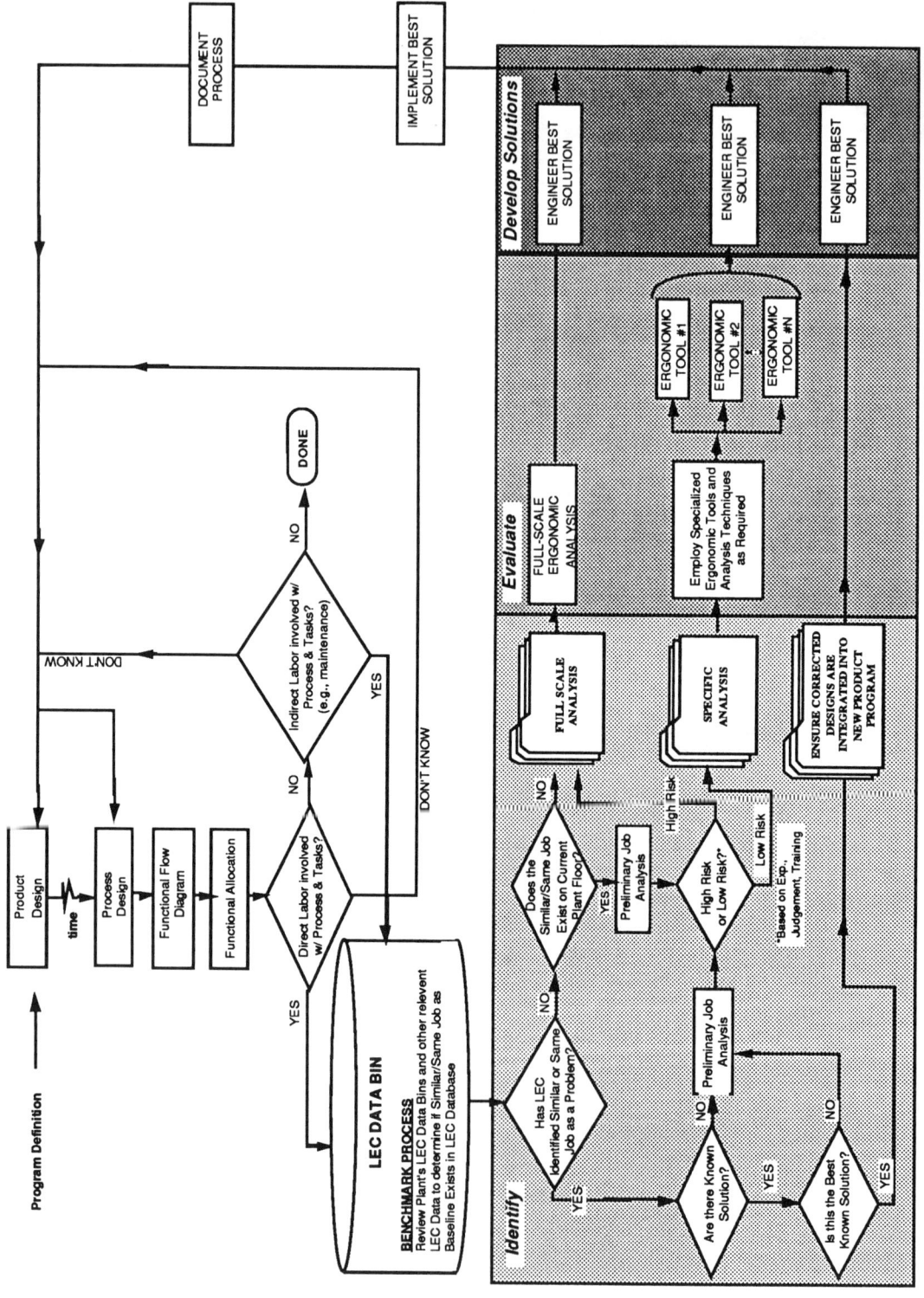

Ergonomics in the Automobile Industry

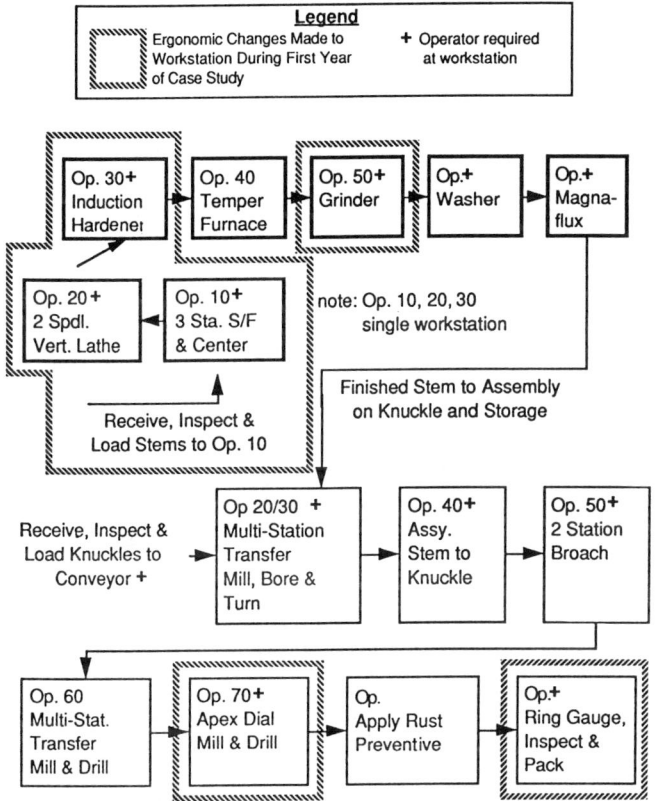

Figure 6 Functional flow of light truck spindle department.

are generally developed in two parts—a functional flow and a functional allocation. The function flow is an operation-by-operation overview of the steps necessary to manufacture and assemble a product. A function allocation is a summary of the tasks necessary to complete each operation. Engineers are required to evaluate each operation to determine if the process is cost-effective and feasible.

These data are valuable to the proactive ergonomics process. Figure 6 shows a summary sheet from a function flow for a front-end chassis component of a small truck. Figure 7 shows a function allocation for one operation within the process. Reviewing these documents can help engineers make reliable decisions regarding potential ergonomics issues. For example, according to the function allocation (Figure 7), operators are required to lift two 17-lb parts to the machine 70 times an hour. These parts are going to be lifted from and to an overhead monorail.

These data can be used to compare or "benchmark" the process with other processes already being carried out within the plant. For example, spindles are a necessary component on almost every light truck vehicle. The local ergonomics committee can review operations that are in production and determine if these are ergonomics issues. These data can be col-

---

Figure 5 The design improvement cycle for process design and build.

PC/HR=(3916(2)/ 123) 1.09=70	
PROJECTED WEIGHT OF PART=17 LBS.	

Process Estimate Sheet

Program or ERC Number	Part Name Spindle & Spindle Assy.-Frt.
For Models Light Truck	Material Iron
Oper. No. — Operation Description	Tool-Machine-Equipment Description-Tool or B.T. No.
50 — Broach Brake Surfaces Operator Removes Assy. From OHM, Loads Into RH Fixture, Cycles. LH Fixture Is Unloaded & Reloaded While RH Ram is Cycling Reload Part Onto OHM	15 Ton 90" Dual Ram

Figure 7 Cover page of function allocation/process sheet.

lected by the LEC through the 8-D process and given to the engineers. These data are referred to as the LEC data bin. If a similar allocation has an issue and the LEC has identified the problem, then the engineers should be informed of the issue before final specifications are made for the new process design. According to the flowchart and depending on whether a feasible solution is available, the engineer may be required to perform a complete or partial analysis or implement the solution already suggested by the LEC.

A complete analysis involves looking at four key activities of the process for potential risk—manual handling, assembly and/or disassembly, inspection, and machine operation. Various ergonomic analysis tools can be used to aid in the analysis. Solutions already developed by the LEC and proven to work can be specified directly to the vendor for the new process. It is important to note that, if possible, at least three separate ergonomics evaluations should occur. Recall that there are three phases during which an engineer can impact upon the design of a new process. Also recall that the first two phases (process design and process build) are where the most impact can occur. It is important that at each phase the responsible group review the specifications and sign off for ergonomics.

3. Case Study—Chassis Components Plant, Spindle Line, Light Truck

Plant Description and Layout. The chassis components plant (Study Plant) was a captive automotive supplier for a major automotive concern in the United States. Located in Michigan, it occupies over 1.75 million square feet; 1.5 million square feet for floor space is devoted to the manufacturing processes necessary to make chassis components. At the time of the case study, the manufacturing area was divided into two separate production areas: the rear drive manufacturing and assembly area (areas A: employing approximately 520 hourly workers) and the front and rear suspension components area (area B: employing approximately

603 hourly workers). In addition, there was a plant central maintenance area exists (area C: employing about 200 hourly workers).

Area A manufactured and assembled most of the parts for rear drive axles. Castings were brought into the plant from outside suppliers and machined to specification. Gear, tubes, brackets, bearings, and castings are subassembled then transferred to the final assembly lines. Axles are assembled at the rate of 4000 per day. These axles varied in size and made for vehicles ranging from midsize cars to medium duty trucks. The finished parts are stored in racks and shipped to the auto/truck assembly plants.

Area B manufactured the components necessary for the suspensions of front and rear drive cars and light duty trucks. A stamping area forms suspension arms that house spring and shock assemblies. Some subassembly is required for the parts before they are packaged and shipped. A machining area machines spindles and knuckles from raw castings. These components are part of the front and rear suspensions and are responsible for steering, ride, handling, and holding the wheels to the car.

Area C existed to service, fabricate, and build equipment to operate machinery in both production areas. It consisted mostly of skilled trades, maintenance, cleaning, and fabrication. Its duties included building platforms and special equipment to assist in machine operations and tool changes, submitting proposals for the redesign and modernization of existing machinery to improve efficiency, and maintaining equipment to ensure high quality and lowest possible downtime.

This case study shows three workstations that were altered during the build and implementation phase of the Design for Ergonomics process in area B of the production facility. Area management anticipated that this approach would have two main effects. First, correction of problems before a process was at full production would increase the effectiveness of the program. However, for this method to be most efficient, problems on the jobs would have to be anticipated before they were in full production. Such anticipation would involve job simulation and still risk not fully experiencing all potential problems. Second, making changes to existing jobs is a drain on area resources. As the process was new, the division responsible for installing the equipment was also responsible for any changes to the equipment until final sign-off at the plant. Therefore, the use of divisional resources in conjunction with plant personnel to make the changes would minimize the drain on area B resources.

Area B combined the resources in one large Design for Ergonomics group because combining the resources would eliminate one communication link that might create unnecessary overhead.

Operations 10, 20, 30 of Stem Line. Operations 10, 20, 30 are all in one workstation, and the redesign reflects changes made to all three operations. In general, there was no risk of low back injury for any job because the parts weighed well under 5 lb and all handling was done at or above waist height. However, workplace characteristics existed that could lead to stressful postures of the upper extremity, and modifications were made to correct them (see Figure 8, * marking), including installation of an adjustable platform and automatic doors on the lathe, redesign of load chutes, and repositioning of palm buttons. In addition, the front edge of the lathe was rounded to reduce the risk of striking the elbows when loading and unloading parts. These changes reduced the amount of upper extremity stress to the operator.

Operation 50 of Stem Line. On operation 50 of the stem line, the major area of ergonomics stressors was again the upper extremities. In particular, the gauging portion of the job was quite stressful to the wrist and hands of the operator because the task required rotation of the stem in the pot gauge. The spindle group reduced the magnitude of the force and virtually eliminated the stressful postures by designing a tool to turn the stem in the gauge.

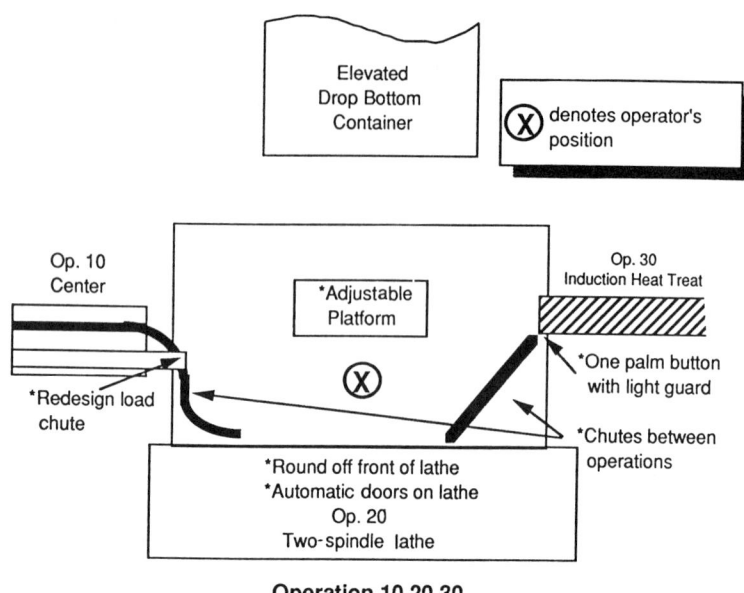

Figure 8 Stem operations 10, 20, 30.

Also, to facilitate using the tool, the vendor redesigned the pot gauge (see Figure 9, * marking). Finally, to improve the general workplace characteristics of the job, an amplifier was lowered to aid the operator in quality control.

Operation 70 of Spindle Line. After operation 40, where the knuckle and stem were assembled, the spindle group made changes for two additional workstations. On operation 70, the operator was required to get left and right spindles, corresponding to the right and left sides of the automobile, from a pallet on a conveyor and load them into the machine. After the machine cycled, the finished spindles were unloaded and moved to the washer conveyor (Figure 10). Because of the weight of the spindle and its location on the conveyor, there was

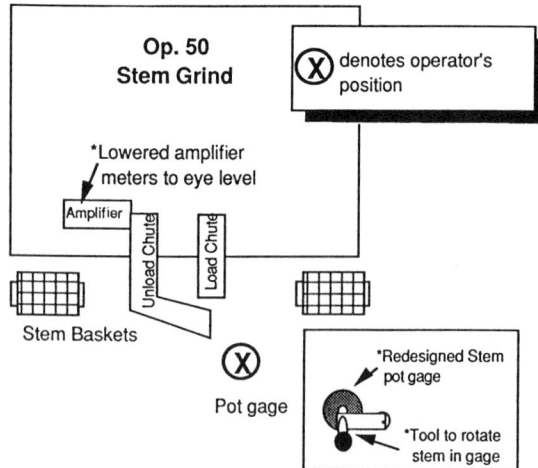

Figure 9 Stem operation 50—grinder.

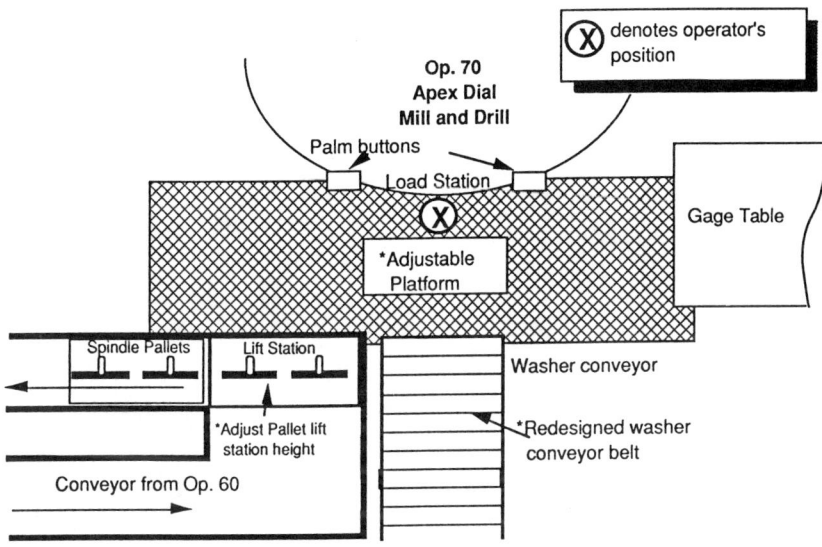

Figure 10 Spindle operation 70—apex mill and drill.

a potential for low back stress. In particular, because of the height at which the spindles were delivered, a source of back stress existed when spindles were being removed from the conveyor and moved to operation 70. To reduce potential stress, a lift table was provided on the conveyor to bring the spindles up to an acceptable height. The spindle group decided on the correct height of this table.

Another area of possible back strain existed when the spindle was removed from the machine and placed on the washer conveyor. Due to the shape of the parts, spindles had to be positioned upright to ensure proper spray dispersion in the washer. Fixtures were provided on the conveyor to aid in holding the spindle upright through the washing. However, these fixtures required the operator to precisely position the spindle on the conveyor. Because of the degree of precision required, the operator was forced to hold the spindle for an extended period of time, leading to back and shoulder stress. The group decided that removing the fixtures and placing the spindles on their side would reduce the strain. To do this required adjusting the aim of the spray nozzles in the washer. This suggestion was completed after a quality control and time study determined its feasibility. Finally, an adjustable platform aided the operator in moving the parts into and out of the machine and onto the washer conveyor. It should be noted that the changes outlined not only affected the low back but also reduced stress to the upper extremities. Note the reductions in stressful upper extremity postures.

Wash, Inspect, and Pack. The last operation on the spindle line were ergonomic changes were made was the packing area. Prior to formation of the ergonomics group, division engineering determined that reaching to the back of the baskets could be excessively stressful to the low back, and rotary tables were installed. However, the height of the tables in relation to the washer conveyor and platform still presented a stressful lifting condition. In response, the spindle group determined that lowering the rotary tables and platform while raising the baskets would dramatically reduce the stress on the back. In addition, these same changes, along with repositioning the spindles on the washer conveyor (see changes to Apex workstation), reduced the amount of stressful postures affecting the upper extremities (see Figure 11, * marking).

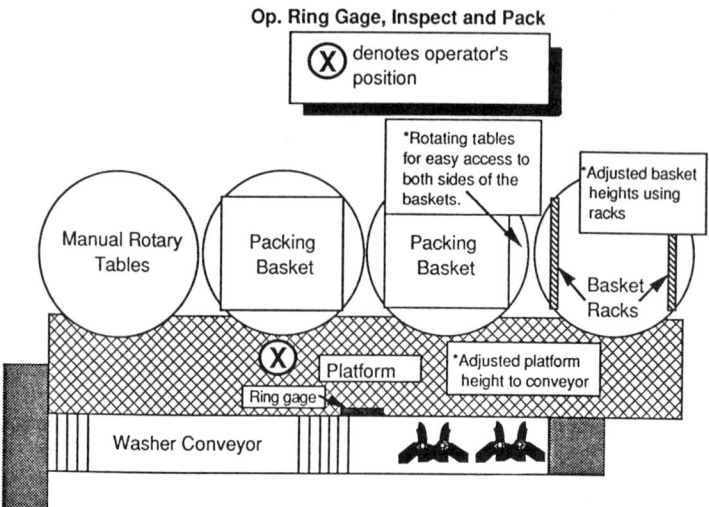

Figure 11 Spindle operation—ring gauge, inspect, and pack.

IV. CRITICAL REVIEW OF CURRENT STATUS OF REACTIVE AND PROACTIVE PROCESSES

A. Reactive Process

Once the program was under way, a process of constant improvement was adopted. After two years, several weaknesses appeared in the process. A summary of key areas of improvement follows.

It is almost impossible to directly measure the success of this process throughout the entire company. Jobs and people are constantly shifting, making it difficult to study the data from a critical baseline. However, it should be noted that in past studies of the joint labor-management ergonomics process, results have been very favorable. Three variables were measured in the study plant: training, group process, and ergonomic outcomes. Overall, these variables led to the following conclusions:

> This single-site project demonstrates that a participative program is an effective way to implement ergonomics changes in the workplace. All of the implemented projects involved both worker and management participation. The top-down approach, however, was unsuccessful at implementing change; none of the changes using this approach were implemented. The affected worker was a valuable data resource and his assistance was essential to enhance change efforts. Obviously, a major problem with the participative approach is that the time spent in meetings is time away from the principal work of the organization. The participative approach can also be more time consuming than the traditional top-down approach. However, in the long run, these negative implications are often more than offset by job changes that produce a healthier and more satisfied workforce [6,7].

Currently, the UAW-Ford process is being evaluated as to how well the local ergonomics committees are performing. This is done by means of an annual audit and a self-report form. The annual audit is completed by an internal team that reviews each plant's progress and makes suggestions for improvement. At the beginning of each calendar year, the LEC is required to submit an action plan to their division coordinator for review. This information, along with the audit, should help direct the LEC along a positive path.

In addition to the internal audit, at least two external groups have reviewed the process. The first reviewed the one plant's data for a one-year period and found that on a case-by-case basis (a case being a job), injury rates dropped dramatically and the financial pay-back, based on injury data, was realized in one and a half to two years. This type of study was very encouraging and will be continued. The second group reviewed the LEC's ability to analyze jobs. The results indicated that, in general, the LECs had a tendency to miss key areas of risk on the job. The cause of this may be twofold—either the LECs were not adequately trained or the training encouraged them to review only risks whose existence was suggested by the medical data. For example, if the medical data indicated shoulder issues, then they would concentrate their evaluation on the shoulder. In any case, the results of the study indicated that a more disciplined approach to the job analysis was required. Currently, the researcher is applying the new methodology in a pilot program, and, if successful, it will be implemented.

These issues resulted in a review of the entire documentation process. The original method of documentation did not allow for the information gathered from the local ergonomics committees to be used by any other group. Therefore, a team was developed to explore better methods of documentation. Two major forms were developed to better document changes. The first is called a concern log. This document records each instance where an ergonomics concern is identified. It does not record much detail; rather it is a listing of the concerns. The second document is referred to as an 8-D form. The "8-D" stands for eight disciplines of problem solving. This system uses eight unique but connected steps to identify and solve problems. Completion of the eight steps ensures complete root cause analysis. Together these documents resulted in a better and more thorough system to communicate ergonomics information within the company. Manufacturing and assembly facilities with similar problems can now share the solutions. In addition, the system resulted in better institutional knowledge for the engineers. When they design and develop new processes, they have a baseline to review to help define the things that went right and those that went wrong (from an ergonomics perspective) with the previous process. This knowledge is essential if ergonomics issues are going to be fixed through the design of new equipment.

B. Proactive Process

Currently, the proactive process is being employed in several pilot programs within the company. It is essential to capture the program at the correct time to have the most impact. Once this is done, it is also important to convince senior program management that the process is important and should be prioritized as highly as other key program objectives. Most important, the analysis tools for ergonomics should be user-friendly and accessible. As stated before, times is the biggest issue in the success of this process. Engineers seem to be willing to try ergonomic analysis if it can be done simply and conveniently. Therefore, the tools should be placed in a simple-to-use toolbox. One method being studied at Ford is the use of computers to do ergonomic analyses. Today, most engineering functions are tied to large computer systems that aid engineers in performing their job functions. These functions include purchasing, computer-aided design, quality, etc. A study has begun at Ford to evaluate the effectiveness of putting the ergonomics analysis tools into the same system. One possible approach is the development of an ergonomics worksheet that will be accessible to engineers during their workplace specification. This worksheet will have most of the state-of-the-art tools available at the push of a button. It is expected that with this system, less time will be required and engineers will be more likely to consider ergonomics design attributes during the early phases of process design.

C. Conclusions and Future Concerns

How effective is the process? At this time, because of various issues noted here, this question cannot be answered. Current trends are encouraging. However, a better data acquisition system is necessary to properly measure performance [8]. Recently, a new health data analysis system was launched. This system is tied to the person, the job, and medical and workers' compensation data. Over the next several years, data will be collected that will enable the analysis of morbidity, cost, quality, and productivity data simultaneously, allowing for a more complete analysis of the ergonomics process.

There are several areas of concern. Currently the reactive part of the program is driven by passive surveillance—reviewing existing data sources to pinpoint areas needing ergonomic intervention. Although this approach has been successful in changing many jobs, it will never be completely successful in primary prevention—preventing the injury before it happens. This can only be accomplished by incorporating active surveillance—the real-time collection of employee complaints and injuries to enable the estimation of the prevalence of various types of injuries—into the process. Active surveillance can be carried out in several ways, ranging from simply using a simplified questionnaire that is taken to the job site to using noninvasive diagnostic examinations involving highly trained medical professionals. In either case, an estimate of risk and employee complaints should be used to estimate exposure.

As one might suspect, even though active surveillance systems are more accurate than passive surveillance, they can be costly to administer. Also, they are time-consuming and disruptive to normal plant operations. In addition, to fully utilize the proactive process, the data generated from the reactive process must be available to persons involved in proactive ergonomics. This institutional knowledge of "things gone right" and "things gone wrong" should help engineers and designers understand where improvements are needed in new program designs. To aid in this process, databases will have to be developed and made accessible to the designers. This will require standardizing data input and output so that the data are easy to interpret and use. Consequently, the reactive and proactive processes are fully dependent on one another, having one without the other will cause the ergonomics process to slow and possibly to fail.

Therefore, the challenge to the manufacturing ergonomics process is to develop an integrated approach that finds and fixes existing issues and prevents future ones from being designed into new equipment. Only then will we get full control of the problem.

ACKNOWLEDGMENTS

The United Auto Workers Union's full name is United Automobile, Aerospace and Agricultural Implement Workers of America. The UAW-Ford National Joint Committee on Health and Safety is jointly chaired by members of the company and the union to oversee certain activities in health and safety for UAW Master Agreement Plants. We thank the UAW-Ford National Joint Committee on Health and Safety for their support and guidance in developing this chapter.

REFERENCES

1. T. O. Kvalseth, Factors influencing the implementation of Ergonomics: An empirical study based on psychophysical scaling technique, *Ergonomics* 23(8):821 (1980).
2. UAW-Ford National Joint Committee on Health and Safety, *The UAW-Ford Ergonomics Process—Implementation Guide*, Univ. Michigan, 1988.
3. UAW-Ford National Joint Committee on Health and Safety, *The UAW-Ford Ergonomics Process—Job Improvement Guide*, Univ. Michigan, 1989.
4. UAW–Ford National Joint Committee Health and Safety, The UAW–Ford Ergonomics Process Leadership Commitment Videotape, 1988.
5. UAW–Ford National Joint Committee on Health and Safety, The UAW–Ford Ergonomics Process, Segments 1–5, 1989.
6. J. K. Liker, B. S. Joseph, and S. S. Ulin, Participatory Ergonomics in two U.S. automotive plants, in *Participatory Ergonomics*, K. Noro and A. Imada, Eds., Taylor & Francis, London, 1991.
7. B. S. Joseph, A participative ergonomics control program in a U.S. automative plant: evaluation and implications, Dissertation, 1986. The University of Michigan; Ann Arbor, Michigan.
8. B. S. Joseph, *In-Plant Ergonomics Programs: Development, Implementation, and Evaluation*, Final Report, Motor Vehicles Manufacturers Association, 1988. Detroit, Michigan.

25
Ergonomic Control Measures in the Health Care Industry

Arthur R. Longmate
Johnson & Johnson, New Brunswick, New Jersey

I. INTRODUCTION

Whether a company is large or small, potential ergonomics-related problems can be found in every workplace where people are an integral component. Johnson & Johnson (J&J) is a highly decentralized corporation composed of many large and small individual companies. The key to effective ergonomics programs at J&J is *proactivity*—identification and resolution of potential problems before the occurrence of serious medical cases. Effective programs to control ergonomics-related disorders require the education of, cooperation of, and input from each person involved in the process. Success in the prevention of serious medical cases depends on the degree of success in the following categories:

Early documentation of ergonomic risk factors and reporting of medical symptoms before the medical cases become chronic and severe

Initial sound ergonomic design of processes and equipment using ergonomic job design guidelines

Prompt intervention steps taken to alleviate identified stresses

Provision of comprehensive medical case management to prevent medical cases from becoming more severe

Ergonomics program components that have been found to be successful at J&J are the focus of this chapter, illustrated by several ergonomics engineering case examples.

II. SETTING REALISTIC ERGONOMICS PROGRAM GOALS

Due to the wide range of personal and work-related ergonomic risk factors, it is virtually impossible to prevent all ergonomics-related medical disorders. Thus, it is critical to set attainable program goals and objectives so that realistic program expectations are maintained. A common mistake is to adopt a "zero defect" mentality with the goal of eliminating all ergonomics-related injuries. An important initial pill to swallow is to accept the fact that some level of ergonomics complaints and medical symptoms will always be present in any business environment—particularly those requiring extensive repetitive motion, manual materials handling, and/or use of computers. A frequent basis for OSHA fines is the failure to report ergonomics-related medical cases when it is readily apparent that ergonomic risk factors are present in the workplace and/or when employee interviews reveal that discomfort and medical symptoms are being experienced. Inaccurate or misleading recording practices are often the result of initially unrealistic program goals. Formulation and acceptance of realistic program expectations, up front, will avoid many problems later on in the process.

It is not uncommon for a well-meaning upper management representative to ask the question, What will it take to make these problems go away? The simple answer to this question is the following. If the repetitive manual work, the related ergonomic risk factors, and the people performing this work go away, so will the related medical cases and discomfort symptoms! Since this does not generally happen, the setting of unrealistic expectations has often led to the unfortunate end result of creative record keeping to create the desired impression that the problems are under control. To prevent this occurrence, performance-oriented goals and objectives should be created rather than those geared to specific incidence rates or to complete elimination of problems. If all elements of a comprehensive ergonomics control program are in place, then improvements in the numbers will be realized in time.

One realistic goal is to attempt to eliminate all *serious* medical cases. This requires the establishment of a sensitive surveillance system, an objective medical case severity rating system, and periodic case reevaluation to allow case tracking. Another realistic goal should be continuous improvement and not the expectation of complete problem eradication. If a manufacturing facility claims that ergonomic injuries or illnesses are not being experienced at any level, one or more of the following factors may be the underlying cause:

Employees (management, professional, and hourly) may not have received sufficient training to recognize ergonomics-related disorders.
Occupational medicine professionals may not have had sufficient training to recognize the work relatedness of ergonomics-related disorders.
The threshold of company ergonomic injury/illness recording criteria may be set too high to detect low-severity cases.
The company may be spending excessive resources to reduce ergonomic stresses to an unreasonably low and financially unjustifiable level.

A key premise to achieving an effective ergonomics control program is to accept the fact that ergonomic medical cases will occur at some baseline level in all environments where ergonomic stresses are present. A key objective for an ergonomics program should be to minimize the incidence and severity of these cases, not necessarily to completely eliminate them. Performance-oriented goals are preferable to the setting of specific incidence rate targets, which may typically lead to creative bookkeeping schemes.

III. ERGONOMICS TEAM STRUCTURE

A typical first step in the establishment of an ergonomics program is the creation of an ergonomics task force composed of team members from the various aspects of the business, such as hourly employees, management, engineering, human resources, medical, and product development representatives. In a simple manufacturing environment, a single ergonomics task force may be effective in identifying and resolving existing problems throughout the entire business. In many of today's complex business environments, it may be necessary to create a network of specialized groups to address problems in various diverse segments of the business. For example, office-related problems may be of little interest to associates working in the manufacturing environment; however, both environments probably do have serious ergonomics issues that need to be addressed.

At one J&J corporate headquarters location, five ergonomics committees have evolved to address issues in different segments of the business. This structure allows a clear focusing of efforts on the problems in a specific area and does not waste the time of a larger group of individuals whose primary interest or responsibility lies in other areas. A brief description of the jurisdiction of the various committees follows.

A. Ergonomics Management Advisory Team

A key component of an ergonomics program is to secure the interest, involvement, and commitment of upper management. The Ergonomics/Safety Management Advisory Team is composed primarily of department directors who represent each functional area of the business and are champions for each of the focused ergonomics teams. In addition, key health professionals such as the ergonomics engineer and medical, safety, industrial hygiene, and human resources personnel are available resources to the committee. The key objectives of this committee are to secure upper management awareness of current ergonomics issues and to ensure that adequate resources are provided to resolve them. The Management Advisory Team also provides a forum to share the efforts being undertaken by each of the other teams.

B. Manufacturing/Engineering Team

The Manufacturing/Engineering Team, informally known as the Safe Workplace Action Team (SWAT team), is made up of hourly associates from each functional manufacturing team. The same group of technical support personnel assist the team as technical support is needed. Each SWAT member receives comprehensive training in the recognition of ergonomics problems and functions as the day-to-day eyes and ears and mouth of the ergonomics program, identifying and reporting problems as they occur. The group also functions as a body of local experts who can provide valuable, practical input in the design of new equipment and processes. This group is also responsible for a variety of other health-related functions, such as

Conducting periodic surveys of all department associates to determine current ergonomic problems and concerns
Conducting monthly departmental safety meetings
Investigation of all illnesses and injuries and completion of related accident reports
Review of all new equipment and/or processes to identify potential problems prior to implementation.
Assisting with the evaluation of the effectiveness of all new ergonomic materials such as chairs, equipment, and tools.

Equipment and process engineers from each product development team provide SWAT members with technical support to ensure that all ergonomics equipment issues are promptly resolved and to ensure that ergonomics equipment design guidelines and specifications have been provided to all equipment suppliers.

C. Product Development Laboratory Team

This organization has many different types of development and testing laboratory environments. Problems experienced with the specialized test equipment found in these laboratories are unlike the typical problems found in other functional areas. As a result, a separate team was formed, with members from each laboratory area, to focus on issues specific to these areas.

D. Office/Computer Ergonomics Team

Office issues are sufficiently unique to warrant a separate, focused ergonomics team. The most frequent source of critical office ergonomics issues is the ever-growing and almost constant user interface with computers. This team adopted the name OSWAT (Office Safe Workplace Action Team) to piggyback off the success of the manufacturing team. The main objective of this team is once again, to be the eyes, ears, and mouth of the office ergonomics program to identify and resolve issues in a preventive mode.

E. Surgical Training Facility Ergonomics Team

A key business objective in the minimally invasive surgery marketplace is the training of surgeons and sales representatives in new endoscopic surgical procedures. To accomplish this strategic objective, a separate training facility was constructed. The ergonomic issues involved with this surgical training facility are once again sufficiently unique to warrant a separate ergonomics team.

F. How Many Teams Are Too Many?

The existence of five ergonomics teams at one primary location is probably quite unusual. In today's competitive and hectic business environment, there is no room for wasted effort. If teams can be effectively combined and remain focused on the specific issues, all the better. The downside of a multiteam structure is that the health professionals who support the focused teams sometimes get stretched somewhat thin attending to the activities of each team. On the other hand, if a group of individuals has a unique set of problems, a separate team is probably justified. Another advantage of multiple teams is that the number of issues facing an individual team will usually be smaller and more manageable than the number that a larger, all-encompassing team must deal with. The smaller focused teams may need to meet only infrequently, such as on a quarterly basis, if the number of issues is small and if all outstanding issues have been resolved. The team meeting may follow a walk-through inspection to determine if any new problems have surfaced. A potential problem for larger teams is to carry too long a list of projects. A smaller team with only a few key issues has a higher probability of success than one whose list is lengthy and frustrating. In either case, issues must be prioritized and addressed within the constraints of resource limitations.

IV. OTHER J&J ERGONOMICS PROGRAM COMPONENTS

Johnson & Johnson initiated its first ergonomics program efforts approximately 12 years ago at Ethicon, Inc. and J&J Products, Inc. (now J&J Consumer Products, Inc.) [1,2]. The goals of these programs have been to establish necessary ergonomics program components. A brief survey of these components follows.

A. Surveillance Techniques

Surveillance techniques are needed to identify existing and potential problems. These techniques consist primarily of passive surveillance methods such as simple review and summarization of available records. In addition, active surveillance techniques such as questionnaires, surveys, and comprehensive medical examinations are often utilized. The most common surveillance techniques used at J&J are

1. Thorough analysis of medical visit data and OSHA logs.
2. Ergonomic review and risk factor documentation of new and existing equipment and workstations.
3. Informal employee surveys—a proactive approach to identifying problems in the early stages before medical symptoms have developed. Why wait for complaints, when you can simply ask the employees which jobs (a) cause discomfort during or after performing the work, (b) are associated with the highest turnover rates, or (c) experience problems meeting expected production goals?
4. Postural discomfort questionnaires [3] to evaluate the effect of specific program interventions before and after implementation.

B. Reporting Systems

Reporting systems are needed to document the incidence and severity of problems and promptly disseminate critical information to parties responsible for problem intervention. The ergonomic teams in each area are responsible for the completion of accident/injury reports that identify causal factors and the formulation of action plans to eliminate the problems.

Medical case incidence and severity data must be maintained in a user-friendly information system to ensure that data are provided to interested parties in the most effective format. It must be easy to customize the system to meet any future information need. Available systems may be purchased or a system can be set up using dBASE III or a similar database software system.

Ergonomic complaints/medical cases are reported to responsible managers and engineers within 48 hr to facilitate formulation of effective action plans.

Monthly injury/illness summary reports, by department, are distributed to each manager and ergonomics team member along with a cover sheet indicating the current monthly trend compared to recent historical data. A summary, with observed trends and currently active intervention projects, is also provided with each monthly report.

Quarterly and year-end summary reports are also provided to indicate trends, successes, and possible failures.

C. Intervention Techniques

Intervention techniques can generally be classified into the following categories:

Comprehensive medical case management.

Conservative medical treatment protocol (surgery is considered only as a last resort).
Physical therapy (most effective when conducted on site)
Effective medical job restriction protocol.
Methods modifications to reduce risk factors.
Case-specific ergonomics engineering interventions (equipment/process modifications).
Work method retraining [methods analysis/modification; possible use of biofeedback monitoring (EMG) to make person aware of least stressful work methods].
Administrative procedures (e.g., job enlargement, job rotation, imposed limitations on overtime).

Complaint/Problem Follow-up.

Team problem resolution (brainstorming) techniques.
Problem resolution tracking—action items with assigned responsibility and target completion dates.

Short-Term and Long-Term Action Plans.

Simple administrative/methods changes made up front.
Long-term job reengineering changes move later.

D. Prevention Techniques

A prevention technique is generally one of the following.

Ergonomics Training. Each and every employee, from the hourly worker to the company president, must receive some level of ergonomics training.

Ergonomics committee members must receive the most extensive training in all phases of ergonomic job assessment, design criteria, prevention, and intervention techniques. The training may be conducted using internal resources or through the use of comprehensive ergonomics training programs offered by academic institutions or private consultants. After initial training, regular follow-up sessions should be conducted to gain critical experience through brainstorming and the sharing of experiences.

Upper management representatives should receive basic training in ergonomics principles and periodic briefings on the extent of problems, goals and objectives of the ergonomics program, and the support required to achieve the objectives. Selected management champions should be identified and invited to attend comprehensive training.

Equipment, process and product development engineers, and designers should receive extensive training in ergonomic design guidelines to prevent future problems. An effective training technique is to get technical support personnel closely involved with medical case follow-ups to allow them to directly interface with persons experiencing the problems. Requiring engineers to work for extended periods actually performing the work tasks they have designed provides an excellent learning experience.

Managers, supervisors, and hourly employees should receive basic ergonomics training to help them understand the types and symptoms of problems that might be experienced and aid in early detection and reporting. The goals of the ergonomics program should be stressed along with specific actions that the company is taking to prevent and manage these problems to minimize the effect on their lives and on the business. Since implementation of ergonomic intervention is a team effort, the team approach should be emphasized along with what is expected of the employees and their managers to address ergonomic issues.

Ergonomic Job/Equipment Design Guidelines. It is critical to establish written job and equipment design guidelines to function as a constant reference and reminder during all phases of the design process. These guidelines are a critical component of the engineer and designer training referenced above.

Ergonomic Equipment Checklists and Reviews. In addition to comprehensive ergonomics guidelines, it is helpful to have simple ergonomics checklists to act as a simple reminder at critical times during the equipment design and development and implementation processes. The complexity and length of the checklists should be minimized to simply encourage the thought process and not to focus on the completion of the checklist itself.

Exercise Programs. To be discussed in detail in the following section.

While all of the above components are critical to a comprehensive ergonomics program, the remainder of this chapter focuses on two critical elements of J&J's ergonomics program strategy: on-the-job ergonomics exercise programs and effective ergonomics engineering intervention efforts.

V. ON-THE-JOB EXERCISE PROGRAMS

A key component of J&J's proactive ergonomics program philosophy is on-the-job exercise. The commitment to on-the-job exercise began over 10 years ago at Ethicon, Inc., one of the corporation's largest domestic operating companies. It should be noted that ergonomic exercises should be considered only after other critical program components (e.g., effective surveillance networks, training programs, problem documentation, and implementation of interventions to alleviate stress) are established. The main objective of the exercise programs is the prevention of ergonomics-related injuries and illnesses. When used as a proactive component of an overall comprehensive ergonomics program, and developed and administered by qualified health professionals, exercise programs can be a valuable tool. On the other hand, when used as primary control strategy in situations where problems are severe and well established, dramatic results may not be achieved [4].

Important benefits of exercise that have been documented are the following:

1. Increased blood flow to active muscle/tendon systems. Many ergonomics-related disorders, such as tendon and muscle strains and inflammation, are partially caused by the inability of the body to transport sufficient nutrients to the site of active tendon/muscle groups. When muscles are contracted or tensed, blood flow to the area is restricted. In the case of static postural situations or highly repetitive tasks, sufficient recovery time is not provided to allow this critical replenishment process to occur to the necessary extent. Each exercise period is, in effect, a break from the repetitious or static postural work regimen. Every break in the work routine allows the muscles to relax and blood flow to be reestablished [3,5]. Whether exercise sessions or simple rest periods are more effective in reducing discomfort and other factors such as mood state is not clear; however, exercise may be more effective in reducing decreases in productivity that occur over the course of the workday. Research focused on selected repetitive tasks has indicated that more widely distributed rest or exercise minibreaks can lead to reduced postural discomfort and increased productivity [6,7].
2. Increased range of motion (flexibility). Periodic muscle/tendon stretching exercises, throughout the worker's entire range of motion, may diminish the risk of developing various types of motion specialization disorders such as trigger finger. It is well known within the sports medicine community that stretching and the resultant increase in range of motion and flexibility are critical factors in the reduction of injury [5].
3. Increased strength. This has been a recognized and well-documented benefit of exercise regimens designed for these purposes. When an individual's maximum strength increases, the relative level of job stress is reduced because the person is working at a lower percentage of maximum strength. This relative reduction in stress lowers the person's risk of injury.
4. Exercise of antagonistic muscle groups. Specific exercises are selected for the program that exercise antagonistic muscles, which act in the opposite direction than those primarily being used while the work is being performed. If work content requires gripping/squeezing motions, it makes little sense to take an exercise break that requires additional gripping and squeezing. A more sensible alternative would be to select an exercise that requires extension of the fingers to exercise the antagonistic muscle groups.
5. Reduction in the level of overall mental and physical tension/stress. While not fully understood, physical exercises are successful in relaxing the tensions that build from performing physical and mental tasks over extended periods.
6. Increase in the level of attention. A person's level of attention becomes degraded by the performance of repetitive tasks over extended periods. Exercise breaks are often referred

to as "energy breaks" because they have the desirable effect of getting the blood flowing and bringing renewed feelings of energy to the individual. This may explain why people have the perception that the exercises are "good" and generally claim that they feel better even while reductions in discomfort are not consistently reported [4].
7. Other benefits, such as reductions in visual strain and improvement in intellectual facilities and sensorimotor skills, have also been documented when exercise breaks are taken during work hours [8,9].

The immediate benefits from identifying and eliminating ergonomic risk factors in the workplace are easy to comprehend and have been clearly documented. In comparison, the benefits of exercise, in tangible terms, are often more difficult to document. Although many important effects of exercise have been documented, the critical scientific correlation between on-the-job exercise and reduced medical case incidence rates has been elusive.

Although preliminary studies have indicated that the productivity lost during exercising can be made up because employees are "energized" and are more highly motivated, it may be overly optimistic to assume that this can always be expected. In today's competitive business environment, the first question often asked by management is how much the implementation of a program will cost and how it will translate into financial savings and/or a reduction in future injuries and illnesses. Since an exercise program performed on company time can potentially result in a loss of 3–4% (approximately 15 min) of productive working time, it is natural for managers to ask what they will get in return. At one J&J facility employing approximately 2000 people, management representatives suggested that 70 new people might have to be hired to make up for the 3.5% conversion of available work time to exercise time. Based upon the assumption that each new employee would cost the company approximately $60,000 (including benefits), the additional annual cost would take $4.2 million for this facility. Whereas the argument could be made that this "lost time" could be made up, it is clear that the potential exists for exercise programs to demand substantial investment to make up for lost production time. It is clear that further research is necessary to explore and establish cost/benefit relationships.

VI. ENGINEERING CONTROL CASE EXAMPLES

A. Trocar Assembly

The following case study illustrates the benefits of many aspects of a comprehensive ergonomics program. During development and production of an endoscopic trocar device, the aggressive project implementation schedule dictated that assembly processes would be primarily manual with the exception of several ultrasonic welding operations. This is a common decision path in the minimally invasive surgery industry due to the requirement of speed to get the product introduced into the marketplace and the critical requirement for frequent quality inspections.

Trocar assembly steps were the following:

1. Get trocar casing bottom half and put into assembly nest.
2. Get knife assembly and compression spring.
3. Slide compression spring over knife stem.
4. Get anti-backup spring.
5. Insert tip of knife shaft through the hole in anti-backup spring while overcoming the pressure of the compression spring by pushing the hands toward each other.
6. Install knife/spring subassembly into trocar casing half to engage slot in knife onto boss in trocar casing while maintaining compression of spring.

7. Get buckling spring and install into trocar assembly.
8. Get two latches and install into trocar assembly.
9. Get trocar casing top half and install onto assembly.
10. Remove completed assembly from nest and place in tray.

During the initial process review, several ergonomic risk factors were noted with this process, particularly the following:

Repetitiveness. Associates were expected to complete approximately 550 assemblies per day.
Force. Substantial pinch force was required to hold the anti-backup spring between the thumb and index finger during compression of the main spring while inserting the knife shaft through the hole in the anti-backup spring.
Static exertion. The pinching of the anti-backup spring was statically maintained from the time the main spring was compressed until the knife shaft was inserted through a hole in the anti-backup spring and installed into the trocar casing.
Contact with sharp edges. The edges of the anti-backup spring were quite sharp and caused local discomfort to the fingers during the spring compression process.

The provision of some type of mechanical assembly assist fixture was considered. However, a feasible design had not been finalized by the time the manufacturing process was launched.

Since all associates had been trained in the recognition of ergonomic disorders and a sensitive surveillance network was in place to detect medical cases in the early stages, the decision was made to proceed in a manual assembly mode and to monitor the process carefully. Efforts also continued to identify an effective mechanical assist fixture.

An additional confounding factor had occurred during the engineering pilot process. For the instrument to function properly, it became necessary to provide a more forceful main compression spring. This incremental increase in the force requirement appeared to be the straw that broke the camel's back.

Over the next 3 months, 14 associates visited the medical department to report upper extremity symptoms, with one interesting pattern—all were left-hand-specific. All cases were relatively minor (most received severity ratings of 1 or 2 on a scale from 1–5, with 1 representing the least serious symptoms and 5 the most severe symptoms). As the cases began to appear, comments from assemblers verified that the problems centered on pinching and statically holding the anti-backup spring in the left hand during the somewhat tedious manual assembly process. It was also felt that if a mechanical fixture could be developed to compress and hold the main spring throughout the remainder of the assembly process, then the related physical stress would be greatly reduced. The project to develop a mechanical assist fixture received even greater attention when it was discovered that the tedious process of squeezing the anti-backup spring was causing damage to the spring and was resulting in occasional instrument failure.

Development of the spring compression fixture was placed at the highest priority, and several prototypes were developed and evaluated. Some of the critical design criteria that were determined through interaction with the instrument assemblers were the following:

1. The main spring should be mechanically compressed and locked into position to allow the rest of the assembly to be completed without the necessity of maintaining this static force.
2. Because different assemblers demonstrated many slightly different methods, the angle of the fixture was made adjustable to more effectively accommodate these individual differences.
3. As some of the assemblers were left-handed, the fixture was designed to be adaptable for convenient use by either left- or right-handed individuals.

The fixtures were designed and built over an approximately 6-week time frame at a cost of approximately $800/per fixture. The relative simplicity of the fixtures is the primary reason for the quick turnaround on custom-designed equipment. It is typical for more complicated fixturing to require several months to design and build.

After implementation of the new assembly fixtures, the stressful static holding of the anti-backup spring was eliminated. The job steps involved with the new Trocar assembly fixture are the following:

1. Get knife assembly and position under fixture.
2. Push in knife assembly retaining knob.
3. Get compression spring and slide over knife stem and into position in front of spring compressor bar.
4. Pull lever to right to compress main spring until latch is engaged.
5. Get anti-backup spring and slide over end of knife stem.
6. Get trocar casing bottom half and position under assembly.
7. Gently squeeze anti-backup spring and position into casing while engaging slot in knife onto boss in trocar casing.
8. Pull out knife assembly retaining knob.
9. Pull assembly down to disengage from spring compressor bar and place on assembly table.
10. Get buckling spring and install into trocar assembly.
11. Get two latches and install into trocar assembly.
12. Get trocar casing top half and install onto assembly.
13. Remove completed assembly from nest and place in tray.

It should be noted that three extra job elements were added through the use of the assembly fixture. However, the stressful job elements associated with compressing the main spring were eliminated. Since the degree of assembly difficulty was reduced significantly, no productivity was lost in conjunction with implementation of the fixtures even though a few simple elements were added.

The fixtures were very positively received by the majority of operators as soon as they were introduced. This highly positive response was better than expected in light of the fact that people had been performing the old method for several months. Usually, a longer adjustment period is required to allow people to acclimate to a new method. Involving the assemblers in the fixture design process clearly played an important role in the feeling of ownership and acceptance of the fixtures.

Within 3 months of implementation of these fixtures, virtually all CTD symptoms and other chronic hand overuse disorders had disappeared. The fixtures also made it possible for all people on the line to perform the job and cleared the way for the elimination of several medical restrictions that had specifically related to this job.

B. Low-Force Activation Buttons

On several ultrasonic welding operations, an elevated incidence of thumb tendinitis was observed. Risk factor assessment and interviews with associates indicated that the 5 lb of force required to fully depress the welder activation buttons in a very dynamic environment might be a primary cause of these problems. The standard activation buttons were positioned in the front plate of the welder base as depicted in Figure 1. The job elements at the welding station were as follows:

Figure 1 Standard welder activation buttons (on front) presence sensing, capacitive activation buttons (at sides).

1. Get assembled stapler from tray.
2. Put the stapler into the welder nest. (Instruments are automatically clamped into position when required).
3. With both hands, fully depress and hold the cycle activation buttons until the vibratory horn contacts the instrument.
4. Get the instrument from the nest and put into a finished instrument tray.

Due to the location of the buttons and the pace of the operation, most welder operators use the thumbs to activate the buttons. To overcome the 5 lb of force quickly, substantially more than 5 lb was actually applied.

Several options were explored to address these issues.

Presence-sensing, capacitive-coupled activation buttons were purchased and installed into auxiliary electrical boxes attached to each side of the welder base (also see Figure 1). This initially appeared to be the perfect solution, because no force on the activation pad was required. Several problems arose during initial evaluations.

The 3 in. of space occupied on each side by the auxiliary electrical enclosures added 3 in. of reach distance each time the associate reached for parts. This was a very undesirable ergonomic consideration in the use of this type of activation control.

Substantial time was required for associated to learn and remember that forceful pressing on these controls was not required. Since the majority of machine activation controls throughout the rest of the facility were push buttons, new operators coming into the area had to be constantly reminded that the controls did not require the application of force. In fact, some associates so expected movement of the "button" that they imparted more force to the stationary sensing pad than would have been required by the original push button control. Also, since the stationary pads were not able to absorb force from quickly moving fingers, more actual force may have actually been imparted by the fingers under dynamic conditions.

The presence-sensing controls also had an inherent problem in that the welder cycle would unexpectedly interrupt for unknown reasons. This required a simple reactivation of the controls after a short pause but was perceived to be a major inconvenience by many operators.

The cost of the presence-sensing controls was also a negative factor compared to the much simpler and less costly push button controls. A project to convert all controls in all facilities to the presence-sensing variety would have been extremely burdensome.

The second option investigated was to evaluate the new photoelectric light beam sensor activation controls. Although these devices are marketed with the claim of eliminating ergonomic stresses by eliminating the pushing of buttons, once again this did not prove to be entirely true, for the following reasons:

When working in a dynamic environment, an operator cannot be expected to quickly extend one finger and calmly place it into the U-shaped opening of the control as is often depicted in the advertisements. Generally, the entire open hand makes dynamic contact, and the fingers generally bump into the relatively sharp edges on the inside edges of the finger cutout area.

When the photoelectric control was oriented upward, in several instances the machine cycle was observed to be accidentally initiated when parts of an operator's clothing were inadvertently dragged over the controls. Safety concern resulted in the necessity to turn the controls 90° to face outward. This modification had three undesirable results:

1. A 6-in. reach was added to the side for components because parts trays had to be located beyond the controls with enough clearance for hands to have access to controls.
2. It was necessary to rotate the hands and forearms 90° from the palm-down posture used to place components to the palm-inward posture used to activate the controls.
3. More complex, two-directional motion (out and back) was needed to move the hands to the controls. It was felt that these motions require both more time and more muscle activity.

Efforts were undertaken to identify other, more traditional button activation controls with reduced activation force and larger button contact surface. A survey of many available push button controls determined that most required approximately the same amount of force—in the 5 lb range. Also, most of these controls were permanently sealed to prevent the removal of internal springs.

One push button switch was identified that could be easily disassembled and the main compression spring removed to reduce the required activation force. An additional spring, built into the contact set, provided sufficient force to return the button to its original position after pressure on it was released. The force required to activate the modified button was approximately 1 lb. This represents a highly significant 80% reduction in required force. When an application called for two contact sets, the required activation force was approximately 2 lb, which still represented in a highly significant 60% reduction in required finger-pressing force. A standard low-force pushbutton switch requiring 2 lb of finger or thumb actuation force was eventually developed by the Allen Bradley Company. This switch was preferred because it did not require in-house modification to reduce the actuation force. Dukane Corporation, manufacturer of the ultrasonic welders, agreed to equip future welders with the low-force Allen Bradley buttons as a standard feature.

C. Knob Twist Fixtures

In departments assembling intestinal staplers, CTD incidence rates were running at 2-3 times the overall plant average. Review of the OSHA log revealed that about 67% of these illnesses were related to various types of elbow tendinitis. Risk factor assessment and interviews with associates indicated that the problem was more than likely related to the repetitive twisting of the adjustment knob at the end of the instrument. Some of the operations required only slight knob adjustment (less than one complete knob rotation), whereas others required extensive twisting, (up to 10 twists, occasionally in both directions). The majority of the twisting required negligible torque but was highly repetitive—occasionally over 10,000 90-110° forearm rotations per 8-hr work shift. On one operation, the operator had to forcefully twist the knob loose to break weld adhesions after the casing halves were ultrasonically welded together. This operation required substantial torque—up to 40 in-lb.

When the probable cause of the illnesses was identified, a crude powered knob twist device was built that used an air motor mounted inside a square aluminum box that sat on top of the work surface. This device was effective in turning the knob but created additional problems for the shoulders and arms because it was necessary for the operator to reach approximately 18 in. above the workstation each time an instrument was placed into the fixture. The device also required the operator to press and hold a button to turn on the motor, which resulted in discomfort to the hands and fingers. Also, it did not have any torque-limiting capability, which resulted in occasional damage to the instrument.

Input was solicited from employees, engineers, and management in the area to determine the most critical features for the next-generation fixture. This was an extremely critical process to maximize the changes that employees would use the fixture, that it would satisfy process constraints, and that production requirements would be met. Based on this input, a more sophisticated fixture was designed and built that had the following features necessary for effective process control:

1. Orientation of the fixture such that the instrument could be inserted into the nest with a lateral move of about 2-3 in. and without instrument reorientation. This was accomplished by using drive belts from the motor to the instrument chuck rather than driving the chuck directly from the motor shaft.
2. Layout of the fixture to consume minimum prime working space. The fixture was designed to be extremely narrow and took up only about 8 in. to the side of the assembler.
3. Selectable number of turns to the nearest quarter turn. The machine would automatically stop when the specified number of turns for a particular application was reached.
4. Selectable autoreverse feature in case assembler had to alternate direction of rotation.
5. Adjustable speed control.
6. An adjustable clutch to prevent instrument overtorquing.

A 300-instrument fixture validation study was conducted to ensure that the fixtures did not affect the functionality of the instruments. Employees were somewhat slow in warming to the new fixture, which took up over 8 in. of their workspace, but after about 2 weeks most people said they didn't know how they ever survived without it.

Within 6 months of implementation of these devices, the incidence of new cases of elbow trauma was essentially eliminated (reduced by approximately 95%). Utilization of the new fixtures also allowed employees with preexisting conditions to work on many jobs from which they were previously restricted due to the requirement of repetitive twisting.

In addition to eliminating the repetitive motions responsible for the occurrence of elbow tendinitis, an additional benefit was noted, during the follow-up evaluation, when the new fixture was used on a calibration torquing operation. The old method required manually twisting the instrument several times and then mounting it in a fixture. A manual torque wrench was then used to prestress the instrument prior to final calibration. Using the new fixture, the twisting was performed automatically and the torquing was also coincidentally achieved through the built-in torque clutch. Once again, a validation study was conducted to ensure that the torque achieved by the fixture was within process limits. Provision of these fixtures at this operation resulted in much lower physical stress for the operator plus an approximately fourfold increase in productivity. Productivity was increased on most other jobs by 10–15% through the use of these fixtures.

D. Instrument Test Fire Fixtures

Hand firing refers to power gripping a handle or trigger instrument configuration and squeezing to provide the power to form surgical staples or ligating clips. For some surgical instrument end users, hand firing of instruments requires a high percentage of their available strength (e.g., some instruments require approximately 50th percentile maximum female grip strength). For the end user, the redeeming factor is that the instrument firing task is performed only once every few minutes over a short period in the operating room and certainly less than 100 times over the course of the day.

During the manufacture of endoscopic surgical instruments, it is critical to ensure that instruments function properly during these critical medical procedures. As a result, each instrument must typically be hand fired several times during the manufacturing process to ensure its functionality in all ranges of future potential use. On production lines where hundreds and perhaps thousands of these instruments are assembled and tested daily, thousands of hand firings may be required.

Surgical instrument assemblers were experiencing a high incidence of upper extremity disorders (tendinitis and other overuse syndromes) that seemed to be related to the highly repetitive and forceful hand firing. Incidence was higher on lines assembling instruments that required higher firing force (e.g., up to 50 lb of grip force). Attempts were initiated to provide alternative powered fixtures to perform this repetitive manual function.

One of the problems was that it was difficult to know how much force a given person was applying to the instrument. The verbal instructions given during training were to squeeze until the trigger of the instrument just bottomed out with the handle. It was found that many of the more highly motivated employees were experiencing problems on a more frequent basis. Upon questioning and observation, it appeared that these individuals were unconsciously squeezing the instrument much harder than the minimum requirement because they were trying to do the best job possible. Training efforts were undertaken to demonstrate that excessive squeezing was a non-value-added function and was indeed increasing the risk of their experiencing a cumulative trauma disorder. To augment these training efforts, training fixtures were also provided that read out how much force an individual was actually applying. These fixtures were modified to have a light indicate when "sufficient" force limit was reached.

An additional argument was offered that powered fixtures would provide a more effective SPC (statistical process control) quality evaluation tool than the hand firing method. Process parameters such as the magnitude of force applied to the trigger and trigger closure speed could be tightly controlled and manipulated in a powered mechanical fixture. Control of these parameters during manual test firing was much more difficult due to the wide range of capabilities within the employee population. Even with training, much greater variability existed with the hand method, particularly over the course of the workday as fatigue began to take its toll. Subsequent tests, using prototype mechanical fixtures, verified the opportunity for the manufacturing process to be controlled to much tighter tolerances.

Efforts to provide powered test fire fixtures were temporarily delayed by the argument that instruments should be test fired by hand because that is the method used in the operating room. In cases where this argument remained an obstacle, a compromise was reached to the effect that at least one of the test firings would be performed manually and the test would be performed on the powered test fixture. For product lines where this argument remained unresolved, efforts were undertaken to provide training aids as described above that trained personnel to limit their exertions.

On instruments designed for multiple firings, such as skin staplers, hernia staplers or ligation clip appliers, quality procedures required that several instruments out of each batch be completely fired out to comprehensively test all instrument functions. Personnel responsible for this function were known to experience an extremely high incidence of upper extremity disorders; in fact, nearly all persons performing this function had experienced some type of medical symptom. Powered test fire fixtures were provided for these operations to greatly reduce the repetition of test firing. In most cases, about 95% of the manual firing could be eliminated because only one or two of the firings were now performed by hand to test the actual "feel" of the instrument. Elimination of the repetitive firedown requirement virtually eliminated all incidence of CTDs.

In the interim, and in situations where firing fixtures were delayed, a job rotation matrix was constructed to guide supervisors in the required frequency of job rotation based upon the actual force requirement and the frequency of firing. Very often, the inconvenience of frequent job rotation provided added incentive to remove obstacles that were delaying implementation of powered test fire fixtures.

An additional benefit to the implementation of the powered test fire fixtures was the elimination of many injuries involving staples being accidentally fired into the fingers and hand during the manual test fire procedure. Limiting manual test firing also greatly limited the occurrence of this unpleasant event.

VII. CRITICAL REVIEW OF CURRENT STATUS

Although many ergonomics problems can be eliminated through engineering redesign of processes and equipment, many others must be addressed through administrative and other control methods due to financial and feasibility constraints. The discussed ergonomic program components will improve the chances of proactively preventing problems, detecting existing problems early, providing a responsive management support environment, and, most important being able to implement solutions within a supportive, involved, and educated workforce. These components are

Employee empowerment to identify and solve problems (focused ergonomics committees)
Realistic, performance-oriented program objectives
Upper management support and involvement
Comprehensive medical case management
Effective medical case reporting and tracking system
Ergonomic job design guidelines
Focus on prevention of problems (critical review of new processes and equipment—don't wait for problems to come to you)

Tailored ergonomics training programs
On-the-job exercise programs

The key to improving the probability of achieving a successful program and maintaining long-term management commitment lies in setting realistic, attainable, performance-oriented goals. Improvements will come, and goals will be accomplished, but only after a lot of hard work has been put into creating a strong foundation upon which the framework of a comprehensive ergonomics program can be built.

Setting goals strictly aimed at reducing the incidence of ergonomic medical cases, in the absence of performance-oriented objectives to initiate and maintain critical program components, often leads to the development of creative bookkeeping schemes that lead to force-fitting incidence rates to meet objectives. The unfortunate side effect of trying to force incidence numbers down is that the legitimacy of each medical case is often questioned. This phenomenon usually leads to the development of mistrust and negative perceptions between employer and employees. The basic management philosophy must be to accept and believe what employees are saying and to work directly with them to identify and resolve problems. If employees are continually scrutinized in an attempt to disprove work relatedness and company responsibility, a company philosophy of basic mistrust will be very quickly understood.

While J&J's ergonomics philosophy includes on-the-job exercise, for which many benefits have been documented in the literature, very limited scientific evidence has linked exercise with documented reductions in the incidence of ergonomics-related medical disorders. Although participants say that they enjoy the exercise programs and that they "feel better," significant reductions in postural discomfort and in the severity of medical symptoms have not been scientifically documented. Although exercise clearly has favorable preventative effects on the body, it has not been shown to be a successful therapeutic intervention when medical problems are well entrenched. Additional research is needed to better document the benefits of on-the-job exercise.

VIII. FUTURE CONCERNS

One of the most significant barriers confronting ergonomics practitioners, in selling the need for ergonomic changes, is the shortage of objective, practical guidelines governing the science. If more precise guidelines were available regarding the number of repetitions or the force or posture deviation levels at which problems will begin to precipitate, the ergonomist would be in a better position to know when to react with strong action plans in the best interests of the company and its employees. With current knowledge, there is always some level of guesswork and risk associated with any proposed action plan. An overly conservative approach may place the company under unfair economic constraints in an already highly competitive marketplace. If the ergonomist is too lenient in allowing stressful work practices to persist, CTDs and other ergonomics-related medical cases and employee complaints will begin to appear. There are always a multitude of possible solutions and very few precise answers when trying to determine to what extent actions must be taken to manage ergonomic risk. As discussed, it is critical to have an educated workforce whose members can recognize problems in the early stages and to create a sensitive surveillance network that facilitates early detection.

It is clear that many of the ergonomic "epidemics" that have occurred in various geographical locations around the world have resulted from lack of education of the workforce, which has allowed emotions to get out of control and myths to run rampant. The more the working population is educated on the recognition of the risk factors and medical symptoms relating to ergonomic problems, the more efficiently these problems can be understood, prevented, identified, and resolved. It is clear that the earlier potential problems are identified, the better the chances are of resolving them with minimal effort and cost to the company along with minimal suffering to the people performing the work.

REFFERENCES

1. A. Longmate and T. Hayes, Making a difference at Johnson & Johnson: some ergonomic intervention case studies, in *Industrial Ergonomics Case Studies*, B. Pulat and D. Alexander, Eds., Industrial Engineering and Management Press, 1991, p. 181.
2. A. Longmate and C. Welker, Components of an industrial ergonomics program: the Johnson & Johnson experience, in *Industrial Ergonomics: A Practitioners Guide*, D. Alexander and B. Pulat, Eds., Industrial Engineering and Management Press, 1985, p. 129.
3. T. Hansford, H. Blood, B. Kent, and G. Lutz, Blood flow changes at the wrist in manual workers after preventive interventions, *J. Hand Surg. 11A*:4 (1986).
4. B. Silverstein, T. Armstrong, A. Longmate, and D. Woody, Can in-plant exercise control musculoskeletal symptoms, *J. Occup. Med. 30*:12 (1988).
5. G. Lutz and T. Hansford, Cumultative trauma disorder controls: the ergonomics program at Ethicon, Inc., *J. Hand Surg. 12A*:863 (1987).
6. N. Swanson and S. Sauter, The effects of exercise on the health and performance of data entry operators, in *Work with Display Units '92*, Luczak, Cakir, and Cakir, Eds., Elsevier, New York, 1993, p. 288.
7. K. Lee and L. Humphrey, Comparison of the relative effectiveness of ergonomic design and physical exercise, in *Ergonomics International '85*, Brown, Ed., Taylor & Francis, London, 1985, p. 556.
8. W. Laporte, The influence of a gymnastic pause upon recovery following post office work, *Ergonomics 9*:6 (1966).
9. D. Thompson, Effect of exercise breaks on musculoskeletal strain among data entry operators: a case study, in *Promoting Health and Productivity in the Computerized Office: Models of Successful Ergonomic Interventions*, Sauter and Dainoff, Eds., Taylor & Francis, London, 1986, p. 118.
10. B. Silverstein, L. Fine, and T. Armstrong, Occupational factors and carpal tunnel syndrome, *Am. J. Ind. Med. 11*:343 (1987).
11. E. Corlett and R. Bishop, A technique for assessing postural discomfort, *Ergonomics 19*:2 (1976).
12. V. Bhatnager, C. Drury, and S. Schiro, Posture, postural discomfort and performance, *Hum. Factors 27*:2 (1985).

26
Ergonomic Case Studies in Industry
Health Care

Roger C. Jensen
UES, Inc., Dayton, Ohio

I. INTRODUCTION

Several sources of data indicate that nursing homes are not particularly safe places to work. In the United States, nursing homes are included in the industrial classification "nursing and personal care facilities." Data indicate that this industrial category had 151,000 occupational injuries and illnesses in 1988 [1]; only five industries had more. The large number of cases in the nursing and personal care industry is attributable to a combination of a large number of employees and high injury rates [1].

Back injuries account for over 40% of the workers' compensation claims from the nursing and personal care industry [2]. Nearly all of these back claims are sprains or strains [2]; only a very small percentage of back claims are for other types of injuries such as fractures, burns, and lacerations. Most of the back sprains and strain are experienced by nursing assistants.

A particularly disturbing concern is that the injury and illness rate for the industry has shown an increasing trend [1,2]. Most of this trend has been attributed to an increase in back injuries [2]. These data strongly suggest that the programs commonly used by the nursing home industry to prevent back injuries are not very successful. Although some nursing homes and/or nursing home chains may have successful prevention programs, the industry as a whole is not having success reducing back injury rates. Consequently, it appears that the common traditional programs for preventing back injuries among nursing home employees are failing to accomplish the objective of prevention.

II. BACKGROUND AND SIGNIFICANCE OF OCCUPATIONAL ERGONOMICS

Traditional prevention programs rely almost exclusively on training staff in body mechanics, resident handling, and personal fitness exercises. These programs can be only partially effective for two reasons. First, consistent use of proper body mechanics and proper resident-handling procedures are not feasible for all resident transfers owing to the sometimes unpredictable behavior of residents and constraints such as too little work area around toilets and beds. Second, it is unrealistic to expect all nursing assistants working in nursing homes to exercise regularly.

The ergonomic approach is being used in other industries. Its underlying principle is that work should be designed to fit the capabilities and limitations of the people who perform it, rather than trying to change all the workers to fit the task requirements.

The ergonomic approach to reducing back injuries in an employment setting has been described as a four-step process [3].

1. First is the identification of jobs with the greatest likelihood of back injuries. This is usually based on past rates of reported back injuries or workers' compensation claims for back injuries. Consideration is also given to total frequency of back injuries among workers performing the various jobs.
2. For the jobs with the greatest back injury likelihood, the second step is the identification of the specific tasks that impose the largest stresses on the workers' backs.
3. Then, for the most back-stressing tasks, the third and most creative steps is to look for ways to change the task demands. This may involve elimination of the stressful task, substitution of lifting equipment for the back muscles of the worker, or control of the exposure level.
4. The fourth step is implementation. Most successful ergonomic interventions have found that the implementation of change works much better if employees actively participated in the earlier steps.

III. CASE STUDY

A. Approach

A project of the National Institute for Occupational Safety and Health (NIOSH) included a contact with the University of Wisconsin to conduct an intervention trial of the ergonomic approach at a county nursing home [4]. The basic steps in the trial were selecting which units would be included and getting support from both the nursing home administration and the nursing assistants, identifying the most back-stressing tasks based on data from the nursing assistants, determining safer ways to perform these tasks, implementing the changes, and assessing the effects. These steps are described more completely below.

For purposes of conducting an objective trial, it was desirable to have two similar units. One would implement the trial program first, and if that program showed no negative side effects, a similar program would be implemented in the second unit. Support of the nursing home administration was obtained by appealing to their interest in reducing the incidence and cost of back injuries among nursing assistants. Getting full support from the nursing assistants was more difficult. Of 57 nursing assistants working the units, 38 volunteered to participate in the initial phase. Apparently, those who had not volunteered initially were gradually convinced that the program was worthwhile; all 57 volunteered to participate in the subsequent phases. Demographic information on the nursing assistants is provided in a descriptive paper from the project [5].

The second step, identifying the most back-stressing tasks, involved the following procedures.

1. Asking nursing assistants to list the tasks they considered most stressful to their backs. This was done in small groups.
2. Based on their lists of most stressing tasks, preparing a list of the 16 tasks most frequently mentioned.
3. Asking each nursing assistant to put the listed tasks in order of most stressful to least stressful for a typical resident.

4. Computing average rank values to produce a rank-ordered list of the most stressful task. The 10 most stressful tasks identified are listed in Table 1 [3,5].

Three other methods for determining the most stressful tasks were also used [3,5]. One of these methods was to have the nursing assistants fill out a form in which they gave a rating of perceived exertion for each task based on how performing the task affects their low back, upper back, shoulders, and whole body [5]. Two other methods used biomechanical analyses to compute compressive force on the L5/S1 vertebral disk and tension force on the erector spinae muscles [5,6]. All four methods showed good correlation and identified the same seven tasks as the top seven [3].

The third step was determining safer ways to perform the most stressful tasks. Initially the standard procedure being used was determined [5]. Then alternative procedures were tested in a laboratory. Alternatives considered included eliminating the transfer, substituting commercially available resident-handling equipment for the back muscles of the nursing assistants; and reducing the intensity of biomechanical stress on the lower back by the use of slings and belts to assist with manual resident transfers.

Results of the laboratory studies were used to select a better procedure for making the transfers [7,8]. Criteria for making this choice were compressive force on the L5/S1 disk, strength required, perceived stress rating, overall preference of the nursing assistants, transfer time, comfort and security of the resident, and applicability of the procedure to a broad range of resident transfers.

The need to manually lift residents who could not bear their own weight was eliminated for several transfers by using a portable hoist. As part of the project, three leading portable hoists were assessed and compared using criteria such as perceived comfort and security of the resident, ease of use, and versatility [9]. The hoist that best satisfied these criteria was

Table 1 Rank-Ordered List of Tasks Considered Most Stressful to the Low Back of Nursing Assistants in One Nursing Home

1.	Transfer from toilet to wheelchair
2.	Transfer from wheelchair to toilet
3.	Transfer from wheelchair to bed
4.	Transfer from bed to wheelchair
5.	Transfer from bathtub to chair[a]
6.	Transfer from bathtub chairlift to chair[b]
7.	Weighing patient
8.	Lifting patient up within bed
9.	Repositioning patient within bed[c]
10.	Repositioning patient within chair

[a]One bathtub had no chairlift, so patients had to be manually lifted for transfers between their wheelchair and the bathtub.
[b]Six bathtubs were equipped with a water-pressure-driven chairlift, so patient transfers were between a wheelchair and the chairlift seat.
[c]A typical within-bed repositioning task is to move the patient in a sideways direction.
Source: Refs. 3-5.

one made by Ambulift [9]. It came with multiple slings to match the needs for particular transfers. Its usefulness included transfers between beds and wheelchairs, wheelchairs and bathtub, wheelchairs and bathtub chairlift, and wheelchairs and toileting/shower chairs. A scale included on one hoist was used for weighing residents, thus eliminating the need to transfer them onto a scale. Some residents were positioned on the toilet by using the hoist with a sling suitable for toilet usage.

For residents who could bear their own weight, procedures were implemented that substantially reduced the level of biomechanical stresses imposed on the lower backs of the nursing assistants who performed the transfers. Most notable of these were the transfers of residents between their wheelchairs and their bed, a toilet, or another chair. Experiments were conducted with several different pulling procedures that used slings or belts. Results led to selection of a technique that makes use of a walking belt with handles on each side and behind [4,9]. It is 12.5 cm (5 in.) wide and comes in different lengths. The standard model has buckles that took too much time to fasten, so the manufacturer (Posey) made a model with the type of fasteners used for personal flotation devices or life vests that required much less time. The procedure was for the nursing assistants to pull the resident from a seated position by holding handles on the belt fastened around the resident's waist. Each nursing assistant would use one hand to hold the handle on his or her side of the resident. One of the nursing assistants would put the free hand on the armrest of the wheelchair and gently push it back as the resident was being pulled from the wheelchair. Then, with the resident supporting his/her own weight, the nursing assistants would help the resident accomplish a 90° rotation and complete the transfer. The standard procedure was to have two nursing assistants make this transfer.

The fourth step was implementing the changes [4]. Two units of the nursing home were involved. They were on different floors. Resident populations were similar, with most needing frequent help with daily activities. The intervention was implemented in unit 1 and monitored for 5 months before changes were introduced in unit 2.

Because the choice of transfer method depended on the abilities and needs of the resident, a system for categorizing the residents was needed. Three classifications were used: (1) dependent, weight-bearing; (2) dependent, non-weight-bearing; and (3) independent.

A hoist was recommended for those who were dependent, non-weight-bearing (class 2). For the dependent, weight-bearing (class 1) residents, the belt-assisted two-person pulling method was recommended. Colored stickers were put on beds to indicate which procedure was to be used for the particular resident. Training all the nursing assistants took about 1 month for each unit. It was carried out in small groups of four to eight. Practice was included by having the nursing assistants take turns acting as the resident. Assistants had to demonstrate competency before they were allowed to use a new procedure with the residents.

The fifth step was assessing the effects. Several measures were used [4].

Acceptability. The extent to which the nursing assistants actually used new procedures was measured by conducting observations. The acceptability rate for a task was the percentage of observations in which the new procedure was used for a task where it was appropriate.

Perceived exertion. Nursing assistants provided ratings of perceived exertions (RPEs) using the Borg scale [10]. These were obtained for the whole body and for the shoulders, lower back, and upper back.

Biomechanical loading. For transfers that used a two-person lift before the intervention and a two-person pull after the intervention, it was possible to compare biomechanical stresses. Specific measures were force on the hands, compressive force on the L5/S1 disk, and the strength required expressed in terms of percentage of the female workforce capable.

Injuries. Injury records were examined before and after the intervention. Measures used for comparison were back injury rate, total injury rate, and back injury severity rate. Data were available for about 4 years before the intervention. After the intervention, data were available for 8 months on unit 1 and for 4 months on unit 2.

B. Changes Made

Commercially available patient-handling equipment was identified in a laboratory study [9]. In addition, a portable chair suitable for both toileting and showering was selected (Figure 1). Arrangements were made for the equipment to be purchased by or loaned to the nursing home. Each unit was provided with four Ambulift portable hoists (Figure 2) and four toileting/shower chairs [4]. A walking belt was provided for each dependent, weight-bearing resident. The belts for some of the female residents were too loose at the top, so a seamstress made modifications to provide a better fit for those individuals.

One set of transfers were those between wheelchair and toilet. The usual procedure had been to push the resident into the bathroom on a wheelchair. Two nursing assistants would then lift the resident from the wheelchair and place him/her on the toilet. A second manual transfer took place after toileting when the resident was lifted from the toilet and returned to the wheelchair. Three alternatives were available for the intervention, with the choice being made to best suit the resident's needs. On shower days residents could be transferred from their bed or wheelchair into a toileting/shower chair. The chairs sat on four casters and had a hole in the seat somewhat smaller than that of a toilet seat. The chair could be pushed into the bathroom and positioned over the toilet, thereby eliminating two transfers inside the bathroom. A number of non-weight-bearing residents who were being transferred with a hoist

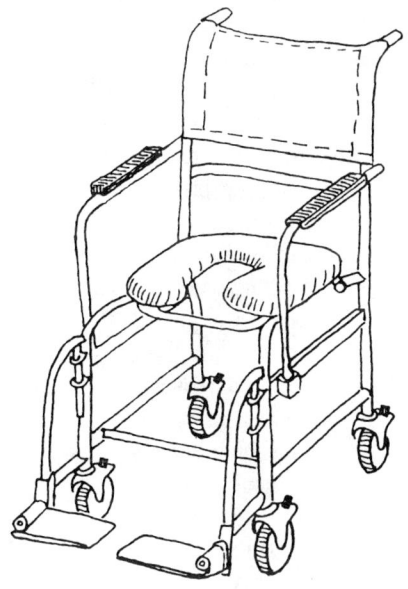

Figure 1 Dual-purpose chair used in the intervention trial for toileting and showering residents.

Figure 2 Portable patient hoist used in the intervention trial for transferring residents.

could be taken into the bathroom while still on the hoist. They were then positioned over the toilet for toileting. This required a sling with an opening and special adaptive clothing. This procedure eliminated the two manual transfers inside the bathroom. The third option for toileting was to use a wheelchair to transport the resident into the bathroom and use a belt-assisted two-person pull method to transfer the resident onto the toilet.

Another pair of eliminated transfers were transfers between a wheelchair and a scale used for weighing residents. On one unit a chair had been bolted on the scale and non-weight-bearing residents were lifted out of their wheelchair and carried to the chair on the scale. The procedure was reversed after weighing. These two transfers were eliminated by replacing the old scale with a ramp scale. The nursing assistants could push the resident onto the scale without lifting him or her out of the wheelchair. Of course, this required subtracting the weight of the wheelchair (which was written on tape stuck to the chair) from the measured weight. Additionally, one of the eight portable hoists had a scale that could be used to measure the weight of residents who were lifted with the hoist.

Additional manual transfers were eliminated in the bathing process. The traditional procedure had been to lift a resident from the wheelchair and place her or him on a seat beside the bathtub. This seat was part of a chairlift apparatus that would raise the resident over the side of the tub and permit lowering into the tub and subsequent raising out of the tub. After bathing, the resident was lifted from this chair and carried back to the wheelchair. The intervention included three options, depending on the resident's needs and preferences. The first option was to take residents from toileting directly to the shower area using the toileting/shower chair. In the shower they were washed using a flexible shower head while remaining in their portable chair. This procedure eliminated the manual transfers required with the bathtub. A second option was to use a portable hoist to transfer from a wheelchair directly to the bathtub or chairlift. The third option was to make the transfer between wheelchair and chairlift by using the belt-assisted two-person pulling method.

C. Effects of Intervention

The extent to which the nursing assistants actually used new procedures was measured by conducting observations. The acceptability rate for a task was the percentage of observations in which the new procedure was used for a task for which it was appropriate. The number of observations differed for each task depending on the frequency of the particular type of transfer. For example, bed-to-wheelchair transfers were observed 140 times, toilet-to-wheelchair 117 times, and wheelchair-to-shower chair 10 times [4]. For the various categories of resident transfers, acceptability rate ranged from 87% to 100% [4].

Nursing assistant perceptions were measured by their ratings of perceived exertion (RPE) using the Borg scale [10]. The scale ranges from 6 to 20. A value of 13 indicates a perception of "somewhat hard" exertion, and a value of 9 indicates "very light" exertion. During the preintervention phase, the lower back RPE values for transfers ranged from 13.3 to 14.3. After the intervention, lower back RPE values ranged from 8.6 to 9.5. Statistically, the decrease was significant for each transfer category at the 0.01 confidence level [4].

For transfers that used a two-person lift before the intervention and the belt-assisted two-person pull after the intervention, it was possible to compare biomechanical stresses. Force on the hands decreased from a mean of 312 N to 122 N after the intervention [4]. Compressive force on the L5/S1 disk decreased from a mean of 4751 N before to 1964 N after the intervention [4]. And the strength required, expressed in terms of percentage of the female workforce capable, improved from 41% to 83% [4]. All three of these were statistically significant at the 0.01 level [4].

Injury records were examined before and after the intervention. Details of comparisons within each unit are reported in the primary research paper [4]. Combined data are presented here. From pre- to postintervention the total injury rate changed from 114 to 84 per 200,000 employee hours [4]. Back injury rate changed from 83 to 47 per 200,000 hr [4]. Severity was expressed as days lost or restricted per 200,000 employee hours. Back injury severity rate changed from 634 to 317 [4]. After the intervention had been implemented on both units, there was a 4-month period of data. During this time the severity rate was zero for back injuries as well as total injuries [4].

IV. CRITICAL REVIEW AND FUTURE DIRECTIONS

The ergonomic approach to reducing back injuries in an employment setting has been described as a four-step process [3]. First is the identification of jobs with the greatest back injury likelihood. The second step is the identification of the specific tasks in those jobs that impose the largest stresses on the workers' backs. Then, for those tasks, the third step is to look for ways to change the task demands. The fourth step is implementation.

The ergonomic intervention described in this chapter illustrates the approach applied to a nursing home. It is important to recognize that the tasks changed in this facility may not be the most appropriate for another nursing home. Consequently, going through the process of asking nursing assistants to identify their most back-stressing tasks is recommended.

If hospital management would like to reduce their back injury rates, the four-step ergonomic approach ought to work, although a well-documented trial has not been reported in the literature. The first step for hospitals, selecting priority jobs, could result in very different choices. Some might decide to focus on dietary staff, others could choose linen services, and others might select nursing. Some might prefer to start with a ward or unit and assess the tasks of all workers on the unit, e.g., nurses, nursing assistants, and unit assistants. There is not one correct answer for the outcome of step 1. Each facility should take into account its own unique needs and culture.

Regardless of whether the facility is a hospital or nursing home, the four-step process would provide a logical approach. Getting affected workers involved in the early stages is strongly recommended because of their knowledge of the work, their potential to generate practical ideas for changing the most stressful tasks, and the value of having their support as changes are being introduced.

This pioneering trial of an ergonomic intervention in a nursing home was a major step in the process of understanding how to reduce musculoskeletal injuries among nursing assistants in nursing homes. However, because the project involved only one nursing home, there is a need to conduct other similar projects in other nursing homes to develop a better understanding of the extent to which different nursing homes will benefit from an ergonomic intervention.

There is also a need for improvements in patient-handling devices and equipment. This project found large differences in the design of hoists and belts, some being much easier to use than others and some being better in terms of patient comfort and security. It does no good for a nursing home to purchase equipment and devices that will not find acceptance among the nursing assistants. Nursing home administrators need access to information on the pros and cons of different commercial products so they can purchase equipment and devices that will be used.

ACKNOWLEDGMENT

The author wishes to express appreciation for the excellent work of Bernice Owen, Ph.D., and Arun Garg, Ph.D.

REFERENCES

1. M. E. Personick, Nursing home aides experience increase in serious injuries, *Mon. Labor Rev.* *113*(2):30 (1990).
2. R. C. Jensen, The increasing occupational injury rate in nursing homes, in *Advances in Industrial Ergonomics and Safety II*, B. Das, Ed., Taylor & Francis, New York, 1990, p. 569.
3. B. D. Owen, A. Garg, and R. C. Jensen, Four methods for identification of most back-stressing tasks performed by nursing assistants in nursing homes, *Int. J. Ind. Ergon.* 9:213 (1992).
4. A. Garg and B. Owen, Reducing back stress to nursing personnel: an ergonomic intervention in a nursing home, *Ergonomics* *35*:1353 (1992).
5. A. Garg, B. Owen, and B. Carlson, An ergonomic evaluation of nursing assistants' job in a nursing home, *Ergonomics* *35*:979 (1992).
6. A. Garg and D. Chaffin, A biomechanical computerized simulation of human strength, *AIIE Trans.* 7:1 (1975).
7. A. Garg, B. Owen, D. Beller, and J. Banaag, A biomechanical and ergonomic evaluation of patient transferring tasks: wheelchair to shower chair and shower chair to wheelchair, *Ergonomics* *34*:407 (1991).
8. A. Garg, B. Owen, D. Beller, and J. Banaag, A biomechanical and ergonomic evaluation of patient transferring tasks: bed to wheelchair and wheelchair to bed, *Ergonomics* *34*:289 (1991).
9. B. D. Owen and A. Garg, Assistive devices for use with patient handling tasks, in *Advances in Industrial Ergonomics and Safety II*, B. Das. Ed., Taylor & Francis, London, 1990, p. 585.
10. G. A. V. Borg, Perceived exertion: a note on "history" and methods, *Med. Sci. Sports* 5:90 (1973).

27
Injuries and Ergonomic Applications in Construction

Hongwei Hsiao and Ronald L. Stanevich
National Institute for Occupational Safety and Health, Morgantown, West Virginia

I. INTRODUCTION

Construction work is very hazardous. Workers frequently perform their tasks at elevations, work with heavy construction machinery, face exposure to various types of hazardous energy such as electrical energy, or manually handle a wide variety of bulky, heavy materials. In addition, at construction sites, the work environment, the work to be done, and the composition of crews are subject to continuous change. These are some of the factors that make construction a high-risk industry.

During the period 1986–1990, the U.S. construction industry had an average annual employment of 5.2 million workers [1–3]. The occupational injury and illness rate per 100 full-time construction workers per year was estimated to be 15.2 cases in 1986 with a steady decline to 14.2 cases in 1990. The corresponding lost workday incidence rate per 100 workers during this period was 6.9 in 1986 and 6.7 in 1990. Although slightly in decline during 1986–1990, the construction industry incidence rates were still the highest among the major industries (Table 1) [1–3]. In addition, during the period 1980–1990, the U.S. construction industry had an overall fatality rate of 25.6 per 100,000 full-time workers. This rate was more than 3.5 times the occupational fatality rate for all other industries in the United States for the same period [4].

In 1987, 21 states reported 180,876 compensable cases of construction injury and illness into the Bureau of Labor Statistics' Supplementary Data System (SDS) (Table 2) [5]. This represented a compensable incidence rate of 9.5 per 100 construction workers in these 21 SDS states. Of these injuries, 25.5% involved overexertion, 20.1% struck by, 15% falls from elevations, 7.8% struck against, 6.6% bodily reaction, and 6.6% falls from the same level. Many of these injuries could have been prevented through ergonomic design of the tools, work procedures, and work environments.

II. BACKGROUND AND SIGNIFICANCE OF THE CHAPTER TO OCCUPATIONAL ERGONOMICS

In 1990, the U.S. Public Health Service (PHS) developed and published *Healthy People 2000: National Health Promotion and Disease Prevention Objectives*. The construction industry was identified as a "special population target" [6]. Ergonomics-related issues such as cumulative trauma, traumatic injuries, and low back disorders were some of the target topics listed as applicable to the construction industry.

Table 1 Incidence Rate[a] per 100 Full-Time Workers in Major Industries

Industry	1986	1987	1988	1989	1990
Construction	15.2	14.7	14.6	14.3	14.2
Manufacturing	10.6	11.9	13.0	13.1	13.2
Agriculture	11.2	11.2	10.9	10.9	11.6
Transportation	8.2	8.4	8.9	9.2	9.6
Mining	7.4	8.5	8.8	8.5	8.3
Trade[b]	7.7	7.7	7.8	8.0	7.9
Service	5.3	5.5	5.4	5.5	6.0
Finance[c]	2.0	2.0	2.0	2.0	2.4

[a]Includes fatalities, injuries, and illnesses.
[b]Wholesale and retail trade.
[c]Finance, insurance, and real estate.
Source: Refs. 1–3.

The focus of most ergonomics research has been on highly repetitive manufacturing jobs, traditionally related to the cause of cumulative trauma disorders. Construction workers face a host of other ergonomic risk factors for both traumatic and musculoskeletal injuries. Their jobs are marked by long work cycles, changing work environments, and varied tasks. The

Table 2 Participating States in the SDS Program in 1987[a]

State	Injury frequency	Percent
Alaska	1,033	0.6
Arizona	4,418	2.4
California	83,860	46.4
Colorado	5,996	3.3
Hawaii	2,414	1.3
Iowa	3,723	2.1
Indiana	8,610	4.8
Kentucky	6,752	3.7
Maryland	12,226	6.8
Maine	6,651	3.7
Missouri	9,303	5.1
Mississippi	2,701	1.5
Nebraska	1,672	0.9
New Mexico	889	0.5
Oklahoma	3,112	1.7
Oregon	4,804	2.7
Tennessee	4,888	2.7
Virginia	5,320	2.9
Washington	6,569	3.6
Wisconsin	5,285	2.9
Wyoming	650	0.4

[a]Twenty-five states/territories participated in the SDS program in 1987. The data from Louisiana, Michigan, and Ohio were excluded from the analysis because their waiting period (7 days) for compensable claims is different from those of other states (0–3 days). Data from the Virgin Islands were also excluded due to lack of occupation-specific population data.
Source: Ref. 5.

demands of construction activities include repetitive and forceful exertions, awkward postures, prolonged static activities, and the stress of being struck by falling objects or falling from an elevated work surface. Despite the importance that has been placed on worksite construction safety and the high injury rate suffered by the workers in this industry, research into construction ergonomics in the United States has been relatively limited.

This chapter focuses on construction work–related nonfatal injuries. It addresses some of the injury sources, their ergonomic solutions, and the research needs for ergonomic applications in the construction industry. First we identify the high-risk U.S. construction occupations with their injury characteristics. Then we present a summary of general ergonomic principles for construction work and their application to two selected construction occupations. Finally, we recommend five areas for ergonomic research emphasis in the future.

III. OCCUPATIONAL INJURY CHARACTERISTICS OF THE CONSTRUCTION INDUSTRY

This section identifies the occupational groups in the U.S. construction industry that experience high rates of injury. Priority-target occupations and the prevalent injury characteristics of these occupations are discussed.

A. Data Resources

Injury data obtained through the U.S. Department of Labor, Bureau of Labor Statistics' (BLS) Supplementary Data System (SDS), and the occupational employment data obtained from the U.S. Department of Labor and the U.S. Department of Commerce construction census reports were used for this investigation [5,7,8].

The SDS program is a federal-state cooperative system that provides information on occupational injuries and illnesses reported to the participating states' workers' compensation programs. The SDS does contain a small percentage of data on occupational illness and fatalities. Because there were so few occupational illnesses and fatalities in these data, results and discussion of the SDS analysis in this chapter focus on work-related nonfatal injuries. The data used for this assessment are based on reports provided by the 21 participating states (Table 2). Both the 1987 and 1988 SDS data tapes were searched to identify the cases that occurred in 1987 [5]. Data on the state-specific employment levels for the major occupational groups in 1987 were obtained through the U.S. Departments of Labor [7] and Commerce [8] industry census reports. These data provided a means for estimating occupation-specific injury rates in the 21 participating states.

B. Analysis Methods

The following procedure was used in developing occupation-specific injury frequencies and injury rates and determining research priorities for the occupational groups.

1. Determination of Occupation-Specific Injury Frequencies

The occupation-specific injury frequencies of workers' compensation cases were established from the SDS database. The SDS database consists of 17 key variables and 13 optional variables (Table 3). Key variables include state code, reference year, occupation, and industry. Optional variables include age of employee, duration of employment, and weekly wages. A frequency analysis was performed using the Statistical Analysis System (SAS) program to

Table 3 Key Variables and Optional Variables in the SDS Program

Key variables	Optional variables
State code	Time of accident
Reference year	Time workday began
Year of occurrence	Hour of shift
Month of occurrence	Associated object or substance
Day of occurrence	Age
Year of receipt	Duration of employment
Occupation	Wage code
Industry	Weekly wage
Ownership	Kind of insurance
Nature of injury or illness	Indemnity compensation code
Part of body affected	Medical payments code
Source of injury or illness	Medical payments
Type of accident or exposure	Case type
Sex	
Extent of disability	
Indemnity compensation	
Sample weight	

Source: Ref. 109.

generate the frequencies of injuries by occupations for all sectors of the construction industry, for incidents occurring in 1987.

2. Determination of Occupation-Specific Employment

Specific occupational employment figures for the 1987 SDS states were derived through the equation

$$E_O = E_{NO} \times (E_C/E_{NC})$$

where

E_O = estimated employment by occupation in the 21 participating SDS states for 1987
E_{NO} = national employment by occupation for 1987 according to Ref. 7
E_C = construction industry employment in the 21 SDS states for 1987 (This was obtained by summing the 21 SDS states' construction employment, as listed in Ref. 8.)
E_{NC} = national construction industry employment for 1987 (This was obtained from Ref. 8.)

3. Determination of Injury Rates

Occupation-specific injury rates for 1987 were determined by dividing the number of reported workers' compensation cases by the corresponding established occupational employment and multiplying by 100 to obtain the number of cases per 100 workers. The overall construction industry injury rate (9.5) for 1987 was determined by dividing the total number of construction injuries (180,876 cases) reported in the 21 SDS states in 1987 by the construction employment (1,897,040 workers) in the same 21 states.

4. Determination of Priority Occupations

A frequency-weighted occupation-specific rating technique was developed to prioritize occupations as targets for injury prevention. The rankings were obtained through the equation

$$FW\_RK = \text{Rank of } [ORDER \times (FREQ_{SO} / FREQ_{AO})]$$

where
FW_RK = frequency-weighted rank of priority occupations (listed from high to low priority)
ORDER = sequence rank of injury rates
(The lowest injury rate was assigned ORDER = 1.)
$FREQ_{SO}$ = injury frequency of a specific occupation
$FREQ_{AO}$ = injury frequency of all occupations in construction

C. Results

The priority ranking was applied to 30 occupations. Each of these 30 occupations was involved in more than 0.3% of the total number of compensation claims (180,876) filed in the 21 SDS states for the construction industry in 1987. The occupations with the highest frequency of injury were construction laborer (25.8%), carpenter (14.8%), electrician (5.9%), and plumber (5.6%). Occupations with the highest injury rate (number of cases per 100 workers) were miscellaneous materials-moving equipment operators (47.6), construction laborer (43.3), miscellaneous construction trades workers (32.2), and carpet installer (32.0) (Table 4).

Seven occupations were identified as priority research targets on the basis of the frequency-weighted occupation-specific rating technique. They were construction laborer, carpenter, roofer, drywall installer, plumber, electrician, and structural metal worker (Table 4). These seven occupations represented an estimated 30.7% of the U.S. construction industry workforce and 65% of the total compensable injuries reported for construction. The combined injury rate of the seven target occupations (20.2 cases per 100 workers) was more than twice that for all construction workers combined (9.5 cases per 100 workers).

A cross-analysis was performed with injuries typed by injury source, injured body part, and nature of injury for the seven occupations (Table 5). This was done to identify problem activities where there was a high rate of injury. In all seven occupations, the sum of overexertion, struck by, and falls from elevation accounted for more than 50% of the injuries. Roofers and drywall installers are most frequently injured as a result of fall-from-elevation and overexertion incidents. For the remaining five occupational groups, overexertion and struck-by incidents caused the greatest number of injuries. Overexertion and fall-from-elevation incidents most frequently result in sprain/strain and fracture injuries to the back; struck-by incidents typically result in cuts and lacerations to the fingers and legs and foreign objects in the eyes.

D. Discussion

Differences in state workers' compensation coverage and reporting requirements have been cited as limitations of the SDS program [9]. This study used data available from 21 states to determine the priority rankings. In spite of these limitations, the broad database and uniform coding format employed by the SDS, combined with employment data from the Department of Labor and the Department of Commerce on construction, allow a more comprehensive analysis of the occupational characteristics associated with work-related injuries in construction than presently exists in the literature. It may not be perfect, but it is the best available.

The censuses of construction industries were performed by the Bureau of the Census, U.S. Department of Commerce every 5 years (for all years ending in 2 and 7) and were available for 1982 and 1987. Occupational employment data in the construction industry were available for 1981, 1984, and 1987. These were separately published by the Bureau of Labor Statistics, U.S. Department of Labor, in 1984, 1986, and 1989. The SDS database con-

Table 4 Potential High-Risk Construction Occupations and Ranking of Safety Research Priorities (1987 SDS, 21 States)

OCC3	Occupation	FREQ	EMP21	RATE	ORDER	FW_RK
869	Construction laborers	46,838	108,116	43.3	29	1
567	Carpenters	26,857	200,188	13.4	18	2
595	Roofers	7,731	31,987	24.2	25	3
573	Drywall installers	7,327	26,688	27.5	26	4
585	Plumbers, pipefitters	10,148	79,312	12.8	16	5
575	Electricians	10,738	102,984	10.4	11	6
597	Structural metal workers	4,187	21,457	19.5	23	7
599	Misc. construction trade workers	3,707	11,495	32.2	28	8
534	Heating & air cond. mechanics	4,470	36,942	12.1	14	9
579	Painters & paperhangers	5,462	55,496	9.8	9	10
596	Sheetmetal duct installers	2,414	13,770	17.5	20	11
593	Insulation workers	2,672	20,238	13.2	17	12
566	Carpet installers	1,654	5,166	32.0	27	13
584	Plasterers	1,783	7,789	22.9	24	14
804	Truck drivers, heavy	3,654	36,578	10.0	10	15
516	Heavy equipment mechanics	1,529	8,342	18.3	22	16
783	Welders & cutters	1,759	10,102	17.4	19	17
653	Sheet metal workers	2,570	21,480	12.0	13	18
550	Construction supervisors	5,901	79,925	7.4	5	19
565	Tile setters	1,128	6,211	18.2	21	20
589	Glaziers	1,298	10,223	12.7	15	21
844	Operating engineers	2,488	33,013	7.5	7	22
859	Misc materials moving equip. operators	573	1,203	47.6	30	23
563	Brickmasons	2,849	38,202	7.5	6	24
588	Concrete finishers	2,411	39,402	6.2	4	25
643	Boilermakers	486	4,424	11.0	12	26
544	Millwrights	611	6,654	9.2	8	27
833	Excavating machine operators	814	18,683	4.4	3	28
019	Managers, nec & manage. occ.	2,013	141,550	1.4	1	29
855	Grader, dozer, scraper operators	745	22,252	3.3	2	30
All occupations		180,876	1,897,040	9.5		

OCC3 = three-digit occupation code
FREQ = frequency (reported SDS cases in the construction industry, 1987)
EMP21 = construction employment in the selected 21 states (1987 SDS)
RATE = FREQ/EMP21*100
ORDER = sequence rank of RATE (from least to greatest)
FW_RK = frequency-weighted rank of research priorities, i.e., the rank of [ORDER*(FREQ of a specific occupation /FREQ of all occupations)]

tains annual workers' compensation data for 1981–1989. The states participating in this program varied from year to year. Based on the available data sets, 1987 was selected as the best data year for this study.

These analyses provide an approach for determining priority occupations for injury prevention research and analysis of the occupational characteristics associated with work-related injuries in the construction industry.

Table 5 Major Injury Characteristics of Eight Selected Construction Occupations

Occupation	Type	Source	Parts	Nature
Contruction laborers	Overexertion [26.2%]	Tools (6.0%) Containers (3.5%) Metal (3.4%) Wood (2.7%)	Back (15.8%)	Sprain/strain (22.4%)
	Struck by [24.8%]	Tools (6.3%) Metal items (6.3%) Wood items (2.8%)	Finger (4.7%) Leg (3.2%) Foot (2.5%) Toe (2.3%)	Cut laceration (8.0%) Fracture (6.5%) Contusion (4.9%)
	Fall from elevation [12.0%]	Working surface (10.8%)	Back (1.6%) Leg (1.6%) Multiple (1.9%)	Fracture (3.7%) Sprain/strain (3.3%)
Carpenter	Struck by [27.5%]	Tools (0.7%) Metal items (6.5%) Wood items (5.6%)	Finger (8.4%) Hand (3.3%) Leg (3.3%) Eye (2.2%)	Cut laceration (14.3%)
	Overexertion [22.4%]	Structure (3.5%) Tool (3.4%) Metal (1.8%) Wood (6.5%)	Back (13.3%)	Sprain/strain (18.8%)
	Fall from elevation [18.6%]	Working surface (16.7%)	Back (2.4%) Leg (1.8%) Ankle (1.8%) Multiple (3.2%)	Fracture (7.1%) Sprain/strain (4.0%)

(continued)

Table 5 Continued

Occupation	Type	Source	Parts	Nature
Roofer	Fall from elevation [24.7%]	Working surface (22.4%)	Back (3%) Leg (2.4%) Ankle (2.4%) Foot (2.3%) Multiple (5.0%)	Fracture (11.1%) Sprain/strain (5.5%)
	Overexertion [23.6%]	Lifting containers (8.6%) Tools (3.0%)	Back (14.8%)	Sprain/strain (20.5%)
	Struck by [13.6%]	Tools (4.3%) Metal item (3.8%)	Finger (2.9%) Leg (2.6%)	Cut laceration (7.6%)
	Contact with extreme temperature [10.4%]	Coal/oil product (9.3%)	Hand (3.2%) Arm (1.9%) Upper ext. (1.1%) Multiple (2.2%)	Burn (heat) (10.1%)
Dry wall installer	Overexertion [28%]	Miscellaneous (14.4%) Containers (4.2%)	Back (15.2%)	Sprain/strain (24.3%)
	Fall from elevation [24.7%]	Working surface (22.7%)	Multiple (3.4%) Back (2.7%) Leg (2.8%) Ankle (2.7%) Wrist (2.7%)	Fracture (8.3%) Sprain/strain (6.8%)
	Struck by [15.1%]	Tools (4.8%) Metal items (4.2%)	Finger (5.4%) Hand (2.4%)	Cut laceration (8.3%)

Occupation	Event [%]	Source (%)	Part (%)	Nature (%)
Plumber	Overexertion [29.2%]	Metal items (8.8%) Tools (6.6%)	Back (17.0%) Shoulder (1.7%)	Sprain/strain (24.4%)
	Struck by [18.6%]	Metal items (7.3%) Tools (4.7%)	Finger (3.8%) Eye (2.3%)	Cut laceration (5.9%)
	Fall from elevation [14.6%]	Working surface (4.7%)	Back (1.7%) Multiple (1.5%)	Fracture (3.0%) Sprain/strain (3.9%)
Electrician	Overexertion [25.7%]	Electric apparatus (4.4%) Metal item (4.3%) Containers (4.3%)	Back (14.7%) Shoulder (1.6%)	Sprain/strain (20.9%)
	Fall from elevation [14.6%]	Working surface (4.7%)	Multiple (2.5%) Back (1.9%) Leg (1.9%) Ankle (1.4%)	Fracture (3.0%) Sprain/strain (3.9%)
Structural metal workers	Overexertion [28.4%]	Metal (17.8%)	Back (18.4%) Shoulder (1.8%)	Sprain/strain (24.0%)
	Struck by [17.7%]	Metal (11.3%)	Finger (2.1%) Foot (2.0%) Eye (1.8%)	Cut laceration (4.7%) Fracture (4.5%) Contusion (3.5%)
	Fall from elevation [16.6%]	Working surface (15.3%)	Back (1.9%) Leg (1.8%) Multiple (4.1%)	Fracture (3.7%) Sprain/strain (2.9%)

[ ] indicates percentage of injury frequency of a specific injury type to overall injury frequency in a specific occupation.
( ) indicates percentage of injury frequency of a specific category (source, parts, and nature) within a specific injury type to overall injury frequency in a specific occupation.

IV. ERGONOMIC PRINCIPLES AND APPLICATIONS FOR WORKER SAFETY

The analysis of the BLS workers' compensation data presented in Section III shows that construction workers are vulnerable to overexertion, fall, and struck-by injuries. Manual handling of construction materials, which are usually heavy and bulky, presents one of the problems. Inappropriate design or operation of hand tools, which are used in almost all construction operations, can result in injury. Heavy equipment, which is used in many construction operations, also presents several problems. Whole body vibration, seat design, operator field of vision, and access systems are some of the potential ergonomic issues. Additionally, many construction workers perform tasks above shoulder level or below knee height and are thereby exposed to situations hazardous to the musculoskeletal system.

Although ergonomics has been increasingly employed in occupational settings and in safety and health disciplines, little ergonomic research has been done in the United States on its application to construction work. Previous studies have concentrated on manufacturing and office workers. Nevertheless, some of the risk factors that have been studied in manufacturing settings and other industries are characteristic of many construction tasks. These factors include (1) forceful exertion, (2) repetitive motions, (3) awkward or static posture, (4) direct external stress over the skin and muscle tissue, (5) vibration, (6) extreme temperature, (7) psychosocial stresses, (8) noise, (9) task difficulty, and (10) illumination. Some of these factors may be more closely associated with overexertion problems in manufacturing settings than with the construction industry. However, in construction, they may result in overexertion, fall, and struck-by incidents or combinations thereof.

A. General Ergonomic Principles

In general, a three-tiered injury control program, using ergonomic principles as outlined in Ref. 10 can be applied to the construction industry. These approaches are

1. Engineering controls—ergonomic design changes to tools, handles, equipment, work methods, or other aspects of the workplace
2. Administrative controls—changes in work practices or organizational and management policies
3. Use of personal protective equipment (PPE)

However, this approach may be more difficult at a construction site because the work environment, the work to be done, and the composition of crews are subject to continuous change. The ergonomic approaches to reducing construction injuries need to be more creative in this setting. Table 6 summarizes the general ergonomic principles in construction.

B. Ergonomic Applications for Selected Construction Occupations

Seven occupations have been identified as having priority for ergonomic research. The most hazardous activities for two of the seven occupations (roofer and drywall installer) and potential ergonomic countermeasures are discussed in this section. These two occupations are presented as examples of potential ergonomic applications in the construction industry.

1. Roofers

Roofers are primarily concerned with covering roofs and exterior walls of structures with slate, asphalt, aluminum, rubber, wood, and related materials. Most roofers work for special trades

Table 6 General Ergonomic Principles for Construction Safety

Suggested action	Benefit	References
1. Engineering controls		
a. Use mechanical aids, torque control devices, or lifting aids to transport construction materials and equipment	Reduce frequency and duration of heavy lifting	11,14,25
b. Automate certain construction tasks	Reduce the chance of overexertion due to repetitive and overreach work	15
c. Optimize the weight of construction tools and containers	Reduce biomechanical stress	11–13,20,21
d. Improve mechanical advantages of tool designs or work procedures (such as hoisting and pulleys) for handling heavy loads	Reduce required reaction force	25
e. Improve visibility during heavy construction equipment operation by modifying vehicle design	Improve work postures; reduce injury incidents due to poor visibility	17–19
f. Improve heavy construction vehicle access systems by cab step redesign or mounting and dismounting procedures design	Reduce fall injuries	22,23,25
g. Optimize size and shape of construction tool handles and strength requirement to operate the tool	Reduce biomechanical stresses due to awkward postures and forceful grip exertions	24,27,28
h. Design hand-held tools and containers for balance movement on the spine during lifting and carrying of construction materials	Reduce rotation on the forearm or lateral bending on the trunk	29,30
i. Select construction tools/equipment with minimum vibration; use isolation/damping for equipment that operates above the point of resonance; adjust tool speed to avoid resonance	Reduce vibration exposure	32,37
j. Select methods to minimize surface and edge finishing	Reduce repetitive motions and static loadings	15
k. Use full hand grip, not just fingertips, when lifting heavy objects	Avoid external mechanical stress	25,38,39
l. Alter position of tools or worker to avoid awkward working postures	Reduce pronounced strains on back, shoulders, and arms	16
m. Provide good couplings between workers and tools/environment, such as texturing handles to prevent hands slipping and using appropriate slip-resistant materials for shoe soles	Reduce overexertion, slipping, and being struck by objects	39,40

(*continued*)

Table 6 Continued

Suggested action	Benefit	References
2. Administrative Controls		
a. Use biomechanical principles such as reducing the distance between the worker and the load, reducing the distance the load must be moved, keeping all movement close to the body, and keeping movements between the knuckles and the shoulder	Reduce biomechanical stress	11,21
b. Remove all obstacles between the worker and the load by practicing good housekeeping procedures	Reduce slipping and tripping hazards and awkward working posture exposures	41,42
c. Keep construction tools in good condition and in a safe place	Reduced hand forces and the chance of tool slipping and striking the worker.	43
d. Seek help when handling large and/or bulky loads	Balance materials being handled and lessen the load distribution during lifting and carrying work	25
e. Provide frequent rest breaks to offset undue fatigue in jobs requiring heavy labor or high performance/production rates	Offset undue fatigue	44
f. Limit overtime work and periodically rotate workers to less stressful jobs	Job stress is spread over a longer time and over different body parts, resulting in less strain	15
g. Vary work tasks or broaden job responsibilities	Offset boredom and sustain worker motivation	15
h. Train workers in techniques to avoid awkward postures and repetitive motions	Reduce construction injuries that result from lack of knowledge and experience	45
3. Use of personal protective equipment		
a. Use safety shoes and protective eye wear	Reduce injuries due to being struck by objects	47,112
b. Use fall arrest devices	Reduce injuries due to falls	47,112
c. Use vibration-attenuating gloves	Reduce vibration exposures	47,112

contractors (Standard Industrial Classification Code SIC 176) at various building construction sites. Roofers represent 1.7% of the construction workforce. Their incidence rate (24.2 cases per 100 workers) was more than 2.5 times that of the average construction worker. The leading exposures for roofers were falls from elevation (24.7%), overexertion (23.6%), being struck by an object (13.6%), and contact with extreme temperature (10.4%).

Figure 1 presents samples of some roofing activities. Studies of roofing-related injuries have shown that the most hazardous activities of roofers are associated with (1) transporting materials or equipment while ascending and descending ladders [49], (2) using unsecured/inadequate ladders and scaffolds [50–52], (3) working at the edge of the roof or slipping/losing balance while on the roofing surface [49], (4) working on roofs that contain holes or thin sheeting [53, 54], and (5) manually handling heavy and/or bulky material or equipment [55–57]. These bulky materials include bundles of shingles (weighing approximately 80 lb), 5-gallon buckets of hot tar, reels/rolls, and containers [58, 59].

Ergonomic approaches to reducing injuries during roofing work have been proposed by some researchers and institutes. One general approach is to use mechanical lifting devices to transport materials and equipment between the ground and the roof. This can reduce the possibility of overexertion and the chance of losing a grip on manually handled materials, which could result in dropped materials. The Occupational Safety and Health Administration (OSHA) and many other safety agencies recommend that workers practice ladder safety to prevent falls and overexertion. The OSHA recommendations include (1) securing the ladder properly at the roof top to prevent horizontal movement, (2) resting the ladder on a solid base, (3) extending the ladder 3 ft above the point where a worker steps onto the roof, (4) placing the base of the ladder a distance of one-fourth of its working length from its vertical line of support, (5) employing the concept of three-point contact with the ladder when ascending and descending ladders, and (6) not carrying objects while ascending and descending ladders [60].

Other ergonomic and safety countermeasures for fall prevention in roofing work include, but are not limited to, (1) using guard rails [61, 62, 110], (2) using fall protection anchoring devices and protective scaffolds [63, 64, 111], (3) covering and securing openings in roofs and roof decks with plywood or other suitable materials [60], and (4) employing a safety line with a shock absorber [65]. Medical examination to detect defects of worker equilibrium may also have a positive impact in preventing roofer fall injuries [66].

Furthermore, roofers should practice good housekeeping [41, 49] and practice the philosophy of "good couplings" to reduce struck-by, slipping, and tripping hazards. This involves such practices as removing all obstacles between the worker and the load, using steel toe safety shoes [40], selecting shoes with appropriate slip-resistant characteristics [40, 68, 69], and providing handles on the items being lifted. The most frequent roofer injury scenarios involve a worker being struck by objects such as rolls, barrels, buckets, cans, asphalt, shingles, or equipment that fell, slipped, or was dropped by another worker; alternatively, the worker may have lost his grip on the material he was handling and injured himself [49].

Tool selection is another important ergonomic concern during roofing work. The principles are to select the correct/best tools for the job, use them correctly, maintain them in good condition, and keep them in a safe place [43]. Tools such as knife blades or chisels should be sharpened regularly. Dull tools require greater hand forces, which can trigger the development of hand, wrist, and upper extremity cumulative trauma disorders and can also result in the tool slipping and striking the worker.

Use of appropriate tools can reduce the biomechanical stress created by some job tasks. Roofers who fasten single-ply roofing systems in place frequently do so in a bent-over posture. A new Dutch device for fastening single-ply roofing has been designed to allow for less bent-over posture during that task, thereby reducing stresses on the musculoskeletal system [70].

Training is also an important countermeasure. Workers should be trained in techniques to avoid awkward postures and repetitive motions and in appropriate procedures for new work methods or equipment [62]. Prior to being assigned work on a roof, new employees should be given safety and health training to control/reduce hazardous exposures [55].

Roofing workers appear to have a high risk of occupation-related burn injuries [71]. In California during 1979, burns and scalds in roofing work occurred 21 times more often than in the construction industry as a whole [57]. Transferring hot asphalt in a can or bucket and filling the melt kettle are hazardous work activities for roofers. Ergonomic design can reduce

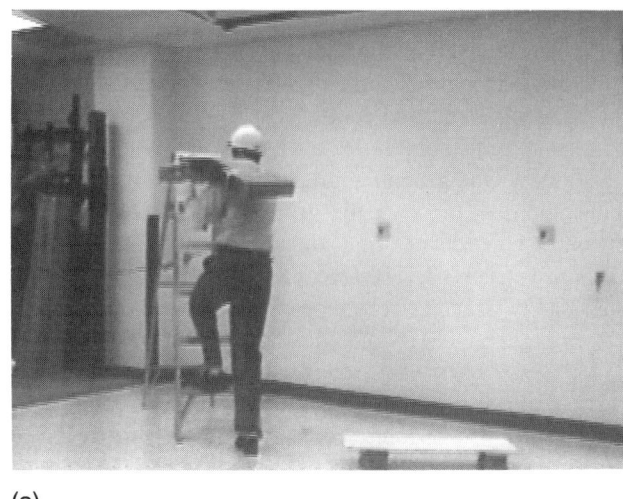

(a)

(b)

Figure 1 Some ergonomics-related at-risk activities of roofing work. (a) Transporting heavy materials; (b) working at the roof edge; (c) prolonged kneeing work; (d) unsafe overreach work.

the hazard of burn injuries. Half-lids installed on cans and buckets can reduce splashing during carrying and pouring tasks. Better designs for placing solid asphalt into the melt kettles can further reduce the incidence of burns and scalds [72]. In addition, improved materials for personal protective equipment must be identified to effectively protect roofers using asphalt. Parsons and Pizatella [59] found that roofers who wore gloves were absent from work longer for recovery from asphalt burn injuries to the hands than workers whose hands were burned but were not protected by gloves. It was surmised that the protective gloves were of the cotton variety and actually caused the workers' hands to be exposed to the hot asphalt for a longer time, resulting in more severe burns.

Ergonomic Applications in Construction 559

(c)

(d)

Figure 1 Continued

2. Drywall Installers

Drywall installers are primarily concerned with installing, taping, and surfacing plasterboard and drywall in the interior of a structure. Typical drywall comes in sheets measuring 4 ft wide by 8, 10, or 12 ft long and 1/4, 1/2, 3/4, or 5/8 in. thick. The sheets weight 55–84 lb each. In 1987, the U.S. construction industry employed 70,510 drywall installers—1.41% of the construction workforce. Their compensable injury rate (27.5 cases per 100 workers) was almost three times the average rate of 9.5 for all construction occupations combined. The injury data revealed that overexertion (28%), fall from elevation (24.7%), and being struck by an object (15.1%) were the leading types of injury. Most drywall installers work for special trades contractors (SIC 174).

The hazardous exposures for drywall installers include manually transporting drywall; applying glue to studs; hanging drywall on ceilings; applying tape to joints and corners; applying skim coats of joint compound to joints and corners; and sanding wall joints, attachment indentions, and corners (see Figure 2). All of these activities involve overhead work, heavy lifting, repetitive motion, forceful cutting, or work at elevated heights. These activities may put considerable biomechanical stress on the worker or expose drywall workers to tripping/falling and struck-by hazards. During taping, compound application, and finishing activities, stilts are frequently used. The stilts are 2-ft-high structures that can be strapped to the workers' legs for quickly moving around while completing the upper part of drywall work (Figure 2). Although no studies have yet been reported, the use of stilts may place workers at increased risk for knee injuries [73] and falls.

(a)

(b)

Figure 2 Some ergonomics-related at-risk activities of drywall installation. (a) Hanging drywall on ceiling; (b) applying tape to joints and corners; (c) finishing wall joints, nail indentions, and corners; (d) use of stilts while taping compound.

Some countermeasures have been developed and used to reduce the potential hazards associated with drywall work. The biomechanical stress associated with transporting the drywall sheets to the wall can be reduced by the use of carts and dollies [74]. Plasterboard lifts can be used to raise the board to the desired height and to hold it in place for fastening the drywall [75]. The biomechanical stress also can be reduced by reducing the width of drywall sheets. Studies have shown that the reduction of drywall board width from just under 4 ft to slightly less than 3 ft did reduce stress on workers' backs and resulted in slower heart rates among workers [76]. This reduction in width makes it considerably easier to handle the sheets of drywall because of the reduction in both weight and size. The increased material consumption, mainly studs and screws, can be compensated for by lower working costs and more rapid fitting of the 3-ft-wide sheets [76].

Screw guns have been developed with a swiveling handle to reduce wrist flexion in fastening drywall to metal studs in overhead and under-knee positions [44]. Using "micro-

(c)

(d)

Figure 2 Continued

pauses" or short breaks after installation of a few screws has also been suggested to reduce stress and increase productivity [44]. Proper training and the use of carrier handles also reduced biomechanical stress among the drywall installers [77]. Finally, sanding high and low areas requires awkward body-part angles for applying proper sanding force. Use of an electric sander may reduce the force required but requires a static hold to support the sander [73].

V. CRITICAL REVIEW OF CURRENT STATUS OF CONSTRUCTION ERGONOMIC RESEARCH AND FUTURE CONCERNS

Ergonomic research in the United States specific to the seven selected construction occupations is limited. Review of construction ergonomics literature related to the seven occupations indicated that most studies presented retrospective analyses of injury frequencies collected from historical records [26, 31, 41, 46, 48, 49, 57, 59, 67, 79, 89, 94, 104–108] and subjective questionnaire/observation surveys of safety-related matters [26, 31, 46, 53, 55, 70, 74, 88, 90, 92, 93, 95, 108]. Occupation-specific injury rates and evaluation of worker tasks associated with resulting injuries were very limited [91]. This limitation might have resulted from the fact that technologies for construction hazard exposure assessment were themselves limited in number. Recently, a few ergonomic-assessment methods such as postural measurement [85], postural balance evaluation [83], biomechanical measurement [76, 85], physiological measurement [76, 80], electromyography [82, 84, 86], sensorimotor behavior test [45], movement performance [82, 84], vibration measurement [81, 87], and audiometric evaluation [83] were applied in some construction ergonomics studies. In addition, syntheses of overexertion problems, metabolic cost, and muscle fatigue during construction work were presented by some researchers [73, 96–103]. Injury prevention approaches such as training [45, 78], tool redesign [85–87], and work design [74, 76] were proposed in some of the studies. A number of these activities were from Bygghalsan of Sweden [97] and the Construction Safety Association of Ontario [25]. It appears that more injury prevention intervention and ergonomic evaluation studies of construction occupations are needed. To effectively reduce U.S. construction worker injuries, the five following areas are recommended for future research.

1. Occupation-specific tasks associated with injuries must be identified and evaluated. Past studies have emphasized retrospective analyses of injury frequency data for the overall construction industry. This chapter has identified injury rates of construction occupations. Active surveillance programs must be initiated to investigate occupation-specific tasks and activities for the priority construction occupations so that effective interventions can be developed.
2. There is a need for more emphasis on in-depth study of injury prevention strategies for high-risk tasks in the seven occupations. Current studies on existing work techniques and injury prevention strategies of these occupations are limited. More in-depth laboratory studies as well as field evaluation to determine and overcome barriers to the use of existing injury prevention technologies are desired.
3. The methods for measuring construction worker exposure to biomechanical and other stressors need to be improved because construction tasks differ from tasks performed by other industrial workers. The environment, the work to be done, and the composition of crews change continually. Studies on measurement methodologies such as telemetry and other technologies would be very useful for researchers and safety professionals in the construction industry.
4. Emphasis should be placed on evaluating and redesigning tools, equipment, assistive devices, and work procedures. Change of workplace, assignment, and crew composi-

tion is the nature of construction work. However, tools, equipment, and work procedures remain fairly consistent. Therefore, redesign of equipment or assistive devices will be more effective and should be emphasized for reducing ergonomics-related injuries in construction.
5. Studies of fall prevention should not just look at passive protective devices; other creative solutions need to be investigated also, continued development of new technologies to evaluate and design better fall prevention solutions is needed.

VI. CONCLUSION

This chapter addressed injury sources, ergonomic solutions, and research needs in the construction industry.

The study has identified construction laborers, carpenters, roofers, drywall installers, plumbers, electricians, and structural metal workers as being at an elevated risk of nonfatal injuries. They are likely to be injured from overexertion, fall from an elevation, or being struck by an object. Some ergonomic solutions have been proposed to reduce these injuries in these selected occupations. Clearly, there is a need to focus additional safety research and ergonomic programs on these occupations and their cause of injuries.

Continued research on the following subjects is recommended to further reduce construction injuries:

1. Identification of occupation-specific risk tasks and activities so that effective interventions can be developed
2. Systematic evaluation of existing injury prevention technologies to overcome the barriers to the use of these technologies
3. Improvement of biomechanical exposure assessment technologies to help identify risk factors
4. Improvement of construction equipment and assistive devices to reduce work-related stress
5. Implementation of new technologies to evaluate and develop fall prevention solutions that better protect construction workers

REFERENCES

1. U.S. Department of Labor, *Occupational Injuries and Illnesses in the United States by Industry, 1987*, Bureau of Labor Statistics Bull. 2328, U.S. Dept. of Labor, Washington, DC, 1989.
2. U.S. Department of Labor, *Occupational Injuries and Illnesses in the United States by Industry, 1989*, Bureau of Labor Statistics Bull. 2379, U.S. Dept. of Labor, Washington, DC, 1991.
3. U.S. Department of Labor, *Occupational Injuries and Illnesses in the United States by Industry, 1990*, Bureau of Labor Statistics Bull. 2399, U.S. Dept. of Labor, Washington, DC, 1992.
4. S. M. Kisner and D. E. Fosbroke, Injury hazards in the construction industry, *J. Occup. Med.* 36(2):137–143 (1994).
5. U.S. Department of Labor, *Supplementary Data System*, Bureau of Labor Statistics, U.S. Dept. of Labor, Washington, DC, 1987–1988.
6. U.S. Department of Health and Human Services, *Healthy People 2000: National Health Promotion and Disease Prevention Objectives*, DHHS Pub. No. (PHS) 91-50212, U.S. Govt. Printing Office, Washington, DC, 1990.
7. U.S. Department of Labor, *Occupational Employment in Mining, Construction, Finance, and Services,* Bureau of Labor Statistics Bull. 2330, U.S. Dept. of Labor, Washington, DC, 1989.
8. U.S. Department of Commerce, *1987 Census of Construction Industries*, Geographic Area Series

CC87-A-1 through CC87-A-9 and U.S. Summary CC87-A-10, Bureau of the Census, U.S. Dept. of Commerce, Washington, DC, 1990.
9. N. Root and D. McCaffrey, Providing more information on work injury and illness, *Mon. Labor Rev.* 101:16-21 (1978).
10. U.S. Department of Labor, *Ergonomics Program Management Guidelines for Meatpacking Plants*, OSHA 3123, U.S. Dept. of Labor, Washington, DC, 1990.
11. NIOSH, *Work Practices Guide for Manual Lifting*, Tech. Rep. DHHS Pub. No. 81-122, U.S. Govt. Printing Office, Washington, DC, 1981.
12. University of Michigan, *2D Static Strength Prediction Program*, Center for Ergonomics, Univ. Michigan, Ann Arbor, MI, 1990.
13. M. M. Ayoub, N. J. Bethea, S. Deivanayagum, S. S. Asfour, G. M. Bakken, D. Liles, A. Mital, and M. Sherif, *Determination and Modeling of Lifting Capacity*, Final Report, HEW, Grant No. 5R01-OH-00545-02, 1978.
14. T. H. Hettinger, Occupational hazards associated with diseases of the skeletal system, *Ergonomics* 28(1):69-75 (1985).
15. S. A. Knoz and A. Mital, Carpal tunnel syndrome, *Int. J. Ind. Ergon.* 5:175-180 (1990).
16. T. J. Armstrong, *Cumulative Trauma Disorders and Forceful Exertions*, occupational ergonomics summer course materials, Univ. Michigan, Ann Arbor, MI, 1990.
17. F. Hella, M. Tisserand, J. F. Schouller, and M. Englert, A new method for checking the driving visibility on hydraulic excavators, *Int. J. Ind. Ergon.* 8(2):135-145 (1991).
18. J. Rouquie, A wide-angle rear-view periscope for dump trucks, *Protect. Civ. Secur. Ind.* 184:33-35 (1970) (French; English abstr.).
19. W. K. Miller, *Analysis of Haulage Truck Visibility Hazards at Metal and Nonmetal Surface Mines*, Mining Enforcement and Safety Administration, U.S. Dept. of the Interior, Inf. Rep. 1038, Washington, DC, 1976.
20. S. H. Snook, The design of manual handling tasks, *Ergonomics* 21(12):963-985 (1978).
21. T. R. Waters, V. P. Anderson, A. Garg, and L. J. Fine, Revised NIOSH equation for the design and evaluation of manual lifting tasks, *Ergonomics* 36(7):749-776 (1993).
22. R. L. Stanevich, A study of earthmoving and highway construction machinery fatalities and injuries, *Ann. Am. Conf. Gov. Ind. Hyg.* 14:703-710 (1986).
23. Construction Safety Association of Ontario, *Mounting and Dismounting Procedures for Construction Equipment*, Toronto, ON, Canada, 1989.
24. D. B. Chaffin and G. B. J. Andersson, *Occupational Biomechanics*, Wiley, New York, 1984.
25. Construction Safety Association of Ontario, *Stand, Lift, Carry*, rev. ed., Construction Safety Association of Ontario, Toronto, ON, 1993.
26. Bureau of Labor Statistics, U.S. Dept. of Labor, *Survey of Power Saw Accidents*, U.S. Govt. Printing Office, Washington, DC, 1980.
27. T. J. Armstrong, *An Ergonomics Guide to Carpal Tunnel Syndrome*, AIHA Ergonomic Guide Series, American Industrial Hygiene Association, Akron, OH, 1983.
28. D. B. Chaffin, Localized muscle fatigue—definition and measurement, *J. Occup. Med.* 15(4):346-354 (1973).
29. A. Mital and M. Ilango, Load characteristics and manual carrying capabilities, *Proc. 27th Annual Meeting Hum. Factors Soc.*, 1983, pp. 274-278.
30. A. Mital and H. F. Fard, Psychophysical and physiological responses to lifting symmetrical and asymmetrical loads symmetrically and asymmetrically, *Ergonomics* 29:1263-1272 (1986).
31. G. R. McCormack, *Work Injuries and Work-Injury Rates in the Highway and Street Construction Industry*, BLS Rep. No. 257, Bureau of Labor Statistics, U.S. Dept. of Labor, Washington, DC, 1961.
32. M. J. Griffin, *Handbook of Human Vibration*, Academic, London, 1990.
33. T. Miwa, Vibration-isolation systems for hand-held vibrating tools, in *Vibration Effects on the Hand and Arm in Industry*, A. J. and W. Taylor, Eds., Wiley, New York, 1981, pp. 303-310.
34. G. A. Hampel and W. J. Hanson, Hand vibration isolation: a study of various materials, *Appl. Occup. Environ. Hyg.* 5(12):859-869 (1990).

35. E. R. Andersson, Design and testing of a vibration attenuating handle, *Int. J. Ind. Ergon.* 6(2):119 (1990).
36. A. M. Steward and D. F. Goda, Vibration syndrome, *Br. J. Ind. Med.* 27:19–27 (1970).
37. D. E. Wasserman, The control aspects of occupational hand-arm vibration, *Appl. Ind. Hyg.* 4(8):22–26 (1989).
38. P. Tadano, A safety/prevention program for VDT operators: one company's approach, *J. Hand Ther.* 3(2):64–71 (1990).
39. C. G. Drury, Handles for manual material handling, *Appl. Ergon.* 11(1):35–42 (1980).
40. Knollma, Safety shoes for roofers, *BAU-Berufsgen. Wuppertal Mitteilungsbl.* 3:147–149 (1988).
41. H. H. Cohen and D. M. J. Compton, Fall accident patterns—characterization of most frequent work surface-related injuries, *Prof. Safety* 27(6):16–22 (1982).
42. Construction Safety Association of Ontario, *Construct. Safety* 4(3), Toronto, ON, Canada, 1993.
43. National Safety Council, *Accident Prevention Manual for Industrial Operations, Engineering, and Technology*, 9th ed., National Safety Council, Chicago, IL, 1988.
44. P. Andersson, Manual screw tightening with and without micro pauses, Bygghalsan Bull. 91-09-16, The Construction Industry's Organization for Working Environment, Occupational Safety and Health, Danderyd, Sweden, 1991, p. 39.
45. K. Prather, R. A. Crisera, and S. Fidell, *Behavior Analysis of Workers and Job Hazards in the High Risk Construction Occupation of Roofing*, NIOSH Contract 099-72-0121, Theodore Barry and Associates, Los Angeles, CA, 1975.
46. E. B. Holmstrom, J. Lindell, and U. Moritz, Low back and neck/shoulder pain in construction workers: occupational workload and psychosocial risk factors. Part 1: Relationship to low back pain, *Spine* 17(6):663–671 (1992).
47. J. B. Moran and R. M. Ronk, Personal protective equipment, in *Handbook of Human Factors*, G. Salvendy, Ed., Wiley, New York, 1987, pp. 876–894.
48. B. Salengro and F. Commandre, Musculoskeletal disorders at work in building constructions: epicondylitis and low back pains, in *Musculoskeletal Disorders at Work*, P. Buckle, Ed., Taylor & Francis, London, 1987, pp. 254–258.
49. T. J. Parsons, T. J. Pizatella, and J. W. Collins, Safety analysis of high risk injury categories within the roofing industry, *Prof. Safety* 31(6):13–17 (1986).
50. State of California, *Work Injuries in Roofing and Sheet Metal Work, 1970*, Div. Labor and Statistics, San Francisco, CA, 1972.
51. Health and Safety Executive, HSE warns builders over standards for timer ladders, *Health and Safety Executive News Release* E64, 1990, pp. 1–2.
52. J. Poirel, On roofs, *Cah. Com. Prevent. Batiment Trav. Public* 17(3):130 (1962).
53. R. A. Crisera, J. P. Martin, and K. L. Prather, *Supervisory Effects on Worker Safety in the Roofing Industry*, Div. Safety Research, NIOSH, Morgantown, WV, Contract No. 099-74-0035, 1977.
54. Sigeru, Repair of flat roofs, *Hochbau* 78(4):110–112 (1990).
55. NBS, *Roofing Accident Data Base*, Rep. No. NBS-GCR76-62, Natl Bur. Standards, Washington, DC, 1975.
56. T. Barry, K. Prather, R. A. Crisera, and S. Fidell, *Behavioral Analysis of Workers and Job Hazards in the Roofing Industry*, NIOSH, Res. Rep., NIOSH Contract No. 099-72-0121, 1975.
57. K. O'Gara, *California Roofing and Sheet Metal Work: Analysis of Work Injuries and Illnesses*, Res. Bull. 6, Div. Labor Stat. Res., San Francisco, CA, 1982.
58. U. Gustafsson and L. Wimnell, *An Ergonomic Study of Rooflayers*, Bygghalsan Bull. 1984-09-01, The Construction Industry's Organization for Working Environment, Occupational Safety and Health, Sweden, 1984, pp. 18–19.
59. T. J. Parsons and T. J. Pizatella, *Safety Analysis of High Risk Activities Within the Roofing Industry*, Tech. Rep., NTIS PB-85163236, 1985.
60. OSHA, *Construction Industry*, OSHA Safety and Health Standards, 29 CFR 1910.29 and 1926.500, Occupational Safety and Health Administration, Washington, DC, 1991.
61. *Health and Safety at Work*, Sloping roofs, *Health Safety Work* 1(11):32–33 (1979).

62. T. Niskanen and E. Seppanen, *Measures for Preventing Falls in the Construction Industry* (Engl. Abstr., Tyoterveyslaitos, Julkaisumyynti, Topeliuksenkatu 41 a A, 00250, Helsinki, Finland, 1987.
63. H. Moser, *Safety Aspects of Roofing and Work on Roofs*, Caisse Nationale Suisse d'Assurance en Cas d'Accidents, Sheet 22024, Luzern, Switzerland, 1982.
64. Federation of Industrial Mutual Accident Insurance Associations (FIMAIA), *Safety Rules for Work on Roofs and Corrugated Roofing,* Carl Heymanns Verlag KG (in German, Engl. Abstr.), 1980.
65. P. O. Axelsson and L. Nilsson, *Hanging in There* (in Swedish). Rep. TRITA-AOG-0041. Royal Inst. Technology, Stockholm, Sweden, 1987. Cited by M. G. Helander, Safety hazards and motivation for safety work in the construction industry, *Int. J. Ind. Ergon.* 8:205–233 (1991).
66. D. Andreoni, in *Encyclopedia of Occupational Health and Safety,* 3rd ed., L. Parmeggiani, Ed., Vol. 2, Int. Labour Office Publications, Geneva, Switzerland, 1983, pp. 1960–1962.
67. E. Broberg, *Ergonomic Injuries at Work*, National Board of Occupational Safety and Health, ISA Information System on Occupational Injuries, Rep. No. 1984:3E, Sweden, 1984.
68. S. M. Szymusiak and J. P. Ryan, Prevention of slip and fall injuries, Part I, *Prof. Safety* 27(6):11–16 (1982).
69. S. M. Szymusiak and J. P. Ryan, Prevention of slip and fall injuries, Part II, *Prof. Safety* 27(7):30–35 (1982).
70. P. Vink, Application problems of a biomechanical model in improving roofwork, *Appl. Ergon.* 23(3):177–180 (1992).
71. W. Inancsi and T. L. Guidotti, Occupation-related burns: five-year experience of an urban burn center, *J. Occup. Med.* 29(9):730–733 (1987).
72. NIOSH, *Health and Safety Guide for the Commercial Roofing Industry*, DHEW Pub. No. 78-194, U.S. Govt. Printing Office, Washington, DC, 1978.
73. S. Schneider and P. Susie, *Ergonomics and Construction: A Review of Potential Hazards in New Building Construction*, The Center to Protect Workers' Rights, Washington, DC, 1993.
74. M. Bjorklund, P. Helmerskog, M. Nordberg/Bohlin, U. Soderman, L. Holmqvist, B. Lindblad, J. Makynen, and S. Ahrman, Bygghalsan Bull. 91-09-16, The Construction Industry's Organization for Working Environment, Occupational Safety and Health, Danderyd, Sweden, 1991, pp. 35–36.
75. E. Rosenlund, B. Maenpaa, H. Nilsson, and I. Persson, *Ergonomic Equipment in Construction Work,* Bygghalsan Bull. 1987-05-01, The Construction Industry's Organization for Working Environment, Occupational Safety and Health, Danderyd, Sweden 1987, pp. 38–39.
76. H. Isakson and J. Kling, *Fitting 90 mm and 1200 mm Plasterboard Sheets—An Ergonomic/Economic Comparison,* Bygghalsan Bull. 91-09-16, The Construction Industry's Organization for Working Environment, Occupational Safety and Health, Danderyd, Sweden 1991, pp. 37–38.
77. M. Wahlin, 1981, *Work Methods for Handling Plasterboard and Chipboard in Building Training,* Bygghalsan Bull. 1981-07-01, The Construction Industry's Organization for Working Environment, Occupational Safety and Health, Sweden, 1981, pp. 20–21.
78. B. Zimolong, Hazard perception and risk estimation in accident causation, in *Trends in Ergonomics/Human Factors II* R. E. Eberts and C. G. Eberts, Eds., North-Holland/Elsevier, Amsterdam, Netherlands, 1985, pp. 463–470.
79. J. R. Myers and R. B. Trent, Hand tool injuries at work: a surveillance perspective, *J. Safety Res.* 19(4):165–176 (1988).
80. I. Astrand, Degree of strain during building work as related to individual aerobic work capacity, *Ergonomics* 10(3):293–303, (1967).
81. E. Hammarskjold, K. Harms-Ringdahl, J. Ekholm, and B. Samuelson, Effect of short-time vibration exposure on work movements with carpenters' hand tools, *Int. J. Ind. Ergon.* 8(2):125–134 (1991).
82. E. Hammarskjold, K. Harms-Ringdahl, and J. Ekholm, Reproducibility of carpenters' work after cold exposure, *Int. J. Ind. Ergon.* 9(3):195–204 (1992).
83. K. H. Kilburn, R. H. Warshaw, and B. Hanscom, Are hearing loss and balance dysfunction linked in construction iron workers?, *Br. J. Ind. Med.* 49(2):138–141 (1992).

84. E. Hammarskjold and K. Harms-Ringdahl, Effects of arm-shoulder fatigue on carpenters at work, *Eur. J. Appl. Physiol. Occup. Physiol. 64*(5):402–409 (1992).
85. R. G. Knowlton and J. C. Gilbert, Ulnar deviation and short-term strength reduction as affected by a curve-handled ripping hammer and a conventional claw hammer, *Ergonomics 26*(2):173–179 (1983).
86. R. Ortengren, T. Cederqvist, M. Lindgberg, and B. Magnusson, Workload in lower arm and shoulder when using manual and powered screwdrivers at different working heights, *Int. J. Ind. Ergon. 8*(3):225–235 (1991).
87. H. Wos, J. Lindberg, R. Jakus, and S. Norlander, Evaluation of impact loading in overhead work using a bolt pistol support, *Ergonomics 35*(9):1069–1079 (1992).
88. P. O. Axelsson, *Development of a Safety Device for Roof Work,* Tekniska Hogskolan, Arbetsolycksfalls-Gruppen, Stockholm, Sweden, 1983.
89. S. Tanaka, A. B. Smith, W. Halperin, and R. Jensen, Carpet-layer's knee, *N. Engl. J. Med. 307*(20):1276–1277 (1982).
90. P. J. Leather, Attitudes towards safety performance on construction work: An investigation of public and private sector differences, *Work Stress 2*(2):155–167 (1988).
91. T. J. Parsons, *Establishment of a Construction Safety Research Plan,* Internal Rep., Div. Safety Research, NIOSH, Morgantown, WV, 1984.
92. H. Bergenudd and B. O. Nilsson, Back pain in middle age; occupational workload and psychological factors, *Spine 13*(1):58–60 (1988).
93. H. Riihimaki, S. Tola, T. Videman, and K. Hanninen, Low-back pain and occupation. A cross-sectional questionnaire study of men in machine operating, dynamic physical work, and sedentary work, *Spine 14*(2):204–209 (1989).
94. D. A. Stubbs and A. S. Nicholson, Manual handling and back injuries in the construction industry: an investigation, *J. Occup. Accidents 2*:179–190 (1979).
95. R. J. Cleveland, Factors that influence safety shoe usage, *Prof. Safety 29*(8):26–29 (1984).
96. University of Iowa, *Work-Related Musculoskeletal Disorders: A Construction Bibliography*, The Center to Protect Workers' Rights, Washington, DC, 1993.
97. Bygghalsan, Bygghalsan Bull. 91-09-16, Bygghalsan, The Construction Industry's Organization for Working Environment, Occupational Safety and Health, Danderyd, Sweden, 1991.
98. T. D. Proctor and F. J. Rowland, Development of standards for industrial safety helmets—the state of the art, *J. Occup. Accidents 8*:181–191 (1986).
99. Health and Safety Commission, *Safety Helmets on Construction Sites: Recommendations of the Construction Industry Advisory Committee on Their Provision and Use,* Health and Safety Commission, London, 1979.
100. M. G. Helander, Safety hazards and motivation for safety work in the construction industry, *Int. J. Ind. Ergon. 8*:205–233 (1991).
101. E. A. P. Koningsveld, Permissible loads for the Dutch construction industry, *Ergonomics 28*(1):359–363 (1985).
102. A. D. F. Price, Calculating relaxation allowances for construction operatives. Part 1: Metabolic cost, *Appl. Ergon. 21*(4):311–317 (1990).
103. A. D. F. Price, Calculating relaxation allowances for construction operatives. Part 2: Local muscle fatigue, *Appl. Ergon. 21*(4):318–324 (1990).
104. S. Kumar, Injury profile of the construction industry in Alberta in Canadian context: a case study of the impact of a non-monetary motivational factor in a multicenter Albertan company, *Int. J. Ind. Ergon. 8*(3):197–204 (1991).
105. T. Niskanen and J. Lauttalammi, Accidents in materials handling at building construction sites, *J. Occup. Accidents 11*(1):1–17 (1989).
106. T. Niskanen and O. Saarsalmi, Accident analysis in the construction of buildings, *J. Occup. Accidents 5*(2):89–98 (1983).
107. P. Barnard, *Injuries in the Erection Industry,* Construction Safety Assoc. Ontario, Appl. Res. Dept., Toronto, ON, Canada, Res. Publ. No. 14, 1970.
108. Bygghalsan, *Protect Your Knees,* Bygghalsan Broschyr 10(1), Stockholm, Sweden, 1983.

109. U. S. Department of Labor, *Supplementary Data System Microdata Files User's Guide, 1985 Edition*, Bureau of Labor Statistics, U.S. Dept. of Labor, 1989.
110. Construction Safety Association of Ontario, *Safety Guidelines for Roofers,* Construction Safety Assoc. Ontario, Toronto, 1993.
111. Construction Safety Association of Ontario, *Safety Tips—Sloped Roofs,* Construction Safety Assoc. Ontario, Toronto, TP004, 1993.
112. Construction Safety Association of Ontario, *Fundamentals of Occupational Health in Construction,* Construction Safety Assoc. Ontario, Toronto, 1993.

28

An Ergonomic Analysis and Abatement Recommendations to Reduce Musculoskeletal Stress in Warehousing Operations

A Case Study

Donald S. Bloswick
University of Utah, Salt Lake City, Utah

Emil Golias
Occupational Safety and Health Administration, Salt Lake City, Utah

I. INTRODUCTION

Warehousing operations require considerable manual handling of loads. One of us was requested to assist OSHA with hazard analysis and abatement recommendations relating to grocery warehouse operations in the eastern United States. Epidemiological data indicated a high musculoskeletal injury rate, which suggested the existence of manual materials handling hazards. A variety of abatement recommendations were made, ranging from those that required a relatively low cost change in work procedure, to those that would require costly facility modification. These abatement recommendations relate to general warehousing operations as well as grocery storage. This chapter describes the actual abatement efforts made in two typical grocery warehouses and discusses their impact on musculoskeletal injury rates.

II. BACKGROUND AND SIGNIFICANCE TO OCCUPATIONAL ERGONOMICS

The development of low back and other musculoskeletal problems is often associated with the magnitude and duration of spinal compressive loads, frequent repetition of manual materials handling activities, awkward postures, and torso rotation during the lift or load movement.

Since the purpose of a warehouse is to store material, the principal product of a warehouse is space. This often results in the requirement that loads be stacked high and be tightly packed together. These operational conditions combined with the fast work pace resulting from the piecework incentive pay programs result in an occupational setting in which the primary risk factors for back and musculoskeletal injuries exist: high loads, awkward postures, and high repetition.

III. CASE STUDY OF A TYPICAL GROCERY WAREHOUSING OPERATION

A. Epidemiological Information

The back injury history for one grocery warehouse is shown in Table 1. The tabulated data may include injuries to some "nonselectors." (A selector is a worker who selects requested items from the warehouse storage space.) If so, the effective rate for the selectors may actually be higher than noted in Table 1.

B. Task Description—Groceries and Perishables

Selectors are required to move cases and objects of varying sizes and weights from the pallet racks to pallet jacks. For one typical warehouse, the average frequency for this task was 1.9 cases/min with an average weight of 23.9 lb (10.8 kg) for groceries and 1.85 cases/min with an average weight of 28.3 lb (12.8 kg) for perishables.

A summary of the Grocery and Perishables workload for one typical grocery warehousing task is shown in Table 2. In the Perishables area, some of the boxes of meat are very heavy, which increases the biomechanical hazard. A sample of 18 boxes of pork had an average weight of 77 lb (34.9 kg) with a box width of 34 in. (86.4 cm), and a sample of 10 boxes of beef had an average weight of 64.6 lb (29.3 kg) with a box width of 24 in. (61 cm). The workers were also required to wear additional clothing, which reduces the range of motion and increases the strain on the body. In some cases the boxes were frozen together, which required considerable force to break the bond.

C. Task Analysis

1. General Discussion of Acceptable Loads and Storage Locations

When a load is lifted from the floor, additional stresses are exerted on the low back due to the body weight moment during stooping to pick the load up. Thus heavy loads should not

Table 1 Musculoskeletal Injury Rate for a Typical Grocery Warehouse Selecting Operation

Year	No. of selector lifting injuries	No. of selectors	Percent selectors injured
Grocery			
1988	21	65	32
1987	26	68	38
1986	10	61	16
1985	26	55	47
1984	35	50	70
Total	118	299	39
Perishables			
1988	24	35	69
1987	28	38	74
1986	12	38	32
1985	17	35	49
1984	36	55	65
Total	117	201	58

Table 2 Grocery and Perishable Workload for a Typical Grocery Warehouse Selecting Operation

Groceries	
Selector workload	863 pcs per person per shift (7.5 hr) = 1.9 cases/min
Average case size	15.9 × 10.9 × 8.7 in. (40.4 × 27.7 × 22.1 cm)
Average case weight	23.9 lb (10.8 kg)
Maximum case weight	Several cases in the range of 40–50 lb (18.1–22.7 kg) were observed
	Company time standards allow "80 lb (36.3 kg) and above."
Perishables (dairy, meat, fish, produce)	
Perishables workload	1027 lifts per person per shift (9.25 hr) = 1.85 cases/min
Average case size	16.7 × 11.9 × 7.8 in. (42.4 × 30.2 × 19.8 cm)
	Boxes of meat up to 34 in. (86.4 cm) wide were observed
Average case weight	28.3 lb (12.8 kg)
Maximum case weight	77 lb (34.9 kg), sample of 18 items on 2 pallets of pork
	64.6 lb (29.3 kg), sample of 10 boxes of beef

be stored on the floor but should be raised to about standing knuckle height [minimum 20 in. (50.8 cm)] to avoid the need to stoop and lift.

Mital [1,2] notes that the average maximum weights for male industrial workers for 8- and 12-hr shifts lifting medium-size boxes at the rate of 1 lift/min are floor to knuckle 8-hr = 33.3 lb (15.1 kg), floor to knuckle 12-hr = 26.5 lb (12 kg), knuckle to shoulder 8-hr = 35.9 lb (16.3 kg), knuckle to shoulder 12-hr = 28.3 lb (12.8 kg). Mital [3] also indicates that lifting capacity may decrease by up to 24% as the vertical lift height increases from 30 in. (76.1 cm) to 65 in. (165.1 cm).

The *Ergonomics Guide to Manual Lifting* [4] indicates that a 50-lb (22.7-kg) weight held 20 in. (50.8 cm) away from the body is acceptable for 75% of the male industrial population. Chaffin and Park [5] suggest that two-handed lifting in the sagittal plane of loads greater than about 35 lb (15.9 kg) is associated with increased incidence of low back pain if the load is close to the body. It should be noted that in many warehousing tasks loads are not lifted in the sagittal plane nor can they generally be held close to the body. Herrin [6] notes that loads in excess of 50 lb (22.7 kg) cause an increase in medical incidents and a significant increase in days lost and days on restricted duty. Snook [7] notes that the average maximum weights for lifting medium-size boxes at the rate of 1 lift/min to accommodate 75% of the male industrial population is approximately 49.2 lb (22.3 kg) for floor to knuckle height and 44.1 lb (20 kg) for knuckle to shoulder height. Snook also lists a maximum carry weight of 62.8 lb (28.5 kg) for an approximate 8-ft (2.4-m) carry. The limit for the lift *and* carry would, of course, be less than for the lift alone.

The International Labor Office [8] suggests a maximum of 55.1 lb (25 kg) for males 20–35 yr old, 46.3 lb (21 kg) for males 35–50, and 35.3 lb (16 kg) for males over 50. Koningsveld [9] suggests that limits for frequent lifts by 21–44-yr-old males during the whole day be 39.7 lb (18 kg) for ideal conditions and 17.6 lb (8 kg) for aggravating conditions, and for males over 45 years, 33.1 lb (15 kg) for ideal conditions and 11 lb (5 kg) for aggravating conditions.

The Ergonomics Training Program by the National-American Wholesale Grocers' Association (NAWGA) [10] provides the following guidelines:

1. Try to eliminate items weighing more than 55 lb (25 kg).
2. Locate cherry picker so that heavier items can be lifted between knuckle and waist height and lighter items can be lifted between knuckle and shoulder height.

3. Limit weight on upper racks to 45 lb (20.4 kg).
4. Arrange picks so light items are at the top of the pallet.
5. Arrange stock so that the heaviest picks are between knuckle and elbow height.

It would appear that research supports a lifting limit of between 30 and 50 lb (13.6 and 22.7 kg) independent of the load location. In light of the fact that the actual lift frequency is frequently higher than the 1 lift/min used in many of the studies noted above and awkward postures are frequently required in handling material in warehouses, an approximate limit of 30 lb (13.6 kg) would be appropriate. It is suggested that consideration be given to the establishment of the guidelines shown in Table 3 for warehouse operations.

Note that the information cited above does not recognize the fatigue caused by highly repetitive lifting or the additional stress resulting from torso twisting and/or constrained working postures. These lifting guidelines do not relate to the net weight of the contents but to the gross weight of the box. In a sample of items weighed at one facility, the gross weight was approximately 40% higher than the net. In addition, these limits generally assume that the workforce is physically fit and do not recognize individual limitations due to age, physical disability, gender, or other personal factors.

2. NIOSH *Work Practices Guide*

Since this analysis was performed in 1989, an analysis based on the 1981 version of the NIOSH *Work Practices Guide for Manual Lifting* (WPG) [11] was performed for the warehousing operation. This version of the NIOSH WPG considers the horizontal distance that the load is held out from the body, the vertical location of the load at the beginning of the lift/lower, the vertical distance that the load is moved, and the frequency of lift/lower. The result of the NIOSH WPG is an action limit (AL) that indicates the load that can be lifted by most people without significant risk of injury and a maximum permissible limit (MPL) that is three times the action limit and indicates the load that will be harmful for most people. The NIOSH WPG applies to populations who are physically fit and accustomed to physical labor.

To facilitate the general application of the NIOSH WPG, the grocery and perishable selection operations were synthesized into the tasks shown in Table 4. The result of the NIOSH WPG analysis is presented in Table 5. This analysis indicates that the movement of objects from the lowest and highest rack levels results in a situation where the load exceeds the action limit and may be hazardous. The following factors suggest that the actual hazard was higher than that noted above:

1. These lifts were often performed in a constrained posture with considerable twisting of the torso.
2. The analysis was performed with the estimated average load. This means that in 50% of the lifts the load was higher than used in the analysis, and these higher loads were often located on the lower levels where the hazard was highest.

Table 3 General Lifting Recommendations for Warehouse Operations

>50 lb (22.7 kg)	Not allowed
30–50 lb (13.6–22.7 kg)	Implementation of control measures
0–30 lb (0–13.6 kg)	Analysis and implementation of control measures if indicated by NIOSH *Work Practices Guide* or biomechanical analysis

Table 4 Synthetic Jobs for NIOSH *Work Practices Guide* Analysis

Lift from pallet racks to waist/thigh
  Horizontal location out from body = 20 in.
  Vertical location at beginning = 4, 36, 68 in. (10.2, 91.4, 172.7 cm)
  Vertical location at end = 30 in. (76.2 cm) (objects moved to waist)
  Frequency = 1.9 cases/min.
Lift/lower to pallet jack from waist/thigh
  Horizontal location out from body = 12 in.
  Vertical location at beginning = 30 in. (76.2 cm) (objects moved from waist)
  Vertical location at end = 4, 36, 68 in. (10.2, 91.4, 172.7 cm)
  Frequency = 1.9 cases/min

3. The loads were carried to the pallet jack in addition to the lifting and lowering included in the task analysis.
4. The population may not be physically fit and accustomed to physical labor.

A revised version of the NIOSH *Work Practices Guide* was published in 1991 that recognizes the effect of twisting the torso and the coupling between the hands and the load [12]. The application of the revised WPG results in a recommended weight limit (RWL) at both the beginning and end of the lift/lower, which is similar to the action limit in the 1981 NIOSH WPG. The revised guide does not contain a maximum permissible limit but does include a limit of three times the recommended weight limit. This indicates the load that will be harmful for most people.

The revised NIOSH WPG can also be applied to the same tasks noted in Table 4 with the assumption that the torso is rotated approximately 60° at the beginning and end of the lift and that the hands are gripping the bottom of the box. The result of the revised NIOSH WPG analysis is presented in Table 6.

This analysis indicates that the average load frequently exceeds the recommended weight limit and approaches 3×RWL. This indicates that workers would be at risk during much of the manual materials handling activity.

3. Biomechanical Analysis

A biomechanical model analysis was run to establish the relationship between load, lift posture, and back stresses [13]. This was done to determine the biomechanical stress and estab-

Table 5 Results of 1981 NIOSH *Work Practices Guide* Analysis[a]

	AL	MPL	Load
Rack (4 in., 10.2 cm) to waist	14.7 (6.7)	44.1 (20.0)	23.9 (10.8)
Rack (36 in., 91.4 cm) to waist	24.4 (11.1)	73.2 (33.2)	23.9 (10.8)
Rack (68 in., 172.7 cm) to waist	13.3 (6.0)	39.9 (18.1)	23.9 (10.8)
Waist to jack (4 in., 10.2 cm)	37.2 (16.9)	111.6 (50.6)	23.9 (10.8)
Waist to jack (36 in., 91.4 cm)	42.7 (19.4)	128.1 (58.1)	23.9 (10.8)
Waist to jack (68 in., 172.7 cm)	33.3 (15.1)	99.9 (45.3)	23.9 (10.8)
Average	20.9 (9.5)	62.7 (28.4)	23.9 (10.8)

[a]Waist is 30 in. above the floor. Weights are in lb (kg).

Table 6 Results of 1991 NIOSH *Work Practices Guide* Analysis Assuming That the Torso Is Rotated 60° and the Boxes Are Lifted from the Bottom

	RWL		3 × RWL		Load
	Begin	End	Begin	End	
Rack (4 in., 10.2 cm) to waist	9.1 (4.1)	9.6 (4.4)	27.3 (12.3)	28.8 (13.2)	23.9 (10.8)
Rack (36 in., 91.4 cm) to waist	12.8 (5.8)	13.4 (6.1)	38.4 (17.4)	40.2 (18.3)	23.9 (10.8)
Rack (68 in., 172.7 cm) to waist	8.3 (3.8)	11.7 (5.3)	24.9 (11.4)	35.1 (15.9)	23.9 (10.8)
Waist to jack (4 in., 10.2 cm)	19.8 (9.0)	15.1 (6.8)	59.4 (27.0)	45.3 (20.4)	23.9 (10.8)
Waist to jack (36 in., 91.4 cm)	22.3 (10.1)	21.2 (9.6)	66.9 (30.3)	63.6 (28.8)	23.9 (10.8)
Waist to jack (68 in., 172.7 cm)	19.4 (8.8)	13.8 (6.3)	58.2 (26.4)	41.4 (18.9)	23.9 (10.8)
Average	15.3 (6.9)	14.1 (6.4)	45.9 (20.7)	42.3 (14.2)	23.9 (10.8)

[a]Waist is 30 in. above the floor. Weights are in lb (kg).

lish acceptable storage locations for loads of different weights and involved the determination of the back compressive force for a 50th percentile male with a height of 70 in. (1.78 m) and a weight of 166 lb (75.3 kg). The results are noted in Table 7.

The back compressive forces noted in bold in Table 7 are above the NIOSH action level of 770 lb (350 kg). This level of back compressive force is thought to cause back problems in a significant portion of the workforce [11]. The additional factors mentioned in the earlier NIOSH WPG analysis would also suggest that the actual hazard is higher than that noted above.

D. Abatement Recommendations

It should be noted that not all recommended controls need necessarily be implemented for each employee in the operations evaluated. It is understood that often more than one operator uses these workstations and often operators have alternate jobs that may also contribute to their problems. The recommendations are not all-inclusive, and a more in-depth study may be

Table 7 Back Compressive Force as a Function of Load Weight and Location for an Average Male[a]

Vertical from floor	Load (lbs)	Back compressive force (lb)		
		Horizontal location (inches from center of ankles)		
		12	20	28
68	10	184	226	327
	20	236	307	416
	30	287	383	499
	40	335	456	577
	50	383	524	649
52	10	151	215	300
	20	206	296	399
	30	258	372	491
	40	309	445	576
	50	359	514	655
36	10	139	399	485
	20	194	486	590
	30	248	569	689
	40	300	648	**778**
	50	351	723	**867**
20	10	433	568	642
	20	503	668	**773**
	30	572	766	**902**
	40	640	**862**	**1029**
	50	706	**956**	**1154**
4	10	591	649	679
	20	674	763	**817**
	30	757	**878**	**956**
	40	**840**	**992**	**1094**
	50	**924**	**1106**	**1232**

[a]Numbers in bold represent back compressive forces exceeding the NIOSH *Work Practices Guide* recommendations.

required to determine safe operating parameters for particular employees. Other factors that should be included for any program to be effective include employee training and participation in planning, ergonomic abatement committees, supervisor training, hazard analysis, administrative controls, and medical monitoring.

The abatement recommendations relating to this typical grocery warehousing operation were divided into the following categories:

1. Abatement recommendations that require primarily changes in work methods or storage procedure. These abatement recommendations require relatively little capital expenditure but may affect time standards and/or storage capacity.
2. Abatement recommendations that require some capital expenditure for lift tables, hoists, and facility modification.
3. Abatement recommendations that require significant changes in storage facility layout and design.

The feasibility of the proposed recommendations is noted in light of state-of-the-art warehouse practices and examples.

Abatement Recommendations That Require Primarily Changes in Work Methods or Storage Procedure. (These abatement recommendations require relatively little capital expenditure for materials handling equipment (MHE) but may affect time standards and/or storage capacity.)

1. Store items weighing more than 30 lb (13.6 kg) between knuckle and shoulder height. The optimum storage location would be between 32.5 in. (82.6 cm), which is the approximate knuckle height of a 95th percentile male, and 51.8 in. (131.6 cm), which is the approximate shoulder height of a 5th percentile male. This can be accomplished by raising the height of the pallet rack "floor" and limiting the material on each pallet. Since this is a lower (as opposed to a lift), it may be appropriate to increase the upper height somewhat. The feasibility of this recommendation is also supported by the Ergonomics Training Program of NAWGA, where it is noted that heavier items should be lifted between knuckle and waist height and lighter items between knuckle and shoulder height [10]
2. Store only light, nonbreakable items on the pallet racks above shoulder height. Be sure that hooks are provided at every pallet location to facilitate selection of items without excessive reaches.
3. Limit the pallet loads so that loads greater than 30 lb (13.6 kg) are stored, or moved to, below shoulder height. The placement of even lighter loads above shoulder height should also be minimized if possible. Also investigate the feasibility (and compliance with regulations) of maintaining the pallet on the pallet jack at a level that minimizes bending while putting the cases or objects on the pallet jack. OSHA regulations relating to transporting a pallet in the raised position must be clarified. The feasibility of this recommendation is supported by the Ergonomics Training Program of NAWGA, where it is noted that heaviest picks should be between knuckle and elbow height and picks should be arranged so that light items are at the top of the pallet [10].
4. Limit the item weight to 50 lb (22.7 kg). The feasibility of this recommendation is supported by the Ergonomics Training Program of NAWGA, where it is noted that items weighing more than 55 lb (24.9 kg) should be eliminated if possible [10].
5. Limit the use of the pallet racks with a 45-in. height to very slow moving items. The lifting of items from these pallets requires a constrained posture that increases the musculoskeletal risk. The feasibility of this is clear. It is simply a change in administrative procedure.

6. Improve access to pallets by facilitating access to the pallet racks from three sides for items weighing more than 30 lb (13.6 kg) to decrease the horizontal distance. The pallet racks are generally 94 in. (2.39 m) wide. The pallets are generally 40 × 48 in. (1.02 × 1.22 m). One way to facilitate access to three sides of the pallet is to place only one pallet "sideways" with the 48 in. (1.22 m) side toward the aisle, instead of two pallets, in each bay for fast-moving and/or heavy items. If placed in this fashion, there would be approximately 23 in. (58 cm) on each side of the pallet to allow access of items on the sides and back of the pallet with a minimum of reaching or bending. This is just above the 22 in. (56 cm) found by Ayoub and Mital [14] to cause a 13% decline in carrying capabilities. Alternatively the pallet could be placed in the "normal" way [with the 40-in. (1.02-m) side toward the aisle] with approximately 27 in. (68.6 cm) allowed on each side to access items. The feasibility of this is clear. It is simply a change in procedure. It might, however, decrease facility storage capacity. In addition, the feasibility of this recommendation is supported by the Ergonomics Training Program of NAWGA [10], where it is noted that far reaches may be avoided by placing the pallets with the long side on the aisle.
7. Improve access to pallets by facilitating access to the pallet racks from two sides for items in excess of 30 lb (11.8 kg). As noted earlier, the pallet racks are generally 94 in. (2.39 m) wide and the pallets are generally 40 × 48 in. (1.02 × 1.22 m). One way to facilitate access to two sides of the pallet is to place the two pallets as far apart as possible within each bay. Although this is not optimum, it would allow 14 in. (35.6 cm) between the pallets to increase the accessibility of items on the sides and back of the pallet. The feasibility of this is clear. It is simply a change in procedure.
8. Specify that suppliers stack boxes on pallets in a "noninterlocking" fashion whenever possible. Although interlocking increases pallet stability, it increases the reaches if selectors reach to the rear of the pallet and increases the force required if they remove an interlocked pallet in the front. The feasibility of this is clear. It is simply a change in procedure. This may require a decrease in the number of items stored on a single pallet.
9. Require that pallets with heavy items be rotated to facilitate access. The feasibility of this recommendation is supported by the Ergonomics Training Program of NAWGA [10], where it is noted that far reaches may be avoided by rotating half-empty pallets of heaviest products 180°.
10. Sequence the pick-up of items to ensure that moderate, sturdy objects are loaded on the bottom of the pallet jacks, heavier in the middle, and lighter on the top. This would minimize the musculoskeletal stress and maintain the stability of the load. The feasibility of this is clear. It is simply a change in procedure.
11. Reduce the importance of incentive-based pay systems, which encourage workers to work at excessive rates.

Abatement Recommendations That Require Some Capital Expenditure for Lift Tables, Hoists, Facility Modification.

1. Redesign racks to allow the placement of two pallets in each bay with a minimum of 20.1 in. (51 cm) separation in the center. (This would allow the shoulder width of a 95th percentile male.) This would require a minimum of 100 in. (2.54 m) bay width in the pallet racks [40 in.(1.02 m) pallet + 20 in. (50.8 cm) gap + 40 in. (1.02 m) pallet = 100 in. (2.54 m)]. The feasibility of this is clear. It would, however, require changes in some pallet rack construction and would slightly decrease overall facility storage capacity.

2. Investigate the feasibility of using pullout or rolling platforms on which the pallets holding items weighing over 30 lb (13.6 kg) can be placed. This would allow the heaviest objects to be easily accessed without twisting and lateral bending. The feasibility of this approach is supported by the variety of devices available for this purpose. Care must be taken to ensure the pull forces are not stressful.
3. Place pallets in scissors lifts or lift tables to keep the lift start height at approximately knuckle height. The best abatement would be the combination of the lift with a turntable to allow the load to be lifted and rotated to obtain the optimum vertical location and keep the horizontal distance to a minimum. The feasibility of this approach is supported by the variety of devices available for this purpose. It would, however, require widening of pallet racks and would slightly decrease overall facility storage capacity.
4. Provide small crane-type lift devices where required for heaviest loads. Optimally this would be for loads of over 30 lb (13.6 kg) but would be needed for loads greater than 50 lb (22.7 kg). The feasibility of this approach is supported by the variety of devices available for this purpose. These devices can be attached to the pallet jack, attached to the pallet rack, or suspended on rails above the aisle and include powered and manual devices.

Abatement Recommendations That Require Significant Changes in Storage Layout and Design.

1. Install a conveyor between the pallet racks to transport the selected items to the palletizing/shipping area. The feasibility of this approach is documented by Johnson, *Material Handling Engineering Magazine,* and *Modern Materials Handling* [15–18]. An increase in efficiency is noted for this type of system in discussed in the *Material Handling Engineering Yearbook* [19].
2. Completely automate the selection and transportation of items from the pallet racks (or similar storage system) to the shipping area. This is obviously a very expensive alternative. Johnson notes that a "fully automated" warehouse may not be the best alternative for a grocery warehouse [15]. It has, however, been accomplished to some extent in at least one grocery warehouse [20].
3. One possible materials assist device is a robot palletizer [21]. This device would assist in palletizing items that have been loaded on a conveyor as discussed in some of the abatements noted above.

IV. CRITICAL REVIEW OF CURRENT STATUS

In the facility reviewed above, the following controls were implemented [22]:

Review and upgrade of lifting training for new hires and refresher training for employees who lift on a regular basis

Daily 10-min warm-up program for all selectors

Strength and endurance testing for all new hires

Consideration of ergonomic design factors when replacing existing material handling equipment

Consultation with an ergonomic expert when performing the above and other feasible, beneficial ergonomic controls

The company indicated that these controls were effective.

In another grocery warehouse the following ergonomic abatements were implemented [23]:

Train selectors and supervisors in back injury prevention.
Annually review potential manual materials handling risks and lifting techniques.
Move heavier items from top racks to lower racks.
Use hooks to pull items closer to the body before lifting.
Limit stack height on pallet jacks to the approximate shoulder height of an average size person.
Modify work procedures so that selectors need not reach past the centerline of the pallet.
Require selectors to work at a steady pace through the shift instead of finishing an 8-hr workload in fewer hours.
Work with suppliers to reduce the weight of product containers.

After these abatements were implemented, the OSHA recordable back injuries decreased from 19 cases with 378 lost days in 1990 to 13 cases with 80 lost days in 1991.

V. FUTURE CONCERNS

It is important that ergonomic controls in warehousing operations focus on the design/redesign of the facility. Selection, training, and conditioning of applicants or workers may be part of an effective program, but primary consideration must be given to the reduction/elimination of ergonomic hazards. Of particular concern is the attractiveness of back-belts as an easy, inexpensive "solution." There is little support for the extensive use of back-belts as an effective abatement measure [24,25].

It is also important that warehouse operators realize that it is not uncommon for there to be an increase in the ergonomic incidence rate on the OSHA 200 log after the implementation of an ergonomics program. This indicates that workers are reporting early symptoms before they become serious and require costly medical care. There is also generally a corresponding decrease in the *cost* of ergonomics-related incidents because the early intervention is less expensive.

REFERENCES

1. A. Mital, Comprehensive maxmium acceptable weights of lift database for regular 8-hour work shifts, *Ergonomics 27*(11):1127–1138 (1984).
2. A. Mital, Maximum weights of lift acceptable to male and female industrial workers for extended work shifts, *Ergonomics 27*(11):1115–1126 (1984).
3. A. Mital, Task variables in manual material handling, *J. Safety Res. 12*(4) (1980).
4. American Industrial Hygiene Association, *Ergonomics Guide to Manual Lifting*, Ergonomics Guide Series, July–August 1970.
5. D. B. Chaffin, and K. S. Park, A longitudinal study of low-back pain as associated with occupational weight lifting factors, *Am. Ind. Assoc. J.,* December 1973, pp. 513–525.
6. G. D. Herrin, *A Taxonomy of Manual Materials Handling Hazards*, Report on International Symposium: Safety in Manual Materals Handling, SUNY Buffalo, July 18–20, 1976, pp. 6–15.
7. S. H. Snook, The design of manual handling tasks, *Ergonomics 21*(12):963–985 (1978).
8. International Labor Office, *Manual Lifting and Carrying,* Inf. Sheet No. 3, March 1962.
9. E. A. P. Koningsveld, Permissible loads for the Dutch construction industry, *Ergonomics 28*(1):359–363 (1985).

10. National American Wholesale Grocers' Association *The Backbone of the Food Industry*, NAWGA, Falls Church, VA, 1989.
11. NIOSH, *NIOSH Work Practices Guide for Manual Lifting*, NIOSH Pub. 81-122, Printed by AIHA, 1983, 1987.
12. T. R. Waters, V. Putz-Anderson, A. Garg, and L. Fine, Revised NIOSH equation for the design and evaluation of manual lifting tasks, *Ergonomics 36*(7):749–776 (1993).
13. Two Dimensional Static Strength Model, Univ. Michigan, Ann Arbor, MI, 1989.
14. M. M. Ayoub and A. Mital, *Manual Materials Handling*, Taylor & Francis, London, 1989, p. 49.
15. J. P. Johnson, "Fully automated" isn't always the best solution to your warehouse operations, *Ind. Eng. 22*(2):30–33 (1990).
16. *Material Handling Engineering*, Twenty-six miles of palletflo live storage (Advertisement), *41*(10):14 (1986).
17. G. F. Schwind, Warehousing 87:redefining the role, *Mater. Handl. Eng. 42*(3):57–65 (1987).
18. L. Beck, What's available for moving loads between levels?, *Modern Mater. Handl. 41*(15):68–70 (1986).
19. Case studies: distributor increases productivity with conveyor system, *Mater. Handl. Eng. Handb. Direct. 41*(13):122–123 (1989).
20. C. E. Witt, Automatic identification highlights distribution system, *Mater. Handl. Eng. 43*(10):49–52 (1988).
21. C. E. Witt, Electrolux: a lesson for American manufacturers, *Mater. Handl. Eng. 43*(1):55–61 (1988).
22. Settlement Agreement, Secretary of Labor v. First National Supermarkets, Inc., OSHRC Docket No. 89-2417, May 14, 1991.
23. Consent Agreement, Secretary of Labor v. Weis Markets, Inc., OSHRC Docket No. 89-3139, July 1991.
24. C. R. Reddell, J. J. Congleton, R. D. Huchingson, and J. F. Montgomery, An evaluation of a weightlifting belt and back injury prevention training class for airline baggage handlers, *Appl. Ergon. 23*(5):319–329 (1993).
25. S. Steers, To belt or not to belt?, *Safety Health,* November 1993, pp. 40–41.

29
Upper Extremity Cumulative Trauma Disorders
Current Trends

Daniel J. Habes
National Institute for Occupational Safety and Health, Cincinnati, Ohio

I. INTRODUCTION

The occurrence of work-related soft tissue disorders, particularly to areas of the upper extremity, is not a new phenomenon. The popular literature attributes the first recording of "use-related" injuries to the Italian physician Ramazzini, who, almost 300 years ago, observed clerks and scribes repeatedly using their hands and sitting in constrained postures while they worked [1]. What *is* new is the nearly universal interest that cumulative trauma disorders (CTDs) have generated, particularly among workers, over the past decade. Since the early 1980s, the occupational incidence of CTDs has more than tripled, costs have increased exponentially, and workers from more and different types of jobs have been affected, most notably those in occupations not previously thought to impose excessive physical demands. The purpose of this chapter is to more fully develop the concept of work-related CTDs and to discuss some of the current trends in the analysis and control of these ailments.

II. BACKGROUND AND SIGNIFICANCE TO OCCUPATIONAL ERGONOMICS

Ergonomics is a holistic approach to work layout that is rooted in achieving the best possible match of worker attributes and capabilities to the design and configuration of a work task. As such, application of ergonomic principles requires many considerations, including human anthropometry, biomechanics, work physiology, vibration, vision, and virtually every aspect of the physical environment [2].

In traditional production plants where the work is physically demanding and many types of tools and powered machinery are in use, the classical approach of applying the many disciplines in the field of ergonomics is still necessary as a means to maximize worker health and efficiency. However, now that the health of workers is an issue in many workplaces that are quiet, pleasantly decorated, and environmentally controlled, the attention of ergonomists (and indeed the perception by many people of what ergonomics is) has shifted from its historical multidisciplinary emphasis to the more limited area of CTD prevention and control.

III. CUMULATIVE TRAUMA DISORDERS

A. Definition

Upper extremity cumulative trauma disorders (CTDs) are regional impairments of the muscles, tendons, ligaments, nerves, and joints that are associated with work-related mechanical trauma [3]. The term *disorder* is used to categorize CTDs as a result of chronic injury to these body parts, as opposed to acute injuries, which are mishaps following the transfer of high energy, resulting in sprains, broken bones, cuts, lacerations, or amputations.

Cumulative trauma disorders are usually classified as sprains, strains, inflammations, and irritations and have names such as tendinitis, tenosynovitis, synovitis, myalgia, bursitis, peritendinitis, epicondylitis, thoracic outlet syndrome, and carpal tunnel syndrome. A more complete list of the types of musculoskeletal disorders that are considered to be related to work can be found in *The International Classification of Diseases* [4].

Also known as repeated trauma illness, repetitive motion injury, regional pain syndrome, repetitive strain injury (Australia), and occupational cervicobrachial disorder (Japan, Germany, and Scandinavia) [5,6], CTDs were once named according to the occupation or body part affected, for example, trigger finger, bricklayer's shoulder, pricer's palsy, mouse elbow, writer's cramp, stitcher's wrist, and carpenter's elbow [7]. These terms were descriptive but tended to minimize, even romanticize, the afflictions suffered by workers because of their jobs. The medical terms that are now more commonly used to describe CTDs (even among nonphysicians) tend to add legitimacy to the symptoms that workers experience, and their use likely had a part in the increased awareness that workers have regarding the work-relatedness of overuse disorders.

B. Common CTDs and Symptoms

1. Soft Tissues of the Upper Extremity

Cumulative trauma disorders most often affect the muscles, ligaments, tendons, tendon sheaths, and nerves of the upper extremity. *Muscles* consist of numerous contractile fibers that move the bones of the upper extremity. *Tendons* are tough, ropelike tissues that connect muscles to bones. In appearance, tendons are smooth, white, and shiny, a result of not having as much blood supply as muscles. Unlike muscles, tendons do not contract; they merely transfer movements from muscles to bones. Tendons in the hand and arm are surrounded by *tendon sheaths*, which are filled with a lubricating synovial fluid (Figure 1). *Ligaments* are bands of strong, ropelike fibers that connect bones to bones and provide stability to joints. Ligaments contribute to the lubrication of joints by forming a sealed *joint capsule* that encases synovial fluid. Ligaments themselves are lubricated at certain joints by fluid-filled sacks called *bursae*. Like tendons, ligaments receive little blood supply. *Nerves* are bundles of fibers that connect the central nervous system to the various parts of the body, primarily providing motor (movement) and sensory (tactile) function [8]. Three main nerves are located in the upper extremity: the median, the ulnar, and the radial. These nerves originate in the cervical region of the spine (upper neck) and traverse the shoulder, arm, wrist, and hand. The radial nerve predominantly innervates the extensor muscles in the upper arm and forearm; the ulnar and median chiefly innervate the flexor muscles in the forearm and hand [9]. Blood vessels frequently accompany the nerve network, and their major branching points are known as *neurovascular bundles* or a *plexus*. One example is the *brachial plexus*, comprising nerves and blood vessels located in the shoulder near the scapula and clavicle (see Figure 4).

Injuries to muscles are usually characterized by aching and swelling and are most severe if muscle fibers are torn or crushed. More common in occupational settings are injuries

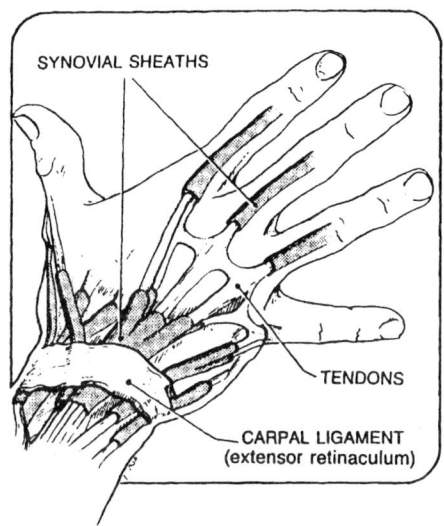

Figure 1 Top view of the hand showing the tendons and tendon sheaths. Also shown is the dorsal (top) side of the carpal ligament.

that occur to tendons and tendon sheaths, nerves, and neurovascular bundles. Pain associated with disorders of the tendons and tendon sheaths (tendinitis, synovitis, tenosynovitis) is primarily due to inflammation and is most intense during movement of the affected body parts [10–12].

Nervous disorders are most often characterized by numbness and tingling along the distribution of the affected nerve. *Carpal tunnel syndrome*, which results from entrapment and/or repeated compression of the median nerve as it passes though the carpal canal, is the most commonly reported nervous disorder associated with occupation. The symptoms are pain and burning, numbness, and tingling of the thumb and the tips of the first three fingers. The carpal canal (or tunnel) is a small, rigid passageway at the wrist, comprising bones on the dorsal (top) side and a ligament called the flexor retinaculum (or transverse carpal ligament) on the palmar side of the wrist. Through it pass the finger flexor muscle tendons and the median nerve (Figure 2). The pathophysiology of CTS is ischemia (lack of blood supply) and/or demyelination (destruction of nerve fiber sheaths) resulting from mechanical trauma [13]. Injuries of these types can occur to the median nerve when it is repeatedly compressed during movements of the hand at the wrist, which are typical in many hand-intensive work activities [14,15]. The swelling of tendons and sheaths during exertions involving high muscle forces (common in industry) can also produce pressure on the median nerve. Compression of the median nerve can be further increased if repetitive, forceful movements result in the development of inflammation of the tendons and tendon sheaths that pass through the carpal tunnel. To summarize, injury to the median nerve can be caused by a reduction in size of the carpal tunnel (wrist movement or arthritis), enlargement of the median nerve (trauma-induced swelling), or an increase in the volume of other structures within the canal (tendinitis or tenosynovitis) [16]. Another example of a nerve entrapment disorder is *cubital tunnel syndrome*, which is compression of the ulnar nerve in the region of the elbow [17]. Figure 3 illustrates the nerves of the hand and arm.

Neurovascular disorders involve nerves and adjacent blood vessels. The most recognized and hence most common occupationally reported CTD of this type is *thoracic outlet syndrome*

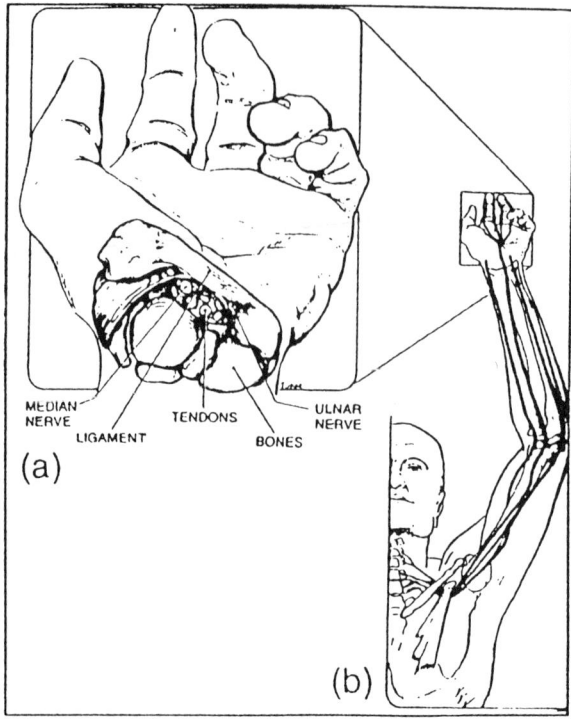

Figure 2 (a) Cross-sectional view of the carpal canal showing the tendons and nerves passing through it. (b) A full view of the three major nerves that originate in the neck and serve the arm and hand: the median, the ulnar, and the radial.

(TOS). TOS is actually an umbrella term that describes a number of syndromes, each of which occurs rarely and has vague diagnostic criteria [18]. TOS can be defined as a compression of the nerves and vessels in the shoulder-neck region of the brachial plexus (Figure 4), characterized by pain in the arms and numbness and tingling in the fingertips, as in carpal tun-

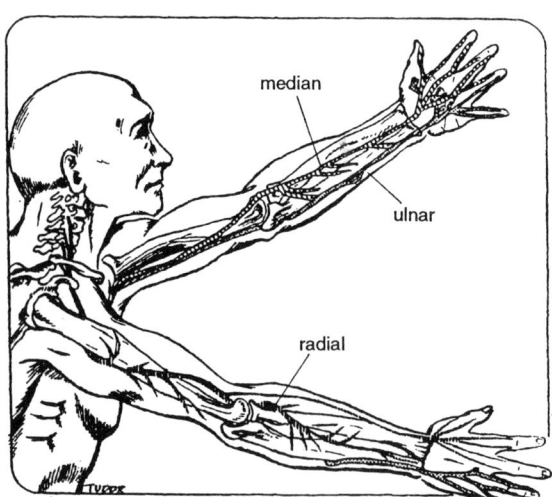

Figure 3 View of the three major nerves of the arm and hand and their areas of innervation.

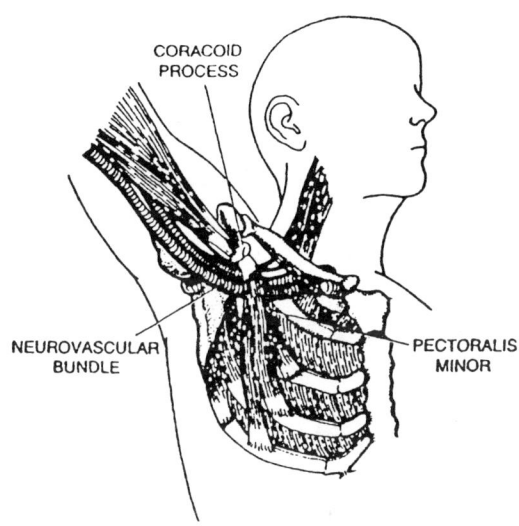

Figure 4 View of the nerves and blood vessels (neurovascular bundle) compressed between the neck and shoulder due to shoulder abduction (movement of the upper arm away from the body).

nel syndrome. Other names for compression or inflammatory disorders that affect the neurovascular bundle in the area of the neck and shoulder are *brachial plexus neuritis* and *cervicobrachial disorder* [19–21].

C. The Development of Cumulative Trauma Disorders

1. Work-Related Factors of CTDs

Because many of the manual activities attributed to CTD development occur with regularity during work tasks, they are known as work-related (or occupational) risk factors for CTDs. This term is not intended to diminish the multifactorial causality of CTDs but rather to emphasize the fact that work or occupation can be a major factor in CTD development. The most common of the occupational risk factors cited in the literature are the upper extremity *posture* (wrist, elbow, shoulder) [22–25] during a manipulation, the amount of *muscular force* associated with the activity [26,27], and the *frequency* or rate of *repetition* [28,29] of the motion. Working in *cold* environments [30,31] and using *vibrating* tools [32,33] are considered to increase the risk of CTDs.

The actual disease mechanisms associated with the development of CTDs are not fully understood, but it is believed that awkward posture, excessive force, and frequent repetition impose mechanical and physiological stress on the soft tissues of the upper extremity. For example, wrist deviations (flexion, extension, ulnar and radial deviation) (Figure 5) stretch the soft tissues across bones and ligaments, causing deformation of tendons and inflammation of tendon sheaths. Wrist deviations result in narrowing passageways that can place mechanical stress on tendons and entrap nerves. Nerve compression due to entrapment in a confined space is thought to be one of the causes of carpal tunnel syndrome [16]. Under severe conditions, muscular force causes tendons and muscles to expand and swell, which can tear tissue and put pressure on nerves. Excessive repetitive movements can overcome the capacity of tendon sheaths to lubricate tendons and that of synovial membranes to lubricate joints. Motions that comprise more than one or all of these occupational factors are more likely to result in the development of CTDs [8].

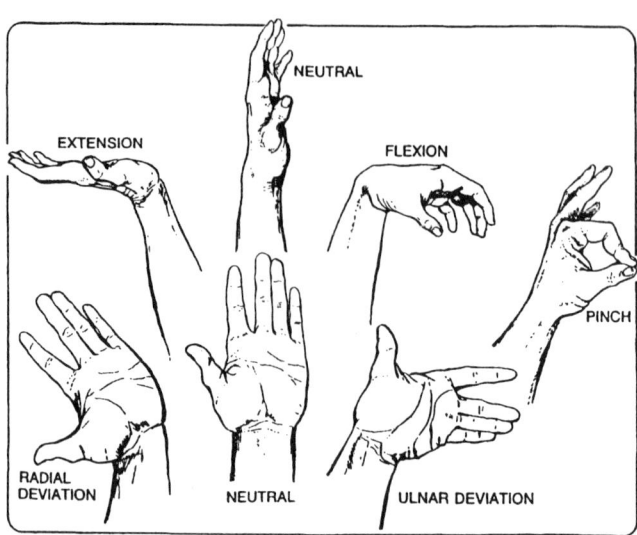

Figure 5 Deviated out postures of the hand and wrist commonly associated with the development of CTDs.

2. Nonoccupational Factors of CTDs

Manual activities and physical conditions that are not related to work but are associated with the occurrence of CTDs are known as *nonoccupational* risk factors or *personal* risk factors. Examples are physically intensive pastimes and hobbies that have many of the same postural and muscular components as work tasks. Some of these are tennis, racquetball, weight-lifting, knitting, sewing, or the playing of video games and musical instruments [24]. Additionally, some physical and medical conditions that are claimed to predispose one to CTDs (particularly carpal tunnel syndrome) are rheumatoid arthritis, gout, acromegaly, diabetes, wrist size and shape, and hormonal factors (menopause, use of oral contraceptives, pregnancy, and gynecological surgery) [16,34,35]. As a number of these conditions are unique to women, women's risk of carpal tunnel syndrome (CTS) may be elevated. Although it is generally reported that the incidence of CTS is greater among women than men [35], studies that have compared the rates of CTS among men and women performing the same work tasks have found no difference in carpal tunnel syndrome risk [75,76].

These nonoccupational and personal/medical conditions, the presence of which has often been sought to account for an outbreak of CTDs in a worker population, are now considered to be minor in the development of work-related CTDs and even more minor in their control. This modification in attitude is rooted in the recognition that few occasionally performed recreational activities can match the day-to-day manual intensity of many industrial tasks. In addition, employers have realized that reducing the work-related risk factors is the most effective way of controlling CTDs [36,37].

3. Industries and Job Tasks Commonly Reporting CTDs

Cumulative trauma disorders have long been associated with the traditional occupations in the United States, in which the work is hand-intensive, assembly-line oriented, and uses many hand and power tools. As such, CTDs have been common in the automotive industry, among meatpackers and garment workers, and among construction workers [38–41]. Although these traditional industries are still among the high-incidence work groups, a new class of high-risk jobs has emerged. Some of the main job titles in this group are cashier, VDT operator, directory assistance operator, and letter-sorting machine operator [42–46]. The latter group is distinct in that increased incidence of CTDs seems to be related to the introduction of a high-technology device or piece of equipment that was intended to make the job easier for the worker and increase productivity for the employer!

D. Incidence Rate and Costs of CTDs

1. Database Reports

For the year 1993, the Bureau of Labor Statistics (BLS) reported that 63% (302,000) of about 482,000 new cases of occupational illness among workers in the private sector were due to repeated or cumulative trauma. About 13.4% of the CTDs reported were designated as carpal tunnel syndrome [47]. Over the years, reported illnesses due to repeated trauma have increased both in number and as a percentage of total illnesses reported. In 1981, there were about 23,000 reports of new cases of illness due to repeated trauma, 18% of total illnesses. The rate of cumulative trauma-related illnesses reached 50% of the total in 1988 and nearly tripled in the decade spanning 1982–1991 (Figure 6). For the year 1992, the most recent year for which industry-specific information is available from BLS, the top five industries for disorders due to repeated trauma were meatpacking, motor vehicle manufacturing, poultry processing, household laundry manufacturing, and textiles: men's and boys' pants.

For some states, particularly in the industrial midwest, the rates of CTDs are even higher than those published by the BLS. In Ohio, for example, cumulative trauma disorders for the year 1991 were 72.6% of the total illnesses reported, up from 48.9% reported in 1985 [48].

The following are some other relevant statistics.

In 1991, CTDs accounted for 2% of all cases reported to Liberty Mutual Insurance Company and 3.5% of workers' compensation paid by the company [49].

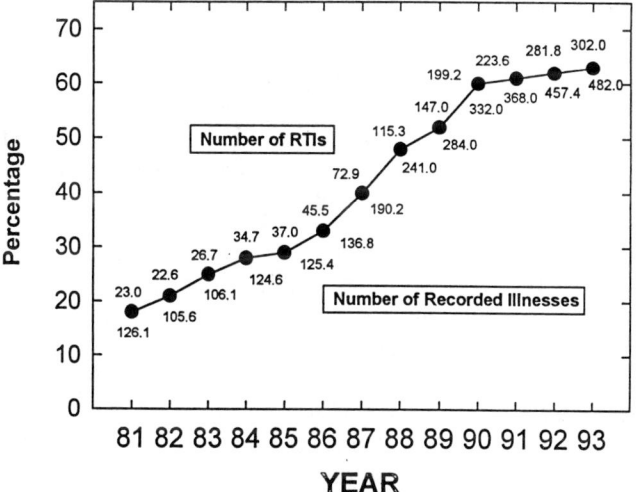

Figure 6 Bureau of Labor Statistics (BLS) data showing the trend in disorders due to "repeated trauma" (number, in thousands, above curve, made up mostly of cumulative trauma disorders) as a percentage of all occupational illnesses reported (number, in thousands, below curve).

In a survey of the frequency and cost of upper extremity disorders [50], it was concluded that each year 16 million injuries occurred, resulting in 90 million days of restricted activities and 16 million lost workdays.

Results of the Occupational Health Supplement of the 1988 National Health Interview Survey (NHIS) indicate that 38% of workers have jobs that require twisting of the hands or wrists, 22% experience hand discomfort, 18% report back pain lasting a week or more, and the most common type of injuries to workers are strains and sprains (26%) [51].

According to the Director of the Occupational Safety and Health Administration's Office of Ergonomic Support [52],

> By the end of the century, about 50 cents of each dollar paid as workers' compensation will go toward musculoskeletal injuries. Currently, spending for musculoskeletal injuries totals about 33% to 40% of compensation spending. That figure will increase as workers age and medical costs continue to increase.

2. Monetary Costs of CTDs

Not surprisingly, the dollar cost of CTDs is increasing and is a major reason for the attention that CTDs have received in recent years. These totals, which can include any combination of the cost of medical treatment and rehabilitation, lost time compensation, pensions for early retirement, or cost to rehire and to retrain new workers, are reason alone to be concerned about the trends in CTD reporting [53].

The following are some statistics of relevance.

The BNA reports that a case of carpal tunnel syndrome can cost up to $30,000; one-third of workers' compensation (more than $10 billion in 1988) is paid out for repetitive motion injuries; in 1989, Liberty Mutual Insurance Company paid out an average of $5670 in workers' compensation for each case of CTD [54]. By 1991, that figure increased to $10,000 [49].

In the state of Ohio, between 1980 and 1983, there were 2581 awarded claims for inflammations of the wrist joint, including carpal tunnel syndrome, with an average cost per claim of $8533 [55].

In 1984, the American Academy of Orthopedic Surgeons conducted a study that included four data sources: (1) the National Health Interview Survey, (2) the National Hospital Discharge Survey, (3) the National Ambulatory Care Survey, and (4) the Health and Nutrition Survey. The Academy estimated that direct costs for musculoskeletal injuries were in excess of $22 billion, indirect costs exceeded $5 billion, and total costs for all musculoskeletal conditions exceeded $65 billion [56].

There are many reasons for the escalating incidence and spiraling cost of CTDs, some of which are not directly related to the actual physical content of work tasks. The emergence of ergonomics as a legitimate discipline and a sound approach to designing work tasks has made workers more aware of the work-relatedness of CTDs. In the past, nonspecific symptoms of hand and arm pain (many of which manifest most dramatically at night after the work shift) were accepted as a consequence of hand-intensive work. Statements like "sure my hands hurt, I use them all day" were typical in field studies where worker comments about their jobs were solicited. What has perhaps reversed this attitude among workers is the knowledge that CTDs are *real* and *preventable*, have specific names with specific symptoms, and are traceable to well-defined work task attributes common to many jobs. Imagine the stigma felt by a worker at finding out that the localized pain being experienced as a result of his/her work could not be labeled or described any more precisely than as "pricer's palsy" or "mouse elbow"!

Despite the changes in attitudes and expectations regarding manual work, the fundamental reason for the increase in CTD incidence and cost has been the physical transformation of jobs and the workforce performing them. Technology advances such as the video display terminal in office environments, the laser scanner in grocery stores, and the bar code sorter in the post office have increased worker productivity, but have also increased the repetitive and stereotyped nature of jobs while depriving workers of the rest periods or microbreaks that were once inherent in most workstation designs. Formerly, an office worker could take a break from the constancy of typing to check a word spelling in the dictionary or go to the printer, or a cashier could pause to check on the correct price of an item. Now there are workplaces where there is no need to stop for breaks or even leave the immediate workstation! The introduction of high tech equipment must be accompanied by work organization changes to accommodate the capacity of workers.

Many production jobs have been fragmented into repetitive, single-task activities that concentrate movements, forces, and mechanical stresses on small areas of the upper extremity. Such job designs are attractive to management and engineers because they simplify workplace layout, increase production output, facilitate work measurement, and minimize the amount of worker skill required. It is not surprising that some of the industries that have fragmented jobs the most are among the highest in CTD rates reported to the BLS, e.g., meatpacking, poultry processing, and motor vehicle manufacturing.

Confounding the physical and mental loads that modern jobs impose is the fact that the workforce is aging and many more women work in traditional jobs than ever before. Cumulative trauma disorders are "wear and tear" injuries, and any limitation in the body's ability to tolerate stress and repair itself will lead to increased risk of CTD development. Women are capable of performing physically intensive work, but many workplaces do not yet have tools and equipment with the size and adjustability features needed to accommodate a diverse workforce with the full spectrum of size and strength capabilities. When job demands cannot be adequately matched to the physical characteristics and abilities of workers, CTDs will probably increase.

E. Controlling Cumulative Trauma Disorders

1. Ergonomics

The most effective and direct remedy for cumulative trauma disorders is to design jobs that are free of ergonomic stress factors such as excessive force, repetition, and awkward postures or to eliminate or minimize these elements in jobs that already exist. The former approach requires a knowledge of ergonomic principles and methods to apply them, and the latter requires a systematic musculoskeletal injury prevention and control program.

In 1990, OSHA published a document detailing the necessary elements of an ergonomics program for the meatpacking industry [57]. The program described in the OSHA document, which emphasizes the need for management commitment and employee involvement, contains the following four key components:

1. *Worksite analysis* to identify existing hazards or conditions where hazards may develop.
2. *Hazard prevention and control* by eliminating job hazards through workstation and tool redesign, work practice controls, and the use of personal protective equipment and by implementing administrative controls.

3. *Medical management* to eliminate or reduce CTD development through early identification and treatment of CTDs to prevent future CTD incidence problems.
4. *Training and education* to enable employees to actively participate in the prevention of CTDs by being able to recognize ergonomic hazards and the early signs of CTDs.

Companies that have adopted plant- or company-wide ergonomics programs consisting of the above elements have reported decreases in musculoskeletal injury incidence rates and turnover and increased productivity [36,37,58–61].

IV. APPLICATION OF ERGONOMICS PRINCIPLES: THE NIOSH HETA PROGRAM

The following case study is presented to illustrate how ergonomic principles can be systematically applied in an industrial setting to evaluate production jobs and formulate recommendations for the control of identified ergonomic hazards. Several years ago, NIOSH was asked to investigate cases of CTDs among workers performing finishing operations in a manufacturer of fiberglass-reinforced plastic products [62]. The OSHA law established the NIOSH Hazard Evaluation and Technical Assistance (HETA) program, a mechanism by which management, union, or employees can request NIOSH assistance in evaluating suspected occupational hazards.

The company produced typewriter housings using a thermoformed molding process. Flashing that resulted from the molding process had to be removed by hand using various types of files. A hand-held pneumatic sander was also used to prepare the typewriter housings for painting.

There were about 25 workers whose primary job activity was filing and sanding the typewriter housings. Medical records indicated that there had been 15 complaints of job-related CTDs among these workers, 13 of which accounted for lost time totaling 185 days. In the previous 2 years, seven workers had undergone the operation for carpal tunnel decompression.

Videotape analysis of workers performing the finishing job indicated that the jobs were highly repetitive and hand-intensive, requiring numerous awkward postures such as wrist flexion and extension, ulnar and radial deviation, and pinching (Figure 7). The ergonomic haz-

(a)

Figure 7 Illustration of wrist postures during typewriter finishing tasks: (a) wrist flexion; (b) wrist extension; (c) ulnar deviation; (d) radial deviation; (e) pinch.

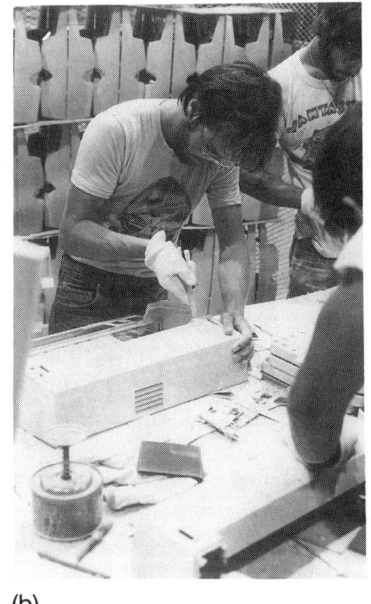
(b)

(c)

(d)　　　　　　　　　　(e)

Figure 7 Continued

ard of these jobs was increased by the use of the vibrating sander. Quantitative analysis of the videotapes indicated that each worker performed about 1000 sanding and filing movements per piece on six pieces per hour, totaling about 6000 repetitive movements per hour, or nearly 50,000 per shift!

Analysis of the patterns of motion of the workers indicated that there were noticeable differences in technique among individuals. Some workers left the typewriter housing flat on the work table and moved their hands to the various surfaces to be finished (Figure 8), whereas others moved the piece around and kept their hands in a standard position (Figure 9). Those workers who repositioned the typewriter often during the work cycle moved their hands more often but were able to maintain their wrists in a neutral position while the forceful filing and sanding motions took place. To remedy the unstructured aspect of the task, an adjustable-height fixture mounted on a ball joint was recommended. This design was intended to allow workers of any size to adjust and rotate a typewriter housing mounted on the fixture to any orientation necessary to permit finishing operations to be performed with a neutral wrist position (Figure 10).

A collection of the files used by the workers is shown in Figure 11a. Standard issue files were the two large ones, which had no handles. The file with the handle is a standard issue file modified by a worker using heater hose from a car. The tool on the far left was also fabricated by a worker using a tongue depressor and sandpaper. The recommended handle for the files is shown in Figure 12a. Pistol-type handles reduce the number of deviated wrist postures required to perform a task when the height of work can be adjusted [63], and a 1.5-in. handle diameter results in the optimum configuration for maximum applied force capability for the least amount of muscular effort [64]. The recommended handle length was about 5 in., angled approximately 60° with respect to the file [65].

The pneumatic orbital sander used by workers is shown in Figure 11b. Most of the sanders in use were worn slick and vibrated, and the air blowing onto the workers' hands

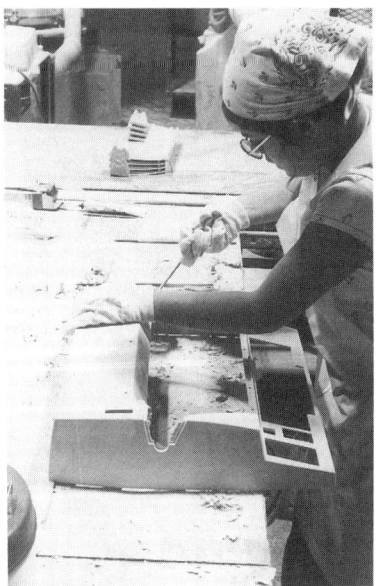
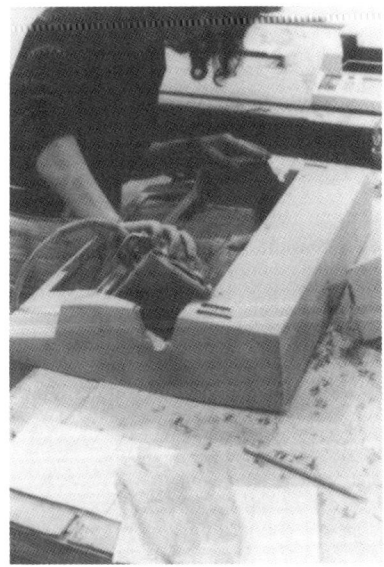

Figure 8 Illustration of wrist extension and deviation needed to file and sand when the typewriter center is left flat on the work table.

Figure 9 Illustration of neutral wrist positions when the typewriter center is tilted.

tended to reduce sensory feedback to the fingertips. The housing of the sander was also too large for some of the women with small hands. All of these factors required the workers to exert large hand forces to hold onto and manipulate the sanders.

The recommended design for the sander is shown in Figure 12b. The same criteria for handle shape and size that were applied to the files were used to develop the sander design. The pistol handle provided the workers using the sanders with the same expected benefits as the file design, with the added feature of isolation from the source of vibration.

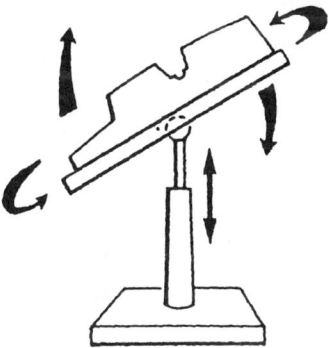

Figure 10 Height- and orientation-adjustable fixture recommended to allow workers to maintain neutral wrist postures while filing and sanding the various openings and surfaces of the typewriter housing.

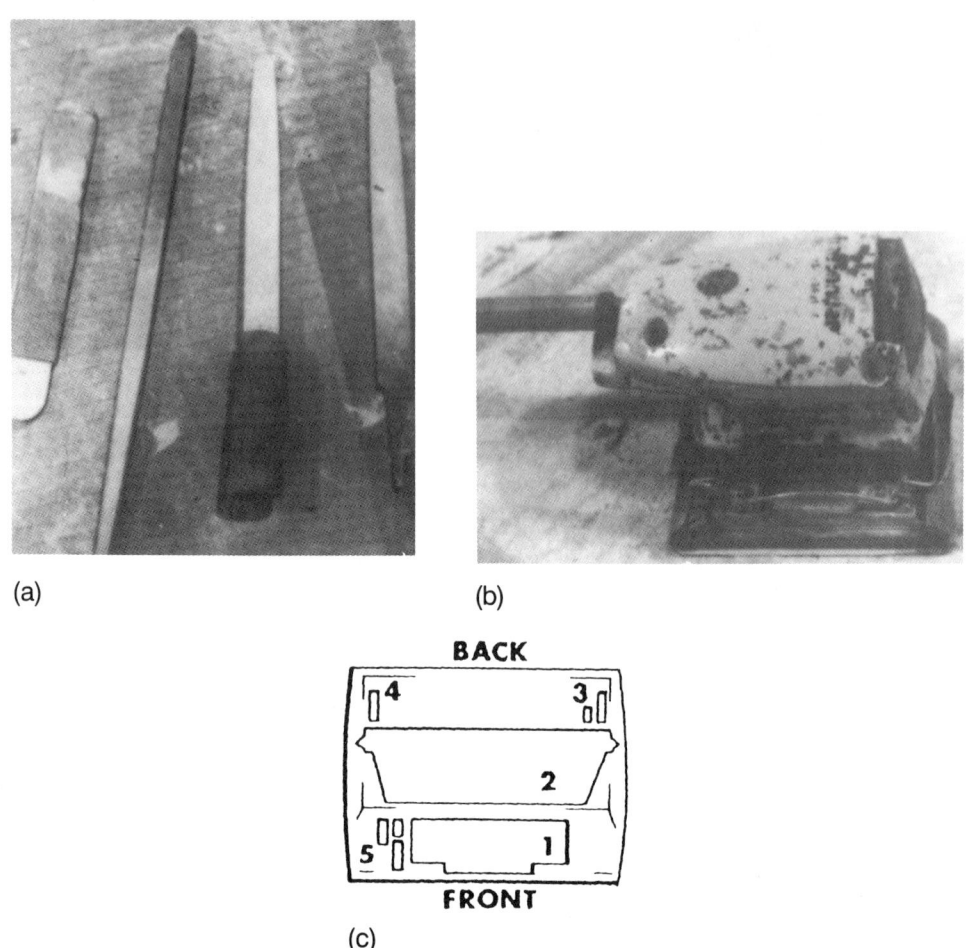

Figure 11 (a) Various types of files used to remove flashing. (b) Hand-held pneumatic sander used to finish the typewriter housing. (c) Top view of housing. Numbered openings are areas where flashing would accumulate during the molding process.

Figure 13 shows a concept drawing of how the job was designed to look using the fixture and modified tools. The use of these components in concert provides the potential to reduce nearly all of the factors that increase risk of CTD development as presented in Sec. III.C.1. The handle orientation and diameter should reduce postures deviated from neutral and minimize hand forces applied to the tool or maximize the shearing force of the file (or sander) if the same muscle forces were applied. If more flashing could be removed from the typewriter housing per filing or sanding stroke, the amount of repetition should decrease. Furthermore, since the handle on the sander is now isolated from the motor, the added risk of CTDs due to vibration exposure is reduced.

A laboratory study was conducted to evaluate the effectiveness of the recommended fixture in controlling the patterns of motion of subjects performing the filing portions of the task. A prototype typewriter housing with only the opening surrounding the keys was constructed of plywood. Three female subjects sanded a pencil line (to simulate the fiberglass flashing) from the four main surfaces of the opening in two ways: with the prototype lying

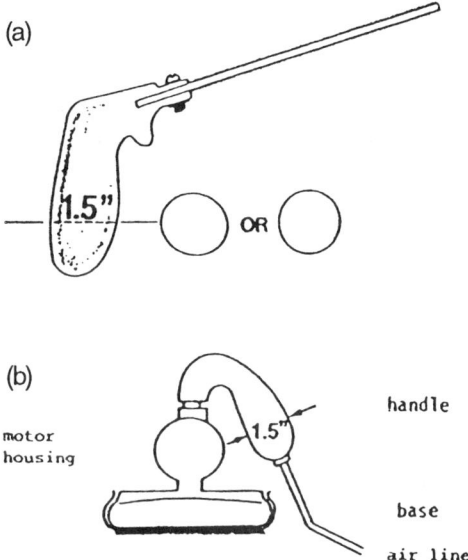

Figure 12 Recommended handles (a) for files and (b) for the pneumatic sander to allow workers to minimize hand forces and deviated postures while filing and sanding.

flat on a table (requiring deviated wrist postures) and in a vertical position and rotated so that each surface was in line (parallel) with the file (to ensure neutral wrist positions). The file used was similar to that used in the plant but had a padded straight handle. The results of the study were that on average the subjects required 2.5 times as many filing strokes to remove the pencil line using the nonneutral wrist method than using the neutral wrist method. In addition, subjects were able to exert over 2.5 times as much force holding the file in a neutral wrist position as when holding it in extended and flexed wrist positions [66]. This study did not evaluate every aspect of the recommendations made to the company, but it was success-

Figure 13 Concept drawing of a worker performing the sanding operation with the recommended handle and the adjustable fixture for holding the typewriter housing.

ful in demonstrating the benefits that could be derived from proper positioning of the typewriter housing with respect to the worker.

The opportunity to perform a prospective study of the effectiveness of the recommended control measures in the work setting was never made available to the NIOSH researchers, but NIOSH was notified by the company that the recommendations were implemented at a cost of about $14,000. A spokesman for the company would only verify that turnover among workers had decreased dramatically. Prior to the tool and fixture modifications, the workforce turned over completely about every 3 weeks. After the interventions, the company experienced what they considered to be an acceptable turnover rate for companies located in that area of the country.

V. FUTURE CONCERNS

There are a number of "activities" currently in progress that will shape the course of occupational ergonomics. Some involve the manner in which OSHA is responding to ergonomics-related worker complaints, and others pertain to the role of proposed legislation such as the OSHA Reform Act and the stated intent of OSHA to develop a standard addressing the occurrence of CTDs in the workplace. Other influential factors include the Americans with Disabilities Act (ADA) and guidelines from major organizations such as the *Control of Cumulative Trauma Disorders* manual from the American National Standards Institute (ANSI).

What eventually happens regarding any or all of these developments will be determined largely by the most powerful factors of all pertaining to occupational ergonomics issues, namely politics and economics. The need to maintain the global competitiveness of American businesses and the ideology of the political party in power may determine most the manner in which the safety and health of American workers are secured in the future. A closer look at each of these legislative issues and their future outlook follows.

A. The Role of OSHA

A question frequently asked by employers and safety and health professionals who have ergonomic concerns in their plants and by the media who report occupational safety and health news is, "What can be done about workplace hazards and CTD incidence rates?" The answer to that question is often situation-specific, but if posed now to OSHA or other legislative officials their answer might be, "Can't tell you just yet, but it will surely be different from what you may have done just a short time ago." The issuance of OSHA general duty clause citations for ergonomic hazards, the recent passage of the Americans with Disabilities Act, the proposed OSHA Reform Bill, OSHA's announcement of the intention to develop a standard to address CTDs, and the formation of a committee of experts by the American National Standards Institute (ANSI Z-365) to develop a technical standard to control CTDs are all indications that the way in which CTDs are dealt with practically and legally is about to change.

Since 1985, OSHA has issued over 350 citations for either lifting or upper extremity CTD hazards. Because there is no standard to address these types of hazards, OSHA uses Section 5(a) (1) of the OSHA Act as the authority to cite. Known as the general duty clause, Section 5(a) (1) states that "each employer shall furnish to each of his employees employment and a place of employment which are free from recognized hazards that are causing or are likely to cause death or serious physical harm to his employees." Many of these citations have been issued to companies in the auto, red meatpacking, poultry, and general food processing industries.

Usually the company and OSHA agree on an abatement plan, but the first case ever to be heard by an administrative law judge of the OSHA Review Commission was decided in 1993 [67]. The case concerned a company that bakes and packages specialty cookies and institutional bulk foods. The company was cited in 1988 for repetitive lifting violations, CTDs from repetitive motions, and records-keeping violations. In March 1993, the judge released his decision, which upheld the records-keeping violations and 21 of 27 lifting violations but vacated all 175 willful violations for CTDs. This "split decision" left both sides claiming victory. The abatement specified by OSHA for the repetitive motion violations was to reduce repetition by reducing conveyor speeds or by adding more workers to the baking lines. The 175 alleged repetitive motion violations were dismissed by the judge because OSHA could not specify exactly how fast each line should run or how many workers would be needed for each different type of cookie produced by the company. OSHA contended that the company could easily determine the correct line speed or staffing levels for each cookie by incrementally reducing conveyor speeds or adding workers until the incidence of CTDs was eliminated or substantially reduced. The company had experienced 35 cases of carpal tunnel syndrome in 2 years among approximately 185 workers. The judge ruled that in the absence of a specific standard, OSHA could not force employers to "experiment" to lessen job safety hazards.

The passage of time will tell how this case affects the course of occupational ergonomics. It may mean that OSHA will no longer use the general duty clause to force abatement of repetitive motion hazards and that a standard for the prevention of musculoskeletal disorders must be passed into law before future activities of OSHA regarding CTDs in industry can take place. No matter what ensues, it must be recognized that the judge did rule favorably on three of the four conditions that must be satisfied in order to uphold a general duty clause citation. The judge ruled that (1) a hazard existed, (2) the hazard was recognized, and (3) the recognized hazard could lead to serious injury (i.e., the carpal tunnel syndrome cases were work-related).

If the judge had ruled that a feasible means of abating the hazard existed at the time of the inspection (i.e., it was known by the company or OSHA what the correct line speed or number of workers should have been), all four conditions would have been met, and the ruling would have been completely in favor of OSHA. Nonetheless, the judge's decision on the first three conditions significantly advances the cause of occupational ergonomics by acknowledging that highly repetitive motions present a risk of injury to workers and that these risks should be recognized by employers.

There was one other important ruling made by the judge in this case relevant to occupational ergonomics. Both sides retained expert medical witnesses who testified regarding the question of whether or not CTDs, particularly carpal tunnel syndrome, could be caused by repetitive manual work, a concept that is controversial among some neurologists and orthopedists in the medical community [68,69]. That CTDs are often work-related and not due to physical attributes or existing medical conditions of workers, or to the hobbies or sports they engage in during off-work hours, is key to the ergonomist's contention that jobs that are associated with a high incidence of CTDs must be redesigned to conform with the capabilities and limitations of people. The judge ruled that those individuals and the body of literature supporting the notion that repetitive motion is *not* a causative factor in upper extremity musculoskeletal disorders [68,69] are "contrarian and in the minority."

In September 1993, the Secretary of Labor filed a petition with the Occupational Safety and Health Review Commission appealing the administrative law judge's decision that the feasibility of the recommended controls (reduce line speeds or add workers) was not demonstrated [70]. In the petition, the Labor Secretary argued that "no case has required the Secretary to show precisely *how many* injuries would be prevented under a given abatement method, in order to sustain a general duty clause violation." The Labor Secretary further argued that there was a wealth of evidence on the feasibility issue and that several ways were proposed to lower the incidence of CTDs, "many of which had been specifically recommended to the employer by its own corporate ergonomists" before OSHA issued its citation. The importance of this appeal is such that both written briefs and oral arguments are being required by the Review Commission before a decision is made. This process will likely delay a decision for several years. Nonetheless, providing that no legislative developments precede the decision, the result of this case will influence OSHA enforcement in the ergonomics area and will possibly affect the direction and timetable of an ergonomics standard for control of CTDs.

The Comprehensive Occupational Safety and Health Reform Act (HR 3160, S 1622) proposed to broaden the coverage of the OSHA Act of 1970 and to increase the role of workers in protecting their own health and safety. The OSHA Reform Act, first introduced in August 1991, called for the Secretary of Labor to issue a "final standard on ergonomic hazards to protect employees from work-related musculoskeletal disorders" not later than 1 year from the adoption of the law. Other requirements influencing ergonomics included increased training and education for workers, the evaluation of jobs for ergonomic hazards, the recording of CTDs as an occupational illness, and the establishment of ergonomics programs having the four elements contained in the OSHA Meatpacking Guidelines [57].

The original OSHA Reform Act, which never reached a vote in the House of Representatives, was reintroduced to the House in April 1993 with some revisions. The major change was the exclusion of federal and postal employees, who will be covered under separate legislation (HR 115). The mandate for OSHA to issue a final standard on ergonomics, still in the bill, underscores the commitment of the federal government to reverse the trend of increasing incidences of CTDs through the rulemaking process.

Subsequent to the reintroduction of the OSHA Reform Act were several alternative variations of the Bill by the opposition political party. One of these (HR 2937) called for voluntary compliance programs offering exemptions from safety and health inspections for an "exemplary safety record" defined as "no employee death caused by occupational injury and fewer lost workdays due to occupational illness than the average for the industry of which the employer is a part." This alternative bill also proposed that the functions and authorities of NIOSH be transferred to the secretary of labor. Developments such as these and the ongoing stalemate in the passage of the OSHA Reform Act led a prominent labor official to conclude that chances for meaningful reform of the OSHA Act in the next two years range from "very bleak to non-existent" [76].

In August 1992, OSHA published an Advance Notice of Proposed Rulemaking (ANPR) in the *Federal Register*, stating its intention to develop an Ergonomic Safety and Health Management Standard. In the ANPR, OSHA invited interested parties to submit information regarding their view on the need for an ergonomics standard, what form it should take, what their experience has been with work-related CTDs, how they have evaluated and controlled CTDs, what works, what doesn't, etc. During the 6-month period allowed, OSHA received a total of 268 responses from all types of individuals and groups: factory management and medical people, lawyers, unions, trade associations, insurance company risk managers, private physicians, university experts, and government agencies, including a response from NIOSH. During that time, OSHA also conducted 3000 telephone interviews soliciting the same type of information as requested in the ANPR.

Throughout its subsequent development and frequent refinement, the major provisions of the proposed OSHA ergonomics standard included: (1) identification of problem jobs through review of medical records or conduct of symptom surveys, (2) use of a risk factor checklist of identify problem jobs and trigger a secondary analysis using more detailed checklists or a professional ergonomist, (3) medical management for workers suffering from work-related musculoskeletal disorders, and (4) employee involvement in all phases of an ergonomics program, and training for employees and supervisors [71].

However, when a draft version of the proposed standard was finally released in March, 1995 (30 months after the ANPR notice), several of these key provisions previously characterized as nearly dogmatic by OSHA had been relaxed. Key changes included: (1) reduction in the coverage of the standard from all 6.1 million U.S. employers to only those where there is employee exposure to "signal" risk factors (2.6 million employers), (2) a "grandparent" provision where employers with existing, effective programs are exempted from certain requirements, (3) use of an alternative effective checklist or other job evaluation method instead of exclusive use of the OSHA checklist, and (4) no requirement for employers to evaluate the effectiveness of their process [77].

The event that changed everything concerning the progress of OSHA reform and the tone of the ergonomics standard was the congressional elections of November, 1994. The result was that the political mood suddenly shifted to budget cutting and antiregulation and to those government agencies associated with regulation and enforcement. An event that provided some closure on the ergonomics standard for now was an announcement by OSHA to abandon plans to pursue the standard in lieu of continuing "to study the problem" [80].

It would be futile to attempt to predict the future course of OSHA reform or the ergonomics standard in view of the current ideological struggle about the role of "big" government and regulation in America today. However, a speech delivered by the President as part of the Administration's "reinventing government" campaign offered a vision for the direction of OSHA that may endure, irrespective of the present or future political climate. The new OSHA would seek to develop worksite health and safety programs in partnership with business firms. Under the plan, employers fully committed to eliminating hazards and reducing injuries and illnesses would be rewarded with fewer inspections and lower or no penalties. The main features of the "reformed OSHA" would be (1) a nationwide implementation of the "Maine 200" program in which companies receive federal assistance to develop safety and health programs using employee participation approaches, (2) inspections focused on only the top hazards in the industry for employers having strong and effective health and safety programs, (3) reduced penalties for employers who provide "quick fixes" for hazards, and (4) a program of information dissemination projects to provide safety and health information to workers and employers through information technology [79].

It is not likely that every aspect of the new plan for OSHA will materialize, but it does appear likely that the philosophy of OSHA will change from one of strict enforcement of regulations to one that provides a choice of partnership to employers who want to develop effective health and safety programs in cooperation with their employees.

B. Other Legislation and Guidelines

The Americans with Disabilities Act (ADA) has already had a profound influence on occupational ergonomics and on how CTDs are dealt with at the workplace, particularly for workers not having the types of disabilities specifically stated in the law. Because the ADA protects those injured at work from losing their jobs, employers will have to look more toward prevention of CTDs [72]. Furthermore, injured workers will have to be given substitute jobs (or the same job with reasonable accommodation) that are within the restrictions of their injuries. During the first year of the ADA, the most common disability alleged in complaints was back injury, either for injuries previously sustained or for lack of accommodation for back injuries occurring after the enactment of ADA [73]. Test cases deciding what is "reasonable" accommodation for injured workers will, over time, determine the actual extent to which the ADA affects the way in which CTDs are managed and controlled at the workplace.

The American National Standards Institute (ANSI) is the clearinghouse and coordinating body for voluntary standards on the national level. It is a federation of standards developing and standards using organizations. In April 1995, ANSI published its most recent draft of a proposed standard on cumulative trauma disorders (ANSI Z-365) [79]. The document was the output of a committee comprised of nearly 275 interested employers, employees, trade associations, professional societies, governmental regulators and researchers, suppliers, insurers, academics, and individuals. The proposed standard and supporting materials contain what the committee believed were the necessary components of a program for effective control of CTDs: surveillance, job analysis and design, medical management, and training.

The significance of this manuscript from ANSI is that it is one more indication of the severity of the cumulative trauma disorder problem in the United States today. As prospects dim for action on the OSHA standard any time before the next presidential elections, the ANSI Z-365 document may well be the best source of consensus information on the control of cumulative trauma disorders available to concerned employers. This is not a new role for ANSI, because when OSHA was formed in 1970, hundreds of the initial OSHA standards were based on existing ANSI standards.

Even though politics and ideology have presently strangled the OSHA ergonomics rule, OSHA is not barred from continuing to work on the standard. The events following the 1994 elections give OSHA and ANSI a "quiet" period to refine their proposed standards and address some of the concerns expressed by critics and other members of the business community. Validation of checklists proposed as screening tools to identify jobs needing closer scrutiny and more data on the effectiveness and cost justification of ergonomic interventions are two controversial topics where current knowledge needs to be strengthened. If better tools for the control of work-related cumulative trauma disorders can be developed and effectively disseminated to the private sector, an ergonomics rule may well be accepted by all affected parties, regardless of the prevailing politics of occupational safety and health.

REFERENCES

1. B. Ramazzini, *De morbis artificum distraba*, 1717. (Translated by W.C. Wright, Univ. Chicago Press, Chicago, 1940.)
2. W. T. Singleton, Introduction, In *The Body at Work: Biological Ergonomics*, Cambridge Univ. Press, London, 1986, pp. 1–28.
3. National Institute for Occupational Safety and Health (NIOSH), Proposed National Strategies for the Prevention of Leading Work-Related Diseases and Injuries, Part 1, published by the Association of Schools of Public Health, 1986.
4. Department of Health and Human Services, *The International Classification of Diseases*, 9th rev. Clinical Modification, 3rd ed., Vol. 1, Diseases/tabular list, U.S. Department of Health and Human Services, Public Health Service, Health Care Financing Administration, DHHS Publ. No. (PHS) 89-1260, Washington DC, 1989.
5. F. T. McDermott, Repetition strain injury: a review of current understanding, *Med. Australia* 144:196–200 (1986).

6. W. E. Stone, Occupational overuse syndrome in other countries, *J. Occup. Health Safety—Aust. NZ 3*(4):397–404 (1986).
7. G. Rosen, The worker's hand, *Ciba Symp. 4*(4):1307–1322 (1942).
8. V. Putz-Anderson, *Cumulative Trauma Disorders: A Manual for Musculoskeletal Diseases of the Upper Limbs*, Taylor & Francis, London, UK, 1988.
9. H. O. Kendall, F. P. Kendall, and G. E. Wadsworth, *Muscles: Testing and Function*, Williams and Wilkins, Baltimore, MD, 1971.
10. K. Kurppa, P. Waris, and P. Rokkanen, Peritendinitis and Tenosynovitis, *Scand J. Work Environ. Health 5*(Suppl. 3):19–24 (1979).
11. T. Jarvinen and I. Kuorinka, Prevalence of tenosynovitis and other occupational injuries of upper extremities in repetitive work, *Arh. Hig. Rada Toksikol. 30*(Suppl.): 1281–1284 (1979).
12. T. J. Armstrong, L. J. Fine, S. A. Goldstein, Y. R. Lifshitz, and B. A. Silverstein, Ergonomics considerations in hand and wrist tendinitis, *J. Hand Surgery 12A*(5):830–837 (1987).
13. D. M. Dawson, Entrapment neuropathies of the upper extremities, *N. Eng. J. Med. 329*(27):2013–2018 (1993).
14. R. G. Feldman, R. Goldman, and W. M. Keyserling, Peripheral nerve entrapment syndromes and ergonomic factors, *Am. J. Ind. Med. 4*:661–681 (1983).
15. R. H. Gelberman, M. D. Hergenroeder, A. R. Hargens, G. N. Lundborg, and W. K. Akeson, The carpal tunnel syndrome: a study of carpal canal pressures, *J. Bone Joint Surg. 63A*(3):380–383 (1981).
16. M. L. Bleecker, Medical surveillance for carpal tunnel syndrome in workers, *J. Hand Surg. 12A*(5, Pt. 2):845–848 (1987).
17. B. T. Harter, Indications for surgery in work-related compression neuropathies of the upper extremity, *Occup. Med. State Art Rev. 4*(3): 485–495. (1989).
18. A. J. Wilbourn and J. M. Porter, Thoracic outlet syndromes, *Spine: State Art Rev. 2*(4):597–626 (1988).
19. J. E. Bateman, Neurovascular syndromes related to the clavicle, *Clin. Orthoped. Rel. Res. 58*: 75–82 (1968).
20. R. R. Tyson and G. F. Kaplan, Modern concepts of diagnosis and treatment of thoracic outlet syndrome, *Orthop. Clin. N. Am. 6*:507–519 (1975).
21. W. A. Dale, Thoracic outlet compression syndrome, *Arch. Surg. 117*:1437–1445 (1982).
22. W. Brain, A. Wright, and M. Wilkinson, Spontaneous compression of both median nerves in the carpal tunnel, *Lancet 1*:277–282 (1947).
23. R. Tanzer, The carpal tunnel syndrome, *J. Bone Joint Surg. 41A*:626–634 (1959).
24. T. J. Armstrong and D. B. Chaffin, Carpal tunnel syndrome and selected personal attributes, *J. Occup. Med. 21*:481–486 (1979).
25. E. R. Tichauer, Some aspects of stress on forearm and hand in industry, *J. Occup. Med. 8*:63–71 (1966).
26. R. Muckart, Stenosing tendovaginitis of abductor pollicis longus and extensor pollicis brevis at the radial styloid (DeQuervain's disease), *Clin. Orthop. 33*:201–208 (1964).
27. B. A. Silverstein, L. J. Fine, and T. J. Armstrong, Occupational factors and carpal tunnel syndrome, *Am. J. Ind. Med. 11*:343–358 (1987).
28. L. Hymovich and M. Lindholm, Hand, wrist, and forearm injuries: the result of repetitive motions, *J. Occup. Med. 8*(1):573–577 (1966).
29. H. Ohara, S. Nakagiri, T. Itani, K. Wake, and H. Aoyama, Occupational health hazards resulting from elevated work rate situations, *J. Human Ergol. 5*:173–182 (1976).
30. W. Fox, Human performance in the cold, *Hum. Factors 9*:203–220 (1967).
31. J. Lockhart and H. Kiess, Auxiliary heating of the hands during cold exposure and manual performance, *Hum. Factors 13*:457–465 (1971).
32. L. J. Cannon, E. J. Bernacki, and S. D. Walter, Personal and occupational factors associated with carpal tunnel syndrome, *J. Occup. Med. 23*(4):225–258 (1981).
33. R. G. Radwin, T. J. Armstrong, and D. B. Chaffin, Power hand tool vibration effects on grip exertions, *Ergonomics 30*:833–855 (1987).

34. C. G. Barnes and H. L. F. Currey, Carpal tunnel syndrome in rheumatoid arthritis, a clinical and electrodiagnostic survey, *Ann. Rheum. Dis. 26*:226-233 (1967).
35. M. S. Sabour and H. H. Fadel, The carpal tunnel syndrome — a new complication ascribed to the pill, *Am. J. Obstet. Gynecol. 107*(3):1265-1267 (1979).
36. F. McKenzie, Storment, P. Van Hook, and T. J. Armstrong, A program for control of repetitive trauma disorders associated with hand tool operations in a telecommunications manufacturing facility, *Am. Ind. Hyg. Assoc. J. 46*(11):674-678 (1985).
37. G. Lutz and T. Hansford, Cumulative trauma disorder controls: the ergonomics program at Ethicon, *J. Hand Surg. 12A*(5, Pt. 2):863-866 (1987).
38. T. J. Armstrong, L. J. Fine, B. Joseph, and B. Silverstein, *Analysis of Selected Jobs for Control of Cumulative Trauma Disorders in Automobile Plants*, Univ. Mich. Tech. Report, Ann Arbor, MI, 1984.
39. M. L. Finkel, The effects of repeated mechanical trauma in the meat industry, *Am. J. Ind. Med. 8*:375-379 (1985).
40. C. Brisson, A. Vinet, and M. Vezina, Disability among female garment workers: a comparison with a national sample, *Scand. J. Work Environ. Health 15*:323-328 (1989).
41. R. C. Jensen, B. P. Klein, and L. M. Sanderson, Motion-related wrist disorders traced to industries, occupational groups, *Mon. Labor Rev.*, September 1983, pp. 13-16.
42. L. Lannersten and K. Harms-Ringdahl, Neck and shoulder muscle activity during work with different cash register systems, *Ergonomics 33*(1):49-65 (1990).
43. A. M. Rossignol, E. P. Morse, M. S. Summers, and L. S. Pagnotto, Video display terminal use and reported health symptoms among Massachusetts clerical workers, *J. Occup. Med. 29*:112-118 (1987).
44. T. Hales and S. Sauter, *U.S. West Communications: Phoenix, Minneapolis, Denver*, NIOSH HETA 89-299-2230, 1992.
45. R. Arndt, *A Prospective Study of the Psychological and Physiological Effects of Machine Paced Work in the United States Postal Service: An Exploratory Study*, USDHHS, Contract Rep. No. 210-79-0072, NTIS PB-87-125-993/A14, 1985.
46. K. Jorgensen, The strain on the shoulder and neck muscles during letter sorting, *Int. J. Ind. Ergon. 3*:243-248 (1989).
47. U.S. Department of Labor, Bureau of Labor Statistics, *Annual Survey of Occupational Injuries and Illnesses in 1992*, Washington, DC, December 1993.
48. Ohio Bureau of Workers' Compensation, *The Number of Compensable Claims Filed for Cumulative Trauma Disorders Compared to all Illnesses and All Injuries and Illnesses in Ohio, 1985-1991*, Prepared by Occupational Health and Safety Research, May 27, 1993.
49. G. E. Brogmus and R. Marko, The proportion of cumulative trauma disorders of the upper extremities in U.S. industry, *Proc. Hum. Factors Soc. 36th Annual Meeting*, Atlanta, GA, 1992.
50. J. Kelsey, H. Pastides, N. Kreiger, C. Hamar, and R. Chernow, *Upper Extremity Disorders: As Survey of Their Frequency and Cost in the United States*, CV Mosby, St. Louis, MO, 1980, pp. 1-70.
51. Department of Health and Human Services, National Health Interview Survey, Vital and Health Statistics: Health Conditions Among the Currently Employed: United States, 1988, Ser. 10, No. 186, DHHS Publ. (PHS) 93-1514, July 1993.
52. BNA Occupational Safety and Health Daily Reporter, *OSHA Official Outlines Future Costs of Musculoskeletal Injuries to Workers*, Bureau of National Affairs, (No. 166) A-1, Washington, DC, Aug. 26, 1992
53. S. Kolare, Strategies for prevention of work-related musculoskeletal disorders: consensus paper, *Int. J. Ind. Ergon. 11*:77-81 (1993).
54. D. E. Elisburg, Ed., *Cumulative Trauma Disorders in the Workplace: Costs, Prevention, and Progress*, Bureau of National Affairs (BNA), Washington, DC, 1991.
55. Industrial Commission of Ohio, Division of Safety & Hygiene, Research and Statistics Section, Columbus, OH, Special Report to NIOSH, prepared May 22, 1989.
56. T. L. Holbrook, K. Grazier, J. L. Kelsey, and R. N. Stauffer, *The Frequency of Occurrence,*

Impact and Cost of Selected Musculoskeletal Conditions in the United States, American Academy of Orthopaedic Surgeons, Chicago, IL, 1984.
57. U.S. Department of Labor, *Ergonomics Program Management Guidelines For Meatpacking Plants*, Occupational Safety and Health Administration (OSHA), Publ. 3123, 1990.
58. J. E. Rigdon, The wrist watch: how a plant handles occupational hazard with common sense, *The Wall Street Journal*, Sept. 28, 1992, p. A1, col. 6.
59. D. T. Geras, C. D. Pepper, and S. H. Rodgers, An integrated ergonomics program at the Goodyear Tire & Rubber Company: the forcing strategy, Goodyear Tech. Rep., 1988, unpublished.
60. G. LaBar, A battle plan for back injury prevention, *Occup. Hazards*, November 1992, pp. 29–33.
61. M. Echard, S. Smolenski, and M. Zamiska, Ergonomic considerations: engineering controls at Volkswagen of America, in *Ergonomic Interventions to Prevent Musculoskeletal Injuries in Industry*, Industrial Hygiene Science Series, ACGIH, Lewis Publishers, Chelsea, MI, 1987, pp. 117–131.
62. J. Boiano, A. Watanabe, and D. Habes, *Armco Composites*, NIOSH HETA 81-143-1041, Hartford City, IN, February 1982.
63. T. J. Armstrong, J. A. Foulke, B. S. Joseph, and S. A. Goldstein, Investigation of cumulative trauma disorders in a poultry processing plant, *Am. Ind. Hyg. J. 43*:103–116 (1982).
64. M. M. Ayoub and P. Lo Presti, The determination of an optimum size cylindrical handle by use of electromyography, *Ergonomics 14*:509–518 (1971).
65. D. B. Chaffin and L. Greenberg, *Workers and Their Tools: A Guide to the Ergonomic Design of Hand Tools and Small Presses*, Pendell, Midland, MI, 1976, p. 67.
66. B. Burnett and A. Bhattacharya, Effect of task design on wrist postures, in *Trends in Ergonomics/Human Factors III*, W. Karwowski, Ed., Elsevier, Amsterdam, 1986, pp. 589–595.
67. Occupational Safety and Health Review Commission, *Secretary of Labor v. Pepperidge Farm, Inc.*, OSHRC Docket No. 89-0265, Washington, DC, April 19, 1993.
68. N. M. Hadler, Cumulative trauma disorders: an iatrogenic concept, (Editorial), *J. Occup. Med. 32*: 1, 38–41 (1990).
69. P. A. Nathan, K. D. Meadows, and L. S. Doyle, Occupation as a risk factor for impaired sensory conduction of the median nerve at the carpal tunnel, *J. Hand Surg. 13-B*(2):167–170 (1988).
70. Bureau of National Affairs, Labor Department appeals ALJ decision vacation 175 violations at Pepperidge Farm, *Occup. Safety Health Reporter, 23*(16):404–405 (1993).
71. BNA Occupational Safety & Health Reporter, *Ergonomics*, Bureau of National Affairs, Inc., Washington, DC, p. 1655 January 11, 1995.
72. F. Swoboda, Motion-injury experts await impact of disabilities act, *Washington Post*, Nov. 3, 1991, p. H2.
73. L. Spayd, The Disabilities Act, *The Washington Post* (LEGI-SLATE Article No. 185189), July 22, 1993.
74. B. A. Silverstein, L. J. Fine, and T. J. Armstrong, Hand-wrist cumulative trauma in industry, *Br. J. Ind. Med. 43*:119–184 (1986).
75. B. A. Silverstein, L. J. Fine, and T. J. Armstrong, Occupational factors and carpal tunnel syndrome, *Am. J. Ind. Med. 11*:343–358 (1987).
76. BNA Occupational Safety & Health Reporter, *AFL-CIO Official Sees Little Chance for Action Soon: Labor's Role Stressed*, Bureau of National Affairs, Inc., Washington, DC, p. 1515, December 7, 1994.
77. Occupational Health & Safety Letter, *OSHA Ergonomic Draft Eases Burden, Allows More Flexibility for Employers*, Business Publishers, Inc., Silver Spring, MD, p. 51, April 3, 1995.
78. The White House, Office of the Press Secretary, *Remarks by the President on Reinventing Worker Safety Regulations*, Stromberg Sheet Metal Works, Inc., Washington, DC, May 16, 1995.
79. American National Standards Institute (ANSI), *Control of Work-Related Cumulative Trauma Disorders, Part1: Upper Extremities*, Document ANSI Z-365, ANSI, New York, April 17, 1995.
80. C. Skrzycki, *OSHA Abandons Rules Effort on Repetitive Injury, Opposition by GOP, Business Cited, The Washington Post* (LEGI-SLATE Article No. 231499), June 13, 1995.

30
Occupational Human Vibration

Michael J. Griffin
University of Southampton, Southampton, England

I. INTRODUCTION

The human body is exposed to vibration in many environments, yet understanding of the effects of oscillatory motion on the body is far from complete. This chapter summarizes human responses relevant to occupational exposures to vibration. More detail may be found elsewhere (e.g., [1]).

Whole-body vibration occurs when the body is supported on a surface that is vibrating. There are three possibilities: sitting on a seat that vibrates, standing on a vibrating floor, or lying on a vibrating surface. Whole-body vibration occurs in all forms of transport.

Hand-transmitted vibration is the vibration that enters the body through the hands. It is caused by various processes in industry, agriculture, mining, and construction when vibrating tools are grasped or pushed by the hands or fingers. Exposure to hand-transmitted vibration can lead to the development of several disorders.

II. BACKGROUND AND SIGNIFICANCE TO OCCUPATIONAL ERGONOMICS

All forms of transport and some industrial machines expose the body to whole-body vibration. Prior to the mechanization of transport, the body was exposed to motion from walking and running, riding on animals, and sailing in boats. There are some similarities between the old and the new: the motions on a horse may have caused difficulty in firing an arrow from a bow just as the motions of a modern tank can cause difficulty for the tasks undertaken by the crew. However, the oscillations caused by traditional forms of transport were very different from those occurring in modern transport, and the tasks undertaken have changed greatly. The people, their postures, and their expectations have also changed. Furthermore, ergonomists can now design seats, controls, and displays to minimize the effects of vibration.

All powered hand tools generate hand-transmitted vibration. Prior to the mechanization of tools, hand-powered hammers exposed the fingers, hands, and arms to vibration. Hand-powered tools mostly produce intermittent shocks, whereas electric, pneumatic, and hydraulic powered tools allow faster work by the production of shocks at a higher rate. This has also greatly extended the range of operations performed by percussive action. Other modern tools cause vibration by high-speed rotation of cutting or polishing surfaces (e.g., grinding tools) or by the oscillation of an engine (e.g., chain saws). The hands of some workers are now exposed to types of vibration never experienced 100 years ago. The design of hand-powered tools has also resulted in increased durations of exposure to vibration. Adverse effects of the vibration from such tools are therefore modern phenomena arising from the use of man-made machines: the effects were unknown before the 20th century.

III. CHARACTERISTICS OF VIBRATION

A. Vibration Magnitude

The magnitude of an oscillation can be expressed as the distance between the extremities reached by the motion (i.e., the peak-to-peak displacement) or the maximum deviation from some central point (i.e., the peak displacement). However, the magnitude of vibration is now most commonly expressed in terms of an average measure of the acceleration of the oscillatory motion, usually the root-mean-square value expressed in meters per second per second (i.e., m/s^2 rms). (The acceleration due to gravity on earth is approximately 9.81 m/s^2; for a sinusoidal motion, the rms value is the peak value divided by $\sqrt{2}$). For rotational vibration, the magnitude is expressed in radians per second per second (i.e., rad/s^2).

When observing vibration, it is sometimes possible to estimate the displacement caused by the motion. For a sinusoidal motion, the acceleration a can be calculated from the frequency f, in hertz (Hz), and the displacement d:

$$a = (2\pi f)^2 d$$

So, for example, a motion with a frequency of 1 Hz and a peak-to-peak displacement of 0.1 m will have an acceleration of 3.95 m/s^2 peak-to-peak, 1.97 m/s^2 peak, and 1.40 m/s^2 rms. The above expression can be used to convert acceleration measurements to corresponding displacements throughout this chapter. However, the conversion is accurate only when the motion has a sinusoidal waveform.

B. Vibration Frequency

The frequency of vibration is expressed in cycles per second using the SI unit hertz (Hz). (1 Hz = 1 cps.) Oscillations at frequencies below about 0.5 Hz can cause motion sickness. The frequencies of greatest significance to whole-body vibration are usually at the lower end of the range from 0.5 to 100 Hz. For hand-transmitted vibration, frequencies as high as 1000 Hz may have detrimental effects.

Mechanical systems have resonance frequencies at which they exhibit a maximum response to vibration. Resonance frequencies for parts of the human body are sometimes suggested but are rarely supported by experimental data. The human body is complex and so has many modes of vibration; it is mostly highly damped so that significant response occurs at frequencies other than at resonance frequencies. Resonance frequencies of the body may also depend on the posture and orientation of the body and differ between individuals. Frequency weightings for both whole-body vibration and hand-transmitted vibration reflect, in a general way, the influence of various modes of vibration in the body, including resonance frequencies.

C. Vibration Direction

The responses of the body differ according to the direction of the motion. The three principal directions for seated and standing persons are fore-and-aft (x axis), lateral (y axis), and vertical (z axis). The vibration is usually measured at the interface between the body and the surface supporting the body (i.e., between the seat and the pelvis, at the ischial tuberosities, for a seated person; beneath the feet for a standing person). The vibration of a backrest, a control, or a display can also be important. Figure 1 illustrates the translational and rotational axes for an origin at the ischial tuberosities of a seated person. A similar set of axes is used for describing the directions of vibration at the back and feet of seated persons and for vibration entering the hand.

D. Vibration Duration

The effects of vibration depend on the total duration of vibration exposure. Additionally, the duration of measurement may affect the measured magnitude of the vibration. The rms acceleration may not provide a good indication of vibration severity if the vibration is intermittent, contains shocks, or otherwise varies in magnitude from time to time.

IV. WHOLE-BODY VIBRATION

Whole-body vibration is produced by machinery and by all forms of transport. The vibration may affect comfort, the performance of activities, and health and cause motion sickness. The

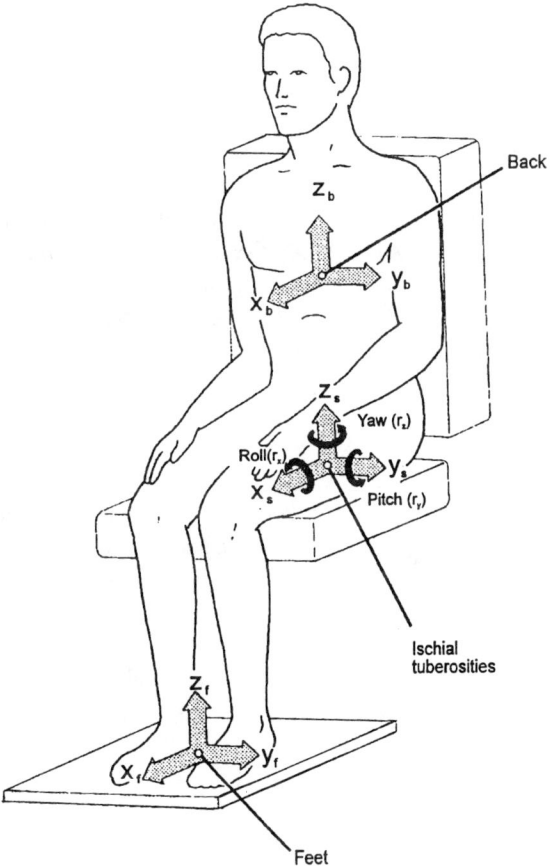

Figure 1 Axes of vibration used for assessing the effects of whole-body vibration.

comments of persons exposed to vibration often derive from the sensations produced by vibration rather than from a knowledge that the vibration is causing harm or interfering with their activities.

A. Effects on Comfort

The human body can detect magnitudes of vibration lower than those that normally cause damage to structures. It is not practical to eliminate the perception of vibration in moving structures (e.g., forms of transport) and often difficult to prevent the perception of vibration in fixed structures (e.g., buildings).

It is now normally possible to estimate the extent of subjective reactions (i.e., discomfort or annoyance) from measurements of vibration. For some types of vibration it is possible to estimate the percentage of persons who will be able to feel vibration. For higher vibration magnitudes, an approximate indication of the extent of subjective reactions is available in a semantic scale of discomfort (e.g., British Standard 6841 [2]).

The design limit to prevent vibration discomfort varies between different environments (e.g., between buildings and transport), between different types of transport (e.g., between cars, motorbikes, trucks), and between types of vehicles (e.g., sports cars and limousines). The design limit depends on external factors (e.g., cost and speed) and the comfort in alternative environments (e.g., competitive vehicles). In recent years there has been a general increase in the expectation of comfort as techniques have become available to make improvements.

1. Effects of Vibration Magnitude

Very approximately, the absolute threshold for the perception of vertical whole-body vibration in the frequency range 1–100 Hz is 0.01 m/s^2 rms; a magnitude of 0.1 m/s^2 will be easily noticeable; magnitudes around 1 m/s^2 rms are usually considered uncomfortable; magnitudes of 10 m/s^2 rms are usually dangerous. The precise values depend on vibration frequency and exposure duration, and they are different for other axes of vibration [1–3].

To a useful approximation, a doubling of vibration magnitude corresponds to a doubling of the discomfort. A halving of vibration magnitude can therefore produce a considerable improvement in comfort.

2. Effects of Vibration Frequency and Direction

The dynamic responses of the body and the relevant physiological and psychological processes dictate that subjective reactions to vibration depend on vibration frequency and vibration direction. Frequency weightings are given in British Standard 6841 [2]; this is currently the most up-to-date published standard giving guidance concerned with vibration discomfort.

3. Effects of Vibration Duration

Vibration discomfort generally increases with increasing duration of vibration. The precise rate of change may depend on many factors, but a simple fourth-power time dependency is sometimes used to approximate how discomfort varies with exposure durations from the shortest possible shock to a full day of vibration exposure (i.e., (acceleration)^4 × duration = constant [1]). This time dependency appears to be more consistent with available information and expectations than either an "energy time dependence" or the ISO 2631 [4] time dependence (see Secs. IV.B and IV.F).

4. Vibration in Buildings

Acceptable magnitudes of vibration in buildings are close to vibration perception thresholds. The effects of vibration in buildings are assumed to depend on the use of the building in addition to the vibration frequency, direction, and duration. Guidance is given in various standards (e.g., American National Standards Institute [5]; International Organization for Standardization [6]). British Standard 6472 [7] defines a procedure that combines the assessment of vibration and shock in buildings by using the "vibration dose value" (see Sec. IV.B.1).

B. Effects on Health

Acute injury to the body can be caused by short exposure to high magnitudes of whole-body vibration or shock. Chronic disease may arise from prolonged exposures to vibration.

Epidemiological studies have reported disorders among persons exposed to vibration from occupational, sport, and leisure activities. The studies do not all agree on either the type or the extent of disorders, and rarely have the findings been related to measurements of vibration exposure. However, it is widely believed that disorders of the back (back pain, displacement of intervertebral disks, degeneration of spinal vertebrae, osteoarthritis, etc.) may be associated with vibration exposure (see Chapter 5 and Appendix 5 of *The Handbook of Human Vibration* [1]).

For persons exposed to vibration, there are often other potential causes of some of the reported disorders of the back (e.g., poor sitting postures, heavy lifting, tobacco smoking). It is not always possible to confidently conclude that a back disorder is solely, or primarily, caused by vibration.

Other disorders claimed to be due to occupational exposures to whole-body vibration include abdominal pain, digestive disorders, urinary frequency, prostatitis, hemorrhoids, balance and visual disorders, headaches, and sleeplessness.

1. Method of Vibration Evaluation

Epidemiological data alone are not sufficient to define how to evaluate whole-body vibration so as to predict the relative risks to health of different types of exposure. A consideration of such data in combination with an understanding of biodynamic responses and subjective responses is used to provide current guidance. The dependence of health effects of vibration and shock on the frequency, direction, and duration of motion is assumed to be the same as, or similar to, that for vibration discomfort. However, it is assumed that the total exposure, rather than average exposure, is important, and so a dose measurement is useful.

International Standard 2631 [4, 8] and American National Standard S3.18 [9] give guidance for the evaluation of whole-body vibration with respect to health or safety. These standards define exposure limits (see Figure 2), which are "set at approximately half the level considered to be the threshold of pain (or limit of voluntary tolerance) for healthy human subjects." Although the latest version of ISO 2631 [8] was published in 1985, it is similar to the 1974 version [4], which was based on research conducted before 1970. British Standard 6841 [2] is more up to date and broadly consistent with a draft revision of the International Standard. Figure 2 also shows an action level for vertical vibration derived from British Standard 6841 [2].

The *vibration dose value* can be considered to be the magnitude of a 1-s duration of vibration that will be equally as severe as the measured vibration. The vibration dose value uses a fourth-power time dependence to accumulate vibration severity over the exposure period from the shortest possible shock to a full day of vibration (see BS 6841, [2]):

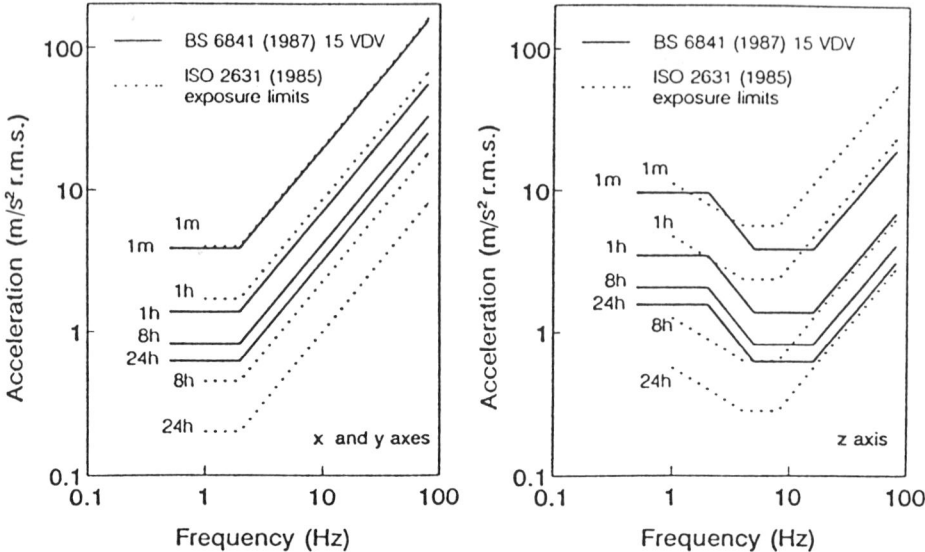

Figure 2 Comparison of ISO 2631 [8] exposure limits with an action level based on a vibration dose value of 15 m/s$^{1.75}$ from BS 6841 [2]. When seated: x axis = fore-and-aft; y axis = lateral; z axis = vertical (From Griffin [1].)

$$\text{Vibration dose value} = \left[\int_{t=0}^{t=\infty} a(t)^4 \, dt\right]^{1/4}$$

The vibration dose value procedure is applicable over a wider range of durations than the exposure limit in ISO 2631 [4] and can be used to evaluate the severity of repetitive shocks. This fourth-power time dependency is also considerably simpler to use than the time dependency in ISO 2631 [4].

C. Effects on Performance

Vibration may interfere with the acquisition of information (e.g., by the eyes), the output of information (e.g., by hand or foot movements), or the complex central processes that relate input to output (e.g., learning, memory, decision making). Effects of oscillatory motion on human performance may impair safety.

Figure 3 illustrates the various component parts of human activity where vibration may affect performance. The greatest effects, and the greatest understanding of the effects, occur with input processes (mainly vision) and output processes (mainly continuous manual control). In both cases there may be disturbance occurring entirely outside the body (e.g., vibration of a viewed display or vibration of a hand-held control), disturbance at the input or output (e.g., movement of the eye or hand), and disturbance affecting the peripheral nervous system (i.e., afferent or efferent system). Central processes may also be affected by vibration, but understanding is currently too limited to make confident generalized statements.

The effects of vibration on vision and manual control are primarily caused by the movement of the affected part of the body (i.e., eye or hand). The effects may be reduced by reducing the transmission of vibration to the eye or to the hand or by making the task less

Occupational Human Vibration

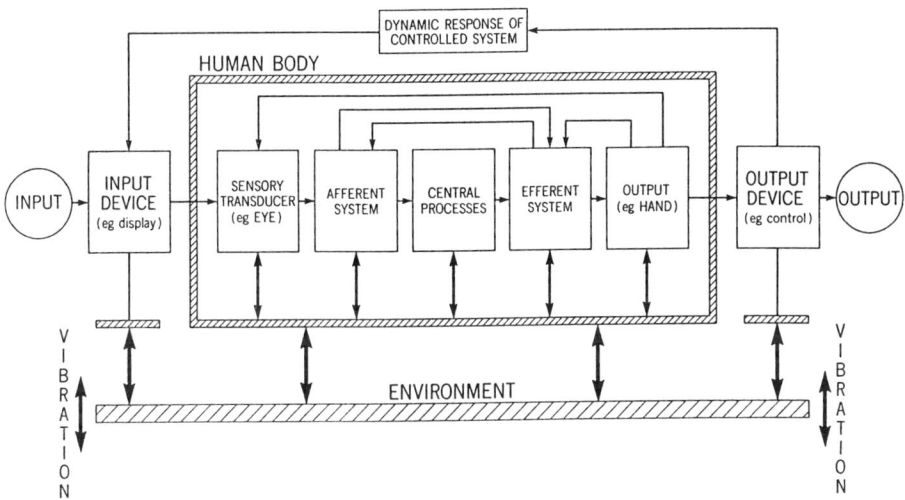

Figure 3 Information flow in a simple system and the areas where vibration may affect human activities. (From Griffin [1].)

susceptible to disturbance (e.g., increasing the size of a display or reducing the sensitivity of a control). Often, the effects of vibration on vision and manual control can be reduced by redesign of the task.

Reading from a book in a moving vehicle may be difficult if there is much vibration. This may arise because the book is vibrating, the eye is vibrating, or both the book and the eye are vibrating. There are many variables that affect visual performance in these conditions, with some variables being difficult to quantify. Published reports of experiments may appear to present a confused picture unless all the relevant variables are taken into account [10].

The most obvious effect of vibration on a control task is the direct mechanical jostling of the hand causing unwanted movement of the control. This is sometimes called breakthrough or feedthrough. Because the errors produced by this means are directly caused by the vibration, they are statistically correlated with the vibration and are also called vibration-correlated error. The inadvertent movement of the pencil caused by jostling while writing in a vehicle is a form of vibration-correlated error. Indeed, an attempt to draw a straight line will result in a graphical display on the paper of the vibration-induced movements of the hand.

Time series analysis techniques can be used to quantify how much tracking error in a control task is correlated with the vibration. In a simple tracking task in which the operator is required to follow movements of a target, some of the error will be correlated with the target movements. This is called input-correlated error and often mainly reflects the inability of an operator to follow the target without delays inherent in visual, cognitive, and motor activity. The part of the tracking error that is not correlated with either the vibration or the tracking task is called the remnant. This includes operator-generated noise and any source of non-linearity: drawing a freehand straight line does not result in a perfectly straight line even in the absence of environmental vibration. The effects of vibration on vision can result in increased remnant with some tracking tasks. There are studies showing that vibration, usually at frequencies above about 20 Hz, can interfere with neuromuscular processes (e.g., [11]) and also increase the remnant [12].

It seems probable that simple cognitive tasks (e.g., simple reaction time) are not affected by vibration other than by changes in arousal or motivation or by direct effects on input and output processes. This may also be true for some complex cognitive tasks. However, the sparsity and diversity of experimental studies allows the possibility of real and significant cognitive effects of vibration [13].

Vibration may influence fatigue, but there is little relevant scientific evidence and none that supports the complex form of the so-called fatigue-decreased proficiency limit offered in International Standard 2631 [4, 8].

D. Motion Sickness

Motion sickness is not an illness but a normal response to motion that adversely affects many fit and healthy people. A variety of motions can cause sickness and reduce the comfort, impede the activities, and degrade the well-being of both those directly affected and those associated with them.

Formulas allow the prediction of the incidence of vomiting and motion illness caused by vertical oscillation over the range of 0.08–0.5 Hz [2,14,15]. However, the probability of sickness in an individual also depends on individual susceptibility, activities, and other aspects of the environment.

E. Control of the Whole-Body Vibration

Whenever possible, the vibration should be reduced at the source. In the case of vehicles, this may involve reducing the undulations of the terrain or reducing the speed of travel. Reduction at the source is not always practicable and sufficient, so ergonomists may be asked to alleviate a problem by other means.

An understanding of the characteristics of the vibration environment and the route for transmission of vibration to the body is important. For example, the magnitude of vibration often varies with location: lower magnitudes will be experienced in some areas.

Seats and beds can be designed to attenuate vibration (see Sec. IV.E.1). There is no satisfactory equivalent method of reducing the transmission of vibration from a floor to standing persons. Table 1 lists some preventive measures that may be considered.

Most seats exhibit a resonance at low frequencies that results in higher magnitudes of vertical vibration occurring on the seat than on the floor! At high frequencies there is usually attenuation of vibration. In use, the resonance frequencies of common seats are in the region of 4 Hz. The amplification at resonance is influenced by the inherent damping in the seat. Increases in the damping of the seat cushioning tend to reduce the amplification at resonance but increase the transmissibility at high frequencies [16].

Large variations in transmissibility between seats result in significant differences in the vibration experienced by people [16]. Seat transmissibility can be measured in the laboratory or in the field.

1. Isolation Efficiency of Seats

The seat effective amplitude transmissibility (SEAT) provides a simple numerical indicator of the isolation efficiency of a seat for a specific application [1]. A SEAT value greater than 100% indicates that, overall, the vibration on the seat is "worse" than the vibration on the floor. Values below 100% indicate that the seat has provided some useful attenuation. Seats should be designed to have the lowest SEAT value.

Table 1 Summary of Preventive Measures to Consider When Persons Are Exposed to Whole-Body Vibration

Group	Action
Management	Seek technical advice.
	Seek medical advice.
	Warn exposed persons
	Train exposed persons.
	Review exposure times.
	Institute policy on removal from work.
Machine manufacturers	Measure vibration.
	Design to minimize whole-body vibration.
	Optimize suspension design.
	Optimize seating dynamics
	Practice ergonomic design to provide good posture, etc.
	Provide guidance on machine maintenance.
	Provide guidance on seat maintenance.
	Provide warning of dangerous vibration.
Technical at workplace	Measure vibration exposure.
	Provide appropriate machines.
	Select seats with good attenuation.
	Maintain machines.
	Inform management.
Medical	Perform preemployment screening.
	Perform routine medical checks.
	Record all signs and reported symptoms.
	Warn workers with predisposition.
	Advise on consequences of exposure.
	Inform management.
Exposed persons	Use machine properly.
	Avoid unnecessary vibration exposure.
	Check that seat is properly adjusted.
	Adopt good sitting posture.
	Check condition of machine.
	Inform supervisor of vibration problems.
	Seek medical advice if symptoms appear.
	Inform employer of relevant disorders.

Source: Adapted from Chap. 5 of Ref. 1.

2. Suspension Seats

Suspension seats have a separate suspension mechanism located beneath the seat pan. These seats, used in some off-road vehicles, trucks, and coaches, have low resonance frequencies (around 2 Hz) and so can attenuate vibration at frequencies above about 3 Hz. The transmissibilities of these seats are usually determined by the seat manufacturer, but their isolation efficiencies vary with operating conditions.

3. Seat Testing Standards

Standards are becoming available for the measurement of seat dynamic performance and the reporting of relevant values. Precautions are required to protect the safety of human subjects if they are used to test seat performance in experimental conditions [17].

F. National and International Standards

No precise limit can be offered to prevent disorders caused by whole-body vibration, but standards define useful methods of quantifying vibration severity. British Standard 6841 [2] offers the following guidance.

> High vibration dose values will cause severe discomfort, pain and injury. Vibration dose values also indicate, in a general way, the severity of the vibration exposures which caused them. However there is currently no consensus of opinion on the precise relation between vibration dose values and the risk of injury. It is known that vibration magnitudes and durations which produce vibration dose values in the region of $15 ms^{-1.75}$ will usually cause severe discomfort. It is reasonable to assume that increased exposure to vibration will be accompanied by increased risk of injury.

At high vibration dose values, prior consideration of the fitness of the exposed persons and the design of adequate safety precautions may be required. The need for regular checks on the health of routinely exposed persons may also be considered.

In the United Kingdom, the value of 15 m/s$^{-1.75}$ has been called a *tentative action level* and not a limit [1]. It may be appropriate for organizations to limit vibration or repeated shock exposures to higher or lower values depending on the situation. The vibration dose value provides a robust measure by which highly variable and complex exposures can be compared. The tentative action level merely serves to indicate the approximate values that might be excessive. Figure 4 illustrates the rms accelerations corresponding to a vibration dose value (VDV) of 15 m/s$^{-1.75}$ for exposures of between 1 s and 24 h.

Figure 4 shows that there are some significant differences between the vibration dose value action level and the exposure limit given in the old ISO 2631 [4]. Unlike the ISO exposure limit, the VDV action level does not allow very high magnitudes at short durations or require very low magnitudes for long duration exposures. (Figure 4 also indicates that an "equal energy" level allows even higher values for short durations yet very low magnitudes for long durations [1].) It would be unwise to exceed the action level without consideration of the possible health effects of an exposure to vibration or shock.

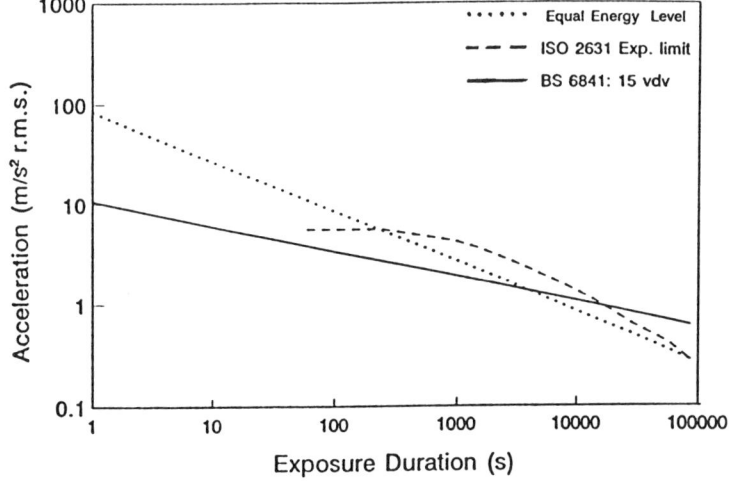

Figure 4 Action level corresponding to 15 m/s$^{1.75}$ (see BS 6841 [2]) compared with ISO 2631 [4] exposure limit and level based on equal energy

The Machinery Safety Directive of the European Economic Community states that machinery must be designed and constructed so that hazards resulting from vibration produced by the machinery are reduced to the lowest practicable level, taking into account technical progress and the availability of means of reducing vibration [18]. The Machinery Safety Directive encourages the reduction of vibration by means additional to reduction at source. There are standards specifying how the attenuation of vibration produced by vehicle seats are to be determined. The provision of good seating (and the dynamic testing of seating) will therefore receive greater attention in the future design of vehicles.

Part 3 of ISO 2631 [4] suggests magnitudes of vertical oscillation in the range of 0.1–0.63 Hz expected to produce a 10% incidence of sickness in sitting or standing fit young men over exposures of 30 min, 2 h, and, tentatively, 8 h [8]. British Standard 6841 [2] defines a motion sickness dose value (MSDV) based on a frequency weighting W_f (i.e., a filter) and a time dependency. The greatest sensitivity to acceleration is in the range 0.125–0.25 Hz, with a rapid reduction in sensitivity at higher frequencies. The exposure duration t (s) and the frequency-weighted rms acceleration a_{rms} (m/s² rms), may be used to compute the motion sickness dose value:

$$\mathrm{MSDV}_z = (a_{rms}^2 t)^{1/2}$$

The percentage of unadapted adults who may vomit is then given by

$$\text{Percentage vomiting} = \mathrm{MSDV}_z/3$$

This relation is based on exposures lasting from about 20 min to 6 h with a incidence of vomiting of up to 70%.

V. HAND-TRANSMITTED VIBRATION

Prolonged regular exposure of the fingers or the hands to vibration or repeated shock can give rise to various signs and symptoms of disorder. The precise extent and interrelation of the signs and symptoms is not fully understood, but five types of disorders can be identified (see Table 2).

The disorders listed in Table 2 may be interconnected: more than one disorder can affect a person at the same time, and it is possible that the presence of one disorder facilitates the appearance of another. The onset of each disorder is dependent on several variables, such as the vibration characteristics, the dynamic response of the fingers or hand, individual susceptibility to damage, and other aspects of the environment. The terms *vibration syndrome* and *hand-arm vibration syndrome* are sometimes used to refer in a vague way to some combination of the effects listed in Table 2.

Table 2 Five Types of Disorders Associated with Hand-Transmitted Vibration Exposures

Type	Disorder
A	Circulatory disorders
B	Bone and joint disorders
C	Neurological disorders
D	Muscle disorders
E	Other general disorders (e.g., central nervous system)

Source: Ref. 32.

A. Vascular Effects (Vibration-Induced White Finger)

The first published cases of vibration-induced white finger are generally acknowledged to be those reported in Italy by Loriga [19]. Not many years later, cases were documented at limestone quarries in Indiana (e.g., Hamilton [20]). Among workers reported to be using pneumatic hammers in the Indiana quarries, there were attacks of numbness and blanching of the fingers that came on suddenly under the influence of cold and then disappeared. The prevalence of the symptoms was 86% among 50 men working on granite, 69% among 38 men working on marble, and 56% among 78 men working on limestone. It was reported that the condition appeared within about 1 year of first using the tools. A survey conducted 60 years later in the same quarry reported no change in the design of the air hammers used for stonecutting and found an 80% prevalence of vibration-induced white finger among stonecutters [21]. Vibration-induced white finger has been reported to occur in many other widely varied occupations in which the fingers are exposed to vibration [1].

1. Signs and Symptoms

The condition now most commonly known as vibration-induced white finger (VWF) is characterized by intermittent blanching of the fingers. The fingertips are first to blanch, but the affected area may extend to all of one or more fingers with continued vibration exposure. Attacks of blanching are precipitated by cold and therefore usually occur in cold conditions or when handling cold objects. The blanching lasts until the fingers are rewarmed and vasodilation allows the return of the blood circulation.

Many years of vibration exposure often occur before the first attack of blanching is noticed. Some persons have other signs and symptoms, such as numbness and tingling; cases of cyanosis and, rarely, gangrene have also been reported. It is not yet clear to what extent these signs are causes of, caused by, or unrelated to, the attacks of white finger.

2. Diagnosis

The diagnosis of VWF must recognize that there are other conditions that can cause similar signs and symptoms. An individual cannot be assumed to have the condition merely because there are attacks of blanching. It is necessary to exclude other known causes of similar symptoms (by medical examination) and also to exclude so-called primary Raynaud's disease (also called constitutional white finger). This exclusion cannot yet be achieved with complete confidence, but if there is no family history of the symptoms, if the symptoms did not occur before the first significant exposure to vibration, and if the symptoms and signs are confined to areas in contact with the vibration (e.g., the fingers, not the ears, etc.), they will often be assumed to indicate VWF.

Diagnostic tests for vibration-induced white finger are being developed, but, at present, most do not provide sufficient agreement with symptoms and other signs to be used as infallible indicators of the disease. The measurement of finger systolic blood pressure during finger cooling is useful [22], but many other tests are also in use.

The severity of the effects of vibration are sometimes recorded by reference to the "stage" of the disorder. The staging of vibration-induced white finger is based on the verbal statements of the affected person. In the Taylor-Pelmear system, the stage of vibration-induced white finger was determined by the presence of numbness and tingling, the areas affected by blanching, the frequency of blanching, the time of year when blanching occurs, and the extent of interference with work and leisure activities (see Griffin [1]). A simpler procedure, the Stockholm Workshop staging system, was subsequently evolved [23] (see Table 3). In this system, the staging only compounds the frequency of attacks of blanching with the areas affected by blanching.

Occupational Human Vibration

Table 3 Stockholm Workshop Scale for the Classification of Vibration-Induced White Finger[a]

Stage	Grade	Description
0	—	No attacks
1	Mild	Occasional attacks affecting only the tips of one or more fingers
2	Moderate	Occasional attacks affecting distal and middle (rarely also proximal) phalanges of one or more fingers
3	Severe	Frequent attacks affecting all phalanges of most fingers
4	Very severe	As in stage 3, with trophic skin changes in the fingertips

[a]If a person has stage 2 in two fingers of the left hand and stage 1 in a finger on the right hand, the condition may be reported as 2L(2)/1R(1). There is no defined means of reporting the condition of digits when the condition varies between digits on the same hand. The scoring system is more helpful when the extent of blanching is to be recorded]
Source: Ref. 23.

A system of recording the areas of the digits affected by blanching is known as the scoring system (see Figure 5). The blanching score for the hands shown in Figure 5 is 01300_{right}, 01366_{left}. The scores correspond to areas of blanching on the digits commencing with the thumb. On the fingers a score of 1 is given for blanching on the distal phalanx, a score of 2 for blanching on the middle phalanx, and a score of 3 for blanching on the proximal phalanx. On the thumbs the scores are 4 for the distal phalanx and 5 for the proximal phalanx. The blanching score may be based on statements from the affected person or on the visual observations of a designated observer (e.g., a nurse).

B. Neurological Effects

In recent years the existence of neurological effects of hand-transmitted vibration (e.g., numbness; tingling; elevated sensory thresholds for touch, vibration, temperature, and pain; and reduced nerve conduction velocity) have been recognized as separate effects of vibration and not merely symptoms of vibration-induced white finger [1]. A method of reporting the extent of vibration-induced neurological effects of vibration has been proposed (see Table 4). This staging system is not directly related to the results of any specific objective test. The sen-

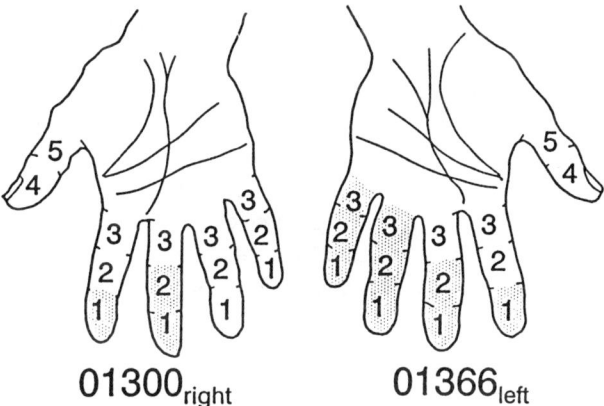

Figure 5 Method of scoring the areas of the digits affected by blanching. (From Griffin [1,32].)

Table 4 Proposed Sensoineural Stages of the Effects of Hand-Transmitted Vibration

Stage	Symptoms
0_{SN}	Exposed to vibration but no symptoms
1_{SN}	Intermittent numbness with or without tingling
2_{SN}	Intermittent or persistent numbness, reduced sensory perception
3_{SN}	Intermittent or persistent numbness, reduced tactile discrimination and/or manipulative dexterity

Source: Ref. 33.

sorineural stage is a subjective impression of a physician that may be based on the statements of the affected person or the results of clinical or scientific testing.

Vibration measurement and evaluation methods evolved to protect against VWF are also used to assess all exposures to hand-transmitted vibration irrespective of the disorder. There is currently no established dose–effect information for disorders other than VWF.

C. Muscular Effects

The research literature includes several reports of muscle atrophy among users of vibrating tools. More commonly, users report difficulty with their grip, including reduced dexterity or reduced grip strength [1]. Many of the reports are derived from symptoms rather than signs and could be a reflection of neurological problems. Measurements of muscle function have rarely been obtained using repeatable tests. Muscle activity may be of great importance to tool users because a secure grip may be essential to the performance of the job and safe control of the tool. The presence of vibration on a handle may encourage the adoption of a tighter grip than would otherwise be used, and a tight grip may increase the transmission of vibration to the hand. An effect of vibration that results in reduced grip may therefore help to protect an operator from further effects of vibration!

D. Articular Effects

Many surveys of the users of hand-held tools have found evidence of bone and joint problems, most often among those operating percussive tools, such as those used in metalworking jobs and mining and quarrying [1]. It is speculated that some characteristic of such tools, probably the low-frequency shocks, is responsible. Some of the reported injuries relate to specific bones and suggest the existence of cysts, vacuoles, decalcification, or other osteolysis, degeneration, or deformity of the carpal, metacarpal, or phalangeal bones. Osteoarthrosis and olecranon spurs at the elbow and other problems at the wrist and shoulder are also documented [1]. Notwithstanding the evidence of many research publications, there is not universal acceptance that vibration is the cause of these problems, and there is currently no dose–effect relation that predicts their occurrence. It seems that adherence to current guidance for the prevention of vibration-induced white finger may provide reasonable protection from articular problems.

E. Other Effects

The reported effects of hand-transmitted vibration on the body are not confined to the fingers, hands, and arms. Many studies have found a high incidence of problems such as headaches and sleeplessness among tool users. Although these are real problems to those affected, these subjective effects are not accepted as real by all researchers. Some current research seeks to establish a physiologial basis for such symptoms.

F. Causes of Hand-Transmitted Vibration

Exposures to vibration from tools vary greatly according to individual tool design and method of use, so it is not possible to categorize individual tool types as "safe" or "dangerous." However, Table 5 lists tools and processes that are sometimes a cause for concern.

The pathogenesis of VWF is not known. Neither is it accurately known what physical conditions (i.e., vibration magnitudes, vibration frequencies, vibration directions, exposure durations, grip forces, etc.) are responsible. Nevertheless, some general information has been used to formulate standardized methods of measuring and evaluating vibration (see Sec. V.H). Some tentative dose–effect guidelines and action levels have been proposed, but there are insufficient epidemiological data to be able to estimate their likely accuracy in any particular application.

Table 5 Tools and Processes Potentially Associated with Vibration Injuries

Type of tool	Examples of tool type
Percussive metalworking tools	Riveting tools
	Caulking tools
	Chipping tools
	Chipping hammers
	Fettling tools
	Hammer drills
	Clinching and flanging tools
	Impact wrenches
	Swaging
	Needle guns
Grinders and other rotary tools	Pedestal grinders
	Hand-held grinders
	Hand-held sanders
	Hand-held polishers
	Flex-driven grinders/polishers
	Rotary burring tools
Percussive hammers and drills used in mining, demolition, and road construction	Hammers
	Rock (etc.) drills
Forest and garden machinering	Chain saws
	Antivibration chain saws
	Brush saws
	Mowers and shears
	Barking machines
Other processes and tools	Nut runners
	Shoe-pounding-up machines
	Concrete vibrothickeners
	Concrete leveling vibrotables
	Motorcycle handle bars

Source: Adapted from Chap. 14 of Ref. 1.

G. Control of Hand-Transmitted Vibration

The protection of workers exposed to hand-transmitted vibration involves actions from management, tool manufacturers, technicians, and medical doctors at the workplace and from the tool user. Table 6 summarizes some of the actions that may be appropriate. The optimum means of reducing vibration varies according to the cause of the vibration exposure. However, the first principles are similar: reduction at the source, reduced transmission, and reduced individual susceptibility.

Measurements of the vibration will determine whether any other tool or process could give a lower vibration severity. Reduction of exposure time may include the provision of breaks from exposure during the day and, if possible, prolonged periods away from vibration exposure.

Table 6 Some Preventive Measures to Considers When Persons Are Exposed to Hand-Transmitted Vibration

Group	Action
Management	Seek technical advice.
	Seek medical advice.
	Warn exposed persons.
	Train exposed persons.
	Review exposure times.
	Institute policy on removal from work.
Tool manufacturers	Measure tool vibration.
	Design tools to minimize vibration.
	Use ergonomic design to reduce grip force, etc.
	Design to keep hands warm.
	Provide guidance on tool maintenance.
	Provide warning of dangerous vibration.
Technical at workplace	Measure vibration exposure.
	Provide appropriate tools.
	Maintain tools.
	Inform management.
Medical	Perform preemployment screening.
	Perform routine medical checks.
	Record all signs and reported symptoms.
	Warn workers with predisposition.
	Advise on consequences of exposure.
	Inform management.
Tool user	Use tool properly.
	Avoid unnecessary vibration exposure.
	Minimize grip and push forces.
	Check condition of tool.
	Inform supervisor of tool problems.
	Keep warm.
	Wear gloves when safe to do so.
	Minimize smoking.
	Seek medical advice if symptoms appear.
	Inform employer of relevant disorders.

Source: Adapted from Chap. 19 of Ref. 1.

Gloves are sometimes recommended as a means of reducing the transmission of vibration to the hands. However, measurements using the frequency weightings in current standards show that commonly available gloves do *not* normally provide effective attenuation of vibration. Gloves and "cushioned" handles may reduce the transmission of high frequencies of vibration, but current standards imply that these are not usually the primary cause of disorders [24]. American National Standard S3.40 [25] provides a guide to the measurement and evaluation of gloves intended to reduce exposure to vibration transmitted to the hand. Other standards are in preparation.

Workers who are exposed to vibration magnitudes sufficient to cause injury should be warned of the possibility of vibration injuries, advised of the symptoms to look out for, and told to seek medical attention if the symptoms appear.

H. National and International Standards

1. International Standard 5349 (1986), British Standard 6842 (1987), and American National Standard S3.34 (1986)

These standards [26-28] specify similar measurement methods for determining the severity of hand-transmitted vibration over the frequency range 8-1000 Hz. They use the same frequency weighting (called W_h in BS 6842 [27]). This weighting is applied to measurements of vibration acceleration in each of the three axes of vibration at the point of entry of vibration to the hand. Frequency-weighted values measured on different tools may be compared. The standards imply that if two tools expose the hand to vibration for the same period of time, the tool having the lowest frequency-weighted acceleration will be least likely to cause injury or disease.

The three standards are not identical. The international and British standards are internally consistent, but BS 6842 expresses caution with regard to dose-effect data and restricts the calculations to a prevalence of 10% (see Sec. V.H.3). The American standard offers different dose-effect guidance and allows different measuring methods in some circumstances.

Threshold limit values for hand-transmitted vibration were proposed by the American Conference of Government Industrial Hygienists (ACGIH) in 1986 [29]. The measurement method is broadly consistent with that defined in International Standard 5349 [26], but the limits fall in a stepwise fashion as the daily exposure duration increases from 1 h to 8 h in a day.

Although the above standards only cover a frequency range up to 1000 Hz, there have been suggestions that higher frequencies might be responsible for some of the observed effects on the fingers and hand (e.g., NIOSH [30]). The frequency weighting and frequency range specified in International Standard 5349 [26] is in widespread use throughout the world. The frequency weighting and the frequency range may be improved with further understanding, but there is no current consensus on what change would be appropriate.

2. Measuring Hand-Transmitted Vibration

When measuring vibration on tools, care is required to obtain representative values for appropriate operating conditions. There can be significant practical difficulties in making valid measurements using common commercial instrumentation (especially when there are high shock levels). It is wise to determine spectra and inspect the acceleration time histories before accepting the validity of any measurements.

3. Exposure Time

Occupational exposures to hand-transmitted vibration can have widely varying daily exposure durations, from a few seconds to many hours. Often, exposures are intermittent. To enable a daily exposure to be reported simply, the standards refer to an equivalent 4-h (ISO 5349 [26]) or an equivalent 8-h (BS 6842 [27]) exposure.

Table 7 shows a relation between years of vibration exposure, 4-h energy-equivalent frequency-weighted acceleration, and the prevalence of finger blanching as proposed in an Annex to ISO 5349 [26]. These relationships are illustrated graphically in Figure 6. The values in Figure 6 and Table 7 refer to frequency-weighted acceleration referenced to the frequency range 8–16 Hz. Figure 7 shows how the magnitudes required for a predicted VWF prevalence of 10% depend on vibration frequency in the range 8–1000 Hz for exposure durations in the range 1–25 years.

4. Prevalence of Vibration-Induced White Finger

It should not be expected that the percentage of affected persons in any group of exposed persons will always closely match the values shown in Table 7 or Figures 6 and 7. Apart from the possible inappropriateness of the assumed frequency weighting, time dependency, and idealized dose–effect information, the numbers affected will depend on the rates at which new persons enter and leave the exposed group. Neither the average exposure time nor the mean latency (i.e., the average period of vibration exposure before those with symptoms of VWF develop the condition) are appropriate measures of the exposure period for this calculation.

British Standard 6842 [27] provides an Annex that summarizes current knowledge and the assumptions concerning dose–effect data for VWF but does not include the detailed table of prevalence rates given in the International Standard. It is stated that vascular symptoms do not normally occur if the frequency-weighted acceleration is below 1 m/s^2 rms.

5. Machinery Safety Directive

The Machinery Safety Directive of the EEC states that machinery must be designed and constructed so that hazards resulting from vibration produced by the machinery are reduced to the lowest practicable level, taking into account technical progress and the availability of means of reducing vibration.

It is proposed that the instruction handbooks for hand-held and hand-guided machinery should specify the equivalent acceleration to which the hands or arms are subjected where this

Table 7 Number of Years Before Blancing Develops in 10–50% of Vibration-Exposed Persons

Weighted acceleration, $a_{hw(eq, 4h)}$ (m/s^2, rms)	Percentage of population affected by finger blancing				
	10%	20%	30%	40%	50%
2	15	23	>25	>25	>25
5	6	9	11	12	14
10	3	4	5	6	7
20	1	2	2	3	3
50	<1	<1	<1	1	1

Source: Ref. 26.

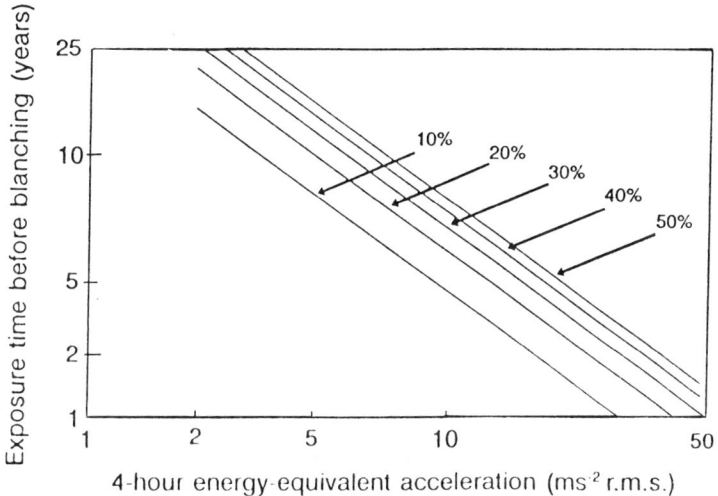

Figure 6 Years of exposure to 4-h energy-equivalent frequency-weighted hand-transmitted vibration required for finger blanching to 10–50% of exposed persons according to ISO 5349 [26].

exceeds some stated value (currently proposed as a frequency-weighted acceleration of 2.5 m/s² rms). The relevance of any such value will depend on the test conditions to be specified in other standards. In normal operation, very many hand-held vibrating tools exceed this value at some time. The specified tests will not encompass all realistic exposures and so may not

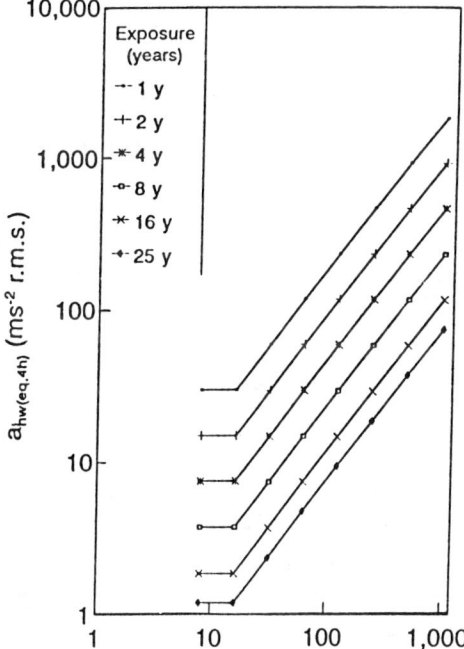

Figure 7 Acceleration magnitudes giving 10% prevalence of VWF for 4-h daily exposures according to International Standard 5349 [26].

represent the exposures received in some jobs. Additionally, the different exposure durations involved in different jobs will carry different degrees of risk.

Standards defining test conditions for the measurement of vibration on chipping and riveting hammers, rotary hammers and rock drills, grinding machines, pavement breakers, and various garden and forestry equipment (including chain saws) are in preparation [31].

VI. CRITICAL REVIEW OF CURRENT STATUS

A. Whole-Body Vibration

The oscillatory motions to which the body is exposed are complex. Vibration may affect comfort, the performance of many different activities, and the physiology and pathology of the body. Although understanding of the interactions between vibration and the body is far from complete, modern standards attempt to predict which types of vibration cause which effect. Some recent research effort has been directed toward standardization rather than toward basic understanding of the mechanisms involved. The complexities of the relevant phenomena are too great to be incorporated within a reasonable standard and sometimes too great to be of interest to those constructing standards.

Ergonomists should be aware of the relevant standards but also aware of the uneven scientific support for such standards. For example, there is no future in designing a hand control or a visual display for a vibration environment based on the information contained in current vibration standards. The design or evaluation process must always include an identification of the relevant variables. A review of the relevant standards may reveal that they do not consider these variables (e.g., control gain, control order, viewing distance, phase between motion of the body and motion of the control or display). Ergonomists considering the effects of vibration should therefore extend their reading beyond standards and into the scientific literature.

B. Hand-Transmitted Vibration

Recognition of the existence of the various effects of hand-transmitted vibration has increased in recent years. Nevertheless, there may be confusion over what is known and what it is desirable to know. It is unlikely that current standards accurately reflect the true influence of vibration magnitude, frequency, direction, or duration on the production of vibration-induced white finger. Current standards fail to provide any quantitative prediction of the factors causing other effects of hand-transmitted vibration.

VII. FUTURE CONCERNS

International agreement on improved standards for evaluating whole-body vibration may allow relevant researchers to turn from the politics and compromise involved in standardization to the clean air of scientific investigation relevant to the true effects of vibration. The complexity of the interactions between oscillatory motion (i.e., vibration and shock) and the functions of the body are great. Systematic multidisciplinary approaches are required to make parallel advances in relevant aspects of biodynamics, physiology, pathology, epidemiology, and psychology (i.e., subjective assessment and performance effects). Knowledge has not, in the past, been advanced by short-term attacks on the subject; there is no substitute for traditional science based on developing an understanding that leads to modeling and hypothesis testing. In this field there are many hypotheses to test en route to establishing the information needed by ergonomics and others.

The output from a better understanding of the effects of whole-body vibration need not be new standards. The complex interactions between variables may be too great for a standard and not amenable to the compromise inherent in standardization. Ergonomic design guides may be a more appropriate means of conveying such information.

Current standards for hand-transmitted vibration provide guidance that cannot be ignored. However, those concerned with the design or evaluation of situations involving hand-transmitted vibration should anticipate that further understanding of the relevant pathology, physiology, and biodynamics may result in significant changes to the methods of assessing the safety of exposures to hand-transmitted vibration.

REFERENCES

1. M. J. Griffin, *Handbook of Human Vibration*, Academic Press, London, 1990.
2. British Standards Institution, *Measurement and Evaluation of Human Exposure to Whole-Body Mechanical Vibration and Repeated Shock*, BS 6841, 1987.
3. K. C. Parsons and M. J. Griffin, Whole-body vibration perception thresholds, *J. Sound Vibration* 121(2):237–258 (1988).
4. International Organization for Standardization, *Guide for the Evaluation of Human Exposure to Whole-Body Vibration*, ISO, 2631, 1974.
5. American National Standards Institute, *Guide to the Evaluation of Human Exposure to Vibration in Buildings*, ANSI S3.29-1983 (ASA 48-1983), 1983.
6. International Organization for Standardization, *Evaluation of Human Exposure to Whole-Body Vibration. Part 2: Continuous and Shock-Induced Vibration in Buildings*, ISO 2631-2, 1989.
7. British Standards Institution, *Evaluation of Human Exposure to Vibration in Buildings (1 Hz to 80 Hz)*, BS 6472, 1992.
8. International Organization for Standardization, *Evaluation of Human Exposure to Whole-Body Vibration. Part 1: General Requirements*, ISO, 2631/1, 1985.
9. American National Standards Institute, *Guide to the Evaluation of Human Exposure to Whole-Body Vibration*, ANSI S3.18-1979, 1979.
10. M. J. Griffin and C. H. Lewis, A review of the effects of vibration on visual acuity and continuous manual control. Part 1: Visual acuity, *J. Sound Vibration* 56(3):383–413 (1978).
11. G. M. Goodwin, D. I. McCloskey, and P. B. C. Matthews, The contribution of muscle afferents to kinaesthesia shown by vibration induced illusions of movement and by the effects of paralysing joint afferents, *Brain* 95:705–748 (1972).
12. R. W. McLeod and M. J. Griffin, A review of the effects of translational whole-body vibration on continuous manual control performance, *J. Sound Vibration* 133(1):55–115 (1989).
13. N. Sherwood and M. J. Griffin, Evidence of impaired learning during whole-body vibration, *J. Sound Vibration* 153(2):219–225 (1992).
14. J. F. O'Hanlon and M. E. McCauley, Motion sickness incidence as a function of the frequency and acceleration of vertical sinusoidal motion, *Aerosp. Med.* 45(4):366–369 (1974).
15. A. Lawther and M. J. Griffin, Prediction of the incidence of motion sickness from the magnitude, frequency, and duration of vertical oscillation, *J. Acoust. Soc. Am.* 82(3):957–966 (1987).
16. C. Corbridge, M. J. Griffin, and P. Harborough, Seat dynamics and passenger comfort, Institute of Mechanical Engineers, Part F, *J. Rail Rapid Transit* 203:57–64 (1989).
17. British Standards Institution, *British Standard Guide to Safety Aspects of Experiments in Which People Are Exposed to Mechanical Vibration and Shock*, BS 7085, 1989.
18. Council of the European Commission, Brussels, Proposal for a Council Directive on the approximation of the laws of the Member States, Proposed Council Directive (88/C29/01), *Off. J. Eur. Communities*, February, 1988.

19. G. Loriga, Il lavoro con i martelli pneumatici (The use of pneumatic hammers), *Boll. Ispett. Lavoro* 2:35-60 (1911).
20. A. Hamilton, A study of spastic anaemia in the hands of stonecutters, *U.S. Bur. Labor Stat. Bull. 236 (Part 19)*: 53-66 (1918).
21. W. Taylor, D. Wasserman, V. Behrens, D. Reynolds, and S. Samueloff, Effect of the air hammer on the hands of stonecutters. The limestone quarries of Bedford, Indiana, revisited, *Br. J. Ind. Med.* 41(3):289-295 (1984).
22. M. Bovenzi, Response of finger systolic pressure to cold provocation in healthy subjects and in Raynaud's phenomenon of occupational origin, in *Proceedings of the 6th International Conference on Hand-Arm Vibration*, Bonn, Germany, H. Dupuis, E. Christ, J. Sandover, W. Taylor, and A. Okada, Eds., HVBG, Sankt Augustin, 1993, pp. 75-80.
23. G. Gemne, I. Pyykko, W. Taylor, and P. Pelmear, The Stockholm Workshop scale for the classification of cold-induced Raynaud's phenomenon in the hand-arm vibration syndrome (revision of the Taylor-Pelmear scale), *Scand. J. Work, Environ. Health* 13(4):275-278 (1987).
24. M. J. Griffin, C. R. Macfarlane, and C. D. Norman, The transmission of vibration to the hand and the influence of gloves, *Proc. 3rd Int. Symp. Hand-Arm Vibration*, Ottawa, May 1981, published as *Vibration Effects on the Hand & Arm in Industry*, A. J. Brammer and W. Taylor, Eds., Wiley, New York, 1982, pp. 103-116.
25. American National Standards Institute, *Guide for the Measurement and Evaluation of Gloves Which Are Used to Reduce Exposure to Vibration Transmitted to the Hand*, ANSI S3.40-1989, 1989.
26. International Organization for Standardization, *Mechanical Vibration Guidelines for the Measurement and the Assessment of Human Exposure to Hand-Transmitted Vibration*, ISO 5349, 1986.
27. British Standards Institution, *Measurement and Evaluation of Human Exposure to Vibration Transmitted to the Hand*, BS 6842, 1987.
28. American National Standards Institute, *Guide for the Measurement and Evaluation of Human Exposure to Vibration Transmitted to the Hand*, ANSI S3.34 (ASA 67), 1986.
29. American Conference of Governmental Industrial Hygienists, *HandArm (Segmental) Vibration. Threshold Limit Values and Biological Exposure Indices for 1986-1987*, 1986.
30. National Instute for Occupational Safety and Health, *Criteria for a Recommended Standard: Occupational Exposure to Hand-Arm Vibration*, U.S. Dept. of Health and Human Services, Natl. Inst. for Occupational Safety and Health, DHHS (NIOSH) Publ. No. 89-106, 1989.
31. International Organization for Standardization, *Hand-Held Portable Tools Measurement of Vibration at the Handle. Part 1: General*, ISO 8662-1, 1988.
32. M. J. Griffin, *The Effects of Vibration on Health*, Univ. Southampton, ISVR Memo. 632, 1982.
33. A. J. Brammer, W. Taylor, and G. Lundborg, Sensorineural stages of the hand-arm vibration syndrome, *Scand. J. Work, Environ. Health* 13(4):279-283 (1987).

31
Revised NIOSH Lifting Equation

Thomas R. Waters and Vern Putz-Anderson
National Institute for Occupational Safety and Health, Cincinnati, Ohio

I. INTRODUCTION

This chapter provides information about a revised equation for assessing the physical demands of certain two-handed manual lifting tasks that was developed by the National Institute for Occupational Safety and Health (NIOSH) and described earlier in an article by Waters et al. [1]. We discuss what factors need to be measured, how they should be measured, what procedures should be used, and how the results can be used to ergonomically design new jobs or make decisions about redesigning existing jobs that may be hazardous. We define all pertinent terms and present the mathematical formulas and procedures needed to properly apply the NIOSH lifting equation. Several example problems are also provided to demonstrate how the equations should be used. An expanded version of this chapter is contained in a NIOSH report [2].

Historically, NIOSH has recognized the problem of work-related back injuries and published the *Work Practices Guide for Manual Lifting* (WPG) in 1981 [3]. WPG contained a summary of the lifting-related literature up to 1981; analytical procedures and a lifting equation for calculating a recommended weight for specific two-handed, symmetrical lifting tasks; and an approach for controlling the hazards of low back injury from manual lifting. The approach to hazard control was coupled to the *action limit* (AL), a term that denoted the recommended weight derived from the lifting equation.

In 1985, NIOSH convened an ad hoc committee of experts who reviewed the current literature on lifting, including the NIOSH WPG.[1] The literature review was summarized in a document containing updated information on the physiological, biomechanical, psychophysical, and epidemiological aspects of manual lifting [4]. Based on the results of the literature review, the ad hoc committee recommended criteria for defining the lifting capacity of healthy workers. The committee used the criteria to formulate the revised lifting equation.[2] Subsequently, NIOSH staff developed the documentation for the equation and played a prominent role in recommending methods for interpreting the results of the lifting equation. *The revised lifting equation reflects new findings and provides methods for evaluating asymmetrical lift-*

---

[1]The ad hoc 1991 NIOSH Lifting Committee members included M. M. Ayoub, Donald B. Chaffin, Colin G. Drury, Arun Garg, and Suzanne Rodgers. NIOSH representatives included Vern Putz-Anderson and Thomas R. Waters.
[2]For this remainder of this chapter, the revised 1991 NIOSH lifting equation will be identified simply as "the revised lifting equation" [1,2]. The abbreviation WPG will continue to be used as the reference to the earlier NIOSH lifting equation, which was documented in the *Work Practices Guide for Manual Lifting* [3].

ing tasks and lifts of objects with less than optimal couplings between the object and the worker's hands. The revised lifting equation also provides guidelines for a more diverse range of lifting tasks than the earlier equation [3].

The rationale and criterion for the development of the revised NIOSH lifting equation are provided in a journal article by Waters et al. [1]. We suggest that those users who wish to achieve a better understanding of the data and decisions that were made in formulating the revised equation consult that article. It provides an explanation of the selection of the biomechanical, physiological, and psychophysical criterion as well as a description of the derivation of the individual components of the revised lifting equation. For those individuals, however, who are primarily concerned with the use and application of the revised lifting equation, this chapter provides a more complete description of the method and its limitations.

Although the revised lifting equation has not been fully validated, the recommended weight limits derived from the revised equation are consistent with, or lower than, those generally reported in the literature [1, Tables 2, 4, 5]. Moreover, the proper application of the revised equation is more likely to protect healthy workers for a wider variety of lifting tasks than methods that rely on only a single task factor or single criterion.

Finally, it should be stressed that the NIOSH lifting equation is only one tool in a comprehensive effort to prevent work-related low back pain and disability. Some examples of other approaches are described elsewhere [5]. Moreover, lifting is only one of the causes of work-related low back pain and disability. Other causes that have been hypothesized or established as risk factors include whole body vibration, static postures, prolonged sitting, and direct trauma to the back. Psychosocial factors, appropriate medical treatment, and job demands may also be particularly important in influencing the transition of acute low back pain to chronic disabling pain. (See Chapter 13, "Manual Materials Handling.")

II. DEFINITION OF TERMS

This section provides the basic technical information needed to properly use the revised lifting equation to evaluate a variety of two-handed manual lifting tasks. Definitions and data requirements for the revised lifting equation are also provided.

A. Recommended Weight Limit (RWL)

The recommended weight limit (RWL) is the principal product of the revised NIOSH lifting equation. The RWL is defined for a specific set of task conditions as the weight of the load that nearly all healthy workers could perform over a substantial period of time (e.g., up to 8 hr) without an increased risk of developing lifting-related low back pain (LBP). By "healthy workers" we mean workers who are free of adverse health conditions that would increase their risk of musculoskeletal injury.

The concept behind the revised NIOSH lifting equation is to start with a recommended weight that is considered safe for an "ideal" lift (i.e., load constant equal to 51 lb) and then reduce the weight as the task becomes more stressful (i.e., as the task-related factors become less favorable). The precise formulation of the revised lifting equation for calculating the recommended weight limit (RWL) is based on a multiplicative model that provides a weighting (multiplier) for each of six task variables:

1. Horizontal distance of the load from the worker (H)
2. Vertical height of the lift (V)

3. Vertical displacement during the lift (D)
4. Angle of asymmetry (A)
5. Frequency (F) and duration of lifting
6. Quality of the hand-to-object coupling (C)

The weightings are expressed as coefficients that serve to decrease the load constant, which represents the maximum recommended load weight to be lifted under ideal conditions. For example, as the horizontal distance between the load and the worker increases from 10 in., the recommended weight limit for that task would be reduced from the ideal starting weight.

The *recommended weight limit* (RWL) is defined as

$$\text{RWL} = \text{LC} \times \text{HM} \times \text{VM} \times \text{DM} \times \text{AM} \times \text{FM} \times \text{CM}$$

where

		Metric	U.S. customary
LC =	load constant	23 kg	51 lb
HM =	horizontal multipler	($25/H$)	($10/H$)
VM =	vertical multipler	$1 - (0.003 \, |V - 75|)$	$1 - (0.0075 \, |V - 30|)$
DM =	distance multipler	$0.82 + (4.5/D)$	$0.82 + (1.8/D)$
AM =	asymmetric multipler	$1 - (0.0032A)$	$1 - (0.0032A)$
FM =	frequency multipler	From Table 5	From Table 5
CM =	coupling multipler	From Table 7	From Table 7

The term *task variables* refers to the measurable task-related measurements that are used as input data for the formula (i.e., $H, V, D, A, F,$ and C); whereas the term *multipliers* refers to the reduction coefficients in the equation (i.e., HM, VM, DM, AM, FM, and CM).

B. Measurement Requirements

The following list briefly describes the measurements required to use the revised NIOSH lifting equation. Details for each of the variables are presented later in this chapter (see Sec. IV).

H = horizontal location of hands from midpoint between the inner ankle bones. Measure at the origin and the destination of the lift (cm or in.).
V = vertical location of the hands from the floor. Measure at the origin and destination of the lift (cm or in.).
D = vertical travel distance between the origin and the destination of the lift (cm or in.).
A = angle of asymmetry—angular displacement of the load from the worker's sagittal plane. Measure at the origin and destination of the lift (degrees).
F = average frequency rate of lifting measured in lifts/min. Duration is defined to be ≤ 1 hr, ≤ 2 hr, or ≤ 8 hr assuming appropriate recovery allowances (see Table 5).
C = quality of hand-to-object coupling (quality of interface between the worker and the load being lifted). The quality of the coupling is categorized as good, fair, or poor, depending upon the type and location of the coupling, the physical characteristics of load, and the vertical height of the lift.

C. Lifting Index (LI)

The *lifting index* (LI) is a term that provides a relative estimate of the level of physical stress associated with a particular manual lifting task. The estimate of the level of physical stress is defined by the relationship between the weight of the load lifted and the recommended weight limit. The LI is defined by the equation

$$\text{LI} = \frac{\text{load weight}}{\text{recommended weight limit}} = \frac{L}{\text{RWL}}$$

where load weight (L) = weight of the object lifted (lb or kg).

D. Miscellaneous Terms

Lifting task   The act of manually grasping an object of definable size and mass with two hands and vertically moving the object without mechanical assistance.

Load weight (L)   Weight of the object to be lifted, in pounds or kilograms, including the container.

Horizontal location (H)   Distance of the hands away from the midpoint between the ankles, in inches or centimeters (measure at the origin and destination of lift). See Figure 1.

Vertical location (V)   Distance of the hands above the floor, in inches or centimeters (measure at the origin and destination of lift). See Figure 1.

Vertical travel distance (D)   Absolute value of the difference between the vertical heights at the destination and origin of the lift, in inches or centimeters.

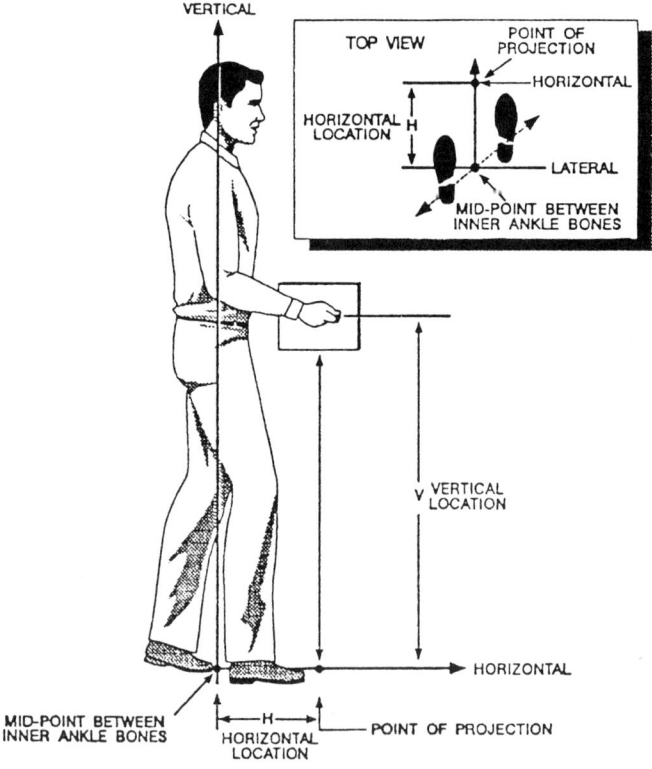

Figure 1   Graphic representation of hand location.

Angle of asymmetry (A) Angular measure of how far the *object* is displaced from the front (midsagittal plane) of the worker's body at the beginning or end of the lift, in degrees (measure at the origin and destination of lift). See Figure 2. The asymmetry angle is defined by the location of the load relative to the worker's midsagittal plane, as defined by the neutral body posture, rather than the position of the feet or the extent of body twist.

Neutral body position Position of the body when the hands are directly in front of the body and there is minimal twisting at the legs, torso, or shoulders.

Frequency of lifting (F) Average number of lifts per minute over a 15-min period.

Duration of lifting Three-tiered classification of lifting duration specified by the distribution of work time and recovery time (work pattern). Duration is classified as either short (1 hr), moderate (1–2 hr), or long (2–8 hr), depending on the work pattern.

Coupling classification Classification of the quality of the hand-to-object coupling (e.g., handle, cutout, or grip). Coupling quality is classified as good, fair, or poor.

Significant control A condition requiring "precision placement" of the load at the destination of the lift. This is usually the case when (1) the worker has to regrasp the load near the destination of the lift, or (2) the worker has to momentarily hold the object at the destination, or (3) the worker has to carefully position or guide the load at the destination.

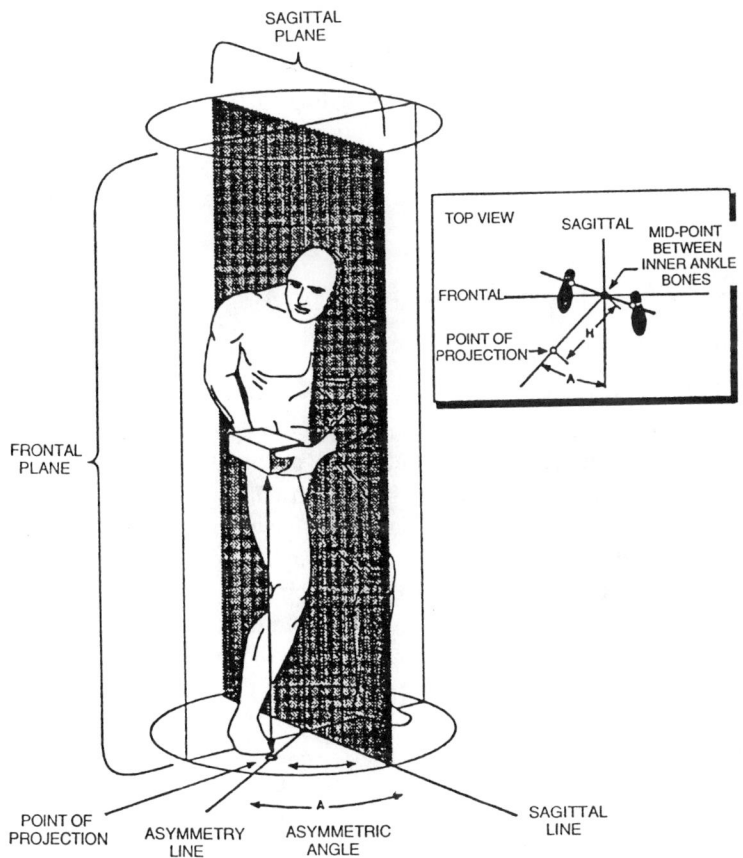

Figure 2 Graphic representation of angle of asymmetry (*A*).

III. LIMITATIONS OF THE EQUATION

The lifting equation is a tool for assessing the physical stress of two-handed manual lifting tasks. As with any tool, its application is limited to those conditions for which it was designed. Specifically, the lifting equation was designed to meet specific lifting-related criteria that encompass biomechanical, physiological, and psychophysical assumptions and data used to develop the equation. To the extent that a given lifting task accurately reflects these underlying conditions and criteria, this lifting equation may be appropriately applied.

The following list identifies a set of work conditions in which the application of the lifting equation could either under- or overestimate the extent of physical stress associated with a particular work-related activity. Each of the following task limitations also highlights research topics in need of further research to extend the application of the lifting equation to a greater range of real-world lifting tasks.

The revised NIOSH lifting equation does not apply if any of the following occur:

Lifting/lowering with one hand
Lifting/lowering for over 8 hr
Lifting/lowering while seated or kneeling
Lifting/lowering in a restricted workspace
Lifting/lowering unstable objects
Lifting/lowering while carrying, pushing, or pulling
Lifting/lowering with wheelbarrows or shovels
Lifting/lowering with "high-speed" motion (faster than about 30 in./sec)
Lifting/lowering with unreasonable foot/floor coupling (<0.4 coefficient of friction between the sole and the floor)
Lifting/lowering an unfavorable environment (temperature significantly outside 66–79°F (19–26°C) range; relative humidity outside 35–50% range)

IV. OBTAINING AND USING THE DATA

A. Horizontal Component

1. Definition and Measurement

Horizontal location (H) is measured from the midpoint of the line joining the inner ankle bones to a point projected on the floor directly below the midpoint of the hand grasps (i.e., load center), as defined by the large middle knuckle of the hand (Figure 1). Typically, the worker's feet are not aligned with the midsagittal plane, as shown in Figure 1, but may be rotated inward or outward. If this is the case, then the midsagittal plane is defined by the worker's neutral body posture as defined above. If significant control is required at the destination (i.e., precision placement), then H should be measured at both the origin and destination of the lift.

Horizontal distance (H) should be measured. In those situations, where the H value cannot be measured, then H may be approximated from the following equations:

Metric (All distances in cm)
$H = 20 + W/2$ for $V \geq 25$ cm
$H = 25 + W/2$ for $V < 25$ cm

U.S. customary (All distances inches)
$H = 8 + W/2$ for $V \geq 10$ in.
$H = 10 + W/2$ for $V < 10$ in.

where W is the width of the container in the sagittal plane and V is the vertical location of the hands from the floor.

2. Horizontal Restrictions

If the horizontal distance is less than 10 in. (25 cm), then H is set to 10 in. (25 cm). Although objects can be carried or held closer than 10 in. from the ankles, most objects that are closer than this cannot be lifted without encountering interference from the abdomen or hyperextending the shoulders. Although 25 in. (63 cm) was chosen as the maximum value for H, it is probably too great a distance for shorter workers, particularly when lifting asymmetrically. Furthermore, objects at a distance of more than 25 in. from the ankles normally cannot be lifted vertically without some loss of balance.

3. Horizontal Multiplier

The horizontal multipler (HM) is $10/H$ for H measured in inches and $25/H$ for H measured in centimeters. If H is less than or equal to 10 in. (25 cm), the multiplier is 1.0. HM decreases with an increase in H value. The multiplier for H is reduced to 0.4 when H is 25 in. (63 cm). If H is greater than 25 in., then HM = 0. The HM value can be computed directly or determined from Table 1.

B. Vertical Component

1. Definition and Measurement

Vertical location (V) is defined as the vertical height of the hands above the floor. V is measured vertically from the floor to the midpoint between the hand grasps, as defined by the large middle knuckle. The coordinate system is illustrated in Figure 1.

Table 1 Horizontal Multiplier

H (in.)	HM	H (cm)	HM
≤10	1.00	≤25	1.00
11	0.91	28	0.89
12	0.83	30	0.83
13	0.77	32	0.78
14	0.71	34	0.74
15	0.67	36	0.69
16	0.63	38	0.66
17	0.59	40	0.63
18	0.56	42	0.60
19	0.53	44	0.57
20	0.50	46	0.54
21	0.48	48	0.52
22	0.46	50	0.50
23	0.44	52	0.48
24	0.42	54	0.46
25	0.40	56	0.45
>25	0.00	58	0.43
		60	0.42
		63	0.40
		>63	0.00

2. Vertical Restrictions

The vertical location (V) is limited by the floor surface and the upper limit of vertical reach for lifting (i.e., 70 in. or 175 cm). The vertical location should be measured at the origin and the destination of the lift to determine the travel distance (D).

3. Vertical Multiplier

To determine the vertical multiplier (VM), the absolute value or deviation of V from an optimum height of 30 in. (75 cm) is calculated. A height of 30 in. above floor level is considered "knuckle height" for a worker of average height (66 in. or 165 cm). The vertical multiplier (VM) is $1-(0.0075\,|V\text{-}30|)$ for V measured in inches $1-(0.003\,|V\text{-}75|)$, for V measured in centimeters.

When V is at 30 in. (75 cm), VM = 1.0. The value of VM decreases linearly with an increase or decrease in height from this position. At floor level, VM = 0.78, and at 70 in. (175 cm) height, VM = 0.7. If V is greater than 70 in., then VM = 0. The VM value can be computed directly or determined from Table 2.

C. Distance Component

1. Definition and Measurement

The distance variable (D) is defined as the vertical travel distance of the hands between the origin and destination of the lift. For lifting, D can be computed by subtracting the vertical location (V) at the origin of the lift from the corresponding V at the destination of the lift (i.e., D is equal to V at the destination minus V at the origin). For a lowering task, D is equal to V at the origin minus V at the destination.

Table 2 Vertical Multiplier

V (in.)	VM	V (cm)	VM
0	0.78	0	0.78
5	0.81	10	0.81
10	0.85	20	0.84
15	0.89	30	0.87
20	0.93	40	0.90
25	0.96	50	0.93
30	1.00	60	0.96
35	0.96	70	0.99
40	0.93	80	0.99
45	0.89	90	0.96
50	0.85	100	0.93
55	0.81	110	0.90
60	0.78	120	0.87
65	0.74	130	0.84
70	0.70	140	0.81
>70	0.00	150	0.78
		160	0.75
		170	0.72
		175	0.70
		>175	0.00

Table 3 Distance Multiplier

D (in.)	DM	D (cm)	DM
≤10	1.00	≤25	1.00
15	0.94	40	0.93
20	0.91	55	0.90
25	0.89	70	0.88
30	0.88	85	0.87
35	0.87	100	0.87
40	0.87	115	0.86
45	0.86	130	0.86
50	0.86	145	0.85
55	0.85	160	0.85
60	0.85	175	0.85
70	0.85	>175	0.00
>70	0.00		

2. Distance Restrictions

The distance variable (D) is assumed to be at least 10 in. (25 cm) and no greater than 70 in. (175 cm). If the vertical travel distance is less than 10 in. (25 cm), then D should be set to the minimum distance of 10 in. (25 cm).

3. Distance Multiplier

The distance multiplier (DM) is $0.82 + 1.8/D$ for D measured in inches $0.82 + 4.5/D$ for D measured in centimeters. For D less than 10 in. (25 cm), D is assumed to be 10 in. (25 cm) and DM = 1.0. The distance multiplier, therefore, decreases gradually with an increase in travel distance. DM = 1.0 when D is set at 10 in. (25 cm); DM = 0.85 when D = 70 in. (175 cm). Thus, DM ranges from 1.0 to 0.85 as D varies from 0 in. (0 cm) to 70 in. (175 cm). The DM value can be computed directly or determined from Table 3.

D. Asymmetry Component

1. Definition and Measurement

Asymmetry refers to a lift that begins or ends outside the midsagittal plane (see Figure 2). In general, asymmetric lifting should be avoided. If asymmetric lifting cannot be avoided, however, the recommended weight limits are significantly less than those limits used for symmetrical lifting.[3]

An asymmetric lift may be required under the following task or workplace conditions:

1. The origin and destination of the lift are oriented at an angle to each other.
2. The lifting motion is across the body, such as occurs in swinging bags or boxes from one location to another.

---

[3]It may not always be clear whether asymmetry is an intrinsic element of the task or just a personal characteristic of the worker's lifting style. Regardless of the reason for the asymmetry, any observed asymmetric lifting should be considered an intrinsic element of the job design and should be considered in the assessment and subsequent redesign. Moreover, the design of the task should not rely on worker compliance but should rather discourage or eliminate the need for asymmetric lifting.

3. The lifting is done to maintain body balance in obstructed workplaces, on rough terrain, or on littered floors.
4. Productivity standards require reduced time per lift.

The asymmetric angle (A), which is depicted graphically in Figure 2, is operationally defined as the angle between the asymmetry line and the midsagittal line. The *asymmetry line* is defined as the line that joins the midpoint between the inner ankle bones and the point projected on the floor directly below the midpoint of the hand grasps, as defined by the large middle knuckle. The *sagittal line* is defined as the line passing through the midpoint between the inner ankle bones and lying in the midsagittal plane, as defined by the neutral body position (i.e., hands directly in front of the body, with no twisting at the legs, torso, or shoulders). *Note*: The asymmetry angle is not defined by foot position or the angle of torso twist, but by the location of the load relative to the worker's midsagittal plane.

In many cases of asymmetric lifting, the worker will pivot or use a step turn to complete the lift. Because this may vary significantly between workers and between lifts, we have assumed that no pivoting or stepping occurs. Although this assumption may overestimate the reduction in acceptable load weight, it will provide the greatest protection for the worker.

The asymmetry angle (A) must always be measured at the origin of the lift. If significant control is required at the destination, however, then angle A should be measured at both the origin and the destination of the lift.

2. Asymmetry Restrictions

The angle A is limited to the range 0–135°. If $A > 135°$, then AM is set equal to zero, which results in an RWL of zero, or no load.

3. Asymmetric Multiplier

The asymmetric multiplier (AM) is $1 - 0.0032A$. AM has a maximum value of 1.0 when the load is lifted directly in front of the body and decreases linearly as the angle of asymmetry (A) increases. The range is from a value of 0.57 at 135° of asymmetry to a value of 1.0 at 0° of asymmetry (i.e., symmetric lift). If A is greater than 135°, then AM = 0, and the load is zero. The AM value can be computed directly or determined from Table 4.

Table 4 Asymmetric Multiplier

A (deg)	AM
0	1.00
15	0.95
30	0.90
45	0.86
60	0.81
75	0.76
90	0.71
105	0.66
120	0.62
135	0.57
>135	0.00

E. Frequency Component

1. Definition and Measurement

The frequency multiplier is defined by (1) the number of lifts per minute (frequency), (2) the amount of time engaged in the lifting activity (duration), and (3) the vertical height of the lift from the floor. Lifting frequency (F) refers to the average number of lifts made per minute as measured over a 15-min period. Because of the potential variation in work patterns, analysts may have difficulty obtaining an accurate or representative 15-min work sample for computing F. If significant variation exists in the frequency of lifting over the course of the day, analysts should employ standard work sampling techniques to obtain a representative work sample for determining the number of lifts per minute. For those jobs where the frequency varies from session to session, each session should be analyzed separately, but the overall work pattern must still be considered. For more information, most standard industrial engineering or ergonomics texts provide guidance for establishing a representative job sampling strategy (e.g., Eastman Kodak Company [6]).

2. Lifting Duration

Lifting duration is classified into three categories based on the pattern of continuous work time and recovery time (i.e., light work) periods. A continuous work time (WT) period is defined as a period of uninterrupted work. Recovery time (RT) is defined as the duration of light work activity following a period of continuous lifting. Examples of light work include activities such as sitting at a desk or table, monitoring operations, and light assembly work. The three categories are short duration, moderate duration, and long duration.

Short Duration. Short duration lifting tasks are those that have a work duration of 1 hr or less followed by a recovery time equal to 1.2 times the work time [i.e., at least a 1.2 recovery time to work time ratio (RT/WT)]. For example, to be classified as short duration, a 45-min lifting job must be followed by at least a 54-min recovery period prior to initiating a subsequent lifting session. If the required recovery time is not met for a job of 1 hr or less and a subsequent lifting session is required, then the total lifting time must be combined to correctly determine the duration category. Moreover, if the recovery period does not meet the time requirement, it is disregarded for purposes of determining the appropriate duration category.

As another example, assume that a worker lifts continuously for 30 min, then performs a light work task for 10 min, and then lifts for an additional 45-min period. In this case, the recovery time between lifting sessions (10 min) is less than 1.2 times the initial 30-min work time (36 min). Thus, the two work times (30 min and 45 min) must be added together to determine the duration. Since the total work time (75 min) exceeds 1 hr, the job is classified as moderate duration. On the other hand, if the recovery period between lifting sessions were increased to 36 min, then the short-duration category would apply, which would result in a larger FM value.

A special procedure has been developed for determining the appropriate lifting frequency (F) for certain repetitive lifting tasks in which workers do not lift continuously during the 15-min sampling period. This occurs when the work pattern is such that the worker lifts repetitively for a short time and then performs light work for a short time before starting another cycle. For work patterns such as this, F may be determined as follows, as long as the actual lifting frequency does not exceed 15 lifts/min:

1. Compute the total number of lifts performed for the 15-min period (i.e., lift rate times work time).

2. Divide the total number of lifts by 15.
3. Use the resulting value as the frequency (F) to determine the frequency multiplier (FM) from Table 5.

For example, if the work pattern for a job consists of a series of cyclic sessions requiring 8 min of lifting followed by 7 min of light work, and the lifting rate during the work sessions is 10 lifts/min, then the frequency rate (F) that is used to determine the frequency multiplier for this job is equal to (10 × 8)/15, or 5.33 lifts/min. If the worker lifted continuously for more than 15 min, however, then the actual lifting frequency (10 lifts/min) would be used.

When using this special procedure, the duration category is based on the magnitude of the recovery periods *between* work sessions, not *within* work sessions. In other words, if the work pattern is intermittent and the special procedure applies, then the intermittent recovery periods that occur during the 15-min sampling period are *not* considered as recovery periods for purposes of determining the duration category. For example, if the work pattern for a manual lifting job were composed of repetitive cycles consisting of 1 min of continuous lifting at a rate of 10 lifts/min, followed by 2 min of recovery, then the correct procedure would be to adjust the frequency according to the special procedure [i.e., F = (10 lifts/min × 5 min)/15 min = 50/15 = 3.4 lifts/min]. The 2-min recovery periods would not count toward the RT/WT ratio, however, and additional recovery periods would have to be provided as described above.

Moderate Duration. Moderate duration lifting tasks are those that have a duration of more than 1 hr but not more than 2 hr, followed by a recovery period of at least 0.3 times the work time [i.e., at least a 0.3 recovery time to work time ratio (RT/WT)].

Table 5 Frequency Multiplier (FM) Table

Frequency[a] lifts/min (F)	Work duration					
	≤ 1 hr		> 1 but ≤ 2 hr		> 2 but ≤ 8 hr	
	V[b] < 30	V ≥ 30	V < 30	V ≥ 30	V < 30	V ≥ 30
≥0.2	1.00	1.00	0.95	0.95	0.85	0.85
0.5	0.97	0.97	0.92	0.92	0.81	0.81
1	0.94	0.94	0.88	0.88	0.75	0.75
2	0.91	0.91	0.84	0.84	0.65	0.65
3	0.88	0.88	0.79	0.79	0.55	0.55
4	0.84	0.84	0.72	0.72	0.45	0.45
5	0.80	0.80	0.60	0.60	0.35	0.35
6	0.75	0.75	0.50	0.50	0.27	0.27
7	0.70	0.70	0.42	0.42	0.22	0.22
8	0.60	0.60	0.35	0.35	0.18	0.18
9	0.52	0.52	0.30	0.30	0.00	0.15
10	0.45	0.45	0.26	0.26	0.00	0.13
11	0.41	0.41	0.00	0.23	0.00	0.00
12	0.37	0.37	0.00	0.21	0.00	0.00
13	0.00	0.34	0.00	0.00	0.00	0.00
14	0.00	0.31	0.00	0.00	0.00	0.00
15	0.00	0.28	0.00	0.00	0.00	0.00
>15	0.00	0.00	0.00	0.00	0.00	0.00

[a]For lifting less frequently than once per 5 min, set F = 0.2 lift/min
[b]V is expressed in inches as measured from the floor.

For example, if a worker continuously lifts for 2 hr, then a recovery period of at least 36 min would be required before initiating a subsequent lifting session. If the recovery time requirement is not met and a subsequent lifting session is required, then the total work time must be added together. If the total work time exceeds 2 hr, then the job must be classified as a long duration lifting task.

Long Duration. Long duration lifting tasks are defined as those that have a duration of 2–8 hr, with standard industrial rest allowances (e.g., morning, lunch, and afternoon rest breaks). *Note*: No weight limits are provided for more than 8 hr of work.

The difference in the required RT/WT ratio for the short (<1 hr) duration category, which is 1.2, and the moderate (1–2 hr) duration category, which is 0.3, is due to the difference in the magnitudes of the frequency multiplier values associated with each of the duration categories. Since the moderate category results in larger reductions in the RWL than the short category, there is less need for a recovery period between sessions than for the short duration category. In other words, the short duration category would result in higher weight limits than the moderate duration category, so larger recovery periods would be needed.

3. Frequency Restrictions

Lifting frequency (F) for repetitive lifting may range from 0.2 lift/min to a maximum frequency that is dependent on the vertical location of the object (V) and the duration of lifting (Table 5). Lifting above the maximum frequency results in RWL = 0.0 (except for the special care of discontinuous lifting discussed above, where the maximum frequency is 15 lifts/min).

4. Frequency Multiplier

The FM value depends upon the average number of lifts per minute (F), the vertical location (V) of the hands at the origin, and the duration of continuous lifting. For lifting tasks with a frequency less than 0.2 lifts/min, set the frequency equal to 0.2 lift/min. Otherwise, FM is determined from Table 5.

F. Coupling Component

1. Definition and Measurement

The nature of the hand-to-object coupling or gripping method can affect not only the maximum force a worker can or must exert on the object, but also the vertical location of the hands during the lift. A "good" coupling will reduce the maximum grasp forces required and increase the acceptable weight for lifting, whereas a "poor" coupling will generally require higher maximum grasp forces and decrease the acceptable weight for lifting.

The effectiveness of the coupling is not static but may vary with the distance of the object from the ground, so that a good coupling could become a poor coupling during a single lift. The entire range of the lift should be considered when classifying hand-to-object couplings, with classification based on overall effectiveness. The analyst must classify the coupling as good, fair, or poor. The three categories are defined in Table 6. If there is any doubt about classifying a particular coupling design, the more stressful classification should be selected.

The decision tree shown in Figure 3 may be helpful in classifying the hand-to-object coupling.

2. Coupling Multiplier

Based on the coupling classification and vertical location of the lift, the coupling multiplier (CM) is determined from Table 7.

Table 6 Hand-to-Container Coupling Classification

Good	Fair	Poor
1. For containers of optimal design, such as some boxes, crates, etc., a "good" hand-to-object coupling would be defined as handles or handhold cutouts of optimal design (see Notes 1–3).	1. For containers of optimal design, a "fair" hand-to-object coupling would be defined as handles or handhold cutouts of less than optimal design (see Notes 1–4).	1. Containers of less than optimal design or loose parts or irregular objects that are bulky, hard to handle, or have sharp edges (see Note 5).
2. For loose parts or irregular objects, which are not usually containerized, such as castings, stock, and supply materials, a "good" hand-to-object coupling would be defined as a comfortable grip in which the hand can be easily wrapped around the object (see Note 6).	2. For containers of optimal design with no handles or handhold cutouts or for loose parts or irregular objects, a "fair" hand-to-object coupling is defined as a grip in which the hand can be flexed about 90° (see Note 4).	2. Lifting nonrigid bags (i.e., bags that sag in the middle).

Notes:
1. An optimal handle design has 0.75–1.5 in. (1.9–3.8 cm) diameter, ≥4.5 in. (11.5 cm) length, 2 in. (5 cm) clearance, cylindrical shape, and a smooth, nonslip surface.
2. An optimal handhold cutout has the following approximate characteristics: ≥1.5 in. (3.8 cm) height, 4.5 in. (11.5 cm) length, semi-oval shape, ≥2 in. (5 cm) clearance, smooth nonslip surface, and ≥0.25 in. (0.60 cm) container thickness (e.g., double thickness cardboard).
3. An optimal container design has ≤16 in. (40 cm) frontal length, ≤12 in. (30 cm) height, and a smooth nonslip surface.
4. A worker should be capable of clamping the fingers at nearly 90° under the container, such as required when lifting a cardboard box from the floor.
5. A container is considered less than optimal if it has a frontal length > 16 in. (40 cm), height > 12 in. (30 cm), rough or slippery surfaces, sharp edges, asymmetric center of mass, unstable contents, or requires the use of gloves.
6. A worker should be able to comfortably wrap the hand around the object without causing excessive wrist deviations or awkward postures, and the grip should not require excessive force.

V. PROCEDURES

Prior to data collection, the analyst must decide (1) if the job should be analyzed as a single-task or multitask manual lifting job and (2) if significant control is required at the destination of the lift. This is necessary because the procedures differ according to the type of analysis required.

A manual lifting job may be analyzed as a single-task job if the task variables do not differ from task to task or if only one task is of interest (e.g., single most stressful task). This may be the case if one of the tasks clearly has a dominant effect on strength demands, localized muscle fatigue, or whole body fatigue. On the other hand, if the task variables differ significantly between tasks, it may be more appropriate to analyze a job as a multitask manual

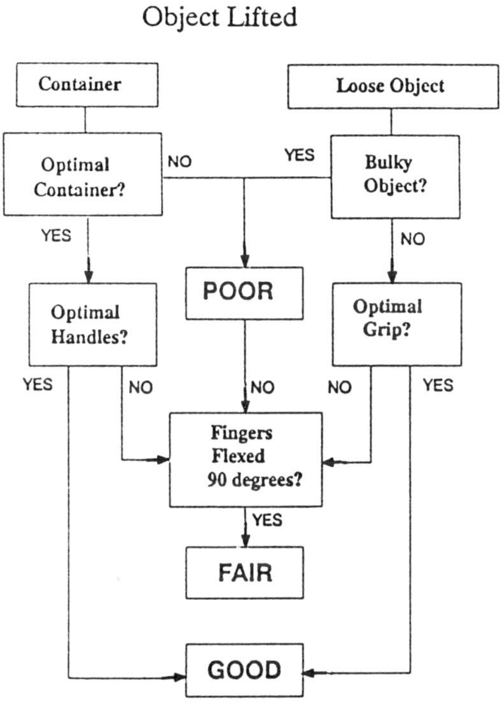

Object Lifted

Figure 3 Decision tree for coupling quality.

lifting job. A multitask analysis is more difficult to perform than a single-task analysis because additional data and computations are required. The multitask approach, however, will provide more detailed information about specific strength and physiological demands.

For many lifting jobs, it may be acceptable to use either the single- or multitask approach. The single-task analysis should be used when possible, but when a job consists of more than one task and detailed information is needed to specify engineering modifications, then the multitask approach provides a reasonable method of assessing the overall physical demands. The multitask procedure is more complicated than the single-task procedure and requires a greater understanding of assessment terminology and mathematical concepts. Therefore, the decision to use the single- or multitask approach should be based on (1) the need for detailed information about all facets of the multitask lifting job, (2) the need for accuracy and com-

Table 7 Coupling Multiplier

Coupling type	Coupling multiplier	
	$V < 30$ in (75 cm)	$V \geq 30$ in. (75 cm)
Good	1.00	1.00
Fair	0.95	1.00
Poor	0.90	0.90

pleteness of data regarding assessment of the physiological demands of the task, and (3) the analyst's level of understanding of the assessment procedures.

The decision about control at the destination is important because the physical demands on the worker may be greater at the destination of the lift than at the origin, especially when significant control is required. When significant control is required at the destination, for example, the physical stress is increased because the load will have to be accelerated upward to slow its descent. This acceleration may be as great as the acceleration at the origin of the lift and may create high loads on the spine. Therefore, if significant control is required, then the RWL and LI should be determined at both locations and the lower of the two values will specify the overall level of physical demand.

To perform a lifting analysis using the revised lifting equation, two steps are taken: (1) data are collected at the worksite as described in step 1 below, and (2) the recommended weight limit and lifting index values are computed using the single-task or multitask analysis procedure described in step 2.

A. Step 1: Collect Data

The relevant task variables must be carefully measured and clearly recorded in a concise format. As mentioned previously, these variables include the horizontal location of the hands (H), vertical location of the hands, (V), vertical displacement (D), asymmetric angle (A), lifting frequency (F), and coupling quality (C). A job analysis worksheet, as shown in Figure 4 for single-task jobs or Figure 5 for multitask jobs, provides a simple form for recording the task variables and the data needed to calculate the RWL and LI values. A thorough job analysis

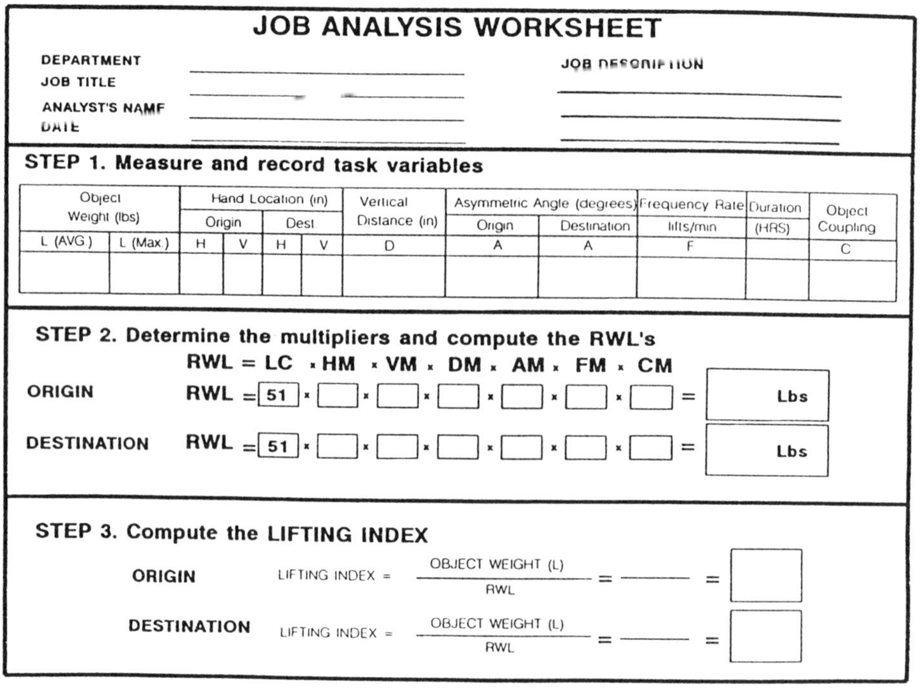

Figure 4 Single-task job analysis worksheet.

| MULTI-TASK JOB ANALYSIS WORKSHEET |

DEPARTMENT _____ JOB DESCRIPTION _____
JOB TITLE _____
ANALYST'S NAME _____
DATE _____

STEP 1. Measure and Record Task Variable Data

Task No.	Object Weight (lbs)		Hand Location (in)				Vertical Distance (in)	Asymmetry Angle (degs)		Frequency Rate lifts/min	Duration Hrs	Coupling
	L (Avg.)	L (Max.)	Origin		Dest.		D	Origin	Dest.	F		C
			H	V	H	V		A	A			

STEP 2. Compute multipliers and FIRWL, STRWL, FILI, and STLI for Each Task

Task No.	LC x HM x VM x DM x AM x CM	FIRWL x FM	STRWL	FILI = L/FIRWL	STLI = L/STRWL	New Task No.	F
51							
51							
51							
51							
51							

STEP 3. Compute the Composite Lifting Index for the Job (After renumbering tasks)

CLI = STLI₁ + △FILI₂ + △FILI₃ + △FILI₄ + △FILI₅

	FILI₂(1/FM₁,₂ - 1/FM₁)	FILI₃(1/FM₁,₂,₃ - 1/FM₁,₂)	FILI₄(1/FM₁,₂,₃,₄ - 1/FM₁,₂,₃)	FILI₅(1/FM₁,₂,₃,₄,₅ - 1/FM₁,₂,₃,₄)

CLI = _____

Figure 5 Mulitask job analysis worksheet.

is required to identify and catalog each independent lifting task in the worker's complete job. For multitask jobs, data must be collected for each task.

B. Step 2: Single-Task Procedure

For a single-task procedure, step 2 consists of computing the recommended weight limit (RWL) and the lifting index (LI). This is accomplished as follows.

Calculate the RWL at the origin for each lift. For lifting tasks that require significant control at the destination, calculate the RWL at *both* the origin and the destination of the lift. The latter procedure is required if (1) the worker has to regrasp the load near the destination of the lift, (2) the worker has to momentarily hold the object at the destination, or (3) the worker has to position or guide the load at the destination. The purpose of calculating the RWL at both the origin and destination of the lift is to identify the most stressful location of the lift. Therefore, the lower of the two RWL values should be used to compute LI for the task, as this value would represent the limiting set of conditions.

The assessment is completed on the single-task worksheet by determining the lifting index (LI) for the task of interest. This is accomplished by comparing the actual weight of the load (L) lifted with the RWL value obtained from the lifting equation.

C. Step 2: Multitask Procedure

For a multitask procedure, step 2 comprises three substeps:

a. Compute the frequency-independent recommended weight limit (FIRWL) and single-task recommended weight limit (STRWL) for each task.
b. Compute the frequency-independent lifting index (FILI) and single-task lifting index (STLI) for each task.
c. Compute the composite lifting index (CLI) for the overall job.

Compute the Frequency-Independent Recommended Weight Limits (FIRWLs). Compute the FIRWL value for each task by using the respective task variables and setting the frequency multiplier (FM) to a value of 1.0. The FIRWL for each task reflects the compressive force and muscle strength demands for a single performance of that task. If significant control is required at the destination for any individual task, the FIRWL must be computed at both the origin and the destination of the lift, as described above for a single-task analysis.

Compute the Single-Task Recommended Weight Limit (STRWL). Compute the STRWL for each task by multiplying its FIRWL by the appropriate FM. The STRWL for a task reflects the overall demands of that task, assuming it was the only task being performed. *Note:* This value does not reflect the overall demands of the task when the other tasks are considered. Nevertheless, it is helpful in determining the extent of excessive physical stress for an individual task.

Compute the Frequency-Independent Lifting Index (FILI). The FILI is computed for each task by dividing the *maximum* load weight (L) for that task by the respective FIRWL. The maximum weight is used to compute the FILI because the maximum weight determines the maximum biomechanical loads to which the body will be exposed, regardless of the frequency of occurrence. Thus, the FILI can identify individual tasks with potential strength problems for infrequent lifts. If any of the FILI values exceeds a value of 1.0, then job design changes may be needed to decrease the strength demands.

Compute the Single-Task Lifting Index (STLI). The STLI is computed for each task by dividing the *average* load weight (L) for that task by the respective STRWL. The average weight is used to compute the STLI because the average weight provides a better representation of the metabolic demands, which are distributed across the tasks rather than being dependent on individual tasks. The STLI can be used to identify individual tasks with excessive physical demands (i.e., tasks that would result in fatigue). The STLI values do not indicate the relative stress of the individual tasks in the context of the whole job, but they can be used to prioritize the individual tasks according to the magnitude of their physical stress. Thus, if any of the STLI values exceeds a value of 1.0, then ergonomic changes may be needed to decrease the overall physical demands of the task. *Note*: It may be possible to have a job in which all of the individual tasks have an STLI less than 1.0 and yet is physically demanding due to the combined demands of the tasks. In cases where the FILI exceeds the STLI for any task, the maximum weights may represent a significant problem, and careful evaluation is necessary.

Compute the Composite Lifting Index (CLI). The assessment is completed on the multitask worksheet by determining the composite lifting index (CLI) for the overall job. The CLI is computed as follows:

1. The tasks are renumbered in order of decreasing physical stress, from the task with the greatest STLI down to the task with the smallest STLI. The tasks are renumbered in this way so that the more difficult tasks are considered first.
2. The CLI for the job is then computed according to the formula

    $$\text{CLI} = \text{STLI}_1 + \Sigma \, \Delta \text{LI}$$

where

$$\sum \Delta \text{LI} = \text{FILI}_2 \times \left(\frac{1}{\text{FM}_{1,2}} - \frac{1}{\text{FM}_1} \right) + \left[\text{FILI}_3 \times \left(\frac{1}{\text{FM}_{1,2,3}} - \frac{1}{\text{FM}_{1,2}} \right) \right]$$

$$+ \left[\text{FILI} \times \left(\frac{1}{\text{FM}_{1,2,3,4}} - \frac{1}{\text{FM}_{1,2,3}} \right) \right]$$

$$+ \ldots + \left[\text{FILI}_n \times \left(\frac{1}{\text{FM}_{1,2,3,4,\ldots,n}} - \frac{1}{\text{FM}_{1,2,3,\ldots,(n-1)}} \right) \right]$$

Note: (1) The numbers in the subscripts refer to the new task numbers and (2) the FM values are determined from Table 5, based on the sum of the frequencies for the tasks listed in the subscripts.

An Example. The following example is provided to demonstrate this step of the multitask procedure. Assume that an analysis of a typical three-task job provided the results shown in Table 8.

To compute the composite lifting index (CLI) for this job, the tasks are renumbered in order of decreasing physical stress, beginning with the task with the greatest STLI. In this case, as shown in Table 8, the task numbers do not change. Next, the CLI is computed according to the formula given above. The task with the greatest CLI is Task 1 (STLI = 1.6). The sum of the frequencies for Tasks 1 and 2 is 1 + 2, or 3, and the sum of the frequencies for Tasks 1, 2, and 3 is 1 + 2 + 4, or 7. Then, from Table 5, $\text{FM}_1 = 0.94$, $\text{FM}_{1,2} = 0.88$, and $\text{FM}_{1,2,3} = 0.70$. Finally,

CLI = 1.6 + 1.0 (1/0.88 − 1/0.94) + 0.67(1/0.70) −1/0.88) = 1.6 + 0.7 + 0.20 = 1.9

Note that the FM values were based on the sum of the frequencies for the subscripts, the vertical height, and the duration of lifting.

VI. APPLYING THE EQUATIONS

A. Using RWL and LI to Guide Ergonomic Design

The recommended weight limit (RWL) and lifting index (LI) can be used to guide ergonomic design in several ways.

1. The individual multipliers can be used to identify specific job-related problems. The relative magnitude of each multiplier indicates the relative contribution of each task factor (e.g., horizontal, vertical, frequency).

Table 8 Computations from Multitask Example

Task No.	Load weight (L)	Task frequency (F)	FIRWL	FM	STRWL	FILI	STLI	New task No.
1	30	1	20	0.94	18.8	1.5	1.6	1
2	20	2	20	0.91	18.2	1.0	1.1	2
3	10	4	15	0.84	12.6	0.67	0.8	3

2. The RWL can be used to guide the redesign of existing manual lifting jobs or to design new manual lifting jobs. For example, if the task variables are fixed, then the maximum weight of the load could be selected so as not to exceed the RWL; if the weight is fixed, then the task variables could be optimized so as not to exceed the RWL.
3. The LI can be used to estimate the relative magnitude of physical stress for a task or job. The greater the LI, the smaller the fraction of workers capable of safely sustaining the level of activity. Thus, two or more job designs could be compared.
4. The LI can be used to prioritize ergonomic redesign. For example, a series of suspected hazardous jobs could be rank-ordered according to LI, and a control strategy could be developed according to the rank ordering (i.e., jobs with lifting indices above 1.0 or higher would benefit the most from redesign).

B. Rationale and Limitations for LI

The NIOSH recommended weight limit (RWL) equation and lifting index (LI) are based on the concept that the risk of lifting-related low back pain increases as the demands of the lifting task increase. In other words, as the magnitude of the LI increases, (1) the level of the risk for a given worker would be increased and (2) a greater percentage of the workforce is likely to be at risk for developing lifting-related low back pain. The shape of the risk function, however, is not known. Without additional data showing the relationship between low back pain and LI, it is impossible to predict the magnitude of the risk for a given individual or the exact percent of the work population who would be at an elevated risk for low back pain.

To gain a better understanding fo the rationale for the development of the RWL and LI, consult Ref. 1, which provides a discussion of the criteria underlying the lifting equation and of the individual multipliers. It also identifies both the assumptions and uncertainties in the scientific studies that associate manual lifting and low back injuries.

C. Job-Related Intervention Strategy

The lifting index may be used to identify potentially hazardous lifting jobs or to compare the relative severity of two jobs for the purpose of evaluating and redesigning them. From the NIOSH perspective, it is likely that lifting tasks with LI > 1.0 pose an increased risk for lifting-related low back pain for some fraction of the workforce [1]. Hence, to the extent possible, lifting jobs should be designed to achieve an LI of 1.0 or less.

Some experts believe, however, that worker selection criteria can be used to identify workers who can perform potentially stressful lifting tasks (i.e., lifting tasks that would exceed an LI of 1.0) without significantly increasing their risk of work-related injury above the baseline level [7,8]. Those who endorse the use of selection criteria believe that the criteria must be based on research studies, empirical observations, or theoretical considerations that include job-related strength testing and/or aerobic capacity testing. Even these experts agree, however, that many workers will be at a significant risk of a work-related injury when performing highly stressful lifting tasks (i.e., lifting tasks with LI > 3.0). Also, "informal" or "natural" selection of workers may occur in many jobs that require repetitive lifting tasks. According to some experts, this may result in a unique workforce that may be able to work above a lifting index of 1.0, at least in theory, without substantially increasing their risk of low back injuries above the baseline rate of injury.

Revised NIOSH Lifting Equation

D. Example Problems

Two example problems are provided to demonstrate the proper application of the lifting equation and procedures. The procedures provide a method for determining the level of physical stress associated with a specific set of lifting conditions and assist in identifying the contribution of each job-related factor. The examples also provide guidance in developing an ergonomic redesign strategy. Specifically, for each example, a job description, job analysis, hazard assessment, redesign suggestion, illustration, and completed worksheet are provided.

To help clarify the discussion of the example problems, and to provide a useful reference for determining the multiplier values, the six multipliers used in the equation have been reprinted in tabular form in Tables 1–5 and 7.

A series of general design/redesign suggestions for each job-related risk factor are provided in Table 9. These suggestions can be used to develop a practical ergonomic design/redesign strategy.

Example 1. Loading Supply Rolls

Job Description. With both hands directly in front of the body, a worker lifts the core of a 35-lb roll of paper from a cart and then shifts the roll in the hands and holds it by the sides to position it on a machine, as shown in Figure 6. Significant control of the roll is required at the destination of the lift. Also, the worker must crouch at the destination of the lift to support the roll in front of the body but does not have to twist.

Job Analysis. The task variable data are measured and recorded on the job analysis worksheet (Figure 7). The vertical location of the hands is 27 in. at the origin and 10 in. at the destination. The horizontal location of the hands is 15 in. at the origin and 20 in. at the destination. The asymmetric angle is 0° at both the origin and the destination, and the frequency is 4 lifts/shift (i.e., less than 0.2 lift/min for less than 1 hr; see Table 5).

Using Table 6, the coupling is classified as poor because the worker must reposition the hands at the destination of the lift and cannot flex the fingers to the desired 90° angle (e.g., hook grip). No asymmetric lifting is involved (i.e., $A = 0$), and significant control of the

Table 9 General Design/Redesign Suggestions

If HM is less than 1.0: Bring the load closer to the worker by removing any horizontal barriers or reducing the size of the object. Lifts near the floor should be avoided; if unavoidable, the object should fit easily between the legs.

If VM is less than 1.0: Raise/lower the origin/destination of the lift. Avoid lifting near the floor or above the shoulders.

If DM is less than 1.0: Reduce the vertical distance between the origin and the destination of the lift.

If AM is less than 1.0: Move the origin and destination of the lift closer together to reduce the angel of twist, or move the origin and destination further apart to force the worker to turn the feet and step, rather than twist the body.

If FM is less than 1.0: Reduce the lifting frequency rate, reduce the lifting duration, or provide longer recovery periods (i.e., light work period).

If CM is less than 1.0: Improve the hand-to-object coupling by providing optimal containers with handles or handhold cutouts, or improve the handholds for irregular objects.

If the RWL at the destination is less than at the origin: Eliminate the need for significant control of the object at the destination by redesigning the job or modifying the container/object characteristics. (See section V.B., p. 643.)

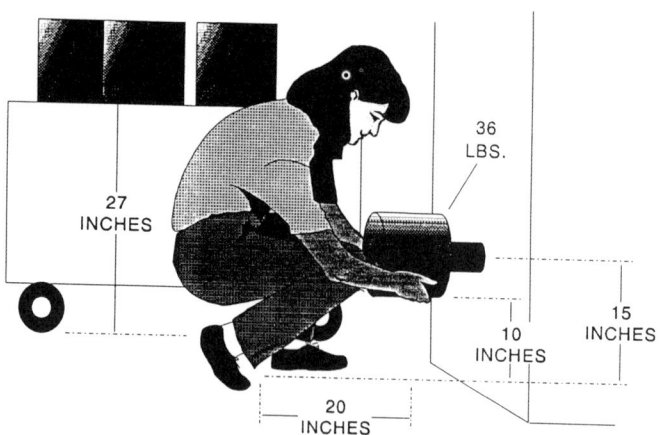

Figure 6 Loading supply rolls, Example 1.

object is required at the destination of the lift. Thus, RWL should be computed at both the origin and the destination of the lift. The multipliers are computed from the lifting equation or determined from the multiplier tables (Tables 1–5 and 7). As shown in Figure 7, for this activity RWL = 28.0 at the origin and RWL = 18.1 lb at the destination.

Hazard Assessment. The weight to be lifted (35 lb) is greater than the RWL at both the origin and destination of the lift (28.0 and 18.1 lb, respectively). At the origin, LI = 35

JOB ANALYSIS WORKSHEET

DEPARTMENT Shipping
JOB TITLE Packager
ANALYST'S NAME
DATE

JOB DESCRIPTION Loading paper supply rolls

Example 2

STEP 1. Measure and record task variables

Object Weight (lbs)		Hand Location (in)				Vertical Distance (in)	Asymmetric Angle (degrees)		Frequency Rate	Duration	Object Coupling
		Origin		Dest.			Origin	Destination	lifts/min	(HRS)	
L (AVG)	L (Max)	H	V	H	V	D	A	A	F		C
35	35	15	27	20	10	17	0	0	<.2	<1	Poor

STEP 2. Determine the multipliers and compute the RWL's

RWL = LC × HM × VM × DM × AM × FM × CM

ORIGIN RWL = 51 × .67 × .98 × .93 × 1.0 × 1.0 × .90 = 28.0 Lbs

DESTINATION RWL = 51 × .50 × .85 × .93 × 1.0 × 1.0 × .90 = 18.1 Lbs

STEP 3. Compute the LIFTING INDEX

ORIGIN LIFTING INDEX = OBJECT WEIGHT (L) / RWL = 35 / 28.0 = 1.3

DESTINATION LIFTING INDEX = OBJECT WEIGHT (L) / RWL = 35 / 18.1 = 1.9

Figure 7 Job analysis worksheet, Example 1.

lb/28.0 lb = 1.3; and at the destination, LI = 35 lb/18.1 lb = 1.9. These values indicate that this job is only slightly stressful at the origin but moderately stressful at the destination of the lift.

Redesign Suggestions. The first choice for reducing the risk of injury for workers performing this task would be to adapt the cart so that the paper rolls could be easily pushed into position on the machine without manually lifting them.

If the cart cannot be modified, then the results of the equation may be used to suggest task modifications. The worksheet displayed in Figure 7 indicates that the multipliers with the smallest magnitude (i.e., those providing the greatest penalties) are 0.50 for HM at the destination, 0.67 for HM at the origin, 0.85 for VM at the destination, and 0.90 for the CM value. Using Table 9, the following job modifications are suggested:

1. Bring the load closer to the worker by making the roll smaller so that the roll can be lifted from between the worker's legs. This will decrease H, which in turn will increase HM.
2. Raise the height of the destination to increase VM.
3. Improve the coupling to increase CM.

If the size of the roll cannot be reduced, then the vertical height (V) of the destination should be increased. Figure 8 shows that if V were increased to about 30 in., then VM would be increased from 0.85 to 1.0; the H value would be decreased from 20 in to 15 in, which would increase HM from 0.50 to 0.67; DM would be increased from 0.93 to 1.0. As shown in Figure 8, the final RWL would be increased from 18.1 lb to 30.8 lb, and LI at the destination would decrease from 1.9 to 1.1.

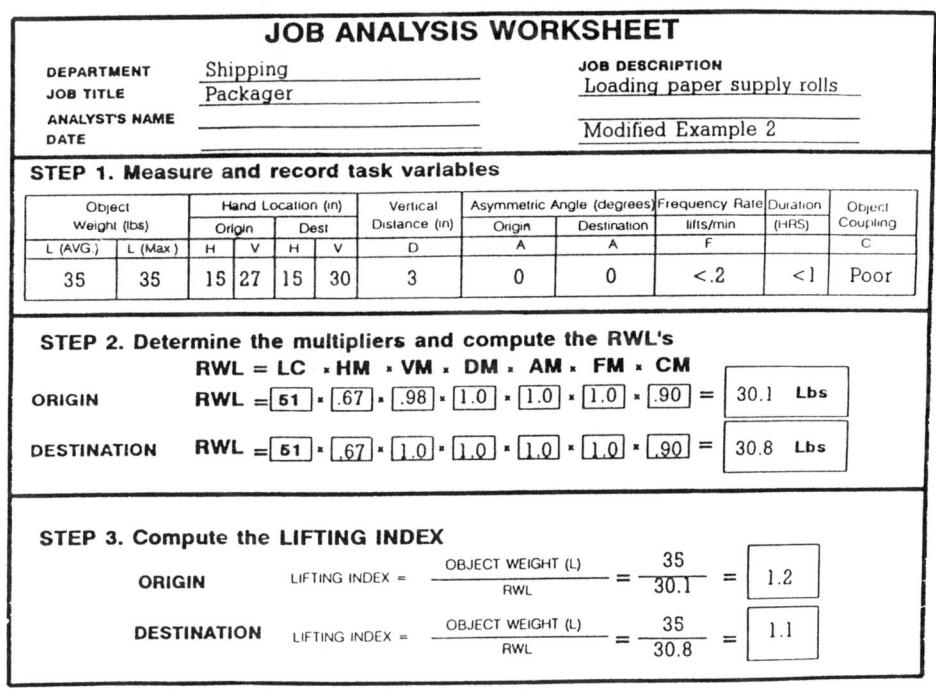

Figure 8 Modified job analysis worksheet, Example 1.

In some cases, redesign may not be feasible. In these cases, the use of a mechanical lift may be more suitable. As an interim control strategy, two or more workers may be assigned to lift the supply roll.

Comments. The horizontal distance (H) is a significant factor that may be difficult to reduce because the size of the paper rolls may be fixed. Moreover, redesign of the machine may not be practical. Therefore, elimination of the manual lifting component of the job may be more appropriate than job redesign.

Example 2. Dishwashing Machine Unloading

Job Description. A worker manually lifts trays of clean dishes from a conveyor at the end of a dishwashing machine and loads them on a cart as shown in Figure 9. The trays are filled with assorted dishes (e.g., glasses, plates, bowls) and silverware. The job takes between 45 min and 1 hr to complete, and the lifting frequency rate averages 5 lifts/min. Workers usually twist their body to one side to lift the trays (i.e., asymmetric lift) and then their body to the other side to lower the trays to the cart in one smooth continuous motion. The maximum amount of asymmetric twist varies between workers and within workers; however, there is usually equal twist to either side. During the lift the worker may take a step toward the cart. The trays have well-designed handhold cutouts and are made of lightweight materials.

Job Analysis. The task variable data are measured and recorded on the job analysis worksheet (Figure 10). At the origin of the lift, the horizontal distance (H) is 20 in., the vertical distance (V) is 44 in., and the angle of asymmetry (A) is 30°. At the destination of the lift, $H = 20$ in., $V = 7$ in., and $A = 30°$. The trays normally weigh between 5 and 20 lb, but for this example, assume that all of the trays weigh 20 lb.

Using Table 6, the coupling is classified as "Good." Significant control is required at the destination of the lift. Using Table 5, the FM is determined to be 0.80. As shown in Figure 10, RWL = 14.4 lb at the origin and 13.3 lb at the destination.

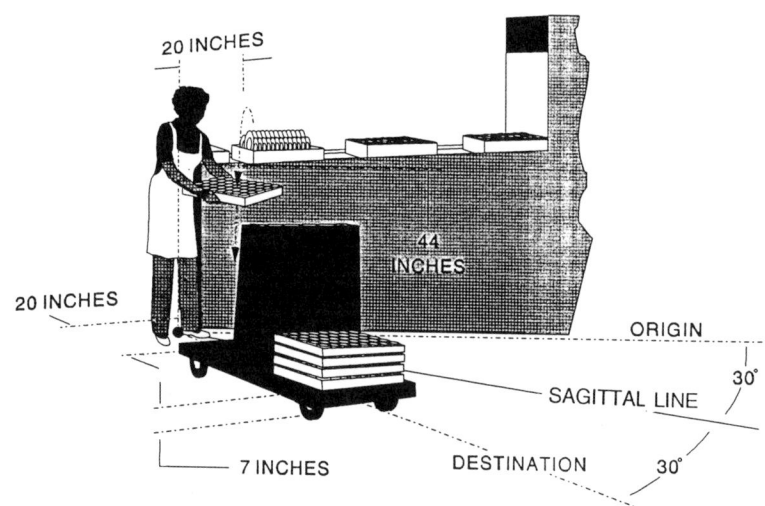

Figure 9 Dishwashing machine unloading, Example 2.

JOB ANALYSIS WORKSHEET

DEPARTMENT	Food Service	JOB DESCRIPTION
JOB TITLE	Cafeteria Worker	Unloading a dish-washing machine
ANALYST'S NAME		Example 5
DATE		

STEP 1. Measure and record task variables

Object Weight (lbs)		Hand Location (in)				Vertical Distance (in)	Asymmetric Angle (degrees)		Frequency Rate lifts/min	Duration (HRS)	Object Coupling
		Origin		Dest			Origin	Destination			
L (AVG)	L (Max)	H	V	H	V	D	A	A	F		C
20	20	20	44	20	7	37	30	30	5	< 1	Good

STEP 2. Determine the multipliers and compute the RWL's

RWL = LC · HM · VM · DM · AM · FM · CM

ORIGIN   RWL = [51] · [.50] · [.90] · [.87] · [.90] · [.80] · [1.0] = 14.4 Lbs

DESTINATION RWL = [51] · [.50] · [.83] · [.87] · [.90] · [.80] · [1.0] = 13.3 Lbs

STEP 3. Compute the LIFTING INDEX

ORIGIN   LIFTING INDEX = $\frac{\text{OBJECT WEIGHT (L)}}{\text{RWL}} = \frac{20}{14.4} = 1.4$

DESTINATION LIFTING INDEX = $\frac{\text{OBJECT WEIGHT (L)}}{\text{RWL}} = \frac{20}{13.3} = 1.5$

Figure 10 Job analysis worksheet, Example 2.

Hazard Assessment. The weight to be lifted (20 lb) is greater than the RWL at either the origin or destination of the lift (14.4 lb and 13.3 lb, respectively). At the origin, LI = 20/14.4 = 1.4, and at the destination, LI = 1.5. These results indicate that this lifting task would be stressful for some workers.

Redesign Suggestions. The worksheet shows that the smallest multipliers (i.e., the greatest penalties) are 0.50 for HM, 0.80 for FM, 0.83 for VM, and 0.90 for AM. Using Table 9, the following job modifications are suggested:

1. Bring the load closer to the worker to increase HM.
2. Reduce the lifting frequency rate to increase FM.
3. Raise the destination of the lift to increase VM.
4. Reduce the angle of twist to increase AM by either moving the origin and destination closer together or moving them further apart.

Since the horizontal distance (*H*) is dependent on the width of the tray in the sagittal plane, this variable can be reduced only by using smaller trays. Both DM and VM, however, can be increased by lowering the height of the origin and increasing the height of the destination. For example, if the height at both the origin and destination is 30 in., then VM and DM are 1.0, as shown in the modified worksheet (Figure 11). Moreover, if the cart is moved so that the twist is eliminated, AM can be increased from 0.90 to 1.00. As shown in Figure

JOB ANALYSIS WORKSHEET

DEPARTMENT	Food Service	JOB DESCRIPTION
JOB TITLE	Cafeteria Worker	Unloading a dish-washing machine
ANALYST'S NAME		
DATE		Modified Example 5

STEP 1. Measure and record task variables

Object Weight (lbs)		Hand Location (in)				Vertical Distance (in)	Asymmetric Angle (degrees)		Frequency Rate lifts/min	Duration (HRS)	Object Coupling
		Origin		Dest			Origin	Destination			
L (AVG)	L (Max)	H	V	H	V	D	A	A	F		C
20	20	20	30	20	30	0	0	0	5	< 1	Good

STEP 2. Determine the multipliers and compute the RWL's

RWL = LC × HM × VM × DM × AM × FM × CM

ORIGIN   RWL = 51 × .50 × 1.0 × 1.0 × 1.0 × .80 × 1.0 = 20.4 Lbs

DESTINATION RWL = 51 × .50 × 1.0 × 1.0 × 1.0 × .80 × 1.0 = 20.4 Lbs

STEP 3. Compute the LIFTING INDEX

ORIGIN   LIFTING INDEX = OBJECT WEIGHT (L) / RWL = 20 / 20.4 = 1.0

DESTINATION LIFTING INDEX = OBJECT WEIGHT (L) / RWL = 20 / 20.4 = 1.0

Figure 11 Modified job analysis worksheet, Example 2.

11, with these redesign suggestions, RWL can be increased from 13.3 lb to 20.4 lb, and the LI values are reduced to 1.0.

Comments. This analysis was based on a 1-hr work session. If a subsequent work session begins before the appropriate recovery period has elapsed (i.e., 1.2 hr), then the 8-hr category would be used to compute the FM value.

REFERENCES

1. T. R. Waters, V. Putz-Anderson, A. Garg, and L. J. Fine, Revised NIOSH equation for the design and evaluation of manual lifting tasks, *Ergonomics* 36(7):749–776 (1993).
2. T. R. Waters, V. Putz-Anderson, and A. Garg, Applications Manual for the Revised NIOSH Lifting Equation, National Institute for Occupational Safety and Health Tech. Rep., DHHS(NIOSH) Pub. No. 94-110, 1994. Available from the National Technical Information Service, NTIS Doc. No. PB94-176930.
3. NIOSH, Work Practices Guide for Manual Lifting, NIOSH Tech. Rep. No. 81-122, 1981, U.S. Dept. of Health and Human Services, Natl. Inst. for Occupational Safety and Health, Cincinnati, OH.
4. NIOSH, Scientific Support Documentation for the Revised 1991 NIOSH Lifting Equation, Technical Contract Reports, May 8, 1991, U.S. Dept. of Health and Human Services, Natl. Inst. for Occupational Safety and Health, Cincinnati, OH. Available from the National Technical Information Service, NTIS No. PB-91-226-274.

5. ASPH/NIOSH, Proposed National Strategies for the Prevention of Leading Work-Related Diseases and Injuries: Part 1, Published by the Association of Schools of Public Health under a cooperative agreement with the National Institute for Occupational Safety and Health, 1986.
6. Eastman Kodak, Ergonomic Design for People at Work, Vol. 2, Van Nostrand Reinhold, New York, 1986.
7. M. M. Ayoub and A. Mital, Manual Materials Handling, Taylor & Francis, London, 1989.
8. D. B. Chaffin and G. B. J. Andersson, Occupational Biomechanics, Wiley, New York, 1984.

32
OSHA's Ergonomic Program

Gregg LaBar
Occupational Hazards, *Cleveland, Ohio*

I. INTRODUCTION

The impact of the Occupational Safety and Health Administration (OSHA) on the field of ergonomics has been significant in recent years. Since the mid-1980s, OSHA has used its enforcement and guidelines-setting authority to address the prevention of cumulative trauma disorders. The agency's recent push to standardize industry's approach to ergonomics helps explain why so many safety and health personnel, consultants, and health professionals are devoting substantial time and resources to this issue. This chapter recaps the history of OSHA's involvement and offers insight into what to expect for the rest of the 1990s and beyond.

Even compared to OSHA's rather short 25-year history, ergonomics is a "new kid on the block" in the field of workplace safety and health (see Table 1). The term *ergonomics*, meaning the study of work, has been around since World War II but until recently was mostly confined to research and military circles. OSHA spent most of its first two decades focusing on acute safety hazards and airborne health hazards.

That started to change in the mid-1980s, as the omnipresent back injuries were found to be responsible for about 30% of all workplace injuries and illnesses, and their costs in financial and human terms continued to rise. Manual materials handlers were at especially high risk of back injuries that disabled them and cost their employers thousands of dollars per case. In addition, the growth of the service sector of the economy and the computerization of all sectors raised concerns about upper extremity cumulative trauma disorders (UECTDs) affecting the fingers, hands, wrists, arms, elbows, shoulders, and neck.

Participants in a 1991 University of Michigan/National Institute for Occupational Safety and Health (NIOSH) conference estimated that musculoskeletal injuries cost American industry as much as $100 billion annually.

In 1993, the last year for which data were available, CTDs accounted for more than 62% of new cases of occupational disease. Most of these involved the upper extremities, although these statistics also include cases of noise-induced hearing loss. Over the years, back injuries have been responsible for up to one-third of occupational injuries.

Since 1990, spurred by the statistics, congressional hearings, research on risk factors and prevention strategies, and pressure from organized labor, ergonomics and the prevention of cumulative trauma disorders have become top regulatory and enforcement priorities for OSHA. As of 1995, OSHA's goal was to promulgate a far-reaching performance-oriented ergonomics standard.

Due to internal review procedures, congressional pressure, and concerns from the business community, OSHA has already encountered several obstacles and delays on the way to

Table 1 History of OSHA's Ergonomic Program and Related Activities

1970	OSH Act passed
1979	First ergonomics-related general duty clause citation
1981	First NIOSH Lifting Guide published
1987	Meatpackers cited
1988	Pepperidge Farm cited
1989	OSHA publishes voluntary safety and health program management guidelines
1990	Americans with Disabilities Act enacted
1990	OSHA publishes meatpacking guidelines, announces plans for general industry standard
1991	Automakers agree to ergonomic settlements
1991	Labor petitions for emergency temporary standard
1992	OSHA publishes advance notice of proposed rulemaking
1993	Second NIOSH Lifting Guide published
1994	Barbara Silverstein leads team developing a standard
1995	Industry and Republican-controlled Congress challenge OSHA
1995	OSHA circulates scaled-down draft ergonomics proposal for comment

creating a standard. Agency officials released a draft of a proposed rule in hopes of getting enough support and insight to move forward with an official proposed rule. They met with various constituents to get their input on if and how OSHA should proceed. An official proposed rule will not be issued until 1996 at the earliest; a final rule is not expected until after the 1996 general election. Most experts predict that eventually there will be an OSHA ergonomics standard for industry, although implementation is at least several years away.

II. BACKGROUND AND SIGNIFICANCE TO OCCUPATIONAL ERGONOMICS

On December 29, 1970, President Richard Nixon signed the Occupational Safety and Health Act of 1970 into law, giving the federal government the power to promulgate and enforce workplace safety and health standards. The law left it up to OSHA to decide what regulations and enforcement procedures were necessary. Regulations addressing "harmful physical agents," which CTD risk factors generally fall under, must, to the extent feasible, ensure that "no employee will suffer material impairment of health or functional capacity."

The law also contains the loosely worded general duty clause (GDC), Sec. 5(a)(1), which states:

> Each employer shall furnish to each of his employees employment and a place of employment which are free from recognized hazards that are causing or are likely to cause death or serious physical harm to his employees.

In the absence of a specific standard, such as with ergonomics, OSHA has used the GDC to issue citations.

In the 1970s, OSHA took direction from the OSH Act and issued and enforced a multitude of safety standards, many simply lifted from consensus standards-setting organizations, on such things as fall protection, personal protective equipment, and machine guarding. During the Carter administration, under the direction of OSHA administrator Eula Bingham, the agency turned its attention to health hazards, implementing standards on substances such as asbestos, benzene, and cotton dust. In the 1980s and early 1990s, OSHA focused on more

III. ERGONOMICS GETS A START

Ergonomics did not really become an OSHA consideration until 1979, when OSHA hired its first ergonomist. Prior to that, the agency was involved in only a couple of pilot projects on back injuries. In the mid-1980s, OSHA considered developing manual lifting guidelines and regulations but never got very far.

OSHA's early foray into the back injury arena was sparked by 1981 lifting guidelines developed by NIOSH, created by the OSH Act as OSHA's research-oriented sister agency. The guidelines recommended a maximum permissible limit (MPL) of 90 lb under ideal lifting conditions. That number would be reduced if the object was bulky, was lifted while bending, must be carried over a distance, or was lifted repeatedly.

These factors could also be used to calculate an action limit (AL), below which lifting jobs were deemed to not pose undue risk. For jobs falling between the MPL and AL, NIOSH said engineering and/or administrative controls could be used. Jobs falling above the MPL were considered high risk, and NIOSH recommended the use of engineering controls.

[A revised lifting guide, published in 1993, established a maximum recommended weight limit (RWL) of 51 lb and added two other variables for evaluating lifting jobs: asymmetrical lifting (twisting and turning) and coupling factors (the person's grip on the object). NIOSH is conducting field studies to evaluate the efficacy of the new RWL and the application of the six variables in the equation.]

Awareness of the cumulative nature of back injuries and repetitive strain injuries developed slowly but steadily, driven by new technology and the higher costs associated with these disorders. In manufacturing and food processing, production lines picked up speed, objects to be moved got heavier, and tools often did not match the required tasks. In addition, the development of the economy's service sector and the computerization of office environments led to a new set of safety and health concerns: office ergonomics, indoor air quality, and electromagnetic radiation.

For the most part, OSHA's early work on back injuries was safety-oriented and focused on the potential for single or a few heavy lifts to cause a back injury, not on the possible hazards of repeated lifting, twisting, bending, standing, or sitting.

In the mid-1980s, however, upper extremity CTDs began showing up in injury and illness logs reviewed by OSHA during recordkeeping inspections. During some reviews, OSHA found that employers had failed to record these instances, and as a result the agency began proposing multimillion-dollar citations for recordkeeping violations.

Bureau of Labor Statistics (BLS) data also pointed to an increasing problem (see Table 2). In 1984, BLS reported 34,700 new cases of CTDs in private industry. By 1993, it found 302,000 cases—a 770% increase—which meant that 62% of all new cases of occupational illness were associated with repetitive trauma, mostly affecting the upper extremities. Furthermore, back injuries continued to be the most common and frustrating of all occupational injuries.

Throughout the 1980s, employers, just like OSHA, were learning a lot about workplace ergonomics. In 1983, for example, only about 25% of major U.S. corporations had human factors experts on staff, and terms like ergonomics, cumulative trauma disorders, and carpal tunnel syndrome were seldom heard in industry. Engineers designed workplaces to maximize

Table 2 Cumulative Trauma Disorders 1984–1993 in Private Industry

1984	34,700
1985	37,000
1986	45,500
1987	72,900
1988	115,400
1989	146,900
1990	185,400
1991	223,600
1992	281,800
1993	302,000

Source: Bureau of Labor Statistics.

the productivity and efficiency of equipment but seldom considered how workers would function in such an environment.

In the late 1980s, budget crunches, corporate downsizing, and total quality management drove interest in ergonomics. Employers became increasingly concerned about rising workers' compensation costs and the impact of workplace and job design on productivity, quality, employee morale, and absenteeism.

Public awareness, even fear, of ergonomic risk factors was heightened by reports that there was an awful new disease called carpal tunnel syndrome. CTS, however, is responsible for only a small percentage of the total of ergonomics-related injuries and illnesses—which led some observers to label the phenomenon "carpal tunnel vision." As word spread that CTS is just the tip of the CTD iceberg, employers, employees, and government increased their investigations into how people work, what tools they use, and how workplaces are designed.

Compliance with the 1990 Americans with Disabilities Act (ADA) also forced many employers to look at matching job tasks and equipment with employee capabilities and making "reasonable accommodations" where necessary. Back injuries are the No. 1 type of ADA claim filed with the Equal Employment Opportunity Commission.

IV. OSHA INVOLVEMENT

Like the cumulative nature of CTDs and many back injuries, it was an accumulation of factors, not a single event, that led to OSHA's interest in this area. Since the mid-1980s, OSHA has responded with steady increases in resources devoted to enforcement of the general duty clause, guidelines, and, ultimately, development of a general industry standard.

In the early 1990s, OSHA identified CTDs as one of the top occupational safety and health problems in the nation. In 1991, OSHA administrator Jerry Scannell said that musculoskeletal injuries were a problem in "every workplace and every institution in this country."

Beginning in 1979 and with increased regularity in the late 1980s, OSHA used the OSH Act's general duty clause in some 500 cases involving alleged ergonomic hazards. The agency targeted the meatpacking and automobile industries, although other types of food processing, nondurable goods manufacturing, warehouses, and nursing homes were also affected. In some cases, OSHA issued large dollar fines and negotiated corporatewide settlements with companies to force them to address ergonomics comprehensively and systematically (see Table 3). Among the major cases were the following.

Table 3 Major Penalties for Alleged Ergonomics-Related Violations

Company	Proposed penalties	Penalties paid
John Morrell, Sioux Falls, S.D.	$4.33 million	$990,000
IBP Inc., Dakota City, Neb.	$3.1 million	$525,760
Ford Electronics & Refrigeration Corp., Lansdale, Pa.	$2 million	$1.95 million
Samsonite Corp., Denver, Co.	$1.6 million	$495,000
Pepperidge Farm, Downingtown, Pa.	$1.4 million	$394,600
General Motors, Detroit, Mich.	$840,000	$420,000
Inland Fisher Guide, Trenton, N.J.	$290,000	$290,000
United Parcel Service, Watertown, Conn.	$279,800	$145,870

Sources: *CTD News*, OSHA's Office of Management Data Systems, and *Occupational Hazards*.

Meatpackers. In a span of 3 months in 1987, OSHA hit meatpacking plants owned by IBP Inc. and John Morrell with proposed penalties of $690,000 and $2.6 million, respectively. The next year, it proposed fines of $3.1 million and $4.3 million, respectively, on other IBP and Morell sites. OSHA cited other companies from the industry, including ConAgra Turkey Co., Empire Kosher Foods, and Sara Lee Corp., and eventually got many of the companies to sign corporatewide settlements that mandate comprehensive programs to identify and reduce CTD risk factors. The settlements' program elements were so similar that OSHA later formally adapted them for use in an industrywide manner, in *Ergonomics Program Management Guidelines for Meatpacking Plants* (For more on the guidelines, refer to Section V.)

The "Big Three" Automakers. Chrysler, Ford, and General Motors were charged large fines for high rates of CTDs, and by early 1991 OSHA had signed all three to corporatewide settlements. These agreements, which cover automobile plants around the country, require facilities to have comprehensive ergonomics programs. The companies were allowed several years of phase-in time, but full compliance is required by 1996.

Pepperidge Farm. In 1988, OSHA fined Pepperidge Farm $1.4 million for alleged violations of recordkeeping requirements and the general duty clause (GDC) as it relates to ergonomics. The case became prominent in 1993 when an administrative law judge (ALJ) ruled that OSHA cannot use the GDC to force abatement of repetitive motion injuries. According to the ALJ, the lack of an ergonomics standard meant that OSHA failed to prove that a feasible means of abating ergonomic hazards exists. The ALJ concluded that there is proof that back injuries and other CTDs are work-related and therefore OSHA should promulgate a standard if it wishes to enforce ergonomic requirements.

In early 1995, an ALJ reached a similar conclusion in a case involving Dayton Tire Co.

According to an evaluation of 1985–1994 federal OSHA data by IBM ergonomist William Kukla, employers in Pennsylvania received 145 of the estimated 500 ergonomics-related citations under the federal general duty clause. Kukla also found that more than 70% of federal OSHA ergonomics inspections were the result of employee complaints and that the number of inspections rose steadily from fiscal 1985 to 1990 and has dropped steadily since then (see Table 4).

In the late 1980s, industry groups demanded that OSHA stop using the broad-based GDC because they said the agency's expectations of employers were inconsistent and were not spelled out. Some even suggested that standards that target specific industries would be preferable because they would know who was covered and what they had to do.

Table 4 Federal OSHA's Ergonomics-Related Inspections (1985–1994)

1985	8
1986	18
1987	26
1988	35
1989	82
1990	93
1991	75
1992	69
1993	65
1994	29

Source: William Kukla, IBM, Feb. 28, 1995.

In the summer of 1991, organized labor turned up the heat on its push for a broad-based standard by petitioning the agency for an emergency temporary standard (ETS). The unions complained that solutions to CTDs are "at hand, but the tools of enforcement are not." In denying the request, OSHA found that ergonomics did not meet the test for an ETS, which, according to Sec. 6(c)(1) of the OSH Act, can be issued only when employees are "exposed to grave danger from exposure to substances or agents determined to be toxic or physically harmful or from new hazards."

V. GUIDELINES

In the late 1980s, OSHA was saying that ergonomics was an important issue but that it was not yet ready for regulation. As a result, the agency responded with the enforcement program described above, as well as a series of booklets, guidelines, and fact sheets.

In January 1989, OSHA published voluntary safety and health program management guidelines (OSHA 3123), which emphasize management commitment and employee involvement as essential components of any successful safety and health program. The guidelines identify four key program elements: worksite analysis, hazard prevention and control, medical management, and training and education. According to OSHA, those elements apply to almost any industry and to any type of program, including ergonomics.

As an outgrowth of its enforcement emphasis for the meatpacking industry, in 1990, OSHA developed the *Ergonomics Program Management Guidelines for Meatpacking Plants*. Although many of the meatpacking citations were issued before the guidelines were published, OSHA announced the release of the guidelines as "the beginning of a nationwide program" to address CTD prevention. Many companies outside the meatpacking industry took the guidelines as reflecting what OSHA expects and implemented programs along these lines.

Reflecting the four key elements in the safety and health guidelines, the meatpacking guidelines provide for:

Worksite analysis. Includes reviewing existing records, conducting surveys, and analyzing jobs systematically such as with a checklist or analysis device.
Hazard prevention and control. Identifies types of engineering controls, work practices controls, personal protective equipment, and administrative controls. As with its other

guidelines and regulations, OSHA states a preference for engineering controls such as redesigned workstations and tools and changes in work methods to reduce awkward postures, repetitive motion, and excessive force.

Medical management. Calls on health care providers to be "part of the ergonomic team." It recommends periodic workplace walk-throughs, symptoms surveys, a list of light-duty jobs, health surveillance, early reporting of symptoms, and protocols for CTD evaluation, treatment, and follow-up.

Training and education. Recommends different levels of training for all affected employees, engineering and maintenance personnel, supervisors, managers, and health care providers. According to the OSHA guidelines, employees who are "potentially exposed to ergonomic hazards should be given formal instruction on the hazards associated with their jobs and with their equipment."

During the development of the meatpacking guidelines, OSHA officials promised that general industry guidelines would be published next. The agency circulated draft guidelines for general industry in 1990 but never formally proposed or published them. In the fall of 1990, Labor Secretary Elizabeth Dole announced the final meatpacking guidelines and announced a shift in policy: OSHA was going to begin rulemaking on a general industry ergonomics standard. Since then, ergonomics has been one of the hottest issues at OSHA and in workplaces around the nation.

VI. ASKING FOR ADVICE

Beginning with Dole's announcement, OSHA has emphasized the enormity of the task at hand and the fact that it would take several years for the agency to move such a far-reaching, controversial standard through the pipeline. In late 1990, for example, OSHA administrator Scannell said that this was not a "fast-track" project and that it would take five to seven years for the agency to complete the rulemaking.

In August 1992, under the direction of acting OSHA administrator Dottie Strunk, OSHA took its first major step by publishing an advance notice of proposed rulemaking on ergonomic safety and health management. The notice contained a series of questions on whether and how OSHA should address workplace ergonomics. Among the questions:

How big a problem are CTDs?
How are jobs analyzed?
What successful control and prevention measures have been implemented?
What kinds of protective equipment and medical devices are available?
What training and education are needed?
What should a health management program include?

The initial request resulted in information from more than 280 commenters, who submitted in excess of 2000 pages and 600 attachments. Responses ranged from urging OSHA to stay out of this area to calling for a strict standard. Many of the comments were from industry sources, who offered wide-ranging information about how they address ergonomics and what OSHA should or should not be doing.

The Bush administration added to the profile of the issue by creating a federal OSHA Office of Ergonomic Support and establishing ergonomic coordinators in each of the 10 regional offices.

VII. A NEW ADMINISTRATION

Bill Clinton's victory in the 1992 general election boosted the chances for an OSHA ergonomics standard. Clinton nominated Joe Dear, former head of Washington State's Department of Labor and Industries, to head the agency. Since officially taking over in November 1993, Dear has made the development of a general industry ergonomics standard a top priority.

He brought in epidemiologist Barbara Silverstein from Washington (who has since left the agency) to head a team developing a standard. At one point, she had 10 people working nearly full time on the standard; a second interagency team reviewed and commented on OSHA's work.

During 1993 and 1994, OSHA reform legislation in Congress would have increased the agency's enforcement and regulatory authority, including requiring issuance of an ergonomics standard. Hearings were held on the labor-backed legislation, introduced by Senator Edward Kennedy (D-Mass.) and Representative William D. Ford (D-Mich.), but it never made it to President Clinton, who had promised to sign it if Congress passed it.

VIII. OSHA'S FIRST TRY

During the first two years of the Clinton administration, the possibilities for an ergonomics standard appeared endless. When the OSHA proposal was in the early stages, Silverstein promised that an OSHA rule would

- Be far-reaching, covering virtually all workplaces in manufacturing, office environments, and construction, and targeting back injuries and cumulative trauma disorders of the upper and lower extremities.
- Be performance-, program-, and system-oriented. OSHA would not tell employers how many repetitions were too many or what lifts were unsafe but would require employers to evaluate jobs and implement improvements that were feasible for that workplace.
- Require employers to do more than just check their OSHA 200 logs and workers' compensation data to determine if they had a CTD problem. OSHA would suggest using ergonomic risk factor checklists, employee symptoms surveys, and other methods to identify early warning signs of CTDs.
- Rely on the safety and health program management and ergonomic meatpacking guidelines, which emphasize worksite analysis, hazard prevention and control, medical management, and training and education. OSHA would point to 1993 BLS data, which show that CTD rates were rising in most industries but declining in meatpacking and automobile manufacturing—where OSHA has already used this approach.
- Be aimed at middle managers and supervisors who had some ergonomic background but did not have to be experts to implement a program. OSHA officials were wary of writing a high-end, state-of-the-art technical standard.
- Be prevention-oriented—"pre-hab instead of rehab"—although the medical management portion of the standard was expected to address treatment.
- Emphasize informing and involving employees. OSHA was considering requiring, for example, that manually handled packages that weigh 25 lb or more be labeled as such.
- Be solutions-oriented. OSHA was not expected to require certain kinds of control measures such as adjustable chairs, lift assist devices, or redesigned tools but to encourage employers to develop and select from a menu of solutions. OSHA was not expected to recommend the use of back supports, wrist splints, or other such devices.

OSHA's Ergonomic Program

During these optimistic times, an OSHA official said that the ergonomics rulemaking could be "the biggest, fastest-developed standard in the history of OSHA. More than ever before, we've done our homework."

IX. A NEW APPROACH

With good reason, OSHA officials are not nearly as confident now (late 1995) as they were in early 1994. With the Republican takeover of Congress in the fall of 1994 and the nation's general antiregulatory mood, OSHA dramatically adjusted its expectations and its proposed requirements.

In March 1995, the agency released a draft of a scaled-down proposed rule in hopes of convincing concerned business leaders and members of Congress that an ergonomics standard does not have to pose an onerous compliance burden.

Agency officials met with stakeholders from industry, labor, professional associations, academia, and government to get their comments and, if OSHA were lucky, their blessing to go ahead with issuance of a proposed rule. Even the most optimistic timetables peg release of a final rule not until well after the 1996 general election.

The revised draft was designed to reduce the burden on employers who already have ergonomics programs and on small businesses that are concerned about the high cost of compliance. Instead of attempting to cover all 6.1 million U.S. workplaces, as it had promised in 1994, for example, OSHA's revision focused on 2.6 million workplaces where there is likely employee exposure.

Instead of advocating a comprehensive program approach to CTD prevention, the draft proposal contained a simpler Look, Learn, Control strategy (see Figure 1). All covered employers would be required to "Look" to see if they have had any work-related CTDs since

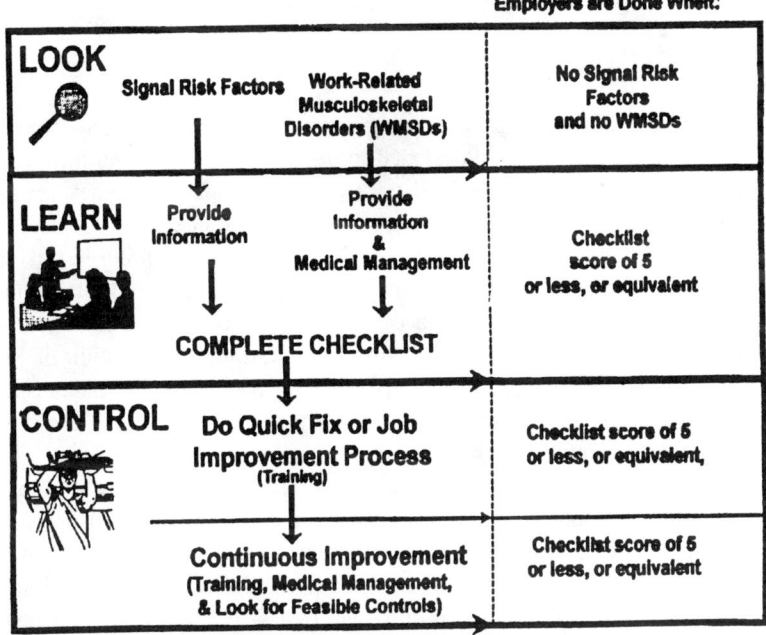

Figure 1 OSHA Look, Learn, Control strategy.

the effective date of the standard or have jobs with "daily exposure" to "signal risk factors." The signal risk factors are

1. Performing the same motion every few seconds for 2, 3, or 4 hr at a time.
2. Maintaining fixed or awkward postures for more than 2, 3, or 4 hr at a time.
3. Using vibrating or impact tools for more than 2, 3, or 4 hr during a work shift.
4. Using forceful hand exertions for more than 2, 3, or 4 hr at a time.
5. Lifting heavy objects or lifting frequently for 1 or 2 hr during a work shift.

OSHA asked constituents and other interested parties which of the numerical thresholds would work best.

At the "Learn" stage, employers would have to provide information and medical management to affected employees. They would use a checklist, either one provided by OSHA or an equally effective alternative, to determine the extent of their ergonomics problems.

At the "Control" stage, employers would implement quick fixes or a longer term job improvement process to reduce risk factors. The proposal calls for administrative or engineering controls to correct problems. Employers would continue to evaluate jobs and make improvements until checklist scores were reduced to 5 on the OSHA form or the equivalent on the other forms. Although the standard calls for continuous improvement in workplace ergonomics, employers would have no further compliance burden if checklist scores were reduced to 5 or the equivalent.

The revised approach includes a "grandparent" exemption, which allows employers who have had effective ergonomics programs in place for at least one year prior to publication of the final rule to continue with their programs as long as they meet the final compliance date.

The agency also included an extra year—five instead of four—for compliance by businesses with fewer than 10 employees. OSHA backed off on requiring medical removal protection for MSD-suffering workers and on reviewing workers' compensation and OSHA 200 log data for the previous two years to identify high-risk workers and jobs.

X. CHALLENGING OSHA

OSHA's proposals for less far-reaching coverage, the "grandparent" exemption, and more phase-in time for small businesses were designed to quell some of the opposition coming from the business community. OSHA also decided not to require employers to pay for an eye care program and corrective lenses for VDT operators with neck/shoulder pain. Nonetheless, many people in the business community are wary of, or opposed to, rulemaking in this area. Small businesses, for example, are worried that the rulemaking could put them out of business and that no amount of extra time will help them comply in a cost-effective manner.

The Coalition on Ergonomics, which includes the National Association of Manufacturers and nearly 200 other trade associations and employers, is concerned that compliance could cost American industry as much as $80 billion. Coalition leaders say they are not automatically opposed to an ergonomics standard but argue that it must be based on "good science and good, hard data on dose–response relationships and risk factors." They are concerned that if as OSHA pushes for continuous improvement in programs, employers will never know when they have done enough to be in compliance.

If there is to be a standard, business leaders say, OSHA must demonstrate why it is necessary, what it will cost, why certain requirements must be met, and what results can be expected. If OSHA cannot meet those tests, business leaders have vowed to oppose the OSHA

OSHA's Ergonomic Program

rulemaking more vehemently than anything before, including labor-backed OSHA reform legislation. Many business leaders have urged OSHA to work instead on a broad-based safety and health program management standard that would include ergonomics, instead of a separate ergonomics rule. OSHA officials, meanwhile, have promised to issue proposals in both areas, which they consider complementary, in 1996.

Industry's concerns about an ergonomics standard are getting a sympathetic hearing in the new Republican-controlled Congress. Congressional leaders, including Senator Nancy Kassebaum (R-Kans.) and Representative Bill Goodling (R-Pa.), promised to hold oversight hearings of the rulemaking if OSHA proposes an ergonomics standard that business leaders feel is too onerous. Congress could also use the appropriations process to prevent OSHA from devoting any resources to the development and enforcement of an ergonomics standard.

XI. CRITICAL REVIEW OF CURRENT STATUS

The Executive Branch's administrative review process ensures that there is never smooth sailing for a far-reaching regulation such as OSHA is proposing for ergonomics. In this case, however, OSHA faces several other hurdles from the business community and Congress that make fast-track rulemaking impossible and any kind of rulemaking a risky proposition at best. In addition, OSHA is faced with the reality that a less far-reaching, legislatively mandated proposal failed in California. As of this writing, the Cal/OSHA Standards Board was considering developing a new proposal, or it could wait to see what OSHA proposes. In 1993, Arizona was developing a standard for its state program, but that is on hold, pending federal action.

When Dear took office in late 1993, he promised that OSHA would have a proposed rule on ergonomics published by September 30, 1994—an ambitious date that most experts thought was too optimistic for such a far-reaching regulation. The agency later revised its timetable to shoot for the end of 1994. OSHA now hopes to have a proposal out by 1996, after which it plans to accept comments for six months and hold hearings across the country. At the earliest, a final rule would not be ready until 1997—more than six years after OSHA announced its intention to develop a standard.

Now, even that timetable appears to be on shaky ground—which would be a major blow to Dear's record, who, in early 1994, said he wanted to "get this job done while I have time to get my signature on the piece of paper." That appears impossible unless Dear stays at OSHA well into the next presidential term, which no previous agency administrator has done.
OSHA acknowledges that some of the requirements it is proposing may appear too stringent for all employers. The strategy of the agency appears to be that it will ask for as much as possible in the early stages and then discuss details, thresholds, and triggers as the project moves toward development of a final rule.

XII. FUTURE CONCERNS

Although some key OSHA personnel are still pushing hard for a standard, an alternative, voluntary compliance-oriented approach is also possible. OSHA is considering, for example, resurrecting the draft guidelines and issuing them to encourage, but not require, employers to have ergonomics programs. There is also a possibility that NIOSH will offer guidelines, which after being evaluated and tested could become a building block for a later OSHA standard.

If either agency publishes guidelines in the near term, they will not help OSHA with enforcement because the agency will still have to prove a violation of the general duty clause—a strict test that the courts have traditionally ruled against. If OSHA hopes to enforce ergonomic requirements, it will have to promulgate a standard and go through all of the political and legal maneuvers associated with the formal rulemaking process.

The consensus among OSHA officials and business and labor leaders is that eventually there will be an OSHA standard. Some observers are predicting that a proposed rule will be issued in 1996 but that a final rule will not be issued until at least 1997. Depending on the political climate and interest in the issue, it could be several years after that before a final rule is adopted and takes effect. Due to the far-reaching nature of the issue, a final standard will probably be greeted by a chorus of lawsuits.

In the short term, the guidelines approach is more likely to produce an official document. But with questions about public input and enforcement, guidelines are unlikely to be more than an interim step on the way to all-out rulemaking.

Until now, many employers who have addressed ergonomics have done so for reasons other than OSHA activity. If the agency goes ahead with its major project, however, OSHA compliance will become the driving influence behind workplace ergonomics programs. OSHA hopes that by the time a standard gets through the pipeline, many employers will have had a positive experience with ergonomics, perhaps as a result of following voluntary guidelines, and will be less likely to challenge the agency over the need to address CTD hazards.

Many observers believe that the Clinton administration will be fortunate just to get a proposed rule out of OSHA during Dear's tenure. No matter how they approach the issue of ergonomics, current and future OSHA officials have no choice but to do something. How and when they proceed will have a dramatic impact on employer programs across the country. It is no wonder, then, that even some six years after OSHA announced its intention to regulate ergonomics, the agency and its constituents are still trying to decide how much of a good thing—designing jobs and workplaces to maximize employee health, productivity, and quality—to mandate and how much to leave open to the instincts and insight of employers and employees.

ACKNOWLEDGMENTS

I thank Dr. Roger Stephens for his forthcoming insight and comments over the years on OSHA's approach to ergonomics. Special thanks also go to Dr. James McGlothlin of NIOSH for his help and support on this chapter and to Steve Minter, editor of *Occupational Hazards* magazine, for providing the time to work on this project.

BIBLIOGRAPHY

Brief Summary of the Key Provisions of the Regulatory Text, Draft Proposed Ergonomic Protection Standard, Mar. 13, 1995.

Control of Work-Related Cumulative Trauma Disorders, Part 1: Upper Extremities, ANSI Z-365, Working Draft, July 14, 1994.

Draft Ergonomic Protection Standard, Summary of Key Provisions, June 17, 1994.

Ergo Inspections Plummet during Standards Formation, *CTD News*, March 1995, Vol. 4, No. 3.

Ergonomics Program Management Guidelines for Meatpacking Plants, OSHA 3123.

Ergonomics Program Management Recommendations for General Industry, OSHA, Sept. 6, 1990.

Ergonomics: The Study of Work, OSHA 3125.

W. Kukla, OSHA ergonomics citations, IBM, Paper presented at conference on Managing Ergonomics in the 1990s: New Challenges, Feb. 28, 1995.
G. LaBar, Building Ergo Land, *Occup Hazards*, October 1991, pp. 29–33.
G. LaBar, Ergonomics: is OSHA's blueprint too hot to handle? *Occup. Hazards*, April 1995, pp. 35–38.
G. LaBar, OSHA: ergonomics can't wait any longer, *Occup. Hazards*, April 1994, pp. 33–36.
G. LaBar, What should OSHA do about ergonomics?, *Occup. Hazards*, April 1993, pp. 31–35.
Occupational Injuries and Illnesses in the United States by Industry, 1983–92, U.S. Department of Labor, Bureau of Labor Statistics, Dec. 21, 1994.
Occupational Safety and Health Act of 1970 (S. 2193), Dec. 29, 1970.
OSHA, *ErgoFacts* Fact Sheets.
OSHA Advance Notice of Proposed Rulemaking for Ergonomic Safety and Health Management, *Federal Register*, Aug. 3, 1992.
OSHA circulates ergonomics draft for support and comments, *Occup. Hazards*, May 1995, pp. 25–26.
OSHA Ergonomics Docket, No. S-777.
Prevention of Cumulative Trauma Disorders, Cal/OSHA Standards Board, September 1994.
Working Safely with Video Display Terminals, OSHA 3092.

33
The Americans with Disabilities Act: Implications for the Use of Ergonomics in Rehabilitation

Jerry A. Olsheski
Ohio University, Athens, Ohio

Robert E. Breslin
University of Cincinnati Medical Center, Cincinnati, Ohio

I. INTRODUCTION

The Americans with Disabilities Act (PL 101-336), or ADA, is regarded by people with disabilities and those in the rehabilitation community to be the most significant piece of civil rights legislation enacted by Congress since the Civil Rights Act of 1964. It is designed to extend civil rights protection similar to that found in existing legislation related to race, sex, age, and ethnicity to individuals with disabilities [1]. Specifically, the ADA is intended to provide a federal mandate prohibiting discrimination based on disability in American society while establishing enforceable standards to address such discrimination and provide redress to those protected individuals who experience discrimination.

In contrast to the medical model, which equates disability with the functional limitations of the individual, ADA addresses the environmental side of disability. The act is specifically designed to remove attitudinal, architectural, and physical barriers that have the effect of excluding individuals with disabilities from places of employment and commerce. As more attention is given to the role that environmental factors play in creating and maintaining disability, the role of ergonomic interventions in disability prevention and management will assume more importance.

In this chapter, the passage of ADA is examined in terms of the impact that this legislation is having on the changing model of disability in American society, the vocational rehabilitation process, and the use of ergonomics in job accommodation.

II. LEGISLATIVE HISTORY OF THE ADA

Although most Americans became aware of the ADA relatively recently, particularly when it became an issue during the presidential election campaign of 1988, the genesis of the law dates back more than a decade to 1983 [2]. At this time the leadership of the National Council on the Handicapped began to explore methods for extending civil rights similar to those found in Title V of the Rehabilitation Act of 1973, which covered federal government agencies

(Section 501), federal contractors (Section 503), and recipients of federal grants (Section 504), to the private sector and state and local governments. An initial draft of the ADA bill was introduced in 1988. This draft provided the 100th Congress with the opportunity to demonstrate the existence of broad bipartisan support for the underlying principles of the act.

After significant revision, the ADA was reintroduced to the 101st Congress in 1989. Throughout 1989 and 1990 a complicated series of negotiations took place among the House of Representatives, the Senate, the White House, and advocates for the disability and business communities. The result was a complex and multifaceted bill developed with extensive input from lobbyists and advocacy organizations representing those with affected interests. The ADA was passed in the Senate and the House of Representatives by overwhelming margins. The final version of the act was signed into law on July 26, 1990 by President George Bush.

A. The Rationale for a New Civil Rights Law

It seems remarkable that a landmark piece of civil rights legislation was conceived and enacted in an era characterized by political conservatism and widespread calls for a diminished federal role in American social and economic life. One needs to look no further than the "Findings and Purposes" [Americans with Disabilities Act of 1991, PL 101-336, Sec. 2(a)] of the ADA, however, to understand why the vast majority of the legislators in Congress felt compelled to support the measure. McMahon and Shaw [3] have called this portion of the ADA "one of the single most eloquent and direct statements of societal culpability ever observed in print." The findings section of the law documents the prevalence of disability in U.S. society, estimating that 45 million Americans have some physical or mental impairment. These individuals have often been isolated, segregated, and discriminated against as the result of erroneous and stereotypical assumptions regarding their abilities. Congress sites data that document that "people with disabilities, as a group, occupy an inferior status in our society, and are severely disadvantaged socially, vocationally, economically, and educationally" [ADA, 1991, PL 101-336, Sec. 2 (a)(7)]. This discrimination has cost the United States billions of dollars as a result of expenses associated with lost productivity and disability support programs.

III. THE ADA: A GUARANTEE OF EQUAL ACCESS FOR AMERICANS WITH DISABILITIES

The ADA contains five separate titles that guarantee equal access to employment, public accommodations, transportation, state and local government operations, and telecommunication services. Equal access to housing and air transportation were not included in the ADA because discrimination in these areas is prohibited by the Fair Housing Amendments Act of 1988 and the Air Carriers Access Act of 1986.

The following section provides a brief description of the scope of the coverage, dates of implementation, and specific protections afforded by the five titles.

A. Title I

Title I prohibits discrimination in *access to employment*. This portion of the law applies to private sector employers, labor unions, employment agencies, and all governmental bodies, with the exception of the federal government, who employ 15 or more employees in 20 or more calendar weeks during a calendar year. Employers who are exempt from ADA include corporations wholly owned by the federal government, Indian tribes, and tax-exempt private membership clubs. Provisions of the law contained in Title I were implemented in stages, with employers having 25 or more employees being subject to the provisions in July 1992. Employers having 15–24 employees were subject to the provisions of Title I in July 1994.

The most significant aspect of Title I is the extension of the "reasonable accommodation" provisions of the Rehabilitation Act of 1973 to the private sector. Under the ADA an employer must consider whether an individual can perform the essential functions of the job with or without reasonable accommodation. Since the employment-related provisions of Title I will directly affect the provision of rehabilitation and ergonomic services in industry, these provisions are the primary concern of this chapter.

B. Title II

Title II prohibits discrimination by any public entity in providing *public services*, including transportation, to individuals with a disability. Public entities include agencies, special-purpose districts, departments, and other instrumentalities of state or local government and commuter authorities of the Rail Passenger Service Act. Title II became effective on January 26, 1992.

C. Title III

Title III prohibits discrimination on the basis of disability in places of *public accommodation*, including places of lodging, establishments for serving food and/or drink, places of exhibition or entertainment, places of public gathering, sales or rental establishments, service establishments, stations used for specified public transportation, places of public display, places of recreation, places of education, social service centers, and places of exercise [4]. Title III also covers all places of commerce. These provisions became effective July 26, 1993, although there are some exceptions for certain provisions that apply to existing business facilities based on size and financial status.

D. Title IV

Title IV contains the *telecommunications* provisions of the ADA. It requires that common carriers of wire or radio communications provide accommodations for individuals with hearing and speech impairments that will serve as the functional equivalent of telephone service for those individuals.

E. Title V

Title V contains *miscellaneous provisions* of the ADA, including those regarding the relationship of the Americans with Disabilities Act to other federal and state laws, those clarifying the effect of ADA on commercial insurance, and the specific prohibition of immunity for states under the eleventh amendment to the Constitution of the United States. Title V also prohibits retaliation against an individual who takes action to exercise his or her rights under the ADA and outlines remedies and procedures as well as the plans for implementation of each of the first four titles. Title V outlines the technical assistance plan and makes some modifications to the definition of handicapped individual in the Rehabilitation Act of 1973 to exclude individuals currently using illegal drugs.

IV. TITLE I EMPLOYMENT: IMPLICATIONS FOR AMERICAN INDUSTRY

As previously described, Title I of the ADA was designed to prohibit discrimination in employment practices against "qualified individuals with disabilities" by employers and other "qualified individuals with disabilities" by employers and other "covered entities." The ADA regulates all aspects of the employment process, including application, testing, hiring and assignments, evaluation, disciplinary actions, training, promotion, and medical examinations, layoff and recall, termination, compensation, leave, and benefits.

A. Definition of Terms

To gain a more accurate understanding of Title I, it is necessary to define several key terms that operationalize the employment-related requirements and compliance process.

1. Individual with a Disability

The definition of disability in the ADA is a derivative of the definition found in the Rehabilitation Act of 1973. For the purposes of the ADA, an *individual with a disability* is any individual with a physical or mental impairment that substantially limits one or more major life activities. These major life activities include walking, speaking, breathing, performing manual tasks, seeing, hearing, learning, caring for oneself, and, perhaps most important from the perspective of both employers and individuals with disabilities, working.

Individuals may also fall under the protection of the act if they have a record of an impairment or if they can demonstrate that discrimination took place because they are regarded as having such an impairment. Finally, individuals may be protected by the act if they are discriminated against because of their relationship with an individual with a disability.

Several conditions that might be considered disabilities under certain circumstances or definitions are specifically excluded from protection under the ADA. These include transvestism, homosexuality and bisexuality, transsexualism, pedophilia, exhibitionism, voyeurism, gender identity disorders, compulsive gambling, pyromania, kleptomania, and current abuse of alcohol and drugs. Individuals with communicable diseases that can be transmitted through food handling may also be denied employment in jobs involving food handling if there is no reasonable accommodation that would eliminate that risk.

2. Qualified Individual with a Disability

In order to be considered a "qualified individual with a disability" under the ADA, the individual must be one who satisfies the requisite skill, experience, education, and other job-related requirements of the employment position that the individual holds or desires and who, with or without reasonable accommodation, can perform the essential functions of such position.

Thus, determining whether an individual with a disability is "qualified" under Title I of the ADA is a two-step process. The first step focuses on the qualities of the individual as a worker. That is, does the individual have the education, skills, work experience, licenses and certifications, and other prerequisites to perform the job successfully? This step should be present in any employment or hiring decision, regardless of the presence of an impairment.

The second step involves the evaluation of whether the individual can perform the "essential functions" of the job with or without reasonable accommodation. This step requires that the employer give attention to the work environment. Typically, a job analysis is conducted to quantify the functional demands and working conditions of a specific job. Job analysis data are used as the basis to identify the essential functions of the job. Second, the employer must evaluate whether the qualified individual with a disability can perform the essential functions of the job, as identified in the job analysis process, with or without reasonable accommodation. This second step, referred to as the job accommodation analysis process, is discussed in more detail later in the chapter.

3. Essential Functions

The ADA does not specifically define the term *essential functions* but rather provides guidelines to employers to aid in their identification. First and foremost, functions identified by an employer to be essential must actually be required of employees performing the job in question. They cannot be arbitrary, e.g., requiring the candidate to type 80 words per minute in a job in which typing is rarely required of the worker.

The employer may also consider whether removing a particular function changes the job in a fundamental way. An employer is not required to accommodate an individual by removing a function if the job exists primarily to perform that function. The employer may also take into consideration the number of employees available to perform a particular function and the degree of specialization or special expertise required to perform the function.

4. Reasonable Accommodation

Perhaps the most significant of the employment-related provisions of the ADA is the legal requirement that employers who are "covered entities" consider reasonable accommodation as a component of employment decisions. *A reasonable accommodation* is defined by the EEOC (1992) as "any modification or adjustment to a job, an employment practice, or the work environment that makes it possible for an individual with a disability to enjoy an equal employment opportunity." Reasonable accommodation is required in order to guarantee equal opportunity in the application process, to enable a qualified individual with a disability to perform the essential functions of a job, and to allow an employee with a disability to enjoy equal "benefits and privileges" of employment. Reasonable accommodations in the application process are intended to guarantee all qualified individuals with disabilities an equal opportunity to be considered for employment. Examples of accommodations in the application process may include modifications that render the interview or application site accessible to an individual who is mobility-impaired or the provision of assistance to an individual who is unable to complete the application process because of visual or manipulative impairments.

Reasonable accommodations to perform the essential functions of a job may include any type of modification of the work environment, including the work devices, the manner in which the job is typically performed, or the administrative policies governing the work and worker. Specific examples of the kinds of accommodations envisioned by Congress are listed in the act [Americans with Disabilities Act of 1991, PL 101-336, Sec. 101(9) (a&b)] and the regulations developed by the EEOC (EEOC-29CFR, 1630, July 1992 (0) (1&2)]. Reasonable accommodations may include

Making facilities readily accessible to and usable by an individual with a disability
Restructuring a job by reallocating or redistributing marginal job functions
Altering when or how an essential job function is performed
Part-time or modified work schedules
Obtaining or modifying equipment or devices
Modifying examinations, training materials, or policies
Providing qualified readers and interpreters
Reassignment to a vacant position

Permitting use of accrued paid leave or unpaid leave for necessary treatment

Providing reserved parking for a person with a mobility impairment

Allowing an employee to provide equipment or devices an employer is not required to provide

There are some limitations on the employer's obligation to provide reasonable accommodation. Only qualified individuals with a disability can request such an accommodation, and the employer is only required to accommodate the *known* limitations of the individual. The ADA does require, however, that covered entities post notices regarding the provisions of ADA including the reasonable accommodation requirement. In addition, the employer may request documentation of the functional limitations claimed by the individual requesting the accommodation. If it can be established that the qualified individual with a disability needs an accommodation, the employer is required to seek input from the individual regarding potential accommodation strategies and to consider his or her preferences for such an accommodation. The employer is not, however, required to select the accommodation chosen by the employee if one can be identified that is equally effective for the individual while better serving the needs of the employer.

5. Undue Hardship

An employer is not required to make an accommodation to a qualified individual with a disability if the employer can demonstrate that the provision of such an accommodation would impose an undue hardship on the business. The employer can use undue hardship as a defense in not making an accommodation if it places an unacceptable financial or other type of burden on the business or employer. This includes modifications that are unrealistic in terms of size, scope, or cost relative to the size and resources of the employer. Accommodations that would alter the fundamental nature or operation of the business or significantly disrupt the business may also be considered to pose an undue hardship.

V. TITLE I OF THE ADA: A SUMMARY OF THE PROHIBITION OF DISCRIMINATION IN EMPLOYMENT

Title I of the ADA prohibits a covered entity from discriminating on the basis of disability against a qualified individual with a disability, i.e., an individual who can perform the essential functions of the employment position with or without reasonable accommodation. Employment discrimination is operationally defined in such a manner that Title I regulates the conduct of employers and their relationship with employees, job applicants, and other aspects of the employment environment.

The reasonable accommodation requirement of the ADA is especially significant in that it recognizes and addresses the environmental aspects of work disability. Although the act clearly addresses attitudinal barriers that unfairly stereotype individuals with disabilities as incapable workers, the reasonable accommodation provisions of the law place emphasis on those aspects of the vocational handicap that are external to the individual with an impairment. This shift, although inconsistent with the medical-legal perception of disability, is compatible with the orientations of the vocational rehabilitationist and the ergonomist.

VI. CRITICAL REVIEW OF CURRENT STATUS

American society has primarily defined disability in medical terms and for the large part has ignored the role that environmental factors play in the creation and maintenance of disability. When disability is essentially defined as a medical condition, the focus of attention centers on the functional limitations of the individual [6, 7]. The medical orientation to disability is personological or individualistic in nature [8]. This approach is based on the assumption that disability is a phenomenon dependent mainly on individual behaviors and individual decisions [9]. Consequently, this paradigm provides an incomplete picture of disability by belittling the political, social, ecological, and environmental variables that play an important role in mediating and determining disability.

The dominance of the medical model in the disability arena has influenced public policy, employment practices, the nature of vocational rehabilitation services, and claims management techniques. Public policy has tended to equate disability with impairment and has restricted disability to clinical criteria. Many employers have refused to reemploy workers with restrictions and have depended on the medical provider to restore the worker to "100%" capacity before the worker returns to work. Vocational rehabilitation in many instances has amounted to little more than an extension of medical interventions as evidenced by the increasing number of registered nurses involved in the "case management" of workers with disabilities and the rapidly escalating costs of medical services in the workers' compensation system. In response to runaway medical costs, insurers who underwrite workers' compensation coverage have developed the concept of *maximum medical improvement* as a short-term cost containment technique. This concept indicates that the worker cannot benefit (i.e., make functional improvements) from additional medical services and the insurer will no longer pay for these types of interventions. Unfortunately, many workers who are evaluated as having achieved maximum medical improvement fail to return to work and instead enter long-term disability support systems [10]. Maximum medical improvement, without reemployment, actually signifies that medical interventions can only go so far in resolving disability. Likewise, the provision of additional medically oriented services under the guise of vocational rehabilitation have resulted in increased criticism of the effectiveness of vocational rehabilitation. A number of states have accordingly repeated mandatory rehabilitation provisions in their workers' compensation systems due to the perception legislators have that rehabilitation services amount to little more than additional medical costs with poor vocational outcomes [11].

A. ADA and the Changing Model of Disability in American Society

The passage of ADA marks a significant departure from the medical model of disability. Disability is defined more in environmental or sociopolitical terms as embodied in the various provisions of the ADA. The primary objective of this legislation is to alter the environmental side of disability. Thus, the emphasis or focus of change as mandated by ADA is on factors that are external to the individual. Whereas the medical model of disability aims at changing the person, the model of disability inherent in the provisions of ADA aims at changing the environment, and this is the essence of the accommodation process. The five titles of the ADA are concerned with removing *environmentally induced disability factors* in such areas as telecommunications, public services, public accommodations, and employment.

VII. FUTURE NEEDS: AN ENVIRONMENTALLY ORIENTED MODEL OF VOCATIONAL REHABILITATION

From a theoretical perspective, the vocational rehabilitation process requires that equal attention be given to characteristics of the person and factors in the environment because disability can originate as much from environmental barriers as from the individual's impairment [12]. Although the environmental side of disability has long been recognized at the theoretical level, the influence of environmenal factors in the rehabilitation process has largely been ignored at the (applied) service delivery level. Rehabilitation interventions that are aimed at altering the characteristics of the person are more widely used than interventions that focus on altering or restructuring the environment. The underutilization of environmentally focused interventions results from a rehabilitation model that continues to view disability as a pathological condition of the person. Pathological elements within the environment are overlooked. In this model the goal of rehabilitation is to increase the functional capacities or in some fashion overcome the occupational liabilities of the person with a disability [6].

The most popular rehabilitation services for workers with disabilities are those that attempt to enhance physical and occupational capacities. The proliferation of work hardening and work capacity centers represents the medical orientation to disability in that such facility-based services focus almost entirely on the characteristics of the person and ignore the importance of environmental variables. Occupational disability is based solely on the capacities of the worker, and the factors in the work environment are not entered into the equation [13]. Thus, although every attempt is made to enhance the capacities of the worker, there is no corresponding attempt to enhance the capacity of the work environment in terms of job accommodation and retention.

A. The Emerging Disability Management Model and the Role of Ergonomics

In response to the continued growth of the personal, social, and economic costs of disability, a revised model of disability and rehabilitation is beginning to emerge. This alternative model has been identified as *disability management* [13–16]. The disability management model recognizes that disability is a complex phenomenon that cannot be defined as a medical condition and resolved by clinical solutions alone [17]. Disability management represents a revitalization of the basic interactional nature of disability, which posits that work disability results from the interaction of the individual's characteristics with the physical, social, and organizational elements in the work environment. Recent research reveals the powerful influence of environmental and organizational variables in the occurrence and outcomes of work disabilities [18–20].

Disability management has been defined [14, p. 2] as

> A workplace prevention and remediation strategy that seeks to prevent disability from occurring or, lacking that, to intervene early following the onset of disability, using coordinated, cost-conscious, quality rehabilitation services that reflect an organizational commitment to continued employment of those experiencing functional work limitations.

A more detailed analysis of this definition highlights the differences between disability management and other types of traditional vocational rehabilitation programs. The differences in these two models are summarized in Table 1.

First, the disability management model contains a primary prevention component that seeks to minimize the possibility of the occurrence of disability. Second, disability management is considered a workplace- or employer-based approach in which labor and management actively participate in both the prevention and rehabilitation processes in the context of the *actual work environment*. Finally, disability management strategies require the organizational commitment to retaining and accommodating workers with permanent restrictions.

Unlike rehabilitation programs that are grounded in the medical model of disability, disability management goes well beyond the mere evaluation of the individual worker and analyzes the environmental or organizational structures within the worker's interactive sys-

Table 1 Comparison of Traditional Vocational Rehabilitation and Disability Management Models

Component	Traditional vocational rehabilitation	Disability management
Concept of disability	Disability defined in medical terms, i.e., functional impairment.	Disability defined as interaction of person and environment.
Focus of rehab	Alter capacity of the person.	Balanced focus on worker capacities and capacities of work environment for accommodation and job retention.
Typical rehab interventions	Physical therapy, work hardening, psychological and vocational testing.	Worksite rehab, ergonomics, job analysis, organizational assessment.
Objective of rehabilitation	Match person to job or no RTW.	Match job to person via job accommodation.
Roles of employer and union	Passive; rely on external provider.	Active participation in rehab and RTW.
Context of rehabilitation	External facility; simulated work.	Actual work environment; real work.
Role of PT/OT	Inclusion/exclusion function; i.e., rule in or out of job.	Negotiate job accommodation and retention.
Source of job accommodation information	Simulated capacities of worker and third party job analysis data.	Direct observation of worker and job.
Role of case manager	External service coordinator; little influence in disability negotiation process.	Empowered as internal negotiator; member of joint labor/management committee.
Disability prevention	Externally generated treatment data are of little use in disability prevention because treatment team does not interact with actual work environment or influence safety policy.	Use of clinical skills at worksite helps identify and resolve ergonomic risks and safety hazards to prevent future injuries.
Role of physician	Typically establishes worker capacities based on type of impairment, limited RTW options.	Provided with precise functional data that document the worker's ability to perform real work tasks; may release worker to RTW with restrictions.

RTW = return to work; PT = physical therapy; OT = occupational therapy.

tem. It is within this interactive arena that the political nature of disability emerges as attempts to reemploy and accommodate workers with disabilities are negotiated.

Ergonomic interventions are of paramount importance in the disability management model. In general terms, *ergonomics* is defined as the scientific study of human work [21]. More specifically, ergonomics is the application of scientific information for the purposes of effectively matching jobs to workers. An effective match is one that maximizes working efficiency (productivity) without compromising the psychological or physical health of the worker. Ergonomic principles, when applied in the disability management model, are valuable tools in the primary prevention of disability and in negotiating the job retention of those with permanent residual limitations.

Ergonomics is a discipline that can help refocus the rehabilitation process so that individual and environmental factors are given equal attention. While medical and psychological interventions are valuable in evaluating the characteristics of the person, the primary value of ergonomic applications is in evaluating environmental factors. Thus, ergonomics and rehabilitation are complementary, for both disciplines strive to maximize the positive work adjustment of the individual [22].

VIII. THE ROLE OF ERGONOMICS IN WORKSITE REHABILITATION PROGRAMS AND JOB ACCOMMODATIONS

The disability management model serves as the philosophical and theoretical foundation for rehabilitation programs that occur in the workplace itself. Worksite rehabilitation programs focus on workers' capacities as they directly interact with the requirements of the job and other elements of the work environment. Transitional work return programs are an example of applying clinical, interdisciplinary rehabilitation skills at the worksite [23, 24]. In transitional work programs, real work activities are used as therapeutic modalities to condition, readjust, and accommodate workers with functional limitations. Transitional work return programs have a unique set of features such as the following.

1. Transitional work programs have established time parameters so long-term or open-ended periods of light duty are avoided. Transitional work is not an occupational goal but an interim step in the physical, psychological, and vocational readjustment of the worker.
2. Workers who participate in transitional work are under the clinical supervision of a licensed physical or occupational therapist. These professionals design and monitor the task progression process with the objective of expeditiously returning the worker to full duty status.
3. Transitional work return programs involve the use of an interdisciplinary rehabilitation team including a vocational rehabilitation counselor, a therapist, and an ergonomist.
4. Since transitional work programs occur at the workplace, the professional rehabilitation team is complemented by the expanded rehabilitation team to include labor and management representatives, work supervisor, the worker, and coworkers.
5. Interaction among the professional and expanded rehabilitation team members facilitates the negotiation process that surrounds job retention and accommodation.
6. Transitional work is not a medically oriented intervention but rather a type of interactive rehabilitation service based on functional capacity and job analysis data. These sets of data form the basis of the job accommodation analysis as required by ADA.
7. The development of transitional work return programs requires a number of organizational interventions including staff development of management and labor representatives, formal policies and procedures that regulate the operational aspects of the program, joint labor/management committees to serve as the steering body for the program, ergonomic surveys of the work environment, and a program evaluation system.

Transitional work return programs provide labor and management with an excellent opportunity to comply with the requirements of ADA. Should the worker have any type of permanent functional limitation at the completion of the transitional work program, information is available to accurately identify specific areas of incompatibility between the worker's capacities and the essential functions of the job.

Certainly, ergonomics can be used as a tool for reducing or removing work–job incompatibilities. Collaboration among the ergonomist (environmental specialist), the therapist (functional capacity specialist), management and labor representatives (organizational specialists), and the worker lies at the core of the interactional model of disability. Information generated from this collaboration of specialists can be presented to joint labor/management committees to initiate the negotiation process and make decisions regarding the reasonableness of specific types of accommodations.

The primary role of the ergonomist in worksite rehabilitation and ADA compliance programs is the application of ergonomic skills in identifying, evaluating, and modifying environmental factors that influence the disability negotiation process. These environmental factors may include labor relations, safety practices, physical job design, job stress factors, coworker relationships, reemployment policies, adversarial claims management techniques, and other characteristics of the physical and psychosocial dimensions of the work environment. Table 2 presents examples of reasonable accommodations achieved through the use of ergonomic interventions in actual cases of workers who participated in transitional work return programs.

Specific job functions of the ergonomist in the disability prevention and management processes include the following.

1. Evaluates the work environment to identify risk factors that contribute to physical injuries or psychosocial stresses.
2. Participates as a member of a professional, interdisciplinary (worksite) rehabilitation team; performs ergonomic job analysis for participants in transitional work return program when needed.
3. As a member of the worksite rehabilitation team, advises physical/occupational therapist, vocational rehabilitation counselor, physician, and others regarding ergonomic issues.
4. Serves as member of joint labor/management steering committee and advises members during job accommodation process; participates in ADA compliance policy development and makes recommendations for specific job accommodations.
5. Establishes and serves as member of ergonomic task force to identify and resolve risk factors; assists in data collection process.
6. Participates in staff development by providing educational programs for management, labor, employees, and rehabilitation professionals.
7. Collaborates with internal engineering staff, supervisors, and workers in designing and implementing job modifications/accommodations.

IX. CONCLUSION

Although many employers initially interpreted the employment-related provisions of ADA to be primarily concerned with preventing discrimination in the *hiring* of people with disabilities, recent EEOC data indicate that the majority of complaints filed thus far have originated from current or former workers with either preexisting or work-related disabilities [25]. ADA is accordingly having a significant impact on the reemployment practices of employers and unions who have traditionally refused to return a worker to the job unless he/she was able to perform all aspects of the pre-injury job. More emphasis must now be given to using rea-

Table 2 Examples of Actual Ergonomic Hazard Prevention Methods and Engineering Controls

Type of industry	Problem	Control	Est. cost
Manufacturer of electronic automobile parts	CTS from using a standard screw driver to "pry" distributor parts for adjustment.	Covered screw driver handles with thermo plastic material that can be custom molded and remolded for each operator.	$25/handle
Manufacturer of electronic automobile parts	Hand/wrist tendinitis, CTS from cutting plastic tabs manually using diagonal cutters	Replaced manual cutters with an automatic trimming machine that uses a pneumatic rotary actuator and four hardened steel blades.	$1300
Manufacturer of electronic automobile parts	CTS from removing automobile distributors from an assembly line, rotating them 90°, placing them in a fixture, and installing two screws using a suspended in-line powered driver.	Replaced in-line screwdriver with a suspended pistol-grip model to install the screws while the distributor remains on the assembly line.	$1500
Manufacturer of fiberglass compression-molded breaker boxes	CTS from manually filing to remove flashing.	Covered all file handles with thermo plastic that can be custom-molded for each operator.	$8/handle
Manufacturer of fiberglass compression-molded breaker boxes	CTS from manually filing to remove flashing.	Purchased pneumatic belt sanders, bench-mounted and foot-controlled.	$400/unit
Manufacturer of fiberglass compression-molded breaker boxes	CTS from manually cutting fiberglass using straight using straight-knives.	Replaced straight knives with 35° curved knives.	$8/knife
Manufacturer of fiberglass compression-molded breaker boxes	CTS from manually cutting fiberglass using straight-handled utility knives.	Installed adjustable-height pneumatic "guillotines."	$3800/unit
Manufacturer of large commercial trucks	Back injury from opening large sliding door with one hand.	Replaced small door handle with a vertical two-hand "grab bar."	$40
Manufacturer of large commercial trucks	Left shoulder injuries from repeatedly lifting and pulling 40-lb lubrication and fuel hoses.	Recommended overhead rail and balances system to suspend hoses and ease handling.	$6000
Manufacturer of large commercial trucks	CTS and low back and shoulder pain from poor seating posture while installing dashboard wiring.	Replaced metal stools with adjustable polyurethane workseat.	$260/stool

CTS = carpal tunnel syndrome.
Source: Courtesy of J. DeWees, Ergo Accommodations, Union, Kentucky.

sonable accommodations not only in the personnel selection process but also as a method of job retention because many workers who sustain work-related injuries are protected by Title I of ADA.

In 1991, total workers' compensation costs in the United States were estimated at $60 billion per year, while medical costs for work-related injuries were increasing 50% faster than non-work-related injuries [26]. Employers and insurers may find it more prudent, from both the economic and legal standpoints, to invest more effort in seeking environmental solutions to disability. Research indicates that altering the characteristics of jobs rather than focusing only on the nature of the worker's impairment can substantially reduce work loss [27]. Many employers are beginning to realize the advantage of developing worksite rehabilitation programs as an alternative to the more costly approach of waiting for the worker to achieve "full recovery" in a nonwork environment. These employers are reporting significant reductions in lost work time and substantially lower compensation and rehabilitation costs [25, 28, 29].

Ergonomic interventions play a central role in the disability management model of prevention and rehabilitation. The science of ergonomics is helping to shift the context of industrial rehabilitation services from medically oriented environments to actual work environments. Work disability must be understood in the context of the employment situation, and the rehabilitation process must be extended as much as possible to the workplace [17].

Worksite rehabilitation programs allow for more accurate assessments of the worker's functional capacities as they interact with the demands of the job and characteristics of the work environment. A primary objective of delivering rehabilitation services in the actual work environment is to prevent or contain the phenomenon known in the disability research literature as "spread" [7, 30]. The delivery of rehabilitation services in the context of the real work environment decreases the chances of inappropriately "spreading" or generalizing a specific functional limitation to all work tasks. Unfortunately, many workers have been deemed vocationally disabled on the basis of third party job analysis data and standardized functional capacity information that are characteristic of facility-based rehabilitation programs. The worksite clinician, however, has the advantage of observing the interaction between a worker's capacities and required work tasks. This type of clinical observation generates a worker–job compatibility profile and pinpoints very precisely what tasks the worker is capable or incapable of performing. This information forms the basis of the job accommodation analysis process and identifies the need for possible ergonomic interventions as a method of reducing or eliminating discrepancies between the worker's abilities and job requirements. Worksite rehabilitation programs mitigate the "all or nothing at all" approach that marks the reemployment practices of many employers and unions.

The use of ergonomic interventions in the rehabilitation process ensures that the environmental side of disability is addressed. As employers attempt to comply with the reasonable accommodation requirements of Title I, more emphasis will be placed on seeking environmental solutions to disability.

The ultimate goal of ADA is the total integration of people with disabilities into all aspects of American society. The primary objective of vocational rehabilitation is the successful integration and retention of people with disabilities in the world of work. If either of these missions is to be fully realized, vocational rehabilitation interventions must reflect the true nature of disability and develop strategies that deal with organizational, environmental, and political variables that contribute to the high rate of unemployment among people with disabilities. The implementation of rehabilitation services at the actual workplace ensures the inclusion of environmental and organizational disability factors in the rehabilitation process. Conversely, industrial rehabilitation services that tend to segregate people with disabilities in facilities that rely on simulated work activities or synthetic work environments may ironically promote the discriminatory stereotype that people with disabilities do not belong in the workplace. Hopefully, the increased use of ergonomic interventions in rehabilitation will result in services that are more congruent with the ecology of the world of work and more effective in preventing environmentally induced disability.

REFERENCES

1. P. Morrissey, *The Americans with Disabilities Act*, Report to the U.S. House of Representatives, Committee of Education and Labor, Washington, DC, 1990.
2. F. Bowe, Development of the ADA, in *The Americans with Disabilities ACT: Access and Accommodation*, N. Hablutzel and B. McMahon, Eds., Paul M. Deutsch Press, Orlando, FL, 1992, pp. 3–10.
3. B. McMahon and L. Shaw, Considerations for the rehabilitation consultant, *The Americans with Disabilities Act: Access and Accommodation*, N. Hablutzel and B. McMahon, Eds., Deutsch, Orlando, FL, 1992, pp. 199–211.
4. H. Gertsman, Environmental barriers: Questions and answers, in *The Americans with Disabilities Act: Access and Accommodations*, N. Hablutzel and B. McMahon, Eds., Deutsch, Orlando, FL, 1992, pp. 225–235.
5. EEOC, *A Technical Assistance Manual of the Employment Provisions (Title I) of the Americans with Disabilities Act*, U.S. Equal Employment Opportunity Commission, Washington, DC, 1992.
6. H. Hahn, A sociopolitical perspective, in *Economics and Equity in Employment of People with Disabilities*, R. Habeck, D. Galvin, W. Frey, D. Tate, and L. Chadderdon, Eds., University Center for International Rehabilitation, Michigan State Univ., East Lansing, MI, 1985, pp. 18–21.
7. C. DeLoach, When attitude begets impairment, in *The Americans with Disabilities Act: Access and Accommodations*, N. Hablutzel and B. McMahon, Eds., Deutsch, Orlando, FL, 1992, pp. 11–33.
8. D. K. Mitchell and S. Leclair, *Strategic Case Resolution, Enhancement, and Management Systems*, International Center for Industry, Labor, and Rehabilitation, Dublin, OH, 1988.
9. C. Castillo-Salgado, Assessing recent developments and opportunities in the promotion of health in the American workplace, *Soc. Sci. Med.* 19(4):349–358 (1984).
10. E. Hester and P. Decelles, *The Worker Who Becomes Physically Disabled: A Handbook of Incidence and Outcomes*, Menninger Foundation, Topeka, KS, 1988.
11. M. Berkowitz, Should rehabilitation be mandatory in workers' compensation programs?, *J. Disabil. Policy Stud.* 1(1):63–80 (1990).
12. D. Hershenson, A theoretical model for rehabilitation counseling, *Rehabil. Counsel. Bull.* 33(4):268–278 (1990).
13. D. K. Mitchell and S. Leclair, Building a working alliance with employers: the politics of work disability, in *Physical Medicine and Rehabilitation: Rehabilitation of the Injured Worker*, E. Johnson, Ed., Saunders, Philadelphia, 1992, pp. 647–664.
14. S. Akabas, L. Gates, and D. Galvin, *Disability Management: A Complete System to Reduce Costs, Increase Productivity, Meet Employee Needs, and Ensure Legal Compliance*, Amacon, New York, 1992.
15. D. Galvin, Health promotion, disability management, and rehabilitation in the workplace, *Rehabil. Lit.* 47(9):218–223 (1986).
16. R. Habeck, *Implementing Disability Management: Programs in Industry*, Paper presented at the meeting of the Center for Advancement of Industrial Rehabilitation and Evaluation, Denver, CO, 1989.
17. R. Habeck, Managing disability in industry, *NARPPS J.* 6(4):141–146 (1992).
18. R. Habeck, M. Leahy, H. Hunt, F. Chan, and E. Welch, Employer factors related to workers; compensation claims and disability management, *Rehabil. Counsel. Bull.* 34(3):210–226 (1991).
19. E. Hester and P. Decelles, *The Effect of Employer Size on Disability Benefits and Cost-Containment Practices*, Menninger Foundation, Topeka, KS, 1990.
20. P. Rousmaniere, Too many cooks, no chefs, *Health Manage. Quart.* 11:2 (1989).

21. S. Pheasant, *Ergonomics: Work and Health*, Aspen, Gaithersburg, MD, 1991.
22. S. Kumar, Rehabilitation: an ergonomic dimension, *Int. J. Ind. Ergon.* 9(2):97–108 (1992).
23. D. Shrey and J. Olsheski, Disability management and industry-based work return transition programs, in *Physical Medicine and Rehabilitation: State of the Art Review*, C. Gordon and P. Kaplan, Eds., Belfus and Hanley, Philadelphia, PA, 1992, pp. 303–314.
24. D. Shrey and R. Breslin, Employer-based disability management strategies and work return transition programs, in *The Americans with Disabilities Act: Access and Accommodation*, N. Hablutzel and B. McMahon, Eds., Detusch, Orlando, FL, 1992, pp. 139–154.
25. R. Groepper, Structured return-to-work and the Americans with Disabilities Act, *Healthcost Monitor,* 2:3 (1993).
26. R. Victor, Workers compensation at a crossroads, Fifth Annual John R. Commons Lecture, Univ. Wisconsin, Madison, WI, 1992.
27. E. Yelin, The myth of malingering: why individuals withdraw from work in the presence of illness, *Milbank Quart.* 64(4):622–647 (1986).
28. M. Weinstein, Many happy returns, *Risk and Insurance,* November 1993.
29. M. Padgett, E. Hollander, L. Warden, R. Coleman, and R. Schwartz, Evaluation of a model return-to-work program in Texas, *Work* 3(3):21–41 (1993).
30. B. Wright, *Physical Disability: A Psychological Approach,* Harper and Row, New York, 1960.

34
Legal Aspects of Ergonomics

James J. Montgomery
Montgomery, Rennie & Jonson, Cincinnati, Ohio

I. INTRODUCTION

On March 10, 1980, Michael Village had no idea that he was changing the course of workers' compensation law in the state of Ohio—all he knew was that his back ached. Mr. Village worked at an automobile assembly plant. He had worked there in various capacities since 1973, but for the last five days he had been working as a battery securer: he chose batteries from a rack, each weighing between 20 and 40 pounds, and installed them in automobiles. On March, 10, 1980 he reported to the plant dispensary with a backache; on March 11, he was unable to get out of bed [1]. Mr. Village was a victim of a common ailment, in a common work setting.

Mr. Village filed a workers' compensation claim alleging that his injury occurred in the course and scope of his employment. Though the Industrial Commission allowed his claim, the employer appealed, the decision was reversed by the trial court, and the reversal was affirmed by the Court of Appeals. To be compensable, the lower courts ruled that "[T]he injury must be accidental in character as a result of [a] sudden mishap and not in the usual course of events." This rationale, imposed by the Supreme Court of Ohio as recently as 1978 [2], was the law in Ohio when Mr. Village developed his backache, and it was the law that the lower courts were bound to follow until it was changed by the Ohio State Supreme Court in *Village v. General Motors*. Reversing a long line of cases, the Court [2] held that

> An injury which develops gradually over time as a result of the performance of
> the injured worker's job-related duties is compensable.

With the release of this decision, Ohio industry was suddenly exposed to an entirely new category of compensable injuries, significant expense, legal involvement, and an affirmative obligation to minimize or eliminate situations that create gradually developing injuries. Workers with such injuries were now able to obtain disability classifications and draw compensation for what used to be considered [2] predictable and expected noncompensable "normal wear and tear" injuries. After *Village v. General Motors,* Ohio industry had a tangible economic incentive to consider ergonomics.

II. BACKGROUND AND SIGNIFICANCE TO OCCUPATIONAL ERGONOMICS

Similar issues have been resolved, or are pending, in other states. Ergonomics, developing as a distinct discipline for decades, rose to prominence in the 1980s, in the field of occupational safety and health. With recent and evolving research on the identification and prevention of cumulative trauma disorders, industry is rapidly becoming sensitive to the need for ergonomic intervention to prevent injuries that occur over time in the normal course of work. This book is an example of that awareness: what constitutes "normal course of work" is being redefined and redesigned, from biomechanical, physiological, and psychophysical perspectives, on subjects as diverse as computer keyboards and materials handling systems.

Lawyers are now heavily involved. Since state courts began to recognize repetitive stress injuries as compensable, state workers' compensation systems have become flooded with claims. Numerous suits pend for repetitive stress injuries caused by the allegedly inadequate design and manufacture of keyboards, keypunches, alphanumeric machines, video display terminals, cash registers, supermarket workstations, stenographic machines, and computer "mouse" devices. Entire industries have been sued for noise-induced hearing loss—industries as diverse as pneumatic air tools, diesel engines, and tire-making equipment. Repetitive stress injury suits were unheard of in the 1970s; they literally did not exist. Now they constitute one of the fastest growing areas of the legal history. Practicing ergonomists and health and safety professionals must be aware of emerging legal precepts in the context of their disciplines. Ergonomists should expect and prepare for additional, significant involvement in administrative and litigation matters and should know and practice the standard of care to which they will be held.

III. WORKERS' COMPENSATION

Historically, state governments developed workers' compensation plans to meet the needs of injured workers who had difficulty recovering for injuries under the adversary system. In order to recover, injured workers had to prove some fault on the part of their employer; employers, in addition, could assert that the worker was negligent, or voluntarily assumed the risk of injury, to thwart these claims. Increasing recognition of the many uncompensated injuries led, in the first few decades of this century, to the widespread adoption of no-fault compensation schemes as the exclusive remedy for workers injured on the job. The workers did not have to establish fault; the employers did not have a defense. The workers received a certain and speedy recovery; the employers received a more limited liability [3].

For most of this century, the compensation systems required that the injury be traceable to a definite time, place, and cause [4], that is, a discernible "accident." This standard, by definition, excludes repetitive stress and cumulative trauma disorders, which, though scientifically recognized, were judicially and legislatively declared to be predictable, expectable, yet noncompensable elements of normal work. The only motivation, then, for ergonomic awareness was either a self-generated concern for the health of employees or the goal of increased production. If a worker simply "wore out," compensation was not available because he could not demonstrate a "precise moment of collapse and dysfunction" [5].

In the 1980s, decisions like that of *Village v. General Motors* became more commonplace. Ergonomic concerns appeared in reported decisions:

> We consider the implication of this rule for all of the employees, factory workers, supervisors, managers, secretaries, salespeople and others, working in Illinois in this technological age. In real life, the erosion of a bodily structure to the point of uselessness translates into arms that cannot lift, legs that cannot walk, knees that cannot bend, lungs that cannot breathe, and eyes chronically irritated or worse. But evidence of such work related injuries alone is not sufficient un-

der the prior interpretations of "accidental injury." Instead, useless limbs, damaged organs and disabled bodies must be pushed to a precise moment of collapse and dysfunction. Then, and only then, according to these interpretations, may a court of this state find an employee eligible for compensation under the Act.

The time has come to abandon an interpretation of "accidental" which fails to address documentable and medically recognizable risks faced by the individuals in connection with their employment. The risk of injury from repeated trauma and exposure endured by truck drivers, CRT operators, chemists and others must be recognized. The judicial interpretation of "accident" must be refined to reflect the purpose of the act and the reality of employees obligated to perform repetitive tasks. [5]

Today, motivated by concern for worker health and safety and the economic implications of enlightened compensation systems, employers and ergonomists have an increased responsibility to eliminate or minimize the risk of cumulative trauma or repetitive stress injuries. A heightened awareness of the role of ergonomics is reflected in OSHA's proposed rule on ergonomic safety and health management [6], which is the federal response to the "significant increase in the reported cases of ergonomic disorders in the work place [which have] more than tripled since 1984" [6]. Purely altruistic motivation, while admirable, is simply not as effective as the prospect of punitive fines and awards, so the evolution in law described here should provide additional motivation. Ergonomists have a job to do. If they fail to perform that job effectively, workers will be at risk and employers will face significantly increased compensation costs, administrative regulation, fines, litigation, and potential punitive damage exposure.

IV. INTENTIONAL TORTS

With ergonomic analysis and intervention, cumulative trauma disorders are quantifiable and predictable. For example, the ergonomic investigation in the John Morrell & Co meatpacking plant disclosed an upper extremity CTD incident rate of an astonishing 41.7 per 100 full-time workers per year and identified intermediate to high risk factors for 171 out of 185 analyzed jobs [7]. One of the conclusions advanced in the NIOSH report was that "ergonomic job analysis revealed the majority of jobs require tasks that are *known risk factors for developing upper extremity CTDs*" (emphasis added) [7]. The legal implications of such conclusions are staggering.

Consider Ohio as an example. As recently as the 1978 decision of *Bowman v. National Graphics* [2], the Ohio courts held that gradually developing disabilities may well be predictable and expected yet were not compensable under the workers' compensation system. From such decisions, industry could justifiably conclude that it was not responsible for normal wear and tear associated with the long-term performance of job activities. However, at the same time the *Bowman* decision was being written, a worker named Blankenship and seven of his coworkers were feeling wronged as a result of their claimed exposure to dimethyltin dichloride at the chemical plant at which they worked. These eight workers filed suit against their employer in 1979, claiming that they were entitled to file such a direct action, outside of the workers' compensation system, by virtue of the employer's allegedly intentional conduct in exposing the workers to hazardous chemicals [8]. The Supreme Court of Ohio held that the exclusive remedy provisions of the workers' compensation act did not preclude a civil suit against an employer for its intentional conduct. Although the *Blankenship* case did not involve ergonomic issues, it did engender a line of cases and legislation affirming the right to sue and defining the standard of proof to establish intent [9], thereby setting, if not springing, the trap for the unwary ergonomist who must evaluate and predict the occurrence of cumulative trauma disorders in an occupational setting.

An intentional tort claim has, as its major component, the allegation that the perpetrator—for our purposes, the employer—intended to harm the employee. With proof of intent comes the major reward of an intentional tort claim: punitive damages—damages awarded over and above the amount necessary to compensate injured parties for their loss, damages that are awarded to punish the offender and serve as an example to other potential offenders that such conduct will not be tolerated. Punitive damages are unpredictable, at best; the large awards generate publicity and provide impetus to otherwise latent claims of a similar nature.

Interpreted in the context of its plain and ordinary meaning, "intent" would not appear to present much of a problem for the ordinary employer–employee relationship. Short of an assault, no one ever actually "intends" to harm an employee. Unfortunately, the law is not accustomed to dealing with ordinary and simple understandings, and definitions of "intent" are very qualified, resulting in widely ranging interpretations.

Again, Ohio is representative of the national trend on these issues. To establish the necessary intent for an intentional tort claim against an employer, the employee must demonstrate

> (1) knowledge by the employer of the existence of a dangerous process, procedure, instrumentality or condition within its business operation; (2) knowledge by the employer that if the employee is subjected by his employment to such dangerous process, procedure, instrumentality or condition, then harm to the employee will be a substantial certainty; and (3) that the employer, under such circumstances, and with such knowledge did act to require the employee to continue to perform the dangerous task. [10]

Mere knowledge, or appreciation of a risk short of substantial certainty, is not intent. However, if employers know that injuries to employees are certain or substantially certain to occur, and they still proceed, the law presumes intent. The qualifier, and perhaps sole comfort, in this equation, is contained in the following hierarchy:

> To establish an intentional tort of an employer, proof beyond that required to prove negligence, and beyond that to prove recklessness must be established. Where the employer acts despite his knowledge of some risk, his conduct may be negligence. As the probability increases that particular consequences may follow, then the employer's conduct may be characterized as recklessness. As the probability that the consequences will follow further increases, and the employer knows that injuries to employees are certain or substantially certain to result from the process, procedure or condition, and he still proceeds, he is treated by the law as if he had in fact desired to produce the result [10].

These quotes should be read carefully: they reflect court-imposed employer standards to determine if conduct is intentional and punitive damages are appropriate.

Consider the following situation. You are the ergonomist for a large insurance company that employs hundreds of clerical workers. Each clerical worker has a standard microcomputer used for data entry and text editing. You are aware of the variety of upper extremity disorders that can occur from the use of conventional keyboards in classic workstations, and you are trying to respond to the individual needs of the keyboard operators as inquiries and complaints come in. As a scientist and ergonomist, you know that, with a population as large as the one you are monitoring, there is a substantial certainty that at least one or some of the keyboard operators will experience a disabling upper extremity disorder as a result of repetitive stress or cumulative trauma. You continue to respond, as responsibly as you can, to the needs of the clerical population. Eventually a member of that population develops and reports a serious disabling condition stemming from the long-term use of the conventional keyboard.

The medical reports confirm that the disability is related to the long-term keyboard operation, just as you predicted. In addition to the claim for a permanent partial disability, the clerical worker sues your company, claiming an intentional tort and punitive damages.

The attorney for the clerical worker has obtained, through normal discovery, copies of your entire file, including those notations or memoranda that reflect your awareness that some portion of the clerical population is at risk for an upper extremity disorder. Using the standards mentioned before, the attorney contends that your actions were intentional because you knew that prolonged use of a conventional microcomputer keyboard had a potential for danger that was substantially certain to occur within the population you were monitoring. With this knowledge, you nonetheless permitted the work to continue.

You, of course, are astonished at these allegations, for you have been striving to do your job properly by being as responsive to the needs of the clerical population as you could be under the circumstances. You believe that you acted reasonably and do not believe that the statistical probabilities of your own science should be the predicate for your and your company's culpability. Unfortunately, the law isn't that clear: you were aware of the problem and allowed the condition to continue even though someone was bound to be affected. Established case law supports the proposition that your actions were beyond reckless, simply because there was a substantial certainty that harm to an employee must result.

This is an uncomfortable example, for it is not a situation where a safety device was disabled or a chemical exposure tolerated, where circumstances would make it certain that injury would result. This is not a situation that was contemplated by the courts when they formulated the standards for determining intent. However, compensable injuries are the responsibility of the employer, and cumulative trauma or repetitive stress disorders are compensable. If you have a population where an injury is substantially certain to occur, then the potential for punitive exposure theoretically exists. Given a theoretical punitive exposure, there is also a substantial certainty that some lawyer is going to jump on that opportunity as soon as it arises. In order to combat such claims, ergonomists need to do their jobs in the professional and responsible manner in which they were trained. The court-imposed standards, admittedly subjective, will govern the analysis regarding responsibility and intent. At a minimum, the ergonomist has the affirmative obligation to act reasonably under the circumstances, for an act of negligence occurring after the employer or ergonomist realizes the substantial certainty of injury may be construed as intent.

The reasoning preserved in another case, *Cantrell v. GAF Corporation* [11], is instructive. This case involved the exposure of certain manufacturing plant workers to asbestos. The plaintiffs presented proof of the employer's awareness of hazards relating to asbestos exposure, including evidence of the employer's failure to heed the recommendations and warnings of a physician-hygienist who was specifically retained to investigate the problem. Instead, the employer terminated the physician-hygienist and failed to advise the employees of the risks associated with their asbestos exposure. The employer argued lack of specific intent to injure. The Sixth Circuit Court of Appeals placed great emphasis on the conduct of the employer:

> [the employer], with knowledge that injury and even death were substantially certain to result from asbestos exposure, continued to expose its employees to asbestos, and followed a policy of not explaining to its employees when medical screening revealed the existence of a disease. [12]

With this focus on the employer's failure to warn and advise, and the employees' continuing exposure, the Court held that "where the employer acts with knowledge that injury is a substantial certainty, the employer has gone beyond negligence, recklessness and wantonness and committed an intentional tort" [11]. The multimillion dollar punitive damage verdict against the employer was upheld.

The *Cantrell* court, when presented with the substantial certainty that a particular condition would result in injury over time, analyzed the *conduct* of the employer and its professional staff. Had the employer heeded the advice of its professionals and followed through with a program of education and prevention, there would have been no punitive verdict. Had the employer acted reasonably, the worst that would be anticipated would be the no-fault compensation claim for injuries sustained at work. If the employer and its professionals had acted in a responsible manner, the potential for punitive exposure would have been minimized.

V. THE STANDARD OF CARE FOR AN ERGONOMICS PROFESSIONAL

The ergonomist, as a professional, has an obligation to act, in the practice of his or her profession, with that degree of care, skill, and diligence commonly exercised by similar professionals under similar circumstances. If the professional does not act in accord with that standard, he or she is negligent, reckless, or worse. If that negligence, recklessness, or "intent" causes harm to a worker, lawyers appear. Simply stated, a cause of action is premised on (1) a duty, (2) which is breached, (3) causing injury.

It is relatively easy for a professional to avoid advancing from negligence to recklessness and beyond in the analysis and treatment of ergonomic problems: Do the job in a professional manner and be able to prove it. Consider, again, the example of the ergonomist charged with the responsibility of monitoring the clerical population. Ergonomics, since it analyzes the manner in which specific individuals work in specific environments, is customization in accord with generalized principles: The workstation for clerical worker No. 1 may not be suitable for worker No. 10, though there will be certain commonalities between them. In such a context, the professional must respond to the entire population with generalized precepts and meet the needs of the specific workers who need ancillary back support, wrist rests, specialized keyboards, trackballs, individually adjustable chairs, etc. Interaction among the professionals and the worker population, essential to the surveillance–cure cycle of ergonomics and the ability to respond to specific needs, is also critical to the legal process. If there is one overwhelmingly significant fundament to the decisions assessing punitive measures against employers, it is in the failure to communicate. The courts look for an exchange of information; they look for advice, cautions, and warnings to the workers from the professionals who are supposed to be watching out for them. Where there is proof of such interaction, it will be difficult for a court to find facts to support a punitive damages award. In *Cantrell*, the federal court of appeals was obviously influenced by the fact that the employer discharged its consultant, ignored his recommendations, and kept the plant population ignorant of the potential risks presented by the asbestos used in the manufacturing process. Recast the facts: The employer, now aware of a problem through interaction with its consultant, follows the recommendations about plant hygiene and surveillance, informs the workers of the risks it has discovered and the program it is establishing, and works with the population to minimize future risk. Will workers who develop asbestos-related problems after a latency period be compensated? Yes; that is precisely what the compensation system was

Legal Aspects of Ergonomics 691

designed to do. Will the workers have a case for an intentional tort and punitive damages? No. The employer responded to the emergent situation with appropriate safeguards and information. The employer acted responsibly. The employer did not conceal information critical to the health and safety of its worker population.

As a profession, ergonomics depends upon employer–worker interaction and worker awareness. Such an exchange, in the legal context, becomes a self-checking process. If ergonomists do their jobs with appropriate interaction, then the information exchange so critical to the legal process is a natural result of a job well done. Proof of that interaction may require some extra effort, for inherent in the nature of injuries that develop over time is the fact that memories fade over that same period of time. Contact records should be retained. Follow-up inquiries not only provide written confirmation of the resolution of a problem but also exist to show that the problem was addressed and resolved. Document all professional, supervisory, and worker involvement in the process. Keep track of announcements, postings, notices, meetings, and other publicity. Since documentation, again, is necessary to the overall ergonomic analysis of the process, the legal end of things is covered by simply doing the job in a professional manner. The standard of care for an ergonomics professional is to perform the job professionally.

VI. THE OSHA ERGONOMIC STANDARD

OSHA's proposed rule on ergonomic safety and health management [6] has generated a significant amount of attention because it is intended to codify ergonomic concerns and establish an administrative process for review, comment, and ultimate enactment. When enacted, the standard will be valuable as a benchmark, or reference, standard for all industry. Employers with existing ergonomics programs will probably not be affected by the compliance requirements of the final rule, for such employers will already be familiar with the science, surveillance, and interaction involved in any ergonomics program. In addition, courts have held that ergonomic considerations are implicit in the OSHA General Duty clause [12], giving rise to the conclusion that specific ergonomic standards are not necessary to enforce ergonomic concerns in any event.

No standard can address every potential situation. It is important to keep in mind that, although compliance with an OSHA ergonomic standard will be mandatory, with fines and enforcement provisions for noncompliance, compliance alone will not guarantee a liability-free existence. The test is the standard of care expected of industry and the ergonomics professionals discussed above. That includes, of course, compliance with all statutory and administrative regulations. If a situation arises that, in the opinion of the professional, requires some action beyond that set forth in any administrative regulation, then the professional opinion, geared to the specific facts under analysis, becomes the standard of care. Consider the following example.

In warehousing operations, there is considerable manual handling of loads (see Chapter 28). If, for example, a rule were promulgated that limited the height from which heavy loads could be lifted to a minimum of 20 in. (50.8 cm), then the law must be followed in its prescriptive sense: it exists as a minimum requirement and if violated subjects the employer to sanctions and fines. As any properly trained ergonomist knows, or should know, the 20-in. minimum height for lifting heavy loads is based on ranges of "standing knuckle height" (see Chapter 28) for some ergonomically significant proportion of the population. However, if the population at your specific warehouse consists of individuals 6 ft 6 in. tall or taller, then mere "compliance" would not be in accord with the standard of care; standing knuckle height for such a tall population is probably something on the order of 32 in. Since a fundamental component of ergonomics consists of fitting the job to the person and "matching the capabilities of specific workers" [6], the 20-in. minimum lift-station height would be ergonomically inconsistent with your specific population. And the provision of a 20-in. lift-station would not be in accord with the applicable standard of care. Because each situation has the potential for its own, individually analyzed standard of care, reference standards can only prescribe where to *begin* an analysis. An ergonomist provides custom services consistent with recommended guidelines.

VII. THE AMERICANS WITH DISABILITIES ACT

The American justice system and the legal profession are frequently derided for the ponderous nature of their proceedings and their obvious preoccupation with economics and wealth redistribution. Bad jokes and egos aside, the Americans with Disabilities Act, discussed elsewhere in this book, will necessarily involve a significant amount of ergonomist, attorney, and court time. An act so fundamentally premised upon individual accommodation must be vague, and therefore subject to interpretation, and case-specific analyses will be required for any disputed failure to comply with the act's provisions. Ergonomists obviously play a major role in the application of the act, for it will fall to them to consider, analyze, discuss, reject, and implement accommodations for the industry and individuals they serve. Ergonomists evaluate and design "facilities, environments, jobs, tasks, tools, equipment, processes and training methods to match the capabilities of specific workers" [6], and the ADA *mandates* such matching for the disabled.

The ADA is enforced by an individual's ability to place employers on trial for violations of civil rights. The enforcement provisions are the same as those used to enforce Title VII of the Civil Rights Act of 1964. The EEOC (Equal Employment Opportunity Commission) receives and investigates charges of discrimination, and if it is unable to mediate or resolve differences, the EEOC might elect to file suit, or it might issue a "right to sue" letter to the person making the charge of discrimination. Available relief includes hiring, reinstatement, accommodation, attorneys' fees, expenses, compensatory damages, and punitive damages according to a scale.

If an ergonomics professional or an employer is placed on trial for a violation of civil rights under the ADA, it is important to remember that neither will be judged by other ergonomists. Judges and juries will assess compliance, and it will be the obligation of the professionals to prove that their actions were reasonable under the circumstances. The greatest weakness—and strength—of the ADA is that it is vague and subjective; ergonomists are relatively free to use their best judgment and will be judged by what is reasonable—that is, whether they complied with the standard of care.

The recently publicized Rodney King cases—one set of facts, two disparate verdicts—illustrate the unpredictable nature of juries and trials. Jurors available for the jury pool will be similar individuals, selected from the voting population at random, who have the time and ability to serve without hardship. Consequently a broad cross section of people will consider any ADA claim. The ergonomists' job, when they defend an employer's actions under the ADA, is to explain the complexities of the situation and the significance of the decisions made.

The plaintiff, or claimant, has the affirmative burden of proof; the claimant has to present facts that, if believed, would indicate that there had been a violation of the provisions of the act. However, the jury will tend to give the claimant the benefit of the doubt and will perceive the ergonomist's role to be that of explaining just why the company doesn't think it owes this claimant a job or at least some money. Sympathy is generally thought to lie with the individual, not the employer.

Read the act: 29 CFR 1630. Read it all. Give it to the company attorney and to management. Read the actual text of the law, the interpretive guidelines, and the EEOC technical assistance manual [13].

The act itself is designed to remove the barriers that prevent qualified individuals with disabilities from enjoying the same employment opportunities as are available to persons without disabilities. It is designed to ensure access to equal employment opportunity based on merit. It accomplishes these goals by legislating that where an individual's disability creates a barrier to employment, the employer must consider whether a reasonable accommodation could remove that barrier. Where an individual's functional limitation impedes job performance, an employer must take steps to reasonably accommodate, and thus overcome, the particular impediment, unless to do so would impose an undue hardship. This is not, however, something to be accomplished by relieving the job applicant of the obligation to perform the essential function of the job itself.

As an illustration of how the process is designed to work, consider the situation of a sight-impaired individual who has applied for a job as an all-purpose executive secretary/administrative assistant. The essential functions of this job are many and varied, including frequent movement about the entire office and plant; out-of-town, overnight travel; and the operation of a large number of office machines, such as a copier, computer, fax, dictaphone, shredder, telephone console, printer, and mail machine. Fully half of the job requires reading, document handling, sorting, abstracting, analyzing, composing, and keyboard entry.

It takes little analysis to conclude that a sight-impaired individual cannot perform many of these essential functions, especially when they are combined into one job. With such a disability, the more frequently considered "accommodation" is the provision of a pair of eyes—a reader or leader. With our theoretical applicant, the question arises, as a matter of law, whether it is a reasonable accommodation to provide a reader or leader to assist the disabled applicant with the job.

With this job description, the answer is undoubtedly no, for the reader, coach, or guide would actually have to perform and duplicate certain of the essential elements of the job: part of any clerical position is the nonlinear, simultaneous analysis of visual input from multiple sources, such as the letter to which the manager is responding, its attachments, the letter written by the manager, its attachments, the keyboard, screen, dictation machinery and controls, and the layout of the desk. A reader is limited to linear output; an executive secretary/administrative assistant requires simultaneous input from a variety of sources. A "reasonable" accommodation does not contemplate hiring an additional individual to actually do the job, or any "essential" portion of it, for the disabled applicant.

By contrast, consider another example. The position available is that of quality control inspector and sealer on the output line in a bulk potato chip operation. The process station is at that portion of the line where the chips are in an unsealed polyethylene bag that has been placed in a 5-gallon can. The essential functions of this job are to look in; examine the contents for contamination, quantity, and breakage, sometimes lifting the bag out of the can to examine the sides and bottom; guide the bag back into the can; twist the bag; apply a wire tie to the bag; place a lid on the can; and go on to the next can. The cycle takes about 6 s. The applicant, otherwise qualified and experienced, has a diagnosed cumulative trauma syndrome, such as carpal tunnel syndrome or thoracic outlet syndrome, and cannot twist a wire tie 2000 times per shift. However, the applicant wants the job and asks for a reasonable accommodation.

The employer is now required to work with the applicant to determine potential reasonable accommodations. The essential functions of the job are inspection, bagging, moving the can, and sealing the bag. The applicant cannot seal the bag with the wire tie currently in use. The first accommodation proposal is a device, similar to that frequently seen in produce departments, that tapes bags shut with a straightforward push of the bag's neck through a slot in the device. Though the proposition sounds reasonable, the sales department quite correctly points out that the container used is the large economy-size can, not an individual portion that is opened only once. The company has to provide, in order to remain competitive, a reusable, simple-to-use, closure device, and the tape used on the proposed device is a one-time, nonreusable closure. Through the dialogue that follows, a second accommodation is proposed: a reusable plastic tab, frequently used to seal bread packages, roughly square, with a serrated notch. The employer agrees, but the applicant points out that use of such a device requires a push-twist motion almost as aggravating as the wire-tie twist. Instead of the low-load, high-repetition twist required by the wire tie, there is an equally destructive high-load, low-repetition motion required by the plastic tab. What is the solution? Can the plastic tab or the wire tie be machine- or tool-applied? The plant ergonomist should be consulted to see if another accommodation is reasonable.

These simplistic examples illustrate the process required by the ADA to determine if an accommodation is possible and whether that accommodation is reasonable. The ergonomist will be an intimate part of this process, and if disagreement results, the ergonomist will be involved in its resolution as well. Whether involved in an accommodation effort or an actual dispute, the ergonomist will need to concentrate on defining the essential functions of a job, analyzing accommodations proposed for specific disabled individuals who are otherwise qualified, analyzing and proving the defense of substantial hardship, and, finally, designing and implementing workplace ergonomic controls to reduce and prevent injuries and to minimize aggravation of preexisting conditions. While this last task sounds like a general job description for a professional ergonomist, it is, in fact, an essential element of compliance under the ADA.

We have already discussed the *Village* decision where a court held that gradually developing injuries arising out of the performance of work are compensable. Courts have also held that a physical injury occasioned solely by mental or emotional stress incurred in the course and scope of employment is also compensable if the injury results from greater strain or tension than that to which all workers are occasionally subjected [14]. Finally, courts have also held that a worker who has proven an aggravation of a preexisting condition is not required to prove that the aggravation was substantial in order to receive compensation [15]. Such decisions are representative of the current evolution of workers' compensation schemes in general: Cumulative trauma cases are compensable; physical or mental conditions that are not related to an *event* are compensable; a minor aggravation of preexisting conditions is compensable. However, under the ADA, an employer may not inquire into preexisting medical conditions or workers' compensation history before offering employment to an otherwise qualified employee. An employer may administer a medical examination or make inquiry into compensation history only if such examinations are required of all applicants in the same job category. An employer may not require a medical examination simply because the results from a medical inquiry disclose a previous job-related injury unless all applicants are required to have the same examination.

Employers will hire people, whether they like it or not, with preexisting conditions or a predisposition to mental or physical injury because the ADA prohibits prescreening on such bases. If the hiring process is conducted in accord with the ADA and the post-offer screening discloses preexisting conditions or predisposition, the applicant can ask for an accommodation. As the working population in a plant ages, or as employees succumb to the effects of job stresses, activities, or even gravity, productivity will diminish. However, increasingly, workers will respond to management's admonition with requests for accommodation. Consequently, the combined effect of the ADA and evolving workers' compensation law has created a serious need for professional ergonomists who will identify the essential elements of a job, design ADA accommodations, and redesign the work practices and stations so that jobs do not cause or aggravate injury. In all of these tasks, the ergonomist is subject to the "reasonableness" standard of care.

VIII. ADA GUIDELINES

The ergonomist should prepare written descriptions of the essential elements of the jobs for which an employer is hiring. Be reasonable. Write these up in advance. The ADA states that advance written descriptions are "relevant." That makes it easier for personnel to defend against a later claim that the description of the essential functions was contrived to rule out the specific applicant or specific disability. Essential functions of a job are those functions that the individual who holds the position must be able to perform unaided or with the assistance of a reasonable accommodation. The interpretive guidelines to Section 1630.2(a) of the ADA outline the factors to be considered in determining if a function is essential. The ADA also requires an employer to consider the terms of applicable collective bargaining agreements.

If a listed job function is not essential, an employer, as a reasonable accommodation, *must* (1) provide an accommodation, (2) transfer the function, or (3) exchange the function. The process of accommodation will require ergonomic supervision or intervention, and the ergonomist who fills this role must work with the otherwise qualified applicant to determine if a particular accommodation is appropriate. Ergonomists must be reasonable and uniform: They may not arbitrarily set production requirements unless those requirements are applied uniformly. The interpretive guidelines suggest a few examples of reasonable accommodations: making employer-provided transportation available; providing personal assistants, such as page turners, for an employee with no hands; providing a travel attendant to act as a sighted guide to assist a blind employee on an occasional business trip; permitting an employee who is blind to bring a guide dog to work, even though the animal is not furnished by the employer.

The benchmark for determining a reasonable accommodation is that it be without undue hardship to the employer. Any accommodation that would be unduly costly, extensive, substantial, or disruptive or that would fundamentally alter the nature or operation of the business could be deemed to be an undue hardship.

The process for determining the appropriate reasonable accommodation is clearly described in the interpretive guidelines relating to Section 1630.09(a). That process includes job analysis, individual consultation with the individual seeking the accommodation, and identification of potential accommodations and the preferences of the individual. If these steps do not result in an agreement, then the guidelines suggest more individualized assessment, including the employer's consultation with the EEOC, state or local rehabilitation agencies, or disability constituent organizations. The guidelines are not prescriptive and contain comments to that effect. However, they do exist as a checklist for items that will be considered as the minimum standard of care for the ergonomist and the employer in determining whether both or either of them have caused or committed any violation of an individual's civil rights under the ADA. What the guidelines really are is a list of the minimum efforts necessary to arrive at an accommodation. Professionals should follow the suggestions, for if they do not, someone will claim that this failure represents a lack of "reasonableness." The ADA was designed to favor the disabled applicant; therefore, to get into the best position possible to avoid a claim, the ergonomist must follow and document every suggestion contained in the act.

Ultimately, occupational ergonomists dealing with the ramifications of the ADA must follow the same standard of care that they apply to the normal performance of the profession: reasonableness. The "reasonableness" requirement suggests a negligence analysis: whether there is a duty, whether that duty was breached, and whether damages resulted. The duty is provided by the ADA statute itself, which also goes a long way to defining the minimum standard of care, that is, what is regarded as reasonable under the circumstances.

Ergonomists should not fear the ADA. It contains a set of nonspecific standards with which industry should comply, but compliance will ultimately be judged on a case-by-case analysis. The courts will look to the specifics of the job and the options, if any, for accommodation. The ergonomist who has acted reasonably has the best chance of prevailing when a claim is presented under the ADA.

IX. CONCLUSION

No brief survey of current trends in workers' compensation law, the ADA, intentional torts, or OSHA proposed rules can substitute for detailed knowledge of the actual cases and laws. The references cited in this survey provide the best available source, or starting point, for fluency in these areas. The actions of industry-employed ergonomists will be imputed to their employers, so it is critical that ergonomists comply with the standard of care for their profession; ergonomists must act in a reasonable and responsible manner and document contacts with management and workers. There is no prescribed standard for any particular situation; whether an ergonomist is reasonable will be judged by what is expected of other practicing ergonomists under similar circumstances.

> There ain't no answer. There ain't going to be any answer. There never has been an answer. That's the answer.
>
> *Gertrude Stein*

REFERENCES

1. *Village v. General Motors* (1984) 15 Ohio St. 3d 129, 472 N.E.2d 1079.
2. *Bowman v. National Graphics Corp.* (1978) 55 Ohio St. 2d 222, 378 N.E.2d 1056.
3. *Van Fossen v. Babcock & Wilcox Co.* (1988) 36 Ohio St. 3d 100, 522 N.E.2d 489.
4. *General Electric v. Industrial Commission* (1982) 89 Ill. 2d 432, 433 N.E.2d 671.
5. *Peoria County Bellwood Nursing Home v. Industrial Commission of Illinois* (1985) 138 Ill. App. 3d 880, 487 N.E.2d 356.
6. 29 CFR 1910; 57 FR 34192-01 (August 3, 1992).
7. See, generally, NIOSH HETA 88-180-1958, John Morrell & Co., Sioux Falls, SD (April 1989).
8. *Blankenship v. Cincinnati Milacron Chemicals* (1982) 69 Ohio St. 2d 608, 433 N.E.2d 572.

9. Section 4121.80(G) *Ohio Revised Code* (1986); *Jones v. VIP Development Co.* (1984) 69 Ohio St. 2d 608, 472 N.E.2d 1046; *Van Fossen v. Babcock & Wilcox Co.* (1988) 36 Ohio St. 3d 100, 522 N.E.2d 489; *Fyffe v. Jeno's Inc.* (1991) 59 Ohio St. 3d 115, 570 N.E.2d 1108.
10. *Fyffe v. Jeno's* (1991) 59 Ohio St. 115, 570 N.E.2d 1108; paragraph 1 of the syllabus.
11. *Cantrell v. GAF Corporation* (1993) 999 F.2d 1007.
12. *In the Matter of the Establishment Inspection of Kelly-Springfield Tire Co.* (1992) 808 F. Supp. 657, which held that even in the absence of specific regulations regarding ergonomics, the general duty provisions of the Occupational Safety and Health Act do cover ergonomic concerns.
13. EEOC, *Title I of the Americans with Disabilities Act,* Technical Assistance Manual, Jan. 26, 1992.
14. *Ryan v. Connor* (1986) 28 Ohio St. 3d 406, 503 N.E.2d 1379.
15. *Schell v. Globe Trucking, Inc* (1990) 48 Ohio St. 3d 1, 548 N.E.2d 920.

35
Real-Time Exposure Assessment and Job Analysis Techniques to Solve Hazardous Workplace Exposures

James D. McGlothlin, Michael G. Gressel, William A. Heitbrink, and Paul A. Jensen
National Institute for Occupational Safety and Health, Cincinnati, Ohio

I. INTRODUCTION

This chapter discusses real-time exposure assessment techniques to solve workplace hazards. Two case studies illustrate useful equipment and techniques for controlling problems of dust and gas exposure in industry. These case studies also show how integrating ergonomic and industrial hygiene principles pinpoints exposure sources and provides effective solutions.

Industrial hygienists often measure a worker's exposure to industrial contaminants by sampling the air he or she breathes. A small, battery-powered pump draws a known flow rate of air through a filter or other collection medium for a measured period of time. The collection medium is analyzed to quantify the contaminant collected and to compute the average exposure for the sampling period. Although these results indicate the extent of exposure, integrated air sampling provides little insight into the specific causes of the worker's exposure. Recommendations for controlling air contaminant exposures are often based upon the industrial hygienist's judgment and can result in control measures that do not address the major air contaminant exposure sources. Direct-reading instruments and data-recording devices can overcome the problem by recording events and exposures in the workplace as a function of time. The data from such a system associates events and exposures, and promotes more effective and focused recommendations for controlling the air contaminant exposures.

Through studies conducted in a variety of industries, researchers with the National Institute for Occupational Safety and Health (NIOSH) have developed a systematic approach to identify the sources of workplace exposures and to provide an effective means for communicating these results to workers and management [2–5]. This system employs:

Direct-reading instruments and data-recording devices to monitor and store data characterizing worker exposures
Video cameras and recorders to document worker activities
Task analyses to evaluate work activities
Statistical techniques to develop predictive models and to summarize the results
Personal computers to perform analyses on the data and to combine the activity data and the exposure data into a presentable form

---

This chapter is largely condensed from a NIOSH technical report [1].

The present system evolved from a series of studies, conducted either to evaluate the effectiveness of engineering controls or to identify characteristics of worker exposures. Direct-reading instruments permitted researchers to monitor exposure changes over short intervals (on the order of seconds). The output from these instruments was stored in an electronic recording device, rather than on a strip chart recorder, so that the data would not require rekeying for statistical analysis. Workers' activities were documented by video-recording systems to determine whether exposures were the result of particular work practices. Work activity data were combined with the real-time exposure data by determining both the exposure and the activity at any given time. Time series analysis of the combined real-time and work activity data set resulted in a model to predict worker exposures. After several studies, however, it became apparent that time series analysis could become a prohibitive task because of processing the tremendous amount of data that can be collected over a very short period. To ease this problem, several simplified analysis techniques were developed. Although these techniques were not as powerful as the time series analysis, they identified those activities that contributed the most to the worker's contaminant exposures.

During completion of the initial studies, a need became obvious: communication of the study results to workers and management. The consensus among the studies' researchers was to provide the facility with a video recording of the work activity, combined with a display of the real-time exposure measurement. The exposure data could be presented in two forms on the video screen: numerically, with the value of the exposure measure displayed, or graphically. Both options were exploited by displaying both the numerical exposure concentration and a bar representing the relative magnitude of the exposure. To place the bar and number on the video screen, a computer program read the exposure data file and generated and updated the bar with time. The system required the use of consumer-quality video and ordinary personal computer equipment; the only specialized equipment required was a special graphics card for the personal computer. The result was a video recording that graphically showed how exposure to a particular substance was affected by activities of the worker.

II. VIDEO EQUIPMENT

Two types of video equipment were used: conventional video equipment for documenting the worker's activities and infrared video equipment for visualizing specific air contaminant plumes. Conventional equipment is used for conducting video exposure monitoring; infrared equipment is used with direct-reading instruments to characterize workplace contaminant concentrations.

A. Conventional Video Equipment

The conventional video recording system consists of a video camera and a videotape recorder. A camcorder, having both capabilities, provides better portability. Mounting the video camera onto a tripod eliminates the need to hold the camera throughout the process. The tape format (Beta, VHS, 8 mm) is not important, and many consumer-quality video recording systems are suitable for video exposure monitoring. There are, however, two important requirements. First, the video system must have a National Television System Committee (NTSC) standard video output signal—a signal used by the video overlay system described in Section V in this chapter. This standard is used by most home video equipment. Second, an on-screen clock or timer is needed—one that can be synchronized with the real-time clock of the data-recording device. Synchronizing the data-recording device with the video camera can be as simple as starting the timer in the camera at the same time the data logger is turned on.

The clock or timer should have a resolution of at least 1 second. The on-screen clock permits an exposure to be coordinated with an associated activity. The video recording of the work cycle or process can then be reviewed while simultaneously tracking the worker's exposure from a printout or plot of the real-time exposure data.

B. Infrared Video Equipment

Effective control of air contaminants depends on understanding the characteristics of their release. It is important to know not only the concentration but also the source and path of the emission. Although some gasses and vapors are visible, most are not. Infrared (IR) imaging is a technique that can provide a real-time picture of some otherwise invisible emissions.

A schematic of such an infrared imaging system is presented in Figure 1. An IR scanner (Thermovision 782) [6] detects changes in absorption of infrared radiation by contaminant gases or vapors. Two versions of the scanner may be used, depending on the range in which the gases absorb infrared radiation: a shortwave band (2–5.6 microns) and a longwave band (8–12 microns). The images received by the scanner are transmitted to a display unit and may be converted from the normal infrared gray-scale image to a colored scale. This image is then simultaneously transmitted to a monitor and video recorder for real-time viewing and recording.

The system uses a flat, black panel as an infrared radiator. The panel is a square, 2-in.-thick, aluminum tank filled with water. A flat-sheet electrical heater is glued to the back surface of the tank; the front surface is painted black. An electronic temperature controller maintains the tank at a constant temperature (120°F). The water in the tank is circulated by a laboratory stirrer to inhibit the formation of a temperature gradient across the panel surface.

The radiant panel and the infrared scanner are positioned so that the emission source is between them. The scanner sees the panel as a constant temperature source and displays it as a uniform image. As a contaminant gas passes between the scanner and the heat source, it absorbs some of the radiated infrared energy. The scanner detects the gas as a lower temperature, which is then displayed as a different color or shade of gray and recorded.

This system is useful for detecting certain process emissions because it provides a real-time image that identifies both the source and path of the emissions. Medical processes, such as the release of nitrous oxide (N_2O) during dental surgery, and industrial processes can be monitored. Also, the IR imaging system, using tracer gas, can determine flow patterns around

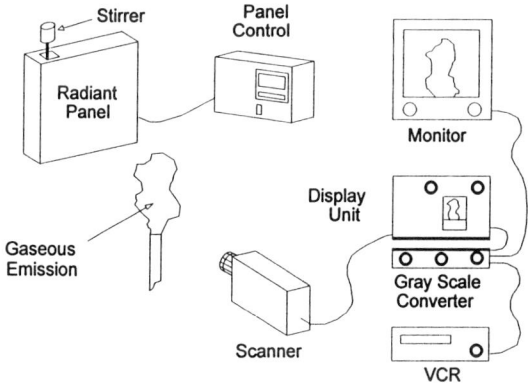

Figure 1 Infrared imaging system.

exhaust openings. This technique has the advantage that the effect of specific work activities or changes in control configuration can be determined immediately.

The most important limitation of this system is sensitivity. The absorption of the emission cloud is related directly to the concentration of the emission and the path length through the cloud. Thus, lower concentrations must be present in greater quantities to be visualized. For example, the sensitivity for N_2O is on the order of 200 ppm meter, i.e., a cloud of nitrous oxide having a concentration of 200 ppm must be more than 1 m in diameter to be detected. System sensitivity can be increased by the use of narrow band pass filters that filter radiation falling outside the narrow band containing the absorption peak of the monitored contaminant. The high concentrations typical of the emission generation point can generally be visualized using this system. Detection of contaminants at lower concentrations typical of the range recommended by occupational health standards is, however, limited. Another limitation of the system is lack of portability. Because the radiant panel is a water-filled tank, it is quite heavy ($\sim$55 lbs) and not easily positioned. Although this is not a severe limitation for laboratory use, it does make field operation difficult.

Some of these limitations are addressed by recent advances in thermal imaging technology. A system that uses a laser in combination with the infrared scanner to detect changes in energy is now available. The laser scans the viewed object, thus eliminating the need for a radiant panel, greatly increasing portability, and making the system much more convenient for field use. This system also has a sensitivity approximately one order of magnitude greater than the one previously described.

III. MONITORING EQUIPMENT

Any air contaminant–monitoring instrument that produces an output signal of the concentration measurements can predict real-time assessments of a worker's exposure to an air contaminant. The usefulness of a specific instrument will vary with the situation. To evaluate the utility of an instrument, consider: (1) the nature of the analog or serial output, (2) the response time of the instrument, (3) specificity for the contaminant of interest, and the instrument's (4) portability and size.

A. Output

The collection of real-time concentration data to evaluate the relationship between events in the workplace and air contaminant concentrations generally must to be recorded automatically. For a monitor to be useful, it should produce a digital or analog output, the latter often being voltage that is proportional to concentration. Techniques for recording analog data are given in Section VI, "Data Acquisition." Some instruments also provide a digital output that is periodically updated. The frequency of these concentration measurements is usually a function of the instrument and normally cannot be adjusted by the user.

B. Response Time

The total system response time (for the monitor and the setting being evaluated) can be defined as the sum of (a) the time required for the air contaminant to be transported to the worker's breathing zone and begin to accumulate and (b) the time required for the instrument to respond to a change in concentration in the worker's breathing zone. To conduct video exposure-monitoring studies of air contaminant concentrations, the total system response time must be less than that of the events of interest. As a result of the response time delays, the instrument output lags behind work events in the workplace.

Monitoring instruments often measure some general parameter that is proportional to concentration. For example, aerosol photometers respond to any aerosol that scatters light. Such a limitation requires either that the monitor be calibrated for the specific air contaminant being measured or that the results be reported as a relative concentration.

C. Portability

To allow for worker acceptance, the monitoring equipment should be light enough to be worn comfortably. The equipment should be battery-operated and should weigh as little as possible. If workers cannot wear the equipment, tubing can transport the air contaminant from a worker's breathing zone to the instrument. This arrangement, however, adds some complications. The monitoring system's response time will increase because of the time needed to transport the air contaminant through the tube to the monitor. In addition, aerosols and contaminants can be lost to the tubing walls and other collecting surfaces. If the tubing is later struck or vibrated, these contaminants can be released and may contribute to the instruments' signal. Organic vapors can be adsorbed onto the tubing walls during periods of high concentration and desorbed during periods of low concentration.

Users need to consider the limitations and capabilities of the direct-reading instruments when designing and conducting studies to yield useful information about exposure sources. Background information on these instruments can be obtained from *Air Sampling Instruments*, by the American Conference of Governmental Industrial Hygienists (ACGIH) and the NIOSH Manual of Analytical Methods (NMAM) [7,8].

IV. PERSONAL COMPUTER SOFTWARE

Several types of software collect and analyze real-time exposure data. Control software operates an analog-to-digital converter card, and communications software downloads portable data loggers. Spreadsheets are valuable for manipulating real-time exposure data, as well as for performing some simple analyses. For more sophisticated data analyses, full-function statistical analysis packages may be required. In addition, if the exposure data are to be combined with a work activity video recording, a customized computer program can be used to generate a graphical representation of the worker's exposure.

Control software operates the analog-to-digital converter that is either a card located in the computer or a stand-alone system with an interface to the computer. These software packages usually require special device drivers for the particular hardware system in use. Many of these control packages can process the real-time data during collection, and some packages provide limited data analysis capability. Besides collecting data from an analog source (a direct-reading instrument, for example), control software can also instruct the computer to send out signals, although this function is beyond the scope of this chapter.

Configuring the analog-to-digital system is normally done by menu-driven software. Many of the software packages allow readings—graphical or tabular—to be displayed on the computer screen as the data are being collected. Once data collection is complete, the readings are stored in a data file. Some programs link directly with a spreadsheet program, making it possible to save the data in a spreadsheet file. For other programs, the data are stored in file formats that can be imported into the spreadsheet. There are several different control programs with many different functions and capabilities. Two specific packages are Labtech

Notebook (Laboratory Technologies Corporation, Wilmington, MA) and ASYST (MacMillan Software Company, New York, NY). Both of these programs work with a variety of analog-to-digital converter cards.

If a portable data-recording device (data logger) records the real-time exposure data, control software is not needed. Instead, a program to download the data logger to a personal computer is required. Downloading software either comes complete with most data loggers or is available for an additional cost. After the data logger is downloaded, some programs allow simple data analysis to be performed. Many of these programs store the data in a file that can then be imported into a spreadsheet program. In addition to the programs supplied with the data loggers, there are communications programs such as Crosstalk (Crosstalk Communications, Roswell, GA) and Procomm (Datastorm Technologies Inc., Columbia, MO) that can download some data loggers through the computer's asynchronous communications port. Communications programs may require nonstandard use, since the format of the data from the data logger may vary with the device.

After the data have been collected and stored in a file, spreadsheet programs can manipulate data and do simple data analysis. Lotus 1-2-3 (Lotus Development Corporation, Cambridge, MA) and Microsoft Excel (Microsoft Corporation, Redmond, WA) are examples of two spreadsheet programs. If the data are to be analyzed by worker activity, a spreadsheet is useful for keying activities with the real-time exposure data: a researcher must determine the time a particular reading was recorded and then observe the worker's activities for that time on the video recording of the work activity. Spreadsheets not only sort data and perform elementary statistical analysis but also format data sets for analysis in a statistical analysis program, or combine the work activities and the real-time exposure data onto videotape.

To combine the real-time exposure data with the video recording of the worker's activities, NIOSH researchers have written a program for IBM-compatible computers that generates a graphical representation of the worker's exposure [1,9,10]. This IBM-compatible program reads a real-time data file, generates a bar to represent the magnitude of the exposure, and then displays the bar on the screen. When this program is run through a video overlay system, a video recording graphically shows how a worker's exposure is influenced by the work activity. The video overlay system is discussed in Section V, "Personal Computer Hardware." The bar is updated with each time interval of readings in the data set. The program allows either one or two bars to be displayed on the screen at one time. Two bars can be displayed if the exposures of two workers are to be compared, or if one worker is monitored with two different instruments. To use the program, the real-time exposure data must be stored in a properly formatted ASCII file. For the program to display one bar, the format of the data file must have three columns of data: two columns for the time the reading was recorded (minutes and seconds) and one column for the exposure measurements. For the program to display two bars, the data file format must have an additional column for the second exposure measurement. The first data set is displayed on the left side of the screen, whereas the second data set is displayed on the right. The time interval between the readings must be constant. The spreadsheet program arranges the data file into the proper format. and generates a bar that is overlaid onto the work activity video recording by a video overlay system.

V. PERSONAL COMPUTER HARDWARE

The computer hardware required for collecting and presenting real-time data is fairly basic. Specialized equipment is required only for combining the graphical exposure bars with the video recordings of the work activity or for running a computer-based analog-to-digital converter system. The basic computer system used by NIOSH researchers is an IBM PC–compatible personal computer. The computer should have sufficient memory (i.e., 4 megabytes) and a hard disk drive. Additional memory (i.e., 4–12 megabytes) may be desirable to improve performance if unusually large data sets (i.e., 1200 readings) are to be manipulated in a spreadsheet. If the real-time exposure data are not going to be combined with the video recording of the work activity, then the type of graphics card is not critical. If data loggers are to be downloaded to the computer, an asynchronous (serial) communications port is required (most computers are sold with this port as standard equipment).

Computer-based analog-to-digital converters are special cards that fit into an expansion slot of the computer. Special software drivers and control programs may be required to operate this board. Section VI, "Data Acquisition," contains more detailed descriptions of the analog-to-digital converter systems.

To overlay the real-time exposure data with the video recording of the work activity, the computer will need either an enhanced graphics adapter (EGA) card and a video overlay board, or a variable graphics array (VGA) card with the overlay features built in. A monitor appropriate for the graphics card also is needed. Both VGA and EGA are high-resolution color graphics adapters, with VGA having slightly higher resolution.

If an EGA card is used, it must be combined with a video overlay system. One such system, the Video Charley (Progressive Image Technology, Folsom, CA), consists of a single computer card. The Video Charley requires the EGA card to have a standard features connector (most EGA cards do). The features connector links the video overlay board with the computer. The video overlay board converts the computer's graphics signal to an NTSC signal and overlay the graphics onto the activity video recording. Besides the features connector, most EGA cards also have a DB9-pin connector for the EGA monitor and two RCA-type connectors. Under normal circumstances (without the Video Charley), the two RCA connectors serve no function. With the Video Charley board installed, however, one RCA connector inputs the activity video signal, and the other outputs the video signal with computer graphics overlaid.

When overlaying computer graphics using the Video Charley, signal differences require the computer display system to operate at a resolution of 640 × 200 pixels, rather than at the typical EGA resolution of 640 × 350 pixels. To combine the activity video signal with the computer graphics signal, the two signals must have the same synchronization frequencies. In the case of the video signal, an NTSC signal, the horizontal sync frequency is 15.7 kHz and the vertical sync frequency is 60 Hz. In the 640 × 200 pixels mode, the horizontal and vertical sync frequencies are also 15.7 kHz and 60 Hz, respectively. In the 640 × 350 pixels mode, the vertical sync frequency is 60 Hz; however, the horizontal sync frequency is 21.8 kHz. To get both signals at the same horizontal sync frequency, the computer graphics card must operate at the lower resolution mode. Depending on the type of EGA card used, either software drivers or hardware switches can set the resolution.

If the VGA option is chosen, an appropriate VGA card is required. Two such cards are the USVideo VGA/NTSC Recordable® graphics card with the Genlock Overlay Module (USVideo, Stamford, CT) and the Willow Peripherals VGA-TV GE/O® (Willow Peripherals, Bronx, NY). These two systems allow computer graphics to be overlaid onto video images at a higher resolution than does the EGA system with the Video Charley. To overlay on VGA systems, only one setting needs to be changed to direct the card's output to the video monitor: on the USVideo card, this is done with a hardware switch; on the Willow Peripherals card, a software program is run. Both cards have two RCA-type ports, one for video-in and one for video-out. The cabling setup, shown in Figure 2, is the same for both the EGA/Video Charley and the VGA systems. Operation of the VGA overlay system is similar to the normal use of the computer, except that the video monitor (connected to the video-out RCA-type port) is the primary monitor.

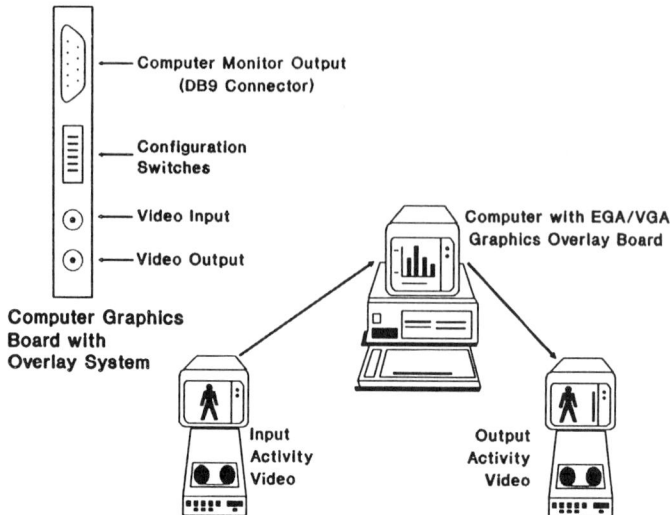

Figure 2 Diagram of personal computer system equipment and connections needed to overlay exposure data with work activities.

VI. DATA ACQUISITION

Many direct-reading monitors have analog output capability, usually in the form of a DC voltage signal, typically on the order of 1–10 V, full scale. Before the proliferation of the personal computer, this analog output typically drove a strip chart recorder. To perform data analysis with a computer, the data from the strip chart was keyed into the computer—a tedious process. With advances in personal computers, the analog output from these monitors can now be stored digitally, allowing the data to be transferred to the computer in just a few easy steps.

Data-recording devices generally fall into two categories: portable data loggers and computer-based analog-to-digital (A/D) converter systems. Both types of devices have a limited resolution over their working voltage range. Depending on the type, the device has either a fixed voltage range (0–2 V, for example) or a working range that can be chosen with hardware switches or control software. This working range is then broken down into intervals. Resolution of a data-recording device is usually given in bits. For example, the working range of 0–2 V for an 8-bit data logger consists of 256 intervals. The number of intervals is determined as follows:

$$\text{Number of intervals} = 2^{\text{bits}} \tag{1}$$

The magnitude of these voltage intervals is calculated from

$$V_I = \frac{V_U - V_L}{N} \tag{2}$$

where
V_I = Interval, V
V_U = Upper working range, V
V_L = Lower working range, V
N = Number of intervals

For the example of an 8-bit, 0- to 2-V working range, there must be a difference of at least 0.008 V for the data-recording device to detect a difference between two voltage readings. In most instances, an 8-bit device should be sufficient. Data loggers are typically 8-bit devices, whereas A/D converters range from 8 to 16 bits; 12-bit boards are very common.

Computer-based systems store the data directly onto the computer's hard drive or onto a disk drive. These systems require software programs to control the parameters. Depending on the program, the exposure measurements can be displayed on the computer screen as the data are being collected. Computer-based A/D converter systems are more flexible than portable data loggers. Computer-based systems are usually more expensive; A/D boards can cost $1000 or more and the control software can cost another $1000.

Portable data loggers store the data in a built-in bank of memory. After data collection, the data logger must be downloaded to the computer, typically through the computer's communication port. A general communications program or a program written specifically for downloading a particular data logger can control the downloading procedure. Most data loggers have parameter-setting programs built in and require no additional control software and most display only limited amounts of data while recording. Since the data logger is likely to be fastened to a worker, observation of the data as it is being generated is not feasible. Portable data loggers, including downloading software, can be purchased for as little as $500.

A hybrid of the A/D systems and the portable data loggers is a telemetry system. Telemetry systems use a transmitter and receiver to transfer data from an instrument to a base unit for storage. The base unit may include a personal computer and may allow the data to be displayed as it is being generated. As with portable data loggers, telemetry systems do not require a worker to be tethered to a computer. Unfortunately, most commercially available telemetry systems tend to be expensive (i.e., >$10,000) [11].

VII. WORKER ACTIVITY ANALYSIS

Activity analysis is an important step in video exposure monitoring because such an analysis helps catalog work activities. This systematic method breaks a complex job into its elements, permitting study of and improvements in a specific task. More importantly, these elements can be sorted so those contributing most to a worker's air contaminant exposure can be dealt with first.

The first phase of activity analysis is a time-and-motion study, which determines the work content of the job. Time-study and production records, as well as timed observations, provide the necessary interval data. Managers, supervisors, and workers can provide job descriptions and demonstrations from which to determine tasks. A job is described as a set of tasks, with each task consisting of a series of steps or elements [12], that is, the fundamental movements or acts (reaching, grasping, moving, positioning, using, etc.) required to perform a job. Groups of elements making up a task (or an activity) are usually performed in the same sequence to accomplish a common end. Examples of tasks might include the following: "turn on machine," "operate machine," and "cleanup." By observing the job or slow-motion video recordings of that job, the researcher identifies the elements composing a task. Gilbreth suggested that formal element definitions are arbitrary in that one can increase or decrease detail as necessary. For example, "get" adequately describe the process of "reach-grasp-move," and "put" works well for "move-position-release" [12].

The second phase of activity analysis is an actual review of the job for recognized occupational risk factors that may cause excess exposure to air contaminants. If a trained investigator can record the risk factors as the worker is performing the job, this analysis can be done at the worksite. A more thorough analysis can be done, however, by viewing the video recording of the worker's activities. The clock or timer in the video camera documents the time it takes the worker to perform the various activities. The clock or timer also allows activities to be studied as changes occur in the air contaminant exposure as measured by direct-reading instruments. When evaluating air contaminant data, the researcher needs not analyze the job in more detail than what the real-time exposure data can reveal. For example, if the response time of the instrument measuring the air contaminant exposure is longer than the time required to complete a set of tasks being video-recorded, then analysis of those individual activities is of little value, and they should be combined into a "principal" activity. This principal activity can then be studied with regard to air contaminant exposure.

VIII. DATA ANALYSIS

To perform data analysis, researchers must combine worker exposure measurements and descriptions of events in the workplace into a single data set. Descriptive statistics describe the contribution of workplace events to a worker's air contaminant exposure. In addition, statistical analysis evaluates whether workplace events significantly affect exposure. The findings of the data analyses help to focus control measures upon actual sources of worker air contaminant exposure.

A. Transportation Lag and Autocorrelation

As a prolog to data analysis, an appreciation is needed of how events in the workplace affect the contaminant concentration measured by an instrument. Consider a worker standing at a workstation. Turbulence in front of the worker transports the air contaminant from a source at the workstation into the worker's breathing zone. If it takes 2 sec for the air contaminants to travel from the source to the worker, the concentration in the worker's breathing zone does not start to change until 2 sec after the event has occurred. In statistical terms, the concentration is said to lag behind the workplace events. This can be referred to as transportation lag. The actual magnitude of this lag can be estimated by observation and measurement, or it can be addressed in the selection of a statistical modeling and data analysis package.

After the air contaminant has been transported into the worker's breathing zone, the direct-reading monitor begins to respond to the changing concentration. A monitor with a time constant of 1 sec would require 3 sec to complete 95% of the change in response to an abrupt change in concentration. Because of the dynamics of the monitor's response, the measured concentration at any moment in time is a function of the concentration in the preceding time intervals. This phenomenon is called autocorrelation.

B. Assembling the Data Set

Concentration measurements from a direct-reading instrument are recorded and stored by data-logging devices. Because the software written for controlling or downloading data-logging devices has limited data analysis capabilities, real-time concentration data can be imported into a spreadsheet program for manipulation and data analysis. Many of the downloading or control programs include utilities for storing the real-time concentration measurements in a *print file* that can be imported into the spreadsheet. (A print file is a text file in ASCII format that can be printed directly by the operating system's print command.) The interval between the concentration measurement readings is set either before the data are recorded by the data-logging device or when the data are stored in the print file.

The real-time exposure data are loaded into the spreadsheet using the "import" command. (The name of this command may vary from program to program, but the command loads a print file.) The print file loaded into the spreadsheet may contain several columns of numbers, depending on the type of data-logging device used. These columns may include several time columns (elapsed time, clock time, etc.), event markers, and concentration measurements. The data can be manipulated in the spreadsheet to create a data set that includes only two columns, one for the real-time concentration measurement and one for the time the readings were recorded. This time reading can be elapsed time or clock time, depending on how the data-logging device was synchronized with the video camera's clock or timer.

After the time and concentration readings have been isolated, work activity variables can be added to the spreadsheet. The video recording of the work activities can be viewed while tracking the worker's exposure in the data set. From this recording, the worker's activities can be defined in two different ways: so that only one activity can occur at any given time, or so that any one of several activities can occur at any given time. For each concentration measurement, the activity can be coded into the data set in one of several ways, depending on how the activities were defined and on the type of data analysis to be conducted. Two methods are frequently used: (1) to enter the activity as a single variable with a different value for each activity, or (2) to enter each activity as a separate variable, with one value if the activity occurs, and another value if it does not occur ("1" and "0," for example). If the activities are defined such that only one activity can occur at a time, the single-variable method is usually more appropriate since it will result in a smaller data set than if each activity were to be entered as a separate variable. If, however, several activities can occur at a time or if data analysis involves using a spreadsheet program to perform multiple regression, then each activity is usually entered as a separate variable. If the activities were entered using the single-variable method, a different value would be needed for every combination of activities.

As discussed earlier, the air contaminant concentration lags behind the causal activities because of the time required to transport the air contaminant from the source to the monitor. If the transportation lag is not addressed by a statistical analysis package, the air contaminant concentration measurements can be "slipped" with respect to the worker activity variables, after the researchers estimates the magnitude of the lag. The lag time matches the worker's activities with the associated air contaminant concentration measurements.

C. Data Analysis Techniques

After the data set is assembled as a time series, it can be analyzed to determine the effect of workplace activities on changes in worker air contaminant exposures. Autocorrelation considerably complicates (1) statistical analysis for modeling worker exposures and (2) examination of whether the worker's activities are affecting the air contaminant exposures. When conducting statistical analysis, researchers should compare the extent of the changes in exposure attributing to a worker's activities with the variability of the exposure data. When the changes in exposure are large with respect to the exposure data variability, one can conclude that the activities significantly affect the exposure. Autocorrelation can cause the variability of the exposure data to be underestimated during regression analysis and analysis of variance. Thus, autocorrelation can cause these two data analysis techniques to overstate the level of confidence in the conclusion that workplace events affect the worker's exposure. Special techniques, called time-series analysis, have been devised to deal with autocorrelated data.

A variety of techniques are available to analyze real-time data and deal with autocorrelation, but because of the time and complexity required to deal with autocorrelation, descriptive statistics are commonly used instead. For a quantitative evaluation of whether activities are causing air contaminant exposures, autocorrelation in the data can be addressed either by censoring the data to remove autocorrelation or by performing time-series analysis. At times, too much information is lost when the data are censored to remove autocorrelation. Time-series analysis methods can evaluate the relationship between the worker's activities and air contaminant concentrations without censoring the data set [13]. Because time-series analysis can be very complicated, the assistance of a statistician may be needed.

Descriptive statistics can aid exploratory data analysis. In such an analysis, the identity of workplace activities causing differences in the worker's exposure is investigated. If there are no differences or if the differences are greater than an order of magnitude, conclusions can usually be based on the findings of the descriptive statistics. However, when the observed differences in concentration are less than an order of magnitude, statistical analysis should be performed. In conducting statistical analysis, the effect of autocorrelation on the analysis must be evaluated.

Real-time data are frequently analyzed to evaluate whether specific workplace activities affect worker exposures. When a workplace activity occurs and the worker's exposure increases, one can conclude that the activity has contributed to the exposure. Because many activities can occur simultaneously in an industrial environment, the change in the worker's exposure may be due to some unrecognized activity. Thus, judgment must be exercised when interpreting the results of the data analysis. After one analyzes the real-time data, control measures can be focused on actual exposure sources.

IX. PRACTICAL APPLICATION OF VIDEO EXPOSURE MONITORING

Video exposure monitoring is effective for identifying those specific activities that contribute most to a worker's exposure to an air contaminant. Integrated monitoring, such as sorbet tube or filter sampling, is normally conducted to determine the extent of the worker's exposure (averaged over the sampling period) before video exposure monitoring. After determining the extent of the exposures, the researcher can apply the techniques for video exposure monitoring. A typical video exposure monitoring evaluation might proceed in the following manner.

1. With the worker activity and contaminant of concern identified, the appropriate direct-reading monitor must be chosen. The monitor should be appropriate for the contaminant, e.g., an aerosol photometer to monitor for aerosols. It should have a minimal time constant so that activities of short duration can be evaluated, and it should be as portable as possible. The monitor should be set to zero and calibrated according to the manufacturer's instructions.
2. In addition to the direct-reading monitor, an IR video system may prove useful, depending on the contaminant being sampled. Such a video system can visualize air contaminant plumes, identify contaminant sources, and identify work practices that may contribute to a worker's exposures.
3. The output of the direct-reading instrument should be recorded by a data acquisition system. Setup of this system consists either of programming the data logger or of running the control software of the analog-to-digital converter system. The clock on the video camera and the data-recording device should also be synchronized at this point.

4. Data collection begins by starting the data-recording device and the video camera, and continues for a period judged to be representative of the process being studied. After the data-collection period, the data must be stored in a data file. If a data logger has been used, the data must be downloaded to a computer for storage to a file.
5. After the data are collected and filed, they are imported into a spreadsheet program. Work activity is analyzed from the video recording of the work activities. The activity variables are entered into the spreadsheet to accompany the air contaminant exposure data. Data analysis can be conducted with the spreadsheet or by statistical analysis programs. The spreadsheet analyses can consist of simple descriptive statistics or regression analysis. Statistical analysis programs are used for more sophisticated analyses, such as time-series analysis.
6. If the exposure data are to be overlaid onto the video recording of the work activities, the video overlay system must be assembled and the exposure data stored in an ASCII file specifically formatted for use by the bar-generating program. To overlay the exposure data onto the video recording of the work activities, the bar-generating program is set up (inputs entered); the work activity videotape then is played back. When the time on the video image reaches the time of the initial reading from the data file, the program's display is started. This synchronizes the exposure data with the video recording. The overlaid signal can be displayed on a video monitor and recorded on a second video recorder.
7. In some situations, the real-time concentration data is useful for evaluating ventilation systems for contaminant dilution and for determining the contaminant generation rate for the process. In these instances, the spreadsheet's regression function or a statistical analysis program can be used to determine the room's mixing factor. With this factor, the generation rate can be estimated, and ventilation systems can be further evaluated.

A. Case Study 1: Manual Material Weigh-out

This plant manufactures a variety of plastic and rubber materials. At the operation studied, powdered acrylic copolymer was weighed into batch lots at a weigh-out booth as diagrammed in Figure 3 [2]. The final manufactured product from this operation was vinyl wall covering. In the weigh-out booth, a hinged segment of the work platform could be raised to allow a drum of raw material to be placed inside the booth. An exhaust plenum formed the back wall of the booth. At the booth, the worker emptied 22.7-kg (50-lb) bags of powder into a fiber drum measuring 84 cm (33 in.) high and 55 cm (21.5 in.) in diameter. Then, using a scoop, the worker transferred the powder from the drum to a small paper bag. The bag was placed on the scale and the weight of powder in the bag adjusted. Usually, two scoops of the powder were required to achieve the proper weight. Finally, the filled bag was closed and placed in a bin behind the worker. This process was repeated until the required number of batches were filled or the fiber drum was emptied.

Methodology

Direct-reading monitors measured the effect of depth of material in the drum and the elements of the job cycle on dust exposure. The worker began with a full drum and weighed the powder into paper bags. An aerosol photometer, the Hand-held Aerosol Monitor (HAM) (PPM Inc., Knoxville, TN), showed the dust concentration in the worker's breathing zone. Every two seconds, the HAM's analog output was recorded by an Apple II Plus computer, equipped with an AI 13 analog-to-digital converter (Interactive Structures Inc., Bala Cynwyd, PA). The evaluation ended when the drum was nearly empty (about 22 minutes).

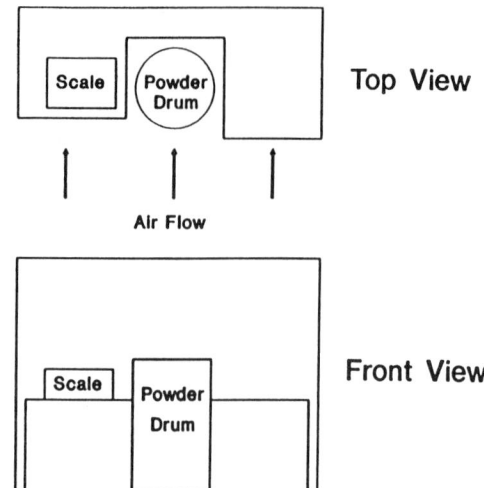

Figure 3 Diagram of a powder weigh-out workstation.

The voltage output was statistically analyzed to determine if the amount of powder in the drum affected worker dust exposure, and if it did, which activities contributed the most to this increase. The strategy for this analysis was to fit a regression model involving the relation of the variable "worker" (a time-dependent measure of dust exposure) to the independent variables "bagcount," "scooping," "weighing," and "turning." "Worker" was the voltage output of the direct-reading monitor mounted on the worker. "Bagcount" was the cumulative number of bags that were weighed. "Scooping" was the cumulative time during each cycle spent scooping material from the drum and into the bag. "Weighing" was the cumulative time during each cycle spent weighing the bag on the scale and adjusting the amount of powder in the bag. "Turning" was the cumulative time during each cycle spent placing the bag in the bin. The worker's exposure was modeled closely enough to provide a fair representation of its relationship to the variables. There was no attempt to continue to add terms to the model until the lack of fit was not statistically significant.

A key assumption in the data analysis was the independence of measurements. Successive readings from the instrument were not independent. When a dust-generating event occurred, dust concentrations did not increase immediately; time was needed for the air to transport the dust cloud from the point of generation to the inlet of the instrument. Also, the HAM was operated with a time constant of 1 second, and it required some time to respond to fluctuating concentrations. The total instrument response time appeared to be 2 to 5 seconds, meaning the instrument responded 2 to 5 seconds after a dust-generating activity occurred. As a result, autoregressive terms were used in the analysis.

The results of the regression analysis are shown in Figure 4, which shows that dust exposure during the scooping activity increased as the bag count increased. Bag count was a surrogate measure for the level of powder in the drum; a bag count of 0 corresponded to a full drum, and a bag count of 55 corresponded to an empty drum. During weighing and turning, the worker's dust exposure either remained constant or failed to increase as fast as the exposures during scooping.

Figure 5 illustrates the effect of job cycle upon dust exposure. During the scooping activity, the dust exposure increased. During the weighing and turning activities, the dust exposure decreased. This suggests that most of the worker's dust exposure was caused when

Assessment of Hazardous Workplace Exposures

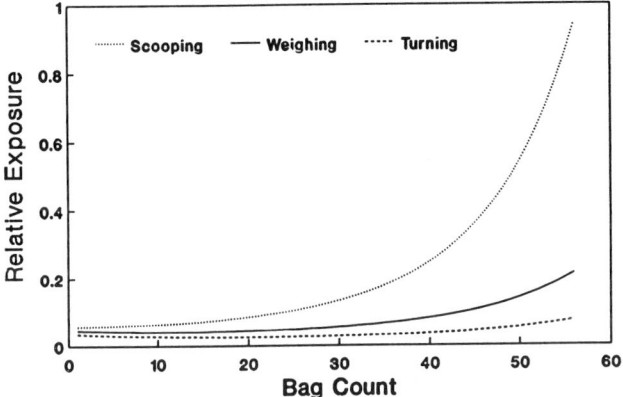

Figure 4 Modeled dust exposure of a worker as a function of bag count for scooping, weighing, and turning.

scooping the powder from the drum. Dust exposures caused by weighing and turning were much smaller than the dust exposures caused by scooping and may have been controlled by the ventilation system. The weighing activity appeared to be associated with higher dust exposure than did the turning activity. This difference, however, may be an artifact caused by the delay of the HAM's response to the high dust exposures during scooping.

Findings

Figure 4 shows that dust exposure increased with bag count, which is a surrogate variable for depth of scooping. The data were collected over approximately 20 minutes. This same conclusion was reached with the use of conventional short-term measurement of dust concentrations with pumps and filters. The filter data, which required three full shifts to collect, however, did not provide any insight into the relationship between job cycle and the worker's dust exposure. Knowledge of the specific task that elevated the worker's dust exposures was crucial to the redesign of the weigh-out booth.

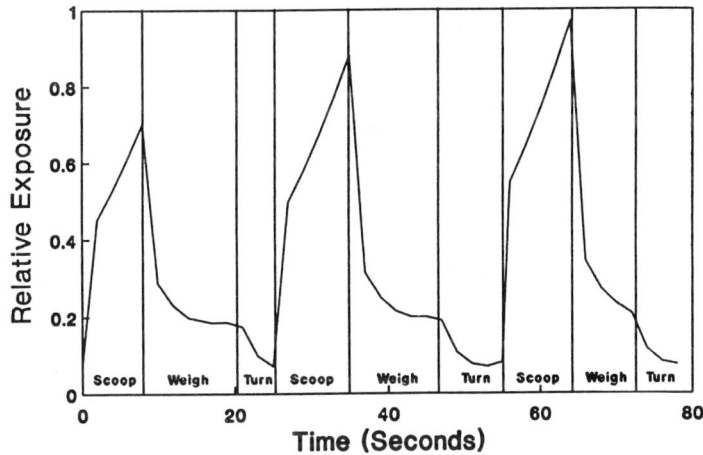

Figure 5 Modeled dust exposure of worker for filling bags 51 through 53.

Recommendations

Based upon the results presented in Figures 4 and 5, researchers recommend that the facility use shorter storage drums for bulk powder to reduce the dust generated by scooping. This case study clearly showed that direct-reading monitors can qualitatively and quantitatively measure sources of dust exposure during the work cycle—exposures too short with regard to time to be studied with integrated air sampling methods.

B. Case Study 2: Dental Administration of Nitrous Oxide

This study was conducted to evaluate how effectively scavenging systems reduce occupational exposure to waste nitrous oxide (N_2O) [14]. For more than 100 years, dentistry has used N_2O as a general anesthetic agent, analgesic, and sedative [15]. Today, N_2O is used primarily for psychosedation, to reduce fear and anxiety in the conscious patient [15]. N_2O scavenging systems typically have three principal components: an N_2O and oxygen (O_2) gas delivery system, a nasal cone for the patient from which to inhale the gases, and an exhaust system that carries the respired gas from the patient out of the building. A schematic of the nasal cone is shown in Figure 6. Although the studies show that scavenging systems significantly reduce N_2O concentrations, the systems do not reduce it to the NIOSH recommended exposure limit (REL) of 25 ppm during the time of administration [16]. In addition to evaluating the effectiveness of scavenging systems, this study was conducted also to determine why exposures exceeded 25 ppm.

Methodology

A dental facility that uses a commercial scavenging system during dental surgery was evaluated by NIOSH researchers. Ten dental operations (e.g., filling, extracting) were monitored by using a combination of sampling strategies: personal breathing zone sampling (dentist and dental assistant), general area sampling, and real-time sampling. A Miran 1A (Foxboro Instruments, Foxboro, MA) monitored the real-time N_2O concentrations. A sampling probe connected to the Miran 1A was placed approximately 12 in. above the patient's head. In addition, video and IR scanning equipment recordings monitored the dental practices (activities). Because N_2O is IR absorbing, it can be "visualized" by using IR thermography. Motion and time measurement techniques were used to document activities of the dentist, dental assistant, and patient during the operation [12]. These activities, listed below, were coded into a computer spreadsheet along with the associated N_2O concentration data.

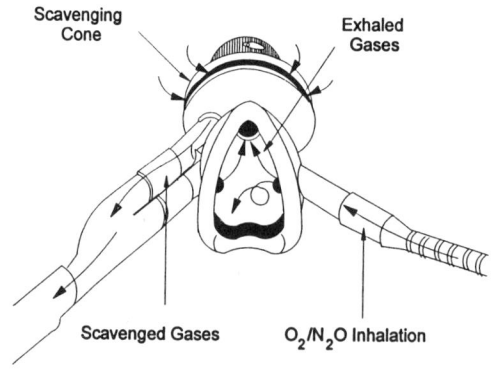

Figure 6 Diagram of a typical dental N_2O scavenging system.

injecting local anesthetic
extracting a tooth
filling a tooth
using the aspirator
using the water and air syringe
using the rubber dam (small rubber sheet that isolates the operative site)
using the curing light for restorative composite resin material
talking, coughing, and yawning of patient
turning on N_2O
turning off N_2O
adjusting N_2O flow rate

Statistical analysis of the N_2O concentration and changes in concentration were modeled as a function of these work elements from the spreadsheet [17].

Findings

Average real-time N_2O concentrations for the 10 operations ranged from 206 to 770 ppm. The average real-time concentration over all 10 operations was 442 ppm. The average personal breathing zone (integrated sample) concentration over all 10 operations for the dentists was 487 ppm. There was no significant difference ($p < 0.68$) between the real-time and personal breathing zone concentrations for dentists. There was, however, a significant difference ($p < 0.014$) between the overall average real-time sampling concentration and the average personal breathing zone concentrations among dental assistants (150 ppm). The differences in dental assistant breathing zone concentrations and the real-time concentrations may have been because the sampling probe was placed closer to the patient's and the dentist's breathing zone than to the dental assistant's breathing zone. Thus, these real-time sampling results may be more representative of the dentists' exposure than that of the dental assistants. It also was determined that the dentists, by nature of the dental surgery, worked closer to the patient's breathing zone than did the dental assistants.

Real-time sampling results and work activities were combined to determine if changes in N_2O concentrations were related to these activities. From the video recordings, several dental surgery activities were selected for analysis. For data analysis, the real-time concentrations were matched with the identified dental activities. A plot of this relationship is shown in Figure 6. Based on this analysis, the only activities that showed significant N_2O concentration changes occurred: (a) when the dentist turned on the N_2O gas, (b) when the dentist adjusted the N_2O concentration during the operation, and (c) when the dentist turned off the N_2O gas, following the operation. Statistical analysis showed that 98% of the changes in N_2O exposure could be accounted for by the N_2O concentration of the gas delivered to the patient as opposed to the specific dental surgery activities (note the "sawtooth" pattern in Figure 7). Thus, the primary source of N_2O exposure was not from the work practices of the dentists, but from N_2O delivery and the inadequacy of the scavenging exhaust system.

During 2 of the 10 dental operations, an IR video camera qualitatively evaluated scavenging mask leakage. The infrared camera revealed N_2O leakage between the mask and face seal, indicating that the scavenging mask did not fit the patient's face properly. The off-gassing of N_2O during patient mouth breathing also affected exposure during these two operations. The IR video camera also revealed that a sudden increase in N_2O exposure could be traced to the patient's expired breath. This increase was corroborated by real-time data. When the patient inspired, the N_2O concentrations decreased. Synchronization of the real-time data with the IR video camera helped to confirm that patient's mouth breathing was also a source of N_2O exposure.

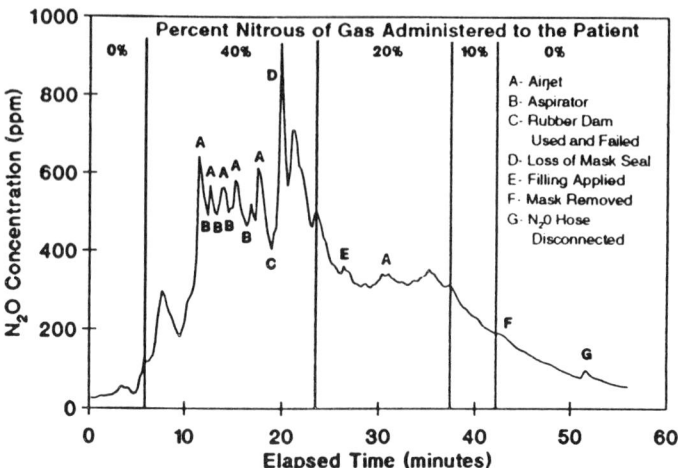

Figure 7 Plot of real-time N$_2$O concentration with activities and supply concentrations.

Recommendations

Scavenging mask leakage and an inadequate scavenging exhaust system caused most of the N$_2$O exposure in this study. Patient mouth breathing was a secondary source of exposure. If the scavenging system had been more efficient, the work practices, such as use of the aspirator, air and water syringes, and patient mouth breathing might have had a greater impact on the N$_2$O exposures of the dentists.

The IR video camera proved to be a valuable tool for detecting N$_2$O leakage from the patient's mask as well as from patient mouth breathing. By following the real-time data patterns, NIOSH researchers discerned when there was a mask leak, when the patient was mouth breathing, or both. This ability to determine these exposure sources helped provide recommendations for improving scavenging system mask design and work practices, and for reducing overall N$_2$O exposures.

X. CRITICAL REVIEW OF CURRENT STATUS

Video exposure monitoring is a set of flexible techniques that can determine specific sources of a worker's exposure to air contaminants. However, there are a number of practical considerations when integrating real-time exposures with the worker's activities. The researcher has to make sure the changes in exposures match worker activities. Problems challenging the researcher are (1) instrument transportation lag, such as natural diffusion of chemical from their source to the detectors, (2) response time, such as the instruments' sensors responding to a contaminant once it has arrived, and (3) autocorrelation of data, which arises from instrument lag and response time. These challenges are equally matched by the difficulty of analyzing videotapes of worker activities and defining the work activity elements so that they correspond to changes in chemical exposure. For example, if the worker is performing several work activities in a short period of time, the changes in chemical exposure shown in the data may not match these activities. Therefore, the researcher has to take special precautions to collect enough data (i.e., videotape and real-time personal sampling) so that patterns of exposure emerge. This will guide the researcher to identify exposure sources, and from this to devise effective controls. The key to successful integration of both real-time exposure assessment and job analysis techniques is to use both methods in such a way that patterns of exposure quickly emerge.

XI. FUTURE CONCERNS

Most of the integration of real-time direct-reading instruments with work analysis has been conducted by Federal and academic researchers [18–20]. For real-time exposure assessment and workplace job analysis techniques to succeed as a useful tool in industry, several things need to occur: (1) real-time instruments need to be more compact, portable, and easy to operate; (2) real-time instrument costs need to be in line with costs of traditional sampling instruments; (3) the availability and reliability of real-time sampling instruments need to be improved; (4) standards and guidelines need to be developed so that real-time data can be compared with traditional integrated sampling results; and (5) industry needs to invest in educating and training of its health and safety professionals to perform real-time sampling and work analysis assessments. The future of these sampling techniques may depend on a number of factors. The first is for government and academic researchers to continue applied research and development in this area; then to transfer this technology to industry to demonstrate the effectiveness of pinpointing exposure sources so that controls can be more effectively applied. The second is for industry to invest in this technology by training its own personnel to use these methods for cost-effective controls. We believe that if these factors occur, real-time sampling and work analysis will become a valued tool in evaluating and controlling hazardous chemical and physical agents in the workplace.

REFERENCES

1. U.S. Department of Health and Human Services, Public Health Service, Centers for Disease Control and Prevetion, National Institute for Occupational Safety and Health (NIOSH), Analyzing workplace exposures using direct reading instruments and video direct reading instruments and video exposure monitoring techniques, M. G. Gressel and W. A. Heitbrink, eds., DHHS (NIOSH) Publ. No. 92-104, 1992.
2. J. D. McGlothlin, W. A. Heitbrink, M. G. Gressel, and T. J. Fischbach, Dust control by ergonomic design, Proceedings of the IXth International Conference on Production Research, Cincinnati, OH, pp. 687–694, 1987.
3. M. G. Gressel, W. A. Heitbrink, J. D. McGlothlin, and T. J. Fishbach, Advantages of real-time data acquisition for exposure assessment, *Appl. Indus. Hyg.* 3(11):316–320 (1988).
4. M. G. Gressel, G. J. Deye, W. A. Heitbrink, and J. D. McGlothlin, Graphical exposure demonstration: Computers and video link together. Presented at the American Industrial Hygiene Conference, St. Louis, MO, May 21–26, 1989.
5. U.S. Department of Health and Human Services, Public Health Service, Centers for Disease Control and Prevention, National Institute for Occupational Safety and Health (NIOSH), control of nitrous oxide in dental operatories, J. D. McGlothlin, K. G. Crouch, and R. L. Mickelsen, DHHS (NIOSH) Publ. No. 94-129, 1994.
6. Operating manual, Thermovision 782, AGA Infrared Systems AB, Secaucus, NJ, 1984.
7. American Conference of Governmental Industrial Hygienist: Air sampling instruments for evaluation of atmospheric conditions, 8th ed., ACGIH, Cincinnati, OH, 1995.
8. U.S. Department of Health and Human Services, Public Health Service, Centers for Disease Control and Prevention, National Institute for Occupational Safety and Health, NIOSH Manual of Ananlytical Methods, 4th ed., DHHS (NIOSH) Publication No. 94-113, 1994.

9. M. G. Gressel, W. A. Heitbrink, and P. A. Jensen, Video exposure monitoring—A means of studying sources of occupational air contaminant exposure, Part I: Video exposure monitoring techniques, Proceedings of the International Symposium on Air Sampling Instrument Performance, *Appl. Occupa. Environ. Hyg.* 8(4):334–338 (1993).
10. W. A. Heitbrink, M. G. Gressel, T. C. Cooper, T. J. Fischbach, D. M. O'Brien, and P. A. Jensen, Video exposure monitoring—A means of studying sources of occupational air contaminant exposure, Part 2: Data interpretation, Proceedings of the International Symposium on Air Sampling Instrument Performance, *Appl. Occupa. Environ. Hyg.* 8(4):339–343 (1993).
11. D. Skoog, and D. West, *Principles of Instrument Analysis*, Holt, Rinehart and Winston, New York, 1971.
12. R. M. Barnes, *Motion and Time Study: Design and Measurement of Work*, 7th ed, John Wiley and Sons, New York, 1980.
13. G. Box, and G. Jenkins, *Time Series Analysis: Forecasting and Control*, Holding-Day, San Francisco, 1981.
14. J. D. McGlothlin, P. A. Jensen, W. F. Todd, T. J. Fishbach, and C. L. Fairfield, Control of anesthetic gases in dental operatories at Children's Hospital Medical Center, Dental Facility, Cincinnati, Ohio, NTIS Pub. No. PB-90-155-946, National Technical Information Service, Springfield, VA, 1990.
15. *Nitrous Oxide/N_2O*, E. I. Eger, ed., Elsevier, New York, 1985.
16. U.S. Department of Health and Human Services, Public Health Service, Centers for Disease Control, National Institute for Occupational Safety and Health, Criteria for a recommended standard, occupational exposure to waste anesthetic gases and vapors, Pub. No. 77-140, NIOSH, Cincinnati, OH, 1977.
17. J. D. McGlothlin, P. A. Jensen, W. F. Todd, and T. J. Fischbach, Study protocol: Control of anesthetic gases in dental operatories, U.S. DHHS (NIOSH) Report No. ECTB 166-03, NIOSH, Cincinnati, OH, 1989.
18. K. Willeke and P. A. Baron, Bridging science and application in aerosol measurement, in *Aerosol Measurement: Principles Techniques and Applications*, K. Willeke and P. A. Baron, Eds., Van Nostrand Reinhold, New York, 1993, pp. 3–7.
19. D. O'Brien, Data acquisition and analysis, in *Aerosol Measurement: Principles, Techniques, and Applications*, K. Willeke and P. A. Baron, Eds., Van Nostrand Reinhold, New York, 1993, pp. 521–532.
20. P. A. Jensen and D. O'Brien, Industrial hygiene, in *Aerosol Measurement: Principles, Techniques, and Applications*, K. Willeke and P. A. Baron, Eds., Van Nostrand Reinhold, New York, 1993, pp. 537–555.

36
Ergonomics and Concurrent Design

Peter M. Budnick, Donald S. Bloswick, and Don R. Brown
University of Utah, Salt Lake City, Utah

I. INTRODUCTION

Many industries are adopting product development strategies intended to streamline the time and financial commitments during the conception-to-market phase of product life cycles. Termed "concurrent design," "design for manufacturing (DFM)," or "simultaneous engineering," for example, these methods strive to integrate traditionally separate engineering disciplines to achieve a more comprehensive design at a lower initial cost. The time saved during this process can also provide significant advantage by beating competitors to the market. As related by Walsh et al. [1],

> In the 1990's . . . the rules of successful competition are changing. Companies not only have to develop and manufacture well-designed, high quality products at a price that offers the customer value for money, but this has to be done at least as fast as, or preferably faster than, the competition.

The later a design change is made, the greater the cost burden. De Fazio et al. [2] estimate that 70% of a product's life-cycle costs are determined when it is designed. The choices made during design determine materials, fabrication, and assembly methods and to some extent the material handling, inspection, and other options available to the process designers. Decisions that affect assembly in particular also affect nearly every other aspect of the manufacture and use of a product [2]. Figure 1 indicates the relative relationships between cost, design change, and development time for a typical product from the point of product conception through production, comparing concurrent design methods with traditional design methods.

Production methods are, of course, a primary concern to ergonomics experts, because assembly, materials handling, and inspection tasks are major contributors to industrial ergonomic problems, including productivity, quality, and health and safety issues. With concurrent design, process design viewpoints are integrated into early product and process design stages. In this way, many potential problems can be eliminated or accommodated early in development. This saves time and money and can positively influence productivity and quality as well. Currently, far too many ergonomic recommendations are denied because they are presented or solicited late in the product life cycle when changes are financially prohibitive. We contend that a proactive ergonomics advocate is crucial in the design synthesis stages.

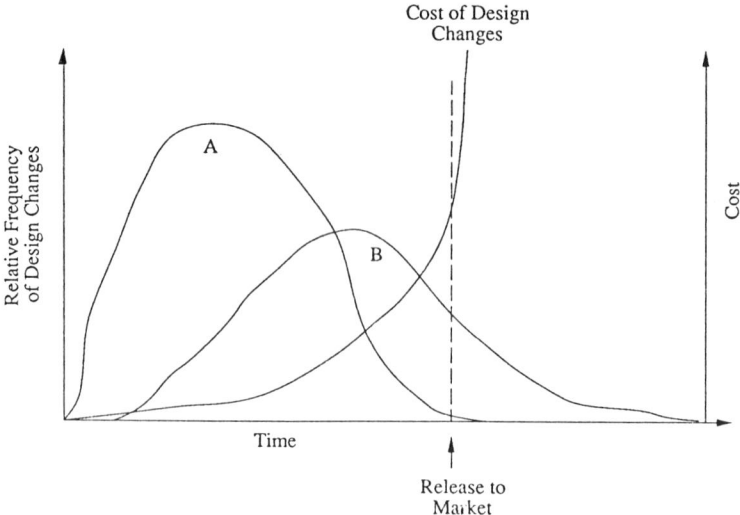

Figure 1 Relative relationships between cost, design changes, and development time for traditional design methods (curve B) and concurrent design methods (curve A).

II. BACKGROUND AND SIGNIFICANCE TO OCCUPATIONAL ERGONOMICS

A. The Design Process

The product development process is, as stated by Rosenthal [3], "inherently a creative, interactive problem-solving process, not just a technological accomplishment." However, not all designs are innovative, and not all arise from R&D. In fact, the vast majority of new products involve variations on existing technology or incremental improvements to existing designs. Many design efforts are actually redesigns focusing on reliability, quality, and economic manufacturing [1].

Arriving at a final design involves narrowing a large set of possible choices, often representing conflicting views, until a satisfactory product can be developed. There is rarely one unique solution to a design quest. The ultimate choice depends upon the way the problem is defined or specified and the nature of the information available to solve it. The viewpoints and knowledge applied by the designers themselves probably have the strongest influence on final design choices. Based on these observations, Stoll [4] proposes the following: (1) In general, many different solutions to a design problem are possible; (2) the solution that is chosen may be selected for one or a combination of many right and wrong reasons; and (3) the design process provides no intrinsic guarantee that the selected solution is, in fact, the best solution or even the right solution. If this is true, then how can the "one best solution" be arrived at in the shortest period of time? This question is addressed in Sec. II.A.2. First, traditional and concurrent design methods are reviewed and compared.

1. Traditional Design Methods

Traditional product life cycles followed an incremental and sometimes linear model, as depicted in Figure 2. In the worst cases, little communication occurred between the traditionally separate departments of design, manufacturing, marketing, distribution, and after-sales service. Each department performed its duties and would then "throw it over the wall" to the next in line. Because decisions made in one area affect choices in others, such methods re-

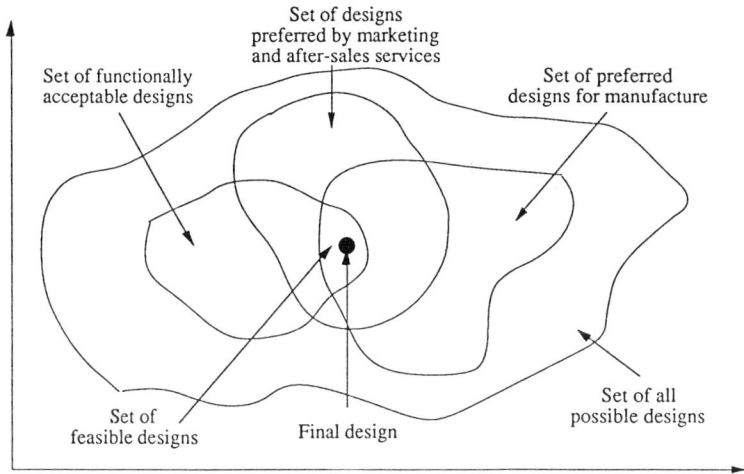

Figure 2 Representation of a multidimensional design space that narrows the final solutions set for a particular design problem.

sulted in a particular product being "thrown" back and forth, losing valuable time and money before it was released to market. Significant compromises in product and process quality were also evident in this turbulent process.

It has become readily evident that traditional compartmental information hoarding and decision making is not competitive in the world market. There must be some sort of communication mechanism between the different concerns that encourages and accommodates varied viewpoints in a timely fashion. Corporate "culture" and organizational structure can either be a barrier or provide the potential for positive interaction among those with different perspectives (see, e.g., Rosenthal [3] for discussion relating to such issues).

2. Concurrent Design Methods

Recommended methodologies and case examples of successful concurrent design abound (see, e.g., Ettlie and Stoll [5], De Fazio et al. [2], Susman [6]). We do not propose or endorse any one method but conceptualize the underlying principles and identify specific methods and types of communication that we believe are important in the effort to infuse ergonomic information into the design process. Returning to the question previously posed: How can the "one best solution" be arrived at in the shortest period of time?

Concurrent design methods approach this problem by superimposing critical product and process life-cycle information on the problem definition. This approach, while using more information than has traditionally been handled during early development stages, quickly narrows the solution set to a manageable set of choices that can be optimized in a shorter period of time (see Figure 2). The greatest opportunity to "get the design right" exists in the earliest stages of design formulation. Thus, it is critical to get as comprehensive as possible a set of product knowledge in the planning and requirements definition design stage. The information must be clearly stated as requirements and constraints. Design is an iterative process, and the greatest flexibility exists in the planning stage. In concurrent design, the range of available choices narrows quickly, and changes become increasingly expensive and difficult [4]. The challenge for the field of ergonomics is to integrate ergonomic knowledge into this process.

As typically represented, concurrent design approaches can be broken down into three general and interrelated categories: concept generation, product engineering, and process engineering (below, this is expanded by considering distribution and after-sales services). Concept generation often reflects the viewpoint of marketing, which provides a *functional* description in terms of target customer desires, expectations, needs, and problems. Product engineering develops a description of the *form* of the product in terms of geometry, tolerance, and material selections. Process engineering develops a description defining the *fabrication* of the product in terms of plant layout, tool and equipment design, work design, and so forth [4]. This presents three distinct viewpoints in which to infuse ergonomic knowledge, to which we will add distribution and after-sales services. This provides a more robust framework in which to advocate and expand ergonomic influences in product life cycles.

Functional knowledge may be viewed as more conceptual in nature, whereas form and fabrication information relies on more objective ergonomic knowledge. For example, a functional constraint may be that the product must be easy to grasp. This problem could be solved in a variety of ways, such as by providing traditional handles or by defining product geometry such that it "invites" a user to grasp in a certain location. This general functional information can be further refined into form criteria by imposing anthropometric, biomechanical, and psychophysical constraints relating to handles. Likewise, fabrication constraints defined by form can be positively influenced by industrial ergonomic principles and criteria. The key is that each design iteration and decision must be accompanied by specific and applicable ergonomic criteria. The type of ergonomic knowledge and its organizational structure is discussed in greater detail below.

III. INFUSING ERGONOMICS INTO CONCURRENT DESIGN METHODS

The "best" methods for getting ergonomic knowledge included in concurrent design is a matter of debate. It would be ideal if an ergonomic expert were part of every design effort, as suggested by Mital and Morse [7], for instance, but for numerous reasons this is rarely the case. These reasons include, for example, (1) ergonomists do not even agree among themselves on who qualifies as an expert; (2) ergonomists have been largely unsuccessful in the past at presenting their knowledge to nonexperts; (3) ergonomic knowledge is often ambiguous, heuristic, and less universally applicable than knowledge in other design domains; (4) ergonomists are generally unable to objectify the need for ergonomic design influences in terms of economics; and (5) ergonomists often have limited knowledge regarding design and engineering practices.

Alternatively, since a designer's knowledge, inspired by culture, world view, and life experiences, inevitably affects design choices, it would be advantageous if engineers received basic ergonomics training as part of their general education. Although efforts are under way to incorporate ergonomic and safety knowledge into the undergraduate engineering curriculum (see, e.g., Kavianian et al. [8]; Bloswick [9]; Talty [10]), ergonomists must not wait for the next generation of engineers. Product and process developers need tools that can be useful and instructive today. There are many viable ways to satisfy this need, including incorporating ergonomic knowledge into existing design tools or developing new tools specific to ergonomics. However, practicing engineers have warned that stand-alone products sit on the shelf [11]. Therefore, it is to the ergonomist's advantage to enhance existing tools by providing an ergonomic viewpoint. Since ergonomists must appeal to designers and engineers, it is to their benefit to work within the framework, language, and formats with which such professionals are already familiar. Computers are increasingly the vehicle for engineering tools and thus will play a major role in efforts to incorporate ergonomic knowledge into concurrent design practices. We believe that computerized methods will be most applicable in the future, especially for large organizations.

A. Product Life-Cycle Model

"Product life cycle" includes every stage of a product, from the time it is merely an idea in someone's mind to the time when it becomes inert (which may not necessarily be the time of disposal).

It has already been noted that traditionally compartmentalized and sequential product life cycles erect barriers to efficient concurrent development strategies. Therefore, rather than encourage and prolong the potential pitfalls associated with the traditional view depicted in Figure 3, we will consider the product life cycle from a higher level mission- or goal-oriented viewpoint of function, form, and fabrication, to which distribution and after-sales services have been added, creating the model shown in Figure 4.

The product life cycle begins with a functional concept, which is then formed, fabricated, marketed, and serviced. Within this model, when, where, and how is ergonomic knowledge best presented? Expanding on Brown et al. [12], we want to encourage and work within a framework that represents the viewpoints of all concerns within one central design space. Most likely this is supported by a computerized framework within which decisions made by any participant are immediately available and evident to all other participants. It is this comprehensive and expedient flow of critical information that streamlines the portion of the product life cycle prior to release to market.

Figure 4 depicts the relationship between differing product viewpoints as each simultaneously monitors and contributes to the design space. Naturally, contributions and viewpoints will be prioritized based upon organizational and functional decisions, but the concept of multiple viewpoints should be clear. Ergonomic knowledge should be organized in such a way that each viewpoint is afforded an additional human-oriented perspective. Furthermore, the knowledge and its presentation should act as an advocate for human interface perspectives, which may be in competition with purely technical or economic perspectives offered by other viewpoints. Our objectives must be to define and demonstrate the need for ergonomics, not to just supply solution-oriented data.

B. Ergonomic Knowledge Base

The knowledge base should consist of information taken from multidisciplinary ergonomics literature, design and analysis tools, expert knowledge from in-house and external experts, and experiential knowledge gleaned from existing processes and methods within a given or-

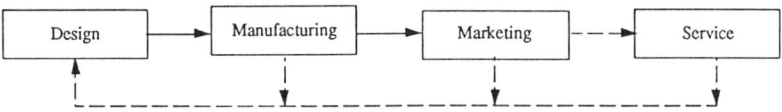

Figure 3 Traditional sequential, compartmentalized product life cycle. The broken lines denote the often poor or nonexistent functions and lines of communication in older product development and life-cycle methods.

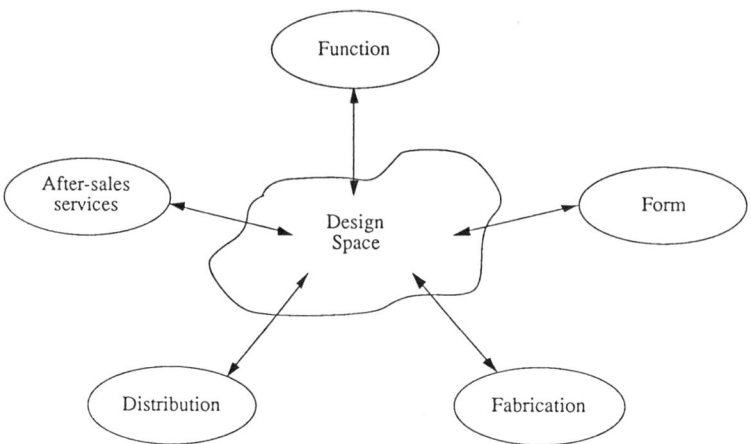

Figure 4 Model of all viewpoints having access to the design space.

ganization (see Figure 5). (Nagamichi and Yamada [13] review participatory ergonomics, providing an example method that can be used as a platform to generate and capture experiential knowledge.)

"Ergonomic knowledge base" is used very broadly to describe a source of well-organized ergonomic concepts and data. The organization and structure of the knowledge should make it easily accessible, interpretable, and applicable. It could reside in an expert, or it could be a hard copy reference source, or it could be an electronically stored and manipulated source.

We put forth ideas and goals for such a knowledge base but leave the actual content and structures to those interested in further research and development in this area.

1. Presentation Format

For an audience of nonexperts, ergonomic knowledge must be presented in a way that advocates its inclusion into areas where it has traditionally been neglected. The knowledge base

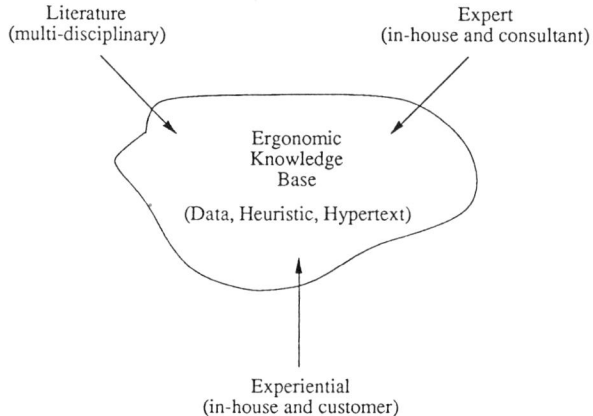

Figure 5 The ergonomic knowledge base should consist of knowledge elicited from multidisciplinary literature sources, in-house and out-of-house experts, and experiences gathered in-house and in the environment of use.

must also be able to defend ergonomic solutions in the context important to product development, which includes technical, organizational, and financial considerations—the so-called bottom line. The nonexpert must not be burdened by a requirement to sift and search through volumes of extraneous materials, yet should be free to navigate easily when interested. In short, the format must present ergonomic knowledge so that it is respectful, useful, and easily applied by nonexperts of varying specializations and backgrounds. Nonexperts often don't understand the need for, let alone the context of, ergonomic information. Therefore, it is up to us to positively and proactively influence developers, which is no small task.

In conceptual stages, problem solutions are very uncertain. Minimizing the uncertainty surrounding design choices is central to concurrent design. As modeled by Stoll [4], this can be viewed as a "funnel process," where the set of choices is wide at first but quickly narrows as iterative decisions are made. Beginning with a set containing all ergonomic knowledge, the set must be quickly narrowed to only applicable information as choices are made. This quick elimination of unnecessary information and the bringing forward of applicable information implies a structured ergonomic knowledge base. (See Lehto [14], for example, for a thorough discussion relating to knowledge base "development with structured application to safety ergonomics.") In general, the ergonomic knowledge should be organized from top-level goal-oriented information that progressively funnels to specific solution-oriented data that can be applied as material and geometric choices are made.

An additional benefit of presenting ergonomics in a goal-oriented fashion is that it can appeal to the problem-solving creativities and abilities of engineers and developers. While the knowledge base must be able to provide specific solution-oriented data based on design choices, it must also be flexible enough to encourage, recognize, and accept new solutions to old problems.

The type of ergonomic knowledge that is applicable to each global perspective is now explored.

2. Topical Knowledge Structure and Organization

While each of the general product life-cycle categories is treated as a separate topical activity in this discussion, an effective concurrent design strategy dictates that they occur together and that much of the knowledge be applied during functional formulations.

Function. Functional ergonomic knowledge and its presentation format should focus on topics applicable early in product conceptualization and development. Product functional characteristics are defined during the conceptual stage for a new product and revised and updated during product redesign. This is a crucial point at which to advocate the human interface in any product or process design. It is important that experiential ergonomic knowledge gleaned from in-house sources and customer experiences monitored by after-sales services be added to the design space for product redesign. For a new product, it is important that the traditional technologically driven parameters be augmented with human-oriented goals. It must be impressed upon decision makers that all products will interface with humans in a variety of ways and that product success requires that ergonomic human interface criteria play a significant role in defining functional criteria.

During functional definition, human-oriented concepts such as product usability, maintainability, serviceability, and safety should be advocated and prioritized. Depending on the product and its intended application and user environment, some ergonomic topics may be given priority over others.

One question that should be answered early on is, Where will humans interfere with a product during its lift cycle? Nearly all products will have some in-house human interface during design, manufacturing, and distribution. After market, depending on the nature of the product, it may have varying degrees of interface with people of different needs and skill levels. A consumer product, for example, might require considerable ergonomic knowledge relating to user operation and training, skilled or unskilled maintenance and service providers, and disposal. On the other hand, a product that serves as a component within a larger machine or system may require ergonomic knowledge geared primarily to assembly, maintenance, and service. By targeting the human interfaces that will occur during the product life cycle, one can determine the ergonomic knowledge that will be most applicable and eliminate the information extraneous to the project at hand.

Those responsible for market research can benefit from human-oriented research methods. Experimental design and statistical methods used in the field of ergonomics can help target user populations and more accurately determine potential user attributes, concerns, and desires. Just as statistically sound experimental and measurement methods have improved product quality control and reliability, ergonomists should advocate appropriate research methods to identify human responses, desires, capabilities, expectations, etc. Crucial topics such as the skill and product knowledge level of the target user population are often overlooked or given low priority by designers. Proactive ergonomic knowledge can and should positively influence designers to consider such parameters.

Another area of market research that is useful in this stage of design relates to market experience with similar products. Information concerning both successes and failures can be very instructive in early product development decisions. In particular, accident and injury data, consumer dissatisfaction, standards, laws, and related liability information can be beneficial in early stages of development. While the ergonomic knowledge base may not contain such specific information (unless it is gleaned from in-house experimental knowledge, which should be continually updated), it should advise as to how and why developers can and should conduct such research.

Unless market research has identified ergonomic parameters, it is up to the ergonomic knowledge base and presentation format to advocate them. At the top level, the advice might be quite heuristic, such as "Portable consumer products must be easy to move, lift, or carry." Getting slightly more detailed, "If a product is meant to be lifted, it should be easy to grasp and hold." Thus, the knowledge first reminds developers to consider ergonomic issues and then immediately supplies applicable data when a choice is made. For instance, if "easy to grasp" survives as a design goal, then ergonomic data pertinent to grasps, such as hand-hold types and the advantages and disadvantages of each, must be immediately provided. If a particular hand-hold is selected, then location, orientation, shape, and clearance data should be immediately presented, and so on as the functional product decisions become product form decisions.

Form. As the functional parameters are set, the design quickly progresses toward form-oriented solutions. Ideally, the ergonomics advocate is able to influence design goals and parameters as functional determination progresses. Either way, it must continue to infuse applicable ergonomic knowledge into the development of form criteria. Choices made here will impact all aspects of the human interface throughout the product life cycle. Therefore, the knowledge base must offer industrial and consumer ergonomic expertise at varying levels of detail (conceptual and detailed) to an audience consisting mostly of design and manufacturing engineering experts.

For product redesign, experiential knowledge will be very valuable. For new products, it is important to learn from similar product and fabrication experiences. Many redesigns focus on product and process quality and reliability issues, driven primarily from a technical standpoint. The role of the human operator is often given the least consideration in fabrication processes, yet without ergonomic considerations it can be the least controllable and least predictable component in the system. Not only must an ergonomic knowledge base impress this fact upon engineers, it must also demonstrate that ergonomics can minimize this uncertainty and improve productivity, quality, and reliability.

Geometric, tolerance, and material choices have direct effects on manufacturing processes and the people who interface with them. It is this relationship between form and fabrication that has received the greatest attention to date and has resulted in the proliferation of DFx (Design for x) tools, where x represents a particular interest or viewpoint. For example, there are DFManufacturability, DFAssembly, DFDisassembly, DFMaintainability, DFEnvironment, etc. Unfortunately, few of these tools are influenced by ergonomic knowledge.

Assembly operations are defined in part by the choices made while defining the form of a product and its components. The type of connection and the assembly procedures directly affect consumer concerns such as product quality, reliability, maintainability, and serviceability. Connection methods also have a dramatic effect on industrial ergonomic concerns, which can translate into productivity, product quality, reliability, and inspection problems. Many engineers are unaware of the role human capabilities and limitations can have on these important product characteristics.

Heuristic and specific ergonomic knowledge is applicable in the determination of form. For example, heuristic knowledge might identify to the product and process developers potential ergonomic hazards related to the use of threaded fasteners as component connectors. If threaded fasteners are nevertheless chosen, then methods to minimize potential hazards, such as recommended tools and assembly workstation design parameters, must be immediately provided. In this way designers can make efficient and informed iterations and decisions.

In short, an effective ergonomic knowledge base must be able to specify geometric, tolerance, and material selections for both product and fabrication processes that will minimize ergonomic hazards and enhance usability, productivity, quality, and reliability.

Design of product support materials, which should begin in this design phase, can also significantly benefit from ergonomic perspectives. The importance of product user and maintenance manuals, product warnings, product packaging, and marketing materials is often neglected or poorly understood. These materials can not only help consumers and users to better use a product, thereby increasing customer satisfaction, but they can also have significant impacts on safety and health issues and related liability.

Fabrication. In concurrent design, determination of design criteria that reflect form and fabrication considerations will occur in parallel. This relationship has received the most attention in industry, resulting in the variety of DFx tools mentioned above. Ergonomic knowledge must become a part of this critical process, because decisions made here will affect every aspect of a product. Industrial ergonomic knowledge, which is most applicable in fabrication, is undoubtedly one of the most complete sections in the body of ergonomic knowledge. To be useful in concurrent design, the knowledge must be structured and directed proactively, rather than reactively, as Corbett [15] notes is the typical mode of use for ergonomics design criteria.

If ergonomics is to assume a role in product and fabrication design, the ergonomics advocate must defend, with bottom-line data, the need for, and consequent benefits of, its application. Grossmith [16], for example, presents design-oriented ergonomic considerations from the viewpoint of reducing manufacturing costs, which is the type of bottom-line knowledge that will influence design decisions. Ergonomic methods that increase productivity or improve quality, reliability, and inspection processes will also receive serious attention from fabrication experts.

Engineers involved in the design of product fabrication processes typically plan assembly sequences and shop floor layouts, select appropriate fabrication machinery, design workstations, and determine the means of transport and the in-house materials handling, storage, and distribution schemes. Each of these activities can be enhanced by the inclusion of ergonomic principles.

Fabrication knowledge should again be organized from a conceptual, goal-oriented level down to specific applicable process design data. Conceptually, ergonomic input can help process designers distribute tasks between humans and machines, taking advantage of distinctly human talents. Processes and methods that have been known to contribute to ergonomic hazards and degradation in productivity, quality, or reliability can be flagged early, at which time they can be replaced with better alternatives or improved from an ergonomic standpoint.

Distribution. In-house employees, employees of distributors, and consumers are all impacted by marketing methods, packaging, layout, and product-handling methods. Commonly, such distribution characteristics are driven by operational and quality (minimizing transport damage) concerns but pay little attention to the human interface with this phase of the product life cycle.

From the perspective of in-house distribution, ergonomic knowledge should be applied to the design of shipping packaging (size, shape, weight, and handling characteristics) and storage and handling methods. Where applicable, packaging should be designed so that units can be handled manually with minimal biomechanical hazards. Where this is not possible, lifting assistance devices should be provided. When products will pass to second- and third-party distributors and outlets, specifying the ergonomic handling and training methods may be beneficial. However, since control over handling methods may not be possible once a product leaves a manufacturing facility, especially once it is passed to the consumer, we encourage that safe handling methods be achieved through initial product and packaging design. As with all perspectives in concurrent design, such principles must be presented and defended in the earliest stages of product functional formulation.

After-Sales Service. Depending on the type of after-sales services provided, varying degrees of ergonomic hazards may exist for in-house employees, consumers, and third-party service providers. Effective after-sales services ideally monitor and provide follow-up services for customers. There are two main roles for effective after-sales services: (1) product and customer support and (2) product and customer monitoring. In many organizations, the functions of product development and after-sales services are far removed from each other, which often negates the inclusion of experiential knowledge in future production and redesign efforts. An effective concurrent design program encourages and provides an infrastructure to bring this critical information to the functional development stage. Ergonomics can enhance this process by providing human-oriented product and consumer research and monitoring methods as well as by advocating design parameters that improve maintenance and serviceability parameters such as accessibility, visibility, tool selection, and biomechanical accommodations. For product maintenance and repair services, the design can either hinder or enhance maintainability and serviceability, which can have a significant impact on customer satisfaction and safety.

IV. CRITICAL REVIEW OF CURRENT STATUS

Significant efforts are under way to integrate ergonomics into design, but no single example has demonstrated great success. Further, most ergonomic tools are evaluation-oriented rather than being synthesis methods that can be applied during concept formulation for new products and processes. Many tools that perform specific ergonomic functions have been proposed or developed, but none are known to have achieved widespread use within the design or engineering community to date. Young [17] suggests that "widespread use of [such tools] has been inhibited due to several factors: cost, complexity, ease of integration, functionality and validity." It must also be understood that until the need for ergonomics can be advocated and defended within the "bottom-line" parameters that drive development, this lack of use will continue.

A. Review of Related Research and Applications

1. Useful References

A number of useful discussions relating to ergonomics in design for manufacturability can be found in the volume edited by Helander and Nagamichi [7, 13, 16, 19]. In particular, Mital and Morse [7] discuss the role of ergonomics in designing for manufacturing and identify some data sources that can be used in the process. Drury [18] discusses a systematic procedure for designing for inspectability. Grossmith [16] presents design-oriented ergonomic considerations from the viewpoint of reducing manufacturing costs, which is the type of bottom-line knowledge that will influence design decisions. He also presents some useful manufacturing ergonomic design parameters relating to materials, fasteners, and product handling. Seidel [19] reviews an assembly sequence planning system from a human factors perspective. Each of these example references represents a portion of ergonomic knowledge that could be valuable in a concurrent design effort; however, none appears to be part of a larger system to capture and disseminate ergonomic principles, concepts, and design data in a proactive manner.

2. Ergonomic Design Tools

Many ergonomic tools described as "design tools" are probably better called "design evaluation tools." That is, they do not provide design synthesis or conceptual development knowledge in the functional definition stages, which is necessary in the concurrent design environment. Further, many of these tools, though computerized, are stand-alone and have not been integrated into the tools and systems to which designers are accustomed (geometric modelers, material databases, etc.). Interested readers should refer to the Mattila Karwowski volume [20,21], for example, for a review of computerized ergonomic applications.

A specific type of computerized ergonomic tools are expert systems. One such system that we have under development is briefly described below. The long-term value of expert systems, an outgrowth of artificial intelligence research, is debatable. However, because ergonomic experts are somewhat scarce, and because ergonomic knowledge is sometimes uncertain and heuristic, such knowledge is a prime candidate for expert systems. A number of ergonomic expert systems have been proposed and prototypes constructed. Brown et al. [23], Jarvinen and Karwowski [20], and Kern and Bauer [21], for instance, provide reviews and references for ergonomic expert systems. In addition to expert systems, we also suggest computerized hypertext methods to present ergonomic knowledge to nonexperts, which is discussed below.

B. Review of CDEEPS

CDEEPS (Concurrent Design and Engineering for the Ergonomics of Production Systems) is a prototype system currently under development within a computational framework for concurrent engineering developed by Brown et al. [12]. This prototype system provides a common computing environment that uses an object-oriented programming language (Common Lisp). The computing environment is designed such that all critical design decision data can be viewed by all design participants. Currently, CDEEPS is geared toward product and process design and is limited to providing advice relating to industrial ergonomic concerns only. The architecture, however, is expandable to accommodate ergonomic concerns throughout the entire product life cycle. In essence, it captures the design space as depicted in Figure 4 and allows product designers and process designers to view and alter design criteria.

CDEEPS uses expert systems and hypertext technology, outgrowths from the field of artificial intelligence. Hypertext can be thought of as an on-line electronic library. Hypertext allows the user to browse through information by selecting keywords from the text, which then immediately opens other sources for review, just as one might use cross-referencing in a library. The hypertext tool is quite useful in the conceptual formulation stages, as it allows the nonexpert to easily peruse background ergonomic knowledge to any level of detail. In this sense, it may serve as an educational tool as well.

Currently, the expert system triggers the hypertext in the design formulation phase and performs more detailed task analysis as assembly processes are developed. The expert system prototype contains over 700 rules and is able to predict manual assembly process times using MTM [22] time measurement units (tmu's) and to predict a general hazard level for a given task.

Interested readers may find more detailed information concerning CDEEPS in Brown et al. [23] and Budnick et al. [24].

V. FUTURE CONCERNS

Due partially to the proliferation of computers and their ability to quickly process and pass on information, product development activities are becoming more organized and accelerated as competitors seek to introduce better products to market faster. Historically, it has been difficult to establish ergonomics in design practices, especially in the design synthesis and functional definition stages. Missed opportunities while setting design parameters translate to form definitions with inferior human interface characteristics throughout the product life cycle, including both workplace and consumer environments. To fully realize its benefits, the inclusion of ergonomics must be advocated and defended early in the design stages. To be useful to product and process developers, ergonomic knowledge must be organized in such a way that each design decision is immediately accompanied by appropriate and applicable concepts and data. This implies a highly organized and structured knowledge base with an interface that presents design information in a format consistent with that of other engineering and development data. Such a knowledge base must also be integrated with the environment and tools used by design experts. Short of an ergonomics expert on every design team, a worthy goal but unrealistic in the near future, we suggest hard copy and computerized tools targeted for proactive use rather than the reactive use that characterizes most ergonomic applications to date. It is strongly suggested that those interested in concurrent engineering applications study and work within the environments in which those applications will be used, which includes advanced computer hardware and software systems.

REFERENCES

1. V. Walsh, R. Roy, M. Bruce, and S. Potter, *Winning by Design: Technology, Product Design and International Competitiveness*, Blackwell, Cambridge, MA, 1992, pp. 4–5, 10.
2. T.L. De Fazio, A. C. Edsall, R. E. Gustavson, R. E. Metzinger, and W. A. Dvorak, in *Concur-*

rent Design of Products and Processes: A Strategy for the Next Generation in Manufacturing, J. L. Nevins and D. E. Whitney, Eds., McGraw-Hill, New York, 1989, p. 2.

3. S. R. Rosenthal, Bridging the cultures of engineers: challenges in organizing for manufacturable product design, in *Managing the Design-Manufacturing Process*, J. E. Ettlie and H. W. Stoll, Eds., McGraw-Hill, New York, 1990, pp. 21-52.
4. H. W. Stoll, Design for life-cycle manufacturing, in *Managing the Design-Manufacturing Process*, J. E. Ettlie and H. W. Stoll, Eds., McGraw-Hill, New York, 1990, pp. 79-113.
5. J. E. Ettlie and H. W. Stoll, Eds., *Managing the Design-Manufacturing Process*, McGraw-Hill, New York, 1990, p. 22.
6. G. I. Susman, Ed., *Integrating Design and Manufacturing for Competitive Advantage*, Oxford Univ. Press, New York, 1992.
7. A. Mital and I. E. Morse, The role of ergonomics in designing for manufacturability, in *Design for Manufacturability: A Systems Approach to Concurrent Engineering and Ergonomics*, M. Helander and M. Nagamachi, Eds., Taylor & Francis, Washington, DC, 1992, pp. 147-159.
8. H. R. Kavianian, N. Meshkati, C. A. Wentz, and J. K. Rao, Should engineering schools address occupational and environmental safety and health issues? *Prof. Safety*, June 1993, pp. 48-49.
9. D. S. Bloswick, Developing and marketing safety and ergonomics courses in traditional engineering curricula: process and content, *Proc. NW Workshop on Occup. Safety and Heath Engineering Curricula*, Corvallis, OR, April 27-28, 1992.
10. J. T. Talty, Project shape—integration of safety and health into engineering school curricula, *Proc. NW Workshop Occup. Safety and Health in Engineering Curricula*, Corvallis, OR, April 27-28, 1992.
11. B. Joseph, Corporate Ergonomist, Ford Motor Company, Dearborn, MI, personal communication, 1992.
12. D. R. Brown, M. R. Cutkosky, and J.M. Tenenbaum, Next-cut: a second generation framework for concurrent engineering, in *Computer Aided Cooperative Product Development*, D. Sririam and R. Logcher, Eds., Springer-Verlag, New York, 1991.
13. M. Nagamachi and Y. Yamada, Design for manufacturability through participatory ergonomics, in *Design for Manufacturability: A Systems Approach to Concurrent Engineering and Ergonomics*, M. Helander and M. Nagamachi, Eds., Taylor & Francis, Washington, DC, 1992, pp. 219-231.
14. M. R. Lehto, *A Structured Methodology for Expert System Development with Application to Safety Ergonomics*, UMI Dissertation Services, Ann Arbor, MI, 1985.
15. J. M. Corbett, Human centered advanced manufacturing systems: from rhetoric to reality, *Int. J. Ind. Ergon.* 5:83-90 (1990).
16. E. J. Grossmith, Product design considerations for the reduction of ergonomically related manufacturing costs, in *Design for Manufacturability: A Systems Approach to Concurrent Engineering and Ergonomics*, M. Helander and M. Nagamachi, Eds., Taylor & Francis, Washington, DC, 1992, pp. 232-243.
17. M. F. Young, Development of an Integrated Computer Aided Ergonomics Toolbox for Industrial Workstation Design, Masters Thesis, Dept. Mech. Eng., Univ. Utah, Salt Lake City, UT, 1994, Unpublished.
18. C. G. Drury, Product design for inspectability, in *Design for Manufacturability: A Systems Approach to Concurrent Engineering and Ergonomics*, M. Helander and M. Nagamachi, Eds., Taylor & Francis, Washington, DC, 1992, pp. 204-216.
19. U. A. Seidel, MI system for assembly sequence planning: human factors considerations, in *Design for Manufacturability: A Systems Approach to Concurrent Engineering and Ergonomics*, M. Helander and M. Nagamachi, Eds., Taylor & Francis, Washington, DC, 1992, pp. 189-203.
20. J. Jarvinen and W. Karwowski, Applications of knowledge-based expert systems in industrial ergonomics: a review and appraisal, in *Computer Applications in Ergonomics, Occupational Safety and Health*, M. Mattila and W. Karwowski, Eds., Elsevier, New York, 1992, pp. 45-54.
21. P. Kern and W. Bauer, Expert systems in ergonomics, perspective applications in industrial practice, in *Computer Applications in Ergonomics, Occupational Safety and Health*, M. Mattila and W. Karwowski, Eds., Elsevier, New York, 1992, pp. 37-44.

22. MTM Association for Standards and Research, 1601 Broadway, Fair Lawn, NJ.
23. D. R. Brown, P.M. Budnick, D. S. Bloswick, and J. Zhou, Design for ergonomics in a concurrent environment, Department of Mechanical Engineering, University of Utah, 1993.
24. P. M. Budnick, D. S. Bloswick, and D. R. Brown, Integrating industrial ergonomics into the design process: accommodating the design engineer, in *Advances in Industrial Ergonomics and Safety*, Vol. IV, S. Kumar, Ed., Taylor & Francis, Wasington, DC, 1992, pp. 11–18.

37
Overview of Ergonomic Research and Some Practical Applications in Sweden

Lennart Dimberg
Volvo Aero Corp., Trollhättan, Sweden

I. INTRODUCTION

The Scandinavian countries, especially Sweden, have a long history of concern for the health of their workers, Sweden being one of the first countries in the world to adopt regulations on the ergonomics of the work environment, The Swedish Ordinance Concerning Work Postures and Working Movements, in 1983 [1].

Sweden has 8.7 million inhabitants [2] and one of the western world's highest proportion of female workers: 86% of women 15-64 years of age with children less than 10 years old had employment in 1988 [3]. It has 2.2 physicians per 1000 inhabitants, and in 1990 about 75% of all workers were covered by an occupational health program [4]. The retirement age is 65 years of age, and in 1993 the unemployment rate was 8.8%. The Swedes have one of the world's highest average life expectancies—78 years for women and 73 for men in 1992—and the second lowest infant mortality rate in the world (second to Japan) [2].

Volvo has used ergonomic designs for the interior of its cars for many years, and a study by Kelsey et al. [5] suggested that its low back support seats, developed in cooperation with orthopedic surgeons, reduced the frequency of lumbar disk surgery. Over the years Volvo has transformed the worker-oriented approach into practical manufacturing examples in the car industry as well, in both the Kalmar (1974-1994) and Uddevalla (1985-1993) plants. However, in spite of all the work in the field of ergonomics, the number of reported work-related musculoskeletal injuries as well as overall sickness absenteeism continued to grow during the 1980s [6], and the Volvo workers were equally affected.

II. HISTORICAL BACKGROUND AND SIGNIFICANCE TO OCCUPATIONAL ERGONOMICS

A. General Aspects

In Sweden, the first law on workers' protection was adopted in 1912. In those days the main problem was long working hours (12-14 hr/day), which was thought to be the main reason for work-related accidents. In 1920 the 8-hr workday was stipulated by law. A new law was adopted in 1949 that called for the introduction of mandatory health and safety committees in companies with more than 50 employees and worker safety representatives in companies with more than five employees.

The National Board of Occupational Health and Safety was also formed. The current Work Environmental Act was adopted in 1977 and emphasizes the importance of prevention in both physical and psychological terms. It applies to all areas of occupational life, private and public sectors alike. Through special ordinances, this board instructs employers as to how the work environment law should be followed.

From 1951 to 1992, agreements between management and the unions regulated the development of occupational health care services. Since 1993 these agreements have been under renegotiation. Occupational health care is not required by law.

1. The Labor Inspectorate

The National Board of Occupational Safety and Health is the central administrative authority in Sweden and is also the authority guiding the field organization, the Labor Inspectorate. The Labor Inspectorate is a law enforcement agency whose inspectors regularly monitor workplaces. In recent years, its actions have been directed more toward system inspection, looking at the structure of how health and safety work is organized and implemented. Traditionally, it has focused on accidents, and the implementation of ergonomic regulations has, relative to all the reported musculoskeletal work injuries, only rarely been enforced.

2. The Work Environment Fund

The work environment fund was founded in 1972 and grants research funding of about US $100 million per year (1990) for work environment–related research.

3. The Fund of Working Life

As of 1989, a special fee (1.5% of salary costs) has been paid to a Work Environment Fund, which distributes grants totaling about US $300 million annually to employers for special projects on rehabilitation and improvement of the work environment. To receive funds, an employer must present a written program describing existing problems and showing plans to deal with them. The fund will pay part of the costs of projects that meet its criteria.

4. Occupational Health

There are, in principle, three types of occupational health services in Sweden: in-plant service, out-of-plant group service, and branch service. In-plant service is the most common in the larger companies (with more than 1000 employees), out-of-plant service is most common in smaller companies, and branch service is representative of, for instance, the construction industry (Bygg-hälsan). In general, in the larger companies there are 1 physician, 1 safety engeineer, 1/2 physiotherapist, and 2 nurses per 2000 employees [7]. Together, these professionals usually form a team to work with ergonomic problems.

B. Time Trends: Only Being Best Is Good Enough

Sweden has, like all other highly industrialized countries, a high cost of labor. Therefore, Swedish industry has found other ways of competing with low-cost countries, manufacturing high-value products with a focus on high technology, robotization, and a highly competent workforce. Swedish managers realize the high potential of a motivated worker, and a good work environment is an important motivator. However, now that high technology is readily available in low-cost countries as well and new computer programs make it easier to learn and use computers, the employers' demand for the best possible team is understandable. This also constitutes a risk that sick or disabled workers will no longer be taken care of by employers. It is important to find a humane solution to this problem.

1. Decentralized Organization and Increased Worker Involvement

Swedish companies like Sandvik, Asea Brown Boveri (ABB), Ericsson, and Volvo all have programs to decentralize their organizations. The hierarchy is flattened (fewer bosses), meaning that the number of decision levels is reduced to hasten the decision-making process. Work groups are being introduced, and supervisory responsibilities for planning, economy, quality, personnel, and production are shared among the workers. The goals of a group are set, and it is up to the group to decide on the means to achieve them. This has meant that health and safety matters are also delegated, which constitutes both an opportunity and a risk. If a worker needs an ergonomic chair or a lifting device, that can be decided on the floor, but there is also a risk that the specialized knowledge of rules and regulations is lost because the responsibility is now shared by several people.

The Japanese have shown that worker participation in safety work is crucial to good results [8]. However, individuals with certain handicaps may need special consideration, which, in such groups, may not necessarily exist. Leif Wallin [9] developed an interesting method to have the group analyze their work situation by means of a questionnaire and subsequent discussion of the results.

1. Job Enlargement and Job Enrichment

Since repetitive work, especially in fixed body positions, has been associated with muscular strain problems [10,11], occupational physicians have applauded when jobs have been enlarged. In the beginning, this often meant that a worker rotated between different, but similarly stressful, production operations—for example, from assembly to inspection of parts. It was clear that this was not enough to eliminate the repetitive strain problems. Also, work technique and its consequences for musculoskeletal disorders were evaluated by Kilbom [12]. Anyone who has played the guitar knows what the right technique means and the muscular fatigue of an incorrect technique.

The recent ideas of enriching the work by introducing completely different types of work such as planning, economics, and quality may be a better way to prevent these problems, but this still remains to be shown.

2. Automation, Robots, and Computers

The fast growth of computer technology has meant that the next generation of industrial robots will be run by workers without extensive assistance from programmers and computer personnel. This clearly represents a breakthrough in the prevention of muscular pain problems. For example, a very high frequency of neck and arm problems was reported on a production line at Volvo Aero Corp., where connecting rods were made (Figure 1). An automatic process using a line of robots was introduced, (Figure 2), and now the workers only have to watch and serve the machines, and orthopedic problems are rare. At the Matsushita Company in Tokyo that makes radios for Volvo, robots have now relieved women of tasks that gave them pains in the assembly of radios.

3. ISO 9000 and Total Quality Management

Quality control is a fashionable term now in Sweden. The ISO 9000 quality standard calls for the documentation of every process and routine. Total quality management is an instrument of wider scope used to define, describe, and measure all processes and their relation to a business.

The quality of work in occupational health services also needs to be looked at. For example, in the field of ergonomics there is often, out at the plants, a lack of prognosis and little or no follow-up of implemented ergonomic solutions to work-related health problems.

Figure 1 Manual work with connecting rods.

4. Process Organization Analysis

Volvo's main process is the manufacture of transport equipment. The role of occupational health services in this process concerns the prevention of injuries and diseases, the treatment of medical problems, and rehabilitation supporting the core process. In some companies many of the peripheral processes such as cleaning and guarding are contracted outside the company. In-plant occupational health services have also been questioned by some companies, but experience of outside contracting is that quality deteriorates, and that is especially true of ergonomic work [7].

Figure 2 A numerically controlled (NC) machine.

5. Kaizen

It is clear that the closer one is to the workplace, the better one understands the problems. Many workers have suggestions not only on how to improve the quality of work but also on ergonomic solutions. The Japanese term *kaizen* means daily improvement, and this must also include the ergonomics of the workplace. It is important to act before the problems get out of hand. In Sweden there has been a tendency to let the worker health and safety representatives (*skyddsombud*) deal with these problems through their organization. However, the law has recently been changed to emphasize that the employer alone is responsible, which has meant that these issues are now dealt with together with other production and quality issues. It is hoped that this will lead to the greater involvement of all workers in ergonomics.

6. The Worker Selection Process

Over the years, much hope has been placed on preventing musculoskeletal problems by selecting the right workers for the job. Unfortunately, the conclusion reached by Kilbom and Hagberg [13] in a major literature survey is that no good screening program exists for the prevention of repetitive strain injuries. The idea that strength and fitness are of importance in the prevention of back problems [14] is contradicted by Battié's prospective study at Boeing in Seattle [15], where only the number of previous back problems was a predictive variable.

7. Health Promotion and Worker Fitness

Saab has developed a questionnaire and physical test that are presented to each participant to obtain a health profile. The general idea is that this profile should be an instrument for behavior change to a sounder lifestyle. Volvo has a similar program called the Life Line. In an evaluation, this has been shown to increase general well-being, but no preventive effect against muscular stain problems has been documented. A proven prophylactic program, however, has been the Volvo Back School, which is a prevention program that was used in a prospective, controlled study of patients with chronic back problems [16].

8. Flexible Working Hours

A system of flexible hours for blue-collar workers has also been implemented in some Volvo plants. It has the advantage of reducing stress associated with rigid shift hours, and workers with rheumatoid problems and morning stiffness may, for example, benefit from starting the workday somewhat later.

9. From Blue- and White-Collar Worker to Fellow Worker

Asea Brown Boveri and Volvo are presently discussing the problems associated with the separation of workers into blue- and white-collar collectives. From the perspective of rehabilitation, it has been shown to be very difficult, especially in a downsizing situation, to transfer, for example, a welder (blue-collar) with a chronic neck problem to a suitable office (white-collar) job. Here, the unions sometimes cannot agree.

C. Examples at Volvo

1. The Kalmar Plant

Attempts to facilitate the production by preassembly of, for example, doors was introduced at the original Torslanda factory as shown in Figure 3. The vision of the Kalmar plant was to break the traditional line system. Involving the workers in more varied assembly work was expected to improve worker satisfaction, avoid repetitive strain injuries, and thereby increase

Figure 3 Traditional car assembly line at Volvo Torslanda.

quality. In the Kalmar plant, which was started in 1974, the mechanical conveyor belt was replaced by self-propelled trolley carriers (Figure 4) on which the car bodies were moved. The carriers were also equipped with tilting equipment, so that work on the underside of the car could be performed in a comfortable posture. The work team became a factor in an endeavor to break down the impersonality of a big operation into sections that were easier to assimilate. The car was still moving between different workstations, but the trolley did not move during the time a given team was working on the car. However, even though the rate of reported muscular strain problems was lower at this plant compared to traditional line work, it was still unacceptably high.

Figure 4 The moving line was broken at Kalmar and assembly performed on a stationary platform.

2. The Uddevalla Plant

In 1985, the opening of the Uddevalla plant signaled a revolution in the process of making cars (Figure 5). The Uddevalla plant's new methods were based on small work teams. Since each work team had a number of different tasks to perform and it was possible to lengthen the work cycles greatly, the individual team member's work could be more varied. The idea was that involving the team in planning, assembly, and quality work would provide so much variation that it would be mentally stimulating to the workers, and with this variation one would also avoid all repetitive strain problems. In Uddevalla the whole car was built at one station by a team of eight persons. Owing to a number of interesting ergonomic solutions, 80% of the time, the car could be manufactured in an upright ergonomically optimum body posture, as shown in Figure 6, by rotating it by 90°, compared to 20% in the old line system. Unfortunately, this did not prove to be the final solution for repetitive strain problems. In fact, the frequency soon increased to parallel the reported work accident frequency of the Kalmar plant. One reason was the high percentage of female assembly operators, whose higher sickness absenteeism (40% higher than for male workers) was a contributing factor. Perhaps the main reason was that the cars were still being built almost completely by hand. Even if each particular assembly task was repeated only a few times per day, some assembly tasks required working with force in awkward positions.

3. The Skövde Plant

The moving platform system, originally implemented at the Kalmar plant, has come into widespread use. In the new engine line at Skövde, the platforms have individually adjustable fixtures (Figure 7), which clearly afford less musculoskeletal strain. In this plant, an almost totally automated production line for gearboxes has shown the way to completely avoid manual assembly and thereby repetitive strain problems.

4. The Volvo Aero Plant

Manual deburring of turbine blades for jet engines (Figure 8) has, over the years, caused large numbers of muscular strain problems in the neck and shoulders. The final solution to this problem turned out to be the introduction of a deburring robot (Figure 9).

III. WHAT WENT WRONG?

A. The Missing Link

Given the fact that improved ergonomics in general has not resolved the problem of repetitive strain injuries, one may speculate as to whether or not the injury mechanism really is clearly understood. Is there a missing link?

1. Theoretical Injury Mechanisms of Repetitive Strain Disorders

In a review article, Hagberg [17] summarizes injury mechanism. The following paragraphs draw strongly on that article.

Degenerative joint disease—osteoarthrosis or osteoarthritis—may be caused by increased stress on the cartilage, such as repetitive impulsive loading. According to some authors, this is sometimes preceded by trabecular microfractures in the subchondral bone caused by trauma. Other authors point to clinical evidence of polyarticular disease and suggest a metabolic abnormality of the articular cartilage. It is claimed that insertion disorders in tendons, ligaments, and articular capsules are caused by local ischemia leading to degeneration and producing inflammation and pain. In particular, the tendons of the supraspinatus, the biceps brachii, and

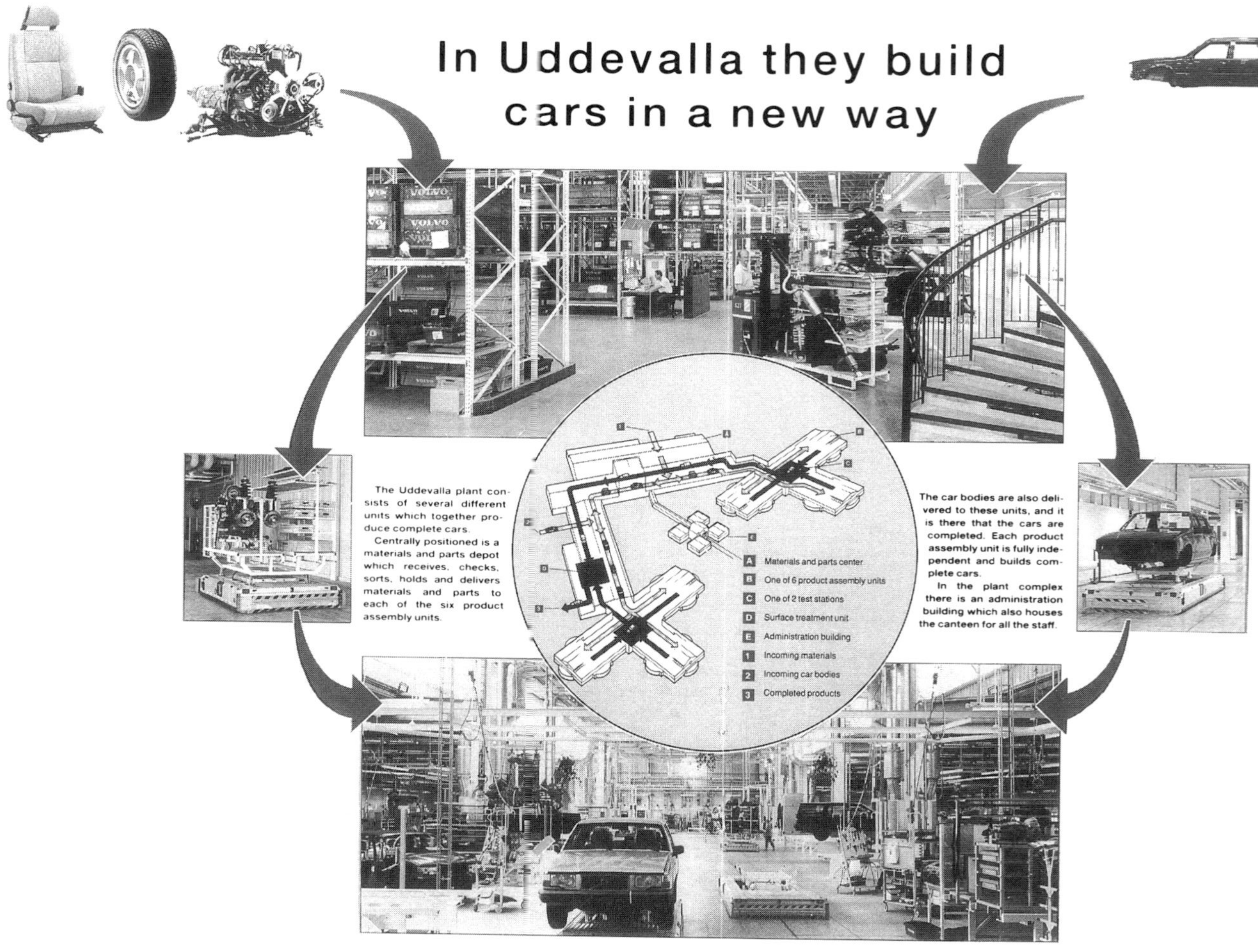

Figure 5 The concept of the Uddevalla plant.

Ergonomic Research and Applications in Sweden 741

Figure 6 At the Uddevalla plant, 80% of the time the car was built in an upright ergonomically optimum body posture.

the upper part of the infraspinatus muscles have a zone of avascularity. This has been found to be the site of microruptures and degeneration that may be accelerated by aging.

Impairment of the venous circulation may occur when the humeral head compresses the tendons (elevated arm) but also when there is increased tension in the tendon. Tendon inflammation has been provoked by repetitive contractions in rabbits.

Figure 7 The engines at the Skövde plant are built on platforms with individually adjustable fixtures.

Figure 8 Manual polishing of turbine blades.

Degenerative tendinitis in the shoulder girdle aroused by exertion, for example, may trigger a foreign body response inflammation.

Tenosynovitis is an inflammation of the tendon sheath and its synovia. In the long biceps tendon, this may be caused by the tendon and its sheath rubbing against the lesser tuberosity during overhead movements.

Postinfective arthritis as well as tendinitis may presumably predispose a person exposed to shoulder stress to a more severe reaction.

Muscle tenderness, myofascial syndrome, trapezius myalgia, and related disorders are obscure conditions because the pain does not originate from the contractile muscle fibers themselves. It may possibly derive from pain fibers within the blood vessels or the connec-

Figure 9 A deburring robot.

tive tissue. Hagberg [17] points to three pathophysiological routes. The first is mechanical failure with ruptures of Z-disks probably caused by temporary high local stress. The second is local ischemia due to the impairment of the circulation by continuous muscular performance, which may already occur at 10–20% of the maximum voluntary contraction. This leads to a fall in pH and reversible enzyme inhibition. It is postulated that the tissue irritation causes extravasation of blood, edema, and fibrositis in some individuals. Highly repetitive work may then possibly cause cumulative trauma to the muscle cell [thence cumulative trauma disorders (CTDs)] affecting both morphology and energy metabolism.

The third pathophysiological route would be energy metabolism disturbance. Energy depletion in the muscle cell has been suggested as one factor in muscle pain. Defects in the energy metabolism are often associated with painful disorders of the muscle. Laboratory experiments involving repetitive shoulder flexions have produced energy depletion as indicated by an increasing serum creatine kinase and accompanying pain. It is hypothesized that this may also be important for static loads. The possibility of certain primary metabolic disturbances in certain individuals has also been proposed.

The carpal tunnel syndrome is a textbook example where the injury mechanism is clearly understood. Friction-caused synovitis of the tendon sheaths in the carpal tunnel causes pressure on the median nerve [18]. Another example is supraspinous tendinitis, where the supraspinous tendon is pressed against the acromion in a space that has been limited by inflammation [19].

Fibrositis/fibromyalgia and generalized muscular pain is a common condition but with a poorly understood pathogenetic mechanism. The theory that static muscular work causes ischemia in the muscle, creating morphological changes in the muscle fibers, has been suggested but remains to be proved [20]. It is easy to prove muscular fatigue and pain after minor static load [21,22], but the pathogenic link from repetitive chronic muscular fatigue to permanent damage and chronic pain remains to be shown.

Microfractures have been suggested by Hansson et al. [23] to be the reason for back pain in certain individuals, but no one knows how common this is. The lumbar disk and its degeneration have been connected to low back pain, and the intradiscal pressure as measured by Nachemson and Elfström [24] was for a long time the theoretical mechanism for ergonomic advice on lifting but has since been abandoned by Nachemson [25].

Spinal shrinkage as measured from height before and after loading provides a method for measuring mechanical load on the spine [26]. The shrinking is dependent on the elasticity of the intervertebral disks. However, no one knows if the shrinkage leads to permanent disk problems or whether this may be a pathogenetic mechanism for back pain.

2. Recent Research

Overview. Recent Swedish research on ergonomics has broadened the concept from physical workplace adjustment to psychological factors [27] and sociological aspects of work groups and leadership [28].

A comprehensive literature survey and analysis of health risks at work has been performed by the National Swedish Commission for the Work Environment [13]. Its conclusion is that the prevalence of and risk factors for many disorders in the locomotor system are still not satisfactorily known.

Case Studies. Evaluations of the use of an ergonomic chair, the Ullman chair, as a secondary prophylaxis for low back problems has shown significantly less pain and fewer sick days for low back pain in this group than in a randomized control group using a traditional office chair in a 1-year prospective study [29]. Also, prophylactic inversion therapy (autotraction) by the Swedish Mastercare Inversion System (Svenska Hälsobänken) with self-training for 10 min daily has been shown to reduce the pain score significantly in chronic low back pain patients in a 1-year study at Volvo Aero Corp. [30].

B. Rules and Regulations

1. The Workers' Compensation System

The Swedish workers' compensation system introduced in 1977 has been one of the most liberal in the world. This may be one explanation why Swedish industry, in spite of substantial improvements in the physical work environment and ergonomic advances, has had about the highest rate of work injuries in the world.

The acceptance of a work injury is based on two steps, the first being that a disorder should be known to be caused by work. If that was the case, in the next step even a very low probability would allow a claim to be recognized as a work injury. This caused an explosion of filed claims for most pain problems involving the locomotor system, and over 90% were accepted as work injuries [6].

The acceptance of a work injury meant full salary during sick leave, and, in addition to this, compensation of pain and suffering and medical disability would provide the claimant with considerably more money than the regular salary and there was no incentive to return to work. There are several persons in Sweden with claimed muscular pain in the neck and shoulders that, at less than 30 years of age, have gone into early retirement in spite of few objective findings! In July 1993, the law was changed so that it now takes a high probability for work to be considered the causal factor of medical complaints.

2. The General Welfare System

Sweden has for many years had a very liberal and generous health insurance paid by the government through equal fees from all employers. Although the Swedes have one of the highest life expectancy rates in the Western world, we also have one of the highest rates of sickness absenteeism. Although the Volvo plants had an average sickness rate of 12% among their blue-collar workers in 1991, it is important to know that two-thirds of these workers take sick leave less than 1 week/year. In 1991, the law was changed so that employers pay directly for only the first 2 weeks of sick leave, and the first 3 days of sickness compensation was reduced from 90% to 75% of the worker's salary. In 1992, the sickness rate decreased by 1.3% for blue-collar workers. When, in April 1993 a one-day waiting period (with no pay) was introduced for health insurance, the number of sick reports per day at Volvo Aero Corp. was reduced by about 50% (see Figure 10). Against this background it is probable that the real benefit of ergonomic applications in the work environment has been shadowed by effects of the health insurance system.

IV. CRITICAL REVIEW OF CURRENT STATUS

A. The Nachemson Back Review

In a major survey of international studies on low back pain, Nachemson [25] claims that very few prospective and controlled studies on work-relatedness exist. He concludes that even if the effect of physical factors cannot be denied, work satisfaction seems to be the most important predictor of future work-limiting back problems. The importance of early and active rehabilitation is emphasized.

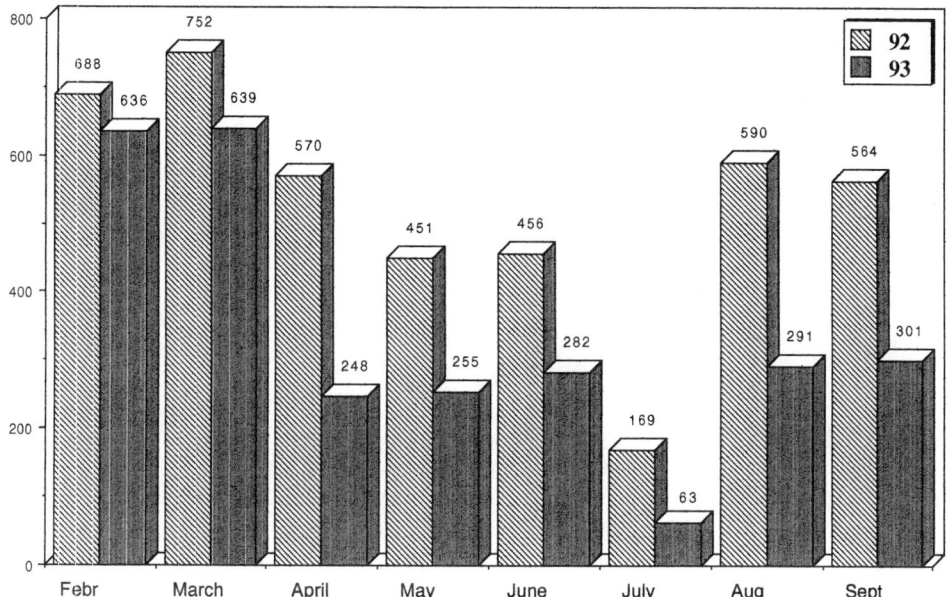

Figure 10 Sick reports by Volvo Flygmotor (3000 employees) during the time of a change in the compensation system. One day without pay was introduced in April 1993.

B. The Svanborg Study of 70-Year-Old People

In one of the largest cross-sectional studies published to date, Svanborg [31] interviewed in great detail and physically examined some 700 70-year-old men and women in Göteborg. He concludes that there was no difference in the prevalence of pain problems in the locomotor system between the workers who had performed heavy manual labor and those who had been office workers. The only difference noted was that the former were physically stronger and had higher bone density. The group has since been followed for 10 years. At age 79, previously sedentary workers were more disabled in activities of daily living than those whose work had been physically strenuous.

C. Methodological Problems in the Evaluation of Implemented Solutions

Hagberg et al. [32] pointed out some methodological aspects that call for brief comment. The evaluation of doses of exposure to repetition injuries and the dose–response correlation is important. The diagnostic criteria should be stringent. Unfortunately, only a few clinical tests have been validated, and these clinical tests may be helpful for confirming a diagnosis but not for excluding a disease [33]. Ideally, an age-, gender-, and education-matched control group should be used to evaluate suggested ergonomic changes because over a period of time job content may change, new work processes may be implemented, some workers may be transferred to other departments, and so on, which may be the real reason for improvements in health rather than the ergonomic solution. Also, time could be a healing factor that coincides with ergonomic change. Other factors such as tobacco use and alcohol consumption should also be checked [34]. The use of visits to physicians as an endpoint for measuring the prevalence of certain conditions such as tennis elbow at the workplace is not a good method. Workers with tennis elbow are more prone to see a doctor for their problem if they have a heavy job even if the real prevalence of tennis elbow is the same among workers in heavy and light jobs [35].

Swärd [36] studied top athletes and found that 85% of male gymnasts experience some low back pain and 54% have severe low back pain. Still they can operate at the top international level. Sickness absenteeism as a measure of the problem is thus seen to be affected by many psychological and motivational factors.

V. FUTURE CONCERNS

A. Finding the Answers

The criticism from Nachemson, the Svanborg study, and my own experience indicated that the irreversible chronic low back pain syndromes, neck/shoulder pain, fibrositis, etc. in many cases have a causative background (metabolic, infectious?) other than work. Work may aggravate aches and pains, but the aches and pains probably had another origin. It is important to distinguish between the pathological process and the provoking factor that produces pain. In such cases, ergonomic solutions may be of some help to the individuals but will usually not solve the problem.

The typical muscle strain work injury should be improved if working conditions are improved by, for instance, ergonomic changes. To be a candidate for ergonomic intervention, an injury should comply with most of the nine criteria stated for causation by Sir Bradford Hill [37]:

1. *Strength of association.* High odds ratios mean a strong association.
2. *Consistency.* Repeatedly observed by different investigators in different countries.
3. *Specificity of the association.* The injury is limited to the exposed persons.
4. *Temporal relationship.* Effects appear *after* the cause.
5. *A dose-response curve.* The higher the dose, the greater the number of cases.
6. *Biological plausibility.* The biological mechanism should be understood.
7. *Coherence of evidence.* Cause and effect should not seriously conflict with the generally known facts of natural history and biology.
8. *Reversible association.* As an example, when smokers stop smoking cigarettes, the rate at which they develop lung cancer should fall.
9. *Analogy.* After having discovered a drug effect of thalidomide on fetuses, we are more ready to accept fetal effects by other drugs.

The missing link of a pathologic process may suddenly appear. The etiologic process is better understood when, as in the case of rheumatoid arthritis, the immunologic mechanism is discovered. New imaging techniques such as magnetic resonance imaging (MRI) make it possible to visualize soft tissue and verify, for instance, changes in the carpal tunnel that explain the pain. The ultimate prerequisite for an ergonomic solution is that the pathogenetic mechanism is understood in detail.

The Swedish National Commission for the Work Environment (Arbetsmiljökommissionen, 1989 [13]) suggests that a systematic follow-up of implemented ergonomic solutions should be performed. Many suggested modifications are based more on individual ideas than on a solid base of knowledge.

B. Implementing the Solutions

Compiling and using collected examples of evaluated ergonomic modifications such as those published by Oxenburgh [38] is one way of implementing ergonomic solutions. A possible vision is that the next generation of intelligent, seeing, and feeling industrial robots will take over all the highly repetitive tasks that continue to provoke discomfort and pain in human workers.

ACKNOWLEDGMENTS

I particularly wish to thank Ulf Jeverstam, AM Volvo, for valuable advice on the manuscript; Mats Ericson, Royal Swedish School of Technology, for critical comments; Dennis Savage for reviewing the language; and Sture Hård, Gothenburg University Medical Library, for help with the references.

REFERENCES

1. Arbetarskyddsstyrelsens författningssamling (Swedish), The Swedish Ordiance Concerning Work Postures and Working Movements, 1983:6.
2. *Statistics Sweden*, 1993.
3. Parliament and Ministries 28 (1991).
4. SAF:s statistik om företagshälsovård (1991).
5. J. L. Kelsey and R. J. Hardy, Driving of motor vehicles as a risk factor for acute herniated lumbar intervertebral disc, *Am. J. Epidemiol. 102*(1):63–73 (1975).
6. The National Board of Occupational Health and Safety (Sweden), ISA, 1993.
7. L. Stenudd, Investigation of the occupational health services of Procordia, manuscript (Swedish), 1993.
8. K. Noro, *Participatory Ergonomics: Concept, Advantages and Japanese Cases in Human Factors in Organizational Design and Management*, North-Holland, Amsterdam, 1990, pp. 83–86.
9. L. Wallin, Modification of work organization, *Ergonomics 30*(2):343–349 (1987).
10. S. Kvarnström, Occurrence of musculoskeletal disorders in a manufacturing industry with special attention to occupational shoulder disorders, *Scand. J. Rehab. Med. Suppl. 8*:1–144 (1983).
11. L. Dimberg, A. Olafsson, E. Stefansson, et al. The correlation between the work environment and the occurrence of cervicobrachial symptoms, *J. Occup. Med. 31*(5):447–453 (1989).
12. Å. Kilbom and J. Persson, Work technique and its consequences for musculoskeletal disorders, *Ergonomics 30*(2):273–279 (1987).
13. Å. Kilbom and M. Hagberg, *Arbeten utsatta för särskilda hälsorisker,* Arbetsmiljökommissionen, (Swedish), 1989.
14. S. H. Snook, Approaches to preplacement testing and selection of workers, *Ergonomics 30*(2):241–247 (1987).
15. M. C. Battié, The reliability of physical factors as predictors of the occurrence of back pain reports. A prospective study within industry, Thesis, Göteborg, 1989.
16. A. Nachemson, I. Lindström, C. Öhlund, et al., Symposium om "ryggproblemens lösning," *Läkartidningen 85*(50):4437 (1988) (in Swedish).
17. M. Hagberg, Occupational musculoskeletal stress and disorders of the neck and shoulder: a review of possible pathophysiology, *Int. Arch. Occup. Environ. Health 53*(3):269–278 (1984).
18. G. Lunborg, *Nerve Injury and Repair*, Churchill Livingstone, 1988.
19. P. Herberts, R. Kadefors, C. Hogfors, et al., Shoulder pain and heavy manual labor, *Clin. Orthop. 191*:166–178 (1984).

20. K. G. Henriksson, Exercise induced myopathies in man, *Opuscula Med.* 28(2):34-37 (1983).
21. M. Hagberg, On evaluation of local muscular load and fatigue by electromyography, *Arb. Hälsa* 24:1-53 (1981).
22. B. G. Jonsson, J. Persson, and Å. Kilbom, Disorders of the cervicobrachial region among female workers in the electronics industry. A two-year follow up, *Int. J. Ind. Ergon.* 3:1-12 (1988).
23. T. Hansson, T. Keller, and R. Jonsson, Fatigue fracture morphology in human lumbar motion segments, *J. Spinal Disord.* 1(1):33-38 (1988).
24. A. Nachemson, and G. Elfström, Intravital dynamic pressure measurements in lumbar discs. A study of common movements, maneuvers and exercises, *Scand. J. Rehab. Med. (Suppl.)* 1:1-40 (1970).
25. A. Nachemson, *Ont i ryggen, orsaker, diagnostik och behandling*, SBU, 1991, (in Swedish).
26. M. Ericson and I. Goldie, Spinal shrinkage with three different types of chair whilst performing video display unit work, *Int. J. Ind. Ergon.* 3:177-183 (1989).
27. T. Sivik, Diagnosis and treatment of patients with idiopathic low back pain, Thesis, University of Göteborg, 1992.
28. L. Dimberg, L. Wallin, and B. Eriksson, Unpleasant atmosphere at work increases the risk of musculoskeletal disorders, *Läkartidningen* 88(11):981-985 (1991) (in Swedish).
29. L. Dimberg, M. Ericson, and J. Ullman, The effects of low back pain and disorders by sitting in non-traditional chairs with forward sloping seats compared to traditional chairs with horizontal seats, *Proc. Swedish Med. Assoc. (Riksstämman)* 46:121 (1992) (in Swedish).
30. L. Dimberg, L. G. Josefsson, and B. Eriksson, Prophylactic inversion therapy in chronic low-back pain patients in the workplace—a prospective, randomized controlled study, *Proc. Swedish Med. Assoc. (Riksstämman)* 9P:114 (1993).
31. A. Svanborg, 70-Year-old people in Gothenburg, Sweden: a population study in an industralized Swedish city. Practical and functional consequences of aging, *Gerontology* 34(Suppl. 1):11-15 (1988).
32. M. Hagberg, L. Jorulf, and Å. Kilbom, *Methodologic Problems in Muscular Strain Epidemiology*, Natl. Inst. of Occupational Health, Sweden, Report 19, 1988 (in Swedish).
33. E. Viikari-Juntura, Interexaminer reliability of observations in physical examinations of the neck, *Phys. Ther.* 67:1526-1532 (1987).
34. L. Dimberg, A. Olafsson, E. Stefansson, et al., Sickness absenteeism in an engineering industry—an anlysis with special reference to absence for neck and upper extremity symptoms, *Scand. J. Soc. Med.* 17(1):77-84 (1989).
35. L. Dimberg, The prevalence and causation of tennis elbow (lateral humeral epicondylitis) in a population of workers in an engineering industry, *Ergonomics* 30(3):573-579 (1987).
36. L. Swärd, The back of the young top athlete. Symptoms, muscle strength, mobility, anthropometric and radiologic findings, Thesis, Univ. Gothenburg, Sweden, 1990.
37. B. Hill, Association is not causation, in *Principles and Practice of Research for Surgical Intervention,* Troidl, Spitzer, McPeek, et al. Eds., Springer-Verlag, New York, 1986.
38. M. Oxenburgh, Increasing production and profit by health and safety, 1991.

38
Overview of Ergonomic Needs and Research in India

Rabindra Nath Sen
Ergonomics Laboratory, University of Calcutta, Calcutta, India

I. INTRODUCTION

Ergonomics has been defined as the science, technology, and art of man at work [1]. Its aim is mainly to optimize human-machine-environment interactions to get the maximum efficiency, productivity, and improvement of the working conditions and working life of the workers. Though the basic principles of ergonomics have been practiced by workers in India from prehistoric times without the specialized knowledge of ergonomics, their formal application started only a few decades ago.

II. HISTORICAL BACKGROUND AND SIGNIFICANCE OF OCCUPATIONAL ERGONOMICS

India is the second most populous country of the world, with about 920 million people, and its unemployment rate is huge even at the present time. A 3% annual population growth rate amounts to a 19-fold increase in a century. India was formerly a predominantly agricultural country, with about 80% of the population directly or indirectly involved in agriculture. Only since its independence from British rule and the initiation of the Second Five Year Plan has it experienced industrial development. Hence, the application of occupational ergonomics is becoming more and more necessary.

Among the many constraints in industrial development, financial constraints are the most important. Hence, in industrially developing countries (IDCs) such as India, the application of ergonomics must in the initial stages consider the implementation of improvements whose cost will be low or negligible. The significance of the use of occupational ergonomics in the progress and development of the country as a whole, not only in industry but also in agriculture and other areas, is unquestionable.

III. SOME ASPECTS OF ERGONOMICS STUDIES IN INDIA

The work done in India regarding several ergonomics studies may also be applicable to other IDCs. Sen [1,2] and Kogi and Sen [3] reviewed the research work done in India in the following areas:

1. Energy expenditure of the workers in different types of jobs in various factories, and classification of heaviness of jobs on the basis of physiological responses.
2. Anthropometric measurements in relation to the design of machines, tools, implements, and consumer products [1,2].
3. Problems of agricultural and other unorganized workers with special reference to manual materials handling [4-15].
4. Ergonomic solutions of problems in small-scale and cottage industries [16].
5. Low-cost improvements of working conditions, safety, health, and welfare of the workers [17].
6. Postural analysis [18].
7. Training programs for workers, supervisors, trade union officials, managers, government executives, etc., to bring awareness of ergonomics to people at all levels including the general populace.

Because of the huge population of India, any slight ergonomic improvement at the individual level would yield in total a very significant quantitative effect. In the unorganized sectors such as in agriculture and among manual laborers in various vocations—e.g., masons, carpenters, blacksmiths, construction workers, household workers, and service workers such as laundry workers, barbers, handloom workers [19], cobblers, and about 8 million handicrafts workers in India [20]—the use and application of ergonomics are much less common than in the organized sectors. This is due mainly to a lack of awareness of the basic principles, poor economic conditions, and reluctance to change existing and traditional work methods and tools.

A. Agricultural Ergonomics

1. Use of Tools and Implements and Their Design

Several studies have been conducted for the improvement of the existing designs of tools, implements, and aids such as the sickle [5], plough [6,7], shovel [8,9], spade [10], combined shovel and hoe [11], and "float-seat" [12] used in agricultural work.

2. Use of Pesticides

In agriculture, the increased use of pesticides, mainly spraying without adequate protection, has increased the health hazards to agricultural workers by severalfold.

3. Use of Work Methods

Sen and coworkers [13-15,21,22] have studied the various traditional methods used in agricultural work.

4. Use of Modern Sophisticated Machines

Traditional agricultural workers are unfamiliar with modern agricultural mechanized systems such as the use of the combine harvester and thresher, and modern work methods [13]. Similarly, traditional industrial workers are not acquainted with modern sophisticated industrial machinery, complex tools, and work methods. Many work-related diseases occur very frequently. There are very few statistical data on accidents in either agriculture or industry, especially from small-scale and medium-sized organizations [23]. Cultural, social, and economic differences make the problems of IDCs more difficult to solve than those of the developed countries. This poses a great challenge to professional ergonomists.

B. Industrial Ergonomics

Like most other IDCs, India is traditionally agriculture-based and has evolved into an IDC only in the last few decades.

1. Ergonomic Design of Factory Buildings

There are several monographs on the design of factory building in cold climates. However, there have been practically no studies on the ergonomic design of factory buildings suitable for a tropical climate characterized by high ambient temperatures and high relative humidity.

In any industry—big, medium-sized, or small—it is very important to plan much in advance the layout, design, and orientation of factory buildings. Sen [24] considered in detail the ergonomic design of factories suitable for the tropical climate of India. He stressed the importance of orienting the building to provide good natural ventilation and daylight, to reduce solar radiation by sun shades and double roofs, to circulate cooler underground air, and other factors. He emphasized the usefulness of a correct layout of the shop floor with respect to the positions of windows, fans, floor fans, materials, machines, workstations for different operations, storage, shipping, etc., so that efficient movement of personnel and a quicker flow of materials and products could be effected. It is imprudent to spend a huge sum of money to change a badly designed factory building at a later stage. It always pays to plan and design factory buildings ergonomically well in advance of the start of production.

There is an acute shortage of housing for industrial workers. Whereas most of our public sector employers have tried to provide some housing for workers, the records of private sectors are not at all encouraging.

2. Reduction of Heat Stress

Engineering control for the reduction of heat stress is very expensive in most cases. Hence, ergonomic designs of low-cost thermal barriers and low-cost personal protective clothing are of considerable importance.

Thermal Barriers. In improving working conditions, various types of thermal barriers, personal protective devices, floor fans, "man coolers" (big, high-speed air circulating fans), etc., have been designed to reduce the effects of high heat stress, especially for workers in the steel industry, engineering work, glass factories, and the like.

To protect against thermal radiation from furnaces, molten metal, slag, etc., in various factories, low-cost barriers against thermal radiation have been fabricated from two oxidized iron sheets, each about 1 square meter in size with a gap of 25 cm between them. The two sheets are held together by horizontal iron rods at four corners and at the middle. These portable thermal barriers are put on low-cost stands with different points of suspension to permit placing them between the sources of high thermal radiation and the workers. The hot air rises in the gap between the two sheets to significantly reduce the effects of thermal radiation [17].

Molten Metal or Slag Channel Cover. To reduce the thermal radiation from red hot molten metal or slag flowing in specific channels, small metal covers hinged on one side have been fabricated for the channels; these covers can be raised during maintenance and cleaning of the channels. The channel covers also act as guards against splashes of metal, sparks, etc. to effectively reduce the risks of burns and other accidents from spatters of molten metal [17].

Personal Protective Devices. The acceptance of personal protective devices, equipment, and protective measures for technical devices depends substantially on their effective functionality and ergonomic design.

Ergonomic designs of low-cost, washable, ventilated (by a "man cooler" at the back), vapor-permeable, flame-retardant, radiant-heat-reflecting type of special work clothing have been made for hot process workers such as furnace tenders and helpers in glass factories, and furnace tenders, teemers, chargers, cleaners, launderers and bull-ladle operators in hot metal factories to reduce heat stress and physiological costs of work. The new special clothing was found to be 3 to 4 times more comfortable than the existing workwear, as revealed by worker polls. The use of this work clothing increased productivity in quality and quantity, especially during the summer months [3,25-27]. An ergonomically designed low-cost transparent face shield and fire-resistant, hard-toe safety boots with a hinged wooden sole are also suitable for workers working on the top of a coke oven battery and for furnacemen to protect them from thermal radiation and from injuries from metal splashes or sparks and to increase efficiency and productivity. Further research is necessary for ergonomic designs of similar low-cost personal protective devices for work situations having much more intense heat stress.

Sen and Das [43] ergonomically designed a manual metal arc (MMA) welders' screen and the protective work clothing for reducing the workload and risks of damage by ultraviolet radiation, shielding also against toxic fumes generated in the welding process, and reducing static local muscular fatigue. The improvement was due to the elimination of hand movements by the hand-held screen of the old design because the lower-jaw-controlled screen window protecting against ultraviolet rays was fixed in the helmet in the new design.

The welder's electro-optical protection filters and the welder's curtain must conform to the needs of labor safety, comfort of wearing, and good ergonomic layout of the workplace.

3. Manual Materials Handling

From time immemorial, one of the main problem areas in industry has been manual materials handling. Use of human labor is very extensive in IDCs like India. Use of machinery reduces the workload considerably for a worker and saves a lot of time, but it involves very high financial costs. It is often so expensive to mechanize and automate a manufacturing system that human adaptability makes it cheaper to use people for the awkward jobs such as materials handling and inspection. In a country with very cheap labor, a medium-sized or small-scale industry always uses workers in manual materials handling in the traditional manner.

4. Use of Ergonomic Designs of Tools and Implements

Another way to bring about improvement is to use ergonomically modified designs of existing tools and implements so as to enhance their efficiency. Sen and coworkers [2,8-11] made attempts to improve the designs of the traditional shovel (Figure 1), spade, and shovel cum hoe to increase productivity and to reduce physiological costs to the workers by minimizing the degree of stooping required and enhancing leverage.

During the use of a standard shovel in lifting and handling materials such as sand and stone chips, the protracted bending or stooping posture causes backache and fatigue. An additional handle fitted at the base of the shovel significantly reduced the bending of the back and simultaneously increased the throwing distance by about 25% due to the improvement in leverage.

Sen [29] similarly improved the existing designs of the beater and the ballast rake used for the manual maintenance of railway tracks.

Efforts were also made to modify the designs of bullock-cart [30] and hand-pulled carts [42] and to use an ergonomically designed special harness [31] for the pullers.

5. Mechanical Materials Handling

Special Cart for Moving Heavy Materials. In a company manufacturing ceramic bricks, an ergonomically designed special cart for transporting the bricks to the drying chamber

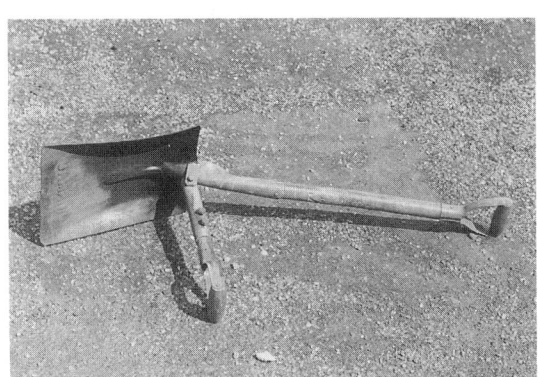

Figure 1 An additional handle fitted at the base of a traditional shovel significantly reduces the bending of the back and simultaneously increases the throwing distance by about 25%. A removable clip designed to temporarily hold the second handle when the shovel is used is not shown.

on a metal plate was fabricated from low-cost bicycle parts and iron angles, which reduced the efforts of the workers by about 25% and increased productivity by about 15% [17].

Crane Design. In mechanical materials handling, improvements were also made in the ergonomic design of a chain pulley system, forklift truck, platform truck, and overhead transport crane [28,43]. In many of the factories there was simultaneous use of different designs of overhead transport cranes that had different types of controls and displays according to their positions, purpose of use, distance from the operator, etc., and there were many control-display incompatibilities. The need for the same worker to use different types of cranes on the same day resulted in an increase in the risk of accidents due to the greater chance of making a mistake in operating the controls. Sen and coworkers [28,43] suggested standardizing the positions of controls and displays and changing their locations so that the normal motion stereotypes are maintained and there are no risks of accident. There is scope for similar improvements in the design of tools, implements, trucks, crane cabins, etc.

C. Application of Ergonomics to the Improvement of the Quality of Working Life

The issue of the quality of working life will become the major concern of the people of India in the coming few decades. A more participative and humanizing approach to labor relations will be helpful only in the presence of a cooperative attitude of top management, who should foresee the benefits of such work redesign.

Ergonomists should undertake some studies relevant to improving the quality of working life. The challenge of change is faced by all big and small undertakings and enterprises.

Much of the information collected and knowledge gained should encourage enterprises who embark on innovations to give due prominence to the human aspects of the proposed changes so as to ensure both the advances that are sought and continued improvement in the quality of working life of the people involved.

1. Cosumer Ergonomics

In most of the industries manufacturing products and consumer goods, much stress has been laid on the economic and aesthetic aspects, with very little or no attention paid to the functional and ergonomic aspects. A change in this attitude is desirable.

2. Participatory Ergonomics

In industry, there are many problems that could easily be solved by collaborative, cooperative, and participatory efforts among ergonomists, factory managers, other officials, supervisors, foremen, and workers. Typical examples are the various low-cost modifications and improvements of existing work conditions, methods, and tools suggested by various experts in IDCs such as India, Sri Lanka, Indonesia, Singapore, Thailand, Malayasia, and the Phillippines, for a project supported by the International Labour Organisation [17].

Studies on participation of workers in the management of problems concerning work methods, machines, tools, and the work environment including the effects of noise [32,33] have shown that, as in Japan, job satisfaction and high morale are possible only in small working units. This is encouraging a growing number of industrial companies to re-design their larger organizations to form a collection of interlocking small groups.

Considerable attention has been paid in recent years, by those interested, to change the attitude to work and restructure with innovations. Ergonomics is being applied to eliminate boring repetitive jobs [4], create more interesting work and thus to avoid physical and mental fatigue.

3. Ergonomics Application in Information Technology

Though the development of modern management with radical improvements in information technology and instant communication has provided effective and dynamic corporate functioning and growth, yet it has also added to stresses and strains due to the operation of VDUs and other components of office automation. The stress includes all manners of pressure—physical, physiological, sociological, and/or a combination of all these and other relevant factors. It is well accepted that only a relaxed and healthy mind in a healthy body can cope with demanding responsibilities. With explosive use of information technology in IDCs like India, it is essential that ergonomically bad design aspects are excluded during the technology transfer from industrially developed countries.

No system can claim to be the best; no organization possesses the one best form. There is a continuing interaction between all parts of the system. All parts of the system must therefore be developed together. The origin of a new philosophy of enterprise is to be found in the dramatic changes in India; these changes are evident in human relations in all aspects of living.

Employers and/or governments must realize that though workers may lack scientific training, knowledge, or familiarity with technical jargon, it is they who experience the hazards at first hand. They must also recognize that it is often the pressure to produce more, and not recklessness or carelessness, that causes accidents and exposes workers to ill health. Impractical goals and unrelenting drives force the workers to work faster than the safe limits and thus experience accidents.

Often trade unions too do not bother very much about shortcomings in enforcement of health and safety regulations. They are busy negotiating for higher wages and other monetary benefits. The trade union leaders should be put through special orientation and educational programs to make them aware of the importance and intricacies of safety and health issues. The message of the safety movement will have to reach every factory, farm, and even the workers on roadways and railways. Of late, there has been an increase in the number of accidents in the railway and road transport industries. Most of them could probably be avoided with better preventive action. Even work on farms, which used to be relatively safe from accidents, is becoming more and more hazardous because of the use of poisonous chemicals and pesticides and unsafe machines.

4. Ergonomics Application for Low-Cost Improvements

In the industrial sector, Sen [17] has suggested low-cost improvements in the work process, work organization, and working conditions of various industries such as textile, soap, food processing, and engineering.

Today many industrial managers believe that in order to achieve optimum efficiency in a plant, it should be kept rather smaller than what has generally been thought to be the optimum size.

5. Ergonomics Application in Transport Systems

Railways Transport. In railway transport Sen and Ganguli [34,35] considered improvements in the design of both diesel and electric railway locomotive drivers' cabins—control display aspects, the positions of primary and secondary controls and displays—the seating arrangements of the drivers and the assistants and provision of their routine requirements. Sen [2,29] also stressed the need for ergonomic modifications of tools for railway track maintenance, existing ticket counters for punching tickets, the ergonomic design of signaling and signal control rooms and arrangements of non-air-conditioned and air-conditioned passenger coaches.

Motor Vehicle Transport System. Sen and Nag [36] considered ergonomic improvements for double-decker public buses that were blind copies of London buses and consequently unsuitable for a tropical climate, having minimum facilities for ventilation and very high thermal radiation due to the metal body being heated by solar radiation. Sen and Nag [37] suggested several improvements such as higher ventilation, greater ease of passenger flow inside and out of the bus, better seating arrangements for the passengers as well as the driver, improved control and display arrangements for the driver, and greater access of the driver and the passengers into the bus.

Public Transport. Similarly, application of ergonomics is needed for better designs of cars, minibuses, trucks, and nonmotorized transport for rural areas such as cycle rickshaws, bullock carts, handcarts, cycle vans, and bicycles. Future research is also required on the design of roads and roadsigns to reduce the high rate of traffic accidents in India.

Nonmotorized Transport Systems. Particularly in rural areas and in some urban and suburban areas in India, most passengers, goods, patients, etc., are transported in three-wheel cycle rickshaws manually operated by pullers [38]. Sen [38] carried out studies on improving the design of hand-pulled and cycle rickshaws.

Cycle rickshaws form one of the most widely used transport systems in both urban and rural areas throughout the country for short journeys. Cycle-rickshaw pulling is an important occupation for millions of people. The main advantages of using cycle rickshaws are

1. It causes almost no environmental pollution.
2. It is the only transport available in some instances and at some places.
3. It is not very costly.
4. It provides self-employment to a large number of people.

In the years to come, research into improvements in the design of cycle rickshaws and the extent of such improvements should be ergonomically evaluated for implementing critical improvements from the anthropometric, physiological, and other relevant points of view. Design improvements in the line of new regenerative brakes, energy-storing devices, noncircular or elliptical or multisprocket drives, directional and positional stability, reduced cycle-rickshaw weight, lower center of gravity, and similar aspects constitute a few of the areas of possible improvement.

IV. INTERNATIONAL SYMPOSIA AND MEETINGS ON ERGONOMICS

In the international symposia on work physiology and ergonomics and on applied physiology and ergonomics held in Calcutta in 1974 and 1983, respectively, various problems faced by the developing countries were discussed [39]. A review of papers from a number of industrially developing countries confirmed that the main ergonomics effects are directed toward the analysis of work problems and the implementation of cost-efficient workplace improvement.

V. ERGONOMICS EDUCATION AND TRAINING PROGRAM

In IDCs, ergonomics education at the school, college, and university levels is as important as bringing awareness to the individual and the public by using mass media such as radio, television, newspapers, and magazines. It is very important to make the contents of each educational program suitable for those for which it is intended. Ergonomic ways of using audiovisual aids, suitable examples through practical demonstrations, and the organization of popular exhibitions would go a long way toward bringing the principles of ergonomics to the persons concerned and elucidating the scope and benefits of its applications in day-to-day activities in all spheres of life.

Invariably, the workers are the worst sufferers in all types of accidents. There has to be health and safety education for industrial workers as well as for supervisors and managers, both separately and jointly, particularly in accident-prone industries, such as the jute [40], mining, chemicals, fertilizer, engineering, and construction industries. There is as yet no such sustained educational program.

Social attitudes and practices cannot be forced to change through strict legislation. What is needed is a genuine change of heart and a scientific approach through awareness and training. These take both time and a lot of effort and cannot be achieved overnight.

VI. ERGONOMIC ORGANIZATIONS IN INDIA AND THEIR ROLES

The pioneering research work on various aspects of work physiology and ergonomics such as anthropometry and energy costs of different activities including basal metablic rate (BMR) of industrial workers was conducted in the Department of Physiology, Presidency College, Calcutta beginning in 1953. Similarly, the Section of Physiological and Industrial Hygiene in the All India Institute of Hygiene and Public Health and the Department of Physiology, University Colleges of Science, Technology, and Agriculture, University of Calcutta started research on work physiology in the 1960s. Various institutions including the Defence Institutes of Physiology and Allied Sciences, Delhi; Industrial Physiology Division of Central Labour Institute, IIT and NITIE, Bombay; and the National Institute of Occupational Health, and NID, Amhedabd, under the Government of India and other institutions supported by the Indian Council of Medical Research (ICMR), Indian Council of Social Research (ICSR), Council of Scientific and Industrial Research (CSIR), Department of Science and Technology (DST), and others carried out projects along similar lines. In 1970–1971, the University of Calcutta started the first postgraduate course in India on ergonomics and work physiology with limited resources. It is still the only university where this postgraduate specialty is taught, though several organizations run short-term orientation and refresher courses on work physiology and ergonomics.

More centers of research and training should be established in the rural and urban sectors to apply the multidisciplinary approach of ergonomics to solve various problems of both the organized and unorganized sectors by optimizing the human-machine-environment interactions to utilize human resources most efficiently without adverse effects.

Scientific and professional societies like the Indian Society of Ergonomics (ISE) and academic institutions should thus play a very important role in exploring science and technology applications to generate awareness of multidisciplinary ergonomics, occupational health, safety, and the environment.

As the interactions and the collaborative activities among the various institutions, industries, and agricultural institutes are very meager, steps must be taken to increase these interactions to have mutual benefits in solving pressing problems.

Ergonomics should be introduced in a very simple form even in the school curricula. Students are capable of enthusiastically applying ergonomics in project work to produce successful designs for various purposes.

An ergonomics survey of India similar in scope and function to the anthropological, zoological, and botanical surveys of India should be constituted to undertake comprehensive and large-scale surveys at the national level for the establishment of ergonomic norms of the Indian people. These should consider the different ethnic groups and habitual activities, static and dynamic body dimensions, abilities, performance and environmental standards, for various working conditions.

VII. ERGONOMIC DESIGN OF HOSPITALS AND HEALTH CARE UNITS

The ergonomic design of hospitals and health care units for a large number of patients, in both rural and urban areas, is one of the most important areas that need the attention of ergonomists to improve the efficiency and productivity of health care personnel and the quality of patient care. It is similarly important to consider ergonomic designs for children and elderly people and for the physically and mentally handicapped and functionally impaired population of developing countries such as India. This includes design of buildings and rooms and of products, such as crockery and cutlery, and crutches and other mobility aids [41].

It is apparent that in the years to come the importance of ergonomics will increase in all spheres of life in the IDCs, depending on present approaches in the application, training, and research needs of ergonomics.

Maintaining a balance between a few very expensive, high technology–oriented enterprises and a large number of low-cost small enterprises—all intending to bring benefits to the people—would be the key point in achieving improvement in the quality of working life in IDCs. The government of India in collaboration with the state governments should assume a supervisory and regulatory role in such matters.

REFERENCES

1. R. N. Sen, Application of ergonomics to industrially developing countries, *Ergonomics* 27:1021–1032 (1984).
2. R. N. Sen, Ergonomics—science and technology of man at work—its role in our national development, *Proc. Indian Science Congress, 66th Session*, Part II, Hyderabad, India, 1979, pp. 53–77.

3. K. Kogi and R. N. Sen, Third world ergonomics, *Int. Rev. Ergon.* *1*:77-118 (1987)
4. R. N. Sen and P. K. Nag, Optimal work load for Indians performing different repetitive heavy manual work, *Ind. J. Physiol. Allied Sci.* *33*:18-25 (1979).
5. R. N. Sen and D. Chakraborti, An ergonomic study on sickle designs for a reaping task in Indian agriculture, *Proc. Ergonomics Society Conf., UK,* published as *Contemporary Ergonomics,* E. D. Megaw, Ed., Taylor & Francis, London, 1989, pp. 313-317.
6. R. N. Sen and D. Chakraborti, A new ergonomic design of a "desi" plough, *Ind. J. Physiol. Allied Sci.* *38*:97-105 (1984).
7. A. De and R. N. Sen, Ergonomic evaluation of ploughing process of paddy cultivation in India, *J. Hum. Ergol.* *15*:103-112 (1986).
8. R. N. Sen and S. N. Bhattacharya, Evaluation of an ergonomically designed shovel, *J. Physiol. Allied Sci.* *38*:150-154 (1984).
9. R. N. Sen and S. Bhattacharya, Development of an ergonomic design of a shovel from the viewpoint of increasing productivity in manual material handling in India, Project Report No. 23, Ergonomics Lab., Dept. of Physiology, Univ. Calcutta, 1976, pp. 1-14.
10. R. N. Sen, C. K. Pradhan, S. Basu, and A. K. Ganguli, An ergonomic evaluation of design of spades ("kodal"), *Pro. Indian Sci. Congr., 66th Session,* Part III, Hyderabad, India, 1979, pp. 6-7.
11. R. N. Sen and S. Sahu, An ergonomic design of a multi-purpose shovel-cum-"kodal" for manual material handling, *Adv. Ind. Ergon. Safety* *6*:561-566 (1994).
12. R. N. Sen and D. Chakraborti, A new ergonomic "float-seat" for improvement of paddy cultivation in India, in *Ergonomics in Developing Countries,* OSH Ser. No. 58, ILO, Geneva, 1987, pp. 419-427.
13. R. N. Sen and A. De, Ergonomics study of sheafing in paddy cultivation in India, *Ind. J. Physiol. Allied Sci.* *38*:85-94 (1984).
14. A. De and R. N. Sen, A work measurement method for application in Indian agriculture, *Int. J. Ind. Ergon.* *10*:285-292 (1992).
15. R. N. Sen, A. K. Ganguli, G. G. Ray, A. De, and D. Chakraborti, Tea leaf plucking—workloads and environmental studies, *Ergonomics* *26*:887-893 (1983).
16. R. N. Sen, Case studies to improve working conditions through technological choices in small and medium-sized (food processing, textiles and metal working) enterprises in West Bengal, in *Improving Working Conditions in Small Enterprises in Developing Asia,* K. Kogi, Ed., International Labour Office, Geneva, 1985, p. 158.
17. R. N. Sen, Examples of low-cost improvements in working conditions and environment, work methods, occupational health and safety at the enterprise level in West Bengal, in *Low-Cost Ways of Improving Working Conditions—100 Examples from Asia,* K. Kogi, W. O. Phoon, and J. E. Thurman, Eds., International Labour Office, Geneva, 1988, p. 179.
18. R. N. Sen and D. Pal, A new ergonomic method for postural analysis, in *Biomechanics,* K. B. Sahay and R. K. Saxena, Eds., Wiley, New Delhi, 1989, pp. 226-270.
19. R. N. Sen and D. Ghoshthakur, Ergonomic study on work-rest cycle, physiological responses and production rate of hand loom weavers, *Ind. J. Physiol. Allied Sci.* *38*:47-53 (1984).
20. R. N. Sen and A. Kar, An ergonomic study on bamboo handicraft workers, *Ind. J. Physiol. Allied Sci.* *38*:69-77 (1984).
21. R. N. Sen, A. K. Ganguli, G. G. Roy, A. De, and D. Chakraborti, Ergonomics studies of tea-leaf plucking operations—criteria for selection and categorization, *Appl. Ergon.* *12*:83-85 (1981).
22. R. N. Sen, A. K. Ganguli, G. G. Roy, A. De, and D. Chakraborti, A preliminary ergonomic study on tea plucking operations: two leaves and a bud, *TRA J.* *27*:74-80 (1980).
23. R. N. Sen and S. Gangopadhyay, Ergonomics study of accidents during manual material handling in factories of West Bengal, *Proc. 11th Asian Conf. Occup. Health,* Bombay, India, 1989, p. 95.
24. R. N. Sen, Certain ergonomic principles in the design of factories in hot climates, *Proc. Int. Symp. Occup. Safety, Health and Working Conditions,* International Labour Office, Geneva, 1982, pp. 123-147.
25. R. N. Sen and N. C. Dutta, An ergonomic design of low-cost protective work clothing for furnace workers, in *Ergonomics in Developing Countries,* Occupational Safety and Health Series No. 58, Int. Labour Office, Geneva, 1987, pp. 222-223.

26. R. N. Sen, Personal protective devices for workers working in radiant and hot environments in India, *Proc. 2nd Int. Symp. Clothing Comfort Studies, in Mount Fuji*, Japan Research Association for Textile End-Uses, Osaka, Japan, 1991, pp. 123–142.
27. R. N. Sen, Special work-wear and protective clothing for Indian furnace workers, *Proc. Int. Conf. on Human-Environment System*, Pergamon, Tokyo, Japan, 1991, pp. 411–415; *J. Therm. Biol.* 18:677–681, 1993.
28. R. N. Sen, A. K. Ganguli, and D. Chakraborti, Some anthropometric considerations related to Indian railways locomotive drivers, *Ind. J. Physiol. Allied Sci.* 38:106–113 (1984).
29. R. N. Sen, Ergonomics design of some tools for manual maintenance of railway tracks in India, *Proc. 10th Int. Ergon. Congr.*, Sydney, Australia, 1988; *Ergonomics in Developing Countries*, A. S. Adams, R. R. Hall, B. J. MePhee and M. S. Oxenburgh, Eds., Ergonomics Society of Australia Inc. Vol. 1, 1988, pp. 227–229.
30. R. N. Sen and P. Chatterjee, An ergonomic approach to the design of Indian bullock-cart used for carrying load in the city of Calcutta, Project Rep. No. 140, Ergonomics Laboratory, Dept. of Physiology, Univ. Calcutta, 1986, pp. 18.
31. R. N. Sen and S. Gangopadhyay, *New Ergonomic Harness Design for Hand-Cart Pullers*, Project Rep. No. 120, Ergonomics Laboratory, Calcutta Univ. 1985, p. 20.
32. R. N. Sen, Some studies on the physiological effects of noise, *Proc. Semin. Higher Productivity Through Building Insulation*, Bombay, India, 1967, pp. 53–69.
33. S. K. Chatterjee, R. N. Sen, and P. N. Saha, Determination of the level of noise originating from room air conditioners, *J. Heating Vent. J. Air Cond.* 38:429–433 (1965).
34. R. N. Sen and A. K. Ganguli, An ergonomics analysis of railway locomotive driver functions in India, *J. Human Ergol.* 11:187–202 (1982).
35. R. N. Sen and A. K. Ganguli, Preliminary investigation into the loco-man factor on the Indian railways, *Appl. Ergon.* 13:107–117 (1982).
36. R. N. Sen and P. K. Nag, Are Calcutta public buses ergonomically designed?, *Ind. J. Physiol. Allied Sci.* 27:156–157 (1957).
37. R. N. Sen and P. K. Nag, Design ergonomics of some public buses in Calcutta, *Proc. Indian Sci. Congr., 61st Session*, Nagpur, India, Part IV, 1974, pp. 126–127.
38. R. N. Sen, S. Basu, and A. Goswami, An ergonomic design of hand-pulled rickshaw, *Proc. Indian Sci. Congr., 66th Session*, Hyderabad, India, 1979, pp. 8–9.
39. R. N. Sen and S. Bannerjee, Eds., *Proceedings of the International Satellite Symposium on Work Physiology and Ergonomics*, held in Calcutta, 1–3 Nov., 1974 (during the 24th World Congress of Physiological Sciences held at New Delhi), *Ind. J. Physiol. Allied Sci.* 33: 1–145 (1979).
40. R. N. Sen and D. Majumdar, Ergonomics in relation to occupational safety and health in jute industries in eastern India, *Proc. 10th Asian Conf. Occup. Health*, Singapore, 1982, pp. 289–298.
41. R. N. Sen and A. Dutta Gupta, Ergonomics study on rehabilitation of a bilateral below-elbow amputee, in *Advances in Industrial Ergonomics and Safety*, Vol. 6, F. Aghazadeh, Ed., Taylor & Francis, London, 1994, pp. 437–440.
42. S. R. Dutta and S. Ganguli, An ergonomic approach to the design of Indian hand-pulled carts ("Thelas"), *Ind. J. Physiol. Allied Sci.* 33:102–106 (1979).
43. R. N. Sen and S. Das, Ergonomics, occupational health and safety improvements for the manual metal arc welders, *Proc. 1st Int. Symp. Ergon., Occup. Health, Safety and Environment*, Bombay, India, 1991, Indian Society of Ergonomics (ISE), Calcutta, India, 1993, pp. 97–102.

39
Overview of Ergonomic Needs and Research in China

Tian Lin Li and Zun Yong Liu
Institute of Occupational Medicine, Chinese Academy of Preventive Medicine, Beijing, People's Republic of China

Xiou Fen Zhang
Harbin Institute of Industrial Hygiene and Occupational Disease, Harbin, People's Republic of China

I. INTRODUCTION

Research in ergonomics in China originated from its application in industry. During the 1950s, a methodological investigation was carried out on standardization of gas masks, gloves, clothes, and the design of aircraft cockpits. In the 1960s some problems in the measurement of human feet were studied, such as the application of two-dimensional normal distribution theory in systematic research for developing shoe sizes. Efforts were also made to develop standard measurements for human body segments. Standards of clothing were developed by the Mathematical Institute of the Chinese Academy of Sciences and the Light Industry Ministry of China. During the 1970s, an investigation of head types of the Chinese people was carried out by the Ministry of Posts and Telecommunications of China. An illumination standard was developed by the Institute of Psychology of the Chinese Academy of Sciences and the Physical Institute of the Chinese Academy of Architecture. The measurement of values of the spectral stimulation ($V\lambda$) curve for Chinese observers of standard chromatics was studied by the Chinese Academy of Measurement and the Institute of Psychology of the Chinese Academy of Sciences. The visual irritability of different wavelengths to the eyes of Chinese subjects was summarized. After the 1980s, ergonomics in China developed more quickly.

The development of ergonomics in China is somewhat lagging compared to other countries. Although some ergonomics experiments of airplane design were made by some institutes during the 1960s, ergonomics did not emerge as an independent academic field until the 1980s. Since 1980, ergonomic laboratories have been gradually established in Chinese universities and institutes. In 1980, the first book on ergonomics was published in China [1]. In the same year, the China National Technical Committee for the Standardization of Ergonomics (CNTCSE) was founded in Beijing. The Chinese Institute of Standardization and Information put CNTCSE in charge of developing classification and a coding system. The thirty members of the CNTCSE include a physiologist, a psychologist, an anthropologist, a hygienist, engineers, and professors and specialists on standardization and labor protection from Chinese research institutions and universities. Each member serves on the committee for about 3 years. The main tasks of CNTCSE are to make recommendations for guidance of policy and technical measures of ergonomic standards for the industries in China and to develop guidelines for the formulation and revision of standards.

The CNTCSE has been given the task of accomplishing the following objectives:

1. To develop ergonomic principles for the design of work systems
2. To develop ergonomic principles for the design of the workplace
3. To develop ergonomic principles of safety distance and measures
4. To develop ergonomic principles of design for living space
5. To develop ergonomic principles for position and working range of machine operators
6. To develop a general demand for ergonomics in work arrangements for machine operators
7. To develop ergonomic parameters for evaluating the quality of products
8. To define ergonomic terms

In order to accomplish these objectives, eight technical subcommittees are set up in CNTCSE. They are responsible for research and academic activities in ergonomics in China. These committees address the issues of (1) anthropometry and biomechanics, (2) signals and controls, (3) color, (4) illumination, (5) the physical environment, (6) labor safety, and (7) working systems and ergonomic demand.

II. HISTORICAL BACKGROUND AND SIGNIFICANCE OF OCCUPATIONAL ERGONOMICS IN CHINA

In the industrialized countries, ergonomics and standardization of work are combined. With the development and use of ergonomics in China, many research achievements on ergonomics have been standardized. The main contents of ergonomics standards in China are as follows:

1. Relevant standards of basic human characteristics (physical, physiological, psychological, social, etc.)
2. Standards related to the physical factors in the environment that affect the human body
3. Standards related to the functions associated with production procedures and systems during human operation
4. Standards for measurement methods in ergonomics and data handling
5. Coordination with International Standards Organization (ISO) technical committees

The combination of ergonomics with standardization leads to research achievements in ergonomics for the benefit of worker health and safety. Ergonomics has contributed to the promotion of product development, increasing production efficiency and ensuring worker safety.

The Chinese Ergonomics Society (CES) was organized in 1988. It consists of several specialized committees including committees on human/machine engineering, environmental ergonomics, management ergonomics, ergonomic standardization, cognitive ergonomics, and ergonomic biomechanics. The Technical Committee on Systematic Engineering of Human-Machine-Environment in China (TCSEHMEC) was founded in 1982.

III. OCCUPATIONAL ERGONOMICS IN CHINA—RESEARCH AND STANDARDS

The research achievements and standards of ergonomics have been used widely in many aspects of industrial production in China. They offer the guiding principle and are an important reference for engineering design and technique. For instance, human characteristics are often considered as a basic design factor in engineering and product design. The standard for the ranking of head types of the Chinese adult (GB2428-81) is the important foundation for the design of protective tools for respiration and the standardization of product. With the development of national standards on body sizes of Chines adults, a database of Chinese adult sizes was set up, including the operational range of sitting and standing postures. At the same time, relevant research was carried out on hand type, foot type, and the body type of the Chinese people. These are very important investigations for the recognition of ergonomic needs in China.

Research on standards of occupational ergonomics on labor safety has also been successful. For example, the national standard on ergonomics for the safety distance for protection of the operator during machine operation and protective clothing for workers (GB 12265) have had a significant impact on labor safety, employee training, workplace design, devising of protective measures, etc. In addition, some other ergonomic standards have also shown good results. These national standards have already produced related social and economic benefits. In the past 10 years, about 35 national standards on occupational ergonomics have been set up in China. The most important standards are designed to address issues related to anthropometry of Chinese adults, the functional sizes of chairs, classification of physical workload, and classification of physical workload in hot environments. Of these standards, about one-third are mandatory national standards and the other two-thirds are recommended national standards.

IV. REVIEW OF CURRENT STATUS

In order to understand the current status of occupational ergonomics in China, a questionnaire survey type of investigation was carried out. The information gathered is summarized below.

1. Many national universities offer courses in ergonomics. In some universities and colleges, teaching materials on ergonomics are offered as part of their major field. The topic of ergonomics has also found its place at the graduate level in Chinese education. These efforts are important to the popularization and development of ergonomics at educational institutions.
2. There have been encouraging developments in the actual application of occupational ergonomics. For instance, the research achievements with respect to ergonomic standards for mechanical vehicles, labor safety, and military supplies and equipment have influenced the development of science and technology and social progress.
3. The weak point of occupational ergonomics in China is that basic ergonomic research is greatly lacking. We should enhance the exploration of fundamental experimental methods and accumulate basic knowledge of ergonomics. We should also apply ergonomic research information to develop and apply industrial standards.
4. We should also strengthen the exchange of information among scientists with the formation of an international ergonomic society and an international standardization organization.

V. FUTURE CONCERNS

There are still areas of concern in occupational ergonomics in China that require further attention. There is a need to form a complete set of ergonomic standards for such items as operators' chairs, seats for workers whose work requires a sitting position, the position of workers working in a standing position, and minimum requirements for hygiene in the workplace. The ergonomic demands of operating different switches, knobs, handwheels, handbrakes, operating stations, and control sticks need to be studied and dealt with.

The modern design and manufacture of machines must be in accordance with the demands of human characteristics and functional capacity and ergonomic principles as well as resolving the relationship between human and machine, to ensure the safety and efficiency of the workplace. For these reasons, systematic ergonomic standards and rules should be formulated. China should rouse itself to catch up with ergonomic developments. The distance in ergonomics between developing countries and China should be shortened gradually. We should also deal with the theoretical study and application of occupational ergonomics more widely and more thoroughly.

REFERENCES

1. C. Qi, A brief history of development of human engineering, in *Human Engineering*, Chen Du, Ed., Science-Technical Publishing House of Si Chuan, 1991, p. 1.
2. Labour Hygiene Department of Nan Jing Medical College, An investigation on human health and air ions in the computer room, *Annu. Rep. Nan Jing Med. College* 8(2):137 (1988).
3. T. L. Li, X. F. Zhang, and Z. G. Li, A brief introduction of ergonomic standardization, *Labour Protect. Water Conservancy Electr. Power 1*:20 (1991).
4. China National Technical Committee for the Standardization of Ergonomics, A brief introduction of the China national technical committee for the standardization of ergonomics, *Proc. CNTCSE'89*, Beijing, 1989, pp. 3–4.
5. China Institute of Standardization and Information (Classification and Coding), Ergonomics and standardization, *Proc. CNTCSE'92*, Beijing, 1992, pp. 1–5.
6. Chinese Institute of Standardization, The research direction of Chinese ergonomics and diagram mark, *Proc. CNTCSE'82*, Beijing, 1982, pp. 3–13.
7. China Institute of Standardization and Information (Classification and Coding), A systematic table on standardization of Chinese ergonomics, *Proc. CNTCSE'88*, Beijing, 1988, pp. 1–4.
8. China National Technical Committee for the Standardization of Ergonomics, Summary minutes of the meeting of China National Technical Committee for the Standardization of Ergonomics in 1990, *Proc. CNTCSE'90*, Beijing, 1990, pp. 1–3.
9. Chinese Ergonomics Society, Summary of minutes of establishing meeting of Chinese Ergonomics Society, *Proc. CES'89*, Shanghai, 1989, pp. 1–4.
10. Technical Committee on Systematic Engineering of Human–Machine–Environment in China, Systematic table on standardization of systematic engineering of human–machine–environment, *Proc. TCSEHMEC'85*, Beijing, 1985, pp. 1–5.

40
Overview of Ergonomic Needs and Research in Taiwan

Chi-Yuang Yu and Eric Min-yang Wang
National Tsing Hua University, Hsinchu, Taiwan, Republic of China

I. HISTORICAL REVIEW

Ergonomics was first introduced to Taiwan in the early 1960s by industrial designers. In 1962, the Chinese Productivity Center (an affiliate agency of the Ministry of Economic Affairs) sponsored a series of introductory industrial design training courses that were aimed at the promotion of the design of industrial products. During these courses, "human engineering" was presented by Professor Ohara Zirou of Ciba University in Japan. A course in human engineering was taught for the first time 3 years later in the industrial design department of Ming-Chi Institute of Technology in Taipei. Within a decade, many universities, colleges, and junior colleges followed this new trend, offering human engineering courses in their industrial design departments. The term *human engineering* was increasingly heard by many people through commercials in the media. However, most people had only a vague impression of ergonomics, which was considered to be the knowledge of fitting products (e.g., cars and chairs) to the physical dimensions of human beings.

Ergonomics has grown rapidly in Taiwan since 1983, becoming increasingly important since the initiation of the course in industrial engineering departments. The industrial engineering department at National Tsing Hua University (NTHU) in Hsinchu was one of the pioneers in offering this course. The course was titled "Human Factors," as in the United States, where a majority of the faculty members had received their graduate degrees. The first ergonomics graduate program in Taiwan was initiated a couple of years later at NTHU. Many other universities and colleges followed suit. More than 20 schools currently offer at least one human factors course for undergraduate studies. Human factors has become an important branch in the field of industrial engineering and is increasingly known for its many applications, e.g., workplace and protective equipment design, traffic and transportation system design, and control room design for nuclear power plants and chemical plants.

Two governmental agencies, the National Science Council (NSC) and the Council of Labor Affairs (CLA), have devoted a substantial amount of energy over the past several years to supporting the promotion of ergonomics. In 1985, NSC invited ergonomics professors from a number of disciplines, e.g., industrial design, industrial engineering, and psychology, to discuss and plan the future development of ergonomics in Taiwan. Those meetings concluded with a directive plan, which was then adopted by the NSC as a guideline to grant research projects thereafter.

In a different approach, CLA, focusing its efforts on the welfare of laborers, has devoted much of its efforts to promoting ergonomics in the industrial sector for practical purposes. In recent years, CLA has sponsored many introductory seminars for its employees and industrial safety and health personnel. In 1992, the Council established the Institute for Occupational Safety and Health (IOSH), one of whose primary missions involves promoting ergonomics research that would lead to the establishment of safety and hygiene standards for the entire nation. This mission is having a significant and positive impact for the future development of ergonomics in Taiwan.

II. ERGONOMICS EDUCATION IN TAIWAN

Ergonomics courses in Taiwan have primarily been taught in the departments of industrial design and industrial engineering in universities and colleges. For industrial design departments, six universities and 20 junior colleges offer ergonomics courses that stress design applications; therefore, anthropometry and biomechanics seem to be the core. Among these universities, two have master's programs in ergonomics. The educational backgrounds of those faculty members are industrial design, industrial engineering, psychology, or related areas. For industrial engineering departments, five universities and 20 junior colleges have general ergonomics courses, and of these, three universities currently offer doctoral and master's programs. The educational backgrounds of those faculty members are mainly ergonomics, industrial engineering, and psychology. Ergonomics education in the industrial engineering departments is more diversified and in depth than that in industrial design departments. The numbers of faculty members and students along with the titles of the ergonomics courses of the three universities that offer doctoral programs are listed in Table 1.

III. ERGONOMICS SOCIETY OF TAIWAN ROC

The Ergonomics Society of Taiwan ROC (EST) was founded in February 1993, with the stated objectives of promoting indigenous research and technological applications as well as encouraging international cooperation and exchange in ergonomics. The EST's executive council tentatively planned to hold one national conference and publish at least one journal each year and also to occasionally sponsor international symposiums. The EST also coordinates research projects and provides training programs for interested parties, such as IOSH and those in the industrial sector. The EST is an independent organization. However, it closely coordinates with local institutes of industrial engineering and the Industrial Design Society for their common interests.

The society had roughly 180 individual members and seven organizational chapters by the end of 1993. In terms of educational background, industrial engineering is the largest group and industrial design the second largest (see Figure 1). In terms of occupations, university and college faculty are the largest group, with students being second (see Figure 2). Three of the seven organizational chapters are governmental institutions, i.e., IOSH, Taiwan Power Company, and the Center for Industrial Safety and Health Technology at the Industrial Technology Research Institute; the remaining four are colleges. The officers of EST are striving diligently to promote their society in all sectors of Taiwan. It is hoped that more individual members will be enrolled from among nonacademic professionals and that more organizational chapters will be formed in the private sector.

Table 1 Ergonomics Education Programs in Universities in Taiwan as of 1993

	National Tsing Hua University	National Chiao Tung University	National Taiwan Institute of Technology
Faculty	4	3	4
Doctoral students	5	4	9
Master's students	13	5	10
Undergraduate courses	Psychology Work study Human factors 1 Human factors 2 Industrial safety	Motion and time study Human factors Industrial safety and hygiene Industrial organization psychology	Industrial psychology Human factors Industrial safety Consumer behavior Time and motion study
Postgraduate courses	Human performance Work physiology Environmental analysis Human–machine system Human–computer interface Industrial safety Safety engineering Biomechanics Vision and eye physiology Color science Design evaluation	Human information process Advanced human factors Measurement and evaluation of human performance Human performance Human and computer Human–machine system Problem solving	Biomechanics Fundamental physiology Fundamental psychology Advanced human factors Cognition Human reliability Human–machine system

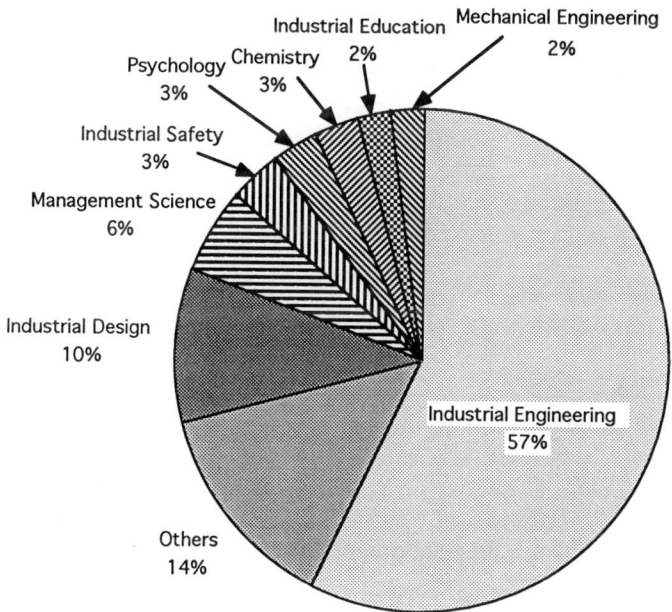

Figure 1 Educational backgrounds of EST's 180 members as of 1993.

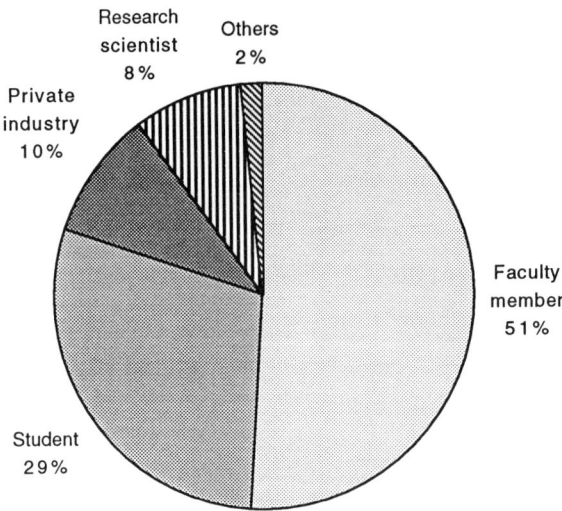

Figure 2 Classification of EST members' occupations as of 1993.

Less than 6 months after its formation, the society had already sponsored an industrial ergonomics and safety workshop and had also conducted a domestic ergonomics manpower survey. On its first anniversary, an international conference on ergonomics and occupational safety and health was held. Around 200 local researchers and practitioners as well as many from Japan and the United States participated. The society is currently coordinating an ergonomic industrial chair development and evaluation research project. The prototype of the chair is presently in the mass production stage and will be released to industry for a large-scale user test within months. Several other nationwide ergonomics projects—e.g., anthropometric, biomechanical, craniofacial, and work physiological databases for the domestic population—are also in the planning stage. These databases are expected to be completed within the next 6 years.

In addition, the society is tentatively planning to place an emphasis on promoting ergonomics to the public in the coming years. Among others, one of the most effective approaches involves presenting introductory films and publishing articles for the mass media, e.g., public television programs and newspapers.

IV. RESEARCH PROJECTS AND FUTURE DEVELOPMENT

A. Research Projects Funded by NSC

Previous ergonomics research was primarily funded by NSC; however, several research projects have recently been funded by other governmental agencies, such as CLA and the Ministry of Transportation. Since 1985, NSC has given grants to more than 70 ergonomics-related research projects, and the total budget has exceeded US $8 million. As ergonomics-related research was comparatively new to NSC, the funded research projects were of small scale and not integrated. Among those projects, some important ones are listed in the following.

1. Taiwanese Static Anthropometric Data bank

The primary purpose of this study focused on establishing a static anthropometric data bank for Chinese residents in Taiwan using a computerized photographic method [1]. Stratified random sampling was applied to determining the sampling site; in addition, a sample size was established by considering standard errors in a pilot study. A photographic method was next employed in addition to direct measurement of selected body dimensions. A cumulative total of 933 subjects were measured. The data were entered from photos via a digitizing tablet into a microcomputer for processing. The resulting anthropometric data bank was established, and recommendations were presented for future research efforts.

2. Ergonomics Research for Chinese Computer Keyboard Operations

The primary purpose of this research effort [2] involved developing ergonomic design guidelines for a Chinese computer keyboard. Some of the subproject titles translate as follows:

1. A questionnaire survey on Chinese VDT keyboard operation behaviors and the causes of discomfort
2. The anthropometry of hands for computer keyboard design
3. A study of the key grouping and layout for computer keyboard design
4. A study of finger movements in keyboard operation
5. A study of table–chair height and arm–shoulder postures in keyboard operation at a VDT workstation
6. A study on readability and legibility of Chinese VDT key legends
7. A study of keyboard design for the domestic population

3. Driver Visibility and Safety for Motor Vehicle Design

The primary purpose of this research effort focused on developing ergonomic design guidelines for cars and motorcycles by emphasizing driver visibility and safety. This project consisted of five subprojects:

1. A questionnaire survey on driver comfort and safety for heavy and light motorcycles
2. A study of the drivers' visual field, dark perception, and signal response time
3. An ergonomic study of motorcycle helmet design
4. The effects of tail light sizes on daytime braking reaction time for motorcycles
5. An evaluation of driving effectiveness on tail light configurations and the visual interference between dashboard displays and the steering wheel

4. Ergonomics in a Nuclear Power Station

This research effort was composed of a series of projects concerning supervisory control, error diagnosis, and problem solving for nuclear power stations. The research projects included

1. Application of queuing theory to quantify information workload in supervisory control systems [3]
2. Stochastic modeling of human errors on system reliability [4]
3. Dynamic hierarchical modeling of problem solving [5]
4. The prevention of human errors in nuclear power plant maintainability through root cause analysis [6]
5. The design strategies of Chinese information display in supervisory control systems [7]

B. Research Projects Funded by IOSH

Other than NSC, IOSH has funded several research projects during the past few years. Among these projects, two are especially important, the development of a high-mobility industrial chair and a computerized manual materials handling assessment system.

1. Development of a High-Mobility Industrial Chair

An anthropometric measurement procedure [8] was developed for investigating the design parameters for a low sit-stand posture, and the data on these parameters were applied to designing a chair for high-mobility industrial tasks.

The anthropometric procedure used ischial tuberosities as the seat reference point. Investigating the design parameters required tracing spinal curves of five postures and a posterior thigh curve via a three-dimensional curve-tracing device. The high-mobility chair was designed with a special seat pan and backrest profiles to accommodate the musculoskeletal geometrical configuration of a low sit-stand posture. The seat pan consists of pelvic support that supports the ischial tuberosities and the areas behind them and a thigh support that maintains the thighs at a 15° inclination angle, resulting in a 105° torso-to-thigh angle. The backrest consists of a lumbar support that preserves lumbar lordosis and a thoracic support that supports the upper back during backward leaning.

2. Computerized Manual Materials Handling Assessment System

The primary purpose of the second project [9] was to develop an automated computer system for evaluating the risk of low back injuries that occur in manual materials handling. The system employs a computer vision technique to assess postures while a worker is performing manual materials handling tasks. Based on one of the six biomechanical models available, e.g., Chaffin's [10], the system can calculate the strain produced in the lower back and some joints on the basis of anthropometric data. The calculated strain values are then compared with the corresponding predicted injury load, indicating the worst posture to take during the job cycle. Recommendations for correct manual materials handling posture are finally provided.

C. Future Development

The National Science Council funded a research project in 1990 to plan ergonomics development for five years. The ergonomics research topics suggested by local researchers are summarized below. This summary includes many of the most essential aspects in this field (excluding military applications), which indicates the enormous requirements and potential areas for ergonomics research.

1. Occupational safety and industrial ergonomics
   a. Design and evaluation of traffic signs and symbols
   b. Human factors applications in product design
   c. Measurement and standards for workplace environments
   d. Computerized biomechanical evaluation for manual materials handling
   e. The effects of workplace environment quality on work performance
2. Anthropometry, biomechanics, and work physiology
   a. Structural and functional anthropometric database of Taiwanese adults
   b. Physical strength data for Taiwanese
   c. Standards for physical work
   d. Epidemiological study of occupational musculoskeletal disorders

3. Human information processing and decision behaviors
   a. Error detecting and diagnostic behaviors
   b. Behavioral models of human errors
   c. Problem-solving behavior model and its applications in automated systems
   d. Human factors standards for the design of a knowledge-based information system
   e. The theory and applications of mental models
   f. The application of human decision-making behavior theory in a decision support system

In addition, IOSH has drawnup a six-year ergonomics research plan that is aimed specifically at establishing a variety of work guidelines and standards for various tasks. The plan has been classified as a long-term directive plan in Executive Yuan (the highest public administration organization in Taiwan). The plan involves establishing indigenous anthropometric, strength, and work physiology databases for enhancing occupational safety and hygiene in the workplace. The main ideas are listed in the following.

Anthropometric Databases. This project will consist of establishing static and dynamic databases for domestic workers. For static measurements, about 300 body dimensions will be measured from 3200 adult subjects. Stratified random sampling on gender, age, and other variables will be employed on the basis of the national demographic data. For dynamic measurement, more than 50 ranges of motion for specific postures will be measured from the same subjects.

This project, with its estimated budget of approximately US $1,200,000, will be initiated in 1994 and reach completion in 1996. A mobile measurement laboratory is proposed, and more than 30 investigators will participate in this project at three measurement centers stationed in major universities nationwide. Specific results of this project will consist of static and dynamic anthropometric databases. Three-dimensional (3-D) and two-dimensional (2-D) manikins will be made according to the actual data.

Craniofacial Anthropometric Database. The purpose of this project involves establishing a head and face anthropometric database for Taiwanese workers as well as constructing a set of 3-D head and face models for various applications such as respiratory protective equipment. It is planned to have about 1000 subjects drawn by stratified random sampling on gender, occupation, age, and area based on national demographic data. The data will then be categorized, by considering the facial length, facial width, and standard deviations, into several groups, and a "typical model" will be recommended for each group.

This project started in 1993 and will be ended in 1995. The estimated budget for this project is about US $200,000. It is expected to yield a craniofacial anthropometric database and a complete set of typical 3-D craniofacial models.

Strength Database. The primary purpose of this project involves collecting basic strength data for various biomechanical applications as well as paving the way for establishing physical work standards. These data will be collected along with anthropometric measurement by the same group of researchers. The strength measurements will include lifting, pushing, pulling, carrying, gripping, and gripping torque. The data will be tabulated according to strength and torque moment with respect to specific conditions both isometrically and isotonically. Roughly 1000 subjects will be randomly selected from those of the anthropometric database project and measured for this purpose.

Work Physiology Database. The primary purpose of this project is to collect physiological work data for setting up physiological work standards. These data will include maximal aerobic power, oxygen uptake at various workload levels, respiratory rate, pulmonary ventilation capacity, and similar information. This project is still in its planning stage.

V. CONCLUSIONS

A significant amount of progress in ergonomics education and research has been achieved in Taiwan over recent decades. Ergonomics education is currently progressing at a good pace in terms of both quantity and quality. The Ergonomics Society of Taiwan ROC, founded in 1993, will play an important role in promoting future research, education, and international exchange in the global village of ergonomics. Ergonomics research is gradually becoming solid and integrated, especially since the establishment of IOSH. Much faster growth is foreseeable in Taiwan over the next decade.

REFERENCES

1. C.-C. Li, S.-L. Hwang, and M.-Y. Wang, Static anthropometry of civilian Chinese in Taiwan using computer-analyzed photography, *Hum. Factors 32*:359 (1990).
2. L. F. Chen, F. K. Wu, S. S. Lai, M. C. Lin, K. S. Chen, M. J. Suon, and S. S. Kuan, The human factors studies on computer keyboards (Proj. NSC75-0415-E006-02), 1987. (In Chinese, available from National Science Council, Taipei, Taiwan, ROC.)
3. C.-C. Her and S.-L. Hwang, Application of queuing theory to quantify information workload in supervisory control systems, *Int. J. Ind. Ergon. 4*: 55 (1989).
4. C.-T. Hwang and S.-L. Hwang, A stochastic model of human errors on system reliability, *Reliability Eng. Sys. Safety* (accepted).
5. C.-H. Wang and S.-L. Hwang, The dynamic hierarchical model of problem solving, *IEEE Trans. Sys., Man. Cybern. 19*:946 (1989).
6. T.-M. Wu and S.L. Hwang, The prevention of human errors in nuclear power plant maintainability through root cause analysis, *Appl. Ergon. 20*:115 (1989).
7. S.-L. Hwang, M.-Y. Wang, C.-C. Her, D.-M. Wu, and C.-D. Hwang, The design strategies of Chinese information display in supervisory control systems, *Int. J. Hum. Comput. Interaction 2*:41 (1990).
8. C. Y. Yu, K. K. Li, and Y. P. Chen, Development of a high mobility industrial chair (Proj. IOSH82-H124), 1993. (In Chinese, available from Institute for Occupational Safety and Health, Council of Labor Affairs, Taipei, Taiwan, ROC.)
9. M. J. Wang and K. J. Hwang, Computerized manual material handling assessment system (Proj. IOSH82-H121), 1993. (In Chinese, available from Institute for Occupational Safety and Health, Council of Labor Affairs, Taipei, Taiwan, ROC.)
10. D. B. Chaffin, Biomechanical modeling of the low back during load lifting, *Ergonomics 32*:685 (1988).

Appendix A
Biomechanical Modelling of Carpet Installation Task

First, a three-dimensional model was developed to estimate the joint stresses associated with static postures commonly assumed by the carpet layers. Second, a two-dimensional model was developed for analyzing the dynamic carpet-stretching task. The results emphasize joint reaction forces and ground reaction forces.

I. DESCRIPTION OF THE MODELS

A. Static 3-D Biomechanical Model

A static link segment model was developed to aid in the estimation of joint reaction forces and muscle moments at six joints on each side of the body. This model has 12 element link segments. These links are: foot, shank, thigh, upper arm, and lower arm of each side of the body, plus one for torso and one for head and neck. During the carpet-stretching activity, three points of the body touch the ground (1). These anatomical sites for a right-legged person are: leg foot, knee, and palm of the hand. In addition, the knee kicker tool used for the carpet stretching task makes ground contact at its front and back ends (Figure 1). For the determination of ground reaction forces at these five contact points, the following equations apply:

The static equations of equilibrium are:

Sum of z-direction forces = 0   (1)

Sum of moments about x axis = 0   (2)

Sum of moments about y axis = 0   (3)

One more constraint equation is obtained by taking moment about the left knee:

$$R_f(x_f - x_k) = W_1(x_1 - x_k) + W_2(x_2 - x_k) \qquad (4)$$

where

R_f = ground reaction force at the left foot (newton)
x_f, y_f = x and y coordinates of R_f (meter)
W_1 = weight of the left foot segment (newton)
W_2 = weight of the left shank (newton)
$x_1 \rightarrow x_{12}$ = x coordinates of segments 1 to 12 (meter)
$y_1 \rightarrow y_{12}$ = y coordinates of segments 1 to 12 (meter)
x_k, y_k = x and y coordinates of left knee (meter)

Developed by Amit Bhattacharya and his staff at the Biomechanics-Ergonomics Laboratory, University of Cincinnati Medical School, through a cooperative agreement with the National Institute for Occupational Safety and Health

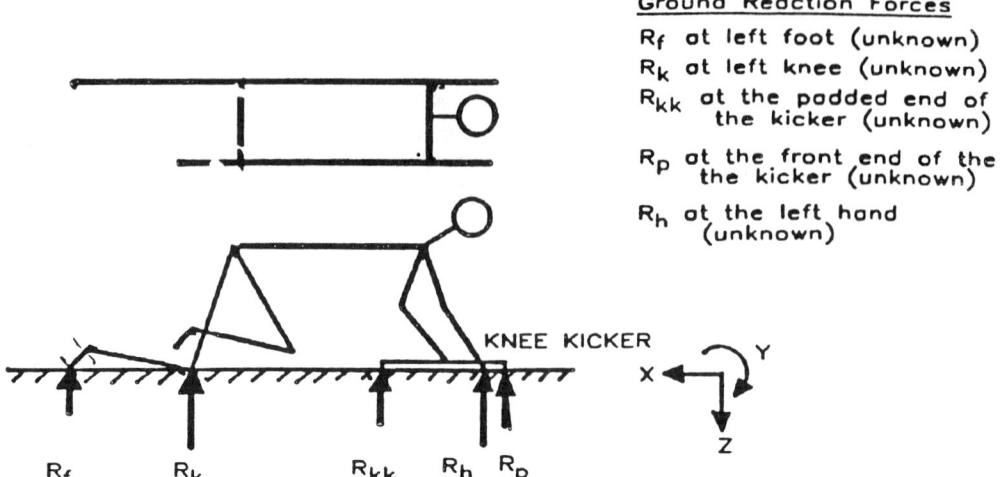

Figure 1 Static-free body diagram for carpet installation task.

Eq. (4) is solved for R_f. The remaining equations are:

$$\sum F_z = \sum_{l=1}^{12} W_l + (W_{kk} - R_p - R_f) - R_k - R_h - R_{kk} = 0 \quad (5)$$

$$\sum M_x = \sum_{l=1}^{12} W_l y_l + W_{kk} y_{kk} - R_p y_p - R_f y_f - R_k y_k - R_h y_h - R_{kk} y_{kk} = 0 \quad (6)$$

$$\sum M_y = \sum_{l=1}^{12} W_l x_l + W_{kk} x_{kk} - R_p x_p - R_f x_f - R_k x_k - R_h x_h - R_{kk} x_{kk} = 0 \quad (7)$$

Equations 5, 6, 7 are solved simultaneously for R_k, R_h, and R_{kk}
where

R_p = ground reaction force at the front end of the kicker (measured by the force platform) (newton)
R_k = ground reaction force at the left knee (newton) (unknown)
R_k, R_{kk}, = ground reaction forces at the left hand and at the padded end of the kicker (newton (unknown)
W_{kk} = weight of the kicker (newton)
$W_1 \rightarrow W_{12}$ = weight of segments 1 to 12 (newton)
X_p, Y_p = x and y coordinates of the front end of the kicker (meter)
X_h, Y_h = x and y coordinates of the left hand (meter)
X_{kk}, Y_{kk} = x and y coordinates of the padded end of the kicker (meter)

Once the ground reaction forces are calculated, the equations of static equilibrium are written for each of the 12 body segments, and the knee kicker and the head/neck segments. These are:

Sum of Z or vertical direction forces = 0   (8)

Sum of moments about y axis = 0   (9)

Biomechanical Modelling of Carpet Installation Task

An example of static free body diagram for the right thigh is shown in Figure 2A. Similar static free body diagrams for the other segments are drawn for the equations of static equilibrium. The static free body diagrams for the knee kicker and the head/neck segment are shown in Figures 2B and 2C.

The above-mentioned generalized equations are used for the calculation of joint reaction force and the muscle moment at each body joint. A computer program written in IBM-PC BASIC for Eq. 4-9 has built-in anthropometric tabular data (Tables 1 and 2) for segments weights as percent of body weight, and the location of segment center of gravity (CG) as percent of segment length (proximal and distal). This program first calculating the joint reaction forces and moments from the left ankle joint to the left hip joint. Then it calculates these values from the right ankle joint to right hip joint. After performing these calculations, the program determines the values of joint forces and moments from the right wrist joint to right shoulder joint and then for the left side of the body.

Finally, for the checking of force (R_1, R_2, R_3, R_4) and moment (M_1, M_2, M_3, M_4) balance at the four corners (right shoulder and hip joint and left shoulder and hip joint) of

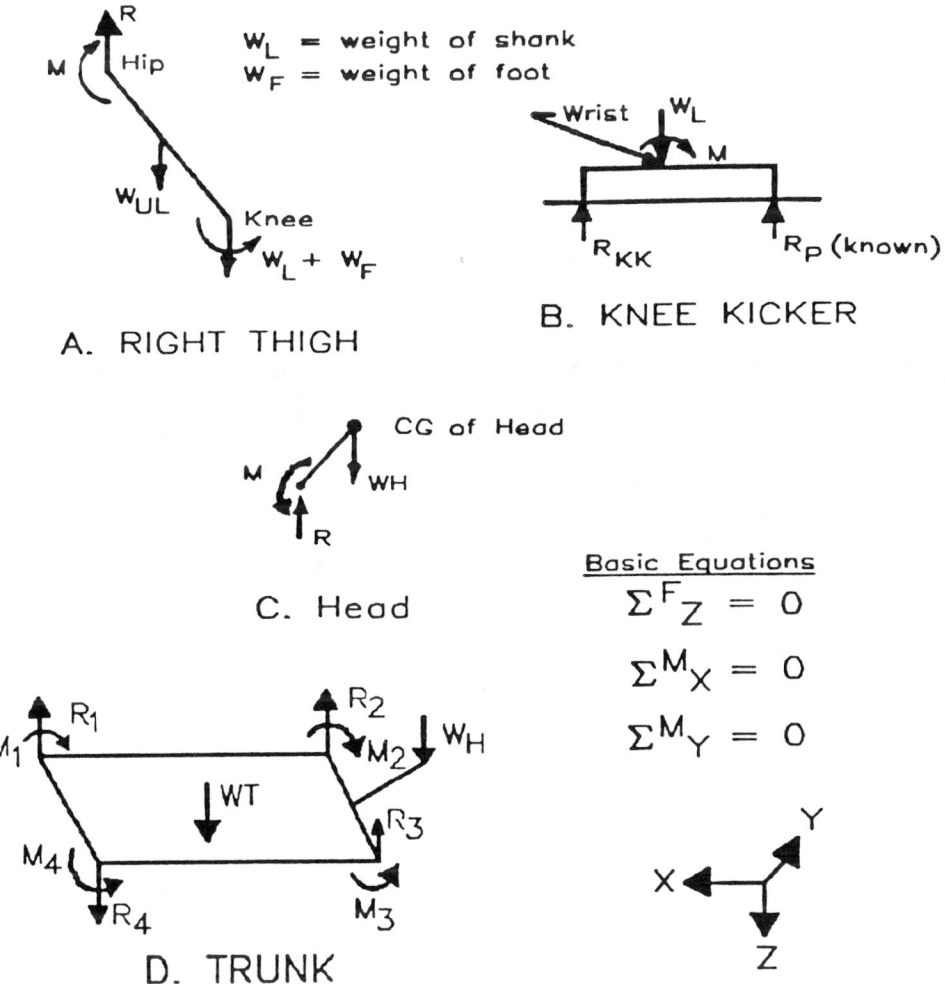

Figure 2 Static free body diagram for the individual link segment.

Table 1 Weight of Body Segments[a]

Segment	Percentage of body weight[b]	Weight (N)	Weight (lb)
Weight of foot = WA	0.014	11.20	2.52
Weight of calf = WK	0.046	36.81	8.28
Weight of thigh = WH	0.097	77.61	17.46
Weight of trunk = WS	0.486	388.87	87.48
Weight of upper arm = WE	0.027	21.60	4.86
Weight of lower arm = WW	0.014	11.20	2.52
Weight of hand = WN	0.006	4.80	1.08

[a] For a subject of BW = 180 lb (800 newton).
[b] Ref. 2.

the trunk in association of the head/neck segment, a static free body diagram is used (Figure 2D). For static equilibrium at the trunk and the head/neck segments, the program checks whether or not the sum of forces at all four joints equals the sum of weights of the trunk and the head/neck. Next, it checks whether all clockwise moments about the CG of the trunk equal the corresponding counterclockwise moments. For static equilibrium, these values are equal to each other.

Model inputs: These are body weight, R_p, $W_1 \to W_{12}$, W_{kk}, $X_1 \to X_{12}$, $Y_1 \to Y_{12}$, Z_1, Z_{12}, X_k, Y_k, X_p, Y_p, X_h, Y_h, X_{kk}, and Y_{kk}.

Model outputs: R_f, R_k, R_h, R_{kk}, X, Y, Z coordinates of segment CG, X, Y, Z coordinates of whole body CG; joint reaction forces (vertical) and moments (M_y) at all body joints, and force and moment check at the trunk segment.

Two-Dimensional Dynamic Model for Carpet-Stretching Task

This phase describes a model to outline the dynamic aspects of the carpet-stretching task, which requires the use of a knee kicker (4). This is a 6-segment, or link, model of the kicking side of the body. The segments are foot, shank, thigh, torso, upper arm, and lower arm. The head and the hand are not included in this model at this stage of the model development. This model assumes that the major body motion is in the sagittal plane (xy) and in the longitudinal di-

Table 2 Location of Segment Center of Gravity as a Function of Segment Length

Segment	Segment length from proximal joint (m)	Definition of segment
Foot	0.50	Lateral malleolus to head metatarsal II
Calf	0.433	Femoral condyles to medial malleolus
Thigh	0.433	Greater trochanter to femoral condyles
Trunk	0.50	Greater trochanter to glenohumeral joint
Upper arm	0.436	Glenohumeral axis to elbow axis
Lower arm	0.430	Elbow axis to ulnar styloid
Hand	0.506	Ulnar styloid to knuckle II middle finger

Source: Ref. 2.

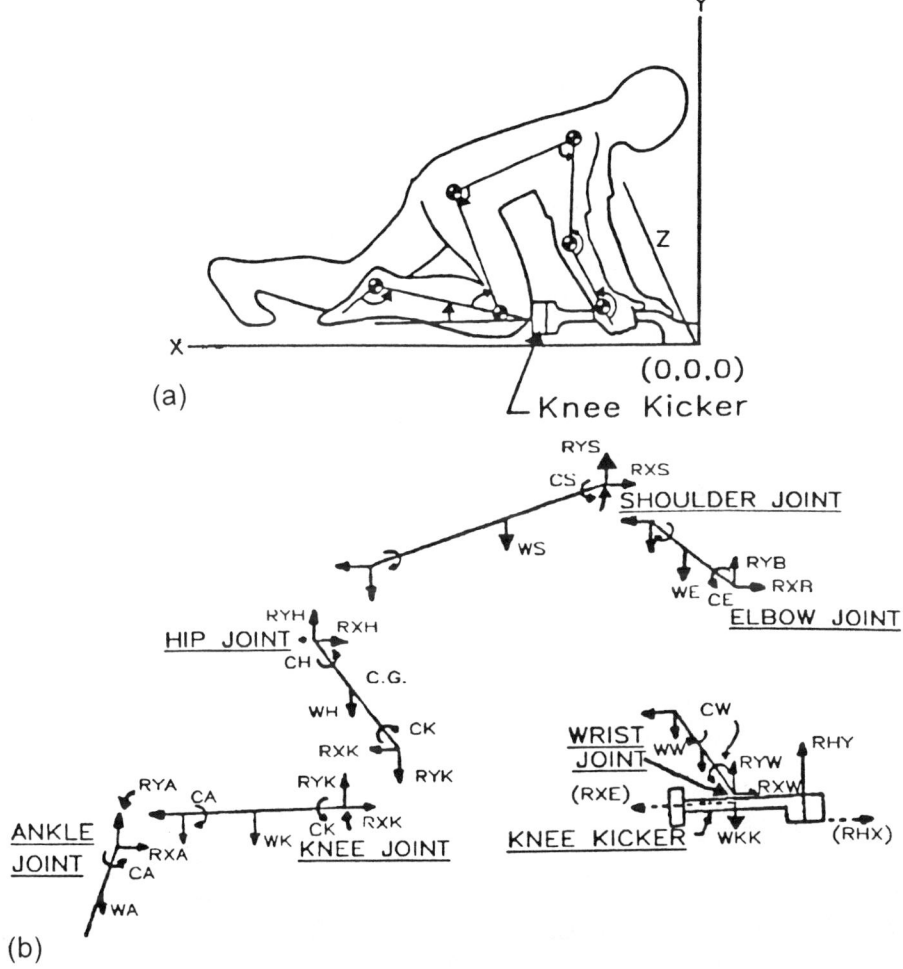

Figure 3 (a) Schematic of coordinate system for this study. (b) Link segment model.

rection (along the x axis) (Figure 3), and the inertial forces in the lateral direction (z axis) are minimal. During the knee-kicking task, the kicking side of the body generally remains airborne until the impact phase, when the suprapatellar region of the knee makes contact with the padded end of the knee kicker. During the entire kicking cycle, the only point on the right side of the body that is in contact with the ground is the front end of the kicker. During dynamic knee-kicking experiments, we found that the padded end produces no (or negligible) ground reaction forces. Therefore, the ground reaction force at the front end of the kicker essentially provides an integrated effect of whole body weight shifts associated with knee kicking. This information is an important input for the determination of joint stresses and muscle moments for the right, or the kicking, side of the body. The lateral forces in the z direction under the front end of the kicker are assumed insignificant in comparison to those in the vertical (y) and the horizontal (x) directions. It is realized that the body weight is also supported at the left or stationary knee and the left or stationary hand. However, it is assumed that the inertial effects along the lateral direction is minimal; therefore, ground reaction forces

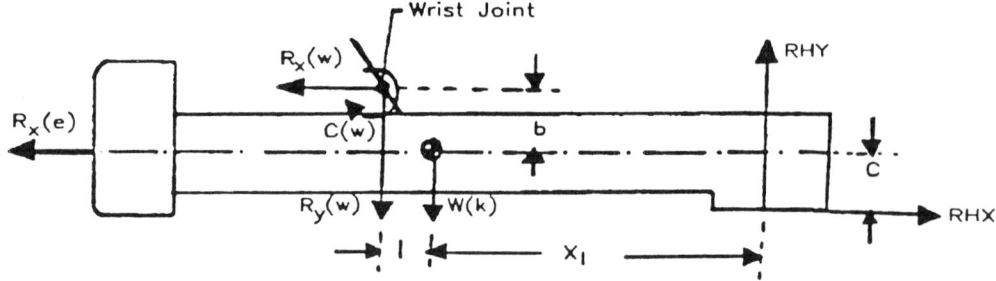

Figure 4 Static free body diagram for the wrist joint–knee kicker segment.

at these body contact points will affect mainly the joint stress calculations of the stationary, or left, side of the body.

As a first attempt in the modelling of dynamic aspects of knee kicking, the above assumptions are reasonable. During the experiment phase, these assumptions were validated.

Figure 4 is a static free body diagram showing the wrist joint–knee kicker segment. In the following, the hand weight is assumed negligible, and the wrist joint and the knee kicker are assumed stationary during each kicking cycle:

$$RH_x - R_x(w) + R_{xe} = 0 \tag{10}$$

$$RH_y - R_y(w) - W(R) = 0 \tag{11}$$

Taking moments around the CG of the kicker gives:

$$-C(w) + R_y(w) \times 1 + RH_y \times X_1 + RH_x \times c + R_x(w) \times b = 0 \tag{12}$$

where

RH_y = vertical ground reaction force at the front end of the kicker (measured with a force platform) (newton)

RH_x = horizontal friction force at the front end of the kicker (measured with a force platform) (newton)

R_{xe} = impact knee force applied at the padded end of the kicker [in Eq. (12), the direction of force applied is assumed to be toward the right] (newton)

$R_x(w)$, $R_y(w)$ = horizontal and vertical joint reaction forces at the wrist joint (newton)

$W(k)$ = weight of the kicker (newton)

$C(w)$ = muscle moment at the wrist joint (newton-meter)

1 = horizontal distance between the CG of the kicker and the wrist joint (constant value, measured manually) (meter)

X_1 = horizontal distance between the CG of the kicker and the middle of the front end of the kicker (point 0) (constant value, measured manually) (meter)

c = vertical distance between the bottom of the front end of the kicker and the CG of the kicker (constant value, measured manually) (meter)

b = vertical distance between the wrist joint and the CG of the kicker (constant value, measured manually) (meter)

For the calculations of joint reaction forces and muscle moments, the following generalized equations are used (3):

Biomechanical Modelling of Carpet Installation Task

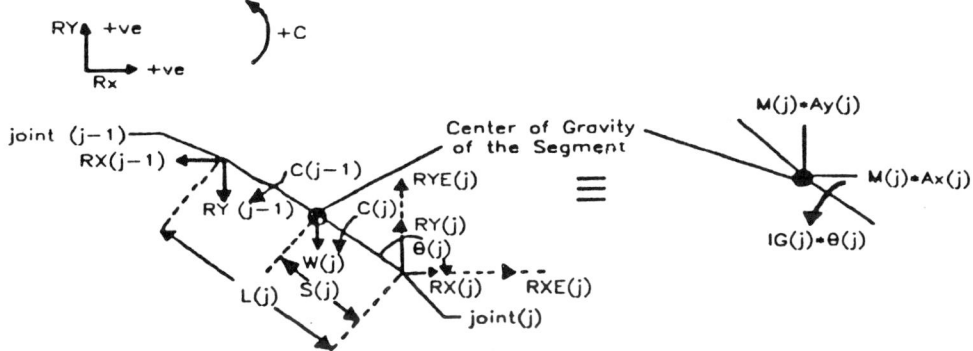

Figure 5 Schematic of generalized segment for force and moment equilibrium.

$$R_x(j) - R_x(j - 1) + R_{xe}(j) = M(j) \times A_x(j) \tag{13}$$

$$R_y(j) - R_y(j - 1) - W(j) + R_{ye}(j) = M(j) \times A_y(j) \tag{14}$$

$$\begin{aligned}C(j) = {} & C(j - 1) - R_x(j - 1) \times \{L(j) - S(j)\} \times \sin \theta(j) + R_y(j) \\ & \times \{L(j) - S(j)\} \times \cos \theta(j) + R_y(j) \times S(j) \times \cos \theta(j) \\ & + R_{ye}(j) \times S(j) \times \cos \theta(j) - R_x(j) \times S(j) \times \sin \theta(j) \\ & - R_{xe}(j) \times S(j) \times \sin \theta(j) + IG(j) \times \ddot{\theta}(j) \end{aligned} \tag{15}$$

where

R_x, R_y: joint reaction forces in the x and y directions, respectively, at (j)th and (j − 1)th joints (newton)

R_{xe}, R_{ye}: externally applied forces in the x and y directions, respectively, at (j)th and (j − 1)th joints (newton)

A_x, A_y: linear acceleration in the x and y direction of the CG of the segment with respect to a fixed reference point (m/sec²)

$\ddot{\theta}(j)$: angular acceleration of CG of the segment (rad/s²)

$\theta(j)$: angular location of segment with respect to right horizontal, degrees

$W(j)$ and $M(j)$: weight (newton) and mass (kg) of the segment

$S(j)$: location of CG of the segment with respect to the joint-understudy (meter)

$L(j)$: segment length (meter)

$IG(j)$: mass moment of inertia of the segment about its CG (N · m.s²)

$J = 1$ to $n = $ number of joints

$R_x(0) = 0$, $R_y(0) = 0$, $C(0) = 0$

For the static case

$A_x(j) = 0$, $A_y(j) = 0$ and $\ddot{\theta}(j) = 0$

Figure 5 presents a generalized segment for force and moment equilibrium.

The linear acceleration of CG of the segment with respect to the jth joint is given by the following generalized equation:

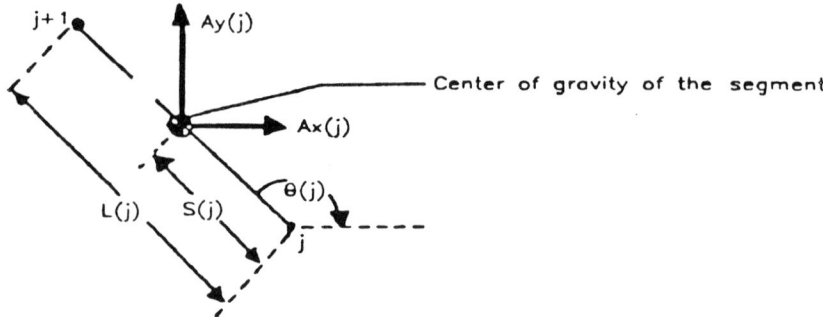

Figure 6 Schematic of generalized segment for linear acceleration calculation.

$$AG_x(j) = S(j)\ \dot{\theta}^2(j)\ \cos\theta(j) + (m)\ S(j)\ \ddot{\theta}(j)\ \sin\theta(j) \tag{16}$$

$$AG_y(j) = S(j)\ \dot{\theta}^2(j)\ \sin\theta(j) - (m)\ S(j)\ \ddot{\theta}(j)\ \cos\theta(j) \tag{17}$$

The linear acceleration equation of CG of the generalized segment (in Figure 6) with respect to a stationary reference point (usually a point of attachment) on the body is given by:

$$A_x(j) = AG_x(j) + \text{linear x - acceleration of the jth joint with respect to the point of attachment} \tag{18}$$

$$A_y(j) = AG_y(j) + \text{y - acceleration of the jth joint with respect to the point of attachment} \tag{19}$$

In the carpet installation modelling study, the wrist joint is considered to be a stationary reference point or point of attachment. The second terms in the above two equations are therefore equal to zero for the first segment

where

J = 1 to n = number of joints
m = +1 for counterclockwise rotation
m = −1 for clockwise rotation
$\ddot{\theta}(j)$ = angular acceleration of the segment CG; it can have either +ve or −ve values, (rad/s²)
  and $\dot{\theta}(j)$ = angular velocity (rad/s)

The remaining parameters have been explained previously.

The above equations (10–19) for the model were programmed in Microsoft FORTRAN for the IBM-PC laboratory computer. This program first starts calculating the values of $A_x(j)$ and $A_y(j)$ from the point of attachment, the wrist joint, and continues on to the foot segment. Next, the program calculates joint reaction forces at the ankle, knee, and hip joints. And, finally, it calculates these values at the wrist, elbow, and shoulder joints.

Model input: X and Y coordinates of body joints as a function of time, RH_x, RH_y, R_{xe}, R_{ye}, W(k), l, X_1, c, b, M(j), subject height, and body weight.

Model output: $R_x(j)$, $R_y(j)$, $A_x(j)$, $A_y(j)$, $\dot{\theta}(j)$ $\ddot{\theta}(j)$, and C(j).

REFERENCES

1. A. Bhattacharya, M. Mueller, and V. Putz-Anderson, Quantification of traumatogenic factors affecting the knee: A worksite study of carpet installation, *Applied Ergonomics* 16(4):243–250 (1985).
2. W. Dempster, Space requirements of the seated operator, *WADS-TR-55-150* Wright Patterson Air Force Base, (1955).
3. D. Winter, Biomechanics of Human Movement, John Wiley and Sons, New York; 1979.
4. A. Bhattacharya, Biomechanical analysis of carpet installation task, ASME Conference, Cincinnati, OH (1987).

Appendix B
Ergonomics Checklists

I. GENERAL WALKTHROUGH ERGONOMICS CHECKLISTS
A. General Workstation Design Principles

Use of the following general workstation design principles will help achieve an optimum match between the work requirements and operator capabilities. This, in turn, will maximize the performance of the total system while maintaining human comfort, well-being, efficiency, and safety.

- [ ] 1. Make the workstation adjustable so that the tall person will fit, and the small person can reach easily.

- [ ] 2. Provide materials and tools in front of the worker to reduce twisting.

- [ ] 3. Avoid static work postures. Avoid jobs requiring operator to:
    - lean to the front or the sides
    - hold an extremity in a bent or extended position
    - bend the torso or head forward/backward more than 15 degrees

- [ ] 4. Set the work height at 2 inches below the elbows for most work and 6 inches above the elbows for precision work.

- [ ] 5. Provide adjustable, properly designed chairs.

- [ ] 6. Allow the workers, at their discretion, to alternate between sitting and standing. Provide floor mats/padded surfaces for prolonged standing.

- [ ] 7. Support the limbs. Provide wrist, arm, foot and back rests as needed.

- [ ] 8. Use gravity to move materials.

- [ ] 9. Design the workstation so arm movements are continuous and curved. Avoid straight-line, jerking arm movements.

- [ ] 10. Keep arm movements in the normal area (limit reaches to 16 inches).

- [ ] 11. Provide dials and displays that are simple, logical, and easy to read and operate.

- [ ] 12. Eliminate excessive noise, heat, humidity, cold, and poor illumination.

The checklists in Section I were developed by Dave Ridyard, CIH, CSP, CPE, Applied Ergonomics Technology, 270 Mather Rd., Jenkintown, Pennsylvania.

B. Lifting and Lowering Tasks: Ergonomic Design

The following checklist should be used to eliminate the need to manually lift heavy or bulky materials, and reduce unnecessary bending, twisting and reaching when lifting materials:

1. Optimize Material Flow Through the Workplace:

    ☐ Reduce Manual Handling of Materials to a Minimum.

    ☐ Establish Adequate Receiving, Storage, and Shipping Areas.

    ☐ Maintain Adequate Aisle and Access Areas.

2. Eliminate the Need to Lift or Lower Manually:

    ☐ Lift Tables and Platforms   ☐ Elevated Pallets
    ☐ Lift Trucks   ☐ Gravity Dump Systems
    ☐ Cranes and Hoists   ☐ Elevating Conveyors
    ☐ Drum and Barrel Dumpers   ☐ Vacuum Systems

3. Increase the Weight so it must be Handled Mechanically:

    ☐ Palletized Handling of Raw Materials and Products.

    ☐ Unit Load Concept (Bulk Handling in Large Bins).

4. Reduce the Weight of the Object:

    ☐ Reduce the Weight and Capacity of the Container.

    ☐ Reduce the Load in the Container.

    ☐ Specify the Quantity per Container to Suppliers.

5. Reduce the Hand Distance From the Body:

    ☐ Change the Shape of the Object or Container.

    ☐ Provide Grips or Handles.

6. Convert Lift/Lower Combined with a Carry to a Push or Pull:

    ☐ Conveyors, Hand Trucks, Carts   ☐ Ball Caster Tables

C. Pushing and Pulling Tasks: Ergonomic Design

The following checklist should be used to eliminate manually pushing or pulling materials, or to reduce the exertion hazard when materials are pushed or pulled:

1. Eliminate the Need to Push or Pull:

   ☐ Conveyors ☐ Lift tables

   ☐ Slides or chutes ☐ Powered trucks

2. Reduce Force Required to Push or Pull:

   ☐ Reduce size and/or weight of load.

   ☐ Utilize four-wheel trucks or dollies.

   ☐ Utilize non-powered conveyors.

   ☐ Require that wheels & casters on hand-trucks & dollies have:
   - Periodic lubrications of bearings
   - Adequate maintenance
   - Proper sizing (provide larger diameter wheels & casters)

   ☐ Maintain floors to eliminate holes & bumps.

   ☐ Require surface treatment of floors to reduce friction.

3. Reduce the Distance of the Push or Pull:

   ☐ Relocate receiving, storage, production, or shipping areas.

   ☐ Improve production to eliminate unnecessary material handling.

4. Optimize Technique of the Push or Pull:

   ☐ Replace pull with a push whenever possible.

   ☐ Use ramps with slope less than 10%.

D. Repetititve Hand Tasks: Ergonomic Design

Workers may experience pain, discomfort, and disabling hand and wrist disorders such as carpal tunnel syndrome if the high number of repetitions are combined with abnormal wrist postures and excessive forces. The following checklist should be used as a guide for designing safe hand and wrist activities:

- ☐ 1. Reduce the number of repetitions per shift. Automated systems should be used whenever possible.

- ☐ 2. Maintain neutral (handshake) wrist positions. Design jobs and tools so that the wrist is not flexed forward, extended backward, or bent from side to side. Avoid inward and outward rotation of the forearm when the wrist is bent to minimize stress to the elbow.

- ☐ 3. Reduce the force or pressure on the wrists and hands. Reduce the weight and size of objects that must be handled repetitively. Avoid using the hand as a "hammer".

- ☐ 4. Design tasks so a power grip rather than a finger pinch grip can be used to grasp materials. Note that a pinch grip is five times more stressful than a power grip.

- ☐ 5. Avoid reaching more than 16 inches in front of the body.

    Avoid reaching above shoulder height, below waist level, or behind the body to minimize shoulder disorders.

- ☐ 6. Provide support devices where awkward body postures (elevated hands, elbows and extended arms) must be maintained

- ☐ 7. Avoid tools & equipment that transmit vibration to the hands.

- ☐ 8. Avoid exposure of the hands to cold, hot, and vibration.

- ☐ 9. Wear gloves that fit properly to enhance grip strength and manual dexterity.

- ☐ 10. Select and use properly designed hand tools.

E. Selection of Ergonomic Hand Tools

Poorly designed hand tools that combine repetitive forceful grip exertions with bent wrist postures can cause carpal tunnel syndrome and other cumulative trauma disorders. The following checklist can be used as a guide for selecting hand tools:

- ☐ 1. Maintain straight wrists. Avoid bending or rotating the wrists. Remember, bend the tool, not the wrist. A variety of bent-handle tools are commercially available.

- ☐ 2. Avoid static muscle loading. Reduce both the weight and size of the tool. Do not raise or extend elbows when working with heavy tools. Provide counter-balanced support devices for heavier tools.

- ☐ 3. Avoid stress on soft tissues. Stresses result from poorly designed tools that exert pressure on the palms or fingers. Examples include short-handled pliers and tools with finger grooves.

- ☐ 4. Reduce grip force requirements. A compressible gripping surface rather than hard plastic is best.

- ☐ 5. Whenever possible, select tools that utilize a full-hand power grip rather than a precision finger grip.

- ☐ 6. Maintain optimal grip span. Optimal grip spans for pliers, scissors, or tongs, measured from the fingers to the base of the thumb, range from 2.5 to 3.5 inches. Recommended handle diameters for circular-handle tools such as screwdrivers are 1.5 to 2 inches when a power grip is required, and 0.3 to 0.6 inches when a precision finger grip is needed.

- ☐ 7. Avoid sharp edges and pinch points. Select tools that will not cut or pinch the hands even when gloves are not worn.

- ☐ 8. Avoid repetitive trigger-finger actions. Select tools with large switches that can be operated with all four fingers. Also, use of the thumb is preferred to using a single finger for trigger action. Proximity switches are the most desirable triggering mechanism.

- ☐ 9. Protect hands from excessive heat, cold, and vibration.

- ☐ 10. Wear gloves that fit properly to enhance strength and dexterity.

F. Computer Workstation: Ergonomic Design

Use of the following set of computer workstation design guidelines will maximize the performance of the computer operator while maintaining human comfort, well-being, efficiency, and safety. Note that <u>adjustability</u> of the workstation is the key.

- [ ] The <u>keyboard and mouse surface</u> should be height adjustable between 23 and 28 inches. Both keyboard and mouse should be positioned at elbow height.

- [ ] <u>Workstation width</u> should be at least 30 inches.

- [ ] <u>Depth of the workstation</u> should allow for screen, keyboard, and approximately 3 inches to serve as a wrist rest area.

- [ ] <u>Edges of the work surface</u> must be rounded and at least one inch thick to prevent stress on the arms and wrists.

- [ ] A <u>wrist rest</u> should be provided to enhance workstation adjustability.

- [ ] <u>Knee room design considerations:</u>
  - Knee room height should be a minimum of 26 inches.
  - Knee room width should be a minimum of 20 inches.
  - Knee room depth should be a minimum of 15 inches.

- [ ] <u>Chair design considerations</u>:
  - Chair base should be five-point, on casters. Chair should swivel.
  - Chair should have an adjustable seat height.
  - Seat size between 15 to 17 inches deep, and 18 inches wide with a "waterfall" front edge.
  - Seat slope should be between 10 forward and 10 degrees backward.
  - Seat pan and backrest should be upholstered. When seated, the seat pan and backrest should not compress more than 3/4 inch.
  - Backrest size should be 7 inches high and 13 inches wide; backrest height should be adjustable between 3 and 6 inches; backrest depth should be adjustable between 14 and 17 inches; and backrest tilt should be adjustable 15 degrees forward and backward.
  - Removable, height-adjustable arm rests should be incorporated.
  - Backrest, if adjustable, should have a locking mechanism.

- [ ] <u>Screens</u> should be located directly in front of the operator at a distance of 18 to 20 inches. Screens should be located at right angle to windows.

- [ ] <u>Task lighting,</u> rather than overhead lighting, should be provided to minimize glare. Work surfaces and walls should be furnished with non-glare materials.

G. Ergonomic Guidelines for Computer Operators

The simple adjustments outlined below may increase the comfort of your computer workstation. Consider the following to prevent musculoskeletal and visual fatigue:

1. <u>Adjust the height of your work surface and the height of your chair</u> so that your keyboard is at elbow height and your feet are flat on the floor. If the work surface is not adjustable, a footrest should be used when the feet do not rest flat on the floor.

2. <u>Adjust the back rest of your chair</u> so that it provides support to your lower back. Do not sit on the edges of the chair: rest your back against the backrest.

3. <u>Position the screen</u> directly in front of you. The distance between your eyes and the screen should be approximately 18 to 20 inches.

4. <u>Adjust the height of the screen so</u> that your eyes are level with the top of the screen.

5. <u>Tilt the screen to minimize glare</u>. Tilting the screen will help reduce glare caused by bright overhead lights.

6. <u>Draw drapes or shades and utilize task lighting</u> rather than bright overhead lighting when working at the computer to reduce glare.

7. <u>Use a document holder</u>. Documents placed flat on the desk will cause you to lean forward and flex your neck, leading to fatigue and discomfort. The document and screen should be located at approximately the same distance to eliminate constant eye refocusing at varying viewing distances.

8. <u>Keep the area under your desk clear</u> for adequate leg and knee room.

9. <u>When keying, rest your wrists and/or forearms, and keep the upper arms nearly vertical</u> to prevent fatigue. Use a wrist rest, if necessary, to maintain your wrists, hands, and arms in a straight horizontal line.

10. <u>Take frequent micro-breaks and stretch periodically</u> to reduce the soreness and stiffness related to fixed, static work postures.

H. Ergonomic Guidelines for Laboratory Employees

1. Position equipment and tools directly in front of your body to reduce twisting motions and reach distances.

2. Eliminate excessive reaches over 16 inches.

3. Set the work height (hand height) at 2 inches below the elbows to minimize static muscle fatigue of the shoulders, neck, and back.

4. Maintain good seated posture. Adjust the backrest so that it provides firm support to your lower back. Do not sit on the edge of the chair.

5. Minimize static loads and fixed work postures. Avoid:
   - leaning to the front or sides
   - holding an extremity in a bent or extended position
   - tilting your head forward more than 15 degrees
   - bending your body forward or backward more than 15 degrees.
   - supporting your weight with one leg

6. Support the limbs. Use elbow, wrist, arm, and back rests when needed.

7. Reduce the number of repetitive hand, wrist, and finger motions. The use of the thumb is preferred to the use of a single finger for trigger action. Motorized pipettes eliminate repetitive, forceful trigger-finger motions.

8. Maintain neutral (handshake) wrist postures. Design experiments so that the wrist does not need to be flexed forward, extended backward, or bent from side to side.

9. Reduce grip force requirements. Select pipettes with a compressible, rather than hard plastic-gripping surface. Use tools that utilize a full-hand power grip, rather than a more forceful precision finger grip.

10. Avoid pounding with the pipette to pick up tips.

11. Select tools which minimize stress on soft tissues. Stress concentrations result from tool handles that exert pressure on the palms or fingers. Examples include short-handle pliers and tools with finger grooves that do not fit the specific employee's hand.

Ergonomics Checklists

I. Audit of Materials-Handling Risk Factors

The following items should be considered when evaluating new or existing materials-handling equipment, processes, work activities, or work stations:

1. Does the task require any of the following activities?:

- ☐ Lifting, lowering, or carrying more than 25 pounds
- ☐ Lifting, lowering, or carrying objects that are too bulky to easily grip and hold close to the body
- ☐ Lifting, lowering, or carrying materials more than 50 times per shift
- ☐ Lifting above shoulder height or below waist level
- ☐ Lifting in cramped areas resulting in bending, reaching, twisting
- ☐ Pushing/pulling carts, etc. that require large forces to get moving
- ☐ Maintaining a fixed or awkward work posture (e.g., overhead work, twisted or bent back, kneeling, stooping or squatting)

2. Do any of the following unsafe practices or conditions exist?:

- ☐ Employees not following procedures (taking short-cuts, etc.)
- ☐ Employees not using appropriate materials-handling equipment
- ☐ Materials-handling equipment inadequate or damaged
- ☐ Materials improperly stacked, loaded, or banded on fork trucks

3. Do any of the following unsafe work area conditions exist?:

- ☐ Unsafe (cracked or broken pavement, etc.) walking surfaces
- ☐ Poor housekeeping (wet, oily floors; debris; clutter)
- ☐ Poor layout of work area --- crowding, congestion, excessive traffic
- ☐ Excessive noise, heat, humidity, cold, or poor illumination

AREA AUDITED: _____

AUDITED BY: _____

J. Audit of Repetitive Hand Tasks

The following items should be evaluated when evaluating new or existing work activities requiring highly repetitive motions with the hands or wrists (i.e., manual packaging or inspection activities, and the use of hand tools:

1. **Does the task require any of the following activities?:**

    - ☐ Performing the same motion every few seconds with the hands or wrists

    - ☐ Repetitively shaking cartons, bottles, or other materials (repetitive bending of the wrists)

    - ☐ Bending the wrists when working. The wrists should be maintained in a neutral (handshake) position. The job task should not require the wrists to be flexed forward, extended backward, or bent from side to side

    - ☐ Working for extended periods of time with awkward body postures (elevated hands and elbows; extended arms; reaching behind body)

    - ☐ Exerting high grip forces with the hands

    - ☐ Using a pinch grip rather than a power (curled-finger) grip

    - ☐ Using the hand as a "hammer"

    - ☐ Repetitively reaching more than 16 inches in front of the body

2. **Do any of the following unsafe hand tool practices or conditions exist?:**

    - ☐ Using vibrating or impact tools or equipment

    - ☐ Using tools that require bending or rotating the wrists

    - ☐ Raising or extending the elbows when working with heavy tools

    - ☐ Using tools requiring repetitive trigger-finger actions, excessive grip forces, or excessive forceful exertions

AREA AUDITED: _____

AUDITED BY: _____

II. WALKTHROUGH ERGONOMICS CHECKLIST FOR CARPENTRY TASKS

Site: _____ ID CODE: _____

Date: _____

Contact: _____ phone: _____

Ergonomist: _____

Job Specialty Observed: _____ JOB CODE: _____

Task Description: _____

Total Time of Observation: _____ (hrs)   START: _____
STOP: _____

A) WORKER INFORMATION:

Subject name: _____ SS# _____

Address: _____

Phone: _____

Age: _____ yrs.   SEX: M / F   ETHNIC CODE: _____

Height: _____ feet   Weight: _____ lbs.

Union Local: _____   Years employed: _____

Toolbelt: Yes / No   Tools used: _____

B) WORK EXPERIENCE: (YES = 2, NO = 1)   SCORE

1. High level of background noise?   YES NO   _____
    ÷
2. Frequent, loud impact noise?   YES NO   _____
    ÷
3. Exposure to vibration?   YES NO   _____
Body parts exposed _____ RANK4 = _____
4. Weather Conditions: Temperature _____ °F
Cloudy/Sunny   Wet/Dry   Working surface: _____

Comments /Observations: _____
_____

The checklist in Section II copyright © 1992, Greater Cincinnati Occupational Health Center, Cincinnati, Ohio. Developed by Amit Bhattacharya and his staff at the Biomechanics-Ergonomics Research Laboratory, University of Cincinnati Medical School, Cincinnati, Ohio, through a cooperative agreement with the National Institute for Occupational Safety and Health.

C) REPETITIVE MOTION:

The following frequency periods and ratings are based on the analysis made during the observation period on the specific day of the ergonomic walkthrough.

Torso   Repetition/Minute   Rating   SCORE

1. Bending at waist (not lifting)   _____ X _____ = _____

Rating: 1 for less than 9 repetitions/minute
        2 for 9 or more repetitions/minute   +

2: Stooping (bending with legs straight)   _____ X _____ = _____

Rating: 1 for less than 6 repetitions/minute
        2 for 6 or more repetitions/minute   +

3. Turning/Twisting of upper torso   _____ X _____ = _____

Rating: 1 for less than 9 repetitions/minute
        2 for 9 and more repetitions/minute   RANK 1 = _____

Lower Extremities   Repetition/Minute   Rating: SCORE

1. Squatting   _____ X _____ = _____

Rating: 1 for less than 6 repetetitions/minute
        2 for 6 and more repetitions/minute   +

2. Kneeling   _____ X _____ = _____

Rating: 1 for less than 6 repetitions/minute
        2 for 6 or more repetitions/minute   RANK 2 = _____

Upper Extremities/Wrist
1. Cycle time: Rating   SCORE

Rating: 1 for cycle time ≥ 30 seconds
        2 for cycle time < 30 seconds   _____ = _____

2. Deviation of wrist   +

Rating: 1 for deviation < 30°
        2 for deviaion ≥ 30   _____ = _____

                                                                      RANK 3 = _____

D. POSTURES

NECK-SHOULDERS

1	Free and relaxed Time spent in this posture:_____	
2	In a natural posture, but limited by the work Time spent in this posture:_____	
3	Tense due to the work Time spent in this posture:_____	
4	Neck twisted or bent and/or the upper part of the arms at or above shoulder level Time spent in this posture:_____	
5	Neck bent backwards, great demand for strength in the arms Time spent in this posture:_____	

BACK

1	In a natural posture and/or well supported in seated and standing position Time spent in this posture:_____	
2	In a good posture, but limited by the work Time spent in this posture:_____	
3	Bent and/or poorly supported Time spent in this posture:_____	
4	Bent and twisted without support Time spent in this posture:_____	
5	In a poor posture during heavy work Time spent in this posture:_____	

ELBOW

1	Free in a posture of choice, demand for strength small Time spent in this posture:_____	
2	Arms in a position required by the work, slightly tense at times Time spent in this posture:_____	
3	Arms tense and/or the joints in an extreme posture Time spent in this posture:_____	
4	Arms maintain static contraction with elbows elevated at an angle $> 30°$ Time spent in this posture:_____	
5	Great demand for strength in the arms Time spent in this posture:_____	

HIPS-LEGS

1	In a free position that can be changed at will Time spent in this posture: _____	
2	In a good posture, but limited by the work Time spent in this posture: _____	
3	Poorly supported, or unstable/soft footing; feet at two different levels Time spent in this posture: _____	
4	Standing on one foot or in a kneeling or stooping position Time spent in this posture: _____	
5	In a poor posture during heavy work Time spent in this posture: _____	

WRIST
(not using tools)

#	Description	Illustration
1	Wrist relaxed and straight no bending, little movement no stress on joint Time spent in this posture:_____	**1** neutral
2	rist bent side to side radial or ulnar deviation. Slight force on joint. Time spent in this posture:_____	radial / ulnar **2**
3	Wrist bent in up/down motion in flexion or extension. Slight force on joint Time spent in this posture:_____	flexion / extension **3**
4	Wrist bent and used to grasp object (other than a tool) involving thumb as a support. Time spent in this posture:_____	**4** pinch grip
5	Wrist in extreme/awkward posture requiring excessive force on joint Time spent in this posture:_____	**5**

TOOL DESIGN

1	Tool is light-weight and fits hand comfortably, wrist not deviated Time spent in this posture: _____	1
2	Tool is powered, requires little exertion to use; little to no vibration Time spent in this posture: _____	2
3	Tool handle pinches or puts pressure on palm; tool is hard to grasp and/or blows cool air currents on hand from tool motor Time spent in this posture: _____	3
4	Tool is heavy and awkward to use, and/or causes vibration/impact Time spent in this posture: _____	4
5	Tool design causes deviation of wrist more than 30^0 from neutral position and/or causes use of pinch grip Time spent in this posture: _____	5

E) LIFTING TASK:

PHYSICAL STRESS JOB ANALYSIS SHEET

DEPARTMENT_____ DATE_____

JOB TITLE_____ ANALYST'S NAME_____

| Task Description | Object Weight | | Hand Location: | | Task Freq | ASYM | Coupling | RWL |
	Ave	Max	Origin H V cm cm	Destination H V cm cm				

F) JOB/TASK REQUIREMENT DATA: Start time:_____
                                    Stop time:_____

Resting heart rate:_____ beats /minute

Maximum Heart rate:_____ beats /minute = (220 - age)

TIME	TASK	HEART RATE	TYPE OF FORCE*	MEASURED FORCE

* TYPES OF FORCE: IMPACT, GRIP, PUSH, PULL and LIFT

NOTE: The times used in the above anlysis should be rounded to the nearest second or minute, whichever the case may be

III. SYMPTOMS SURVEY: ERGONOMICS PROGRAM CHECKLIST

A. Symptoms Survey Checklist

Symptoms Survey: *Ergonomics Program*

DATE ___/___/___

_____ _____ _____ _____
Plant   Dept #   Job #   Job Name

_____ _____ _____ ___years___months
Shift   Supervisor   Hours worked/week Time on THIS job

Other jobs you have done in the last year (for more than 2 weeks)

_____ _____ _____ _____ ___months___weeks
Plant   Dept #   Job #   Job Name   Time on THIS job

_____ _____ _____ _____ ___months___weeks
Plant   Dept #   Job #   Job Name   Time on THIS job

(If more than 2 jobs, include those you worked on the most)

Have you had any pain or discomfort during the last year?
☐ Yes   ☐ No   (If NO, stop here)

If YES, carefully shade in the area of the drawing which bothers you the MOST.

Front   Back

(Continued)

Developed by Thomas R. Hales, of the National Institute for Occupational Safety and Health, and Patricia K. Bertsche, of the Occupational Safety and Health Administration. (Reprinted with permission from the *American Association of Occupational Health Nursing Journal* 40(3): 1992.)

B. Physical Examination Recording Form for Health Care Providers

(Complete a separate page for each area that bothers you)

Check Area: ☐ Neck   ☐ Shoulder   ☐ Elbow/Forearm   ☐ Hand/Wrist   ☐ Fingers
☐ Upper Back   ☐ Low Back   ☐ Thigh/Knee   ☐ Low Leg   ☐ Ankle/Foot

1. Please put a check by the word(s) that best describe your problem

   ☐ Aching
   ☐ Burning   ☐ Numbness (asleep)   ☐ Tingling
   ☐ Cramping   ☐ Pain   ☐ Weakness
   ☐ Loss of Color   ☐ Swelling   ☐ Other
                       ☐ Stiffness

2. When did you first notice the problem? _____(month) _____(year)

3. How long does each episode last? (Mark an X along the line)

   _____/_____/_____/_____/_____
    1 hour   1 day   1 week   1 month   6 months

4. How many separate episodes have you had in the last year?_____

5. What do you think caused the problem?_____
   _____

6. Have you had this problem in the last 7 days?   ☐ Yes   ☐ No

   7. How would you rate this problem (mark an X on the line)
   NOW

   _____
   None   Unbearable

   When it was the WORST

   _____
   None   Unbearable

8. Have you had medical treatment for this problem?   ☐ Yes   ☐ No
   8a. If NO, why not_____

   8b. If YES, where did you receive treament?_____
      1. Company Medical ☐   Times in past year_____
      2. Personal doctor ☐   Times in past year_____
      3. Other ☐   Times in past year_____
   8c. If YES, did the treatment help?   ☐ Yes   ☐ No

9. How much time have you lost in the last year because of this problem?_____days

10. How many days in the last year were you on restricted or light duty because of this problem?
    _____days

11. Please comment on what you think would improve your symptoms
    _____
    _____
    _____
    _____

Name: _____   Current Job: _____
Examiner: _____   Date: ___/___/___

Discomfort Scale: 1=no discomfort, 2=mild, 3=moderate, 4=severe, 5=worst ever

NECK:
Inspection: Inflammation (red, swollen, warm) ___Yes   ___No
Palpation: <u>Right</u>   <u>Left</u>
    Trapezius Trigger Point   ___   ___
    Trapezius Spasm   ___   ___
Manevuers:
  Resisted Flexion   ___   ___
  Resisted Extension   ___   ___
  Resisted Rotation   ___   ___

SHOULDER
Inspection: Acromium Inflammation?   ___Yes(R or L)   ___No
Maneuvers: <u>Right</u>   <u>Left</u>
  Passive Abduction   ___   ___
  Active Abduction   ___   ___
  Resisted Abduction   ___   ___
  Deltoid Palpation   ___   ___

ELBOW
Inspection: Olecranon Inflammation: ___Yes(R or L)   ___No
Palpation: <u>Right</u>   <u>Left</u>
    Medial Epicondyle   ___   ___
    Lateral Epicondyle   ___   ___

FOREARM
Inspection: Forearm Inflammation?   ___Yes(R or L)   ___No
Maneuvers: <u>Right</u>   <u>Left</u>
  Passive Wrist Flexion   ___   ___
  Passive Wrist Extension   ___   ___
  Resisted Wrist Flexion   ___   ___
  Resisted Wrist Extention   ___   ___
  Resisted Finger Flexion   ___   ___
  Resisted Finger Extention   ___   ___
  3rd digit resisted Extention   ___   ___

WRIST
Inspection: Inflammation   ___Yes(R or L)   ___No
    Extensor ganglion cyst   ___Yes(R or L)   ___No
    Flexor ganglion cyst   ___Yes(R or L)   ___No
Maneuvers: <u>Right</u>   <u>Left</u>
  Guyon Tinel's   ___   ___
  Carpal Tinel's   ___   ___
  Phalen's   ___   ___

HANDS AND FINGERS
Inspection: Inflammation   ___Yes(R or L)   ___No
Maneuvers: <u>Right</u>   <u>Left</u>
  Trigger Finger   ___   ___
  Finkelstein's   ___   ___

Appendix C
Electronic Sources of Information

BIOMCH-L

BIOMCH-L is a listserver that provides a forum for discussion among experts/practitioners in the fields of biomechanics, bioengineering, and ergonomics. The wide variety of membership generates a multidisciplinary platform for discussing relevant topics of interest.

To subscribe to the list, send e-mail to:

Bitnet users: listserv@hearn.bitnet
Other users: listserv@nic.surfnet.nl

Mail the following one-line message: subscribe biomch-1 YourFirstName YourLastName (YourAffiliation). Follow the instructions sent to you by the listserver to complete the process. World Wide Web (WWW) archives are located at http://www.kin.ucalgary.ca/isd/biomch_1.html

BIOMECHANICS YELLOW PAGES

This is an electronic database of products and services related to the field maintained by Pierre Baudin at the following WWW address:

http://www.orst.edu/~bowenk/byp.html

ERGOWEB

This is a WWW site for useful information and services related to the field. It provides a comprehensive source of information, software, references, and simple software programs for practicing ergonomists and biomechanists:

http://ergoweb.mech.utah.edu/

BIOMECHANICS WORLDWIDE (BWW)

Another comprehensive and updated source maintained by Dr. Pierre Baudin, for navigating the WWW sites dealing with biomechanics and ergonomics:

http://dragon.acadiau.ca/~pbaudin/biomch.html

COMPUTER SYSTEMS TECHNICAL GROUP

This is a forum for discussing topics related to computer applications in human factors. Send a one-line e-mail message to listserv@vtvml.cc.vt.edu. The body of the mail should contain the message:

subscribe CSTG-L "Your Full Name in Quotes"

HUMAN FACTORS TECHNICAL GROUP ON AGING

This is a forum for discussing topics related to human factors and aging. Send a one-line e-mail message to listserv@UCF1VM.cc.ucf.edu. The body of the mail should contain the message:

subscribe HFTGA "Your Full Name in Quotes"

HUMAN FACTORS COMMUNICATIONS TECHNICAL GROUP

This is a forum for discussing topics related to technical discussions in communications in human factors. Send a one-line e-mail message to listserv@VTVM1.cc.vt.edu. The body of the mail should contain the message.

subscribe COMMS-L "Your Full Name in Quotes"

BIOMECHANICS-ERGONOMICS RESOURCE TANK

This is a source of information related to biomechanics and ergonomics, with useful links to WWW sites serving a parallel purpose; maintained by the authors:

http://www.uc.edu/~bhattatt/welcome.html

GOVERNMENT AND ORGANIZATIONAL ELECTRONIC SOURCES OF ERGONOMICS INFORMATION

http://www.cdc.gov/niosh/homepage.html	NIOSH home page
http://www.osha.gov/index.html	Occupational Safety & Health Administration
http://thomas.loc.gov	Thomas: legislative information on the internet
http://www.ccohs.ca/	Canadian Centre for Occupational Health and Safety
http://www.who.ch/	World Health Organization (WHO) home page
http://www.osha-slc.gov/	OSHA Salt Lake City home page
http://ergo.human.cornell.edu	Cornell Ergonomics
http://www2.ncsu.edu/CIL/NCERC/index.html	North Carolina Ergonomics Resource Center
http://www.os.dhhs.gov/	Department of Health and Human Services home page

Appendix D
Ergonomics Software Sources

Package	Description	Hardware	Contact
Vision 3000, by Promatek	Video-based system for upper and lower body posture analysis and biomechanical lifting assessments. Will perform NIOSH lifting equation calculations and interface with Static Strength Prediction Program.	DOS Compatible 386 or 486 w/2MB RAM, 200 MB hard disk, ATI compatible VGA graphics card and monitor. Sony 8-mm camcorder or VCR with remote-controlled frame advance.	Promatek Medical Systems, Inc. 1851 Black Rd. Joliet, IL 60435
OCS TOOLS, by Triangle Research Collaborative Inc.	Video-based time and motion study system. Models also available without video-based system.	IBM or compatible, VCR (dual-audio lines, external advanced sync is preferred). Video time character generator or recordable video overlay.	Triangle Research Collaborative Inc. P.O. Box 12167 100 Park, Suite 115 Research Triangle Park, NC 27709 (919) 549-9093
3D Static Strength Prediction Program, by University of Michigan Software	Predicts human static strength requirements of manual material handling tasks (lifts, presses, pushes, pulls). Applicable to worker motions in 3-D space.	IBM PC or compatible with Intel 486SX or 486DX running at 33 MHz or greater and 80387 math coprocessor. Windows 3.1 or greater and 4 MB RAM.	University of Michigan Software Wolverine Tower Room 3003 S. State St., #2071 Ann Arbor, MI 48109-1280 (313) 936-0435 http://www.tmo.umich.edu
Jack™ by University of Pennsylvania	Human factors design tool for use with CAD systems.		University of Pennsylvania Center for Technology Transfer 133 South 36th St., Suite 419 Philadelphia, PA 191-4-3246 (215) 898-1488
LifTrak, by Motion Analysis Corporation (also, GaitTrak, SpineTrak, and FootTrak)	Determines lumbar isometric compression forces during lifting, pushing, pulling, or setting down activities and compares to NIOSH lifting standards.	Video-based system.	Motional Analysis Corporation 3650 North Laughlin Road Santa Rosa, CA 95403 (707) 579-6500 fax (707) 526-0629

Micro Saint Animation Simulation software, by Micro Analysis and Design	Graphical task network simulation modeling software with animation features to create, for example, moving a pictures of the operator and the controls/displays being used. Used for manpower estimation, workload analysis and human performance prediction.	PC and PC-compatibles, Macintosh (animation feature available).	Micro Analysis and Design 3300 Mitchell Lane, Suite 175 Boulder, CO 80301 (303) 442-6947 fax (303) 442-8274
Dynalift, by Institute for Ergonomics Research, Texas Tech University	A two-dimensional dynamic biomechanical model that calculates L5/S1 compression, L5/S1 shear, and reaction forces at the joints for lifting tasks. Graphics available that plot forces over time to lift.	PC and PC-compatibles.	Institute for Ergonomics Research Dept. of Industrial Engineering Texas Tech University Lubbock, TX 79409 (806) 742-3543 fax (806) 742-3411
Manual Materials Handling, by Institute for Ergonomics Research Texas Tech University	Determines psychophysical and NIOSH calculated load limits for manual materials tasks, such as lifts, prediction of individual capacity.	PC and PC-compatibles.	Institute for Research Dept. of Industrial Engineering, Texas Tech University Lubbock, TX 79409 (806) 742-3543 fax (806) 742-3411
2D Static Strength Prediction Program, by University of Michigan Software	Predicts human static strength requirements of manual material handling tasks, such as lifts, presses, pushes, and pulls, applicable to worker motions in the sagittal plane. (A metabolic load model is also available.)	IBM PC (AT, XT, PS-2) or compatible with 3.5: or 5.25" disk drive, 288 K RAM, Dos 3.0, IBM EGA (color) or CGA (monochrome), high-quality dot matrix or graphics-capable printer for printed output.	University of Michigan Software Intellectual Properties Office 475 E. Jefferson, Room 2354 Ann Arbor, MI 48109-1248 (313) 936-0435

Package	Description	Hardware	Contact
CASHE 1.0 Computer Aided Systems Human Engineering, by CSERIAC	An interactive multimedia computer database for crew system design for prototyping which allows access to and simulation of the information contained in a complete implementation of Engineering Data Compendium and MIL-STD-1472D, with hyperlinks between these two human engineering reference documents.	Macintosh II family of computers equipped with at least a 13"-grayscale or color monitor, a CD-ROM drive, and a hard disk for storage of user-defined files. Some auditory phenomena (e.g., stereo effects) will require use of dual speakers or headphones	Mr. Don Monk Armstrong Laboratory Crew Systems Directorate Human Engineering Division Wright-Patterson AFB OH 45433-6573 (513) 255-8814 fax (513) 255-9198 e-mail dmonk@al.wpafb.af.mil
Seated Workplace Analysis and Design (SWAD) Expert System, by Techman Inc.	An expert system specifically created for analyzing and/or designing regular as well as CDT-seated workplaces.	IBM PC or compatibles.	Techman, Inc. P.O. Box 842 1300 Washington Avenue Miami Beach, FL 33139
CREW CHIEF, by CSERIAC	An interactive computer-aided design (CAD) human factors evaluation system which is interfaced to existing commercial CAD systems. The computerized human-model used to analyze maintenance situations: technician anthropometry; allowances for clothing, personal protective equipment and tools; simulation of posture, and physical and visual accessibility; and strength capabilities of manual material handling tasks and tool use.	Interfaces to: CADAM V.20 with 3-D Interactive Design and Geometry Interface Modules for MVS/SP operating system, FORTRAN 66 CADAM V.21 with Interface User Exit for MVS operating system, VS FORTRAN 66 and 77, and for VM operating system ComputerVision version CADDStation for CADDS 4X Rev. 3.3, UNIX 4.2-based Rev. 5.0 operating system ComputerVision CADS 4001 with an analytical processing unit and CADDS 4X software, Rev. 5B and later System Independent FORTRAN 66 and 77 Interfaces to CATIA V.3 and I-DEAS	Cindy D. Martin CSERIAC Program Office AAMRL/HE/CSERIAC Wright-Patterson AFB, OH 45433-6573 (513) 255-4842 fax (513) 255-4823

Ergonomics Software Sources

COMBIMAN by CSERIAC	An interactive computer graphics human factors evaluation system for human-crewstation interface. Three-dimensional modeling system which creates a computerized man-model with Air Force pilot antropometry allowances for clothing and personal protective equipment, vidual accessibility, strength for operating controls, reach capability with the arms and legs, and fit limitations.	IBM 370 OS/VS-compatible environment using an IBM 2250-3-compatible display tube and an on-line Versatec 22-inc electrostatic plotter. Requires 650 K computer memory and minimum of 20 K graphics buffer control area. Future interfaces to the following CAD systems are in progress: CADAM V.20, CADAM V.21, CATIA V.3 and System Independent	Cindy D. Martin CSERIAC Program Office AAMRL/HE/CSERIAC Wright-Patterson AFB, OH 45433-6573 (513) 255-4842 fax (513) 255-4823
Computer Health Break, by Escape Ergonomics	Program that has 70 exercises that pop up automatically, based on VDT user's workload. Has a learning system that teaches health, business, and computer topics, and a glossary that explains medical terms and computer jargon.	IBM PCs and compatibles, Dps 2.1.	Escape Ergonomics 1111 W. El Camino Real, Suite 109 M/S 403 Sunnyvale, CA 94087
SAMMI CAD	Computerized three-dimensional anthropometric model for workstation analysis.		SAMMI C.A.D. Ltd. 22 Station Rd. Quorn Loughborough Leicestershire LE12 8BS UK

PARTIAL LIST OF SOFTWARE SOURCES FOR THE 1991 REVISED NIOSH LIFTING GUIDELINES

Source	Contact person/phone	Program name	Program type
Continental Insurance Corporate Loss Control One Continental Drive Cranbury, NJ 08570-001	Dave Mahone (609) 395-5734 (312) 822-4447	Smart-Lift	DOS
Kansas State University Dept. of Industrial and Manufacturing System Engineering 237 Durland Hall Manhatten, KS 66506	Stephan Konz (913) 532-5606		DOS
SAFECO Insurance 3637 Geyer Rd. Sunset Hills, MO 63127	Ira Robinson (314) 957-4569	LIFTCALC	DOS
Safetysoft Inc. P.O. Box 45050 Westlake, OH 44145	John DeArmon (216) 871-0115	Lift Guide	DOS
TNO-Netherlands Wassenaarseweg 56 P.O. Box 124 2300 Leiden The Netherlands	Peter Vink 31 71 18 17 73	Lifting Advisor	DOS
UES Inc. 4401 Dayton-Zenia Rd. 900 Central Park West Atlanta, GA 30328	Roger Jensen (404) 390-5500	Lift Limit	DOS
Univ. of Wisconsin-Milwaukee Industrial & Systems Eng. P.O. Box 784 Milwaukee, WI 53201	Arun Garg (404) 229-6240	Revised NIOSH Guide Program	Windows
USF&G Co. Suite 600 9000 Central Park West Atlanta, GA 30328	Bill Boyd (770) 390-5836	ERGO	DOS
SAIF Corp. 2285 Stewart Parkway Roseburg, OR 97470	John Ratliff (503) 672-4440	NIOSH Equations	Apple Hypercard

Appendix E
Information Sources from the National Institute for Occupational Safety and Health

Telephone: 1-800-35-NIOSH (9:00–4:00 EST). Technical information service number that provides access to NIOSH and its information resources.
(513) 533-8573 (fax)
(513) 533-8328 (outside U.S.)

For further information, write:

NIOSH
Education & Information Division
4676 Columbia Parkway
Cincinnati, OH 45226-1998

http://www.cdc.gov/NIOSH/homepage.html (NIOSH internet address)
pubstaft@niosdt1.em.cdc.gov (e-mail address request for NIOSH publications)

SELECTED NIOSH PUBLICATIONS IN ERGONOMICS

95-119	Cumulative Trauma Disorders in the Workplace: Bibliography
95-102	Effective Interventions to Reduce Ergonomic Injuries in Meatpacking
95-114	A Strategy for Industrial Power Hand Tool Ergonomic Research Design, Selection, Installation, and Use in Automotive Manufacturing: Proceedings of a NIOSH Workshop
94-124	Participatory Ergonomic Interventions in Meatpacking Plants
Biblio.	Carpal Tunnel Syndrome: Selected References
Reprint	Revised NIOSH Equation for the Design and Evaluation of Manual Lifting Tasks
94-127	Back Belts: Do They Prevent Injury?
94-122	Workplace Use of Back Belts
91-100	Selected Topics in Surface Electromyography for Use in the Occupational Setting: Expert Perspectives
89-106	Occupational Exposure to Hand–Arm Vibration
83-110	Vibration Syndrome

NIOSH EDUCATIONAL RESOURCE CENTERS OFFERING ERGONOMICS COURSES

Deep South Center for Occupational Health and Safety
University of Alabama at Birmingham; Auburn University
(205) 934-7178, fax (205) 975-7179

Northern California Educational Resource Center
University of California, Berkeley, Davis and San Francisco
(510) 231-5645, fax (510) 231-5648

Southern California Educational Resource Center
University of Southern California; University of California, Los Angeles
(213) 740-3995, fax (213) 740-8789

The Great Lakes Center for Occupational and Environmental Health and Safety
University of Illinois at Chicago
(312) 413-1113, fax (312) 413-7369

Johns Hopkins Educational Resource Center
Johns Hopkins University, Maryland
(410) 955-3602, fax (410) 955-9334

Harvard Educational Resource Center
Harvard University
(617) 432-3314, fax (617) 432-0219

Michigan Education Resource Center
University of Michigan
(313) 936-0148, fax (313) 764-3451

Minnesota Educational Resource Center
University of Minnesota School of Public Health, St. Paul-Ramsey Medical Center
(612) 221-3992, fax (612) 292-4773

New York/New Jersey Educational Resource Center
Mt. Sinai School of Medicine, University of Medicine and Dentistry of New Jersey—Robert Wood Johnson Medical School; Hunter College School of Health Sciences, New York University Medical Center, New Jersey Institute of Technology Consortium
(908) 235-5062, fax (908) 235 5133

North Carolina Educational Resource Center
University of North Carolina, Duke University Medical Center
(919) 962-2101, fax (919) 966-7579

University of Cincinnati Educational Resource Center
University of Cincinnati
(513) 558-1729, fax (513) 558-1756

Southwest Center for Occupational and Environmental Health
University of Texas, Houston
(713) 792-4648, fax (713) 792-4407

Rocky Mountain Center for Occupational and Environmental Health
University of Utah
(801) 581-4055, fax (801) 585-5275

Northwest Center for Occupational and Environmental Health
University of Washington
(206) 543-1069, fax (206) 685-3872

Appendix F
Ergonomics Journals

JOURNALS IN WHICH ARTICLES RELATED TO ERGONOMICS ARE LIKELY TO APPEAR

American Industrial Hygiene Association Journal
American Industrial Hygiene Association, 2700 Prosperity Ave., Ste. 250, Fairfax, VA 22031-4307

American Journal of Industrial Medicine
John Wiley & Sons, Inc., Journals, 605 Third Ave., New York, NY 10158. TEL 212-850-6000. FAX 212-850-6088

Applied Ergonomics
Butterworth-Heinemann, Part of the Reed Elsevier group, Linacre House, Jordan Hill, Oxford OX2 8DP, England. TEL 8065-310366. FAX 0865-310898.

Applied Occupation & Environmental Hygiene
Elsevier Science Inc., 655 Avenue of the Americas, New York, NY 10010

Clinical Biomechanics
Butterworth-Heinemann, Part of the Reed Elsevier group, Linacre House, Jordan Hill, Oxford OX2 8DP, England. TEL 0865-310366. FAX 0865-310898.

Ergonomics
Taylor & Francis Ltd., Rankine Rd., Basingstoke, Hants, RG24 8PR, England. TEL 4-1256-840366. FAX 44-1256-479438

Human Factors
Human Factors and Ergonomics Society, Box 1369, Santa Monica, CA 90406-1369. TEL 310-394-1811. FAX 310-394-2410.

International Journal of Industrial Ergonomics
Elsevier Science B.V., P.O. Box 211, 1000 AE Amsterdam, Netherlands. TEL 31-20-4853911. FAX 31-20-4853598.

Journal of Biomechanics
Elsevier Science Ltd., Pergamon, P.O. Box 800, Kidington, Oxford OX5 IDX, England. TEL 44-1865-843000. FAX 44-1865-843010.

Journal of Hand Surgery: American Volume
Churchill Livingstone International 650 Ave. of the Americas, New York, NY 10011. TEL 212-206-5040. FAX 212-727-7808.

Journal of Neurosurgery
American Association of Neurological Surgeons, 1224 W. Main St., Ste. 450, Charlottesville, VA 22903. TEL 804-924-5503. FAX 804-924-2702.

Muscle and Nerve
John Wiley & Sons, Inc., Journals, 605 Third Ave., New York, NY 10158. TEL 212-850-6645. FAX 212-850-6021.

Neurology
Advanstar Communications, Inc., 7500 Old Oak Blvd., Cleveland, OH 44130. TEL 216-826-2839. FAX 216-891-2726.

*Occupational Ergonomics**
Chapman and Hall, Inc., 29 W. 35th St., New York, NY 10001.
TEL 212-244-3336.
FAX 212-563-2269.

Occupational Medicine
Butterworth-Heinemann, Part of the Reed Elsevier group, Linacre House, Jordan Hill, Oxford OX2 8DP, England.
TEL 0865-310366. FAX 0865-310898.

Orthopedics
Slack, Inc., 6900 Grove Rd., Thorofare, NJ 08086. TEL 609-848-1000. FAX 609-853-5991.

Rheumatology
S. Karger AG, Allschwilerstr. 10. P.O. Box, CH-4009 Basel, Switzerland.
TEL 061-3061111. FAX 061-3061234.

Scandinavian Journal of Work, Environment and Health
Topeliuksenkatu 41 aA, FIN-00250 Helsinki, Finland. TEL (+ 358 0) 474 7694 and (+ 358 31) 260 8644.
FAX (+ 358 0) 878 3326.

Spine
(Philadelphia, 1986); state of the art reviews. 1986. 3/yr. $90 (foreign $99). Hanley & Belfus, Inc., 210 S. 13th St., Philadelphia, PA 19107.
TEL 215-546-7293.
FAX 215-790-9330.

Work
Elsevier Science Ireland Ltd., P.O. Box 85, Limerick, Ireland.
TEL 353-61-471944.
FAX 353-61-472144.

---

*Available by late 1996.

Glossary

Abduction. Motion away from the midline. Increases the angle between a limb and the sagittal plane.
Acceleration. The change in velocity of a body divided by the time over which change occurs.
Achilles tendon. Connects the principal plantar flexor muscles of the foot with the heel of the skeleton of the foot.
Action limit. A term from the 1981 Work Practices Guide that denotes the weight limit that nearly all workers can perform safely. The term has been replaced in the 1991 equation with the term **Recommended weight limit**.
Action potential. For example, motor unit action potential is the nerve impulse propagating down a mononeuron and activating all muscle fibers of a motor unit (see **Electromyography**).
Abduction. Motion toward the midline. Decreases the angle between a limb and the sagittal plane.
Agonistic muscles. Muscles that initiate and carry out motion.
Angle of asymmetry. The angle between the **Asymmetry line** and the **Sagittal line** of the worker's body, as defined by the worker's neutral body position; measure at the origin and destination of lift and use to compute the **Asymmetric multiplier** (see **Asymmetry line**, and **Asymmetric multiplier**)
Anisotropy. The quality whereby a material exhibits unlike mechanical properties when loaded in different directions.
Antagonistic muscles. Muscles that oppose the actions of the agonistic muscles (oppose a movement).
Anthropometry. An empirical science defining the physical measures of a persons size and form.
Area moment of inertia. Quantity that takes into account the cross-sectional area and distribution of material around an axis during bending.
Arthritis. Degenerative joint disease (also inflammation of a joint).
Asymmetric multiplier. A reduction coefficient defined as $(1 - (.0032A))$; has a maximum value of 1.0 when the load is lifted directly in front of the body and decreases linearly as the **angle of asymmetry** (A) increases.
Asymmetry line. The auxiliary line that connects the mid-point of the line drawn between the inner ankle bones and the point projected down to the floor directly below the center of the hand grasps.
Axial rotation. Rotation about an axis.
Axis of motion. Line about which all points move in a body in motion.
Axis of rotation. Line about which all points in a rotating body describe circles.
Axon. The long process of a nerve cell. Conducts impulse.

Bending. A loading mode in which a load is applied to a structure in a manner that causes it to bend about an axis, subjecting the structure to a combination of tension and compression.

Bending moment. A quantity at a point in a structure equal to the product of the applied force and the perpendicular distance from the point to the force line, usually measured in newton meters.

Biceps. Long twin-bellied muscle going from the shoulder blade to the proximal end of the radius, thus crossing and acting on both the shoulder and the elbow joints.

Bone remodeling. The ability of bone to adapt, by changing its size, shape, and structure, to the mechanical demands placed upon it.

Branchialis muscle. Short muscle originating int he lower third of anterior surface of humerus and inserting into the anterior part of the ulna close to elbow joint.

Bursa. Small bag filled with fluid, reducing friction between moving structures.

Bursitis. Inflammation of a bursa.

Capillary. Very small blood vessel that connects the smallest branches of the arteries with those of the veins.

Carpal tunnel. Channel on the palmar side of the wrist formed by the irregular small bones of the wrist and a tough ligament stretched across it. Through the carpal tunnel pass the flexor tendons of the fingers, the median nerve, and some blood vessels.

Carpus. The aggregate of eight small irregular bones forming the wrist.

Center of gravity. Equilibrium point of a supported body where all its weight is concentrated (see **Center of mass**).

Center of mass. That point at the exact center of an object's mass; often called the center of gravity (see **Center of gravity**).

Center of rotation. A point around which circular motion is described.

Chondrocyte. A cartilage cell.

Combined loading. Application of two or more loading modes to a structure.

Composite lifting index. The term that denotes the overall lifting index for a multi-task manual lifting job.

Compression. A loading mode in which equal and opposite loads are applied normal to surface of the structure, resulting in shortening and widening of structure.

Concentric contraction. Increase of tension within a muscle, producing shortening. For example, the brachialis shortens when a weight is lifted by flexing the forearm.

Coronal plan. A vertical plane perpendicular to the median plane that divides the body into anterior (front) and posterior (back) segments.

Coupling classification. The three-tiered classification of the quality of the coupling between the worker's hands and the object (either good, fair, or poor); used in the **Coupling multiplier**.

Coupling multiplier. A reduction coefficient based on the **Coupling classification** and **Vertical location** of the lift.

Creep. Progressive deformation of soft tissues due to constant low loading over an extended period of time.

Cross-sectional area. Measure of the area of a piece of material cut at right angles to its longitudinal axis.

Deformation rate. The speed at which an applied load deforms a structure (see **Speed of loading**).

Degrees of freedom. The number of ways in which a body can move.

Deltoid. Large muscle of the shoulder that abducts and otherwise moves the upper arm about the shoulder joint against external loads.

Distal. In a limb: further away from the body. Elsewhere: further away from the central axis of the body.
Distance multiplier. A reduction coefficient defined as $(.82 + (1.8/D))$, for D measured in inches, and $(.82 + (4.5/D))$, for D measured in centimeters.
Distance variable. The vertical travel distance of the hands between the origin and destination of the lift measured in inches or centimeters; used in the **Distance multiplier**.
Dorsiflexion. Bending upwards around an axis.
Duration of lifting. The three-tiered classification (either short, moderate, or long) of lifting duration specified by the distribution of work time and recovery time (work pattern).
Dynamic effort. Rhythmic alteration of contraction and extension, tension, and relaxation.
Dynamic work. "Work" according to the definition in mechanics. Defined as the product of a force multiplied by the distance through which its point of application moves.
Dynamics. The study of forces acting on a body in motion.
Elasticity. Property of a material which allows the material to return to its original shape and size after being deformed.
Electrogoniometer. Device to quantify in analog fashion or digitally an angle and changes of angle between body segments connected by a joint.
Electromyography. The recording of action potentials emitted from contracting muscles, EMG. Also used to measure action potentials from nerves.
Epicondylitis. Technical term for "tennis elbow."
Equilibrium. State of a body at rest in which the sum of all forces and moments is zero.
Ergonomics. A multidisciplinary activity dealing with the interactions between man and his total working environment, plus such traditional and environmental aspects as atmosphere, heat, light, and sun, as well as of tools and equipment of the workplace. (From American National Standard ANSI Z794.1-1972.)
Excentric contraction. Increase of tension within a muscle while lengthening. For example, the brachialis exerts a force resisting the pull of gravity when extending a flexed forearm slowly.
Extension. The position of the joints of the extremities and back when one stands at rest, or the direction of motion that tends to restore this position; the opposite of **Flexion**.
External forces. Forces applied to the body by outside objects (e.g., weight of a box being held or carried).
Fascia. Layer or sheet of connective tissue.
Fatigue fracture. A fracture typically produced by either low repetition of high loads or high repetition of relatively normal loads.
Flexion. Movement involving the bending of a joint whereby the angle between the bones is diminished; the opposite of **Extension** (except at shoulder).
Force. An action that changes the state of rest or motion of a body to which it is applied.
Force couple. Two parallel forces of equal magnitude but opposite direction applied to a structure.
Free body. A structure considered in isolation for the purpose of studying the effect of forces acting on it.
Free body diagram. Diagram of an isolated portion of a structure used during free body analysis for the purpose of studying the effect of forces acting on the free body.
Frequency-independent lifting index. Defined as (L)/(FIRWL), identifies individual tasks with potential strength problems; values exceeding 1.0 suggest that ergonomic changes may be needed to decrease the strength demands.
Frequency-independent recommended weight limits. A value used in a multi-task assessment; product of all the reduction coefficients and the LC, holding FM equal to unity;

reflects the overall strength demands for a single repetition of that task; used in **Frequency-independent lifting index**.

Frequency of lifting. The average number of lifts per minute over a 15 minute period; used in the **Frequency multiplier**.

Frequency multiplier. A reduction coefficient that depends upon the **Frequency of lifting** (F), the **Vertical location** (V) at the origin, and the **Duration of lifting**.

Friction force. A tangential force opposing motion which acts between two bodies in contact.

Frontal plane. The plane that passes through the longitudinal axis of the body.

Glycosaminoglycan (GAG). A long, flexible chain of repeating disaccharide units that are the building blocks of proteoglycans.

Goniometer. Device measuring the angle and range of angular movement between two body segments connected by a joint.

Gravitational force. A force produced by gravitational attraction by the Earth on a body.

Ground reaction force. A gravitational force produced by the weight of an object against the surface on which it lies.

Horizontal location. The horizontal distance between the mid-point of the hand grasps projected down to the floor and the mid-point of the line between the inner ankle bones; used in the **Horizontal multiplier**.

Horizontal multiplier. A reduction coefficient defined as 10/H, for H measured in inches, and 25/H, for H measured in centimeters.

Instantaneous center of motion. The immovable point existing at an instant in time created by one segment (link) of a body rotating about an adjacent segment; all other points on the body rotate about this immovable point.

Internal forces. Those forces generated by the muscles as a result of muscle tendon action (e.g., holding or carrying a weight).

Interphalangeal joints. Joints connecting finger bones.

Ischial tuberosities. Two bony prominences forming the lowest point of the pelvis. On them rests the weight of the body when seated.

Isometric work. A muscle exerts a force (i.e., contracts) against resistance without producing any motion; for example, to hold a weight still with the extended arm. Isometric work, which results in increased demand for calories, is different from work in mechanics, defined as force multiplied by the distance an object moves.

Joint lubrication. A design feature of the joint which maintains the continuity of the thin film of synovial fluid between the joint surfaces, minimizing the contact and wear of the cartilaginous surfaces.

Joint reaction force. The internal reaction force acting at the contact surfaces when a joint in the body is subjected to external loads.

Kinematics. The branch of mechanics that deals with motion of a body without reference to force or mass.

Kinesiology. The study of human movements as a function of the construction of the musculoskeletal system.

Kinetics. The branch of mechanics that deals with the motion of a body under the action of given forces.

Kyphosis. Convexity of the spine. Normally observed in the thoracic region.

Lateral. Structures farther to the sides, away from the midline.

Lever arm. The perpendicular distance from the line of application of a force to the center of motion in a rigid structure, also known as the moment arm of the force.

Lifting index. Defined as L/RWL; generally relates the level of physical stress associated with a particular manual lifting task to the number of workers who should be able to per-

Glossary

form the task (see **Load weight**). A value of 1.0 or more denotes that the task is hazardous for some fraction of the population.

Lifting task. Denoting the act of manually grasping an object of definable size and mass with two hands, and vertically moving the object without mechanical assistance.

Ligament. Connective tissue attaching bone to bone.

Load constant. A constant term in the RWL equation defined as a fixed weight of 23 kg or 51 lb; generally considered the maximum load nearly all healthy workers should be able to lift under optimal conditions (i.e., all the reduction coefficients are unity).

Load-deformation curve. A curve that plots the deformation of a structure when the structure is loaded in a known direction.

Load weight. The weight of an object to be lifted, in pounds or newtons, including the container; used in the **Lifting index**.

Locomotion. The act of moving the human body from place to place using the musculoskeletal system.

Long duration. Referring to lifting tasks that have a duration of between two and eight hours with standard industrial rest allowances (e.g., morning, lunch, and afternoon rest breaks).

Longitudinal axis. A line through the longer part of a body about which a body rotates.

Lordosis. Concave curvature of the spine. Exists in the neck and in the lumbar region.

Lumbosacral joint. Joint between fifth lumbar vertebra and sacrum.

Mathematical model. A set of mathematical equations that quantitatively describes the behavior of a given system.

Matrix. The intercellular substance of a tissue.

Medial. Reference to that side of an anatomical structure that is closest to the midsagittal plane.

Median nerve. Large important nerve. Activates muscles that pronate the forearm and flex forearm, wrist, and fingers. the sensory part of the nerve provides feedback information from the thumb and the first two and one-half fingers.

Moderate duration. Referring to lifting tasks that have a duration of between one and two hours, followed by a recovery period of at least 0.3 times the work time [i.e., at least a 0.3 recovery time to work time ratio (RT/WT)].

Moment arm. See **Lever arm**.

Moment. A quantity necessary to cause or resist rotation of a body, usually expressed in newton meters. (special case is torque, which is a moment about a longitudinal axis.)

Motion segment. A unit of the spine representing inherent biomechanical characteristics of the ligamentous spine. Physically, it consists of two adjacent vertebrae and the interconnecting disc and ligament tissue, devoid of musculature.

Motor unit. The body of a nerve cell and its axon and all muscle fibers supplied by branches of one axon, that is, the functional "unit" of muscle.

Myoelectric signal. Electrical potential produced during contraction of muscle (see Electromyography).

Nerve root entrapment syndrome. Technical term for "pinched nerve."

Neutral axis. The central plane on which the tensile and compressive stresses and strains due to bending equal zero.

Newton. The unit of force in the Systéme International d'Unités. One newton is the amount of force required to give a 1-kg mass an acceleration of 1 meter per second per second.

Palpate. To locate by touch.

Pathology. The discipline dealing with the development and description of disease in terms of altered structure and function of the body.

Pectoralis major. Large triangular muscle. The base forms the origin, running parallel to the entire length of the breast bone. The apex inserts into the medial side of the humerus. Essentially an abductor of the upper arm.

Physiological cross-sectional area. The amount of muscle fiber in a given cross section of a muscle.

Physiological response. The normal response and adaptation of the living organ and its parts to stress.

Physiology. The science that deals with the normal function of the living organ and its parts.

Plantar flexion. Bending about the ankle joint in the direction of the sole of the foot.

Poor coupling. Referring poor hand-to-object coupling that generally requires higher maximum grasp forces and thus specified a decreased acceptable weight for lifting.

Poplitea. The hollow at the back of the knee.

Pressure. The surface stress acting perpendicular to a unit area.

Pronation. The action of rotating the flexed forearm toward the midsagittal plane, so that the hands become prone, with palms down, back of hand up.

Proteoglycan. A macromolecule composed of glycosaminoglycans forming a hydrated gel; one of the primary structural components of cartilage.

Proximal. In a limb, closer to the body. Elsewhere, closer to the central axis of the body.

Radial deviation. Bending the hand at the wrist in the direction of the thumb.

Range of motion. The range of translation and rotation of a joint for each of its degrees of freedom.

Recommended weight limit. The product of the lifting equation; the load that nearly all healthy workers could perform over a substantial period of time for a specific set of task conditions.

Repetitive loading. Repeated application of a load to a structure.

Rigid body. A collection of particles joined together rigidly. For practical purposes, a body is said to be rigid if its deformation as compared with the other bodies in the system is small within a given range of the loads applied.

Rotation. Motion in which all points describe circular arcs about an immovable line or axis.

Sagittal line. The line passing through the mid-point between the inner ankle bones and lying in the sagittal plane, as defined by the neutral body position.

Scoliosis. Lateral curvature of the spine.

Shear. A loading mode in which a load is applied parallel to the surface of the structure, causing internal angular deformation or slip.

Short duration. Referring to lifting tasks that have a work duration of one hour or less, followed by a recovery time equal to 1.2 times the work time [i.e., at least a 1.2 recovery time to work time ratio (RT/WT)].

Significant control. A term defining a condition requiring "precision placement" of the load at the destination of the lift (e.g., 1. the worker has to re-grasp the load near the destination of the lift, 2. the worker has to momentarily hold the object at the destination, or 3. the worker has to position or guide the load at the destination).

Single-task lifting index. (L)/(STRWL); identifies individual tasks with potentially excessive physical demands and can prioritize the individual tasks according to the magnitude of their physical stress; values exceeding 1.0, suggest that ergonomic changes may be needed to decrease the overall physical demands of the task.

Single-task recommended weight limit. A value used in a multi-task assessment; the product of FIRWL and the appropriate FM; reflects the overall demands of that task, assuming it was the only task being performed. May be used to help determine if an individual task represents excessive physical demand; used in **Single-task lifting index**.

Speed of loading. The rate at which load is applied to a structure (see **Deformation rate**).
Static effort. Prolonged state of muscle contraction. Muscles remain in a state of heightened tension; blood does not flow through the muscle, and no useful work is visible.
Static work. Work that occurs when effect must be made to hold the body in a certain position instead of directing that effect to a task such as grasping.
Statics. The study of forces acting on a body in equilibrium.
Stiffness. A measure of resistance offered to external loads by a specimen or structure as it deforms.
Strain. Deformation (lengthening or shortening) of a body divided by its original length.
Strain gauge. A device that permits strain to be measured.
Strain rate. The speed at which a strain occurs.
Stress. Load per unit area which develops ona plane surface within a structure in response to externally applied loads.
Stress concentration. Any localized stress peak that cannot be predicted by simple strength of material theory.
Stress–strain curve. A curve generated by plotting the stress and the strain during compressive, tensile, or shear loading of a structure.
Suplination. Process of rotating the flexed forearm outward so that hand becomes "supine," that is, "palms up."
Synovia. Membranes lining the inside of joint capsules and moving surfaces of joints. They secrete the synovial fluid, which lubricates joints.
Tangential. Relating to a straight line that is the limiting position of a secant of a curve through a fixed point.
Tendinitis. Also **Tendonitis**. Inflammation of tendon (including tendon sheath).
Tendon. Connective tissue attaching muscle to bone.
Tendonitis. See **Tendinitis**.
Tendon sheaths. Tubular structures through which tendons run. They are lined with a synovial membrane and, therefore, not only guide but also lubricate the tendons.
Tenosynovitis. Inflammation of the tendon sheaths.
Tension. A loading mode in which equal and opposite loads are applied away from the surface of a structure, resulting in lengthening and narrowing.
Torsion. A loading mode in which a load is applied to a structure in a manner that causes it to twist about an axis, subjecting the structure to a combination of shear, tension, and compressive loads.
Translation. Parallel motion of one surface across another.
Transverse. Crosswise; in a horizontal direction.
Transverse plane. A plane at right angles to both the sagittal and coronal planes dividing the body into superior and inferior portions.
Triceps. Three-headed large extensor muscle of the forearm. Originates from the back of the humerus and the shoulder blade and inserts into the proximal tip of the ulna.
Ulnar deviation. Bending the hand at the wrist in the direction of the little finger.
Ultimate failure point. The point on the load-deformation curve past which complete failure of the structure occurs due to continued loading int he nonelastic region.
Vector. A quantity that has magnitude, direction, line of application, and point of application, commonly represented by a directed line segment.
Velocity. The displacement of a body divided by the time over which displacement occurs.
Vertical location. The distance of the hands above the flow measured at the origin and destination of the lift in inches or centimeters; used in the **Vertical multiplier**.

Vertical multiplier. A reduction coefficient defined as $(1 - (.0075 |V - 30|))$, for V measured in inches, and $(1 - (.003 |V - 75|))$, for V measured in centimeters.

Viscoelasticity. The property of a material to show sensitivity to rate of loading or deformation.

Width. Referring to width of a container in the sagittal plane.

Wolff's law. A law which states that bone is laid down where needed and resorbed where not needed.

Work. The amount of energy required to move a body from one position to another. Mechanical work is defined as the product of force applied to the distance moved in the direction of the force.

Index

Abduction, 69
Accelerometers, 263
Accommodation, 673
Acetylcholine, 66
Actin, 67
Action potential, 66
ACTMET, 47
ADA, 669
Adenosine triphosphate, 73
Afferent, 64
Age, 14
    variation, 16
    workers, 409
Anatomic, 2
Anatomical position, 2
Annulus fibrosus, 102
ANSUR, 7
Anterior/posterior, 3
Anthropometer, 6
Anthropometric, 1, 4
    criteria, 15
    dimensions, 4
Anthropometry, 1
    clothed, 18
Applied mechanics, 78
ARBAN, 266
Arc, 5
ARP, 13
Assist devices, 351
Average person, defined, 9

Back,
    injury, 104
    pain, 104

Biomechanical risk factors, 118
    models, 115
Biomechanics, 77, 464
Body,
    function of the, 2
    movement, 47
    size descriptors, 14
    size variability, 15
    temperature, 48
Bones, 69
Book, 47
Borg scales, 139
Breadth, 5
    biacromial, 22
BSLMET, 47
BSMET, 48

Calipers, 6
Cardiac, 66
Cardiovascular system, 47
Carrying, 50
CDEEPS, 729
Center(s)
    of gravity, 105
    of pressure, 106
    subcortical, 64
Central, 63
Cerebellum, 64
Cerebral cortex, 64
Checklists, 273
China, 761
Chronobiology, 405
Circumference, 5
Civilian population, 12

Coefficient of friction, 357
Compressive, 102
Concentric, 67
Construction, 434
    occupations, 554
Controls,
    administrative, 209
    engineering, 209
Contraction, 67, 70
Core temperature, 206
Coronal, 3
Costs
    of CTDs, 587
    of activity, 49

Data,
    military data, 12
    normative data, 222
    specialized data, 17
Degenerated disk, 103
Depolarized, 65
Depth, 5
Design
    concurrent, 721
    improvement cycle, 507
DIGMET, 47
Dimensions, 4
Disability, 672
Discomfort, 155
    assessment, 494

Eccentric, 67
Effects,
    of performance, 610
    on comfort, 608
    on health, 609
Efferent, 64
Elbow, 93
    rest height, 23
Electrogoniometer, 263
Electromyogram, 188
Electromyograph, 263
Electromyography, 71
Endplate, 65
Endurance time, 75

Energy
    expenditure, 190
    requirements, 47
Equilibrium, 81
ErgoMOST, 417
    case study, 422
Ergonomic, 279
    agricultural, 750
    assessment tools, 333
    definition, 2, 279
    education, 755
    standard, 691
    tables and figures, 24
Ergonomically "safe," 63
Ethnicity, 14
Evoked potentials, 72
exercise programs, 230, 525
Exposure
    monitors, 339
    time, 622
Extension, 69
Eyestrain, 387, 394

Fatigue, 75
Fechner's law, 138
Fiber types, 73
Filters, 179
Finger, 622
    flexion, 69
Force, 80
    external, 80
    internal, 80
    measurements, 186
    platform, 106
    torsional, 103

Gait, 73
Gender, 309
    differences, 408
Gloves, 316
Golgi tendon organs, 64, 68
Goniometer, 73
Guidelines, 660

Hammers, 317
Hand-transmitted vibration, 615

Index

Handedness, 311
Handling,
    information , 287
    manual materials (MMH), 329
    materials, 287
Hand trucks, 356
HANES (III), 12
Hardware, 704
Health care, 537
Heart rate monitor, 264
Heat strain, 195, 198
Height, 4
    clearance, 23
    popliteal, 23
HES, 12
HETA program, 590
Human,
    bodyspace, 2
    kinematics, 185
    –machine interface model, 280

India, 749
Inductive transducers, 172
Industrial manipulators, 352
Inertial, 72
Injuries, 436
Input, 283
    process-output model, 281
Instrument, 167
    virtual, 184
Intentional torts, 687
Interfaces, 183
Intervention techniques, 523
Intervertebral disk, 100
Intradiscal pressure, 104
ISO 9000, 735
Isoinertial, 72
Isokinetic, 72
Isometric, 72

Job
    analysis, 259
    analysis example, 271
    improvement cycle, 504
    related risk factors, 331

Joint, 69
    forces, 64

Keyboards, 390
Kinematic, 77, 98
Kinesiology, 72
Kinetic, 77, 99
Knee, 96
Knives, 319

Labor–management committees, 238
Landmarks, 4
Length, 5
    acromiale-radiale, 22
    resting, 67
Lift tables, 354
Lifting,
    dynamics, 121
    index (LI), 629
    isokinetic, 121
Ligaments, 69
Low back
    pain (LBP), 329
    disorders, 116
Lumbosacral angle, 105

Maximum aerobic capacity, 268
Mean, 9
Measure, 6
Measurement, 5, 166
    contour, 4
    direct, 263
    dynamic, 70, 78, 81, 97
Mechanical advantage, 84
Mechanical aids, 342
Mechanical stress, 391
Medial/lateral, 3
Metabolic, 73
    basal rate, 48
    requirement, 47
Metabolism, 47
    aerobic, 64
    anaerobic, 74
Model, 188, 196
    computer, 19
    of the hand, 127

[Model]
  link-segment, 107
  whole body, 133
  of the wrist, 127
Moments, 81
  of inertia, 2
Monitoring equipment, 702
Motion,
  Newton's laws of, 80
  Range of joint (ROJM), 2, 5
Motion sickness, 612
Motivation, 249
Motor
  control, 70
  neurons, 65
  units, 66
Multimedia, 192
Muscle
  agonist, 67
  antagonist, 67
  deltoid, 95
  forces, 64
  length, 64
  spindles, 64
  stretch, 64
Musculoskeletal, 115
  symptoms, 436
Myofibrll, 66
Myosin, 67

NASA, 13
Neurological effects, 617
Neuromuscular, 63
Normal distribution, 8
NTIS, 13
Nucleus pulposus, 102

Occupational biomechanics, 77
Occupational ergonomics, 2
Occupational injury, 547
OSHA, 658
Output, 283
OXUPTK, 50
Oxygen
  debt, 74
  uptake monitor, 264

Percentile, 9
Peripheral, 63
Personal,
  protection, 210
  protective equipment (PPE), 345
  risk factors, 331
Phosphocreatinine, 73
Photosensors, 175
Physical agility tester (PAT), 370
Physical anthropologists, 2
Physical qualifications, 367
Physical stressors 283
Physical work, 47
Piezoelectric, 172
Piezoresistive, 172
Pliers, 318
Postural stability, 72
Posture, 73
  awkward, 389
  working, 18
Power law, 138
Power tools, 319
Prevention techniques, 524
Proactive
  ergonomics, 502
  process, 515
  program, 505
Process, 283
Pronation, 69
Proprioception, 64
Proximal/distal, 3
Psychophysics, 137
PWC, 219

Quality circles, 237

Racial/ethnic variation, 16
Race, 14
Rate
  incidence, 587
  repetition, 267
Reach, 5
Reactive, 502
  process, 514
  program, 503
Recovery periods, 267

Recruitment, 71
Rehabilitation, 678
Risk factors, 118, 331
    environment, 286
Role of OSHA, 596
Rotating shifts, 411
RWL, 628

Sagittal, 3
Sampling theorem, 181
Saws, 318
Scalars, 78
Scanning, 14
Screwdrivers, 318
Seated work, 22
Seating, 393
Sensor, 170
    electrocapacitive, 172
    electrochemical, 173
    electromechanical, 170
    electrothermal, 174
    force, 263
Sensory receptors, 68
Sex, 14
Sexual variation, 15
Shape, 2
Shift schedules, 404
Shoulder, 94
Shovels, 317
Signal, 177
    processing, 180
Size, 2
Skeletal, 66
    muscle, 66
Sleep, 406
Slip resistance, 466
    application, 471
    guidelines, 470
Slips, 464
Smooth, 66
Software, 703
Spectrum analysis, 182
Spinal unit, 99
Spine, 116
Standards, 614, 621
Static, 4, 70, 78, 81, 82, 118

[Static]
    postures, 74
Stevens' power law, 138
Strength, 2
    isometric, 121
    isotonic, 5
    muscular, 72
Stress(es), 80
    contact, 388
    psychosocial, 388, 391
    shear, 103
Superior/inferior, 3
Supination, 69
Surveillance
    active, 489, 492
    passive, 477, 491
    record-based, 478, 481
Sweating, 205
Sweden, 733

System
    lever, 83, 84
    muscular, 47
    nervous, 63
    Ovako Working Posture Analysis
        (OWAS), 266
    reporting, 523
    somatosensory, 63

Taiwan, 765
Tape, 6
Task
    analysis, 570
    demands, 259
Team,
    advisory, 521
    manufacturing/engineering, 521
    office/computer ergonomics, 522
    product development laboratory, 521
Tendons, 69
Tenosynovitis, 304
Tensile, 102
Tension-monitoring receptors, 64
Terminology, 47
Tool, 303
    design, 305

TOTMET, 47
Training and education, 343
Transverse, 3
Tremor, 73
   physiological, 73
Trunk
   kinematics, 126
   mechanics, 122

Univariate, 9
Upright stability, 105

VDTs, 387
Vector, 78
Vertebral column, 100
Vestibular, 64, 68
Vibration, 191
   characteristics of, 606
   -induced white finger, 616
   measuring hand-transmitted, 621
   whole body, 607
Video
   display terminal (VDT), 22
   equipment, 700
   motion analysis, 264
Videotape, 261
Visual, 64, 68
   analog scales, 141

Volvo, 737

Walking, 464
Warehousing operation, 570
WBGT, 202
Weber's law, 138
WLKMET, 48
WMDs, 433, 490
Work
   history, 437
   performance, 407
   teams, 238
Worker
   compensation, 686
   participation, 235
   safety, 554
Workspace layout, 13
Workstation, 283
   sit/stand, 292
   sitting, 290
   standing, 291
Wrist
   activity monitors, 264
   and hand, 89
   position, 307
WT, 48
WTL, 48, 49